T0206912

Generalized Principal Component Analysis

Interdisciplinary Applied Mathmatics

Volume 40

Problems in engineering, computational science, and the physical and biological sciences are using increasingly sophisticated mathematical techniques. Thus, the bridge between the mathematical sciences and other disciplines is heavily traveled. The correspondingly increased dialog between the disciplines has led to the establishment of the series: *Interdisciplinary Applied Mathematics.*

The purpose of this series is to meet the current and future needs for the interaction between various science and technology areas on the one hand and mathematics on the other. This is done, firstly, by encouraging the ways that mathematics may be applied in traditional areas, as well as point towards new and innovative areas of applications; and, secondly, by encouraging other scientific disciplines to engage in a dialog with mathematicians outlining their problems to both access new methods and suggest innovative developments within mathematics itself.

The series will consist of monographs and high-level texts from researchersworking on the interplay between mathematics and other fields of science and technology.

More information about this series at http://www.springer.com/series/1390

René Vidal • Yi Ma • S. Shankar Sastry

Generalized Principal Component Analysis

Springer

René Vidal
Center for Imaging Science
Department of Biomedical Engineering
Johns Hopkins University
Baltimore, MD, USA

Yi Ma
School of Information Science
and Technology
ShanghaiTech University
Shanghai, China

S. Shankar Sastry
Department of Electrical Engineering
and Computer Science
University of California Berkeley
Berkeley, CA, USA

ISSN 0939-6047 ISSN 2196-9973 (electronic)
Interdisciplinary Applied Mathematics
ISBN 978-1-4939-7912-7 ISBN 978-0-387-87811-9 (eBook)
DOI 10.1007/978-0-387-87811-9

Mathematics Subject Classification (2010): 30C10, 30C40, 62-XX, 62-07, 62-08, 62B10, 62Fxx, 62H12, 62H25, 62H35, 62Jxx, 62J05, 62J07, 14-XX, 14N20, 15-XX

Springer New York Heidelberg Dordrecht London

Softcover reprint of the hardcover 1st edition 2016

Printed on acid-free paper

Springer Science+Business Media LLC New York is part of Springer Science+Business Media (www.springer.com)

To Camille, Margot, and Romeo Vidal (R. V.)
To Diana Xiaoyuan Zhu, Barron, and Henry Ma (Y. M.)
To Claire Tomlin, Samuel, and Lucy Sastry (S. S.)

Preface

We are not very pleased when we are forced to accept a mathematical truth by virtue of a complicated chain of formal conclusions and computations, which we traverse blindly, link by link, feeling our way by touch. We want first an overview of the aim and of the road; we want to understand the idea of the proof, the deeper context.

—Hermann Weyl

Classical theory and methods for the analysis of data were established mainly for engineering and scientific problems that arose five or six decades ago. In these classical settings, engineers or scientists usually had full control of the data acquisition process. As a result, the data to be processed and analyzed were typically *clean* and *complete*: they contained only moderate amounts of noise and were often adequately collected for the specific task or problem of interest. In that regime, many data analysis methods were based on the assumption that most data sets have fewer effective degrees of freedom than the dimension of the ambient space. For example, the number of pixels in an image can be rather large, yet most computer vision models used only a few parameters to describe the appearance, geometry, and dynamics of a scene. This assumption motivated the development of a number of techniques for identifying low-dimensional structures in high-dimensional data, a problem that is important not only for understanding the data, but also for many practical purposes such as data compression and transmission. A popular technique for discovering low-dimensional structure in data is principal component analysis (PCA), which assumes that the data are drawn from a *single* low-dimensional affine subspace of a high-dimensional space (Jolliffe 1986, 2002). PCA is arguably the simplest and most popular dimensionality reduction tool, and it has found widespread applications in many fields such as computer vision (Turk and Pentland 1991).

However, in the past decade or so, there has been a fundamental regime shift in data analysis. Currently, scientists and engineers often must deal with data whose dimension, size, and complexity expand at an explosive rate. Moreover, they have pretty much lost control of the data acquisition process. For instance, in 2012, 350 million photos were uploaded to Facebook every day, and 100 hours of video were uploaded to YouTube each minute. Moreover, it is estimated that 3.8 trillion photos had been taken by 2012, 10% of them in the last 12 months.[1] This and other forms of massive amounts of data on the Internet and mobile networks are being produced by billions of independent consumers and businesses. How to extract useful information from such massive amounts of data for numerous tasks (such as search, advertisement, scientific analysis) has become one of the biggest engineering endeavors of mankind. Many call it the era of *Big Data*. Obviously, such a regime shift demands a fundamental paradigm shift in data analysis, since classical theory and methods for data analysis were simply not designed to work under such conditions. The website of Theoretical Foundations of Big Data Analysis[2] puts things into perspective:

> *The Big Data phenomenon presents opportunities and perils. On the optimistic side of the coin, massive data may amplify the inferential power of algorithms that have been shown to be successful on modest-sized data sets. The challenge is to develop the theoretical principles needed to scale inference and learning algorithms to massive, even arbitrary, scale. On the pessimistic side of the coin, massive data may amplify the error rates that are part and parcel of any inferential algorithm. The challenge is to control such errors even in the face of the heterogeneity and uncontrolled sampling processes underlying many massive data sets.*

Since the data acquisition process is no longer under the data gatherer's control, the structure of the data to be processed or analyzed can no longer be assumed to be relatively simple or clean: very often, the data contain significant amounts of noise, corrupted entries, and outliers; or the data could be incomplete or inadequate for a task that arises only after the data have been collected; or the data could even have some degree of unknown nonlinearity due to lack of calibration in the data acquisition. In the past decade, these challenges have led to many revolutionary discoveries and much progress in which many of the classical models and methods for data analysis have been systematically generalized or improved to make them robust to such bad nuisances in the data. In the context of identifying low-dimensional structures in the data, classical PCA is *generalized* so that it can robustly find the correct subspace structure of the data despite such nuisances. The forms of progress include entirely new methods for low-rank matrix completion, robust PCA, kernel PCA, and manifold learning.

Another challenge that arises in the new regime is that we can no longer assume that the data lie on a single low-dimensional subspace or submanifold. This is because many modern data sets are not collected for any specific task. Instead, the data may have already been collected, and the task emerges only afterward. Hence a data set can be mixed with multiple classes of data of different natures, and the intrinsic structure of the data set may be *inhomogeneous* or *hybrid*. In this case, the data set may be better represented or approximated by not one, but multiple low-dimensional subspaces or manifolds. Figure 1 gives an example of face images

[1] http://www.buzzfeed.com/hunterschwarz/how-many-photos-have-been-taken-ever-6zgv.

[2] http://simons.berkeley.edu/programs/bigdata2013.

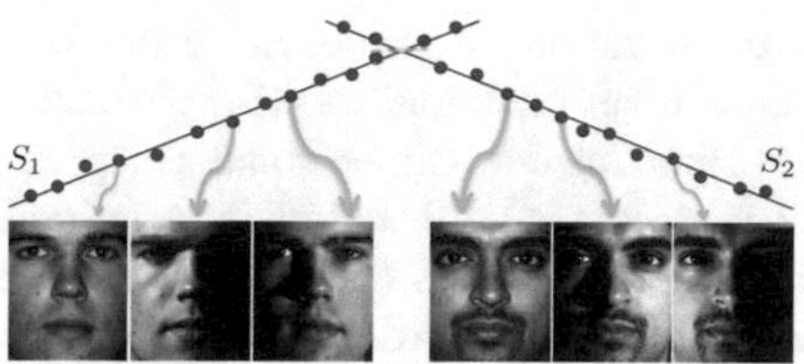

Fig. 1 Face images from multiple individuals can be well approximated by multiple low-dimensional subspaces.

under varying illumination conditions, where each affine subspace corresponds to face images of a different individual. This leads to a general problem: Given a set of data points from a *mixture* of affine subspaces, how does one automatically learn or infer those subspaces from the data? A solution to this problem requires one to *cluster* or *segment* the data into multiple groups, each belonging to one subspace, and then *identify* the parameters of each subspace. To model data with such mixed subspace structures, the classical PCA method needs to be *generalized* so that it can simultaneously identify multiple subspaces from the data. This leads to the so-called *subspace clustering* problem, which has received great attention in the last decade and has found widespread applications in computer vision, image processing, pattern recognition, and system identification.

Purpose of This Book

The purpose of this book is to provide a comprehensive introduction to the latest advances in the mathematical theory and computational tools for modeling high-dimensional data drawn from one or more low-dimensional subspaces (or manifolds) and corrupted by noise, missing entries, corrupted entries, and outliers. This will require the development of new algebraic, geometric, statistical, and computational theory and methods for efficient and robust estimation of one or more subspaces. To distinguish this theory and these methods from classical PCA, we call all such advanced approaches as *generalized principal component analysis* or GPCA for short.[3]

As we will see in this book, in order to generalize classical PCA to the case of corrupted and mixed data, we need to resort to a body of more advanced mathematical tools from estimation theory, algebraic geometry, high-dimensional

[3]In the literature, the word *generalized* is sometimes used to indicate any particular extension to classical PCA (Jolliffe 1986, 2002). In our opinion, each of these extensions is a particular generalization rather than the more systematic generalization that we present in this book. In addition, for the case in which we want PCA to handle large amounts of corruptions or outliers, we may use the special name *robust PCA* (RPCA); for the *nonlinear* case in which each component is an algebraic variety of higher degree such as a quadratic surface or a more complicated manifold, we may use the name *nonlinear PCA* or *manifold learning*; for the case of multiple subspaces or manifolds, names such as *mixtures of probabilistic PCA* (MPPCA), *subspace clustering* (SC) and *hybrid component analysis* (HCA) have been suggested and would also be appropriate.

statistics, and convex optimization. In particular, in this book and its appendices, we will give a systematic introduction to effective and scalable optimization techniques tailored to estimating low-dimensional subspace structures from high-dimensional data (see Appendix A), all the related statistical theory and methods for robust estimation of mixture models (see Appendix B), as well as a complete characterization of the algebraic properties of a union of multiple subspaces as an algebraic set (see Appendix C). As we will see throughout this book, the statistical, algebraic-geometric, and computational aspects of GPCA are highly complementary to each other. Each of them leads to solutions and algorithms of their own that hold certain conceptual or computational advantages against other approaches under certain assumptions about the data and/or the subspaces.

There are several reasons why we feel that the time is now ripe to write a book about GPCA:

1. The limitations of classical PCA have been well known to engineers and practitioners of modern data analysis. However, PCA remains the method of choice by many field engineers simply because they do not have a systematic body of theory and methods for handling different types of nuisances in the data. In the past few years, with advances in algebraic geometry, high-dimensional statistics, and convex optimization, our understanding of the problem of estimating a low-dimensional subspace has gone well beyond classical settings: we have not only a better understanding of the geometric, statistical, and probabilistic nature of PCA, but also computationally efficient algorithms for PCA with missing and corrupted data that give provably correct solutions under broad conditions. In addition, the field of estimating mixture models, in particular a mixture of subspaces, has also gone through revolutionary developments in the past few years. The statistical, algebraic, geometric, and computational properties of this class of models have been reasonably well understood. As result, many effective and efficient algorithms have been developed for this problem.
2. These new developments obviously come at a very good time, since both science and engineering are entering the era of *Big Data*. Many of the new algorithms and techniques have already demonstrated great success and potential in many important practical problems of image processing and pattern analysis, as we will demonstrate with some concrete applications and examples in this book. We anticipate that these new theoretical results and the associated computational methods will provide scientists and engineers with a new set of models, principles, and tools that can be readily applied to a broad range of practical problems and real-world data, far beyond the applications and data illustrated in this book.

Intended Audience of This Book

We have written this book with the idea that it will have both research and pedagogical value. From a research perspective, the topics covered in this book are of great relevance and importance to both theoreticians and practitioners in such areas as data science, machine learning, pattern recognition, computer vision, signal and image processing, and system identification. The motivating examples and applications given in this book are purposely biased by our own research interests in image processing and computer vision, because we believe that from a pedagogical perspective, visual data and examples can best illustrate some of the abstract models and properties introduced. Nevertheless, the basic theory and algorithms are established in fairly general terms, and are obviously applicable to many practical engineering and scientific problems well beyond those described in this book.

We believe that the material of the book is ideal for an introductory graduate course for students in data science, machine learning, and signal processing, or an advanced course for students in computer vision, estimation theory, and systems theory. Through arguably the simplest class of models, the low-dimensional linear models, the book introduces to students some of the most fundamental principles in data modeling, statistical inference, optimization, and computation. Knowledge about these basic models and their properties is absolutely necessary for anyone who strives to study more sophisticated classes of models in which low-dimensional linear models are the key building blocks, such as the sparse models in compressive sensing and the deep neural networks in machine learning (see Chapter 13 for further discussion).

The book is written to be friendly to beginning graduate students and instructors. At the end of each chapter, we have provided many basic exercises and programs from which students may gain hands-on experience with the material covered in the chapters as well as an extensive survey of related literature for research purposes. Additional information, resources, and sample code for most of the examples, algorithms, and applications featured in this book will be made available at the book's website: http://www.vision.jhu.edu/gpca.

We have used material from this book many times to teach a one-semester graduate course at the Johns Hopkins University, the University of Illinois at Urbana-Champaign, the University of California at Berkeley, and the ShanghaiTech University in China. As the reader will see, GPCA is a very unique subject that touches on many fundamental concepts, facts, and principles across engineering, computation, statistics, and mathematics. Therefore, this is a great topic that can shepherd researchers and students to systematically establish some of the most fundamental and useful knowledge for modern data science and machine learning. We also believe that the reader will learn to appreciate the complementary nature of different perspectives and approaches presented in this book, and in the end develop a deep and comprehensive understanding of the subject.

Organization of This Book

Chapter 1 gives a nontechnical introduction to the basic problems, ideas, and principles studied in this book. The remainder of the book is organized into four *parts*:

Part I covers classical and modern theory and methods for modeling data with a single low-dimensional linear or affine subspace (or a nonlinear submanifold). More specifically, **Chapter 2** gives a review of classical PCA theory and methods for subspace estimation, including its statistical, geometric, and rank minimization interpretations. The chapter also covers a simple generative model for PCA, called probabilistic PCA, as well as model selection issues for PCA. **Chapter 3** shows how to estimate a subspace when the data are incomplete or corrupted. The chapter discusses statistical and alternating minimization methods for robust PCA, as well as some advanced tools from compressive sensing for sparse and low-rank recovery. Since complete proofs for these results are beyond the scope of this book, we will simply discuss their implications and show how to use them to develop effective algorithms for robust PCA. **Chapter 4** shows how to extend the methods for learning linear subspaces to nonlinear submanifolds. In particular, the chapter introduces both parametric and nonparametric methods for manifold learning, including nonlinear PCA, kernel PCA, locally linear embedding, and Laplacian eigenmaps. The chapter also introduces the basic K-means algorithm for clustering data distributed around a few cluster centers, as well as the more advanced spectral clustering algorithm, which combines manifold learning methods with K-means to cluster mixed data that have more complex nonlinear structures.

Part II covers three complementary approaches and methods for modeling data with a mixture of multiple subspaces. More specifically, **Chapter 5** studies the algebraic-geometric properties of a mixture of subspaces, also known in modern algebra as a subspace arrangement. The chapter introduces a basic noniterative algebraic method for estimating multiple subspaces, which works effectively and efficiently when the data are relatively clean and the ambient dimension is low. **Chapter 6** introduces several statistical methods for estimating mixture subspace models. They are based on different but related statistical principles, including the minimax principle (the K-subspaces method), the maximum likelihood principle (the EM algorithm), and the minimum description/coding length principle (the compression-based agglomerative clustering method). **Chapter 7** explores the nonparametric spectral clustering method for subspace clustering and introduces many different ways to establish affinity matrices for data points in a mixture of subspaces, based on local, semilocal, and global geometric information. **Chapter 8** develops principled ways to establish affinity matrices for subspace clustering via self-expressive low-rank or sparse representations. It introduces modern convex optimization techniques to find such representations. It also studies under what conditions this approach gives provably correct solutions.

Part III demonstrates a few representative applications of the methods and algorithms introduced in earlier chapters. More specifically, **Chapter 9** shows how to cluster image patches into multiple subspaces and learn a hybrid linear model from them for the purpose of building highly compact and sparse representations of natural images. **Chapter 10** shows how to segment natural images into multiple regions corresponding to different colors and textures based on data compression and subspace clustering techniques introduced in this book. **Chapter 11** shows how to segment multiple moving objects in an image sequence using many of the subspace clustering algorithms presented in this book. The chapter also provides an empirical comparison of these methods on motion segmentation data, and discusses their strengths and weaknesses. The chapter also shows how to extend subspace clustering algorithms to a special class of nonlinear manifolds

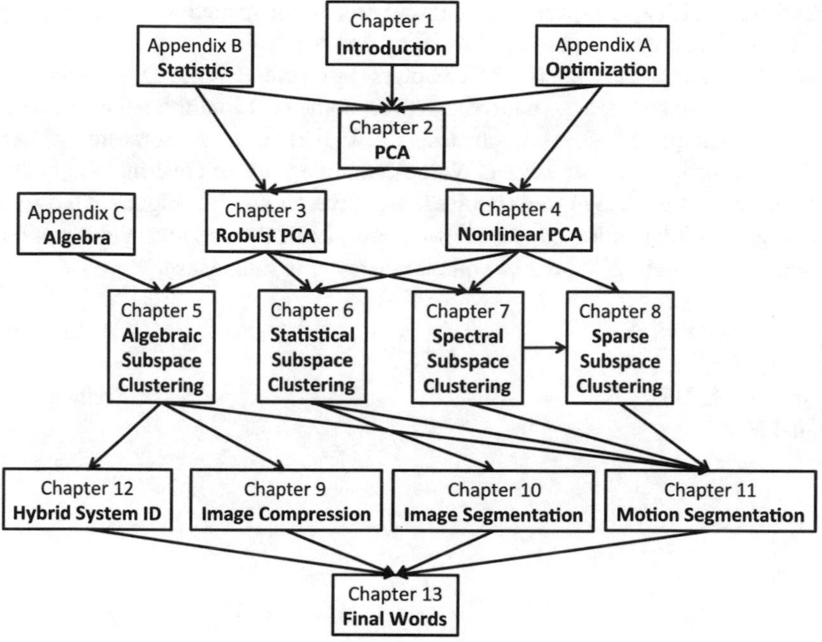

Fig. 2 Organization of the Book—logical dependency among all the chapters and the appendices.

arising in the motion segmentation problem. The chapter also shows how subspace clustering algorithms can be used to segment video and time series into multiple events or actions. **Chapter 12** studies the temporal segmentation problem more systematically. The algebraic subspace clustering method is modified and extended to segment observations that are generated by a hybrid linear dynamical system and to subsequently identify all the underlying dynamical models.

Part IV covers relevant concepts and results in optimization, mathematical statistics, and algebraic geometry in order to make the book self-contained. More specifically, **Appendix A** covers basics notions from optimization, such as first- and second-order conditions for optimality, convexity, gradient descent methods, alternating minimization methods, constrained optimization, duality, Lagrange methods, augmented Lagrange methods, and the alternating direction method of multipliers. **Appendix B** covers basic notions from statistics, such as sufficient statistics, unbiased estimators, maximum likelihood estimation, expectation maximization, mixture models, model selection, and robust statistics. **Appendix C** covers basic notions from algebraic geometry, including polynomial rings, ideals, algebraic sets, subspace arrangements, ideals of subspace arrangements, and Hilbert functions of subspace arrangements. All these concepts and results may come in handy for readers who are not so familiar with certain mathematical facts used in the book, especially for the early chapters.

Last but not least, **Chapter 13** discusses some of the related open research topics and future directions that are not covered by this book.

We have taught the material of Chapters 1–8 several times in a one-semester course, and have covered the entire book with some of the additional proofs for the material in Chapters 3–8 and applications in Part III in a two-semester sequence. We invite instructors to experiment with alternative ways of covering this material. To help instructors design their courses, we have outlined in Figure 2 the overall book organization and logical dependency among all the chapters and appendices. We would be delighted to hear of your experiences in this regard.

Baltimore, MD, USA — René Vidal
Shanghai, China — Yi Ma
Berkeley, CA, USA — S. Shankar Sastry
August 2015

Acknowledgments

As we express our gratitude, we must never forget that the highest appreciation is not to utter words, but to live by them.

—John F. Kennedy

Our initial motivation for trying to generalize principal component analysis to multiple subspaces can be traced back to early 2001, when René, Shankar, and colleagues were developing methods for having a team of robots pursue another team of robots using visual information. For this purpose, we needed to develop methods for estimating the pose of multiple moving objects in a video taken by a moving camera. At the time, methods for estimating the pose of an object relative to a moving camera were well understood, including many methods developed by Yi in his PhD thesis. However, the problem of estimating the pose of multiple moving objects in a video was not as well understood. In particular, the main challenge was that we often do not know which pieces of the video correspond to the same moving object; hence we needed both to segment the video and to estimate the pose of each object, i.e., we needed to solve the *motion segmentation problem*.

To address these issues, René and Yi began to work on a polynomial-based method for solving the motion segmentation problem. The approach was based on fitting a high-order polynomial to the image data and factorizing it into multiple bilinear factors, each one encoding the pose of each one of the moving objects. Interestingly, we observed that the bilinear factorization problem could be reduced to the problem of factorizing a polynomial into a product of linear factors, which in turn provided a solution to the problem of clustering data drawn from a union of planes in three-dimensional space, i.e., the *plane clustering problem*. We soon realized that this polynomial-based method for solving the plane clustering problem could be extended to subspaces of arbitrary dimensions, a problem that was common and fundamental to many data modeling, clustering, and classification problems in pattern recognition, computer vision, signal/image processing, and systems theory. However, at the time there was a serious lack of systematic study and understanding of this very important class of models and problems, and many algorithms at the time were heuristic or ad hoc. This inspired us to work very actively during the next few years to develop a more complete theoretical and algorithmic foundation for this class of models. As of August 2003, René had summarized much of the algebraic-geometric method in his PhD dissertation at Berkeley (see Chapter 5), including some initial applications to motion segmentation.

In early 2004, on the day after Yi's wedding reception, we decided to formalize our findings with a manuscript and sketched an early outline of this book at Café Kopi in downtown Champaign, Illinois. Our initial plan was to extend the algebraic-geometric approach by developing more efficient and robust techniques for estimating low-dimensional subspace structures from imperfect mixed high-dimensional data, and to apply these techniques to a broad class of engineering and scientific problems. Following René's PhD dissertation and some of our earlier papers, many of our graduate students, postdocs, and colleagues enriched and extended the theory, algorithms, and applications of the algebraic-geometric approach to many new settings and problems in computer vision, image processing, and system identification. We especially thank our former students Laurent Bako, Yasmin Hashambhoy, Wei Hong, Kun Huang, Jacopo Piazzi, Dheeraj Singaraju, Roberto Tron, Allen Yang, and John Wright for their development of robust algebraic-geometric approaches to subspace clustering and their applications to image compression, image segmentation, motion segmentation and hybrid system identification, which are featured in Chapters 6, 9, 10, 11, and 12. We are also greatly indebted to Professor Robert Fossum and Professor Harm Derksen. They painstakingly taught Yi algebraic geometry and have helped develop a rather complete characterization of the algebraic properties of subspace arrangements. Their work has helped to provide a rigorous mathematical foundation for the algebraic subspace clustering algorithms developed in this book (Chapter 5 and Appendix C). Professor Harm Derksen was also the first to suggest to Yi the use of compression for subspace clustering. He and John Wright helped develop the compression-based subspace clustering work (featured in Chapter 6), which is very complementary to other existing methods. We also thank Professor Richard Hartley and Professor Brian Anderson for their contributions to the applications of the algebraic-geometric methods to motion segmentation (featured in Chapter 11) and hybrid system identification (featured in Chapter 12), respectively.

As it turned out, around 2007 we realized that our initial plan based on extending the algebraic-geometric approach to subspace clustering was a little premature: We did not fully realize at the time how much and how fast this topic was to evolve in years to come. In fact, during the next few years, even classical PCA for learning a single subspace went through unprecedented development with many new algorithms based on more solid statistical and mathematical principles that are much more effective and robust than classical techniques. We especially thank our former students Arvind Ganesh, Hossein Mobahi, Shankar Rao, Allen Yang, Andrew Wagner, and John Wright of UIUC for their development of robust approaches to classical PCA based on sparse and low-rank minimization, and their applications to face recognition. We are also extremely grateful to Professor Emmanuel Candès for his pioneering and inspiring work in this area. His collaboration with Yi on robust

PCA is featured in Chapter 3 together with his elegant treatment of the low-rank matrix completion problem.

Such exciting developments in the theory of classical PCA led us to revisit the subspace clustering problem with much more advanced mathematical tools and entirely new perspectives. In particular, we are greatly indebted to our former students Ehsan Elhamifar, Alvina Goh, Shankar Rao, Guangcan Liu, and Roberto Tron for the development of manifold learning and spectral clustering approaches to subspace clustering, which we have featured in Chapter 7 and Chapter 8. Special thanks go to Ehsan Elhamifar, who was a key contributor to the development of subspace clustering methods based on sparse representation, and to Guangcan Liu and Professor Paolo Favaro for the development of subspace clustering methods based on low-rank representation, both featured in Chapter 8. Their work extended not only the theory of subspace clustering, but also their applicability to midsize data sets, including face and digit clustering. We also thank Mahdi Soltanolkotabi and Professor Emmanuel Candès for their recent elegant theoretical analysis of the sparse subspace clustering algorithm, which is also featured in Chapter 8.

While we were witnessing such exciting developments, our book plan had been delayed repeatedly. Only very recently did we all become convinced that this topic had become stable and mature enough for us to fulfill our ten-year-old commitment. The final version of this book would have not been completed without the help of René's students Chong You and Manolis Tsakiris. Chong was kind enough to help generate most of the wonderful running examples for the algorithms presented this book, especially those on face images and motion capture, while Manolis generated many of the synthetic examples, especially those based on algebraic methods. In return, this book has inspired their own PhD work: Chong is currently developing theory and algorithms to scale up sparse and low-rank subspace clustering methods to the big data domain, while Manolis is revisiting the algebraic-geometric approach to make it robust. Overall, it looks like this book has come full circle, and we may have robust algebraic-geometric algorithms in the near future thanks to Manolis's work. We also thank Chong, Manolis, and Ben Haeffele for proofreading the final version of this book and giving us fantastic comments on how to improve the presentation of the material.

We thank Professor Martin Vetterli of École Polytechnique Fédérale de Lausanne, Professor David Donoho of Stanford University, and Professor Guillermo Sapiro of Duke University for their encouragement for us to apply generalized PCA models to sparse image representation. Yi would like to thank Professor David Donoho in particular for making the early suggestion about the strong connection between subspace arrangements and sparse representation. One could say that the story of GPCA would never be so profound and complete without the advanced theory and computational tools from compressive sensing and sparse representation.

We thank many of our colleagues for valuable collaborations, discussions, suggestions, and moral support. They are Professor Stefano Soatto of the University of California at Los Angeles, Professor Jana Kosecka of George Mason University, Professor Richard Hartley of the Australian National University, Professors Jitendra Malik and Ruzena Bajcsy of the University of California at Berkeley, Dr. Harry Shum, Dr. Yasuyuki Matsushita, and Dr. David Wipf of Microsoft, Professor Zhouchen Lin of Peking University (or Microsoft Research at the time of collaboration), Professor Shuicheng Yan of National University of Singapore, Professors Robert Fossum, Minh Do, Thomas Huang, Narendra Ahuja, Daniel Liberzon, and Yizhou Yu of the University of Illinois at Urbana-Champaign, Professor Rama Chellapa from the University of Maryland at College Park, and Professors Don Geman, Gregory Hager, Michael Miller, Daniel Robinson, Laurent Younes, of The Johns Hopkins University. In particular, Professor Robert Fossum helped with proofreading an early version of the manuscript (containing essentially Chapters 1–6), which Yi used to teach an earlier course on GPCA at UIUC in 2006. We also thank Professors Alvaro Soto and Domingo Mery from the Catholic University of Chile, Professors Jean Ponce and Francis Bach from INRIA, and Professor Emmanuel Candès from Stanford University for hosting René during his sabbatical in 2012, when many chapters of this book were written.

We are obviously grateful for all the funding agencies and institutes that have supported us through all these years. In particular, we would like to thank our funders and program managers, Dr. Daniel DeMenthon, Dr. Helen Gill, Dr. Haesun Park, Dr. John Cozzens, and Dr. Jie Yang of the National Science Foundation, and Dr. Behzad Kamgar-Parsi of the Office of Naval Research. They have generously supported our research in this direction even when much of the theory and results were still in their infancy. Without their vision and trust, the area of GPCA let alone this book, would not have been possible. We would like to acknowledge the research funding of NSF under grants IIS-0347456, CAREER-IIS-0447739, CNS-EHS-0509101, CRS-EHS-0509151, CCF-TF-0514955, ECCS-0701676, IIS-0703756, CNS-0834470, CCF-0964215, ECCS-0941463, CSN-0931805, OIA-0941362, IIS-0964416, IIS-1116012, IIS-1218709, IIS-1335035, and IIS-1447822, ONR under grants N00014-00-10621, N00014-05-10633, N00014-05-10836, N00014-09-10084, N00014-09-10230, N00014-09-10839, and N00014-13-10116, and DARPA under grants F33615-98-C-3614 and KECoM 10036-100471 for their support of our work. René would like to thank the Sloan Research Fellowship for partially supporting his sabbatical. Yi would like to pay special thanks to the generous startup support from ShanghaiTech and moral support from Dean Cher Wang and President Mianheng Jiang. Their vision and determination to reform Chinese higher education and research has encouraged Yi to focus on writing this book till its completion over the past two years at ShanghaiTech. GPCA has now become part of the regular curricula for the areas of data science, signal processing, and machine learning for the School of Information Science and Technology of ShanghaiTech.

Finally and most importantly, our families, including the six little ones who were born during the gestation of this book, have provided us with a huge amount of love, encouragement, and support in the writing of this book.

Baltimore, MD, USA	René Vidal
Shanghai, China	Yi Ma
Berkeley, CA, USA	S. Shankar Sastry
August 2015	

Contents

Glossary of Notation

Frequently used mathematical symbols are defined and listed according to the following categories:

0. Set theory and logic symbols
1. Sets and linear spaces
2. Transformation groups
3. Vector and matrix operations
4. Geometric primitives in space
5. Probability and statistics
6. Graph theory
7. Image formation

Throughout the book, **every vector is a column vector unless stated otherwise!**

0. Set theory and logic symbols

$\cap$	$S_1 \cap S_2$ is the intersection of two sets
$\cup$	$S_1 \cup S_2$ is the union of two sets
$\doteq$	Definition of a symbol
$\exists$	$\exists s \in S, P(s)$ means there exists an element s of set S such that proposition $P(s)$ is true
$\forall$	$\forall s \in S, P(s)$ means for every element s of set S, proposition $P(s)$ is true
$\in$	$s \in S$ means s is an element of set S
$\lvert S \rvert$	The number of elements in set S
$\setminus$	$S_1 \setminus S_2$ is the difference of set S_1 minus set S_2
$\subset$	$S_1 \subset S_2$ means S_1 is a proper subset of S_2
$\{s\}$	A set consists of elements like s
$\to$	$f : D \to R$ means a map f from domain D to range R
$\mapsto$	$f : x \mapsto y$ means f maps an element x in the domain to an element y in the range
$\circ$	$f \circ g$ means composition of map f with map g
$\vee$	$\mathcal{P} \vee \mathcal{Q}$ is true if either proposition $\mathcal{P}$ or proposition $\mathcal{Q}$ is true
$\wedge$	$\mathcal{P} \wedge \mathcal{Q}$ is true if both proposition $\mathcal{P}$ and proposition $\mathcal{Q}$ are true
$\Longrightarrow$	$\mathcal{P} \Longrightarrow \mathcal{Q}$ means proposition $\mathcal{P}$ implies proposition $\mathcal{Q}$
$\Longleftrightarrow$	$\mathcal{P} \Longleftrightarrow \mathcal{Q}$ means propositions $\mathcal{P}$ and $\mathcal{Q}$ imply each other
$\mid$	$\mathcal{P} \mid \mathcal{Q}$ means proposition $\mathcal{P}$ holds given the condition $\mathcal{Q}$

1. Sets and linear spaces

$\mathbb{C}$	The set of all complex numbers

$\mathbb{C}^n$	The n-dimensional complex linear space
$\mathbb{P}^n = \mathbb{RP}^n$	The n-dimensional real projective space
$\mathbb{R}$	The set of all real numbers
$\mathbb{R}^n$	The n-dimensional real linear space
$\mathbb{R}_+$	The set of all nonnegative real numbers
$\mathbb{Z}$	The set of all integers
$\mathbb{Z}_+$	The set of all nonnegative integers
L	A generic 1-D line in space
S	Typically represents a generic linear or affine subspace
P	A generic 2-D plane in space

2. Geometric primitives in space

$x \in \mathbb{R}$	A lower-case letter normally represents a scalar
$\boldsymbol{x} \in \mathbb{R}^D$	A bold lower-case letter represents a vector or a random vector
$\boldsymbol{x}_j \in \mathbb{R}^D$	The jth sample vector in a data set
$\mathcal{X} \subset \mathbb{R}^D$	Represents a set of data points: $\mathcal{X} = \{\boldsymbol{x}_1, \boldsymbol{x}_2, \ldots, \boldsymbol{x}_N\}$
$X \in \mathbb{R}^{D \times N}$	A capital letter represents a matrix, very often representing the data matrix with the data points as its columns: $X = [\boldsymbol{x}_1, \boldsymbol{x}_2, \ldots, \boldsymbol{x}_N]$
$\mathcal{X}_i \subset \mathcal{X}$	The ith subset or cluster of the dataset $\mathcal{X}$
X_i	The submatrix of X associated with the ith cluster $\mathcal{X}_i$

3. Vector and matrix operations

$\|\|\boldsymbol{x}\|\|_2$	The 2-norm of a vector $\boldsymbol{x} \in \mathbb{R}^n$: $\sqrt{x_1^2 + x_2^2 + \cdots + x_n^2}$
$\|\|\boldsymbol{x}\|\|_1$	The 1-norm of a vector $\boldsymbol{x} \in \mathbb{R}^n$: $\|x_1\| + \|x_2\| + \cdots + \|x_n\|$
$\|\|\boldsymbol{x}\|\|_0$	The 0-norm of a vector $\boldsymbol{x} \in \mathbb{R}^n$: the number of nonzero values
$\langle \boldsymbol{x}, \boldsymbol{y} \rangle \in \mathbb{R}$	The inner product of two vectors: $\langle \boldsymbol{x}, \boldsymbol{y} \rangle = \boldsymbol{x}^\top \boldsymbol{y}$
$\boldsymbol{x} \sim \boldsymbol{y}$	Homogeneous equality: two vectors or matrices $\boldsymbol{x}$ and $\boldsymbol{y}$ are equal up to a nonzero scalar factor
$\boldsymbol{x} \times \boldsymbol{y} \in \mathbb{R}^3$	The cross product of two 3-D vectors: $\boldsymbol{x} \times \boldsymbol{y} = \widehat{\boldsymbol{x}}\boldsymbol{y}$
$\boldsymbol{x} \otimes \boldsymbol{y}$	The Kronecker (tensor) product of $\boldsymbol{x}$ and $\boldsymbol{y}$
$\operatorname{span}(M)$	The range or subspace spanned by the columns of a matrix M
$\operatorname{rank}(M)$	The rank of a matrix M
$\operatorname{null}(M)$	The null space or kernel of a matrix M
$\det(M)$	The determinant of a square matrix M
$M^\top \in \mathbb{R}^{n \times m}$	Transpose of a matrix $M \in \mathbb{R}^{m \times n}$ (or a vector)
$\operatorname{trace}(M)$	The trace of a square matrix M, i.e., the sum of all its diagonal entries, sometimes shorthand as $\operatorname{tr}(M)$
$M = U \Sigma V^\top$	The singular value decomposition of a matrix M
$\|\|M\|\|_*$	The nuclear norm of a matrix M: the sum of all its singular values
$\|\|M\|\|_0$	The 0-norm of a matrix M: the number of nonzero values

$\\|M\\|_F$	The Frobenius norm of a matrix M: the square root of the sum of the square of its entries
$S_1 \oplus S_2$	The direct sum of two linear subspaces S_1 and S_2
$S^{\perp}$	The orthogonal complement of a subspace S
$P_S(\boldsymbol{x})$	Projecting a vector $\boldsymbol{x}$ onto the subspace S

4. Transformation groups

$GL(n) = GL(n, \mathbb{R})$	The real general linear group on $\mathbb{R}^n$; it can be identified as the set of $n \times n$ invertible real matrices
$SL(n) = SL(n, \mathbb{R})$	The real special linear group on $\mathbb{R}^n$; it can be identified as the set of $n \times n$ real matrices of determinant 1
$A(n) = A(n, \mathbb{R})$	The real affine group on $\mathbb{R}^n$; an element in $A(n)$ is a pair $(A, \boldsymbol{b})$ with $A \in GL(n)$ and $\boldsymbol{b} \in \mathbb{R}^n$ and it acts on a point $\boldsymbol{x} \in \mathbb{R}^n$ as $A\boldsymbol{x} + \boldsymbol{b}$
$O(n) = O(n, \mathbb{R})$	The real orthogonal group on $\mathbb{R}^n$; if $U \in O(n)$, then $U^{\top} U = I$
$SO(n) = SO(n, \mathbb{R})$	The real special orthogonal group on $\mathbb{R}^n$; if $R \in SO(n)$, then $R^{\top} R = I$ and $\det(R) = 1$
$SE(n) = SE(n, \mathbb{R})$	The real special Euclidean group on $\mathbb{R}^n$; an element in $SE(n)$ is a pair $(R, \boldsymbol{t})$ with $R \in SO(n)$ and $\boldsymbol{t} \in \mathbb{R}^n$ and it acts on a point $\boldsymbol{x} \in \mathbb{R}^n$ as $R\boldsymbol{x} + \boldsymbol{t}$

5. Probability and statistics

$p_{\theta}(\boldsymbol{x})$	The probability density function of the random variable or vector $\boldsymbol{x}$ with θ as parameters of the distribution, sometimes also written as $p(\boldsymbol{x}, \theta)$
$p(\boldsymbol{y} \mid \boldsymbol{x})$	The conditional probability density function of the random variable $\boldsymbol{y}$ given $\boldsymbol{x}$
$P(\cdot)$	The probability of a random event
$\boldsymbol{\mu} = \mathbb{E}[\boldsymbol{x}]$	The expectation (or mean) of a random variable or vector $\boldsymbol{x}$
$\Sigma_{\boldsymbol{x}} = \mathrm{Cov}(\boldsymbol{x})$	The covariance matrix of a random vector $\boldsymbol{x}$
$\mathcal{N}(\boldsymbol{\mu}, \Sigma)$	The normal (Gaussian) distribution with mean $\boldsymbol{\mu}$ and covariance Σ

6. Graph theory

$\mathcal{G} = \{\mathcal{V}, \mathcal{E}\}$	An (undirected) graph consisting of a set of vertices $\mathcal{V}$ and (weighted) edges $\mathcal{E}$
$\mathcal{V} = \{1, \ldots, N\}$	The set of N vertices of a graph $\mathcal{G}$, where in this book a vertex typically represents one data point
$\mathcal{E} = \{(i, j)\}$	The set of (weighted) edges of a graph $\mathcal{G}$, where in this book an edge typically represents two data points belonging to the same cluster
$w_{ij} \in \mathbb{R}_+$	A weight associated with the edge $(i, j) \in \mathcal{E}$, where in this book the weight value represents the affinity between two data points

W	Weight matrix of a graph $\mathcal{G}$, with w_{ij} as its entries
$\mathcal{D}$	Degree matrix of a graph $\mathcal{G}$, a diagonal matrix whose diagonal entries are the degree $d_{ii} = \sum_j w_{ij}$ of each vertex $i \in \mathcal{V}$
$\mathcal{L}$	Laplacian matrix of a graph $\mathcal{G}$, defined as $\mathcal{L} = \mathcal{D} - W$

7. Image formation

$(R_i, \boldsymbol{T}_i)$	Relative motion (rotation and translation) from the ith camera frame to the (default) first camera frame: $X_i = R_i X + \boldsymbol{T}_i$
$(R_{ij}, \boldsymbol{T}_{ij})$	Relative motion (rotation and translation) from the ith camera frame to the jth camera frame: $X_i = R_{ij} X_j + \boldsymbol{T}_{ij}$
$H \in \mathbb{R}^{3\times 3}$	The homography matrix, and it usually represents an element in the general linear group $GL(3)$

Chapter 1
Introduction

The sciences do not try to explain, they hardly even try to interpret, they mainly make models. By a model is meant a mathematical construct which, with the addition of certain verbal interpretations, describes observed phenomena. The justification of such a mathematical construct is solely and precisely that it is expected to work.

—John von Neumann

The primary goal of this book is to study theory and methods for modeling high-dimensional data with one or more low-dimensional subspaces or manifolds. To a large extent, the methods presented in this book aim to generalize the classical principal component analysis (PCA) method (Jolliffe 1986, 2002) to address two major challenges presented by current applications.

One challenge is to generalize the classical PCA method to data with significant amounts of missing entries, errors, outliers, or even a certain level of nonlinearity. Since the very beginning of PCA nearly a century ago (Pearson 1901; Hotelling 1933), researchers have been aware of PCA's vulnerability to missing data and corruption. Strictly speaking, estimating a subspace from incomplete or corrupted data is an inherently difficult problem, which is generally NP-hard. Nevertheless, due to the practical importance of this problem, many extensions to PCA have been proposed throughout the years in different practical domains to handle imperfect data, even though many of these methods have been largely heuristic, greedy, or even ad hoc. Recent advances in high-dimensional statistics and convex optimization have begun to provide provably correct[1] and efficient methods for finding the optimal subspace from highly incomplete or corrupted data.

Another challenge is to generalize the classical PCA method to a data set that consists of multiple subsets, each subset belonging to a different subspace. In various contexts, such a data set is referred to as "mixed," "multimodal," "piecewise linear," "heterogeneous," or "hybrid." In this book, to be more consistent, we will

[1] Under fairly broad conditions that we will elaborate in this book.

R. Vidal et al., *Generalized Principal Component Analysis*, Interdisciplinary Applied Mathematics 40, DOI 10.1007/978-0-387-87811-9_1

typically refer to such data as "mixed data" and the model used to fit the data as a "mixture model." However, we will not completely exclude other names that have been conventionally used in different application domains.[2] A mixture model typically consists of multiple constituent primitive models (say subspaces). Modeling mixed data with a mixture model implies partitioning the data into multiple (mainly disjoint) subsets and fitting each subset with one of the constituent models. In the literature, the words "cluster," "group," "partition," "segment," and "decompose" are often used interchangeably. In this book, for consistency, we will primarily use the word "cluster," but again, in special application domains, we will use words that have been conventionally used in the literature. For instance, for images, we typically say "image segmentation."

In this chapter, we give a brief introduction to some fundamental concepts and problems involved in modeling incomplete, corrupted, or mixed data. First, we discuss some basic concepts associated with data modeling in general, such as the choice of model class. Next, we motivate the problem of modeling mixed data with mixture models using several examples from computer vision, image processing, pattern recognition, system identification, and system biology. We then give a brief account of geometric, statistical, and algebraic methods for estimating mixture models from data, with an emphasis on the particular case of modeling data with a *union of subspaces*,[3] also known as hybrid linear models in systems theory. We finish the chapter with some discussion about how noise and outliers make the estimation problem extremely challenging, especially when the complexity of the model to be estimated is not known.

1.1 Modeling Data with a Parametric Model

In science and engineering, one is frequently called upon to infer (or learn) a quantitative model M for a given set of sample points $\mathcal{X} = \{\boldsymbol{x}_1, \boldsymbol{x}_2, \ldots, \boldsymbol{x}_N\} \subset \mathbb{R}^D$. For instance, Figure 1.1 shows a simple example in which one is given a set of four sample points in a two-dimensional plane. Obviously, these points can be fit perfectly by a (one-dimensional) straight line L. The line can then be called a "model" for the given points. The reason for inferring such a model is that it serves many useful purposes. On the one hand, the model can reveal information encoded in the data or underlying mechanisms from which the data were generated. In addition, it can simplify the representation of the given data set and help

[2]In the statistical learning literature, the most commonly used term is "mixture model." In systems theory, the typical term is "hybrid model." In algebraic geometry, for the case of subspaces, the typical term is a "subspace arrangement."

[3]In this book, we will use interchangeably "mixture," "collection," "union," and "arrangement" of subspaces or models. But be aware that in the case of subspaces, the formal terminology in algebraic geometry is a "subspace arrangement."

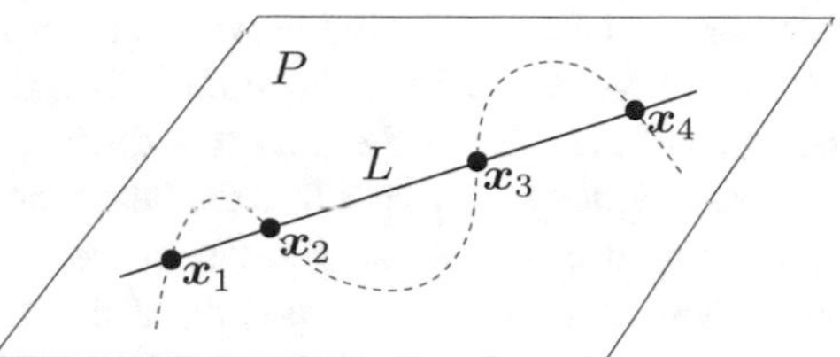

Fig. 1.1 Four sample points on a plane are fit by a straight line. However, they can also be fit by many other smooth curves, for example the one indicated by the dashed curve.

predict future samples. In the case of the four points shown in Figure 1.1, the line model gives a more compact one-dimensional representation than the original two-dimensional plane P. It also suggests that any new point (if generated with a similar mechanism as the existing points) will likely fall on the same line.

1.1.1 The Choice of a Model Class

A first important consideration to keep in mind is that inferring the "correct" model for a given data set is an elusive, if not impossible, task. The fundamental difficulty is that if we are not specific about what we mean by a "correct" model, there could easily be many different models that fit the given data set "equally well." For instance, in the example shown in Figure 1.1, any smooth curve that passes through the sample points would seem to be as valid a model as the straight line. Furthermore, if there were noise in the given sample points, then any curve, including the line, passing through the points exactly would unlikely be the "true model."

The question now is this: in what sense can we say that a model is correct or optimal for a given data set? To make the model inference problem well posed, i.e., to guarantee that there is a unique optimal model for the given data, we need to impose additional assumptions or restrictions on the class of models considered. To this end, we should not be looking for just any model that can describe the data. Instead, we should look for a model M^* that is the *best* among a *restricted* class of models $\mathcal{M}$.[4] In addition, to make the model inference problem computationally tractable, we need to specify how restricted the class of models needs to be. A common strategy, known as the principle of Occam's razor,[5] is to try to get away

---

[4]Or equivalently, we may impose a nonuniform prior distribution over all models.

[5]Occam's (or Ockham's) razor is a principle attributed to the fourteenth-century logician and Franciscan friar William of Occam: "*Pluralitas non est ponenda sine neccesitate,*" which translates literally as "*entities should not be multiplied unnecessarily.*" In science, this principle is often interpreted thus: "*when you have two competing theories that make exactly the same predictions, the simpler one is better.*"

with the simplest possible class of models that is *just necessary* to describe the data or solve the problem at hand. More precisely, the model class should be rich enough to contain at least one model that can fit the data to a desired accuracy and yet be restricted enough that it is relatively simple to find the best model for the given data.

Thus, in engineering practice, the most popular strategy is to start from the simplest class of models and increase the complexity of the models only when the simpler models become inadequate. For instance, to fit a set of sample points, one may first try the simplest class of models, namely linear models, followed by the class of hybrid (piecewise) linear models (subspaces), and then followed by the class of (piecewise) nonlinear models (submanifolds). One of the goals of this book is to demonstrate that among them, piecewise linear models can already achieve an excellent balance between expressiveness and simplicity for many important practical data sets and problems.

1.1.2 Statistical Models versus Geometric Models

There are essentially two main categories of models and approaches for modeling a data set. Methods of the first category model the data as random samples from a probability distribution and try to learn this distribution from the data. We call such models *statistical models*. Models of the second category model the overall geometric shape of the data set with deterministic models such as subspaces, smooth manifolds, or topological spaces.[6] We call such models *geometric models*.

Statistical Learning

In the statistical paradigm, one typically assumes that each data point $\boldsymbol{x}_j$ in the data set $\mathcal{X}$ is drawn independently from a common probability distribution $p(\boldsymbol{x})$. Such a probability distribution gives a *generative* description of the samples and can be used to generate new samples or predict the outcome of new observations. Within this context, the task of learning a model from the data becomes one of inferring the most likely probability distribution within a family of distributions of interest (for example, the Gaussian distributions). Normally, the family of distributions is parameterized and denoted by $\mathcal{M} \doteq \{p(\boldsymbol{x} \mid \theta) : \theta \in \Theta\}$, where $p(\boldsymbol{x} \mid \theta)$ is a probability density function parameterized by $\theta \in \Theta$, and Θ is the space of parameters. Consequently, one popular criterion for choosing a statistical model $p(\boldsymbol{x} \mid \theta^*)$ is the maximum likelihood (ML) estimate given by[7]

[6]Roughly speaking, a smooth manifold is a special topological space that is locally homeomorphic to a Euclidean space and has the same dimension everywhere. A general topological space may have singularities and consist of components of different dimensions.

[7]If the true distribution from which the data are drawn is $q(\boldsymbol{x})$, then the maximum likelihood estimate $p(\boldsymbol{x} \mid \theta^*)$ minimizes the Kullback–Leibler (KL) divergence $KL(q\|p) = \int q(\boldsymbol{x}) \log \frac{q(\boldsymbol{x})}{p(\boldsymbol{x})} \, d\boldsymbol{x}$ among the given class of distributions (see Appendix B.)

$$\theta^*_{ML} \doteq \arg\max_{\theta \in \Theta} \prod_{j=1}^{N} p(\boldsymbol{x}_j \mid \theta). \tag{1.1}$$

If a prior distribution (density) $p(\theta)$ of the parameter θ is also given, then, following the Bayesian rule, the maximum a posteriori (MAP) estimate is given by

$$\theta^*_{MAP} \doteq \arg\max_{\theta \in \Theta} \prod_{j=1}^{N} p(\boldsymbol{x}_j \mid \theta) p(\theta). \tag{1.2}$$

Many effective methods and algorithms have been developed in the statistics and machine learning literature to find the optimal distribution $p(\boldsymbol{x} \mid \theta^*)$ or a good approximation of it if the exact solution is computationally prohibitive. A brief review is given in Appendix B.

Geometric Modeling
In many practical scenarios, we may not know a priori the statistical process that generated the data. Also, the amount of data may not be sufficient to determine a unique optimal distribution within a large class of possible distributions. In such cases, we may exploit the fact that the data points are often subject to topological or geometric constraints, e.g., they must lie in a low-dimensional subspace or submanifold. This implies that the data can be represented only with a probability distribution that is close to being singular.[8]

In general, it is very ineffective to learn such a singular or approximately singular distribution via statistical means (Vapnik 1995). Thus, an alternative data-modeling paradigm is to learn the overall geometric shape of the given data set directly. Typical methods include fitting one or more geometric primitives such as points[9], lines, subspaces, and submanifolds to the data set. For instance, the classical principal component analysis (PCA) method is essentially equivalent to fitting a low-dimensional subspace, say $S \doteq \text{span}\{\boldsymbol{u}_1, \boldsymbol{u}_2, \dots, \boldsymbol{u}_d\}$, to a data set in a high-dimensional space, say $S \subset \mathbb{R}^D$. That is, we try to represent the data points as

$$\boldsymbol{x}_j = y_{1j}\boldsymbol{u}_1 + y_{2j}\boldsymbol{u}_2 + \cdots + y_{dj}\boldsymbol{u}_d + \varepsilon_j, \quad \forall\, \boldsymbol{x}_j \in \mathcal{X}, \tag{1.3}$$

where $d < D$, $y_{ij} \in \mathbb{R}$ with $i = 1, \dots d$, and $\boldsymbol{u}_1, \boldsymbol{u}_2, \dots, \boldsymbol{u}_d \in \mathbb{R}^D$ are unknown model parameters that need to be determined, playing the role of the parameters θ in the foregoing statistical model. The line model in Figure 1.1 can be viewed as an example of applying PCA to the four points in the plane. In the above equation, the term $\varepsilon_j \in \mathbb{R}^D$ denotes the error between the jth sample and the model. As we will

[8] Singular distributions are probability distributions concentrated on a set of Lebesgue measure zero. Such distributions are not absolutely continuous with respect to the Lebesgue measure. The Cantor distribution is one example of a singular distribution.

[9] As the cluster centers.

see in Chapter 2, PCA finds a set of model parameters $\{\boldsymbol{u}_i\}$ and $\{y_{ij}\}$ that minimize the error $\sum_j \|\varepsilon_j\|^2$ in (1.3). When the errors ε_j are independent samples drawn from a zero-mean Gaussian distribution, the geometric formulation of PCA is equivalent to the classical statistical formulation (Jolliffe 1986, 2002). In general, a geometric model gives an intuitive description of the samples, and it is often preferred to a statistical one as a "first-cut" description of the given data set. Its main purpose is to capture global geometric, topological, or algebraic characteristics of the data set, such as the number of clusters and their dimensions. Geometric models can also provide more compact representations of the original data set, making them useful for data compression and dimensionality reduction.

As two competing data-modeling paradigms, the statistical modeling techniques in general are more effective in the high-noise regime when the generating distribution is nonsingular, while the geometric techniques are more effective in the low-noise regime when the underlying geometric space is (at least locally) smooth. The two paradigms thus complement each other in many ways. On the one hand, once the overall geometric shape, the clusters, and their dimensions are obtained from geometric modeling, one can choose the class of probability distributions more properly for further statistical inference. On the other hand, since samples are often corrupted by noise and sometimes contaminated by outliers, in order to robustly estimate the optimal geometric model, one often resorts to statistical techniques. Thus, this book will provide thorough coverage of both geometric and statistical modeling techniques.

1.2 Modeling Mixed Data with a Mixture Model

As we alluded to earlier, many data sets $\mathcal{X}$ cannot be modeled well by a single primitive model M in a pre-chosen or preferred model class $\mathcal{M}$. Nevertheless, it is often the case that if we *group* such a data set $\mathcal{X}$ into multiple disjoint subsets,

$$\mathcal{X} = \mathcal{X}_1 \cup \mathcal{X}_2 \cup \cdots \cup \mathcal{X}_n, \quad \text{with } \mathcal{X}_l \cap \mathcal{X}_m = \emptyset, \text{ for } l \neq m, \tag{1.4}$$

then each subset $\mathcal{X}_i$ can be modeled sufficiently well by a model in the chosen model class:

$$M_i^* = \arg\min_{M \in \mathcal{M}} \ \text{Error}(\mathcal{X}_i, M), \quad i = 1, 2, \ldots, n, \tag{1.5}$$

where $\text{Error}(\mathcal{X}_i, M)$ represents some measure of the error incurred by using the model M to fit the data set $\mathcal{X}_i$. Each model M_i^* is called a *primitive* or a *component* model. Precisely in this sense, we call the data set $\mathcal{X}$ *mixed* (with respect to the chosen model class $\mathcal{M}$) and call the collection of primitive models $\{M_i^*\}_{i=1}^n$ a *mixture model* for $\mathcal{X}$. For instance, suppose we are given a set of sample points as shown in Figure 1.2. These points obviously cannot be fit well by any single

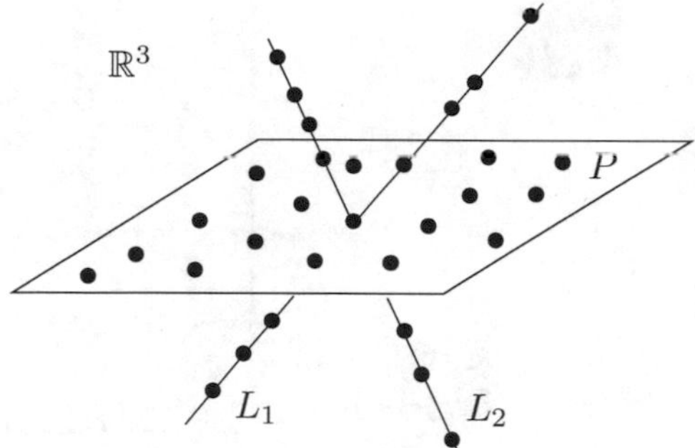

Fig. 1.2 A set of sample points in $\mathbb{R}^3$ are well fit by a mixture model with two straight lines and a plane.

line, plane, or smooth surface in $\mathbb{R}^3$; however, once they are grouped into three subsets, each subset can be fit well by a line or a plane. Note that in this example, the topology of the data is "hybrid": two of the subspaces are of dimension one, and the other is of dimension two.

1.2.1 Examples of Mixed Data Modeling

The problem of modeling mixed data is quite representative of many data sets that one often encounters in practical applications. To further motivate the importance of modeling mixed data, we give below a few real-world problems that arise in image processing and computer vision. Most of these problems will be revisited later in this book, and more detailed and principled solutions will be given.

Face Clustering under Varying Illumination
The first example arises in the context of image-based *face clustering*. Given a collection of unlabeled images $\{I_j\}_{j=1}^N$ of several different faces taken under varying illumination, we would like to cluster the images corresponding to the face of the same person. For a Lambertian object,[10] it has been shown that the set of all images taken under all lighting conditions forms a cone in the image space, which can be well approximated by a low-dimensional subspace called the "illumination subspace" (Belhumeur and Kriegman 1998; Basri and Jacobs 2003).[11] For example, if I_j is the jth image of a face and d is the dimension of the illumination subspace associated with that face, then there exists a *mean face* $\boldsymbol{\mu}$ and d *eigenfaces* $\boldsymbol{u}_1, \boldsymbol{u}_2, \ldots, \boldsymbol{u}_d$ such that $I_j \approx \boldsymbol{\mu} + \boldsymbol{u}_1 y_{1j} + \boldsymbol{u}_2 y_{2j} + \cdots + \boldsymbol{u}_d y_{dj}$. Now, since the images

[10]An object is called Lambertian if its apparent brightness is the same from any viewpoint.

[11]Depending on the illumination model, the illumination space can be approximately three- or nine-dimensional.

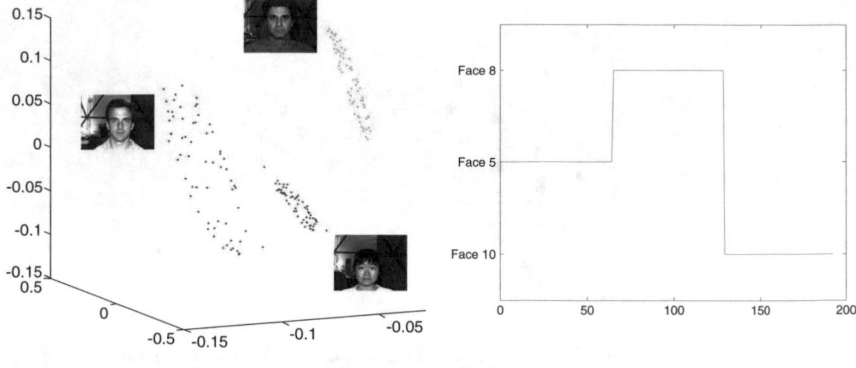

(a) Several images of three faces projected onto the first three principal components

(b) Classification of the images according to the three different faces

Fig. 1.3 Clustering a subset of the Yale Face Database B consisting of 64 frontal views under varying lighting conditions for subjects 5, 8 and 10.

of different faces will live in different "illumination subspaces," we can cluster the collection of images by estimating a basis for each one of those subspaces. As we will see later, this is a special case of the *subspace clustering problem* addressed in Part II of this book. In the example shown in Figure 1.3, we use a subset of the Yale Face Database B consisting of $n = 64 \times 3$ frontal views of three faces (subjects 5, 8 and 10) under 64 varying lighting conditions. For computational efficiency, we first down-sample each image to a size of 30×40 pixels. We then project the data onto their first three principal components using PCA, as shown in Figure 1.3(a).[12] By modeling the projected data with a mixture model of linear subspaces in $\mathbb{R}^3$, we obtain three affine subspaces of dimension 2, 1, and 1, respectively. Despite the series of down-sampling and projection, the subspaces lead to a perfect clustering of the face images, as shown in Figure 1.3(b).

Since face images are rather intuitive real data and have good subspace structures, we will use them to produce many running examples in the book to help demonstrate certain abstract concepts or to evaluate certain methods.

Image Representation and Segmentation

The next set of examples arises in the context of image processing, especially *image representation and segmentation*. It is commonplace that in an image, pixels in different regions have significantly different local color/texture profiles (normally in an $N \times N$ window around a pixel). Conventional image representation/compression schemes, such as JPEG and JPEG2000, often ignore such differences and use the same linear filters or bases (for example the Fourier transform, discrete cosine transform, wavelets, or curvelets) to represent the entire set of local profiles. For example, if I_j is the jth image patch and $\boldsymbol{u}_1, \boldsymbol{u}_2, \ldots, \boldsymbol{u}_d$ are the basis elements, then

[12] The legitimacy of the projection process will be addressed in Chapter 5.

(a) Input image (b) First segment (c) Second segment (d) Third segment

Fig. 1.4 Image segmentation based on fitting different linear subspaces (and bases) to regions of different textures. The three segments (or subspaces) correspond to the ground, the clouds, and the sky.

all image patches are approximated as a linear combination of these basis elements as $I_j \approx \boldsymbol{u}_1 y_{1j} + \boldsymbol{u}_2 y_{2j} + \cdots + \boldsymbol{u}_d y_{dj}$. Nevertheless, modeling the set of local profiles as a mixed data set allows us to segment the image into different regions and represent each region differently. Each region consists of only those pixels whose local profiles span the same low-dimensional linear subspace.[13] Specifically, if the jth image patch belongs to the ith region, then $I_j \approx \boldsymbol{u}_1^i y_{1j} + \boldsymbol{u}_2^i y_{2j} + \cdots + \boldsymbol{u}_d^i y_{dj}$, where the subspace basis $\{\boldsymbol{u}_j^i\}_{j=1}^{d_i}$ can be viewed as a bank of adaptive filters for the ith image region. Figure 1.4 shows regions of an image segmented by such a mixed representation. The obtained subspaces (and their bases) normally provide a very compact representation of the image, often more compact than any of the aforementioned fixed-basis schemes.[14] Hence they are very useful for applications such as image compression, classification, and retrieval. More details on the application of subspace clustering to image representation and segmentation can be found in Chapters 9 and 10, respectively.

Segmentation of Moving Objects in Video

The next example is the *motion segmentation* problem that arises in the field of computer vision: given a sequence (or sometimes only a pair) of images of multiple moving objects in a scene, how does one segment the images so that each segment corresponds to only one moving object? This is a very important problem in applications such as motion capture, vision-based navigation, target tracking, and surveillance.

One way of solving this problem is to extract a set of feature points in the first image and track these points through the video sequence. As a result, one obtains a set of point trajectories such that each trajectory corresponds to one of the moving objects in the video. It is well known from the computer vision literature (Hartley and Zisserman 2004; Ma et al. 2003) that feature points from

[13]In contrast to the previous face example, there is no rigorous mathematical justification for why local profiles from a region of similar texture must span a low-dimensional linear subspace. However, there is strong empirical evidence that a linear subspace normally gives a very good approximation.

[14]That is, the number d_i of basis elements needed to represent the ith region is typically much smaller than the number d of basis elements needed to represent the whole image.

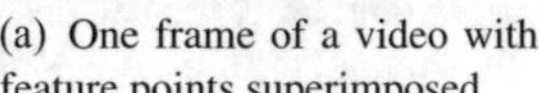

(a) One frame of a video with feature points superimposed

(b) One frame of a video with feature points superimposed

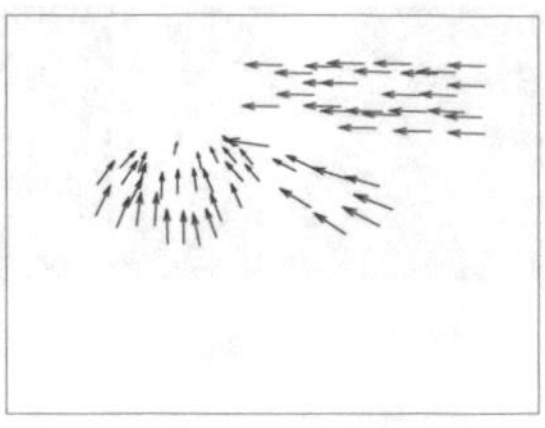

(c) Displacement of the feature points between the two frames

Fig. 1.5 Clustering the relative motion of a collection of feature points between two views of a scene where both the camera and the car are moving, hence there are two different 3-dimensional motions in the scene.

two corresponding views of the same object are related by either linear or quadratic constraints depending on the type of motions and camera projection models (see Chapter 11). Therefore, mathematically, the problem of motion segmentation is equivalent to clustering point trajectories into different linear subspaces (in a certain high-dimensional feature space). Figure 1.5 shows two frames of a video sequence of a moving car. Feature points on both the car and the background are detected and tracked through the sequence. These points undergo different three-dimensional (3D) motions in space as both the car and the camera move; in this sequence, the camera is mainly panning and zooming. The image on the left shows the starting positions of the car and the camera view, and the image in the middle shows the final positions. The image on the right shows the displacement of these feature points from the first to the second image as well as the segmentation of these displacement vectors using a mixture model of two linear subspaces. We will describe in detail the motion segmentation method used to achieve this result in Chapter 11.

Temporal Video Segmentation and Event Detection
Another example arises in the context of *detecting events from video sequences*. A typical video sequence contains multiple activities or events separated in time. For instance, Figure 1.6(a) shows a news sequence in which the host is interviewing a guest and the camera is switching between the frames containing the host, the guest, or both the host and the guest. The problem is to separate the video sequence into subsequences, so that each subsequence corresponds to one of the three events. For this purpose, we assume that all the frames associated with the same event live in a low-dimensional subspace of the space spanned by all the images in the video, and that different events correspond to different subspaces. The problem of segmenting the video into multiple events is then equivalent to a subspace clustering problem. Since the image data live in a very high-dimensional space ($\sim 10^5$, the number of pixels), we first project the image data onto a low-dimensional subspace (~ 10) using principal component analysis (PCA) and then fit a mixture model of multiple subspaces to the projected data to identify the different events. Figure 1.6 shows the segmentation results for two video sequences. In both cases, a perfect segmentation is obtained. We will describe in detail the segmentation method used to achieve these results in Chapter 11.

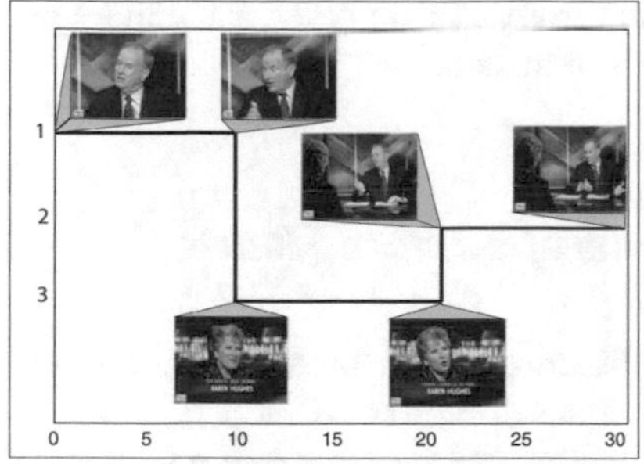

(a) Thirty frames of a video sequence of a television show clustered into three groups: host, guest, and both of them

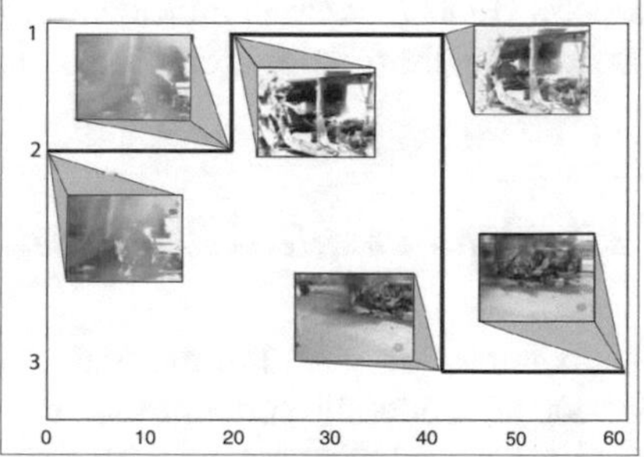

(b) Sixty frames of a news video sequence clustered into three groups: car with a burning wheel, burnt car with people, and burning car

Fig. 1.6 Clustering frames of a news video sequence into groups of scenes by modeling each group with a linear subspace.

Identification of Hybrid Dynamical Models

The last, somewhat more abstract, example arises in the context of modeling time series data with linear dynamical models. A popular dynamical model used to analyze a time series $\{y_t \in \mathbb{R}\}_{t\in\mathbb{Z}}$ is the linear autoregressive (AR) model

$$y_t = a_1 y_{t-1} + a_2 y_{t-2} + \cdots + a_n y_{t-n} + \varepsilon_t, \quad \forall t \in \mathbb{Z}, \tag{1.6}$$

where $\{a_i\}$ are the parameters of the AR model and $\varepsilon_t \in \mathbb{R}$ represents the modeling error or noise, which is often assumed to be a white-noise random process. In order to capture more complex dynamics in the data, one can assume that y_t is the output of a piecewise AR model, where the output at each time instant is drawn from one out of finitely many AR models. Notice that at each time instant, the vector $\boldsymbol{x}_t = [y_t, y_{t-1}, \ldots, y_{t-n}]^\top$ lies on an n-dimensional hyperplane in $\mathbb{R}^{n+1}$. Therefore, the vectors $\boldsymbol{x}_t$ for all t lie in a collection of hyperplanes. As a consequence, the identification of the parameters of a piecewise AR model can be viewed as another subspace clustering problem. We will discuss this and more general classes of hybrid dynamical models, together with algorithms for identifying the parameters of such models, in Chapter 12.

As we can see from the foregoing examples, there are many practical applications whereby one can rigorously show that a given data set belongs to a collection of linear or quadratic surfaces of the same or possibly different dimensions (motion segmentation example). In many other cases, one can use piecewise linear structures to approximate the data set and obtain a more compact and meaningful geometric representation of the data, including segments, dimensions, and bases (image representation, face classification, and video segmentation examples). As we will see in Part II of this book, subspace (or surface) clustering is a natural abstraction of all these problems and thus merits systematic investigation. From a practical standpoint, the analysis of such problems has led to many general and powerful modeling tools that are applicable to a wide variety of data types, including image, video, audio, time series, genomic, and proteomic data.

1.2.2 Mathematical Representations of Mixture Models

The examples presented in the previous subsection argue forcefully for the development of modeling and estimation techniques for mixture models. Obviously, whether the model associated with a given data set is mixed depends on the class of primitive models considered. In this book, the primitives are normally chosen to be simple classes of geometric models or probabilistic distributions.

For instance, one may choose the primitive models to be linear subspaces. Then one can use an arrangement of linear subspaces $\{S_i\}_{i=1}^n \subset \mathbb{R}^D$,

$$Z \doteq S_1 \cup S_2 \cup \cdots \cup S_n, \tag{1.7}$$

also called a piecewise linear model, to approximate many nonlinear manifolds or piecewise smooth topological spaces. This is the standard model considered in geometric approaches to generalized principal component analysis (GPCA), which will be studied in Part II of this book.

The statistical counterpart to the geometric model in (1.7) is to assume instead that the sample points are drawn independently from a mixture of (near singular) Gaussian distributions $\{p_{\theta_i}(\boldsymbol{x})\}_{i=1}^n$, where $\boldsymbol{x} \in \mathbb{R}^D$ but each distribution has mass concentrated near a subspace. The overall probability density function can be expressed as a sum:

$$q_\theta(\boldsymbol{x}) \doteq \pi_1 p_{\theta_1}(\boldsymbol{x}) + \pi_2 p_{\theta_2}(\boldsymbol{x}) + \cdots + \pi_n p_{\theta_n}(\boldsymbol{x}), \tag{1.8}$$

where $\theta = (\theta_1, \ldots, \theta_n, \pi_1, \ldots, \pi_n)$ are the model parameters and $\pi_i > 0$ are mixing weights with $\pi_1 + \pi_2 + \cdots + \pi_n = 1$. This is the typical model studied in mixtures of probabilistic principal component analysis (PPCA) (Tipping and Bishop 1999a), where each component distribution $p_{\theta_i}(\boldsymbol{x})$ is a nearly degenerate Gaussian distribution. A classical way of estimating such a mixture model is the expectation maximization (EM) algorithm, where the membership of each sample is represented as a hidden random variable. Appendix B reviews the general EM method, and Chapter 6 shows how to apply it to the case of multiple subspaces.

In the special case that there is only one subspace or one component distribution (i.e., $n = 1$), the model reduces to the classical (probabilistic) PCA, and we will see that the geometric and statistical formulations are equivalent in the sense that they both give very much the same solution (see Chapter 2). However, in the case of incomplete or corrupted data, or in the general case of a mixture of multiple components, the two formulations can be very different, and their optimal solutions need to be found by very different techniques. In this book, we will study and clarify the similarities and differences between these geometric models and statistical models in Chapters 5 and 6.

Difficulties with Conventional Data Modeling Methods
The reader may have been wondering why we should not simply enlarge the class of primitive models to include mixture models so that we can deal with them by the conventional single-model estimation paradigm to be reviewed in Appendix B. While this is an appealing idea in principle, conventional model estimation methods are applicable mostly to data sampled from smooth manifolds and/or nonsingular distributions. As shown by the examples above, many practical data sets are better modeled by nonsmooth manifolds or singular distributions. As we will see, the underlying topological space of a mixed data set may contain multiple (likely intersecting) manifolds of different dimensions, and conventional manifold learning techniques such as (Tenenbaum et al. 2000; Roweis and Saul 2000) do not apply to such mixtures of smooth manifolds. Also, if one tries to model mixed data sets with a single probability distribution, then the distribution will typically have singularities, and conventional statistical techniques become rather tricky or ineffective in inferring such singular distributions (Vapnik 1995).

An alternative approach to modeling mixed data is to first cluster the data set into coherent subsets and then model each subset using classical single-model methods. This is a popular approach adopted by many practitioners in the field. However, a fundamental difficulty with this approach is that without knowing which subset of sample points belongs to which constituent model, there is seemingly a "chicken-and-egg" relationship between data clustering and model estimation: If the partition of the data were known, one could fit a model to each subset of samples using classical model estimation techniques; and conversely, if the constituent models were known, one could easily find the subset of samples that best fits each model. This relationship has been the rationale behind many *alternating minimization* techniques for mixed data, such as the well-known EM and K-means algorithms widely used in machine learning (see Appendix B) as well as their counterparts for multiple subspaces (see Chapter 6). These alternating methods, however, share several drawbacks:

- The iteration needs to start with a good initial guess of the solution; otherwise, the iteration is likely to converge to a local minimum.
- Without knowing a priori the number of models and the dimension of each model, the algorithm may diverge if it starts with a wrong guess for these key model parameters.
- There are cases or applications in which we may care about only the joint model rather than the individual clusters; or at the opposite end of the spectrum, we may care more about the clusters of the data than their parametric models. In such cases, it might be more efficient to look for a direct solution to what is needed.

In this book, we will see a few representative approaches that lead to effective and efficient solutions without resorting to such alternating schemes. These new methods are global, noniterative, and in many cases provably optimal for the clustering and modeling problem.

Mixture Models as Algebraic Sets

In this book, in addition to manifolds or distributions, we will also view mixed data as algebraic sets, study their algebraic properties, and seek solutions via algebraic-geometric means. Roughly speaking, an algebraic set is the common zero-level set of a family of polynomial equations (see Appendix C). To see the merit of such a representation for a mixture model, let us consider a simple example in which the data corresponding to the ith constituent model belong to a hyperplane of $\mathbb{R}^D$ of the form

$$Z_i = \{\boldsymbol{x} : \boldsymbol{b}_i^\top \boldsymbol{x} = 0\} \quad \text{for} \quad i = 1, 2, \ldots, n. \tag{1.9}$$

In other words, the set Z_i is the zero-level set of the polynomial $p_i(\boldsymbol{x}) = \boldsymbol{b}_i^\top \boldsymbol{x}$. Therefore, we can interpret a mixed data set drawn from a union of n hyperplanes as the zero-level set of the polynomial $p(\boldsymbol{x}) = (\boldsymbol{b}_1^\top \boldsymbol{x})(\boldsymbol{b}_2^\top \boldsymbol{x}) \cdots (\boldsymbol{b}_n^\top \boldsymbol{x})$, i.e.,

$$Z \doteq Z_1 \cup Z_2 \cup \cdots \cup Z_n = \big\{\boldsymbol{x} : p_1(\boldsymbol{x}) p_2(\boldsymbol{x}) \cdots p_n(\boldsymbol{x}) = 0\big\}. \tag{1.10}$$

This polynomial can be determined from a number of (random) sample points on the algebraic set $\mathcal{X} \doteq \{\boldsymbol{x}_j \in Z\}$ using techniques analogous to those used for fitting a circle to three points in $\mathbb{R}^2$. Given the polynomial $p(\boldsymbol{x})$, we can use polynomial factorization techniques to obtain the factors $p_i(\boldsymbol{x}) = \boldsymbol{b}_i^\top \boldsymbol{x}$, and hence the parameters for each constituent model, namely the vector $\boldsymbol{b}_i$ normal to the hyperplane.

This simple example of modeling the data with a union of hyperplanes can be immediately generalized to modeling the data with a union of algebraic varieties.[15] More specifically, let us suppose that the data corresponding to the ith constituent model can be described as the zero-level set of some polynomials in a prime ideal $\mathfrak{p}_i$,[16]

$$Z_i \doteq \{\boldsymbol{x} : p(\boldsymbol{x}) = 0, p \in \mathfrak{p}_i\} \subset \mathbb{R}^D, \quad i = 1, 2, \ldots, n. \tag{1.11}$$

The (mixed) data from a union of n such models then belong to an algebraic set:[17]

$$\begin{aligned} Z &\doteq Z_1 \cup Z_2 \cup \cdots \cup Z_n \\ &= \big\{\boldsymbol{x} : p_1(\boldsymbol{x}) p_2(\boldsymbol{x}) \cdots p_n(\boldsymbol{x}) = 0, \quad \forall p_i \in \mathfrak{p}_i, \quad i = 1, 2, \ldots, n\big\}. \end{aligned} \tag{1.12}$$

From a number of (random) sample points on the algebraic set $\mathcal{X} \doteq \{\boldsymbol{x}_j \in Z\}$, one can determine the (radical) ideal of polynomials that vanish on the set Z:[18]

$$\mathcal{X} \quad \rightarrow \quad \mathfrak{q}(Z) \doteq \big\{q : q(\boldsymbol{x}_j) = 0, \quad \forall \boldsymbol{x}_j \in Z\big\}. \tag{1.13}$$

[15] An algebraic variety is an irreducible algebraic set. An algebraic set is called irreducible if it cannot be written as the union of two proper algebraic subsets. A subspace is one such example.

[16] A prime ideal is an ideal that cannot be decomposed further as the intersection of two other ideals (see Appendix C). The zero-level set of a prime ideal is an irreducible algebraic set, i.e., an algebraic variety.

[17] Notice the correspondence between a "union" of algebraic varieties and the "multiplication" of the polynomials associated with the varieties.

[18] According to Hilbert's Nullstellensatz (see Appendix C), there is a one-to-one correspondence between algebraic sets and radical ideals (Eisenbud 1996).

While the ideal $\mathfrak{q}$ is no longer a prime ideal, once $\mathfrak{q}$ has been obtained, the constituent models $\mathfrak{p}_i$ (or Z_i) can be subsequently retrieved by *decomposing* the ideal $\mathfrak{q}$ into irreducible prime ideals via algebraic means, i.e.,[19]

$$\mathfrak{q} \quad \rightarrow \quad \mathfrak{q} = \mathfrak{p}_1 \cap \mathfrak{p}_2 \cap \cdots \cap \mathfrak{p}_n. \tag{1.14}$$

Clearly, the above representation establishes a natural correspondence between terminologies developed in algebraic geometry and the heuristic languages used in modeling mixed data: the constituent models become algebraic varieties, the mixture model becomes an algebraic set, the mixed data become samples from an algebraic set, and the estimation of mixture models becomes the estimation and decomposition of a radical ideal. Although this nomenclature may seem abstract and challenging at first, we will see in Chapter 5 how to make this very concrete for the case of a subspace arrangement.

Despite its purely algebraic nature, the above algebraic representation is closely related to and complements well the two aforementioned geometric and statistical data modeling paradigms.

From the geometric viewpoint, unlike a smooth manifold M that sometimes can be implicitly represented as the level set of a single function, an algebraic set Z is the zero-level set of a *family* of polynomials. As a result, an algebraic set Z allows components with different dimensions as well as singularities that the zero-level set of a single smooth function cannot describe.

From the statistical viewpoint, one can also view the irreducible components $\{Z_i\}$ of Z as the "means" of a collection of probability distributions $\{p_i(\cdot)\}$ and the overall set Z as the "skeleton" of their mixture distribution $q(\cdot)$. For instance, a piecewise linear structure can be viewed as the skeleton of a mixture of Gaussian distributions (see Figure 1.7). Therefore, mixture models represented by algebraic sets can be interpreted as a special class of *generative models* such that the random variables have small variance outside the algebraic sets, but large variance inside.

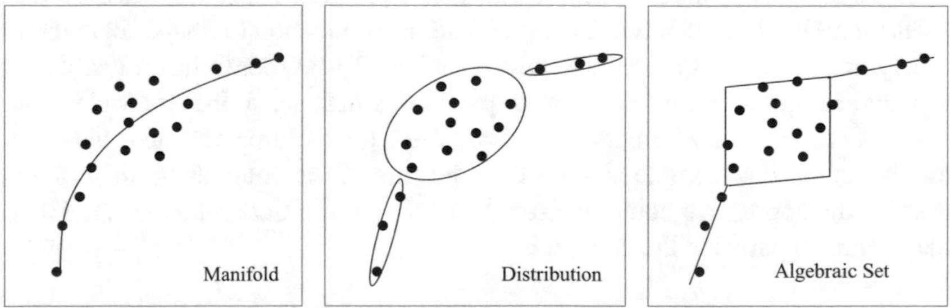

Fig. 1.7 Comparison of three representations of the same data set: a (nonlinear) manifold, a (mixed Gaussian) distribution, or a (piecewise linear) algebraic set.

[19]For the special case in which the ideal is generated by a single polynomial, the decomposition is equivalent to factoring the polynomial into factors.

As we will show in this book, if the primitive models are simple models such as linear subspaces (or quadratic surfaces), then in principle, the problem of segmenting mixed data and estimating a mixture model can be solved *noniteratively* (see Chapter 5). Moreover, the correct number of models and their dimensions can also be correctly determined via purely algebraic means, at least in the noise-free case (see Chapter 5).

1.3 Clustering via Discriminative or Nonparametric Methods

The previous section argued for the importance of identifying a mixture model for clustering mixed data. As a result, we often obtain a parametric (either geometric or statistical) model that best describes how the given sample data are generated. However, there are applications for which there might be no need to obtain parametric and generative models behind the given data. We might be interested only in seeking a more compact representation of the samples themselves as long as certain important information or structure (such as topology) of the data is preserved.

Clustering as a Compression Problem
For instance, for the image segmentation problem, we might be interested only in grouping the pixels into several homogeneous segments, but not necessarily in a generative model that best describes the texture in each segment. Hence, it suffices for our purpose to have a method that directly gives rules (or classifiers) that separate a collection of data points into different segments or clusters. Such methods are often referred to as "discriminative" methods, popular in areas such as object classification and pattern recognition.

However, this does not mean that for discriminative methods one does not need to understand intrinsic structures of the data. In this book, we will see that in order to arrive at an effective discriminative method for classifying subspace-like data, it is very crucial to have precise information about intrinsic geometric properties of the clusters, such as its dimension and volume. The compression-based clustering method described in Chapter 6 and its application to image segmentation in Chapter 10 clearly support this point of view. To be more illustrative, a rather pragmatic reason why we may want to partition a data set $\mathcal{X}$ into multiple subsets $\mathcal{X} = \mathcal{X}_1 \cup \mathcal{X}_2 \cup \cdots \cup \mathcal{X}_n$ might be because the total "volume of space" we need to store the data set as a whole is more than the sum of the volumes of the individual subsets. So, suppose we could measure the volume of a data set as $L(\mathcal{X})$. Then, it makes sense to partition the data set if

$$L(\mathcal{X}) > L(\mathcal{X}_1) + L(\mathcal{X}_2) + \cdots + L(\mathcal{X}_n). \tag{1.15}$$

For the data set shown in Figure 1.2, the whole data set spans a nontrivial volume in 3D space, yet each of the three subsets (on the two lines and in the plane) spans a nearly zero-volume set in three dimensions. In this sense, the data set is separable because it has "compressible" low-dimensional parts.

Clustering as a Graph-Partitioning Problem
In many modern data-driven machine learning tasks, we are very often interested not in each sample data point as a signal defined over space and time, but in a certain high-level semantic label that the signal shares with other similar signals in the same class—say sounds of the same word, or images of the same object. Since the original data could contain a large quantity of irrelevant information or nuisance factors, we often need to find a much more compact representation of the data that extracts and highlights what is relevant but suppresses what is irrelevant.

More formally, we could consider mapping the given sample data $\mathcal{X}$ to another domain (typically of much lower dimension):

$$f: \ \mathcal{X} \subset \mathbb{R}^D \to \mathcal{Y} \subset \mathbb{R}^d, \tag{1.16}$$

$$\boldsymbol{x} \in \mathbb{R}^D \mapsto \boldsymbol{y} \in \mathbb{R}^d. \tag{1.17}$$

The image $\mathcal{Y} = f(\mathcal{X})$ of $\mathcal{X}$ under such a mapping can be considered a *nonparametric representation* for $\mathcal{X}$, and ideally, such a $\mathcal{Y}$ should preserve some key structural information about $\mathcal{X}$, such as its intrinsic dimension, topology, and neighborhood (e.g., in the manifold learning problem studied in Chapter 4). Although $\mathcal{Y}$ is not a parametric model in the conventional sense that we have discussed in previous sections, it could better serve the task at hand (whether it is to cluster the data or to infer some high-level semantic information). In the computer vision or pattern recognition literature, such a representation is loosely called a "feature." More formally, features serve the same role as "sufficient statistics" for the inference tasks of interest (see Appendix B for a definition).

In this book, we will see a representative example of these methods in the context of manifold learning and data clustering. As we will describe in Chapters 4 and 7, if we are interested only in clustering (not modeling) the mixed data, then *spectral embedding* serves as a great example for such a feature map $f(\cdot)$. The basic idea is rather simple: instead of seeking a parametric model for the data set, we view each data point $\boldsymbol{x}_i$ of $\mathcal{X} = \{\boldsymbol{x}_1, \ldots, \boldsymbol{x}_N\}$ as a vertex of a graph $\mathcal{G}$ in which each pair of vertices $\boldsymbol{x}_i, \boldsymbol{x}_j$ are connected by an edge e_{ij} with a weight

$$w_{ij} = \exp\big(-\operatorname{dist}(\boldsymbol{x}_i, \boldsymbol{x}_j)\big), \tag{1.18}$$

where $d(\boldsymbol{x}_i, \boldsymbol{x}_j)$ is a "distance" between the points according to some norm. Ideally, we hope that the weight will be 1 when the two points belong to the same cluster (subspace) and 0 when they do not. Such a weight is often referred to as an *affinity measure* between pairs of data points. With such an affinity measure, the problem of clustering the data set $\mathcal{X}$ becomes one of identifying the connected components of such a graph. As we will see in Chapter 4, the null space of the Laplacian matrix $\mathcal{L} \in \mathbb{R}^{N \times N}$ of the affinity graph $\mathcal{G}$,

$$\mathcal{L}Y = \mathbf{0}, \quad Y \in \mathbb{R}^{N \times d}, \tag{1.19}$$

reveals how the data set should be clustered. More precisely, two rows $\boldsymbol{y}_i, \boldsymbol{y}_j \in \mathbb{R}^d$ of Y are the same if and only if the two vertices $\boldsymbol{x}_i, \boldsymbol{x}_j$ are connected in the graph. Hence, the null space of the Laplacian can be viewed as a nonparametric representation $\mathcal{Y}$ that captures the clustering information of $\mathcal{X}$. For subspace-like clusters, we will see many different ways of defining the graph affinity in Chapter 7.

Clustering as a Sparse Representation Problem
However, again, taking the nonparametric approach for a problem does not mean that one could largely ignore the intrinsic structures of the data. Quite to the contrary, as we will see in Chapter 7 and Chapter 8, for the subspace clustering problem, in order to build an effective and correct affinity measure for application of the spectral method, we need to exploit the local or global *low-dimensional* structures of the subspaces to their fullest extent.

From the graph-partitioning perspective, we hope to establish an affinity measure such that data points are connected only to points that belong to the same subspace. As we will see in Chapter 8, one effective way to obtain such an affinity is to make use of an important property of subspace-like data: Each point can be represented as a linear combination of other points in the same subspace; and in general, this representation, though not necessarily unique, is the most compact one in the sense that it represents each point with the minimum number of points. For instance, in Figure 1.2, a point in $\mathbb{R}^3$ typically can be written as a linear combination of three other points, but for a point on one of the lines, it can be represented as a scaled version of any other point on the same line.

Hence, we could represent all points in the data set $\mathcal{X}$ as linear combinations of other points in the same data set. More specifically, let X be the matrix with the data points as columns $X = [\boldsymbol{x}_1, \boldsymbol{x}_2, \ldots, \boldsymbol{x}_N]$. We have

$$X = XC, \tag{1.20}$$

where $C \in \mathbb{R}^{N \times N}$ is the matrix of coefficients with zeros on its diagonal so as to exclude the trivial representation $C = I$ in which each point equals itself. Among all possible representations of this kind for X, if we can find the one that contains the fewest nonzero coefficients, the matrix C may help us to construct an affinity matrix that has the desired graph-connectivity property. In Chapter 8, we will see that under rather broad conditions, such a sparse representation can be found effectively and efficiently, and the resulting affinity graph indeed respects all the subspace structures.

1.4 Noise, Errors, Outliers, and Model Selection

In many real-world applications, the given data samples may be corrupted by noise or gross errors, or contaminated by outliers. Figure 1.8 shows one such example. In contrast to the noiseless or high signal-to-noise ratio (SNR) scenario, the problem

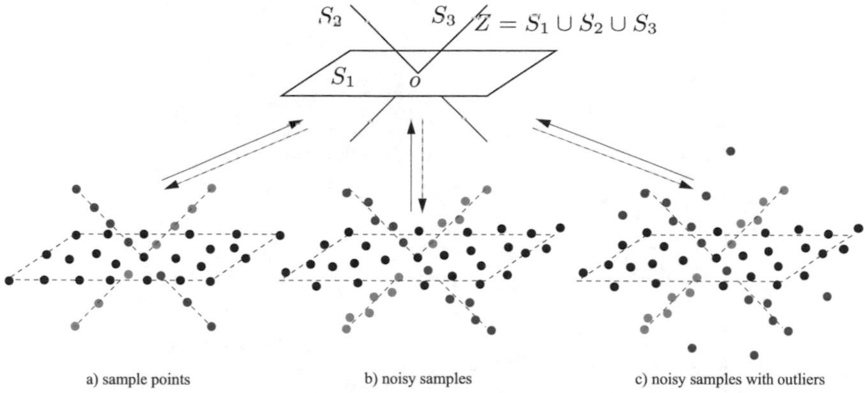

Fig. 1.8 Inferring a mixture model of multiple subspaces Z, consisting of one plane (S_1) and two lines (S_2, S_3), from a set of mixed data, which can be (a) noiseless samples from the plane and lines, (b) noisy samples, (c) noisy samples with outliers.

of finding the "correct" model becomes much more challenging in the presence of a significant amount of noise, errors, or outliers. Proper statistical and robust statistical techniques therefore need to be developed for model estimation and data clustering. These issues will be carefully treated in Chapter 3 for the single-subspace case and Chapter 6 and Chapter 8 for the multiple-subspace case.

Another important observation is that the class of piecewise linear models is very expressive. In the presence of noise and outliers, a mixture model of linear subspaces is not necessarily the best even if it achieves the highest fidelity to the given data. This is especially the case when the number of subspaces and their dimensions are not known a priori. In fact, for every point in the data set, one can fit a separate line to it, which results in no modeling error at all. But such a model is not very appealing, since it has exactly the same complexity as the original data.

In general, the higher the model complexity, the smaller the modeling error.[20] In statistics, this is known as "overfitting." A good (statistical or geometric) model M should strike a good balance between the complexity of the model and its fidelity to the data $\mathcal{X}$.[21] Many general model selection criteria have been proposed in the statistics and machine learning literatures, including the Akaike information criterion (AIC), the Bayesian information criterion (BIC), the minimum description length (MDL), and the minimum message length (MML). (See Appendix B for a brief review.) Despite some small differences, these criteria all make a tradeoff

[20]For example, every function can be approximated arbitrarily well by a piecewise linear function with a sufficient number of pieces.

[21]For instance, the complexity of a model can be measured as the minimum number of bits needed to fully describe the model, and the data fidelity can be measured by the distance from the sample points to the model.

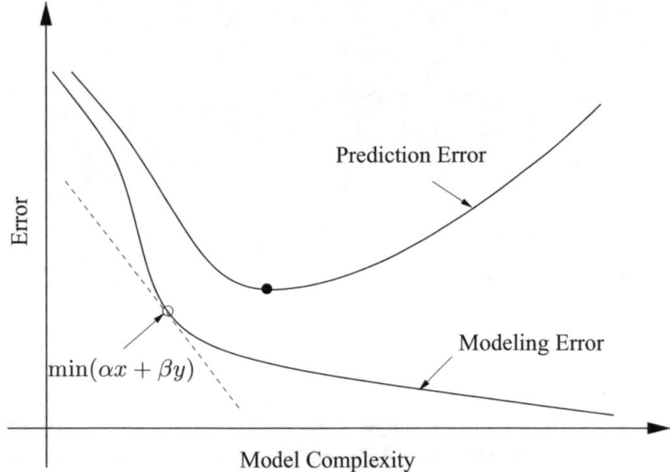

Fig. 1.9 Modeling and prediction error versus model complexity. The optimal model that minimizes prediction error, represented by the black dot, can be different from the optimal model that trades off modeling error and model complexity, represented by the circle.

between modeling error and model complexity and minimize an objective of the following form:

$$\min_{M \in \mathcal{M}} \left[J(M) \doteq \alpha \cdot \text{Complexity}(M) + \beta \cdot \text{Error}(\mathcal{X}, M) \right]. \tag{1.21}$$

In this book, we will introduce a model complexity measure that is specially designed for an arrangement of linear subspaces of arbitrary dimensions, namely the *effective dimension* (see Chapter 5). In Chapter 6, we show how to measure model complexity and data fidelity based on the minimum description length principle, which leads to a compression-based data clustering algorithm. We will also see how such a principle can be effectively applied to the image compression and segmentation problems in Chapter 9 and Chapter 10, respectively.

There is yet another fundamental tradeoff, known as the *bias versus variance* tradeoff in statistics, which is often exploited for model selection. When the model complexity is too high, the model tends to overfit the given data, including the noise in it. Such a model does not generalize well in the sense that it is unlikely to predict well the outcome of new samples. When the model complexity is too low, the model underfits the data, which results in a large prediction error. Therefore, a good model should minimize the prediction error. The typical relationship between modeling error and prediction error as a function of model complexity is plotted in Figure 1.9. Unfortunately, the "optimal" models obtained by trading off modeling error and prediction error can be different, as illustrated in the figure. In such a case, a choice between the two objectives has to be made. In the unsupervised learning setting, it is often difficult to obtain the prediction error curve unless one does cross-validation within the given data set itself. For purposes such as data compression, the prediction error is of less concern than the modeling error. Hence in these cases, we often choose a tradeoff between the modeling error and the model complexity (see Chapter 6 and Chapter 10). However, if the purpose of data modeling is to correctly classify future new samples, e.g., in face recognition, the typical model selection criterion is to minimize prediction error solely.

In the remainder of this book, we will show with great technical detail and depth how to apply many of the mathematical modeling principles discussed in this chapter to two classes of models that are central to many applications: subspace models and union of subspaces models. As we will see, although most of the book will study problems that generalize principal component analysis well beyond its classical setting, the fundamental mathematical, statistical, and computational principles that lead us to good solutions to those generalized settings remain very much the same as those already being utilized in the classical theory.

Part I
Modeling Data with a Single Subspace

Chapter 2
Principal Component Analysis

Principal component analysis is probably the oldest and best known of the techniques of multivariate analysis.

—Ian T. Jolliffe

Principal component analysis (PCA) is the problem of fitting a low-dimensional affine subspace to a set of data points in a high-dimensional space. PCA is, by now, well established in the literature, and has become one of the most useful tools for data modeling, compression, and visualization.

In this chapter, we will give a brief review of the classical theory of PCA, but with some modern twists and enrichment. When the dimension of the subspace is known, we will introduce both the statistical and geometric formulations of the PCA problem and establish their mathematical equivalence. Specifically, we will show that the singular value decomposition provides an optimal solution to the PCA problem and provides an interpretation of it as a rank minimization problem. We will also establish the similarities and differences between PCA and a probabilistic generative subspace model called probabilistic PCA. Finally, when the dimension of the subspace is unknown, we will introduce some conventional and modern model selection methods to determine the number of principal components.

2.1 Classical Principal Component Analysis (PCA)

Principal component analysis (PCA) refers to the problem of fitting a low-dimensional affine subspace S of dimension $d \ll D$ to a set of points $\{\boldsymbol{x}_1, \boldsymbol{x}_2, \ldots, \boldsymbol{x}_N\}$ in a high-dimensional space $\mathbb{R}^D$. Mathematically, this problem can be formulated as either a statistical problem or a geometric one. In this section, we will discuss both formulations and show that they lead to the same solution. We will also formulate PCA as a low-rank matrix approximation problem.

R. Vidal et al., *Generalized Principal Component Analysis*, Interdisciplinary Applied Mathematics 40, DOI 10.1007/978-0-387-87811-9_2

2.1.1 A Statistical View of PCA

Historically, PCA was first formulated in a statistical setting to estimate the principal components of a multivariate random variable $\boldsymbol{x}$ (Pearson 1901; Hotelling 1933). Specifically, given a zero-mean multivariate random variable $\boldsymbol{x} \in \mathbb{R}^D$ and an integer $d < D$, the d "principal components" of $\boldsymbol{x}$, $\boldsymbol{y} \in \mathbb{R}^d$, are defined as the d *uncorrelated* linear components of $\boldsymbol{x}$,

$$y_i = \boldsymbol{u}_i^\top \boldsymbol{x} \in \mathbb{R}, \quad \boldsymbol{u}_i \in \mathbb{R}^D, \quad i = 1, 2, \dots, d, \tag{2.1}$$

such that the variance of y_i is maximized subject to

$$\boldsymbol{u}_i^\top \boldsymbol{u}_i = 1 \quad \text{and} \quad \mathrm{Var}(y_1) \geq \mathrm{Var}(y_2) \geq \cdots \geq \mathrm{Var}(y_d) > 0. \tag{2.2}$$

For example, to find the first principal component y_1, we seek a vector $\boldsymbol{u}_1^* \in \mathbb{R}^D$ such that

$$\boldsymbol{u}_1^* = \arg\max_{\boldsymbol{u}_1 \in \mathbb{R}^D} \mathrm{Var}(\boldsymbol{u}_1^\top \boldsymbol{x}) \quad \text{s.t.} \quad \boldsymbol{u}_1^\top \boldsymbol{u}_1 = 1. \tag{2.3}$$

The following theorem shows that the principal components of $\boldsymbol{x}$ can be computed from the eigenvectors of its covariance matrix $\Sigma_{\boldsymbol{x}} \doteq \mathbb{E}[\boldsymbol{x}\boldsymbol{x}^\top]$.

Theorem 2.1 (Principal Components of a Random Variable). *Assume that* $\mathrm{rank}(\Sigma_{\boldsymbol{x}}) \geq d$. *Then the first d principal components of a zero-mean multivariate random variable $\boldsymbol{x}$, denoted by y_i for $i = 1, 2, \dots, d$, are given by*

$$y_i = \boldsymbol{u}_i^\top \boldsymbol{x}, \tag{2.4}$$

where $\{\boldsymbol{u}_i\}_{i=1}^d$ are d orthonormal eigenvectors of $\Sigma_{\boldsymbol{x}} \doteq \mathbb{E}[\boldsymbol{x}\boldsymbol{x}^\top]$ associated with its d largest eigenvalues $\{\lambda_i\}_{i=1}^d$. Moreover, $\lambda_i = \mathrm{Var}(y_i)$ for $i = 1, 2, \dots, d$.

Proof. For the sake of simplicity, let us first assume that $\Sigma_{\boldsymbol{x}}$ does not have repeated eigenvalues. In this case, since the matrix $\Sigma_{\boldsymbol{x}}$ is real and symmetric, its eigenvalues are real and its eigenvectors form a basis of $\mathbb{R}^D$. Moreover, the eigenvectors are unique (up to sign), and the eigenvectors corresponding to different eigenvalues are orthogonal to each other (see Exercise 2.1).

Now notice that for every $\boldsymbol{u} \in \mathbb{R}^D$, we have that

$$\mathrm{Var}(\boldsymbol{u}^\top \boldsymbol{x}) = \mathbb{E}[(\boldsymbol{u}^\top \boldsymbol{x})^2] = \mathbb{E}[\boldsymbol{u}^\top \boldsymbol{x}\boldsymbol{x}^\top \boldsymbol{u}] = \boldsymbol{u}^\top \Sigma_{\boldsymbol{x}} \boldsymbol{u}. \tag{2.5}$$

Therefore, the optimization problem in (2.3) is equivalent to

$$\max_{\boldsymbol{u}_1 \in \mathbb{R}^D} \boldsymbol{u}_1^\top \Sigma_{\boldsymbol{x}} \boldsymbol{u}_1 \quad \text{s.t.} \quad \boldsymbol{u}_1^\top \boldsymbol{u}_1 = 1. \tag{2.6}$$

To solve the above constrained optimization problem, we use the method of Lagrange multipliers (see Appendix A). The Lagrangian function is given by

$$\mathscr{L} = \boldsymbol{u}_1^\top \Sigma_x \boldsymbol{u}_1 + \lambda_1(1 - \boldsymbol{u}_1^\top \boldsymbol{u}_1), \tag{2.7}$$

where $\lambda_1 \in \mathbb{R}$ is the Lagrange multiplier. From computing the derivatives of $\mathscr{L}$ with respect to $(\boldsymbol{u}_1, \lambda_1)$ and setting them to zero, we obtain the following necessary conditions for $(\boldsymbol{u}_1, \lambda_1)$ to be an extremum of $\mathscr{L}$:

$$\Sigma_x \boldsymbol{u}_1 = \lambda_1 \boldsymbol{u}_1 \quad \text{and} \quad \boldsymbol{u}_1^\top \boldsymbol{u}_1 = 1. \tag{2.8}$$

This means that $\boldsymbol{u}_1$ is an eigenvector of Σ_x with associated eigenvalue λ_1. Since the extremum value is $\boldsymbol{u}_1^\top \Sigma_x \boldsymbol{u}_1 = \lambda_1 \boldsymbol{u}_1^\top \boldsymbol{u}_1 = \lambda_1$, the optimal solution for $\boldsymbol{u}_1$ is given by the eigenvector of Σ_x associated with its largest eigenvalue $\lambda_1 = \text{Var}(y_1) > 0$.

To find the second principal component, $\boldsymbol{u}_2$, we use the fact that $\boldsymbol{u}_1^\top \boldsymbol{x}$ and $\boldsymbol{u}_2^\top \boldsymbol{x}$ need to be uncorrelated. This implies that $\boldsymbol{u}_2$ is orthogonal to $\boldsymbol{u}_1$. Indeed, from

$$\mathbb{E}[(\boldsymbol{u}_1^\top \boldsymbol{x})(\boldsymbol{u}_2^\top \boldsymbol{x})] = \mathbb{E}[\boldsymbol{u}_1^\top \boldsymbol{x}\boldsymbol{x}^\top \boldsymbol{u}_2] = \boldsymbol{u}_1^\top \Sigma_x \boldsymbol{u}_2 = \lambda_1 \boldsymbol{u}_1^\top \boldsymbol{u}_2 = 0 \tag{2.9}$$

and $\lambda_1 \neq 0$, we have $\boldsymbol{u}_1^\top \boldsymbol{u}_2 = 0$. Thus, to find $\boldsymbol{u}_2$, we need to solve the following optimization problem:

$$\max_{\boldsymbol{u}_2 \in \mathbb{R}^D} \boldsymbol{u}_2^\top \Sigma_x \boldsymbol{u}_2 \quad \text{s.t.} \quad \boldsymbol{u}_2^\top \boldsymbol{u}_2 = 1 \quad \text{and} \quad \boldsymbol{u}_1^\top \boldsymbol{u}_2 = 0. \tag{2.10}$$

As before, with an abuse of notation, we define the Lagrangian

$$\mathscr{L} = \boldsymbol{u}_2^\top \Sigma_x \boldsymbol{u}_2 + \lambda_2(1 - \boldsymbol{u}_2^\top \boldsymbol{u}_2) + \gamma \boldsymbol{u}_1^\top \boldsymbol{u}_2. \tag{2.11}$$

The necessary conditions for $(\boldsymbol{u}_2, \lambda_2, \gamma)$ to be an extremum are

$$\Sigma_x \boldsymbol{u}_2 + \frac{\gamma}{2} \boldsymbol{u}_1 = \lambda_2 \boldsymbol{u}_2, \quad \boldsymbol{u}_2^\top \boldsymbol{u}_2 = 1 \quad \text{and} \quad \boldsymbol{u}_1^\top \boldsymbol{u}_2 = 0, \tag{2.12}$$

from which it follows that $\boldsymbol{u}_1^\top \Sigma_x \boldsymbol{u}_2 + \frac{\gamma}{2} \boldsymbol{u}_1^\top \boldsymbol{u}_1 = \lambda_1 \boldsymbol{u}_1^\top \boldsymbol{u}_2 + \frac{\gamma}{2} = \lambda_2 \boldsymbol{u}_1^\top \boldsymbol{u}_2$, and so $\gamma = 2(\lambda_2 - \lambda_1)\boldsymbol{u}_1^\top \boldsymbol{u}_2 = 0$. This implies that $\Sigma_x \boldsymbol{u}_2 = \lambda_2 \boldsymbol{u}_2$ and that the extremum value is $\boldsymbol{u}_2^\top \Sigma_x \boldsymbol{u}_2 = \lambda_2 = \text{Var}(y_2)$. Therefore, $\boldsymbol{u}_2$ is the leading eigenvector of Σ_x restricted to the orthogonal complement of $\boldsymbol{u}_1$.[1] Since the eigenvalues of Σ_x are distinct, $\boldsymbol{u}_2$ is the eigenvector of Σ_x associated with its second-largest eigenvalue.

---

[1]The reason for this is that both $\boldsymbol{u}_1$ and its orthogonal complement $\boldsymbol{u}_1^\perp$ are invariant subspaces of Σ_x.

To find the remaining principal components, we use that fact that for all for $i \neq j$, $y_i = \boldsymbol{u}_i^\top \boldsymbol{x}$ and $y_j = \boldsymbol{u}_j^\top \boldsymbol{x}$ need to be uncorrelated, whence

$$\mathrm{Var}(y_i y_j) = \mathbb{E}[\boldsymbol{u}_i^\top \boldsymbol{x}\boldsymbol{x}^\top \boldsymbol{u}_j] = \boldsymbol{u}_i^\top \Sigma_{\boldsymbol{x}} \boldsymbol{u}_j = 0.$$

Using induction, assume that $\boldsymbol{u}_1, \dots, \boldsymbol{u}_{i-1}$ are the unit-length eigenvectors of $\Sigma_{\boldsymbol{x}}$ associated with its top $i-1$ eigenvalues, and let $\boldsymbol{u}_i$ be the vector defining the ith principal component, y_i. Then, $\Sigma_{\boldsymbol{x}} \boldsymbol{u}_j = \lambda_j \boldsymbol{u}_j$ for $j = 1, \dots, i-1$ and $\boldsymbol{u}_i^\top \Sigma_{\boldsymbol{x}} \boldsymbol{u}_j = \lambda_j \boldsymbol{u}_i^\top \boldsymbol{u}_j = 0$ for all $j = 1, \dots, i-1$. Since $\lambda_j > 0$, we have that $\boldsymbol{u}_i^\top \boldsymbol{u}_j = 0$ for all $j = 1, \dots, i-1$. To compute $\boldsymbol{u}_i$, we build the Lagrangian

$$\mathscr{L} = \boldsymbol{u}_i^\top \Sigma_{\boldsymbol{x}} \boldsymbol{u}_i + \lambda_i (1 - \boldsymbol{u}_i^\top \boldsymbol{u}_i) + \sum_{j=1}^{i-1} \gamma_j \boldsymbol{u}_i^\top \boldsymbol{u}_j. \tag{2.13}$$

The necessary conditions for $(\boldsymbol{u}_i, \lambda_i, \gamma_1, \dots, \gamma_{j-1})$ to be an extremum are

$$\Sigma_{\boldsymbol{x}} \boldsymbol{u}_i + \sum_{j=1}^{i-1} \frac{\gamma_j}{2} \boldsymbol{u}_j = \lambda_i \boldsymbol{u}_i, \ \boldsymbol{u}_i^\top \boldsymbol{u}_i = 1 \text{ and } \boldsymbol{u}_i^\top \boldsymbol{u}_j = 0, j = 1, \dots, i-1, \tag{2.14}$$

from which it follows that for all $j = 1, \dots, i-1$, we have $\boldsymbol{u}_j^\top \Sigma_{\boldsymbol{x}} \boldsymbol{u}_i + \frac{\gamma_j}{2} = \lambda_j \boldsymbol{u}_j^\top \boldsymbol{u}_i + \frac{\gamma_j}{2} = \lambda_i \boldsymbol{u}_j^\top \boldsymbol{u}_i$, and so $\gamma_j = 2(\lambda_j - \lambda_i) \boldsymbol{u}_j^\top \boldsymbol{u}_i = 0$. Since the associated extremum value is $\boldsymbol{u}_i^\top \Sigma_{\boldsymbol{x}} \boldsymbol{u}_i = \lambda_i = \mathrm{Var}(y_i)$, $\boldsymbol{u}_i$ is the leading eigenvector of $\Sigma_{\boldsymbol{x}}$ restricted to the orthogonal complement of the span of $\boldsymbol{u}_1, \dots, \boldsymbol{u}_{i-1}$. Since the eigenvalues of $\Sigma_{\boldsymbol{x}}$ are distinct, $\boldsymbol{u}_i$ is the eigenvector of $\Sigma_{\boldsymbol{x}}$ associated with the ith-largest eigenvalue. Therefore, when the eigenvalues of $\Sigma_{\boldsymbol{x}}$ are distinct, each eigenvector $\boldsymbol{u}_i$ is unique (up to sign), and hence so are the principal components of $\boldsymbol{x}$.

Let us now consider the case in which $\Sigma_{\boldsymbol{x}}$ has repeated eigenvalues. In this case, $\Sigma_{\boldsymbol{x}}$ still admits a basis of orthonormal eigenvectors. Specifically, the eigenvectors of $\Sigma_{\boldsymbol{x}}$ associated with different eigenvalues are still orthogonal, while the eigenvectors associated with a repeated eigenvalue form an eigensubspace, and every orthonormal basis for this eigensubspace gives a valid set of eigenvectors (see Exercise 2.1). As a consequence, the principal directions $\{\boldsymbol{u}_i\}_{i=1}^d$ are not uniquely defined. For example, if λ_1 is repeated, every eigenvector associated with λ_1 can be chosen as $\boldsymbol{u}_1$ and any other eigenvector associated with λ_1 and orthogonal to $\boldsymbol{u}_1$ can be chosen as $\boldsymbol{u}_2$. Nonetheless, the principal components can still be obtained from any set of the top d eigenvectors of $\Sigma_{\boldsymbol{x}}$, as claimed. □

The solution to PCA provided by Theorem 2.1 suggests that we may find the d principal components of $\boldsymbol{x}$ simultaneously, rather than one by one. Specifically, if we define a random vector $\boldsymbol{y} = [y_1, y_2, \dots, y_d]^\top \in \mathbb{R}^d$ and a matrix $U = [\boldsymbol{u}_1, \boldsymbol{u}_2, \dots, \boldsymbol{u}_d] \in \mathbb{R}^{D \times d}$, then, since $\boldsymbol{y} = U^\top \boldsymbol{x}$, we have that

$$\Sigma_{\boldsymbol{y}} \doteq \mathbb{E}[\boldsymbol{y}\boldsymbol{y}^\top] = U^\top \mathbb{E}[\boldsymbol{x}\boldsymbol{x}^\top] U = U^\top \Sigma_{\boldsymbol{x}} U. \tag{2.15}$$

From the definition of principal components, the entries of $\boldsymbol{y}$ are uncorrelated. As a result, the matrix $\Sigma_{\boldsymbol{y}}$ must be diagonal, and from the proof of Theorem 2.1, we showed that the matrix U must be orthonormal, i.e., $U^\top U = I_d$.

Recall that every diagonalizable matrix A can be transformed into a diagonal matrix $\Lambda = V^{-1}AV$, where the columns of V are the eigenvectors of A and the diagonal entries of Λ are the corresponding eigenvalues. Recall also that if A is real, symmetric and positive semi-definite, its eigenvalues are real and nonnegative, i.e., $\lambda_i \geq 0$, and its eigenvectors can be chosen to be orthonormal, so that $V^{-1} = V^\top$ (see Exercise 2.1). Since the matrix $\Sigma_{\boldsymbol{x}}$ is real, symmetric, and positive semidefinite, one solution to the equation $\Sigma_{\boldsymbol{y}} = U^\top \Sigma_{\boldsymbol{x}} U$ is obtained by choosing the columns of U as d eigenvectors of $\Sigma_{\boldsymbol{x}}$ and the diagonal entries of $\Sigma_{\boldsymbol{y}}$ as the corresponding d eigenvalues. Moreover, since our goal is to maximize the variance of each y_i and $\lambda_i = \text{Var}(y_i)$, we conclude that the columns of U are the top d eigenvectors of $\Sigma_{\boldsymbol{x}}$ and the entries of $\Sigma_{\boldsymbol{y}}$ are the corresponding top d eigenvalues.

Principal Components of a Nonzero-Mean Random Variable

When $\boldsymbol{x}$ does not have zero mean, then the d principal components of $\boldsymbol{x}$ are defined as the d uncorrelated affine components

$$y_i = \boldsymbol{u}_i^\top \boldsymbol{x} + a_i \in \mathbb{R}, \quad \boldsymbol{u}_i \in \mathbb{R}^D, \quad i = 1, 2, \ldots, d, \tag{2.16}$$

of $\boldsymbol{x}$ such that the variance of y_i is maximized subject to

$$\boldsymbol{u}_i^\top \boldsymbol{u}_i = 1 \quad \text{and} \quad \text{Var}(y_1) \geq \text{Var}(y_2) \geq \cdots \geq \text{Var}(y_d) > 0. \tag{2.17}$$

As shown in Exercise 2.6, the principal directions $\{\boldsymbol{u}_i\}_{i=1}^d$ are the d eigenvectors of $\Sigma_{\boldsymbol{x}} \doteq \mathbb{E}[(\boldsymbol{x}-\boldsymbol{\mu})(\boldsymbol{x}-\boldsymbol{\mu})^\top]$, where $\boldsymbol{\mu} = \mathbb{E}(\boldsymbol{x})$, associated with its d largest eigenvalues $\{\lambda_i\}_{i=1}^d$. Moreover, $\lambda_i = \text{Var}(y_i)$ and $a_i = -\boldsymbol{u}_i^\top \boldsymbol{\mu}$ for $i = 1, 2, \ldots, d$.

Sample Principal Components of a Zero-Mean Random Variable

In practice, we may not know the population covariance matrix $\Sigma_{\boldsymbol{x}}$. Instead, we may be given N i.i.d. samples of the zero-mean random variable $\boldsymbol{x}$, $\{\boldsymbol{x}_j\}_{j=1}^N$, which we collect into a data matrix $X = [\boldsymbol{x}_1, \boldsymbol{x}_2, \ldots, \boldsymbol{x}_N]$. It is well known from statistics (see Exercise B.1) that the maximum likelihood estimate of $\Sigma_{\boldsymbol{x}}$ is given by

$$\hat{\Sigma}_N \doteq \frac{1}{N} \sum_{j=1}^{N} \boldsymbol{x}_j \boldsymbol{x}_j^\top = \frac{1}{N} X X^\top. \tag{2.18}$$

We define the d "sample principal components" of $\boldsymbol{x}$ as

$$\hat{y}_i = \hat{\boldsymbol{u}}_i^\top \boldsymbol{x}, \quad i = 1, 2, \ldots, d, \tag{2.19}$$

where $\{\hat{\boldsymbol{u}}_i\}_{i=1}^d$ are the top d eigenvectors of $\hat{\Sigma}_N$, or equivalently those of $XX^\top$.

Notice that when the dimension D of the data is very high, we can avoid computing the eigenvectors of a large matrix $XX^\top$ by exploiting the fact that the top eigenvectors of $XX^\top$ are the same as the top singular vectors of X. Therefore, the sample principal components of $\boldsymbol{x}$ may be computed from the singular value decomposition (SVD) of $X = U_X \Sigma_X V_X^\top$ as $\boldsymbol{y} = U^\top \boldsymbol{x}$, where the columns of U are the first d columns of U_X.

Remark 2.2 (Relationship between principal components and sample principal components)**.** *Even though the principal components of $\boldsymbol{x}$ and the sample principal components of $\boldsymbol{x}$ are different notions, under certain assumptions on the distribution of $\boldsymbol{x}$, they can be related to each other. Specifically, one can show that if $\boldsymbol{x}$ is Gaussian, then every eigenvector $\hat{\boldsymbol{u}}$ of $\hat{\Sigma}_N$ is an asymptotically consistent unbiased estimate (see Appendix B) for the corresponding eigenvector $\boldsymbol{u}$ of $\Sigma_{\boldsymbol{x}}$. Interested readers may find a more detailed proof in (Jolliffe 1986, 2002).*

2.1.2 A Geometric View of PCA

An alternative geometric view of PCA, which is very much related to the SVD (Beltrami 1873; Jordan 1874), assumes that we are given a set of points $\{\boldsymbol{x}_j\}_{j=1}^N$ in $\mathbb{R}^D$ and seeks to find an (affine) subspace $S \subset \mathbb{R}^D$ of dimension d that best fits these points. Each point $\boldsymbol{x}_j \in S$ can be represented as

$$\boldsymbol{x}_j = \boldsymbol{\mu} + U\boldsymbol{y}_j, \quad j = 1, 2, \ldots, N, \tag{2.20}$$

where $\boldsymbol{\mu} \in S$ is a point in the subspace, U is a $D \times d$ matrix whose columns form a basis for the subspace, and $\boldsymbol{y}_j \in \mathbb{R}^d$ is simply the vector of new coordinates of $\boldsymbol{x}_j$ in the subspace.

Notice that there is some redundancy in the above representation due to the arbitrariness in the choice of $\boldsymbol{\mu}$ and U. More precisely, for every $\boldsymbol{y}_0 \in \mathbb{R}^d$, we can re-represent $\boldsymbol{x}_j$ as $\boldsymbol{x}_j = (\boldsymbol{\mu} + U\boldsymbol{y}_0) + U(\boldsymbol{y}_j - \boldsymbol{y}_0)$. We call this ambiguity the *translational ambiguity*. Also, for every invertible $A \in \mathbb{R}^{d\times d}$, we can re-represent $\boldsymbol{x}_j$ as $\boldsymbol{x}_j = \boldsymbol{\mu} + (UA)(A^{-1}\boldsymbol{y}_j)$. We call this ambiguity the *change of basis ambiguity*. Therefore, we need some additional constraints in order to end up with a unique solution to the problem of finding an affine subspace for the data.

A common constraint used to resolve the translational ambiguity is to require that the average of the $\boldsymbol{y}_j$ be zero,[2] i.e.,

$$\frac{1}{N}\sum_{j=1}^{N} \boldsymbol{y}_j = \boldsymbol{0}, \tag{2.21}$$

[2]In the statistical setting, $\boldsymbol{x}_j$ and $\boldsymbol{y}_j$ will be samples of two random variables $\boldsymbol{x}$ and $\boldsymbol{y}$, respectively. Then this constraint is equivalent to setting their means to zero.

where $\mathbf{0} \in \mathbb{R}^d$ is the vector of all zeros, while a common constraint used to resolve the change of basis ambiguity is to require that the columns of U be orthonormal, i.e., $U^\top U = I$. This last constraint eliminates the change of basis ambiguity only up to a rotation, because we can still re-represent $\boldsymbol{x}_j$ as $\boldsymbol{x}_j = \boldsymbol{\mu} + (UR)(R^\top \boldsymbol{y}_j)$ for some rotation R in $\mathbb{R}^d$. However, this *rotational ambiguity* can easily be dealt with during optimization, as we shall soon see.

The model in (2.20) now assumes that each point $\boldsymbol{x}_j$ lies perfectly in an affine subspace S. In practice, the given points are imperfect and have noise. For example, if point $\boldsymbol{x}_j$ is contaminated by additive noise ε_j, we have

$$\boldsymbol{x}_j = \boldsymbol{\mu} + U\boldsymbol{y}_j + \varepsilon_j, \quad j = 1, 2, \ldots, N. \tag{2.22}$$

In this case, we define the "optimal" affine subspace to be the one that minimizes the sum of squared errors, i.e.,

$$\min_{\boldsymbol{\mu}, U, \{y_j\}} \sum_{j=1}^{N} \left\| \boldsymbol{x}_j - \boldsymbol{\mu} - U\boldsymbol{y}_j \right\|^2, \ \text{ s.t. } \ U^\top U = I_d \ \text{ and } \ \sum_{j=1}^{N} \boldsymbol{y}_j = \mathbf{0}. \tag{2.23}$$

In order to solve this optimization problem, we define the Lagrangian function

$$\mathscr{L} = \sum_{j=1}^{N} \left\| \boldsymbol{x}_j - \boldsymbol{\mu} - U\boldsymbol{y}_j \right\|^2 + \gamma^\top \sum_{j=1}^{N} \boldsymbol{y}_j + \operatorname{trace}\left((I_d - U^\top U)\Lambda \right), \tag{2.24}$$

where $\gamma \in \mathbb{R}^d$ and $\Lambda = \Lambda^\top \in \mathbb{R}^{d \times d}$ are, respectively, a vector and a matrix of Lagrange multipliers. A necessary condition for $\boldsymbol{\mu}$ to be an extremum is

$$-2 \sum_{j=1}^{N} (\boldsymbol{x}_j - \boldsymbol{\mu} - U\boldsymbol{y}_j) = \mathbf{0} \implies \hat{\boldsymbol{\mu}} = \hat{\boldsymbol{\mu}}_N \doteq \frac{1}{N} \sum_{j=1}^{N} \boldsymbol{x}_j. \tag{2.25}$$

A necessary condition for $\boldsymbol{y}_j$ to be an extremum is

$$-2U^\top (\boldsymbol{x}_j - \boldsymbol{\mu} - U\boldsymbol{y}_j) + \gamma = \mathbf{0}. \tag{2.26}$$

Summing over j yields $\gamma = \mathbf{0}$, from which we obtain

$$\hat{\boldsymbol{y}}_j = U^\top (\boldsymbol{x}_j - \hat{\boldsymbol{\mu}}_N). \tag{2.27}$$

The vector $\hat{\boldsymbol{y}}_j \in \mathbb{R}^d$ is simply the coordinates of the projection of $\boldsymbol{x}_j \in \mathbb{R}^D$ onto the subspace S. We may call such a $\hat{\boldsymbol{y}}$ the "geometric principal components" of $\boldsymbol{x}$.

Before optimizing over U, we can replace the optimal values for $\boldsymbol{\mu}$ and $\boldsymbol{y}_j$ in the objective function. This leads to the following optimization problem:

$$\min_U \sum_{j=1}^{N} \left\| (\boldsymbol{x}_j - \hat{\boldsymbol{\mu}}_N) - UU^\top (\boldsymbol{x}_j - \hat{\boldsymbol{\mu}}_N) \right\|^2 \quad \text{s.t.} \quad U^\top U = I_d. \tag{2.28}$$

Note that this is a restatement of the original problem with the mean $\hat{\boldsymbol{\mu}}_N$ subtracted from each of the sample points. Therefore, from now on, we will consider only the case in which the data points have zero mean. If such is not the case, simply subtract the mean from each point before computing U.

The following theorem gives a constructive solution for finding an optimal U.

Theorem 2.3 (PCA via SVD). *Let $X = [\boldsymbol{x}_1, \boldsymbol{x}_2, \ldots, \boldsymbol{x}_N] \in \mathbb{R}^{D\times N}$ be the matrix formed by stacking the (zero-mean) data points as its column vectors. Let $X = U_X \Sigma_X V_X^\top$ be the SVD of the matrix X. Then for a given $d < D$, an optimal solution for U is given by the first d columns of U_X, an optimal solution for $\boldsymbol{y}_j$ is given by the jth column of the top $d \times N$ submatrix of $\Sigma_X V_X^\top$, and the optimal objective value is given by $\sum_{i=d+1}^{D} \sigma_i^2$, where σ_i is the ith singular value of X.*

Proof. Since $U^\top U = I$, we have $(I - UU^\top)(I - UU^\top) = (I - UU^\top)$. Then, recalling that $\boldsymbol{x}^\top A \boldsymbol{x} = \text{trace}(A\boldsymbol{x}\boldsymbol{x}^\top)$, we can rewrite the least-squares error

$$\sum_{j=1}^{N} \left\| \boldsymbol{x}_j - UU^\top \boldsymbol{x}_j \right\|^2 = \sum_{j=1}^{N} \boldsymbol{x}_j^\top (I_D - UU^\top) \boldsymbol{x}_j \tag{2.29}$$

as $\text{trace}((I_D - UU^\top)XX^\top)$. The first term $\text{trace}(XX^\top)$ does not depend on U. Therefore, we can transform the minimization of (2.29) to

$$\max_U \ \text{trace}(UU^\top XX^\top) \quad \text{s.t.} \quad U^\top U = I_d. \tag{2.30}$$

Since $\text{trace}(AB) = \text{trace}(BA)$, the Lagrangian for this problem can be written as

$$\mathscr{L} = \text{trace}(U^\top XX^\top U) + \text{trace}((I_d - U^\top U)\Lambda), \tag{2.31}$$

where $\Lambda = \Lambda^\top \in \mathbb{R}^{d\times d}$. The conditions for an extremum are given by

$$XX^\top U = U\Lambda. \tag{2.32}$$

Therefore, $\Lambda = U^\top XX^\top U$, and the objective function reduces to $\text{trace}(\Lambda)$. Recall now that U is defined only up to a rotation, i.e., $U' = UR$ is also a valid solution, hence so is $\Lambda' = R\Lambda R^\top$. Since Λ is symmetric, it has an orthogonal matrix of eigenvectors. Thus, if we choose R to be the matrix of eigenvectors of Λ, then Λ' is a diagonal matrix. As a consequence, we can choose Λ to be diagonal without loss of generality. It follows from (2.32) that the columns of U must be d eigenvectors of $XX^\top$ with the corresponding eigenvalues in the diagonal entries of Λ. Since the goal is to maximize $\text{trace}(\Lambda)$, an optimal solution is given by the top d eigenvectors of $XX^\top$, i.e., the top d singular vectors of $X = U_X \Sigma_X V_X^\top$, which are the first d columns of U_X. It then follows from (2.27) that $Y = [\boldsymbol{y}_1, \boldsymbol{y}_2, \ldots, \boldsymbol{y}_N] = U^\top X =$

$U^\top U_X \Sigma_X V_X^\top = \Sigma V^\top$, where Σ is a diagonal matrix whose diagonal entries are the top d singular values of X and V a matrix whose columns are the top d right singular vectors of X. Finally, since $\Lambda = U^\top U_X \Sigma_X^2 U_X^\top U = \Sigma^2$, the optimal least-squares error is given by $\text{trace}(\Sigma_X^2) - \text{trace}(\Sigma^2) = \sum_{i=d+1}^{D} \sigma_i^2$, where σ_i is the ith singular value of X. □

According to the theorem, the SVD gives an optimal solution to the PCA problem. The resulting matrix U, together with the mean $\boldsymbol{\mu}$ if the data do not have zero mean, provides a geometric description of the dominant subspace structure for all the points;[3] and the columns of the matrix $\Sigma V^\top = [\hat{\boldsymbol{y}}_1, \hat{\boldsymbol{y}}_2, \ldots, \hat{\boldsymbol{y}}_N] \in \mathbb{R}^{d \times N}$, i.e., the principal components, give a more compact representation for the points $X = [\boldsymbol{x}_1, \boldsymbol{x}_2, \ldots, \boldsymbol{x}_N] \in \mathbb{R}^{D \times N}$, since d is typically much smaller than D.

Theorem 2.4 (Equivalence of Geometric and Sample Principal Components)**.** *Let $X = [\boldsymbol{x}_1, \boldsymbol{x}_2, \ldots, \boldsymbol{x}_N] \in \mathbb{R}^{D \times N}$ be the mean-subtracted data matrix. The vectors $\hat{\boldsymbol{u}}_1, \hat{\boldsymbol{u}}_2, \ldots, \hat{\boldsymbol{u}}_d \in \mathbb{R}^D$ associated with the d sample principal components of X are exactly the columns of the matrix $U \in \mathbb{R}^{D \times d}$ that minimizes the least-squares error (2.29).*

Proof. The proof is simple. Notice that if X has the singular value decomposition $X = U_X \Sigma_X V_X^\top$, then $XX^\top = U_X \Sigma_X^2 U_X^\top$ is the eigenvalue decomposition of $XX^\top$. If Σ_X is ordered, then the first d columns of U_X are exactly the leading d eigenvectors of $XX^\top$, which give the d sample principal components. □

The above theorem shows that both the geometric and statistical formulations of PCA lead to exactly the same solution/estimate of the sample principal components. This equivalence is part of the reason why PCA has become the tool of choice for dimensionality reduction, since the optimality of the solution can be interpreted either statistically or geometrically in different application contexts.

Figure 2.1 gives an example of a two-dimensional data set and its two principal components.

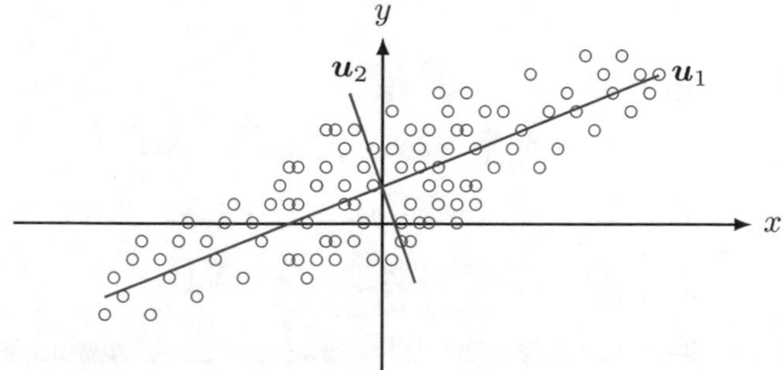

Fig. 2.1 Example showing a two-dimensional data set and its two principal components.

[3]From a statistical standpoint, the column vectors of U give the directions in which the data X has the largest variance, whence the name "principal components."

2.1.3 A Rank Minimization View of PCA

Notice that the geometric PCA problem in (2.23) can be rewritten as

$$\min_{\boldsymbol{\mu},U,Y} \left\|X - \boldsymbol{\mu}\mathbf{1}^\top - UY\right\|_F^2, \quad \text{s.t. } U^\top U = I_d \text{ and } Y\mathbf{1} = \mathbf{0}, \tag{2.33}$$

where $X = [\boldsymbol{x}_1, \ldots, \boldsymbol{x}_N]$, $Y = [\boldsymbol{y}_1, \ldots, \boldsymbol{y}_N]$, $\mathbf{1} \in \mathbb{R}^N$ is the vector of all ones, and $\|X\|_F^2 = \sum_{ij} X_{ij}^2$ is the Frobenius norm of X. Therefore, another interpretation of PCA is to see it as the problem of finding a vector $\boldsymbol{\mu}$ and rank-d matrix that best approximate the data matrix X. This problem can be formulated as

$$\min_{\boldsymbol{\mu},A} \|X - \boldsymbol{\mu}\mathbf{1}^\top - A\|_F^2 \quad \text{s.t.} \quad \operatorname{rank}(A) = d \text{ and } A\mathbf{1} = \mathbf{0}. \tag{2.34}$$

Notice that this formulation is identical to that in (2.23), except that we have now replaced the subspace basis U and the matrix of principal components Y by their product $A = UY$. The constraint $A\mathbf{1} = \mathbf{0}$ comes from the requirement that the principal components be centered, i.e., $\sum \boldsymbol{y}_j = 0$; hence $Y\mathbf{1} = \mathbf{0}$.

Since the problem in (2.34) is the same as that in (2.23), we already know that the optimal solution for $\boldsymbol{\mu}$ is $\frac{1}{N}\sum_j \boldsymbol{x}_j = \frac{1}{N}X\mathbf{1}$. Therefore, after we have centered the data matrix by subtracting $\boldsymbol{\mu}$ from each column, the optimization problem in (2.34) can be reduced to

$$\min_{A} \|X - A\|_F^2 \quad \text{s.t.} \quad \operatorname{rank}(A) = d. \tag{2.35}$$

Notice that we have dropped the constraint $A\mathbf{1} = \mathbf{0}$. This is because this constraint is not needed when the data matrix is centered, i.e., when $X\mathbf{1} = \mathbf{0}$. To see this, let A^* be the optimal solution to (2.35), i.e., the optimal solution without the constraint $A\mathbf{1} = \mathbf{0}$. Suppose $\boldsymbol{a} = \frac{1}{N}A^*\mathbf{1}$ is not zero, and let $\hat{A} = A^* - \boldsymbol{a}\mathbf{1}^\top$. Notice that $\operatorname{rank}(\hat{A}) \leq d$ and $\hat{A}\mathbf{1} = 0$. So $\hat{A}$ satisfies the constraints of the program (2.34). However,

$$\|X - \hat{A}\|_F^2 = \|X - A^* + \boldsymbol{a}\mathbf{1}^\top\|_F^2 \tag{2.36}$$

$$= \|X - A^*\|_F^2 + 2\langle X - A^*, \boldsymbol{a}\mathbf{1}^\top\rangle + \|\boldsymbol{a}\mathbf{1}^\top\|_F^2 \tag{2.37}$$

$$= \|X - A^*\|_F^2 + 2\boldsymbol{a}^\top(X - A^*)\mathbf{1} + N\|\boldsymbol{a}\|_2^2 \tag{2.38}$$

$$= \|X - A^*\|_F^2 - N\|\boldsymbol{a}\|_2^2 < \|X - A^*\|_F^2, \tag{2.39}$$

which contradicts the optimality of A^* to the program (2.35) without the constraint $A\mathbf{1} = \mathbf{0}$.

To solve the problem in (2.35), let $X = U_X\Sigma_X V_X^\top$ and $A = U_A\Sigma_A V_A^\top$ be the SVDs of X and A, respectively. Then, letting $U = U_X^\top U_A$ and $V = V_X^\top V_A$, we have

$$\|X - A\|_F^2 = \|U_X \Sigma_X V_X^\top - U_A \Sigma_A V_A^\top\|_F^2 = \|\Sigma_X - U \Sigma_A V^\top\|_F^2 \tag{2.40}$$

$$= \|\Sigma_X\|_F^2 - 2\langle \Sigma_X, U\Sigma_A V^\top \rangle + \|\Sigma_A\|_F^2. \tag{2.41}$$

Therefore, minimizing $\|X - A\|_F^2$ with respect to A is equivalent to minimizing the above expression with respect to U, V, and Σ_A. We will solve this problem in two steps.

In the first step, we will minimize with respect to U and V only. Notice that this is equivalent to

$$\max_{U,V} \langle \Sigma_X, U\Sigma_A V^\top \rangle, \tag{2.42}$$

where U and V are orthonormal. The solution to this problem can be found from Von Neumann's inequality, which is stated next.

Lemma 2.5 (Von Neumann's Inequality). *For any $m \times n$ real-valued matrices F and G, let $\sigma_1(F) \geq \sigma_2(F) \geq \cdots \geq 0$ and $\sigma_1(G) \geq \sigma_2(G) \geq \cdots \geq 0$ be the descending singular values of F and G respectively. Then*

$$\langle F, G \rangle = \text{trace}(F^\top G) \leq \sum_{i=1}^{n} \sigma_i(F)\sigma_i(G). \tag{2.43}$$

The case of equality occurs if and only if it is possible to find orthonormal matrices U_F and V_F that simultaneously singular value decompose *F and G in the sense that*

$$F = U_F \Sigma_F V_F^\top \quad \text{and} \quad G = U_F \Sigma_G V_F^\top, \tag{2.44}$$

where Σ_F and Σ_G denote the $m \times n$ diagonal matrices with the singular values of F and G, respectively, down the diagonal.

Proof. See (Mirsky 1975). □

Applying this lemma to $F = \Sigma_X$ and $G = U\Sigma_A V^\top$, we obtain

$$\langle \Sigma_X, U\Sigma_A V^\top \rangle \leq \sum_{i=1}^{d} \sigma_i(X)\sigma_i(A), \tag{2.45}$$

because $\sigma_i(A) = 0$ for $i > d$. Notice also that equality can be achieved by taking $U = I$ and $V = I$; hence $U_A = U_X$ and $V_A = V_X$.

In the second step, we will substitute the above solutions for U and V into the objective function $\|X - A\|_F^2$ and optimize over $\{\sigma_i(A)\}_{i=1}^d$. This gives the following optimization problem:

$$\min_{\{\sigma_i(A)\}_{i=1}^d} \sum_{i=1}^{d} \sigma_i(A)^2 - 2\sum_{i=1}^{d} \sigma_i(X)\sigma_i(A). \tag{2.46}$$

Taking the derivatives with respect to $\sigma_i(A)$ and setting them to zero gives us $\sigma_i(A) = \sigma_i(X)$ for $i = 1, \ldots, d$. The value of the objective function is $-\sum_{i=1}^{d} \sigma_i(X)^2$, which reaches the minimum when the d singular values are the largest. We thus have the following result.

Theorem 2.6 (PCA via Rank Minimization). *Let $X = U_X \Sigma_X V_X^\top$ be the singular value decomposition of the mean-subtracted data matrix. An optimal solution for the optimization problem*

$$\min_{A} \|X - A\|_F^2 \quad s.t. \quad \text{rank}(A) = d \tag{2.47}$$

is given by $A = U\Sigma V^\top$, where $U \in \mathbb{R}^{D\times d}$, $\Sigma \in \mathbb{R}^{d\times d}$ and $V \in \mathbb{R}^{N\times d}$ are matrices corresponding to the top d singular vectors and singular values in U_X, Σ_X and V_X, respectively.

Notice that this theorem is essentially equivalent to Theorem 2.3 and that the above derivation based on Von Neumann's inequality provides an alternative proof for the theorem.

In summary, we can view the PCA problem as a statistical problem, a geometric problem, or a rank minimization problem; and all three interpretations lead to the same solution.

Example 2.7 (PCA for Modeling Face Images under Varying Illumination) Face recognition is an area of computer vision in which low-dimensional linear models such as PCA and its variations have been extremely popular tools for capturing the variability of face images. In particular, it has been shown that under certain idealized circumstances (such as Lambertian reflectance), images of the same face under varying illumination lie near an approximately nine-dimensional linear subspace known as the *harmonic plane* (Basri and Jacobs 2003). The principal bases U estimated by PCA are also known as the *eigenfaces* in the computer vision literature (Turk and Pentland 1991), and they capture the principal modes of variation in the face images. The principal components Y estimated by PCA are often less dependent on illumination variations and are hence used for recognition purposes.

In this example, we show how PCA can be used to capture the illumination variations in a data set of face images of a person taken under different lighting conditions. The extended Yale B data set (Georghiades et al. 2001) is a popular data set used to study face recognition under varying lighting conditions. From this data set, we take as input for PCA $N = 10$ frontal face images of one individual (subject 20), shown in Figure 2.2. Each image is of size $D = 192 \times 168$ pixels. We apply PCA to the ten input images and compute the first $d = 2$ principal components and the first two eigenvectors $\boldsymbol{u}_1$ and $\boldsymbol{u}_2$, i.e., the first two eigenfaces.

Fig. 2.2 Face images of subject 20 under 10 different illumination conditions in the extended Yale B data set. All images are frontal faces cropped to size 192×168.

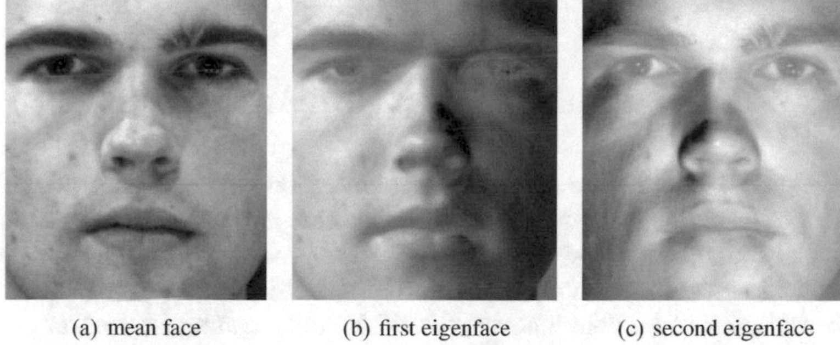

(a) mean face (b) first eigenface (c) second eigenface

Fig. 2.3 Mean face and the first two eigenfaces by applying PCA to the ten images in Figure 2.2.

Figure 2.3 shows the obtained mean face $\boldsymbol{\mu}$ and the first two eigenfaces $\boldsymbol{u}_1$ and $\boldsymbol{u}_2$.[4] Figure 2.3 demonstrates that the first two eigenfaces capture lighting from the right and up directions, respectively. This is visualized more clearly in Figure 2.4. The first row plots how the appearance of the face changes along the direction of the first eigenface: $\boldsymbol{\mu} + y_1\boldsymbol{u}_1$ for $y_1 = -\sigma_1 : \frac{\sigma_1}{3} : \sigma_1$; and the second row plots variations along the second eigenface: $\boldsymbol{\mu} + y_2\boldsymbol{u}_2$ for $y_2 = -\sigma_2 : \frac{\sigma_2}{3} : \sigma_2$, in which σ_1 and σ_2 are the standard deviations of the first and second principal components, respectively. Now it is clear from this experiment that for this data set, the first two eigenfaces mainly encode how the appearance of the face varies along the horizontal and vertical lighting directions, respectively.

[4]In Section 1.2.1, we have seen an example in which a similar process can be applied to an ensemble of face images from multiple subspaces, where the first $d = 3$ principal components are calculated and visualized.

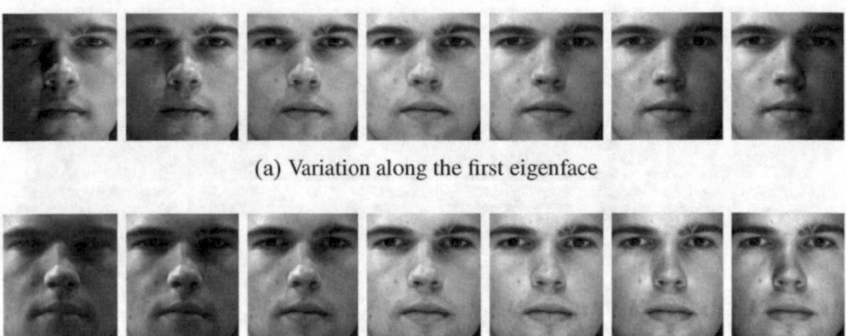

(a) Variation along the first eigenface

(b) Variation along the second eigenface

Fig. 2.4 Variation of the face images along the two eigenfaces given by PCA. Each row plots $\boldsymbol{\mu} + y_i \boldsymbol{u}_i$ for $y_i = -\sigma_i : \frac{\sigma_i}{3} : \sigma_i, i = 1, 2$, where σ_i is the standard deviation of the first or second principal component.

2.2 Probabilistic Principal Component Analysis (PPCA)

The PCA model described so far allows us to find a low-dimensional representation $\{\boldsymbol{y}_j \in \mathbb{R}^d\}$ of a set of sample points $\{\boldsymbol{x}_j \in \mathbb{R}^D\}$, where $d \ll D$ is the desired number of principal components. However, the PCA model is not a proper generative model, because the low-dimensional representation $\{\boldsymbol{y}_j\}$ and the error $\{\varepsilon_j\}$ are not treated as random variables. As a consequence, the PCA model cannot be used to generate new samples of the random variable $\boldsymbol{x}$.

To address this issue, we assume that the low-dimensional representation $\boldsymbol{y}$ and the error ε are independent random variables with probability density functions $p(\boldsymbol{y})$ and $p(\varepsilon)$, respectively. This allows us to generate a new sample of $\boldsymbol{x}$ from samples of $\boldsymbol{y}$ and ε as

$$\boldsymbol{x} = \boldsymbol{\mu} + B\boldsymbol{y} + \varepsilon, \tag{2.48}$$

where $\boldsymbol{\mu} \in \mathbb{R}^D$ and $B \in \mathbb{R}^{D\times d}$ represent a point and a basis for affine subspace S, respectively. Let the mean and covariance of $\boldsymbol{y}$ be denoted by $\boldsymbol{\mu}_y$ and Σ_y, respectively. If we assume that ε has zero mean and covariance Σ_ε, then the mean and covariance of the observations are given by

$$\boldsymbol{\mu}_x = \boldsymbol{\mu} + B\boldsymbol{\mu}_y \quad \text{and} \quad \Sigma_x = B\Sigma_y B^\top + \Sigma_\varepsilon. \tag{2.49}$$

Notice that in contrast to the PCA problem studied in the previous section, here we assume only that B is a rank-d matrix, but we no longer need to assume that B is orthonormal. This is because if we enforce a specific type of probability distribution for $\boldsymbol{y}$, we can then estimate an optimal model from the observations $\boldsymbol{x}$ without any additional constraints on the matrix B via the maximum likelihood (ML) principle (see Appendix B.1.4). The remainder of this section discusses two methods for estimating the parameters of this model, $\boldsymbol{\mu}$, B, $\boldsymbol{\mu}_y$, Σ_y, and Σ_ε, from the mean and covariance of the population, $\boldsymbol{\mu}_x$ and Σ_x, and alternatively from i.i.d. samples $\{\boldsymbol{x}_j\}_{j=1}^N$ of $\boldsymbol{x}$.

2.2.1 *PPCA from Population Mean and Covariance*

Observe that in general, we cannot uniquely recover the model parameters from $\boldsymbol{\mu}_x$ and Σ_x by solving the equations in (2.49). For instance, notice that $\boldsymbol{\mu}$ and $\boldsymbol{\mu}_y$ cannot be uniquely recovered from $\boldsymbol{\mu}_x$. Similarly to what we did in the case of PCA, this issue can be easily resolved by assuming that $\boldsymbol{\mu}_y = \mathbf{0}$. This leads to the following estimate of $\boldsymbol{\mu}$:

$$\hat{\boldsymbol{\mu}} = \boldsymbol{\mu}_x, \tag{2.50}$$

which is the same estimate as that of PCA (see Exercise 2.6). Another ambiguity that cannot be resolved in a straightforward manner is that Σ_y and Σ_ε cannot be uniquely recovered from Σ_x. For instance, $\Sigma_y = 0$ and $\Sigma_\varepsilon = \Sigma_x$ is a valid solution. However, this solution is not meaningful, because it assigns all the information in Σ_x to the error, rather than to the low-dimensional representation.

To resolve this ambiguity, we need to make some additional assumptions. Intuitively, we would like $B\Sigma_y B^\top$ to capture as much information about Σ_x as possible. Thus it makes sense for Σ_y to be of full rank and for Σ_ε to be as close to zero as possible. More specifically, the assumptions made in PPCA are the following:

1. The low-dimensional representation has unit covariance, i.e., $\Sigma_y = I_d \in \mathbb{R}^{d\times d}$.
2. The noise covariance matrix $\Sigma_\varepsilon \in \mathbb{R}^{D\times D}$ is isotropic, i.e., $\Sigma_\varepsilon = \sigma^2 I_D$.

Under these assumptions, the covariance of the observations must be of the form

$$\Sigma_x = BB^\top + \sigma^2 I_D. \tag{2.51}$$

It follows from this relationship that the eigenvalues of Σ_x must be equal to the eigenvalues of $BB^\top$ plus σ^2. Since $BB^\top$ has rank d and is positive semidefinite, $D-d$ eigenvalues of $BB^\top$ must be equal to zero. Therefore, the smallest $D-d$ eigenvalues of Σ_x must be equal to each other and equal to σ^2. In addition, the off-diagonal entries of Σ_x are equal to the off-diagonal entries of $BB^\top$. As a consequence, even though both PPCA and PCA try to capture as much information as possible from Σ_x into Σ_y, the information they attempt to capture is not the same. On the one hand, PPCA tries to find a matrix B such that the covariances are preserved, i.e., the off-diagonal entries of Σ_x. On the other hand, PCA tries to preserve the variances, i.e., the diagonal entries of Σ_x.

As it turns out, the parameters B and σ of the PPCA model can be computed in closed form from the SVD of the population covariance $\Sigma_{\boldsymbol{x}}$, as stated by the following theorem. Again, we emphasize that in the PPCA model, the matrix B can be an arbitrary matrix and does not need to be orthonormal.

Theorem 2.8 (PPCA from Population Mean and Covariance). *The parameters $\boldsymbol{\mu}$, B and σ of the PPCA model can be estimated from the population mean and covariance, $\boldsymbol{\mu}_{\boldsymbol{x}}$ and $\Sigma_{\boldsymbol{x}}$, respectively, as*

$$\hat{\boldsymbol{\mu}} = \boldsymbol{\mu}_{\boldsymbol{x}}, \ \ \widehat{B} = U(\Lambda - \hat{\sigma}^2 I)^{1/2} R, \ \ \hat{\sigma}^2 = \lambda_{d+1} = \lambda_{d+2} = \cdots = \lambda_D, \tag{2.52}$$

where U is the matrix with the top d eigenvectors of $\Sigma_{\boldsymbol{x}}$, Λ is the diagonal matrix in $\mathbb{R}^{d\times d}$ of the corresponding top d eigenvalues, $R \in \mathbb{R}^{d\times d}$ is an arbitrary orthogonal matrix, and λ_i is the ith eigenvalue of $\Sigma_{\boldsymbol{x}}$.

Proof. We have already shown in (2.50) that $\hat{\boldsymbol{\mu}} = \boldsymbol{\mu}_{\boldsymbol{x}}$. We have also shown that σ^2 is equal to the smallest $D - d$ eigenvalues of $\Sigma_{\boldsymbol{x}}$. To find B, let

$$\Sigma_x = \begin{bmatrix} U & V \end{bmatrix} \begin{bmatrix} \Lambda_1 & \mathbf{0} \\ \mathbf{0} & \sigma^2 I_{D-d} \end{bmatrix} \begin{bmatrix} U & V \end{bmatrix}^\top \tag{2.53}$$

be the eigenvalue decomposition of $\Sigma_{\boldsymbol{x}}$, where the columns of U are the top d eigenvectors of Σ_x and the entries of Λ are the corresponding eigenvalues. Then,

$$BB^\top = \Sigma_x - \sigma^2 I_D = \begin{bmatrix} U & V \end{bmatrix} \begin{bmatrix} \Lambda - \sigma^2 I_d & \mathbf{0} \\ \mathbf{0} & \mathbf{0} \end{bmatrix} \begin{bmatrix} U & V \end{bmatrix}^\top \tag{2.54}$$

$$= U(\Lambda - \sigma^2 I_d) U^\top. \tag{2.55}$$

Since both B and U are of rank d, all the solutions for B must be of the form $B = U(\Lambda - \sigma^2 I_d)^{1/2} R$, where R is an arbitrary orthogonal matrix. □

2.2.2 *PPCA by Maximum Likelihood*

In practice, we may not know the population mean and covariance, $\boldsymbol{\mu}_{\boldsymbol{x}}$ and $\Sigma_{\boldsymbol{x}}$. Instead, we are given N i.i.d. samples, $\{\boldsymbol{x}_j\}_{j=1}^N$, from which we wish to estimate the PPCA model parameters $\boldsymbol{\mu}$, B, and σ. In this section, we show that the ML estimates (see Appendix B.1.4) of these parameters can be computed in closed form from the ML estimates of the mean and covariance.

To that end, assume that $\boldsymbol{y}$ and ε are zero-mean Gaussian random variables with covariances I_d and $\sigma^2 I_D$, respectively, i.e., $\boldsymbol{y} \sim \mathcal{N}(\mathbf{0}, I)$ and $\varepsilon \sim \mathcal{N}(\mathbf{0}, \sigma^2 I)$. Then $\boldsymbol{x} \sim \mathcal{N}(\boldsymbol{\mu}_{\boldsymbol{x}}, \Sigma_x)$, where $\boldsymbol{\mu}_{\boldsymbol{x}} = \boldsymbol{\mu}$ and $\Sigma_x = BB^\top + \sigma^2 I_D$. Therefore, the log-likelihood of $\boldsymbol{x}$ is given by

$$
\begin{aligned}
\mathscr{L} &= \sum_{j=1}^{N} \log\Big(\frac{1}{(2\pi)^{D/2}\det(\Sigma_x)^{1/2}} \exp\big(-\frac{(\boldsymbol{x}_j-\boldsymbol{\mu}_x)^\top \Sigma_x^{-1}(\boldsymbol{x}_j-\boldsymbol{\mu}_x)}{2}\big)\Big) \\
&= -\frac{ND}{2}\log(2\pi) - \frac{N}{2}\log\det(\Sigma_x) - \frac{1}{2}\sum_{j=1}^{N}(\boldsymbol{x}_j-\boldsymbol{\mu})^\top \Sigma_x^{-1}(\boldsymbol{x}_j-\boldsymbol{\mu}).
\end{aligned} \tag{2.56}
$$

We obtain the ML estimate for $\boldsymbol{\mu}$ from the derivatives of $\mathscr{L}$ with respect to $\boldsymbol{\mu}$ as

$$
\frac{\partial \mathscr{L}}{\partial \boldsymbol{\mu}} = -\sum_{j=1}^{N} \Sigma_x^{-1}(\boldsymbol{x}_j - \boldsymbol{\mu}) = \boldsymbol{0} \implies \hat{\boldsymbol{\mu}} = \hat{\boldsymbol{\mu}}_N \doteq \frac{1}{N}\sum_{j=1}^{N}\boldsymbol{x}_j. \tag{2.57}
$$

After replacing $\hat{\boldsymbol{\mu}}$ in the log-likelihood, we obtain

$$
\mathscr{L} = -\frac{ND}{2}\log(2\pi) - \frac{N}{2}\log\det(\Sigma_x) - \frac{N}{2}\operatorname{trace}(\Sigma_x^{-1}\hat{\Sigma}_N), \tag{2.58}
$$

where

$$
\hat{\Sigma}_N \doteq \frac{1}{N}\sum_{j=1}^{N}(\boldsymbol{x}_j - \hat{\boldsymbol{\mu}}_N)(\boldsymbol{x}_j - \hat{\boldsymbol{\mu}}_N)^\top. \tag{2.59}
$$

The answer to the question whether B and σ can be estimated as in Theorem 2.8 after replacing Σ_x by $\hat{\Sigma}_N$ is given by the following theorem.

Theorem 2.9 (PPCA by Maximum Likelihood). *The ML estimates for the parameters of the PPCA model $\boldsymbol{\mu}$, B, and σ can be obtained from the ML estimates of the mean and covariance of the data, $\hat{\boldsymbol{\mu}}_N$ and $\hat{\Sigma}_N$, respectively, as*

$$
\hat{\boldsymbol{\mu}} = \hat{\boldsymbol{\mu}}_N, \quad \widehat{B} = U_1(\Lambda_1 - \hat{\sigma}^2 I)^{1/2}R \quad \textit{and} \quad \hat{\sigma}^2 = \frac{1}{D-d}\sum_{i=d+1}^{D}\lambda_i, \tag{2.60}
$$

where U_1 is the matrix with the top d eigenvectors of $\hat{\Sigma}_N$, Λ_1 is the matrix with the corresponding top d eigenvalues, $R \in \mathbb{R}^{d\times d}$ is an arbitrary orthogonal matrix, and λ_i is the ith-largest eigenvalue of $\hat{\Sigma}_N$.

Proof. We have already shown that $\hat{\boldsymbol{\mu}} = \hat{\boldsymbol{\mu}}_N$. To find B, we need to compute the derivatives of $\mathscr{L}$ with respect to B. It follows from Exercise 2.4 that $\frac{\partial}{\partial X}\log(|\det(X)|) = (X^{-1})^\top$, $\frac{\partial}{\partial X}\operatorname{trace}(AX^{-1}B) = -(X^{-1}BAX^{-1})^\top$ and $\frac{\partial}{\partial X}\operatorname{trace}(XBX^\top) = XB^\top + XB$. Therefore,

$$
\frac{\partial \mathscr{L}}{\partial B} = -N\Sigma_x^{-1}B + N\Sigma_x^{-1}\hat{\Sigma}_N\Sigma_x^{-1}B = 0 \implies \hat{\Sigma}_N\Sigma_x^{-1}B = B. \tag{2.61}
$$

One possible solution is $B = 0$, which leads to a minimum of the log-likelihood and violates our assumption that B should be of full rank d. Another possible solution is $\Sigma_x = \hat{\Sigma}_N$, where the covariance model is exact. This corresponds to the case discussed in the previous section, after replacing Σ_x by $\hat{\Sigma}_N$. Thus, the model parameters can be computed as in Theorem 2.8, since equation (2.60) reduces to equation (2.52). A third solution is obtained when $B \neq 0$ and $\Sigma_x \neq \hat{\Sigma}_N$. In this case, let $B = Z\Gamma V^\top$ be the compact SVD of B, where $Z \in \mathbb{R}^{D\times d}$ is a matrix with orthonormal columns, $\Gamma \in \mathbb{R}^{d\times d}$ is an invertible diagonal matrix, and $V \in \mathbb{R}^{d\times d}$ is an orthogonal matrix. Let $Z^\perp \in \mathbb{R}^{D\times(D-d)}$ be an orthonormal matrix such that $Z^\top Z^\perp = \mathbf{0}$, so that the matrix $\left[Z\ Z^\perp\right]$ is orthonormal and $ZZ^\top + Z^\perp Z^{\perp\top} = I$. Then

$$\Sigma_x = Z\Gamma^2 Z^\top + \sigma^2 I_D = Z(\Gamma^2 + \sigma^2 I_d)Z^\top + \sigma^2 Z^\perp Z^{\perp\top}. \tag{2.62}$$

Combining this with (2.61) gives

$$\hat{\Sigma}_N \Sigma_x^{-1} B = \hat{\Sigma}_N (Z(\Gamma^2 + \sigma^2 I_d)^{-1} Z^\top + \sigma^{-2} Z^\perp Z^{\perp\top}) Z\Gamma V^\top \tag{2.63}$$

$$= \hat{\Sigma}_N Z(\Gamma^2 + \sigma^2 I_d)^{-1}\Gamma V^\top = Z\Gamma V^\top, \tag{2.64}$$

whence

$$\hat{\Sigma}_N Z = Z(\Gamma^2 + \sigma^2 I_d). \tag{2.65}$$

Letting $Z = [z_1, \ldots, z_d]$ and $\Gamma = \mathrm{diag}\{\gamma_1, \ldots, \gamma_d\}$, we obtain

$$\hat{\Sigma}_N z_i = (\gamma_i^2 + \sigma^2) z_i \quad \forall i = 1, \ldots, d. \tag{2.66}$$

Hence, Z is a matrix containing d eigenvectors of $\hat{\Sigma}_N$ with corresponding eigenvalues $\gamma_i^2 + \sigma^2$. Let $\hat{\Sigma}_N = U\Lambda U^\top = [U_1, U_2]\mathrm{diag}\{\Lambda_1, \Lambda_2\}[U_1, U_2]^\top$ be the eigenvalue decomposition of $\hat{\Sigma}_N$, where we partition U and Λ so that the d chosen eigenvectors and eigenvalues are in U_1 and Λ_1, respectively. Then all optimal solutions for B are of the form

$$B = Z\Gamma V^\top = U_1(\Lambda_1 - \sigma^2 I_d)^{1/2} V^\top. \tag{2.67}$$

To determine σ, we replace the solution for B in the likelihood in (2.58). Noticing that

$$\det(\Sigma_x) = \det\left(BB^\top + \sigma^2 I_D\right) \tag{2.68}$$

$$= \det\left(U_1(\Lambda_1 - \sigma^2 I_d)U_1^\top + \sigma^2(U_1 U_1^\top + U_2 U_2^\top)\right) \tag{2.69}$$

$$= \det(U_1\Lambda_1 U_1^\top + \sigma^2 U_2 U_2^\top) = \det(\Lambda_1)\sigma^{2(D-d)} \tag{2.70}$$

and that

$$\operatorname{trace}(\Sigma_x^{-1}\hat{\Sigma}_N) = \operatorname{trace}\big((U_1\Lambda_1^{-1}U_1^\top + \sigma^{-2}U_2U_2^\top)(U_1\Lambda_1U_1^\top + U_2\Lambda_2U_2^\top)\big) \tag{2.71}$$

$$= \operatorname{trace}\big(U_1U_1^\top + \sigma^{-2}U_2\Lambda_2U_2^\top\big) = d + \sigma^{-2}\operatorname{trace}\big(\Lambda_2\big), \tag{2.72}$$

we obtain

$$\mathscr{L} = -\frac{N}{2}\big(D\log(2\pi) + \log\det(\Lambda_1) + (D-d)\log\sigma^2 + d + \sigma^{-2}\operatorname{trace}(\Lambda_2)\big). \tag{2.73}$$

The condition for an extremum in σ^2 is given by

$$\frac{\partial\mathscr{L}}{\partial\sigma^2} = -\frac{N}{2}\left(\frac{D-d}{\sigma^2} - \frac{\operatorname{trace}(\Lambda_2)}{\sigma^4}\right) = 0 \implies \sigma^2 = \frac{\operatorname{trace}(\Lambda_2)}{D-d}. \tag{2.74}$$

Therefore, σ^2 is the average of the discarded eigenvalues of $\hat{\Sigma}_N$.

To determine which d eigenvectors and eigenvalues of $\hat{\Sigma}_N$ should be discarded, notice that $\det(\Lambda_1) = \frac{\det(\Lambda)}{\det(\Lambda_2)}$. Hence, after substituting the optimal σ^2 in (2.74) into $\mathscr{L}$, we can see that the maximization of $\mathscr{L}$ is equivalent to the minimization of

$$\mathcal{M} = \log\left(\frac{\sum_{i=d+1}^{D}\lambda_{\pi[i]}}{D-d}\right) - \frac{\sum_{i=d+1}^{D}\log\lambda_{\pi[i]}}{D-d}, \tag{2.75}$$

with respect to a permutation π of all the eigenvalues such that $\lambda_{\pi[1]},\dots,\lambda_{\pi[d]}$ are the chosen eigenvalues and $\lambda_{\pi[d+1]},\dots,\lambda_{\pi[D]}$ are the discarded ones. Since the log function is concave, by Jensen's inequality, $\mathcal{M}$ is nonnegative, and the reader can verify (see Exercise 2.13) that $\mathcal{M}$ is minimized when the discarded eigenvalues are contiguous within the spectrum of the ordered eigenvalues of $\hat{\Sigma}_N$. Further, since the chosen eigenvalues must be such that $\lambda_{\pi[i]} \geq \sigma^2$ for $i = 1,\dots,d$, the discarded eigenvalues must be the $D-d$ smallest eigenvalues. Indeed, if such were not the case, then $\lambda_{\min} = \min_{i=1,\dots,D} \lambda_i$ would be one of the chosen eigenvalues, and we would have $\lambda_{\min} < \sigma^2$, which would be a contradiction to equation (2.66). Therefore, the optimal solutions for B and σ are given by (2.60). Finally, the optimal log-likelihood is given by

$$\mathscr{L} = -\frac{N}{2}\Big(D\log(2\pi) + \sum_{i=1}^{d}\log\lambda_i + (D-d)\log\left(\frac{\sum_{i=d+1}^{D}\lambda_i}{D-d}\right) + D\Big). \tag{2.76}$$

□

Once the parameters of the PPCA model have been identified, one question that arises is how to find the principal components of a specific data point, say $\boldsymbol{x}$. Recall that in both statistical and geometric PCA, the principal components $\boldsymbol{y}$ of $\boldsymbol{x}$ are found as $\boldsymbol{y} = U^\top(\boldsymbol{x} - \boldsymbol{\mu})$, where $(\boldsymbol{\mu}, U)$ are the parameters of the PCA model.

The fundamental difference in PPCA is that we have a proper generative model for the joint distribution of $\boldsymbol{x}$ and $\boldsymbol{y}$. As a consequence, it doesn't make sense to ask for a specific vector $\boldsymbol{y}$ associated with $\boldsymbol{x}$. Instead, we can ask for the entire distribution of $\boldsymbol{y}$ given $\boldsymbol{x}$.

It is easy to show (see Exercise 2.14) that the conditional distribution of $\boldsymbol{y}$ given $\boldsymbol{x}$ is Gaussian, i.e., $\boldsymbol{y} \mid \boldsymbol{x} \sim \mathcal{N}(\boldsymbol{\mu}_{y|x}, \Sigma_{y|x})$, where

$$\boldsymbol{\mu}_{y|x} = (B^\top B + \sigma^2 I)^{-1} B^\top (\boldsymbol{x} - \boldsymbol{\mu}) \quad \text{and} \quad \Sigma_{y|x} = \sigma^2 (B^\top B + \sigma^2 I)^{-1}. \tag{2.77}$$

Therefore, given $\boldsymbol{x}$, we can sample from this distribution in order to obtain its principal components. In practice, however, if the goal is dimensionality reduction, we may be interested in finding only one set of principal components. In this case, we can choose, for example, the mean of the distribution and define the *probabilistic principal components* of $\boldsymbol{x}$ as

$$\boldsymbol{y} = (B^\top B + \sigma^2 I)^{-1} B^\top (\boldsymbol{x} - \boldsymbol{\mu}). \tag{2.78}$$

Now, using the maximum likelihood estimates given in Theorem 2.9, we have

$$B^\top B + \sigma^2 I = R^\top (\Lambda_1 - \sigma^2 I)^{1/2} U_1^\top U_1 (\Lambda_1 - \sigma^2 I)^{1/2} R + \sigma^2 I \tag{2.79}$$

$$= R^\top (\Lambda_1 - \sigma^2 I) R + \sigma^2 I = R^\top \Lambda_1 R. \tag{2.80}$$

Thus, $\boldsymbol{y} = R^\top \Lambda_1^{-1} (\Lambda_1 - \sigma^2 I)^{1/2} U_1^\top (\boldsymbol{x} - \boldsymbol{\mu})$. Therefore, we can see that the main difference between the principal components and the probabilistic principal components is that the latter are a scaled and rotated version of the former, where the scales are given by the diagonal entries of the diagonal matrix $\Lambda_1^{-1}(\Lambda_1 - \sigma^2 I)^{1/2}$, and the rotation is given by R.

Example 2.10 (PPCA for Modeling Faces Images under Varying Illumination) In this example, we apply the PPCA algorithm with $d = 2$ to the same ten face images used in Example 2.7, as shown in Figure 2.2. The obtained mean face and eigenfaces are shown in Figure 2.5. Notice that if we compare to the results in Figure 2.3, the mean face and the first two eigenfaces computed from PCA and PPCA are rather similar: the mean should be the same by construction; and the eigenfaces differ only by a scale factor but appear the same when plotted as images. We also plot the variation along the two eigenfaces by computing $\boldsymbol{\mu} + y_i \boldsymbol{u}_i$ for $y_1 = -1 : \frac{1}{3} : 1$, $i = 1, 2$. The results are shown in Figure 2.6. Observe that the first principal component captures variations in illumination in the horizontal direction, while the second principal component captures variations in brightness (from dark to bright images). In Exercise 2.17, we ask the reader to implement PCA and PPCA and compare the results of the two methods on other data sets.

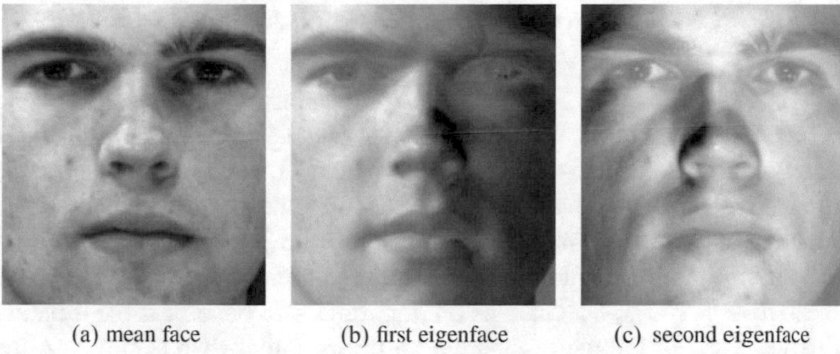

(a) mean face (b) first eigenface (c) second eigenface

Fig. 2.5 Mean face and the first two eigenfaces by applying PPCA to the ten images in Figure 2.2.

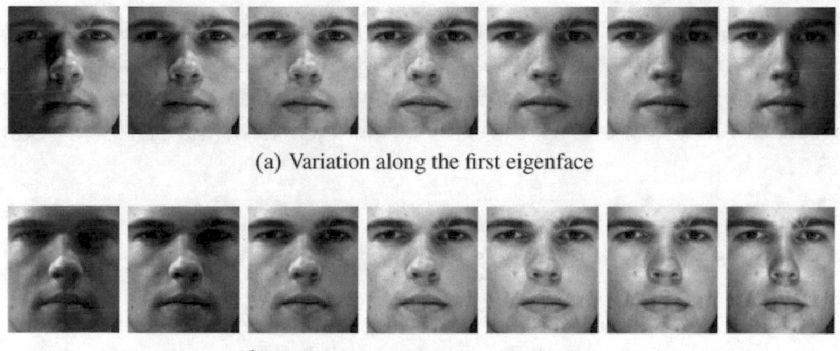

(a) Variation along the first eigenface

(b) Variation along the second eigenface

Fig. 2.6 Variation of the face images along the two eigenfaces given by PPCA. Each row plots $\boldsymbol{\mu} + y_i\boldsymbol{u}_i$ for $y_i = -1 : \frac{1}{3} : 1, i = 1, 2$.

2.3 Model Selection for Principal Component Analysis

One of the main goals of both PCA and PPCA is to reduce the data to a small number of principal components that capture as much information about the data as possible. So far, we have assumed that the number d of principal components or the dimension d of the subspace S is known. In practice, however, we may not know the intrinsic dimension of the data. In this section, we review a few methods (several of them heuristic) for estimating the number of principal components. Some of them are based on the model selection criteria described in Appendix B, while others rely on more modern rank minimization techniques. However, we would like to emphasize that model selection is in general a difficult problem, especially when the amount of noise in the data is unknown.

2.3.1 Model Selection by Information-Theoretic Criteria

Let $X = [\boldsymbol{x}_1, \boldsymbol{x}_2, \dots, \boldsymbol{x}_N] \in \mathbb{R}^{D\times N}$ be the mean-subtracted data matrix. When the data points are noise-free, they lie exactly in a subspace of dimension d. Hence, we can estimate d as the rank of X, i.e., $d = \text{rank}(X)$. However, when the data are contaminated by noise, the matrix X will be of full rank in general; hence we cannot use its rank to estimate d. Nonetheless, notice that the SVD of the noisy data matrix X gives a solution to PCA not only for a particular dimension d of the subspace, but also for all $d = 1, 2, \dots, D$. This has an important side benefit: if the dimension of the subspace S is *not* known or specified a priori, rather than optimizing for both d and S simultaneously, we can easily look at the entire spectrum of solutions for different values of d to decide on the "best" estimate $\hat{d}$ for the dimension of the subspace d given the data X.

One possible criterion is to chose d as the dimension that minimizes the least-squares error between the given data X and its projection $\widehat{X}^d = [\hat{\boldsymbol{x}}_1^d, \hat{\boldsymbol{x}}_2^d, \dots, \hat{\boldsymbol{x}}_N^d]$ onto the subspace S of dimension d. As shown in the proof of Theorem 2.3, the least-squares error is given by the sum of the squares of the remaining singular values of X, i.e.,

$$J(d) \doteq \|X - \widehat{X}^d\|_F^2 = \sum_{j=1}^{N} \|\boldsymbol{x}_j - \hat{\boldsymbol{x}}_j^d\|^2 = \sum_{i=d+1}^{D} \sigma_i^2. \tag{2.81}$$

However, this is not a good criterion, because $J(d)$ is a nonincreasing function of d. In fact, the best solution is obtained when $d = \text{rank}(X)$, because $J(d) = 0$.

The problem of determining the optimal dimension $\hat{d}$ is in fact a "model selection" problem. As we discussed in the introduction of the book, the conventional wisdom is to strike a good balance between the *complexity* of the chosen model and the *fidelity* of the data to the model. The dimension d of the subspace S is a natural measure of model complexity, while the least-squares error $\|X - \widehat{X}^d\|_F^2 = \sum_{i=d+1}^{D} \sigma_i^2$ or its leading term, σ_{d+1}^2, are natural measures of the data fidelity. Perhaps the simplest model selection criterion is to minimize the complexity subject to a bound on the fidelity. For example, we can choose d as the smallest number such that the fidelity is less than a threshold $\tau > 0$, i.e.,

$$\hat{d} = \min_{d} \left\{ d : \sum_{i=d+1}^{D} \sigma_i^2 < \tau \right\} \quad \text{or} \quad \hat{d} = \min_{d} \left\{ d : \sigma_{d+1}^2 < \tau \right\}. \tag{2.82}$$

The second criterion in (2.82) is illustrated in Figure 2.7.

In practice, however, it is very hard to choose an appropriate τ, because the singular values of X are not invariant with respect to transformations of the data, such as scaling. One possible solution is to normalize the singular values by $\|X\|_F^2 = \sum_{i=1}^{D} \sigma_i^2$ and estimate d as

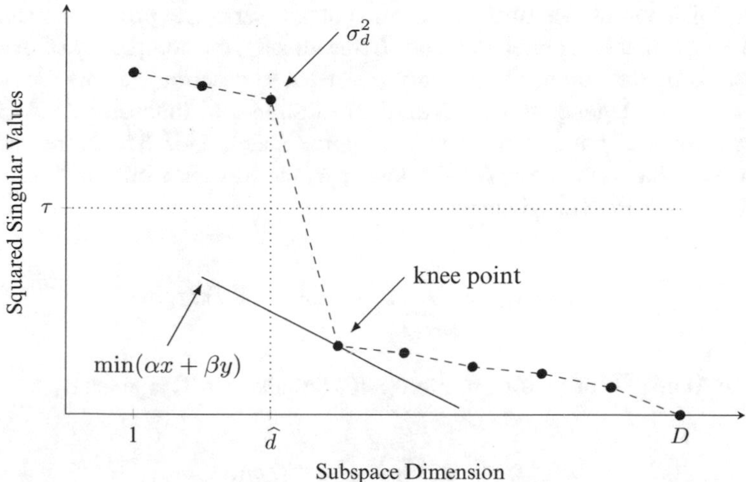

Fig. 2.7 Singular value as a function of the dimension of the subspace. Two model selection criteria are illustrated in this picture: choosing a subspace dimension based on a threshold and choosing a subspace dimension that trades off the model complexity (x-axis) and fidelity (y-axis), both linearly weighted.

$$\hat{d} = \min_d \left\{ d : \frac{\sum_{i=d+1}^{D} \sigma_i^2}{\sum_{i=1}^{D} \sigma_i^2} < \tau \right\} \quad \text{or} \quad \hat{d} = \min_d \left\{ d : \frac{\sigma_{d+1}^2}{\sum_{i=1}^{D} \sigma_i^2} < \tau \right\}. \tag{2.83}$$

The first criterion in (2.83) is widely used, because it has an intuitive interpretation: the number of principal components is chosen as the smallest number such that the fraction of information being discarded is less than a threshold τ. Typical values for τ are in the range 10%–20%.

Yet another model selection criterion seeks a balance between d and σ_{d+1}^2 by minimizing an objective function of the form

$$\hat{d} = \arg\min \tilde{J}(d) \doteq \alpha \cdot \sigma_{d+1}^2 + \beta \cdot d \tag{2.84}$$

for some proper weights $\alpha, \beta > 0$. In general, the graph of the ordered squared singular values of the data matrix X versus the dimension d of the subspace resembles a plot similar to that shown in Figure 2.7. In the statistics literature, this is known as the "scree graph," which was discussed and named by (Cattell 1966). Note that we should expect to see a significant drop in the singular values right after the "correct" dimension $\widehat{d}$, which is sometimes called the "knee" or "elbow" point of the plot. Such a point is a stable minimum, since it optimizes the above objective function (2.84) for a range of values for α and β.

A more principled approach to finding the optimal dimension $\widehat{d}$ of the subspace is to use some of the model selection criteria described in Appendix B. Such criteria rely on a different choice of the model complexity term and provide an automatic way of choosing the parameters α and β. Specifically, the complexity of the model is measured by the number of parameters needed to describe the data distribution. In the case of a degenerate Gaussian distribution in a d-dimensional subspace of $\mathbb{R}^D$, the number of parameters needed is approximately Dd.[5] Therefore, assuming that the noise has variance $\sigma^2 I_D$ with known σ, the Bayesian information criterion (BIC) (Rissanen 1978) is given by

$$\mathrm{BIC}(d) \doteq \sum_{i=d+1}^{D} \sigma_i^2 + (\log N)(Dd)\sigma^2, \tag{2.85}$$

while the Akaike information criterion (AIC) (Akaike 1977) is given by

$$\mathrm{AIC}(d) \doteq \sum_{i=d+1}^{D} \sigma_i^2 + 2(Dd)\sigma^2. \tag{2.86}$$

More recently, a geometric version of the Akaike information criterion has been proposed by (Kanatani 1998). The geometric AIC is given by

$$\text{G-AIC}(d) \doteq \sum_{i=d+1}^{D} \sigma_i^2 + 2(Dd + Nd)\sigma^2, \tag{2.87}$$

where the extra term Nd accounts for the number of coordinates needed to represent (the closest projection of) the given N data points in the estimated d-dimensional subspace. From an information-theoretic viewpoint, the additional Nd coordinates are necessary if we are interested in encoding not only the model but also the data themselves. This is often the case when we use PCA for purposes such as data compression and dimension reduction.

All the above criteria can be loosely referred to as *information-theoretic* model selection criteria, in the sense that most of these criteria can be interpreted as variations to minimizing the optimal code length for both the model and the data with respect to certain classes of distributions and coding schemes (Hansen and Yu 2001).[6] There are many other methods for determining the number of principal

[5]We leave as an exercise to the reader to calculate the number of parameters needed to specify a d-dimensional subspace in $\mathbb{R}^D$ and then the additional parameters needed to specify a Gaussian distribution inside the subspace.

[6]Even if one chooses to compare models by their algorithmic complexity, such as the minimum message length (MML) criterion (Wallace and Boulton 1968) (an extension of the Kolmogrov complexity to model selection), a strong connection with the above information-theoretic criteria, such as minimum description length (MDL), can be readily established via Shannon's optimal coding theory (see (Wallace and Dowe 1999)).

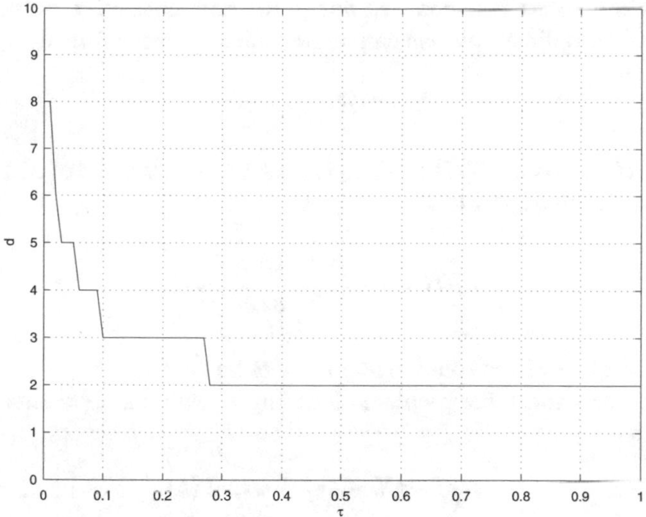

Fig. 2.8 Model selection for the ten face images in Figure 2.2.

components. The interested reader may find more references in (Jolliffe 1986, 2002). We note here that a more recent treatment of the automatic dimension selection for PCA was given through a Bayesian approach (Minka 2000), which assuming a basic noise model, derives an accurate approximation to the probability $p(X|d)$ from which the optimal dimension d^* can be determined.

Example 2.11 (Model Selection for Face Images) As an example, we apply the model selection criteria to the same face data set used in our previous experiments with PCA (Example 2.7) and PPCA (Example 2.10). More specifically, we apply the first dimension selection criterion in (2.83) to ten frontal face images of subject 20 from the extended Yale B data set. We vary the threshold τ in the range $(0, 1)$ and compute the corresponding dimension d. The result is shown in Figure 2.8. As we can see, the first three principal components already capture roughly 90% of the energy in the data.

2.3.2 Model Selection by Rank Minimization

In this section, we present an alternative view of model selection based on the rank minimization approach to PCA introduced in Section 2.1.3. In this approach, the PCA problem is posed as one of finding a rank-d matrix A that best approximates the mean-subtracted data matrix X, i.e.,

$$\min_{A} \quad \|X - A\|_F^2 \quad \text{s.t.} \quad \text{rank}(A) = d. \tag{2.88}$$

Although this problem is nonconvex due to the rank constraint, as we showed in Section 2.1.3, its optimal solution can be computed in closed form as

$$A = U\mathcal{H}_{\sigma_{d+1}}(\Sigma)V^\top, \tag{2.89}$$

where $X = U\Sigma V^\top$ is the SVD of X, σ_k is the kth singular value of X, and $\mathcal{H}_\varepsilon(x)$ is the *hard thresholding operator*:

$$\mathcal{H}_\varepsilon(x) = \begin{cases} x & |x| > \varepsilon \\ 0 & \text{else} \end{cases}. \tag{2.90}$$

However, this closed-form solution requires d to be known.

When d is unknown, the problem of finding a low-rank approximation can be formulated as

$$\min_A \quad \|X - A\|_F^2 + \tau \operatorname{rank}(A), \tag{2.91}$$

where $\tau > 0$ is a parameter. Since the optimal solution of (2.88) for a fixed rank $d = \operatorname{rank}(A)$ is $A = U\mathcal{H}_{\sigma_{d+1}}(\Sigma)V^\top$, the problem in (2.91) reduces to

$$\min_d \quad \sum_{k>d} \sigma_k^2 + \tau d. \tag{2.92}$$

The optimal solution is the smallest d such that $\sigma_{d+1}^2 \leq \tau$. Notice that this model selection criterion is the same as that in (2.82). Therefore, the optimization problem in (2.91) provides a justification for the criterion in (2.82). Under this criterion, and with the notation introduced in this section, the optimal A is given by

$$A = U\mathcal{H}_{\sqrt{\tau}}(\Sigma)V^\top. \tag{2.93}$$

Therefore, the optimal A can still be computed in closed form from the SVD of X, in spite of the fact that the optimization problem in (2.91) is nonconvex.

Most rank minimization problems are, however, NP-hard and cannot be solved as easily as the one in (2.91). This has motivated the development of convex relaxations, which lead to more efficient solutions. A commonly used relaxation (see, e.g., (Cai et al. 2008; Recht et al. 2010)) is to replace the rank of A by its nuclear norm $\|A\|_* = \sum \sigma_k(A)$, i.e., the sum of its singular values.[7] As it turns out, this relaxation leads to a slightly different model selection criterion for PCA. More specifically, the relaxation of (2.91) (modulo the $1/2$ factor) is given by

$$\min_A \quad \frac{1}{2}\|X - A\|_F^2 + \tau \|A\|_*. \tag{2.94}$$

[7]It can be shown that the nuclear norm is a convex envelope of the rank function for matrices.

While this function is not differentiable, one can use the convex hull of all directional derivatives of the function, referred to as the subgradient for optimization. The subgradient of this function with respect to A is given by $A - X + \partial\|A\|_*$, where $\partial\|A\|_*$ is the subgradient of the nuclear norm of A (see Exercise 2.16). Therefore, as shown in (Cai et al. 2008) (see also Exercise 2.16), the optimal solution for A is given by

$$A = \mathcal{D}_\tau(X) = U\mathcal{S}_\tau(\Sigma)V^\top, \tag{2.95}$$

where $\mathcal{D}_\varepsilon$ is the *singular value thresholding operator* and $\mathcal{S}_\varepsilon$ is the *soft thresholding operator*, which is defined as

$$\mathcal{S}_\varepsilon(x) = \operatorname{sign}(x)\max(|x| - \varepsilon, 0) = \begin{cases} x - \varepsilon & x > \varepsilon \\ x + \varepsilon & x < -\varepsilon \\ 0 & \text{else} \end{cases}. \tag{2.96}$$

Notice that the latter solution does not coincide with the one given by PCA, which performs hard-thresholding of the singular values of X without shrinking them by τ. However, the model selection criterion is the same as before: choose d as the smallest integer such that $\sigma_{d+1}^2 \leq \tau$.

2.3.3 Model Selection by Asymptotic Mean Square Error

From the above two sections, we see that by following different model selection criteria or objectives, we essentially have three different types of estimators $\hat{X}$ for a low-rank matrix X_0 from its noisy measurements: $X = X_0 + \sigma E$. If we denote the SVD of X by $X = U\Sigma V^\top$, the three estimators are of the following forms, respectively:

1. If the rank d is known, the optimal estimate $\hat{X}$ subject to $\operatorname{rank}(\hat{X}) = d$ is the *truncated SVD* solution:

$$\hat{X}_1 = U\mathcal{H}_{\sigma_{d+1}}(\Sigma)V^\top. \tag{2.97}$$

   Alternatively, if the rank d is not known and one uses one of the information-theoretic criteria given in Section 2.3.1 to estimate the dimension $\hat{d}$, then we have only to replace the d in the above solution with the estimated $\hat{d}$.
2. If we try to balance the mean squared error and the dimension as in equation (2.91), the optimal estimate is given by the *SVD hard thresholding*:

$$\hat{X}_2 = U\mathcal{H}_{\sqrt{\tau}}(\Sigma)V^\top \tag{2.98}$$

   for some threshold $\tau > 0$.

3. If we try to balance the mean squared error and the nuclear norm as in equation (2.94), then the optimal estimate is given by the *SVD soft thresholding*:

$$\hat{X}_3 = U\mathcal{S}_\tau(\Sigma)V^\top. \tag{2.99}$$

for some threshold $\tau > 0$.

Naturally, this may lead to a certain degree of confusion for practitioners: Which estimate is "the best"? What is the optimal threshold τ^* to use in case we need to threshold the singular values? Which thresholding is better, hard or soft? The short answer to these questions is that none of the above estimators is always better than the others, since each is optimal in its own way and thus useful under different conditions. However, if we all agree on a common objective based on a common noise model, it might be meaningful and even insightful to examine which estimator is better than others.

One such setting was recently proposed by (Donoho and Gavish 2014) to study the different estimators in terms of their mean square errors (MSE) in an asymptotic setting as the size of the matrix $X \in \mathbb{R}^{D\times N}$ becomes large. For simplicity, we first assume that the matrix X is a square matrix of size $N = D$. In the asymptotic setting (as $N \to \infty$), we assume the following noise model:

$$X = X_0 + \sigma E, \tag{2.100}$$

where E is a matrix whose entries are i.i.d. drawn from a probability (say Gaussian) distribution with zero mean and variance $1/\sqrt{N}$. It is easy to see that the noise level in the singular values of X is σ. Among all estimates of X_0 obtained by a hard thresholding of the singular values of X, we are interested in finding the one that minimizes the asymptotic mean square error:

$$\text{AMSE} = \lim_{N\to\infty} \|\hat{X} - X_0\|_F^2. \tag{2.101}$$

The work of (Donoho and Gavish 2014) gives the following answer to this question.

Proposition 2.12 (Optimal Hard Thresholding for Minimizing AMSE). *Given a low-rank matrix $X_0 \in \mathbb{R}^{D\times N}$ and noisy measurements $X = X_0 + \sigma E$ with E zero mean and variance $1/\sqrt{N}$, if the matrix is square, i.e., $D = N$, then the optimal hard threshold estimate $\hat{X} = U\mathcal{H}_{\tau^*}(\Sigma)V^\top$ that minimizes the asymptotic mean square error $\|\hat{X} - X_0\|_F^2$ is given by*

$$\tau^* = 4/\sqrt{3}\sigma \approx 2.309\sigma. \tag{2.102}$$

In the more general case of a nonsquare matrix with $D/N \to \beta$, the optimal threshold is given by

$$\tau^*(\beta) = \sigma \sqrt{2(\beta+1) + \frac{8\beta}{(\beta+1) + \sqrt{\beta^2 + 14\beta + 1}}}. \tag{2.103}$$

The proof of this statement is beyond the scope of this book. However, it is useful to discuss its implications in our context.

In can be shown that under the same noise model, the distribution of the singular values of the matrix $X = X_0 + \sigma E$ forms a quarter-circle bulk, whose radius lies approximately at $(1 + \sqrt{\beta})\sigma$. This is the place where we would normally expect to see a "knee point" in the distribution of singular values (as shown in Figure 2.7). The information-theoretic criteria or the rank-minimization objectives are most likely to choose this value to threshold the singular values. For a square matrix, this gives the threshold $\tau = 2\sigma$, which is close to but not quite at the optimal value 2.309σ. As shown in the work of (Donoho and Gavish 2014), this small difference in the choice of the threshold can result in a $5/3$-fold increase in AMSE.

Interestingly, even if we knew the correct rank d of the matrix X_0 and took the truncated SVD solution $\hat{X} = U\mathcal{H}_{\sigma_{d+1}}(\Sigma)V^\top$, the resulting AMSE would also be $5/3$ the size of the optimal hard thresholding solution given above. In general, soft thresholding does not work as well as hard thresholding in the high signal-to-noise ratio regime, and the AMSE for the optimal soft thresholding solution $\hat{X}_3 = U\mathcal{S}_{\tau^*}(\Sigma)V^\top$ is twice as large as that of hard thresholding. In fact, even if one is allowed to use any singular value shrinkage function instead of merely a hard or soft thresholding, compared to the above optimal hard thresholding solution (2.102), one can at best reduce the AMSE by another $1/3$ (see the work of (Shabalin and Nobel 2010) for more details).

2.4 Bibliographic Notes

As a matrix decomposition tool, SVD was initially developed independently from PCA in the numerical linear algebra literature, also known as the Eckart and Young decomposition (Eckart and Young 1936; Hubert et al. 2000). The result regarding the least-squares optimality of SVD given in Theorem 2.3 can be traced back to (Householder and Young 1938; Gabriel 1978). While principal components were initially defined exclusively in a statistical sense (Pearson 1901; Hotelling 1933), one can show that the algebraic solution given by SVD gives asymptotically unbiased estimates of the true parameters in the case of Gaussian distributions. A more detailed analysis of the statistical properties of PCA can be found in (Jolliffe 2002).

Note that PCA infers only the principal subspace (or components), but not a probabilistic distribution of the data in the subspace. Probabilistic PCA was developed to infer an explicit probabilistic distribution from the data (Tipping and Bishop 1999b). The data are assumed to be independent samples drawn from an unknown distribution, and the problem becomes one of identifying the subspace and the parameters of the distribution in a maximum likelihood or maximum a posteriori sense. When the underlying noise distribution is Gaussian, the geometric and probabilistic interpretations of PCA coincide (Collins et al. 2001). However, when the underlying distribution is non-Gaussian, the optimal solution to PPCA may no longer be linear. For example, in (Collins et al. 2001), PCA is generalized to arbitrary distributions in the exponential family.

2.5 Exercises

Exercise 2.1 (Properties of Symmetric Matrices). Let $S \in \mathbb{R}^{n\times n}$ be a real symmetric matrix. Prove the following:

1. All the eigenvalues of S are real, i.e., $\sigma(S) \subset \mathbb{R}$.
2. Let $(\lambda, \boldsymbol{u})$ be an eigenvalue–eigenvector pair. If $\lambda_i \neq \lambda_j$, then $\boldsymbol{u}_i \perp \boldsymbol{u}_j$; i.e., eigenvectors corresponding to distinct eigenvalues are orthogonal.
3. There always exist n orthonormal eigenvectors of S, which form a basis of $\mathbb{R}^n$.
4. The matrix S is positive definite (positive semidefinite) if and only if all of its eigenvalues are positive (nonnegative), i.e., $S \succ 0$ ($S \succeq 0$) iff $\forall i = 1, 2, \ldots, n$, $\lambda_i > 0$ ($\lambda_i \geq 0$).
5. If $\lambda_1 \geq \lambda_2 \geq \cdots \geq \lambda_n$ are the sorted eigenvalues of S, then $\max_{\|\boldsymbol{x}\|_2=1} \boldsymbol{x}^\top S\boldsymbol{x} = \lambda_1$ and $\min_{\|\boldsymbol{x}\|_2=1} \boldsymbol{x}^\top S\boldsymbol{x} = \lambda_n$.

Exercise 2.2 (Pseudoinverse of a Matrix).

1. Let $A = U_r\Sigma_r V_r^\top$ be the compact SVD of a matrix A of rank r. Show that a pseudoinverse of A is given by $A^\dagger = V_r\Sigma_r^{-1}U_r^\top$.
2. Consider the linear system of equations $A\boldsymbol{x} = \boldsymbol{b}$, where the matrix $A \in \mathbb{R}^{m\times n}$ is of rank $r = \operatorname{rank}(A) = \min\{m, n\}$. Show that $\boldsymbol{x}^* = A^\dagger \boldsymbol{b}$ minimizes $\|A\boldsymbol{x} - \boldsymbol{b}\|_2^2$, where $A^\dagger$ is a pseudoinverse of A defined in part 1. When is $\boldsymbol{x}^*$ the unique solution?

Exercise 2.3 (Convex Sets and Functions). Show the following:

1. The intersection of two convex sets is convex.
2. Let $f : \mathcal{X} \to \mathbb{R}$ be a convex function defined over a convex domain $\mathcal{X} \subseteq \mathbb{R}^n$. Show that for every $c \in \mathbb{R}$, the set $\{\boldsymbol{x} \in \mathcal{X} : f(\boldsymbol{x}) \leq c\}$ is convex.
3. A convex function is pseudoconvex and quasiconvex.

Exercise 2.4 (Derivatives of Traces and Logarithms). Show that

1. $\frac{\partial}{\partial X} \operatorname{trace}(AXB) = A^\top B^\top$
2. $\frac{\partial}{\partial X} \operatorname{trace}(AX^{-1}B) = -(X^{-1}BAX^{-1})^\top$.
3. $\frac{\partial}{\partial X} \operatorname{trace}(A \otimes X) = \operatorname{trace}(A)I$.
4. $\frac{\partial}{\partial X} \operatorname{trace}(X^\top BX) = BX + B^\top X$.
5. $\frac{\partial}{\partial X} \operatorname{trace}(XBX^\top) = XB^\top + XB$.
6. $\frac{\partial}{\partial X} \log|\det(X)| = (X^{-1})^\top$.

Exercise 2.5 (Maximum Likelihood Estimates of the Parameters of a Gaussian). Let $\boldsymbol{x} \in \mathbb{R}^D$ be a random vector. Let $\boldsymbol{\mu}_{\boldsymbol{x}} = \mathbb{E}(\boldsymbol{x}) \in \mathbb{R}^D$ and $\Sigma_{\boldsymbol{x}} = \mathbb{E}(\boldsymbol{x} - \boldsymbol{\mu})(\boldsymbol{x} - \boldsymbol{\mu})^\top \in \mathbb{R}^{D \times D}$ be, respectively, the mean and the covariance of $\boldsymbol{x}$. Given N i.i.d. samples $\{\boldsymbol{x}_j\}_{j=1}^N$, show that the maximum likelihood estimates of $\boldsymbol{\mu}_{\boldsymbol{x}}$ and $\Sigma_{\boldsymbol{x}}$ are, respectively, given by

$$\hat{\boldsymbol{\mu}}_N \doteq \frac{1}{N}\sum_{j=1}^N \boldsymbol{x}_j \quad \text{and} \quad \hat{\Sigma}_N \doteq \frac{1}{N}\sum_{j=1}^N (\boldsymbol{x}_j - \hat{\boldsymbol{\mu}}_N)(\boldsymbol{x}_j - \hat{\boldsymbol{\mu}}_N)^\top. \tag{2.104}$$

Exercise 2.6 (Statistical PCA for Nonzero-Mean Random Variables). Let $\boldsymbol{x} \in \mathbb{R}^D$ be a random vector. Let $\boldsymbol{\mu}_{\boldsymbol{x}} = \mathbb{E}(\boldsymbol{x}) \in \mathbb{R}^D$ and $\Sigma_{\boldsymbol{x}} = \mathbb{E}(\boldsymbol{x} - \boldsymbol{\mu})(\boldsymbol{x} - \boldsymbol{\mu})^\top \in \mathbb{R}^{D \times D}$ be, respectively, the mean and the covariance of $\boldsymbol{x}$. Define the principal components of $\boldsymbol{x}$ as the random variables $y_i = \boldsymbol{u}_i^\top \boldsymbol{x} + a_i \in \mathbb{R}$, $i = 1, \dots, d \le D$, where $\boldsymbol{u}_i \in \mathbb{R}^D$ is a unit-norm vector, $a_i \in \mathbb{R}$, and $\{y_i\}_{i=1}^n$ are zero-mean uncorrelated random variables whose variances are such that $\mathrm{Var}(y_1) \ge \mathrm{Var}(y_2) \ge \cdots \ge \mathrm{Var}(y_d)$. Assuming that the eigenvalues of $\Sigma_{\boldsymbol{x}}$ are distinct, show that

1. $a_i = -\boldsymbol{u}_i^\top \boldsymbol{\mu}_{\boldsymbol{x}}$, $i = 1, \dots, d$.
2. $\boldsymbol{u}_1$ is the eigenvector of $\Sigma_{\boldsymbol{x}}$ corresponding to its largest eigenvalue.
3. $\boldsymbol{u}_2^\top \boldsymbol{u}_1 = 0$, and $\boldsymbol{u}_2$ is the eigenvector of Σ corresponding to its second-largest eigenvalue.
4. $\boldsymbol{u}_i^\top \boldsymbol{u}_j = 0$ for all $i \ne j$ and $\boldsymbol{u}_i$ is the eigenvector of $\Sigma_{\boldsymbol{x}}$ corresponding to its ith-largest eigenvalue.

Exercise 2.7 (Properties of PCA). Let $\boldsymbol{x} \in \mathbb{R}^D$ be a random vector with covariance matrix $\Sigma_{\boldsymbol{x}} \in \mathbb{R}^{D \times D}$. Consider a linear transformation

$$\boldsymbol{y} = U^\top \boldsymbol{x}, \tag{2.105}$$

of $\boldsymbol{x}$, where $\boldsymbol{y} \in \mathbb{R}^d$ and $U \in \mathbb{R}^{D \times d}$ has orthonormal columns. Let $\Sigma_{\boldsymbol{y}} = U^\top \Sigma_{\boldsymbol{x}} U$ be the covariance matrix for $\boldsymbol{y}$. Show that

1. The trace of $\Sigma_{\boldsymbol{y}}$ is maximized by a matrix U whose columns are the first d unit eigenvectors of $\Sigma_{\boldsymbol{x}}$.
2. The trace of $\Sigma_{\boldsymbol{y}}$ is minimized by a matrix U whose columns are the last d unit eigenvectors of $\Sigma_{\boldsymbol{x}}$.

Exercise 2.8 (Principal Angles between Two Subspaces). Given two subspaces S_1 and S_2 of $\mathbb{R}^D$ with $d = \dim(S_1) \le \dim(S_2)$, the principal angles between the subspaces are defined recursively for $m = 1, \dots, d$ as

$$\cos(\theta^1) = \max_{\boldsymbol{u}_1 \in S_1} \max_{\boldsymbol{u}_2 \in S_2} \left\{ \langle \boldsymbol{u}_1, \boldsymbol{u}_2 \rangle : \|\boldsymbol{u}_1\| = \|\boldsymbol{u}_2\| = 1 \right\} = \angle(\boldsymbol{u}_1^1, \boldsymbol{u}_2^1) \tag{2.106}$$

$$\cos(\theta^m) = \max_{u_1 \in S_1} \max_{u_2 \in S_2} \Big\{ \langle u_1, u_2 \rangle : \|u_1\| = \|u_2\| = 1, u_1 \perp u_1^i, u_2 \perp u_2^i, \quad \forall i \in \{1, \ldots, m-1\} \Big\} = \angle(u_1^m, u_2^m). \tag{2.107}$$

Let $U_1 \in \mathbb{R}^{D \times d}$ be an orthogonal matrix whose columns form a basis for S_1 and similarly U_2 for S_2. Show that

$$\cos(\theta^m) = \sigma_m, \quad m = 1, \ldots, d, \tag{2.108}$$

where σ_m is the mth-largest singular value of the matrix $W = U_1^\top U_2$. Show also that the number of angles equal to zero is equal to $\dim(S_1 \cap S_2)$.

Hint: Following the derivation of statistical PCA, find first the smallest angle (largest cosine = largest variance) and then find the second-smallest angle all the way to the largest angle (smallest variance). As you proceed, the vectors that achieve the second-smallest angle need to be chosen to be orthogonal to the vectors that achieve the smallest angle, and so forth, as we did in statistical PCA. Also, let $u_1 = U_1 c_1$ and $u_2 = U_2 c_2$. Show that you need to optimize $\cos(\theta) = c_1^\top U_1^\top U_2 c_2$ subject to $\|c_1\| = \|c_2\| = 1$. Show (using Lagrange multipliers) that a necessary condition for optimality is

$$\begin{bmatrix} 0 & U_1^\top U_2 \\ U_2^\top U_1 & 0 \end{bmatrix} \begin{bmatrix} c_1 \\ c_2 \end{bmatrix} = \lambda \begin{bmatrix} c_1 \\ c_2 \end{bmatrix}. \tag{2.109}$$

Deduce that $\sigma = \lambda^2$ is a singular value of $U_1^\top U_2$ with c_2 as singular vector.

Exercise 2.9 (Fixed-Rank Approximation of a Matrix). Let $A = U \Sigma V^\top$ be the SVD of A. Let $B = U \Sigma_p V^\top$, where Σ_p denotes the matrix obtained from Σ by setting to zero its elements on the diagonal after the pth entry. Show that $\|A - B\|_F^2 = \sigma_{p+1}^2 + \cdots + \sigma_r^2$, where $\|\cdot\|_F$ indicates the Frobenius norm. Furthermore, show that such a norm is the minimum achievable over all matrices $B \in \mathbb{R}^{m \times n}$ of rank p, i.e.,

$$\min_{B:\operatorname{rank}(B)=p} \|A - B\|_F^2 = \sigma_{p+1}^2 + \cdots + \sigma_r^2. \tag{2.110}$$

Exercise 2.10 (Identification of Autoregressive (AR) Systems). A popular model that is often used to analyze a time series $\{y_t\}_{t \in \mathbb{Z}}$ is the linear autoregressive model

$$y_t = a_1 y_{t-1} + a_2 y_{t-2} + \cdots + a_n y_{t-n} + \varepsilon_t, \quad \forall t, y_t \in \mathbb{R}, \tag{2.111}$$

where $\varepsilon_t \in \mathbb{R}$ models the modeling error or noise and is often assumed to be a white-noise random process. Now suppose that you are given the values of y_t for a sufficiently long period of time.

1. Show that in the noise-free case, i.e., $\varepsilon_t \equiv 0$, regardless of the initial conditions, the vectors $x_t = [y_t, y_{t-1}, \ldots, y_{t-n}]^\top$ for all t lie on an n-dimensional hyperplane in $\mathbb{R}^{n+1}$. What is the normal vector to this hyperplane?

2. Now consider the case with noise. Describe how you may use PCA to identify the unknown model parameters $(a_1, a_2, \ldots, a_n)$.

Exercise 2.11 (Basis for an Image). Given a gray-level image $\boldsymbol{I}$, consider all of its $b \times b$ blocks, denoted by $\{B_i \in \mathbb{R}^{b \times b}\}$. We would like to approximate each block as a superposition of d base blocks, say $\{\hat{B}_j \in \mathbb{R}^{b \times b}\}_{j=1}^{d}$. That is,

$$B_i = \sum_{j=1}^{d} a_{ij} \hat{B}_j + E_i, \tag{2.112}$$

where $E_i \in \mathbb{R}^{b \times b}$ is the possible residual from the approximation. Describe how you can use PCA to identify an optimal set of d base blocks so that the residual is minimized.

Exercise 2.12 (Ranking of Webpages). PCA is actually used to rank webpages on the Internet by many popular search engines. One way to see this is to view the Internet as a directed graph $\mathcal{G} = (\mathcal{V}, \mathcal{E})$, where the ith webpage, denoted by p_i, is a node in $\mathcal{V}$, and every hyperlink from p_i to p_j, denoted by e_{ij}, is a directed edge in $\mathcal{E}$. We can assign each webpage p_i an "authority" score x_i and a "hub" score y_i. The "authority" score x_i is a scaled sum of the "hub" scores of other webpages pointing to webpage p_i. The "hub" score is the scaled sum of the "authority" scores of other webpages to which webpage p_i is pointing. Let $\boldsymbol{x}$ and $\boldsymbol{y}$ be the vectors of authority scores and hub scores, respectively. Also, let A be the adjacent matrix of the graph $\mathcal{G}$, i.e., $A_{ij} = 1$ if $e_{ij} \in \mathcal{E}$ and $A_{ij} = 0$ otherwise, and consider the following algorithm:

Answer the following questions.

1. Given the definitions of hubs and authorities, justify the "ranking webpages" Algorithm 2.1.
2. Show that unit-norm eigenvectors of $AA^\top$ (for $\boldsymbol{y}$) and $A^\top A$ (for $\boldsymbol{x}$) give fixed points of Algorithm 2.1.
3. Show that in general, $\boldsymbol{y}$ and $\boldsymbol{x}$ converge to the unit-norm eigenvectors associated with the maximum eigenvalues of $AA^\top$ and $A^\top A$, respectively. Explain why no other eigenvector is possible and why the normalization steps in the algorithm are necessary.

Algorithm 2.1 (Ranking Webpages)

Input: A matrix A and a random vector $\boldsymbol{x}$

while ($\boldsymbol{x}$ not converged) **do**

 $\boldsymbol{y}' \leftarrow A\boldsymbol{x}, \boldsymbol{y} \leftarrow \frac{\boldsymbol{y}'}{\|\boldsymbol{y}'\|}$

 $\boldsymbol{x}' \leftarrow A^\top \boldsymbol{y}, \boldsymbol{x} \leftarrow \frac{\boldsymbol{x}'}{\|\boldsymbol{x}'\|}$

end while

Output: $\boldsymbol{x}$

4. Explain how $\boldsymbol{y}$ and $\boldsymbol{x}$ can be computed from the singular value decomposition of A. Under what circumstances would the given algorithm be preferable to using the SVD?

In the literature, this is known as the *hypertext-induced topic-selection* (HITS) algorithm (Kleinberg 1999; Ding et al. 2004). The same algorithm can also be used to rank competitive sports such as football teams and chess players.

Exercise 2.13 (PPCA by Maximum Likelihood). Study the proof of Theorem 2.8 in great detail and show the missing piece that is left as an exercise to the reader. More specifically, let $\lambda_1, \ldots, \lambda_D$ be the eigenvalues of a covariance matrix $\Sigma \in \mathbb{R}^{D\times D}$. Let $\pi : \{1, \ldots, D\} \to \{1, \ldots, D\}$ be a permutation of the first D integers. We would like to choose d eigenvalues $\lambda_{\pi[1]}, \ldots, \lambda_{\pi[d]}$ such that the discarded ones $\lambda_{\pi[d+1]}, \ldots, \lambda_{\pi[D]}$ minimize

$$\mathcal{M}(\pi) = \log\left(\frac{\sum_{i=d+1}^{D} \lambda_{\pi[i]}}{D-d}\right) - \frac{\sum_{i=d+1}^{D} \log \lambda_{\pi[i]}}{D-d}. \tag{2.113}$$

Use Jensen's inequality to show that $\mathcal{M}$ is nonnegative and use the concavity of the log function to prove that $\mathcal{M}$ is minimized by choosing $\lambda_{\pi[i]}$, $i = d+1, \ldots, D$ to be contiguous in magnitude.

Exercise 2.14 Show that for the PPCA model, $\boldsymbol{x} = \boldsymbol{\mu} + B\boldsymbol{y} + \varepsilon$, where $\boldsymbol{\mu} \in \mathbb{R}^D$, $B \in \mathbb{R}^{D\times d}$, $\boldsymbol{y} \sim \mathcal{N}(0, I_d)$ and $\varepsilon \sim \mathcal{N}(0, \sigma^2 I_D)$, the conditional distribution of $\boldsymbol{y}$ given $\boldsymbol{x}$ is given by $\boldsymbol{y} \mid \boldsymbol{x} \sim \mathcal{N}(\boldsymbol{\mu}_{y|x}, \Sigma_{y|x})$, where

$$\boldsymbol{\mu}_{y|x} = (B^\top B + \sigma^2 I_d)^{-1} B^\top (\boldsymbol{x} - \boldsymbol{\mu}) \text{ and } \Sigma_{y|x} = \sigma^2 (B^\top B + \sigma^2 I_D)^{-1}. \tag{2.114}$$

Exercise 2.15 (An EM Algorithm for PPCA). In Section 2.2.2, we showed that the ML estimate of the parameter $\theta = (\boldsymbol{\mu}, B, \sigma)$ of the PPCA model $\boldsymbol{x} = \boldsymbol{\mu} + B\boldsymbol{y} + \varepsilon$, where $\boldsymbol{\mu} \in \mathbb{R}^D$, $B \in \mathbb{R}^{D\times d}$, $\boldsymbol{y} \sim \mathcal{N}(0, I_d)$ and $\varepsilon \sim \mathcal{N}(0, \sigma^2 I_D)$, can be found in closed form, as shown in Theorem 2.9. An alternative approach, which can be advantageous for large D, is to view $\boldsymbol{y}$ as a hidden variable and use the EM algorithm described in Appendix B.2.1 to find the ML estimate. In this exercise, you will derive an EM algorithm for PPCA.

1. Let $w_j^k(\boldsymbol{y}) = p_{\theta^k}(\boldsymbol{y} \mid \boldsymbol{x}_j)$ be the posterior distribution of the hidden variables with parameters $\theta^k = (\boldsymbol{\mu}^k, B^k, \sigma^k)$ at iteration k of the EM algorithm. Show that the expected complete log-likelihood $Q(\theta \mid \theta^k) = \mathbb{E}_{w^k}[\log p_\theta(\{\boldsymbol{x}_j\}_{j=1}^N, \{\boldsymbol{y}_j\}_{j=1}^N)]$ is given by

$$\begin{aligned} -\sum_{j=1}^{N}\Big(\frac{D}{2}\log(2\pi\sigma^2) + \frac{1}{2\sigma^2}\big(\|\boldsymbol{x}_j - \boldsymbol{\mu}\|^2 - 2(\boldsymbol{x}_j - \boldsymbol{\mu})^\top B\langle \boldsymbol{y}_j\rangle^k \\ + \operatorname{trace} B^\top B\langle \boldsymbol{y}_j\boldsymbol{y}_j^\top\rangle^k\big) + \frac{1}{2}\operatorname{trace}\langle \boldsymbol{y}_j\boldsymbol{y}_j^\top\rangle^k\Big), \end{aligned} \tag{2.115}$$

where

$$\langle y_j \rangle^k = \int_y w_j^k(y) y \, dy = (B^{k\top} B^k + \sigma^2 I_d)^{-1} B^{k\top} (x_j - \mu^k), \tag{2.116}$$

$$\langle y_j y_j^\top \rangle^k = \int_y w_j^k(y) y y^\top dy = (\sigma^k)^2 (B^{k\top} B^k + \sigma^2 I_d)^{-1} + \langle y_j \rangle^k \langle y_j \rangle^{k\top}.$$

Hint: See Exercise 2.14.

2. Show that the parameters $\theta = (\mu, B, \sigma)$ that maximize $Q(\theta \mid \theta^k)$ are given by

$$[B \ \mu] = \begin{bmatrix} \sum_{j=1}^N x_j \langle y_j \rangle^{k\top} & \sum_{i=1}^N x_j \end{bmatrix} \begin{bmatrix} \sum_{j=1}^N \langle y_j y_j^\top \rangle^k & \sum_{j=1}^N \langle y_j \rangle^k \\ \sum_{j=1}^N \langle y_j \rangle^{k\top} & N \end{bmatrix}^{-1}, \tag{2.117}$$

$$\sigma^2 = \frac{1}{ND} \sum_{j=1}^N \|x_j - \mu\|^2 - 2(x_j - \mu)^\top B \langle y_j \rangle^k + \operatorname{trace} B^\top B \langle y_j y_j^\top \rangle^k. \tag{2.118}$$

3. In practice, we know that the ML estimator for μ is $\hat{\mu} = \frac{1}{N} \sum_{j=1}^N x_j$. Therefore, a more efficient approach is to maximize $Q(\theta \mid \theta^k)$ only over the parameters (B, σ). Show that the optimal parameters are given by

$$B^{k+1} = \sum_{j=1}^N (x_j - \hat{\mu}) \langle y_j \rangle^{k\top} \Big(\sum_{j=1}^N \langle y_j y_j^\top \rangle^k \Big)^{-1}, \tag{2.119}$$

$$\sigma^{k+1} = \sqrt{\frac{1}{ND} \sum_{j=1}^N \|x_j - \hat{\mu}\|^2 - 2(x_j - \hat{\mu})^\top B^{k+1} \langle y_j \rangle^k + \operatorname{trace} B^{(k+1\top} B^{k+1} \langle y_j y_j^\top \rangle^k)},$$

where $\langle y_j \rangle^k$ is computed with $\mu^k = \hat{\mu}$. Show also that the above iterations can be rewritten as

$$B^{k+1} = \hat{\Sigma}_N B^k \big[(\sigma^k)^2 I_d + \Sigma_x^{k^{-1}} B^{k\top} \hat{\Sigma}_N B^k \big]^{-1}, \tag{2.120}$$

$$\sigma^{k+1} = \sqrt{\frac{1}{D} \operatorname{trace}(\widehat{\Sigma}_N - \widehat{\Sigma}_N B^k \Sigma_x^{k^{-1}} B^{k+1\top})}, \tag{2.121}$$

where $\hat{\Sigma}_N = \frac{1}{N} \sum_{j=1}^N (x_j - \hat{\mu})(x_j - \hat{\mu})^\top$.

Exercise 2.16 (Properties of the Nuclear Norm). Let X be a matrix of rank r.

1. Show that the nuclear norm $f(X) = \|X\|_* = \sum_{i=1}^r \sigma_i(X)$ of X is a convex function of X.
2. Show that the subgradient of the nuclear norm is given by

$$\partial\|X\|_* = UV^\top + W, \tag{2.122}$$

where $X = U\Sigma V^\top$ is the compact (rank r) SVD of X, and W is a matrix such that $U^\top W = 0$, $WV = 0$, and $\|W\|_2 \leq 1$.

3. Show that the optimal solution of

$$\min_A \quad \frac{1}{2}\|X - A\|_F^2 + \tau\|A\|_* \tag{2.123}$$

is given by $A = \mathcal{D}_\tau(X) = U\mathcal{S}_\tau(\Sigma)V^\top$, where $\mathcal{D}_\tau$ is called the singular-value thresholding operator.

Exercise 2.17 (Face Recognition with Varying Illumination).

1. **Implementation of PCA, PPCA and model selection techniques.** Implement the following functions using at most five lines of MATLAB code per function.

Function `[mu,U,Y]=pca(X,d)`

Parameters

`X`	$D \times N$ data matrix.
`d`	Number of principal components.

Returned values

`mu`	Mean of the data.
`U`	Orthonormal basis for the subspace.
`Y`	Low-dimensional representation (or principal components).

Description

Finds the d principal components of a set of points from the SVD of the data matrix X.

Function `[mu,W,sigma]=ppca(X,d)`

Parameters

`X`	$D \times N$ data matrix.
`d`	Number of principal components.

Returned values

`mu`	Mean of the data.
`W`	Basis for the subspace (does not need to be orthonormal).
`sigma`	Standard deviation of the noise.

Description

Finds the parameters of the PPCA model μ and $\Sigma = WW^\top + \sigma^2 I$.

Function `d=pca_model_selection(X,tau)`

Parameters

`X`	$D \times N$ data matrix.
`tau`	Threshold

Returned values

`d`	Number of principal components.

Description

Finds the number of principal components for PCA as $\hat{d} = \min_d \{d : \sigma_{d+1}^2 < \tau\}$.

Function `d=ppca_model_selection(X,method)`

Parameters

`X`	$D \times N$ data matrix.
`method`	BIC, AIC, G-AIC

Returned values

`d`	Number of principal components.

Description

Finds the number of principal components using different model selection methods.

2. **Face recognition using PCA and PPCA.** In this exercise you will use a small subset of the Yale B data set[8], that contains photos of ten individuals under various illumination conditions. Specifically, you will use only images from the first three individuals under ten different illumination conditions.

   Download the file YaleB-Dataset.zip. This file contains the image database along with the MATLAB function `loadimage.m`. Decompress the file and type `help loadimage` at the MATLAB prompt to see how to use this function. The function operates as follows:

Function `img=loadimage(individual,condition)`

Parameters

`individual`	Number of the individual.
`condition`	Number of the image for that individual.

Returned values

`img`	The pixel image loaded from the database.

Description

Read and resize an image from the data set. The database (directory `images`) must be in the same directory as this file.

   (a) Apply PCA with $d = 2$ to all 10 images from individual 1. Plot the mean face $\boldsymbol{\mu}$ and the first two eigenfaces $\boldsymbol{u}_1$ and $\boldsymbol{u}_2$. What do you observe? Plot $\boldsymbol{\mu} + y_1\boldsymbol{u}_1$ for $y_1 = -\sigma_1 : 0.2\sigma_1 : \sigma_1$ and $\boldsymbol{\mu} + y_2\boldsymbol{u}_2$ for $y_2 = -\sigma_2 : 0.1\sigma_2 : \sigma_2$. What do the first two principal components capture? Repeat for individuals 2 and 3.

[8]http://cvc.yale.edu/projects/yalefacesB/yalefacesB.html.

(b) Apply PPCA with $d = 2$ to all 10 images from individual 1. Plot the mean face $\boldsymbol{\mu}$ and the first two eigenfaces $\boldsymbol{u}_1$ and $\boldsymbol{u}_2$. What differences do you observe between the eigenfaces of PCA and those of PPCA? Plot $\boldsymbol{\mu} + y_1\boldsymbol{u}_1$ for $y_1 = -1 : 0.2 : 1$ and $\boldsymbol{\mu} + y_2\boldsymbol{u}_2$ for $y_2 = -1 : 0.2 : 1$. What differences do you observe between the principal components of PCA and those of PPCA? Repeat for individuals 2 and 3.

(c) Divide all the images into two sets: *Training Set* (images from individuals 1 to 3 and images 1 to 5) and *Test Set* (images from individuals 1 to 3 and images 6 to 10). Apply PCA to the *Training Set* using $d = 10$. Plot the mean face and the eigenfaces. Plot also the singular values of the data matrix. Project the *Test Set* onto the face subspace given by PCA, i.e., $Y_{test} = W^\top(X_{test} - \boldsymbol{\mu}\mathbf{1}^\top)$. Plot the projected faces, i.e., $\text{Proj}(X_{test}) = \boldsymbol{\mu}\mathbf{1}^\top + WY_{test}$. Classify these faces using 1-nearest-neighbor, that is, label an image $\boldsymbol{x}$ as corresponding to individual i if its projected image $\boldsymbol{y}$ is closest to a projected image $\boldsymbol{y}_j$ of individual i. Report the percentage of correctly classified face images for $d = 1, \ldots, 10$. Which value of d gives the best recognition performance? Compare this result with the those obtained using model selection to determine the number of principal components for some threshold τ as well as with the estimates of BIC, AIC, and G-AIC for PPCA.

Chapter 3
Robust Principal Component Analysis

> *... any statistical procedure ... should be robust in the sense that small deviations from the model assumptions should impair its performance only slightly ... Somewhat larger deviations from the model should not cause a catastrophe.*
>
> —Peter J. Huber

In the previous chapter, we considered the PCA problem under the assumption that all the sample points are drawn from the same statistical or geometric model: a low-dimensional subspace. In practical applications, it is often the case that some entries of the data points can be missing or incomplete. For example, the 2-dimensional trajectories of an object moving in a video may become incomplete when the object becomes occluded. Sometimes, it could be the case that some entries of the data points are corrupted by gross errors and we do not know a priori which entries are corrupted. For instance, the intensities of some pixels of the face image of a person can be corrupted when the person is wearing glasses. Sometimes it could also be the case that a small subset of the data points are outliers. For instance, if we are trying to distinguish face images from non-face images, then we can model all face images as samples from a low-dimensional subspace, but non-face images will not follow the same model. Such data points that do not follow the model of interest are often called *sample outliers* and should be distinguished from the case of samples with some corrupted entries, also referred to as *intrasample outliers*. The main distinction to be made is that in the latter case, we do not want to discard the entire data point, but only the atypical entries.

In this chapter, we will introduce several techniques for recovering a low-dimensional subspace from missing or corrupted data. We will first consider the PCA problem with missing entries, also known as *incomplete PCA* or *low-rank matrix completion* (for linear subspaces). In Section 3.1, we will describe several representative methods for solving this problem based on maximum likelihood estimation, convex optimization, and alternating minimization. Such methods are featured due to their simplicity, optimality, or scalability, respectively. In Section 3.2,

R. Vidal et al., *Generalized Principal Component Analysis*, Interdisciplinary Applied Mathematics 40, DOI 10.1007/978-0-387-87811-9_3

we will consider the PCA problem with corrupted entries, also known as the *robust PCA* (RPCA) problem. We will introduce classical alternating minimization methods for addressing this problem as well as convex optimization methods that offer theoretical guarantees of correctness. Finally, in Section 3.3, we will consider the PCA problem with sample outliers and describe methods for solving this problem based on classical robust statistical estimation techniques as well as techniques based on convex relaxations. Face images will be used as examples to demonstrate the effectiveness of these algorithms.

3.1 PCA with Robustness to Missing Entries

Recall from Section 2.1.2 that in the PCA problem, we are given N data points $\mathcal{X} \doteq \{\boldsymbol{x}_j \in \mathbb{R}^D\}_{j=1}^N$ drawn (approximately) from a d-dimensional affine subspace $S \doteq \{\boldsymbol{x} = \boldsymbol{\mu} + U\boldsymbol{y}\}$, where $\boldsymbol{\mu} \in \mathbb{R}^D$ is an arbitrary point in S, $U \in \mathbb{R}^{D\times d}$ is a basis for S, and $\mathcal{Y} = \{\boldsymbol{y}_j \in \mathbb{R}^d\}_{j=1}^N$ are the principal components.

In this section, we consider the PCA problem in the case that some of the given data points are *incomplete*. A data point $\boldsymbol{x} = [x_1, x_2, \ldots, x_D]^\top$ is said to be incomplete when some of its entries are missing or unspecified. For instance, if the ith entry x_i, of $\boldsymbol{x}$ is missing, then $\boldsymbol{x}$ is known only up to a line in $\mathbb{R}^D$, i.e.,

$$\begin{aligned} \boldsymbol{x} \in L &\doteq \left\{[x_1, \ldots, x_{i-1}, x_i, x_{i+1}, \ldots, x_D]^\top, x_i \in \mathbb{R}\right\} \\ &= \left\{\boldsymbol{x}_{-i} + x_i \boldsymbol{e}_i, x_i \in \mathbb{R}\right\}, \end{aligned} \tag{3.1}$$

where $\boldsymbol{x}_{-i} = [x_1, \ldots, x_{i-1}, 0, x_{i+1}, \ldots, x_D]^\top \in \mathbb{R}^D$ is the vector $\boldsymbol{x}$ with its ith entry zeroed out and $\boldsymbol{e}_i = [0, \ldots, 0, 1, 0, \ldots, 0]^\top \in \mathbb{R}^D$ is the ith basis vector. More generally, if the point $\boldsymbol{x}$ has M missing entries, without loss of generality we can partition it as $\begin{bmatrix}\boldsymbol{x}_U \\ \boldsymbol{x}_O\end{bmatrix}$, where $\boldsymbol{x}_U \in \mathbb{R}^M$ denotes the unobserved entries and $\boldsymbol{x}_O \in \mathbb{R}^{D-M}$ denotes the observed entries. Thus, $\boldsymbol{x}$ is known only up to the following M-dimensional affine subspace:

$$\boldsymbol{x} \in L \doteq \left\{ \begin{bmatrix} \mathbf{0} \\ \boldsymbol{x}_O \end{bmatrix} + \begin{bmatrix} I_M \\ \mathbf{0} \end{bmatrix} \boldsymbol{x}_U, \boldsymbol{x}_U \in \mathbb{R}^M \right\}. \tag{3.2}$$

Incomplete PCA When the Subspace Is Known

Let us first consider the simplest case, in which the subspace S is known. Then we know that the point $\boldsymbol{x}$ belongs to both L and S. Therefore, given the parameters $\boldsymbol{\mu}$ and U of the subspace S, we can compute the principal components $\boldsymbol{y}$ and the missing entries $\boldsymbol{x}_U$ by intersecting L and S. In the case of one missing entry (illustrated in Figure 3.1), the intersection point can be computed from

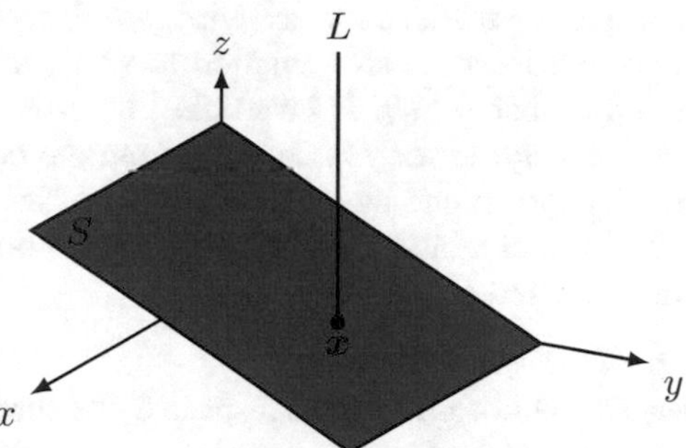

Fig. 3.1 Given a point $\boldsymbol{x} \in \mathbb{R}^D$ with one unknown entry x_i, the point $\boldsymbol{x}$ is known only up to a line L. However, if we also know that $\boldsymbol{x}$ belongs to a subspace S, we can find the unknown entry by intersecting L and S, provided that L is not parallel to S.

$$\boldsymbol{x} = \boldsymbol{x}_{-i} + x_i \boldsymbol{e}_i = \boldsymbol{\mu} + U\boldsymbol{y} \implies \begin{bmatrix} U & -\boldsymbol{e}_i \end{bmatrix} \begin{bmatrix} \boldsymbol{y} \\ x_i \end{bmatrix} = \boldsymbol{x}_{-i} - \boldsymbol{\mu}. \tag{3.3}$$

Note that a necessary condition for this linear system to have a unique solution is that the line L is not parallel to the principal subspace, i.e., $\boldsymbol{e}_i \notin \operatorname{span}(U)$.

In the case of M missing entries, we can partition the point $\boldsymbol{\mu} = \begin{bmatrix} \boldsymbol{\mu}_U \\ \boldsymbol{\mu}_O \end{bmatrix}$ and the subspace basis $U = \begin{bmatrix} U_U \\ U_O \end{bmatrix}$ according to $\boldsymbol{x} = \begin{bmatrix} \boldsymbol{x}_U \\ \boldsymbol{x}_O \end{bmatrix}$. Then, the intersection of L and S can be computed from

$$\begin{bmatrix} \boldsymbol{x}_U \\ \boldsymbol{x}_O \end{bmatrix} = \begin{bmatrix} \boldsymbol{\mu}_U \\ \boldsymbol{\mu}_O \end{bmatrix} + \begin{bmatrix} U_U \\ U_O \end{bmatrix} \boldsymbol{y} \implies \begin{bmatrix} U_U & -I_M \\ U_O & \mathbf{0} \end{bmatrix} \begin{bmatrix} \boldsymbol{y} \\ \boldsymbol{x}_U \end{bmatrix} = \begin{bmatrix} -\boldsymbol{\mu}_U \\ \boldsymbol{x}_O - \boldsymbol{\mu}_O \end{bmatrix}. \tag{3.4}$$

A necessary condition for the linear system in (3.4) to have a unique solution is that the matrix on the left-hand side be of full column rank $d + M \leq D$. This implies that $\boldsymbol{e}_i \notin \operatorname{span}(U)$ for each missing entry i. This also implies that $M \leq D - d$; hence we need to have at least d observed entries in order to complete a data point. When the data point $\boldsymbol{x}$ is not precise and has some noise, we can compute $\boldsymbol{y}$ and $\boldsymbol{x}_U$ as the solution to the following optimization problem:

$$\min_{\boldsymbol{y}, \boldsymbol{x}_U} \|\boldsymbol{x} - \boldsymbol{\mu} - U\boldsymbol{y}\|^2. \tag{3.5}$$

It is easy to derive that the closed-form solution to the unknowns $\boldsymbol{y}$ and $\boldsymbol{x}_U$ is given by

$$\begin{aligned} \boldsymbol{y} &= (I - U_U^\top U_U)^{-1} U_O^\top (\boldsymbol{x}_O - \boldsymbol{\mu}_O) = (U_O^\top U_O)^{-1} U_O^\top (\boldsymbol{x}_O - \boldsymbol{\mu}_O), \\ \boldsymbol{x}_U &= \boldsymbol{\mu}_U + U_U \boldsymbol{y} = \boldsymbol{\mu}_U + U_U (U_O^\top U_O)^{-1} U_O^\top (\boldsymbol{x}_O - \boldsymbol{\mu}_O). \end{aligned} \tag{3.6}$$

We leave the derivation to the reader as an exercise (see Exercise 3.1). Notice that this solution is simply the least squares solution to (3.4), and that in order for U_O to be of full rank (so that $U_O^\top U_O$ is invertible), we need to know at least d entries. Interestingly, the solution for $\boldsymbol{y}$ is obtained from the observed entries ($\boldsymbol{x}_O$) and the part of the model corresponding to the observed entries ($\boldsymbol{\mu}_O$ and U_O). Then the missing entries ($\boldsymbol{x}_U$) are obtained from the part of the model corresponding to the unobserved entries ($\boldsymbol{\mu}_U$ and U_U) and $\boldsymbol{y}$.

Incomplete PCA as a Well-Posed Problem

In practice, however, we do not know the subspace S (neither $\boldsymbol{\mu}$ nor U) a priori. Instead, we are given only N incomplete samples, which we can arrange as the columns of an incomplete data matrix $X = [\boldsymbol{x}_1, \boldsymbol{x}_2, \dots, \boldsymbol{x}_N] \in \mathbb{R}^{D\times N}$. Let $W \in \mathbb{R}^{D\times N}$ be the matrix whose entries $\{w_{ij}\}$ encode the locations of the missing entries, i.e.,

$$w_{ij} = \begin{cases} 1 & \text{if } x_{ij} \text{ is known,} \\ 0 & \text{if } x_{ij} \text{ is missing,} \end{cases} \tag{3.7}$$

and let $W \odot X$ denote the Hadamard product of two matrices, which is defined as the entrywise product $(W \odot X)_{ij} = w_{ij}x_{ij}$. The goal of *PCA with missing data*, also known as *matrix completion*, is to find the missing entries $(\mathbf{1}\mathbf{1}^\top - W) \odot X$, the point $\boldsymbol{\mu}$, the basis U, and the matrix of low-dimensional coordinates $Y = [\boldsymbol{y}_1, \boldsymbol{y}_2, \dots, \boldsymbol{y}_N] \in \mathbb{R}^{d\times N}$ from the known entries $W \odot X$.

Obviously, we cannot expect to always be able to find the correct solution to this problem. Whether the correct complete matrix X can be recovered depends on:

1. Which entries are missing or observed;
2. How many entries are missing or observed.

To see why the location of missing entries matters, suppose the first entry of all data points is missing. Then we cannot hope to be able to recover the first row of X at all. Likewise, suppose that all the entries of one data point are missing. While in this case we can hope to find the subspace from the other data points, we cannot recover the low-dimensional representation of the missing point. These two examples suggest that the location of missing entries should not have any conspicuous patterns.

Now suppose that the matrix X is

$$X = \boldsymbol{e}_1\boldsymbol{e}_1^\top = \begin{bmatrix} 1 & 0 & \cdots & 0 \\ 0 & 0 & \cdots & 0 \\ \vdots & & \ddots & \\ 0 & 0 & \cdots & 0 \end{bmatrix}, \tag{3.8}$$

which is a rank-one matrix. In this case, we cannot hope to recover X even if a relatively large percentage of its entries are given, because most entries are equal to zero, and we will not be able to distinguish X from the zero matrix from many observed entries. This suggests that if we want to recover a low-rank data matrix from a small portion of its entries, the matrix itself should not be too sparse.

Thus, to avoid ambiguous solutions due to the above situations, we must require that the locations of the missing entries be random enough so that the chance that they form a conspicuous pattern is very low; and in addition, we must restrict our low-rank matrices to those that are not particularly sparse. The following definition gives a set of technical conditions to impose on a matrix so that its singular vectors are not too spiky, and hence the matrix itself is not too sparse.

Definition 3.1 (Matrix Incoherence with Respect to Sparse Matrices). *A matrix $X \in \mathbb{R}^{D\times N}$ is said to be* ν-incoherent with respect to the set of sparse matrices *if*

$$\max_i \|\boldsymbol{u}_i\|_2 \leq \frac{\nu\sqrt{d}}{\sqrt{D}}, \quad \max_j \|\boldsymbol{v}_j\|_2 \leq \frac{\nu\sqrt{d}}{\sqrt{N}}, \quad \|UV^\top\|_\infty \leq \frac{\nu\sqrt{d}}{\sqrt{DN}}, \tag{3.9}$$

where d is the rank of X, $X = U\Sigma V^\top$ is the compact SVD of X, and $\boldsymbol{u}_i$, and $\boldsymbol{v}_j$ are the ith row of U and jth row V, respectively.

Notice that since U is orthonormal, the largest absolute value of the entries of $U \in \mathbb{R}^{D\times d}$ is equal to 1, which happens when a column of U is 1-sparse, i.e., when a column of U has only one nonzero entry. On the other hand, if each column of U is so dense that all its entries are equal to each other up to sign, then each entry is equal to $\pm 1/\sqrt{D}$, and the norm of each row is $\sqrt{d/D}$. Therefore, when $\nu < 1$, the first condition above controls the level of sparsity of U. Similarly, the other two conditions control the levels of sparsity of V and $UV^\top$, respectively. From a probabilistic perspective, these conditions are rather mild in the sense that they hold for almost all generic matrices—a random (say Gaussian) matrix satisfies these conditions with high probability when the dimension of the matrix is large enough. As we will see, incoherence is indeed a very useful technical condition to ensure that low-rank matrix completion is a meaningful problem.

Regarding the number of entries required, notice that in order to specify a d-dimensional subspace S in $\mathbb{R}^D$ together with N points on it, we need to specify $D + dD + dN - d^2$ independent entries in $\boldsymbol{\mu}$, U, and Y.[1] That is, it is necessary to observe at least this number of entries of X in order to have a unique solution for X. However, the sufficient conditions for ensuring a unique and correct solution highly depend on the approach and method one uses to recover X.

Incomplete PCA Algorithms

In what follows, we discuss a few approaches for solving the PCA problem with missing entries. The first approach (described in Section 3.1.1) is a simple extension

[1]If $U \in \mathbb{R}^{D\times d}$ and $V \in \mathbb{R}^{N\times d}$, then U and V have $dD + dN$ degrees of freedom in general. However, to specify the subspace, it suffices to specify $UV^\top$, which is equal to $UAA^{-1}V^\top$ for every invertible matrix $A \in \mathbb{R}^{d\times d}$; hence the matrix $UV^\top$ has $dD + dN - d^2$ degrees of freedom.

of geometric PCA (see Section 2.1) in which the sample mean and covariance are directly computed from the incomplete data matrix. However, this approach has a number of disadvantages, as we shall see. The second approach (described in Section 3.1.2) is a direct extension of probabilistic PCA (see Section 2.2) and uses the expectation maximization (EM) algorithm (see Appendix B.2.1) to complete the missing entries. While this approach is guaranteed to converge, the solution it finds is not always guaranteed to be the global optimum, and hence it is not necessarily the correct solution. The third approach (described in Section 3.1.3) uses convex relaxation and optimization techniques to find the missing entries of the low-rank data matrix X. Under the above incoherent conditions and with almost minimal observations, this approach is guaranteed to return a perfect completion of the low-rank matrix. However, this approach may not be scalable to large matrices, since it requires solving for as many variables as the number of entries in the data matrix. The fourth and final approach (described in Section 3.1.4) alternates between solving for $\boldsymbol{\mu}$, U, and Y given a completion of X, and solving for the missing entries of X given $\boldsymbol{\mu}$, U, and Y. Since this method uses a minimal parameterization of the unknowns, it is more scalable. While in general, this approach is not guaranteed to converge to the correct solution, we present a variant of this method that is guaranteed to recover the missing entries correctly under conditions similar to those for the convex relaxation method.

3.1.1 Incomplete PCA by Mean and Covariance Completion

Recall from Section 2.1.2 that the optimization problem associated with geometric PCA is

$$\min_{\boldsymbol{\mu},U,\{\boldsymbol{y}_j\}} \sum_{j=1}^{N} \left\|\boldsymbol{x}_j - \boldsymbol{\mu} - U\boldsymbol{y}_j\right\|^2 \quad \text{s.t.} \quad U^\top U = I_d \quad \text{and} \quad \sum_{j=1}^{N} \boldsymbol{y}_j = \mathbf{0}. \tag{3.10}$$

We already know that the solution to this problem can be obtained from the mean and covariance of the data points,

$$\hat{\boldsymbol{\mu}}_N = \frac{1}{N}\sum_{j=1}^{N} \boldsymbol{x}_j \quad \text{and} \quad \hat{\Sigma}_N = \frac{1}{N}\sum_{j=1}^{N} (\boldsymbol{x}_j - \hat{\boldsymbol{\mu}}_N)(\boldsymbol{x}_j - \hat{\boldsymbol{\mu}}_N)^\top, \tag{3.11}$$

respectively. Specifically, $\boldsymbol{\mu}$ is given by the sample mean $\hat{\boldsymbol{\mu}}_N$, U is given by the top d eigenvectors of the covariance matrix $\hat{\Sigma}_N$, and $\boldsymbol{y}_j = U^\top(\boldsymbol{x}_j - \boldsymbol{\mu})$. Alternatively, an optimal solution can be found from the rank-d SVD of the mean-subtracted data matrix $[\boldsymbol{x}_1 - \hat{\boldsymbol{\mu}}_N, \ldots, \boldsymbol{x}_N - \hat{\boldsymbol{\mu}}_N]$, as shown in Theorem 2.3.

When some entries of each $\boldsymbol{x}_j$ are missing, we cannot directly compute $\hat{\boldsymbol{\mu}}_N$ or $\hat{\Sigma}_N$ as in (3.11). A straightforward method for dealing with missing entries was introduced in (Jolliffe 2002). It basically proposes to compute the sample mean and covariance from the known entries of X. Specifically, the entries of the incomplete mean and covariance can be computed as

$$\hat{\mu}_i = \frac{\sum_{j=1}^{N} w_{ij}x_{ij}}{\sum_{j=1}^{N} w_{ij}} \quad \text{and} \quad \hat{\sigma}_{ik} = \frac{\sum_{j=1}^{N} w_{ij}w_{kj}(x_{ij} - \hat{\mu}_i)(x_{kj} - \hat{\mu}_k)}{\sum_{j=1}^{N} w_{ij}w_{kj}}, \tag{3.12}$$

where $i, k = 1, \ldots, D$. However, as discussed in (Jolliffe 2002), this simple approach has several disadvantages. First, the estimated covariance matrix need not be positive semidefinite. Second, these estimates are not obtained by optimizing any statistically or geometrically meaningful objective function (least squares, maximum likelihood, etc.) Nonetheless, estimates $\hat{\boldsymbol{\mu}}_N$ and $\hat{\Sigma}_N$ obtained from the naive approach in (3.12) may be used to initialize the methods discussed in the next two sections, which are iterative in nature. For example, we may initialize the columns of U as the eigenvectors of $\hat{\Sigma}_N$ associated with its d largest eigenvalues. Then given $\hat{\boldsymbol{\mu}}_N$ and $\hat{U}$, we can complete each missing entry as described in (3.6).

3.1.2 Incomplete PPCA by Expectation Maximization

In this section, we derive an EM algorithm (see Appendix B.2.1) for solving the PPCA problem with missing data. Recall from Section 2.2 that in the PPCA model, each data point is drawn as $\boldsymbol{x} \sim \mathcal{N}(\boldsymbol{\mu}_x, \Sigma_x)$, where $\boldsymbol{\mu}_x = \boldsymbol{\mu}$ and $\Sigma_x = BB^\top + \sigma^2 I_D$, where $\boldsymbol{\mu} \in \mathbb{R}^D$, $B \in \mathbb{R}^{D\times d}$, and $\sigma > 0$. Recall also from (2.56) that the log-likelihood of the PPCA model is given by

$$\mathscr{L} = -\frac{ND}{2}\log(2\pi) - \frac{N}{2}\log\det(\Sigma_x) - \frac{1}{2}\sum_{j=1}^{N}\operatorname{trace}(\Sigma_x^{-1}(\boldsymbol{x}_j - \boldsymbol{\mu})(\boldsymbol{x}_j - \boldsymbol{\mu})^\top), \tag{3.13}$$

where $\{\boldsymbol{x}_j\}_{j=1}^N$ are N i.i.d. samples of $\boldsymbol{x}$. Since the samples are incomplete, we can partition each point $\boldsymbol{x}$ and the parameters $\boldsymbol{\mu}_x$ and Σ_x as

$$\begin{bmatrix} \boldsymbol{x}_U \\ \boldsymbol{x}_O \end{bmatrix} = P\boldsymbol{x}, \quad \begin{bmatrix} \boldsymbol{\mu}_U \\ \boldsymbol{\mu}_O \end{bmatrix} = P\boldsymbol{\mu}, \quad \text{and} \quad \begin{bmatrix} \Sigma_{UU} & \Sigma_{UO} \\ \Sigma_{OU} & \Sigma_{OO} \end{bmatrix} = P\Sigma_x P^\top. \tag{3.14}$$

Here $\boldsymbol{x}_O$ is the observed part of $\boldsymbol{x}$, $\boldsymbol{x}_U$ is the unobserved part of $\boldsymbol{x}$, and P is any permutation matrix that reorders the entries of $\boldsymbol{x}$ so that the unobserved entries appear first. Notice that P is not unique, but we can use any such P. Notice also that the above partition of $\boldsymbol{x}$, $\boldsymbol{\mu}_x$, and Σ_x could be different for each data point, because the missing entries could be different for different data points. When strictly necessary, we will use $\boldsymbol{x}_{jU}$ and $\boldsymbol{x}_{jO}$ to denote the unobserved and observed parts of point $\boldsymbol{x}_j$, respectively, and P_j to denote the permutation matrix. Otherwise, we will avoid using the index j in referring to a generic point.

In what follows, we derive two variants of the EM algorithm for learning the parameters $\theta = (\boldsymbol{\mu}, B, \sigma)$ of the PPCA model from incomplete samples $\{\boldsymbol{x}_j\}_{j=1}^N$. The first variant, called Maximum a Posteriori Expectation Maximization (MAP-EM), is an approximate EM method whereby the unobserved variables are given by their MAP estimates (see Appendix B.2.2). The second variant is the exact EM algorithm (see Appendix B.2.1), where we take the conditional expectation of $\mathscr{L}$ over the incomplete entries. Interestingly, both variants lead to the same estimate for $\boldsymbol{\mu}_x$, though the estimates for Σ_x are slightly different. In our derivations, we will use the fact that the conditional distribution of $\boldsymbol{x}_U$ given $\boldsymbol{x}_O$ is Gaussian. More specifically, $\boldsymbol{x}_U \mid \boldsymbol{x}_O \sim \mathcal{N}(\boldsymbol{\mu}_{U|O}, \Sigma_{U|O})$, where

$$\boldsymbol{\mu}_{U|O} = \boldsymbol{\mu}_U + \Sigma_{UO}\Sigma_{OO}^{-1}(\boldsymbol{x}_O - \boldsymbol{\mu}_O) \text{ and } \Sigma_{U|O} = \Sigma_{UU} - \Sigma_{UO}\Sigma_{OO}^{-1}\Sigma_{OU}.$$

We leave this fact as an exercise to the reader (see Exercise 3.2).

Maximum a Posteriori Expectation Maximization (MAP-EM)

The MAP-EM algorithm (see Appendix B.2.2) is a simplified version of the EM algorithm (see Appendix B.2.1) that alternates between the following two steps:

MAP-step: Complete each data point $\boldsymbol{x}$ by replacing the unobserved variables $\boldsymbol{x}_U$ with their MAP estimates, $\arg\max_{\boldsymbol{x}_U} p_{\theta^k}(\boldsymbol{x}_U \mid \boldsymbol{x}_O)$, where θ^k is an estimate for the model parameters at iteration k.

M-step: Maximize the complete log-likelihood with respect to θ, with $\boldsymbol{x}_U$ given as in the MAP-step.

During the MAP step, the MAP estimate of the unobserved variables can be computed in closed form as

$$\arg\max_{\boldsymbol{x}_U} p_{\theta^k}(\boldsymbol{x}_U \mid \boldsymbol{x}_O) = \boldsymbol{\mu}_{U|O}^k = \boldsymbol{\mu}_U^k + \Sigma_{UO}^k(\Sigma_{OO}^k)^{-1}(\boldsymbol{x}_O - \boldsymbol{\mu}_O^k). \tag{3.15}$$

Therefore, we can complete each data point as $\boldsymbol{x}^k = P^\top \begin{bmatrix} \boldsymbol{\mu}_{U|O}^k \\ \boldsymbol{x}_O \end{bmatrix}$. Letting $\boldsymbol{x}_j^k$ be the completion of $\boldsymbol{x}_j$ at iteration k, we obtain the complete log-likelihood as

$$\mathscr{L} = -\frac{ND}{2}\log(2\pi) - \frac{N}{2}\log\det(\Sigma_x) - \frac{1}{2}\sum_{j=1}^N (\boldsymbol{x}_j^k - \boldsymbol{\mu})^\top \Sigma_x^{-1}(\boldsymbol{x}_j^k - \boldsymbol{\mu}). \tag{3.16}$$

During the M-step, we need to maximize $\mathscr{L}$ with respect to θ. Since the data are already complete, we can update the model parameters as described in Theorem 2.9, i.e.,

$$\boldsymbol{\mu}^{k+1} = \frac{1}{N}\sum_{j=1}^N \boldsymbol{x}_j^k, \;\; B^{k+1} = U_1\big(\Lambda_1 - (\sigma^k)^2 I\big)^{1/2} R, \;\text{ and }\; (\sigma^k)^2 = \frac{\sum_{i=d+1}^{D} \lambda_i}{D-d},$$

where $U_1 \in \mathbb{R}^{D\times d}$ is the matrix whose columns are the top d eigenvectors of the *complete sample covariance* matrix

$$\hat{\Sigma}_N^{k+1} = \frac{1}{N}\sum_{j=1}^{N}(\boldsymbol{x}_j^k - \boldsymbol{\mu}^{k+1})(\boldsymbol{x}_j^k - \boldsymbol{\mu}^{k+1})^\top, \tag{3.17}$$

$\Lambda_1 \in \mathbb{R}^{d\times d}$ is a diagonal matrix with the top d eigenvalues of $\hat{\Sigma}_N^{k+1}$, $R \in \mathbb{R}^{d\times d}$ is an arbitrary orthogonal matrix, and λ_i is the ith-largest eigenvalue of $\hat{\Sigma}_N^{k+1}$. We can then update the covariance matrix as $\Sigma_{\boldsymbol{x}}^{k+1} = B^{k+1}(B^{k+1})^\top + (\sigma^k)^2 I$.

Expectation Maximization (EM)
The EM algorithm (see Appendix B.2.1) alternates between the following steps:

E-step: Compute the expectation $Q(\theta \mid \theta^k) \doteq \mathbb{E}_{\boldsymbol{x}_U}[\mathscr{L} \mid \boldsymbol{x}_O, \theta^k]$ of the complete log-likelihood $\mathscr{L}$ with respect to the missing entries $\boldsymbol{x}_U$ given the observed entries $\boldsymbol{x}_O$ and an estimate θ^k of the parameters at iteration k.

M-step: Maximize the expected completed log-likelihood $\mathbb{E}_{\boldsymbol{x}_U}[\mathscr{L} \mid \boldsymbol{x}_O, \theta^k]$ with respect to θ.

Observe from (3.13) that to compute the expectation of $\mathscr{L}$, it suffices to compute the following matrix for each incomplete data point $\boldsymbol{x}$:

$$S^k = \mathbb{E}_{\boldsymbol{x}_U}[(\boldsymbol{x}-\boldsymbol{\mu})(\boldsymbol{x}-\boldsymbol{\mu})^\top \mid \boldsymbol{x}_O, \theta^k] = P^\top \begin{bmatrix} S_{UU}^k & S_{UO}^k \\ S_{OU}^k & S_{OO}^k \end{bmatrix} P. \tag{3.18}$$

Each block of this matrix can be computed as

$$\begin{aligned}
S_{OO}^k &= \mathbb{E}[(\boldsymbol{x}_O-\boldsymbol{\mu}_O)(\boldsymbol{x}_O-\boldsymbol{\mu}_O)^\top \mid \boldsymbol{x}_O, \theta^k] = (\boldsymbol{x}_O-\boldsymbol{\mu}_O^k)(\boldsymbol{x}_O-\boldsymbol{\mu}_O^k)^\top, \\
S_{UO}^k &= \mathbb{E}[(\boldsymbol{x}_U-\boldsymbol{\mu}_U)(\boldsymbol{x}_O-\boldsymbol{\mu}_O)^\top \mid \boldsymbol{x}_O, \theta^k] = (\boldsymbol{\mu}_{U|O}^k-\boldsymbol{\mu}_U^k)(\boldsymbol{x}_O-\boldsymbol{\mu}_O^k)^\top = (S_{OU}^k)^\top, \\
S_{UU}^k &= \mathbb{E}[(\boldsymbol{x}_U-\boldsymbol{\mu}_U)(\boldsymbol{x}_U-\boldsymbol{\mu}_U)^\top \mid \boldsymbol{x}_O, \theta^k] \\
&= \mathbb{E}[(\boldsymbol{x}_U-\boldsymbol{\mu}_{U|O}^k)(\boldsymbol{x}_U-\boldsymbol{\mu}_{U|O}^k)^\top \mid \boldsymbol{x}_O, \theta^k] + \\
&\quad 2\mathbb{E}[(\boldsymbol{\mu}_{U|O}^k-\boldsymbol{\mu}_U)(\boldsymbol{x}_U-\boldsymbol{\mu}_{U|O}^k)^\top \mid \boldsymbol{x}_O, \theta^k] + (\boldsymbol{\mu}_{U|O}^k-\boldsymbol{\mu}_U)(\boldsymbol{\mu}_{U|O}^k-\boldsymbol{\mu}_U)^\top \\
&= \Sigma_{U|O}^k + (\boldsymbol{\mu}_{U|O}^k-\boldsymbol{\mu}_U)(\boldsymbol{\mu}_{U|O}^k-\boldsymbol{\mu}_U)^\top.
\end{aligned}$$

Let S_j^k denote the matrix S^k associated with point $\boldsymbol{x}_j$ and let $\hat{\Sigma}_N^k = \frac{1}{N}\sum_{j=1}^N S_j^k$. Then the expected complete log-likelihood is given by

$$Q(\theta \mid \theta^k) = -\frac{ND}{2}\log(2\pi) - \frac{N}{2}\log\det(\Sigma_{\boldsymbol{x}}) - \frac{N}{2}\operatorname{trace}(\Sigma_{\boldsymbol{x}}^{-1}\hat{\Sigma}_N^k). \tag{3.19}$$

In the M-step, we need to maximize this quantity with respect to θ. Notice that this quantity is almost identical to that in (2.56), except that the sample covariance matrix $\hat{\Sigma}_N$ is replaced by $\hat{\Sigma}_N^k$. Thus, if $\hat{\Sigma}_N^k$ did not depend on the unknown parameter $\boldsymbol{\mu}$, we could immediately compute B and σ from Theorem 2.9. Therefore, all we need to do is to show how to compute $\boldsymbol{\mu}$. To this end, notice that

$$\frac{\partial}{\partial \boldsymbol{\mu}} \operatorname{trace}(\Sigma_{\boldsymbol{x}}^{-1} S^k) = \frac{\partial}{\partial \boldsymbol{\mu}} \mathbb{E}[(\boldsymbol{x} - \boldsymbol{\mu})^\top \Sigma_{\boldsymbol{x}}^{-1} (\boldsymbol{x} - \boldsymbol{\mu}) \mid \boldsymbol{x}_O, \theta^k] \tag{3.20}$$

$$= -2\Sigma_{\boldsymbol{x}}^{-1} \mathbb{E}[\boldsymbol{x} - \boldsymbol{\mu} \mid \boldsymbol{x}_O, \theta^k] = -2\Sigma_{\boldsymbol{x}}^{-1} (\boldsymbol{x}^k - \boldsymbol{\mu}), \tag{3.21}$$

where $\boldsymbol{x}^k = P^\top \begin{bmatrix} \boldsymbol{\mu}_{U|O}^k \\ \boldsymbol{x}_O \end{bmatrix}$ is the complete data point. Therefore,

$$\frac{\partial}{\partial \boldsymbol{\mu}} Q(\theta \mid \theta^k) = -\frac{1}{2} \frac{\partial}{\partial \boldsymbol{\mu}} \sum_{j=1}^N \operatorname{trace}(\Sigma_{\boldsymbol{x}}^{-1} S_j^k) = \sum_{j=1}^N \Sigma_{\boldsymbol{x}}^{-1} (\boldsymbol{x}_j^k - \boldsymbol{\mu}) = \mathbf{0}, \tag{3.22}$$

and so the optimal $\boldsymbol{\mu}$ is

$$\boldsymbol{\mu}^{k+1} = \frac{1}{N} \sum_{j=1}^N \boldsymbol{x}_j^k. \tag{3.23}$$

Notice that this solution is the same as that of the MAP-EM algorithm. That is, the optimal solution for $\boldsymbol{\mu}$ is the average of the complete data. We can then form the matrix $\hat{\Sigma}_N^k$ and compute B^{k+1} and σ^{k+1} as before. Notice, however, that $\hat{\Sigma}_N^k$ is not the covariance of the complete data. The key difference is in the term S_{UU}^k, which contains an additional term $\Sigma_{U|O}^k$.

The EM algorithm for PPCA with missing data is summarized in Algorithm 3.1. In step 2 of the algorithm, the missing entries of X are filled in with zeros, and an initial estimate of $\boldsymbol{\mu}$ and $\Sigma_{\boldsymbol{x}}$ is obtained from the zero-filled X. Alternatively, one may use other initialization methods, such as the mean and covariance completion method described in Section 3.1.1. In step 7, the missing entries of each $\boldsymbol{x}_j$ are filled in according to the initial estimates of mean and covariance in step 2, while the observed entries are kept intact. This corresponds to the MAP step of the MAP-EM algorithm, and is an intermediate calculation for the E-step of the EM algorithm. Next, steps 9 and 10 update the mean and covariance of the PPCA model. Step 9 is common to both the MAP-EM and EM algorithms, while step 10 is slightly different: the MAP-EM algorithm uses only the first term on the right-hand side of step 10, while the EM algorithm uses both terms. Steps 11–14 update the parameters of the PPCA model and correspond to the M-step of both the MAP-EM and EM algorithms. Finally, step 16 computes the probabilistic principal components. Recall from Section 2.2, equation (2.78), that given the parameters of the PPCA model $(\boldsymbol{\mu}, B, \sigma)$, the probabilistic principal components of a vector $\boldsymbol{x}$ are given by $\boldsymbol{y} = (B^\top B + \sigma^2 I)^{-1} B^\top (\boldsymbol{x} - \boldsymbol{\mu})$.

Algorithm 3.1 (Incomplete PPCA by Expectation Maximization)

Input: Entries x_{ij} of a matrix $X \in \mathbb{R}^{D \times N}$ for $(i,j) \in \Omega$ and dimension d.

initialize

$x_{ij} \leftarrow 0$ for $(i,j) \notin \Omega$, $\boldsymbol{\mu} \leftarrow \frac{1}{N}\sum_{j=1}^{N} \boldsymbol{x}_j$, and $\Sigma \leftarrow \frac{1}{N}\sum_{j=1}^{N}(\boldsymbol{x}_j - \boldsymbol{\mu})(\boldsymbol{x}_j - \boldsymbol{\mu})^\top$.

$P_j \leftarrow$ any permutation matrix that sorts the entries of the jth column of X, $\boldsymbol{x}_j$, so that its unobserved entries (as specified in Ω) appear first.

$\begin{bmatrix} \boldsymbol{x}_U^j \\ \boldsymbol{x}_O^j \end{bmatrix} \leftarrow P_j \boldsymbol{x}_j$, $\begin{bmatrix} \boldsymbol{\mu}_U^j \\ \boldsymbol{\mu}_O^j \end{bmatrix} \leftarrow P_j \boldsymbol{\mu}$, and $\begin{bmatrix} \Sigma_{UU}^j & \Sigma_{UO}^j \\ \Sigma_{OU}^j & \Sigma_{OO}^j \end{bmatrix} \leftarrow P_j \Sigma P_j^\top$.

repeat

  for all $j = 1, \ldots, N$ **do**

    $\boldsymbol{x}_j \leftarrow P_j^\top \begin{bmatrix} \boldsymbol{\mu}_U^j + \Sigma_{UO}^j (\Sigma_{OO}^j)^{-1} (\boldsymbol{x}_O^j - \boldsymbol{\mu}_O^j) \\ \boldsymbol{x}_O^j \end{bmatrix}$.

  end for

  $\boldsymbol{\mu} \leftarrow \frac{1}{N}\sum_{j=1}^{N} \boldsymbol{x}_j$ and $\Sigma \leftarrow \frac{1}{N}\sum_{j=1}^{N}(\boldsymbol{x}_j - \boldsymbol{\mu})(\boldsymbol{x}_j - \boldsymbol{\mu})^\top$.

  $S \leftarrow \Sigma + P_j^\top \begin{bmatrix} \Sigma_{UU}^j - \Sigma_{UO}^j (\Sigma_{OO}^j)^{-1} \Sigma_{OU}^j & \mathbf{0} \\ \mathbf{0} & \mathbf{0} \end{bmatrix} P_j$.

  $U_1 \leftarrow$ top d eigenvectors of S.

  $\Lambda_1 \leftarrow$ top d eigenvalues of S.

  $\sigma^2 \leftarrow \frac{1}{D-d}\sum_{i=d+1}^{D} \lambda_i(S)$.

  $B \leftarrow U_1(\Lambda_1 - \sigma^2 I)^{1/2} R$, where $R \in \mathbb{R}^{d \times d}$ is an arbitrary orthogonal matrix.

until convergence of $\boldsymbol{\mu}$ and S.

$Y \leftarrow (B^\top B + \sigma^2 I)^{-1} B^\top (X - \boldsymbol{\mu}\mathbf{1}^\top)$.

Output: $\boldsymbol{\mu}$, B, and Y.

3.1.3 Matrix Completion by Convex Optimization

The EM-based approaches to incomplete PPCA discussed in the previous section rely on (a) explicit parameterizations of the low-rank factors and (b) minimization of a nonconvex cost function in an alternating minimization fashion. Specifically, such approaches alternate between completing the missing entries given the parameters of a PPCA model for the data and estimating the parameters of the model from complete data. While simple and intuitive, such approaches suffer from two important disadvantages. First, the desired rank of the matrix needs to be known in advance. Second, due to the greedy nature of the EM algorithm, it is difficult to ensure convergence to the globally optimal solution. Therefore, a good initialization of the EM-based algorithm is critical for converging to a good solution.

In this section, we introduce an alternative approach that solves the low-rank matrix completion problem via a convex relaxation. As we will see, this approach allows us to complete a low-rank matrix by minimizing a convex objective function, which is guaranteed to have a globally optimal minimizer. Moreover, under rather benign conditions on the missing entries, the global minimizer is guaranteed to be the correct low-rank matrix, even without knowing the rank of the matrix in advance.

A rigorous justification for the correctness of the convex relaxation approach requires a deep knowledge of high-dimensional statistics and geometry that is beyond the scope of this book. However, this does not prevent us from introducing and summarizing here the main ideas and results, as well as the basic algorithms offered by this approach. Practitioners can apply the useful algorithm to their data and problems, whereas researchers who are more interested in the advanced theory behind the algorithm may find further details in (Cai et al. 2008; Candès and Recht 2009; Candès and Tao 2010; Gross 2011; Keshavan et al. 2010a; Zhou et al. 2010a).

Compressive Sensing of Low-Rank Matrices

The matrix completion problem can be considered a special case of the more general class of problems of recovering a high-dimensional low-rank matrix X from highly compressive linear measurements $B = \mathcal{P}(X)$, where $\mathcal{P}$ is a linear operator that returns a set of linear measurements B of the matrix X. It is known from high-dimensional statistics that if the linear operator $\mathcal{P}$ satisfies certain conditions, then the rank minimization problem

$$\min_{A} \quad \operatorname{rank}(A) \quad \text{s.t.} \quad \mathcal{P}(A) = B \tag{3.24}$$

is well defined, and its solution is unique (Candès and Recht 2009). However, it is also known that under general conditions, the task of finding such a minimal-rank solution is in general an NP-hard problem.

To alleviate the computational difficulty, instead of directly minimizing the discontinuous rank function, we could try to relax the objective and minimize its convex surrogate instead. More precisely, we could try to solve the following relaxed convex optimization problem

$$\min_{A} \quad \|A\|_* \quad \text{s.t.} \quad \mathcal{P}(A) = B, \tag{3.25}$$

where $\|A\|_*$ is the nuclear norm of the matrix A (i.e., the sum of all singular values of A). The theory of high-dimensional statistics (Candès and Recht 2009; Gross 2011) shows that when X is high-dimensional and the measurement operator $\mathcal{P}(\cdot)$ satisfies certain benign conditions,[2] the solution to the convex optimization problem (3.25) coincides with that of the rank minimization problem in (3.24).

In what follows, we illustrate how to apply this general approach to the low-rank matrix completion problem, derive a simple algorithm, and give precise conditions under which the algorithm gives the correct solution.

Exact Low-Rank Matrix Completion with Minimum Number of Measurements

---

[2]Such conditions typically require that the linear measurements and the matrix X be in some sense *incoherent*.

Let $X \in \mathbb{R}^{D \times N}$ be a matrix whose columns are drawn from a low-dimensional subspace of $\mathbb{R}^D$ of dimension $d \ll D$. Assume that we observe only a subset of the entries of X indexed by a set Ω, i.e.,

$$\Omega = \{(i,j) : x_{ij} \text{ is observed}\}. \tag{3.26}$$

Let $\mathcal{P}_\Omega : \mathbb{R}^{D \times N} \to \mathbb{R}^{D \times N}$ be the orthogonal projector onto the span of all matrices vanishing outside of Ω so that the (i,j)th component of $\mathcal{P}_\Omega(X)$ is equal to x_{ij} if $(i,j) \in \Omega$ and zero otherwise. As proposed in (Candès and Recht 2009), we may complete the missing entries in X by searching for a complete matrix $A \in \mathbb{R}^{D \times N}$ that is of low rank and coincides with X in Ω. This leads to the following optimization problem:

$$\min_A \quad \text{rank}(A) \quad \text{s.t.} \quad \mathcal{P}_\Omega(A) = \mathcal{P}_\Omega(X). \tag{3.27}$$

As we have discussed before in Section 3.1, in order for this problem to have a unique solution, we must require that the matrix X be nonsparse, or incoherent according to Definition 3.1. In addition, the missing entries should be random enough and should not fall into any special pattern.

Regarding the minimal number of entries needed, let us assume $D = N$ for simplicity. An $N \times N$ matrix X of rank d has $2Nd - d^2$ degrees of freedom.[3] Therefore, one should not expect to complete or recover a rank-d matrix uniquely with fewer than $O(dN)$ entries, since in general, there will be infinitely many rank-d matrices that have the same given entries.

The question is how many more entries are needed in order for the above problem to have a unique solution and, even more importantly, for the solution to be found efficiently. Since the above rank-minimization problem is NP-hard (even if the solution exists and is unique), inspired by the compressive sensing story, we consider the following convex relaxation:

$$\min_A \quad \|A\|_* \quad \text{s.t.} \quad \mathcal{P}_\Omega(A) = \mathcal{P}_\Omega(X), \tag{3.28}$$

where $\|A\|_* = \sum \sigma_i(A)$ is the sum of the singular values of A, which is the convex envelope of the rank function rank(A).

The seminal work of (Candès and Recht 2009; Candès and Tao 2010; Gross 2011) has established that when the low-rank matrix X is incoherent and the locations of the known entries are sampled uniformly at random, the minimizer to the problem (3.28) is unique and equal to the correct matrix X even if the number of given entries is barely above the minimum. More specifically, the minimum number of measurements that are needed in order for the convex optimization to give the

[3] X can be factorized as $X = UAA^{-1}V^\top$, where $U, V \in \mathbb{R}^{N \times d}$ have Nd entries each, and $A \in \mathbb{R}^{d \times d}$ is an invertible matrix.

correct solution with high probability is very close to the number of degrees of freedom of the unknowns. The following theorem summarizes the results.

Theorem 3.2 (Low-Rank Matrix Completion by Convex Optimization). *Let X be a $D \times N$ matrix of rank d, with $N \geq D$. Assume that X is ν-incoherent with respect to the set of sparse matrices according to Definition 3.1. Let M be the expected number of observed entries, whose locations are sampled independently and uniformly at random.*[4] *Then there is a numerical constant c such that if*

$$M \geq c\, \nu^4 d\, N(\log(N))^2, \tag{3.29}$$

then X is the unique solution to the problem in (3.28) *with probability at least* $1 - N^{-3}$*; that is, the program* (3.28) *recovers all the entries of X with no error.*

Notice that for a general rank-d matrix, this bound is already very tight. To see this, recall from our previous discussion that the minimum number of required measurements is $O(d\,N)$. In essence, the theorem states that with only a polylog factor[5] of extra measurements, i.e., $O(d\,N\ \text{polylog}(N))$, we can obtain the unique correct solution via convex optimization. This bound can be strengthened under additional assumptions. For instance, if $d = O(1)$ (i.e., if X is a matrix whose rank does not increase with its dimension), then the minimum number of entries needed to guarantee the exact completion of X reduces to $M \geq N\ \log(N)$ (Keshavan et al. 2010a). It is worth mentioning that the above statement is not limited to matrix completion. As shown in (Gross 2011), the same bound and statement hold for the compressive sensing of low-rank matrices with general linear observations $\mathcal{P}(X)$, i.e., for the problem (3.25), as long as the linear operator $\mathcal{P}$ is "incoherent" with the matrix X.

Low-Rank Matrix Completion via Proximal Gradient

The work of (Cai et al. 2008) proposes to find the solution to the optimization problem in (3.28) by solving the following problem:

$$\min_{A} \quad \tau\|A\|_* + \frac{1}{2}\|A\|_F^2 \quad \text{s.t.} \quad \mathcal{P}_\Omega(A) = \mathcal{P}_\Omega(X), \tag{3.30}$$

[4]Previously, we have used M to denote the number of observed entries in a specific matrix X. Notice that here, M is the expected number of observed entries under a random model in which the locations are sampled independently and uniformly at random. Thus, if p is the probability that an entry is observed, then the expected number of observed entries is pDN. Therefore, one can state the result either in terms of p or in terms of the expected number of observed entries, as we have done. For ease of exposition, we will continue to refer to M as the number of observed entries in the main text, but the reader is reminded that all the theoretical results refer to the expected number of observed entries, because the model for the observed entries is random.

[5]A polylog factor means a polynomial in the log function, i.e., $O(\text{polylog}(N))$ means $O(\log(N)^k)$ for some integer k.

in which the nuclear norm is augmented with a quadratic penalty term on A. As we will see, the additional quadratic term leads to a very simple algorithm. Furthermore, one can show that as the weight $\tau > 0$ increases, the solution of this regularized program converges to that of (3.28) (Cai et al. 2008).

More specifically, using the method of Lagrange multipliers described in Appendix A, we can write the Lagrangian function of (3.30) as

$$\mathcal{L}(A, Z) = \tau\|A\|_* + \frac{1}{2}\|A\|_F^2 + \langle Z, \mathcal{P}_\Omega(X) - \mathcal{P}_\Omega(A)\rangle, \tag{3.31}$$

where $Z \in \mathbb{R}^{D\times N}$ is a matrix of Lagrange multipliers. The optimal solution is given by the saddle point of the Lagrangian, i.e., the solution to the problem $\max_Z \min_A \mathcal{L}(A, Z)$, which can be found by iterating the following two steps:

$$\begin{cases} A_k &= \arg\min_A \mathcal{L}(A, Z_{k-1}), \\ Z_k &= Z_{k-1} + \beta \frac{\partial \mathcal{L}}{\partial Z}(A_k, Z_{k-1}), \end{cases} \tag{3.32}$$

where $\beta > 0$ is the step size. It is very easy to see that $\frac{\partial \mathcal{L}}{\partial Z}(A_k, Z_{k-1}) = \mathcal{P}_\Omega(X) - \mathcal{P}_\Omega(A_k)$. To compute the optimal A given Z_{k-1}, notice that $\langle Z, \mathcal{P}_\Omega(X) - \mathcal{P}_\Omega(A)\rangle = \langle \mathcal{P}_\Omega(Z), X - A\rangle$, and by completing squares, we have

$$\arg\min_A \mathcal{L}(A, Z) = \arg\min_A \tau\|A\|_* + \frac{1}{2}\|A - \mathcal{P}_\Omega(Z)\|_F^2. \tag{3.33}$$

The minimizer to this problem is given by the so-called *proximal operator* of the nuclear norm: $A^* = \mathcal{D}_\tau(\mathcal{P}_\Omega(Z))$, where $\mathcal{D}_\tau$ is the singular value thresholding operator defined in (2.95). We have left the derivation as Exercise 2.16.

Hence, starting from $Z_0 = 0$, the Lagrangian objective $\max_Z \min_A \mathcal{L}(A, Z)$ can be optimized via Algorithm 3.2. This is also known as the *proximal gradient descent* method. Even though the objective function (3.31) is not smooth, this method is known to converge as fast as the regular gradient descent method for smooth functions, with a rate of $O(1/k)$. If one wants to obtain the solution to the problem (3.28), one can repeat the algorithm with an increasing sequence of τ's and at each run, initialize A with the value previously obtained.

Algorithm 3.2 (Low-Rank Matrix Completion by Proximal Gradient)

Input: Entries x_{ij} of a matrix $X \in \mathbb{R}^{D\times N}$ for $(i, j) \in \Omega$ and parameter $\tau > 0$.

1: Initialize $Z \leftarrow \mathbf{0}$.
2: **repeat**
3: $\quad A \leftarrow \mathcal{D}_\tau(\mathcal{P}_\Omega(Z))$.
4: $\quad Z \leftarrow Z + \beta(\mathcal{P}_\Omega(X) - \mathcal{P}_\Omega(A))$.
5: **until** convergence of Z.

Output: Matrix A.

Example 3.3 (Completing Face Images with Missing Pixels by Convex Optimization) As we have seen in Chapter 2, under certain idealized circumstances (such as Lambertian reflectance), images of the same object taken under different illumination conditions lie near an approximately nine-dimensional linear subspace known as the *harmonic plane* (Basri and Jacobs 2003). In this example, we exploit such a low-dimensional structure to recover face images from the extended Yale B data set that have been corrupted so that the intensity values of some pixels are missing. The data matrix is formed by taking frontal face images of subject 20 under all 64 different illumination conditions. Each image is down-sampled to size 96×84. To synthesize a matrix with missing entries, a fraction of pixels from each image is randomly selected as the missing entries. We apply the proximal gradient algorithm described in Algorithm 3.2 to complete such "missing" entries. Figure 3.2 shows the results of image completion for different parameters τ for varying levels of missing entries (from 30% missing entries to 90%). Notice that with a proper choice of the parameter τ (around $\tau = 4 \times 10^5$ in this case), the convex optimization method is able to recover up to 80% of missing entries.

3.1.4 *Incomplete PCA by Alternating Minimization*

Although the convex-optimization-based approach can ensure correctness of the low-rank solution for the matrix completion problem, it requires solving a convex program of the same size as the matrix. When the data matrix X is very large, parameterizing the low-rank solution A and Lagrange multipliers Z with two matrices of the same size as X seems rather demanding, actually redundant. At least the low-rank solution A could be parameterized more economically with its low-rank factors. Hence, if scalability of the algorithm is a serious concern, it makes sense to look for the low-rank factors of the solution matrix directly.

To this end, we introduce in this section an alternating minimization algorithm for solving the geometric PCA problem with missing data. The main idea behind this approach, which was probably first proposed in (Wiberg 1976), is to find $\boldsymbol{\mu}$, U, and Y that minimize the error $\|X - \boldsymbol{\mu}\mathbf{1}^\top - UY\|_F^2$ considering only the known entries of X in the set $\Omega = \{(i,j) : w_{ij} = 1\}$, i.e.,

$$\begin{aligned}\|\mathcal{P}_\Omega(X - \boldsymbol{\mu}\mathbf{1}^\top - UY)\|_F^2 &= \|W \odot (X - \boldsymbol{\mu}\mathbf{1}^\top - UY)\|_F^2 \\ &= \sum_{i=1}^{D}\sum_{j=1}^{N} w_{ij}(x_{ij} - \mu_i - \boldsymbol{u}_i^\top \boldsymbol{y}_j)^2,\end{aligned} \tag{3.34}$$

where x_{ij} is the (i,j)th entry of X, μ_i is the ith entry of $\boldsymbol{\mu}$, $\boldsymbol{u}_i^\top$ is the ith row of U, and $\boldsymbol{y}_j$ is the jth column of Y. Notice that this cost function is the same as that in (3.10), except that the errors $\varepsilon_{ij} = x_{ij} - \boldsymbol{u}_i^\top \boldsymbol{y}_j$ associated with the missing entries ($w_{ij} = 0$) are removed.

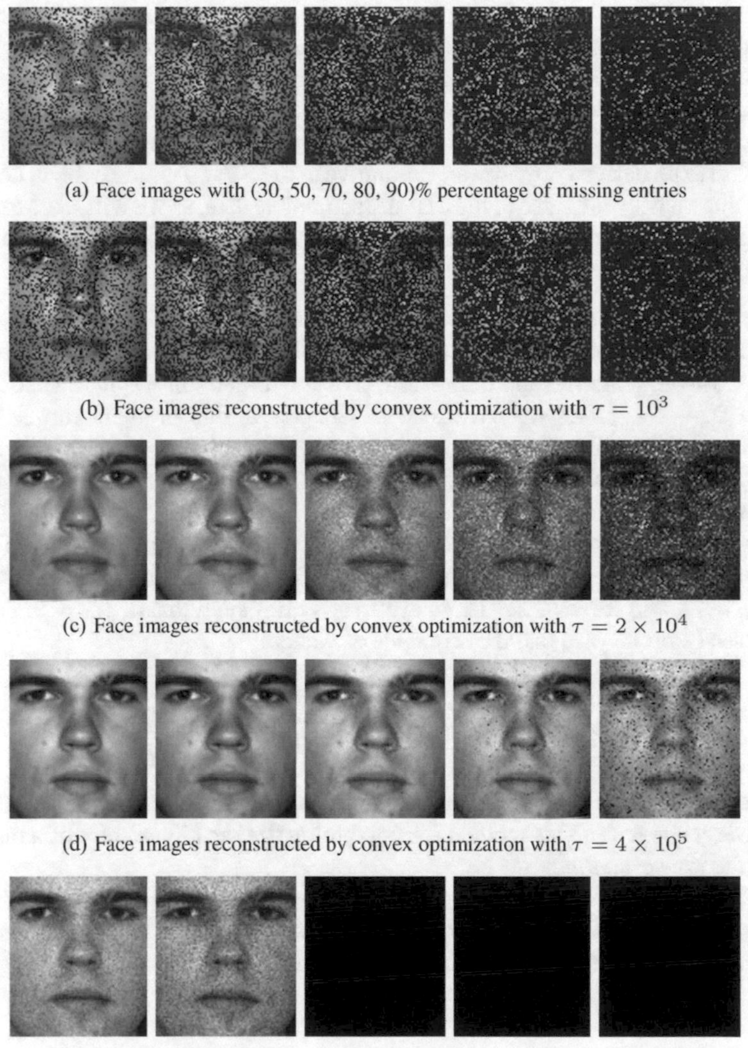

(a) Face images with (30, 50, 70, 80, 90)% percentage of missing entries

(b) Face images reconstructed by convex optimization with $\tau = 10^3$

(c) Face images reconstructed by convex optimization with $\tau = 2 \times 10^4$

(d) Face images reconstructed by convex optimization with $\tau = 4 \times 10^5$

(e) Face images reconstructed by convex optimization with $\tau = 8 \times 10^6$

Fig. 3.2 Matrix completion via convex optimization for face image completion. We take frontal face images (size 96 × 84) of subject 20 from the extended Yale B data set and randomly select a fraction of pixels as missing entries. Each column corresponds to input or result under a different percentage of missing entries. The first row is the input images, and other rows show the completion results by convex optimization with different values of τ for the algorithm. Each image shows one typical example of the recovered 64 images.

In what follows, we will derive an alternating minimization algorithm for minimizing the cost function in (3.34). For the sake of simplicity, we will first derive the algorithm in the case of zero-mean and complete data. In this case, the problem in (3.34) reduces to a low-rank matrix approximation problem, which can be solved using the SVD, as described in Theorem 2.3. The alternating minimization algorithm to be derived provides an alternative to the SVD solution, which, however, can be more easily extended to the case of incomplete data, as we will see. Moreover, the algorithm can also be extended to the more challenging PCA problem with missing entries, as we will see.

Matrix Factorization by Alternating Minimization
In the case of complete, zero-mean data, the optimization problem in (3.34) reduces to the low-rank matrix approximation problem based on explicit factorization $\min_{U,Y} \|X - UY\|_F^2$. As we have seen in Chapter 2, this problem can be solved from the SVD of X. Here, we consider an alternative method based on the *orthogonal power iteration* method (Golub and Loan 1996) for computing the top d eigenvectors of a square matrix.

Suppose that $A \in \mathbb{R}^{N \times N}$ is a symmetric positive semidefinite matrix with eigenvectors $\{\boldsymbol{u}_i\}_{i=1}^N$ and eigenvalues $\{\lambda_i\}_{i=1}^N$ sorted in decreasing order. Suppose that $\lambda_1 > \lambda_2$ and let $\boldsymbol{u}^0 \in \mathbb{R}^N$ be an arbitrary vector such that $\boldsymbol{u}_1^\top \boldsymbol{u}^0 \neq 0$. One can show (see Exercise 3.3) that the sequence of vectors

$$\boldsymbol{u}^{k+1} = \frac{A\boldsymbol{u}^k}{\|A\boldsymbol{u}^k\|} \tag{3.35}$$

converges to the top eigenvector of A up to sign, i.e., $\boldsymbol{u}^k \to \pm \boldsymbol{u}_1$, and that the rate of convergence is $\frac{\lambda_2}{\lambda_1}$. This method for computing the top eigenvector of a matrix is called the *power method*.

More generally, assume that $\lambda_d > \lambda_{d+1}$ and let $U^0 \in \mathbb{R}^{N \times d}$ be an arbitrary matrix whose column space is not orthogonal to the subspace $\{\boldsymbol{u}_i\}_{i=1}^d$ spanned by the top d eigenvectors. One can show (see Exercise 3.3) that the sequence of matrices

$$U^{k+1} = AU^k(R^k)^{-1}, \tag{3.36}$$

where $Q^k R^k = AU^k$ is the QR decomposition of AU^k, converges to a matrix U whose columns are the top d eigenvectors of A and that the rate of convergence is $\frac{\lambda_{d+1}}{\lambda_d}$. This method for computing the top d eigenvectors of a matrix is called the *orthogonal power iteration* method or Lanczos method (Lanczos 1950).

Power Factorization (PF) (Hartley and Schaffalitzky 2003) is a generalization of the orthogonal power iteration approach for computing the top d singular vectors of a (possibly) nonsquare matrix X. The main idea behind PF is that given $Y \in \mathbb{R}^{d \times N}$, an optimal solution for $U \in \mathbb{R}^{D \times d}$ that minimizes $\|X - UY\|_F^2$ is given by $XY^\top(YY^\top)^{-1}$. As before, such a matrix can be made orthogonal by

Algorithm 3.3 (Complete Matrix Factorization by Power Factorization)

Input: Matrices $X \in \mathbb{R}^{D\times N}$ and $Y^0 \in \mathbb{R}^{d\times N}$.

initialize $Y \leftarrow Y^0$.
repeat
  Given Y, find $U \leftarrow Q$, where $QR = XY^\top(YY^\top)^{-1}$.
  Given U, find $Y \leftarrow U^\top X$.
until convergence of the product UY.

Output: Matrices U and Y.

replacing U by the Q factor of the QR decomposition of $XY^\top(YY^\top)^{-1}$. Then, given an orthogonal U, the optimal Y that minimizes $\|X - UY\|_F^2$ is $U^\top X$. The PF algorithm (see Algorithm 3.3) then iterates between these two steps till convergence is achieved. The method is guaranteed to converge to the rank-d approximation of X, as stated in the following theorem, whose proof is left as an exercise to the reader (see Exercise 3.4).

Theorem 3.4 (Power Factorization). *Let X_d be the best rank-d approximation of X according to the Frobenius norm. Let σ_i be the ith singular value of X. If $\sigma_d > \sigma_{d+1}$, then there exists a constant $c > 0$ such that for all $k \geq 0$,*

$$\|X_d - U^k Y^k\|_F^2 \leq c\left(\frac{\sigma_{d+1}}{\sigma_d}\right)^{2k}, \tag{3.37}$$

where U^k and Y^k are the values at iteration k of the matrices U and Y in Algorithm 3.3.

Matrix Completion by Alternating Minimization

Let us now consider the matrix factorization problem with incomplete, zero-mean data, i.e., the problem in (3.34) with $\boldsymbol{\mu} = \mathbf{0}$. Taking the derivatives of the cost function in (3.34) with respect to $\boldsymbol{u}_i$ and $\boldsymbol{y}_j$ and setting them to zero leads to

$$\Big(\sum_{j=1}^N w_{ij}\boldsymbol{y}_j\boldsymbol{y}_j^\top\Big)\boldsymbol{u}_i = \sum_{j=1}^N w_{ij}x_{ij}\boldsymbol{y}_j, \qquad i = 1,\ldots,D, \tag{3.38}$$

$$\Big(\sum_{i=1}^D w_{ij}\boldsymbol{u}_i\boldsymbol{u}_i^\top\Big)\boldsymbol{y}_j = \sum_{i=1}^D w_{ij}x_{ij}\boldsymbol{u}_i, \qquad j = 1,\ldots,N. \tag{3.39}$$

Therefore, given Y, the optimal U can be computed linearly from (3.38). As before, the constraint $U^\top U = I$ can be enforced by replacing U by the Q factor of the QR decomposition of $U = QR$. Then, given U, the optimal Y can be computed linearly from (3.39). This leads to the PF algorithm for matrix factorization with missing entries summarized in Algorithm 3.4.

Algorithm 3.4 (Matrix Completion by Power Factorization)

Input: Matrices $W \odot X \in \mathbb{R}^{D\times N}$ and $Y^0 \in \mathbb{R}^{d\times N}$.

1: **initialize** $Y \leftarrow Y^0$.
2: **repeat**
3: Given $Y = [\boldsymbol{y}_1, \ldots, \boldsymbol{y}_N]$, solve $\min_U \|W \odot (X - UY)\|_F^2$ as

$$U = \begin{bmatrix} \boldsymbol{u}_1^\top \\ \vdots \\ \boldsymbol{u}_D^\top \end{bmatrix}, \quad \boldsymbol{u}_i \leftarrow \Big(\sum_{j=1}^N w_{ij}\boldsymbol{y}_j\boldsymbol{y}_j^\top\Big)^{-1} \sum_{j=1}^N w_{ij}x_{ij}\boldsymbol{y}_j, \quad i = 1, \ldots, D.$$

4: Normalize $U \leftarrow UR^{-1}$, where $QR = U$.
5: Given $U = \begin{bmatrix} \boldsymbol{u}_1^\top \\ \vdots \\ \boldsymbol{u}_D^\top \end{bmatrix}$, solve $\min_Y \|W \odot (X - UY)\|_F^2$ as

$$Y = [\boldsymbol{y}_1, \ldots, \boldsymbol{y}_N], \ \boldsymbol{y}_j \leftarrow \Big(\sum_{i=1}^D w_{ij}\boldsymbol{u}_i\boldsymbol{u}_i^\top\Big)^{-1} \sum_{i=1}^D w_{ij}x_{ij}\boldsymbol{u}_i, \ j = 1, \ldots, N.$$

6: **until** convergence of the sequence UY.

Output: U and Y.

Incomplete PCA by Alternating Minimization

Let us now consider the PCA problem in the case of incomplete data, i.e., the problem in (3.34), where we want to recover both the mean $\boldsymbol{\mu}$ and the subspace basis U. As in the case of complete data, the solution to this problem need not be unique, because if $(\boldsymbol{\mu}, U, Y)$ is an optimal solution, then so is $(\boldsymbol{\mu} - U\boldsymbol{b}, UA, A^{-1}Y)$ for all $\boldsymbol{b} \in \mathbb{R}^d$ and $A \in \mathbb{R}^{d\times d}$. To handle this issue, we usually enforce the constraints $U^\top U = I$ and $Y\mathbf{1} = \mathbf{0}$. For the sake of simplicity, we will forgo these constraints for a moment, derive an algorithm for solving the unconstrained problem, and then find a solution that satisfies the constraints.

To solve the unconstrained problem, let us take the derivatives of the cost function in (3.34) with respect to μ_i, $\boldsymbol{u}_i$, and $\boldsymbol{y}_j$ and set them to zero. This leads to

$$\Big(\sum_{j=1}^N w_{ij}\Big)\mu_i = \sum_{j=1}^N w_{ij}(x_{ij} - \boldsymbol{u}_i^\top \boldsymbol{y}_j), \qquad i = 1, \ldots, D, \tag{3.40}$$

$$\Big(\sum_{j=1}^N w_{ij}\boldsymbol{y}_j\boldsymbol{y}_j^\top\Big)\boldsymbol{u}_i = \sum_{j=1}^N w_{ij}(x_{ij} - \mu_i)\boldsymbol{y}_j, \qquad i = 1, \ldots, D, \tag{3.41}$$

$$\Big(\sum_{i=1}^D w_{ij}\boldsymbol{u}_i\boldsymbol{u}_i^\top\Big)\boldsymbol{y}_j = \sum_{i=1}^D w_{ij}(x_{ij} - \mu_i)\boldsymbol{u}_i, \qquad j = 1, \ldots, N. \tag{3.42}$$

Algorithm 3.5 (Incomplete PCA by Power Factorization)

Input: Matrix W, entries x_{ij} of (i,j) such that $w_{ij} = 1$, and dimension d.

initialize $\begin{bmatrix} \boldsymbol{u}_1^\top \\ \vdots \\ \boldsymbol{u}_D^\top \end{bmatrix} \leftarrow U^0 \in \mathbb{R}^{D\times d}$ and $\big[\boldsymbol{y}_1,\ldots,\boldsymbol{y}_N\big] \leftarrow Y^0 \in \mathbb{R}^{d\times N}$.

repeat

$\quad \mu_i \leftarrow \frac{\sum_{j=1}^N w_{ij}(x_{ij}-\boldsymbol{u}_i^\top \boldsymbol{y}_j)}{\sum_{j=1}^N w_{ij}}$.

$\quad \boldsymbol{u}_i \leftarrow \Big(\sum_{j=1}^N w_{ij}\boldsymbol{y}_j\boldsymbol{y}_j^\top\Big)^{-1} \sum_{j=1}^N w_{ij}(x_{ij}-\mu_i)\boldsymbol{y}_j$.

$\quad U = \begin{bmatrix} \boldsymbol{u}_1^\top \\ \vdots \\ \boldsymbol{u}_D^\top \end{bmatrix} \leftarrow UR^{-1}$, where $QR = \begin{bmatrix} \boldsymbol{u}_1^\top \\ \vdots \\ \boldsymbol{u}_D^\top \end{bmatrix}$.

$\quad Y = \big[\boldsymbol{y}_1,\ldots,\boldsymbol{y}_N\big]$ where $\boldsymbol{y}_j \leftarrow \Big(\sum_{i=1}^D w_{ij}\boldsymbol{u}_i\boldsymbol{u}_i^\top\Big)^{-1} \sum_{i=1}^D w_{ij}(x_{ij}-\mu_i)\boldsymbol{u}_i$.

until convergence of $\boldsymbol{\mu}\mathbf{1}^\top + UY$.

Output: $\boldsymbol{\mu} + \frac{1}{N}UY\mathbf{1}$, U and $Y(I - \frac{1}{N}\mathbf{1}\mathbf{1}^\top)$.

Therefore, given U and Y, the optimal $\boldsymbol{\mu}$ can be computed from (3.40). Likewise, given $\boldsymbol{\mu}$ and Y, the optimal U can be computed linearly from (3.41). Also, given $\boldsymbol{\mu}$ and U, the optimal Y can be computed linearly from (3.42).

As before, we can enforce the constraint $U^\top U = I$ by replacing U by the Q factor of the compact QR decomposition of $U = QR$. Also, we can enforce the constraint $Y\mathbf{1} = \mathbf{0}$ by replacing $\boldsymbol{\mu}$ by $\boldsymbol{\mu} + \frac{1}{N}UY\mathbf{1}$, and Y by $Y(I - \frac{1}{N}\mathbf{1}\mathbf{1}^\top)$. This leads to the alternating minimization approach for PCA with missing entries summarized in Algorithm 3.5.

A similar alternating minimization approach was proposed in (Shum et al. 1995), in which the steps in (3.40) and (3.41) are combined into a single step

$$\sum_{j=1}^N w_{ij}\begin{bmatrix}\boldsymbol{y}_j\\1\end{bmatrix}\begin{bmatrix}\boldsymbol{y}_j\\1\end{bmatrix}^\top\begin{bmatrix}\boldsymbol{u}_i\\\mu_i\end{bmatrix} = \sum_{j=1}^N w_{ij}x_{ij}\begin{bmatrix}\boldsymbol{y}_j\\1\end{bmatrix}, \quad i = 1,\ldots,D. \tag{3.43}$$

This leads to an alternating minimization scheme whereby given Y, one solves for $\boldsymbol{\mu}$ and U from (3.43), and given $\boldsymbol{\mu}$ and U, one solves for Y from (3.42).

Ensuring Global Optimality of Alternating Minimization for Matrix Completion
According to Theorem 3.4, when the data matrix $X \in \mathbb{R}^{D\times N}$ is complete, the alternating minimization method in Algorithm 3.3 is guaranteed to converge exponentially to the optimal rank-d approximation of X as long as $\sigma_{d+1}/\sigma_d < 1$. In the case of incomplete data, the alternating procedure in Algorithm 3.4 is perhaps the simplest and most natural extension of Algorithm 3.3. However, since the objective function is nonconvex, there is no guarantee that the algorithm will

converge. Thus, a natural question is whether there are conditions on the rank of X and the number of observed entries M under which the alternating minimization approach is guaranteed to converge. Now, even if the algorithm were to converge, there is no guarantee that it would converge to the globally optimal low-rank factors, or that the product of the factors would give the optimal rank-d approximation of X. Thus, another natural question is whether there are conditions on the rank of X and the number of observed entries M under which the alternating minimization approach is guaranteed to converge to the globally optimal rank-d approximation of X, and hence perfectly complete X when it has rank d. According to Theorem 3.2, the nuclear norm minimization approach in (3.28) is able to complete most rank-d matrices from $M \geq O(d\, N \log(N)^2)$ entries. Thus, a natural conjecture is that the alternating minimization approach should be able to complete a rank-d matrix from a number of entries that depends on d, Npolylog(N), and some ratio of the singular values of X. However, while alternating minimization methods for matrix completion have been used for many years, theoretical guarantees for the convergence and optimality of such methods have remained elusive.

Nonetheless, recent progress in low-rank matrix factorization (Burer and Monteiro 2005; Bach 2013; Haeffele et al. 2014) has shown that under certain conditions, local minimizers for certain classes of matrix factorization problems are global minimizers. Moreover, recent progress in low-rank matrix completion (Jain et al. 2012; Keshavan 2012; Hardt 2014; Jain and Netrapalli 2014) has shown that under certain benign conditions, certain alternating minimization methods do converge to the globally optimal solution with high probability when the matrix is of sufficiently high dimension. While a detailed explanation of such results is far beyond the scope of this book, we provide here a brief introduction with two purposes in mind. First, the analytical conditions required for optimality provide good intuition as to when we should expect low-rank matrix completion to work well in general. Second, some of the proposed algorithms introduce some modifications to the above alternating minimization methods, which may inspire readers to develop even better algorithms in the future.

As before, we are interested in finding a rank-d factorization UY, with factors $U \in \mathbb{R}^{D\times d}$ and $Y \in \mathbb{R}^{d\times N}$, that best approximates the data matrix $X \in \mathbb{R}^{D\times N}$ given the observed entries $W \odot X$ specified by the matrix $W \in \{0, 1\}^{D\times N}$, i.e.,

$$\min_{U,Y} \|W \odot (X - UY)\|_F^2. \tag{3.44}$$

The alternating minimization algorithm for solving this problem (Algorithm 3.4) uses all of the observed entries of X at each iteration in order to update the factors. In contrast, the work of (Jain et al. 2012) proposes a modified alternating minimization algorithm (see Algorithm 3.6) that uses only a *partition* of the observed entries at each iteration, whence the name *partition alternating minimization*. Specifically, the set of observed entries W is partitioned into $2K+1$ randomly chosen nonoverlapping and equally sized subsets, denoted by $W_0, W_1, \ldots, W_{2K}$. Then the updates of the original alternating minimization algorithm, Algorithm 3.4, are applied using

Algorithm 3.6 (Matrix Completion by Partition Alternating Minimization)

Input: Observed matrix $W \odot X$ and partition matrices $W_1, \dots, W_{2K}$.

```
initialization
    U^0 ← top d left singular vectors of the matrix (1/p) W_0 ⊙ X.
    U^0 ← Q, where QR = U_0 − H_{2ν√d/√N}(U_0).
end initialization
for k = 0, 1, ..., K − 1 do
    Y^{k+1} ← arg min_Y ||W_{k+1} ⊙ (U^k Y − X)||_F^2.
    U^{k+1} ← arg min_U ||W_{K+k+1} ⊙ (U Y^{k+1} − X)||_F^2.
end for
```

Output: Matrix $U^K Y^K$.

the observed entries specified by W_{k+1} to update Y and the observed entries specified by W_{K+k+1} to update U, for each $k = 0, \dots, K-1$, instead of those specified by W. The second main difference between Algorithm 3.6 and the original alternating minimization algorithm, Algorithm 3.4, is the way in which the factor U is initialized. While in Algorithm 3.4, U is typically initialized at random, in Algorithm 3.6, the factor U is initialized using the observed entries. Specifically, let p be the probability that an entry is observed, and let $M = pDN$ be the expected number of observed entries. Let U be the top d singular vectors of $\frac{1}{p} W_0 \odot X$, and $\nu > 0$ the incoherence parameter for X according to Definition 3.1. We clip entries of U that have magnitude greater than $\frac{2\nu\sqrt{d}}{\sqrt{N}}$ to be zero and let the initial U^0 be the orthonormalized version of such U obtained via QR decomposition.

In short, there are two major differences between Algorithm 3.6 and Algorithm 3.4: the initialization based on the singular vectors of $\frac{1}{p} W_0 \odot X$ and the update in each iteration using only a subset of the observations. It is surprising that these small modifications to the basic alternating minimization method can ensure that the new procedure approximates the globally optimal solution as described by the following theorem. A complete proof and explanation of this theorem is beyond the scope of this book. We refer interested readers to (Jain et al. 2012).

Theorem 3.5 (Partition Alternating Minimization for Matrix Completion). *Let X be a $D \times N$ matrix of rank d, with $N \geq D$. Assume that X is ν-incoherent with respect to the set of sparse matrices according to Definition 3.1. Let M be the expected number of observed entries, whose locations are sampled independently and uniformly at random. If there exists a constant $c > 0$ such that*

$$M \geq c\,\nu^2 \left(\frac{\sigma_1}{\sigma_d}\right)^4 d^{4.5} N \log(N) \log\left(\frac{d\|X\|_F}{\varepsilon}\right), \tag{3.45}$$

then with high probability, for $K = C' \log(\|X\|_F/\varepsilon)$ with some constant $C' > 0$, the outputs of Algorithm 3.6 satisfy $\|X - U^K Y^K\|_F \leq \varepsilon$.

In words, the alternating minimization procedure guarantees to recover X up to precision ε in $O(\log(1/\varepsilon))$ steps given that the number of observations is of order $O(d^{4.5}N\log(N)\log(d))$. This result is in perfect agreement with our conjecture that the sample complexity of alternating minimization for matrix completion should depend on d, Npolylog(N), and some ratio of the singular values of X. However, by comparing this result with the one for the convex optimization approach, $M \geq O(\nu^2 dN\log(N)^2)$, we see that this comes at the cost of an increase of the sample complexity as a function of d from linear to polynomial. This has motivated the development of modified versions of Algorithm 3.6 that are guaranteed to recover X up to precision ε under either incomparable or weaker conditions. For example, the method proposed in (Keshavan 2012) requires the expected number of observed entries to satisfy (for some constant c)

$$M \geq c\,\nu\Big(\frac{\sigma_1}{\sigma_d}\Big)^8 dN\log\Big(\frac{N}{\varepsilon}\Big), \tag{3.46}$$

which is superior when the matrix has a small condition number, while the method in (Hardt 2014) requires the expected number of observed entries to satisfy (for some constant c)

$$M \geq c\,\nu\Big(\frac{\sigma_1}{\sigma_d}\Big)^2 d^2\Big(d + \log\Big(\frac{N}{\varepsilon}\Big)\Big)N, \tag{3.47}$$

which reduces the exponent of both the ratio of the singular values as well as the subspace dimension.

Observe also that the results of (Jain et al. 2012; Hardt 2014) are of a slightly different flavor from that of results for convex optimization-based methods, since the minimum number of observed entries depends not only on the dimension of the subspace d, but also on the condition number σ_1/σ_d, which could be arbitrarily large, and the desired accuracy ε. In particular, to achieve perfect completion ($\varepsilon = 0$), we would need to observe the whole matrix. To address this issue, the work of (Jain and Netrapalli 2014) proposes a factorized version of the singular value projection algorithm of (Jain et al. 2010), called stagewise singular value projection, which is guaranteed to complete a rank-d matrix X exactly, provided that the expected number of observed entries satisfies (for some constant c)

$$M \geq c\,\nu^4 d^5 N(\log(N))^3. \tag{3.48}$$

Evidently, this result is worse than that for the nuclear norm minimization approach, which has sample complexity $O(\nu^2 dN\log(N)^2)$. But this comes at the advantage of improving the computational complexity from $O(N^3\log(\frac{1}{\varepsilon}))$ for the nuclear norm minimization approach to $O(\nu^4 d^7 N\log^3(N)\log(\frac{1}{\varepsilon}))$ for the stagewise singular value projection.

In summary, there is currently great interest in trying to develop alternating minimization algorithms for matrix completion with theoretical guarantees of convergence to the optimal rank-d matrix. Such algorithms are computationally less expensive that the nuclear minimization approach, but this comes at the cost of tolerating a smaller number of missing entries. However, as of the writing of this book, existing results do not directly apply to the basic alternating minimization procedure given in Algorithm 3.4. We conjecture that this procedure should be able to correctly complete a matrix under conditions similar to those presented in this section. Having such a result would be important, because in practice, it may be preferable to use Algorithm 3.4 because it is simpler and easier to implement.

Example 3.6 (Completing Face Images with Missing Pixels by Power Factorization) In Example 3.3, we applied the convex optimization approach (Algorithm 3.2) to complete face images in the extended Yale B data set with missing pixels. In this example, we apply the PF method for incomplete PCA (Algorithm 3.5) to the same images. Figure 3.3 shows the results for different values of the subspace dimension d. We see that for a proper choice of d (in this case from 2 to 9), the PF method works rather well up to 70% of random missing entries. However, PF fails completely for higher percentages of missing entries. This is because PF can become numerically unstable when some of the matrices are not invertible. Specifically, since there are only $N = 64$ face images, it is likely that for some rows of the data matrix, the number of observed entries is less than d; thus the matrix $\sum_{j=1}^{N} w_{ij} \boldsymbol{y}_j \boldsymbol{y}_j^\top$ in line 4 of Algorithm 3.5 becomes rank-deficient. We also observed that as expected, PF is faster than the convex approach. Specifically, in this example, PF took 1.48 seconds in MATLAB, while the convex optimization approach took 10.15 seconds.

3.2 PCA with Robustness to Corrupted Entries

In the previous section, we considered the PCA problem in the case that some entries of the data points are missing. In this section, we consider the PCA problem in the case that some of the entries of the data points have been corrupted by gross errors, known as intrasample outliers. The additional challenge is that we do not know which entries have been corrupted. Thus, the problem is to simultaneously detect which entries have been corrupted and replace them by their uncorrupted values. In some literature, this problem is referred to as the *robust PCA* problem (De la Torre and Black 2004; Candès et al. 2011).

Let us first recall the PCA problem (see Section 2.1.2) in which we are given N data points $\mathcal{X} = \{\boldsymbol{x}_j \in \mathbb{R}^D\}_{j=1}^N$ drawn (approximately) from a d-dimensional affine subspace $S = \{\boldsymbol{x} = \boldsymbol{\mu} + U\boldsymbol{y}\}$, where $\boldsymbol{\mu} \in \mathbb{R}^D$ is an arbitrary point in S, $U \in \mathbb{R}^{D\times d}$ is a basis for S, and $\{\boldsymbol{y}_j \in \mathbb{R}^d\}_{j=1}^N$ are the principal components. In the robust PCA problem, we assume that the ith entry x_{ij} of a data point $\boldsymbol{x}_j$ is obtained by corrupting the ith entry ℓ_{ij} of a point $\boldsymbol{\ell}_j$ lying perfectly on the subspace S by an error e_{ij}, i.e.,

$$x_{ij} = \ell_{ij} + e_{ij}, \quad \text{or} \quad \boldsymbol{x}_j = \boldsymbol{\ell}_j + \boldsymbol{e}_j, \quad \text{or} \quad X = L + E, \tag{3.49}$$

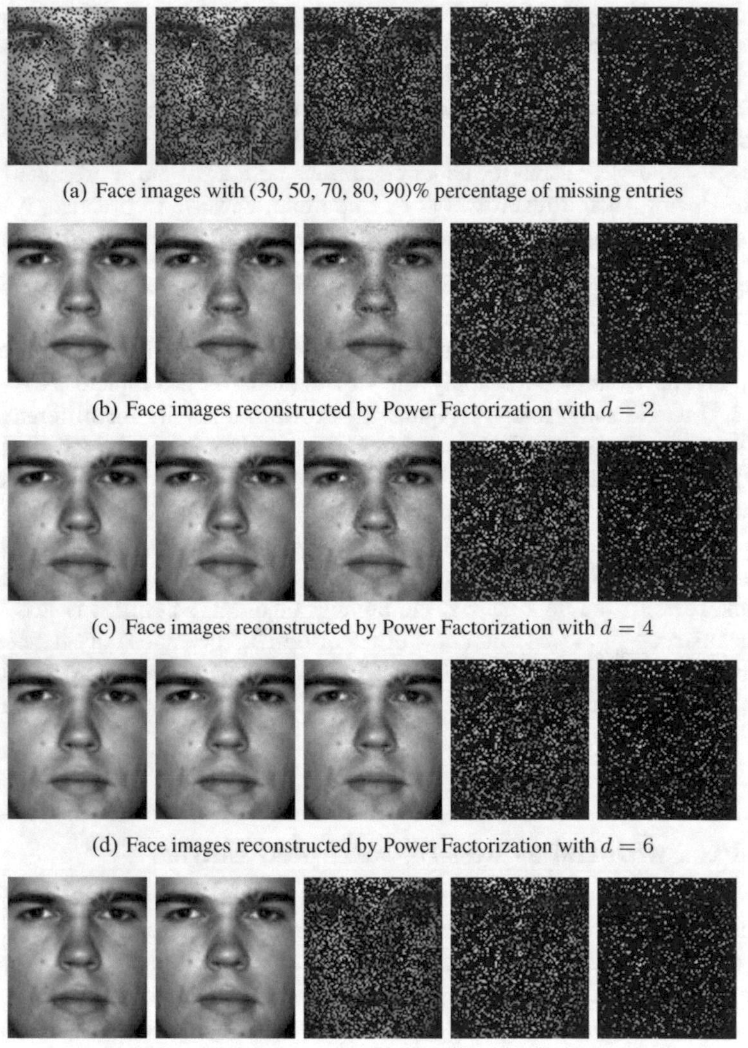

(a) Face images with (30, 50, 70, 80, 90)% percentage of missing entries

(b) Face images reconstructed by Power Factorization with $d = 2$

(c) Face images reconstructed by Power Factorization with $d = 4$

(d) Face images reconstructed by Power Factorization with $d = 6$

(e) Face images reconstructed by Power Factorization with $d = 9$

Fig. 3.3 Power factorization for recovering face images. We take frontal face images (size 96×84) of subject 20 from the extended Yale B data set and randomly select a fraction of pixels as missing entries. Each column corresponds to input or result under a different percentage of missing entries. The first row is the input images, and other rows are the results obtained by power factorization with different values of d used. Each image shows one typical example of the recovered 64 images.

where $X, L, E \in \mathbb{R}^{D\times N}$ are matrices with entries x_{ij}, ℓ_{ij}, and e_{ij}, respectively. Such errors can have a huge impact on the estimation of the subspace. Thus it is very important to be able to detect the locations of those errors,

$$\Omega = \{(i,j) : e_{ij} \neq 0\}, \tag{3.50}$$

as well as correct the erroneous entries before applying PCA to the given data.

As discussed before, a key difference between the robust PCA problem and the incomplete PCA problem is that we do not know the location of the corrupted entries. This makes the robust PCA problem harder, since we need to simultaneously detect and correct the errors. Nonetheless, when the number of corrupted entries is a small enough fraction of the total number of entries, i.e., when $|\Omega| < \rho \cdot DN$ for some $\rho < 1$, we may still hope to be able to detect and correct such errors. In the remainder of this section, we describe methods from robust statistics and convex optimization for addressing this problem.

3.2.1 Robust PCA by Iteratively Reweighted Least Squares

One of the simplest algorithms for dealing with corrupted entries is the iteratively reweighted least squares (IRLS) approach proposed in (De la Torre and Black 2004). In this approach, a subspace is fit to the corrupted data points using standard PCA. The corrupted entries are detected as those that have a large residual with respect to the identified subspace. A new subspace is estimated with the detected corruptions down-weighted. This process is then repeated until the estimated model stabilizes.

The first step is to apply standard PCA to the given data. Recall from Section 2.1.2 that when the data points $\{\boldsymbol{x}_j \in \mathbb{R}^D\}_{j=1}^N$ have no gross corruptions, an optimal solution to PCA can be obtained as

$$\hat{\boldsymbol{\mu}} = \frac{1}{N}\sum_{j=1}^{N} \boldsymbol{x}_j \quad \text{and} \quad \hat{\boldsymbol{y}}_j = \hat{U}^\top(\boldsymbol{x}_j - \boldsymbol{\mu}), \tag{3.51}$$

where $\hat{U}$ is a $D \times d$ matrix whose columns are the top d eigenvectors of

$$\hat{\Sigma}_N = \frac{1}{N}\sum_{j=1}^{N}(\boldsymbol{x}_j - \hat{\boldsymbol{\mu}})(\boldsymbol{x}_j - \hat{\boldsymbol{\mu}})^\top. \tag{3.52}$$

When the data points are corrupted by gross errors, we may improve the estimation of the subspace by recomputing the model parameters after down-weighting samples that have large residuals. More specifically, let $w_{ij} \in [0, 1]$ be a weight assigned to the ith entry of $\boldsymbol{x}_j$ such that $w_{ij} \approx 1$ if x_{ij} is not corrupted, and $w_{ij} \approx 0$ otherwise. Then a new estimate of the subspace can be obtained by minimizing the weighted sum of the least-squares errors between a point $\boldsymbol{x}_j$ and its projection $\boldsymbol{\mu} + U\boldsymbol{y}_j$ onto the subspace S, i.e.,

$$\sum_{i=1}^{D}\sum_{j=1}^{N} w_{ij}(x_{ij} - \mu_i - \boldsymbol{u}_i^\top \boldsymbol{y}_j)^2, \tag{3.53}$$

where μ_i is the ith entry of $\boldsymbol{\mu}$, $\boldsymbol{u}_i^\top$ is the ith row of U, and $\boldsymbol{y}_j$ is the vector of coordinates of the point $\boldsymbol{x}_j$ in the subspace S.

Notice that the above objective function is identical to the objective function in (3.34), which we used for incomplete PCA. The only difference is that in incomplete PCA, $w_{ij} \in \{0, 1\}$ denotes whether x_{ij} is observed or unobserved, while here $w_{ij} \in [0, 1]$ denotes whether x_{ij} is corrupted or uncorrupted. Other than that, the iterative procedure for computing $\boldsymbol{\mu}$, U, and Y given W is the same as that outlined in Algorithm 3.5.

Given $\boldsymbol{\mu}$, U, and Y, the main question is how to update the weights. A simple approach is to set the weights depending on the residual $\varepsilon_{ij} = x_{ij} - \mu_i - \boldsymbol{u}_i^\top \boldsymbol{y}_j$. Our expectation is that when the residual is small, x_{ij} is not corrupted, and so we should set $w_{ij} \approx 1$. Conversely, when the residual is large, x_{ij} is corrupted, and so we should set $w_{ij} \approx 0$. *Maximum-likelihood-type estimators* (M-Estimators) define the weights to be

$$w_{ij} = \rho(\varepsilon_{ij})/\varepsilon_{ij}^2 \tag{3.54}$$

for some robust loss function $\rho(\cdot)$. The objective function then becomes

$$\sum_{i=1}^{D}\sum_{j=1}^{N} \rho(\varepsilon_{ij}). \tag{3.55}$$

Many loss functions $\rho(\cdot)$ have been proposed in the statistics literature (Huber 1981; Barnett and Lewis 1983). When $\rho(\varepsilon) = \varepsilon^2$, all weights are equal to 1, and we obtain the standard least-squares solution, which is not robust. Other robust loss functions include the following:

1. L_1 loss: $\rho(\varepsilon) = |\varepsilon|$;
2. Cauchy loss: $\rho(\varepsilon) = \varepsilon_0^2 \log(1 + \varepsilon^2/\varepsilon_0^2)$;
3. Huber loss (Huber 1981): $\rho(\varepsilon) = \begin{cases} \varepsilon^2 & \text{if} \quad |\varepsilon| < \varepsilon_0, \\ 2\varepsilon_0|\varepsilon| - \varepsilon_0^2 & \text{otherwise;} \end{cases}$
4. Geman–McClure loss (Geman and McClure 1987): $\rho(\varepsilon) = \frac{\varepsilon^2}{\varepsilon^2+\varepsilon_0^2}$,

where $\varepsilon_0 > 0$ is a parameter. Following the work of (De la Torre and Black 2004), we use the Geman–McClure loss scaled by ε_0^2, which gives

Algorithm 3.7 (Robust PCA by Iteratively Reweighted Least Squares)

Input: Data matrix X, dimension d, and parameter $\varepsilon_0 > 0$.

initialize $[\boldsymbol{\mu}, U, Y] = \text{PCA}(X)$ using PCA from Chapter 2.
repeat
$\varepsilon_{ij} \leftarrow x_{ij} - \mu_i - \boldsymbol{u}_i^\top \boldsymbol{y}_j$.
$w_{ij} \leftarrow \frac{\varepsilon_0^2}{\varepsilon_{ij}^2 + \varepsilon_0^2}$.
$\mu_i \leftarrow \frac{\sum_{j=1}^N w_{ij}(x_{ij} - \boldsymbol{u}_i^\top \boldsymbol{y}_j)}{\sum_{j=1}^N w_{ij}}$.
$\boldsymbol{u}_i \leftarrow \left(\sum_{j=1}^N w_{ij}\boldsymbol{y}_j\boldsymbol{y}_j^\top\right)^{-1} \sum_{j=1}^N w_{ij}(x_{ij} - \mu_i)\boldsymbol{y}_j$.
$U = \begin{bmatrix} \boldsymbol{u}_1^\top \\ \vdots \\ \boldsymbol{u}_D^\top \end{bmatrix} \leftarrow \begin{bmatrix} \boldsymbol{u}_1^\top \\ \vdots \\ \boldsymbol{u}_D^\top \end{bmatrix} R^{-1}$, where $QR = \begin{bmatrix} \boldsymbol{u}_1^\top \\ \vdots \\ \boldsymbol{u}_D^\top \end{bmatrix}$.
$Y = [\boldsymbol{y}_1, \ldots, \boldsymbol{y}_N]$ where $\boldsymbol{y}_j \leftarrow \left(\sum_{i=1}^D w_{ij}\boldsymbol{u}_i\boldsymbol{u}_i^\top\right)^{-1} \sum_{i=1}^D w_{ij}(x_{ij} - \mu_i)\boldsymbol{u}_i$.
until convergence of $\boldsymbol{\mu}\mathbf{1}^\top + UY$.
$\boldsymbol{\mu} \leftarrow \boldsymbol{\mu} + \frac{1}{N}UY\mathbf{1}$, $Y \leftarrow Y(I - \frac{1}{N}\mathbf{1}\mathbf{1}^\top)$, $L \leftarrow UY$, and $E \leftarrow X - L$.

Output: $\boldsymbol{\mu}$, U, Y, L and E.

$$w_{ij} = \frac{\varepsilon_0^2}{\varepsilon_{ij}^2 + \varepsilon_0^2}. \tag{3.56}$$

The overall algorithm for PCA with corruptions is summarized in Algorithm 3.7. This algorithm initializes all the weights to $w_{ij} = 1$. This gives an initial estimate for the subspace, which is the same as that given by PCA. Given this initial estimate of the subspace, the weights w_{ij} are computed from the residuals as in (3.56). Given these weights, one can reestimate the subspace using the steps of Algorithm 3.5. One can then iterate between computing the weights given the subspace and computing the subspace given the weights.

Example 3.7 (Face Shadow Removal by Iteratively Reweighted Least Squares) As we have seen in Chapter 2, the set of images of a convex Lambertian object obtained under different lighting conditions lies close to a nine-dimensional linear subspace known as the *harmonic plane* (Basri and Jacobs 2003). However, since faces are neither perfectly convex nor Lambertian, face images taken under different illuminations often suffer from several nuances such as self-shadowing, specularities, and saturations in brightness. Under the assumption that the images of a person's face are aligned, the above robust PCA algorithm offers a principled way of removing the shadows and specularities, because such artifacts are concentrated on small portions of the face images, i.e., they are sparse in the image domain. In this example, we use the frontal face images of subject 20 under 64 different

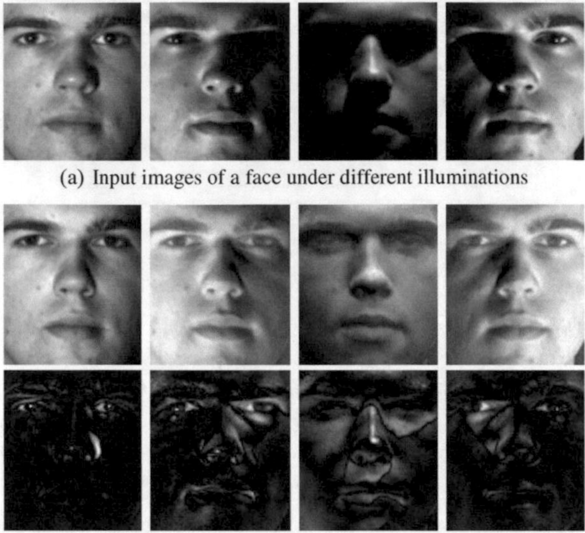

(a) Input images of a face under different illuminations

(b) Low rank and sparse components by IRLS

Fig. 3.4 Removing shadows and specularities from face images using IRLS for PCA with corrupted data. We apply Algorithm 3.7 to 64 frontal face images of subject 20 from the extended Yale B data set. Each image is of size 96×84. (a) Four out of 64 representative input face images. (b) Recovered images from the low-rank component L (first row) and sparse errors E (second row).

illumination conditions. Each image is down-sampled to size 96×84. We then apply the IRLS method (Algorithm 3.7) with $\varepsilon_0 = 1$ and $d = 4$ to remove the shadows and specularities in the face images. The results in Figure 3.4 show that the IRLS method is able to do a reasonably good job of removing some of the shadows and specularities around the nose and eyes area. However, the error image in the third column shows that the recovered errors are not very sparse, and the method could confuse valid image signal due to darkness with true errors (caused by shadows, etc.)

3.2.2 *Robust PCA by Convex Optimization*

Although the IRLS scheme for robust PCA is very simple and efficient to implement, and widely used in practice, there is no immediate guarantee that the method converges. Moreover, even if the method were to converge, there is no guarantee that the solution to which it converges corresponds to the correct low-rank matrix. As we have seen in the low-rank matrix completion problem, we should not even expect the problem to have a meaningful solution unless proper conditions are imposed on the low-rank matrix and the matrix of errors.

In this section, we will derive conditions under which the robust PCA problem is well posed and admits an efficient solution. To this end, we will formulate the robust PCA problem as a (nonconvex and nonsmooth) rank minimization problem in which we seek to decompose the data matrix X as the sum of a low-rank matrix L and a matrix of errors E. Similar to the matrix completion case, we will study convex relaxations of the rank minimization problem and resort to advanced tools from high-dimensional statistics to show that under certain conditions, the convex relaxations can effectively and efficiently recover a low-rank matrix with intrasample outliers as long as the outliers are sparse enough. Although the mathematical theory that supports the correctness of these methods is far beyond the scope of this book, we will introduce the key ideas and results of this approach to PCA with intrasample outliers.

More specifically, we assume that the given data matrix X is generated as the sum of two matrices

$$X = L_0 + E_0. \tag{3.57}$$

The matrix L_0 represents the ideal low-rank data matrix, while the matrix E_0 represents the intrasample outliers. Since many entries of X are not corrupted (otherwise, the problem would not be well posed), many entries of E_0 should be zero. As a consequence, we can pose the robust PCA problem as one of decomposing a given matrix X as the sum of two matrices $L + E$, where L is of low rank and E is sparse. This problem can be formulated as

$$\min_{L,E} \quad \text{rank}(L) + \lambda \|E\|_0 \quad \text{s.t.} \quad X = L + E, \tag{3.58}$$

where $\|E\|_0$ is the number of nonzero entries in E, and $\lambda > 0$ is a tradeoff parameter.

Robust PCA as a Well-Posed Problem

At first sight, it may seem that solving the problem in (3.58) is impossible. First of all, we have an underdetermined system of DN linear equations in $2DN$ unknowns. Among the many possible solutions, we are searching for a solution (L, E) such that L is of low rank and E is sparse. However, such a solution may not be unique. For instance, if $x_{11} = 1$ and $x_{ij} = 0$ for all $(i,j) \neq (1, 1)$, then the matrix X is both of rank 1 and sparse. Thus, if $\lambda = 1$, we can choose $(L, E) = (X, \mathbf{0})$ or $(L, E) = (\mathbf{0}, X)$ as valid solutions. To avoid such an ambiguity, as suggested by the results for matrix completion, the low-rank matrix L_0 should be in some sense "incoherent" with the sparse corruption matrix E_0. That is, the low-rank matrix L_0 itself should not be sparse. To capture this, we will assume that L_0 is an incoherent matrix according to Definition 3.1. Second of all, as suggested also by results for matrix completion, if we want to recover the low-rank matrix L_0 correctly, the locations of the corrupted entries should not fall into any conspicuous pattern. Therefore, as in the matrix completion problem, we will assume that the locations of the corrupted entries are distributed uniformly at random so that the chance that they form any conspicuous pattern is very low.

As we will see, under the above condition of incoherence and random corruptions, the problem in (3.58) will become well posed for most matrices X. However, to be able to state the precise conditions under which the solution to (3.58) coincides with (L_0, E_0), we first need to study the question of how to efficiently solve the problem in (3.58).

Recovering a Low-Rank Matrix or a Sparse Vector by Convex Relaxation
Observe that even if the conditions above could guarantee that the problem in (3.58) has a unique globally optimal solution, another challenge is that the cost function to be minimized is nonconvex and nondifferentiable. In fact, it is well known that the problem of recovering either a low-rank matrix C or a sparse signal $\boldsymbol{c}$ from undersampled linear measurements B or $\boldsymbol{b}$, i.e.,

$$\min_C \ \operatorname{rank}(C) \ \text{ s.t. } \ \mathcal{P}(C) = B, \quad \text{or} \quad \min_{\boldsymbol{c}} \ \|\boldsymbol{c}\|_0 \ \text{ s.t. } \ A\boldsymbol{c} = \boldsymbol{b}, \tag{3.59}$$

is in general NP-hard (Amaldi and Kann 1998).

As we have seen for the low-rank matrix completion problem, the difficulty of solving the rank minimization on the left-hand side of (3.59) can be alleviated by minimizing the convex envelope of the rank function, which is given by the matrix nuclear norm $\|C\|_*$ and gives rise to the following optimization problem:

$$\min_C \ \|C\|_* \ \text{ s.t. } \ \mathcal{P}(C) = B. \tag{3.60}$$

As it turns out, convex relaxation works equally well for finding the sparsest solution to a highly underdetermined system of linear equations $A\boldsymbol{c} = \boldsymbol{b}$, which is the problem on the right-hand side of (3.59). This class of problems is known in the literature as *compressed or compressive sensing* (Candès 2006). Since this linear system is underdetermined, in general there could be many solutions $\boldsymbol{c}$ to the equation $A\boldsymbol{c} = \boldsymbol{b}$. This mimics the matrix completion problem, where the number of given measurements is much less than the number of variables to be estimated or recovered (all the entries of the matrix). Hence we want to know under what conditions the sparsest solution to $A\boldsymbol{c} = \boldsymbol{b}$ is unique and can be found efficiently.

To this end, we briefly survey results from the compressive sensing literature (see (Candès and Tao 2005; Candès 2008) and others). Without loss of generality, let us assume that A is an $m \times n$ matrix with $m \ll n$ whose columns have unit norm. Let $\boldsymbol{b} = A\boldsymbol{c}_0$, where $\boldsymbol{c}_0$ is k-sparse, i.e., $\boldsymbol{c}_0$ has at most k nonzero entries. Our goal is to recover $\boldsymbol{c}_0$ by solving the optimization problem on the right-hand side of (3.59). Notice that if A has two identical columns, say columns 1 and 2, then $\tilde{\boldsymbol{c}}_0 = [1, -1, 0, \dots, 0]^\top$ satisfies $A\tilde{\boldsymbol{c}}_0 = \boldsymbol{0}$. Thus, if $\boldsymbol{c}_0$ is a sparse solution to $A\boldsymbol{c} = \boldsymbol{b}$, then so is $\boldsymbol{c}_0 + \tilde{\boldsymbol{c}}_0$. More generally, if A has very sparse vectors in its (right) null space, then sparse solutions to $A\boldsymbol{c} = \boldsymbol{b}$ are less likely to be unique. Hence, to ensure the uniqueness of the sparsest solution, we typically need the measurement matrix A to be *mutually incoherent*, as defined next.

Definition 3.8 (Mutual Coherence). *The* mutual coherence *of a matrix $A \subset \mathbb{R}^{m \times n}$ is defined as*

$$\nu(A) = \max_{i \neq j = 1,\ldots,n} |\boldsymbol{a}_i^\top \boldsymbol{a}_j|. \tag{3.61}$$

A matrix is said to be mutually incoherent with parameter ν if $\nu(A) < \nu$.

This definition of incoherence is not to be confused with that in Definition 3.1. Definition 3.8 tries to capture whether each column of A is incoherent with other columns so that no sparse number of columns can be linearly independent, whence the name mutual coherence. This notion is useful for finding a sparse solution to a set of linear equations, as we will see in Theorem 3.10. On the other hand, Definition 3.1 tries to capture whether the matrix as a whole is incoherent with respect to sparse missing entries or sparse corruptions, whence the name incoherence with respect to sparse matrices. This notion of incoherence is useful for solving the matrix completion problem, as we saw in Theorem 3.2, and will be useful for solving the robust PCA problem, as we will see in Theorem 3.66.

Another property of a matrix that is typically used to characterize the conditions under which it is possible to solve a linear system is the notion of restricted isometry, as defined next.

Definition 3.9. *Given an integer k, the* restricted isometry constant *of a matrix A is the smallest number $\delta_k(A)$ such that for all $\boldsymbol{c}$ with $\|\boldsymbol{c}\|_0 \leq k$, we have*

$$(1 - \delta_k(A))\|\boldsymbol{c}\|_2^2 \leq \|A\boldsymbol{c}\|_2^2 \leq (1 + \delta_k(A))\|\boldsymbol{c}\|_2^2. \tag{3.62}$$

The remarkable results from compressive sensing have shown that if the measurement matrix A is sufficiently incoherent or the restricted isometry constant is small enough, then to find the correct sparsest solution to $A\boldsymbol{c} = \boldsymbol{b}$, we can replace the ℓ_0 norm in (3.59) by its convex envelope, the ℓ_1 norm, which gives rise to the following optimization problem:

$$\min_{\boldsymbol{c}} \|\boldsymbol{c}\|_1 \quad \text{s.t.} \quad \boldsymbol{b} = A\boldsymbol{c}. \tag{3.63}$$

More precisely, we have the following result:

Theorem 3.10 (Sparse Recovery under Incoherence or Restricted Isometry). *If the matrix A is incoherent, i.e., if $\nu(A) < \frac{1}{2k-1}$, or if it satisfies the restricted isometry property (RIP) $\delta_{2k}(A) < \sqrt{2}-1$, then the optimal solution $\boldsymbol{c}^*$ to the ℓ_1-minimization problem in (3.63) is the correct sparsest solution, i.e., $\boldsymbol{c}^* = \boldsymbol{c}_0$.*

In other words, when the matrix A is incoherent enough, the sparsest solution to the linear system $A\boldsymbol{c} = \boldsymbol{b}$ can be obtained by solving a convex ℓ_1-minimization problem as opposed to an NP-hard ℓ_0-minimization problem.

Robust PCA by Convex Relaxation
Inspired by the above convex relaxation techniques, for the robust PCA problem in (3.58) we would expect that under certain conditions on L_0 and E_0, we can decompose X as $L_0 + E_0$ by solving the following convex optimization problem:

$$\min_{L,E} \quad \|L\|_* + \lambda\|E\|_1 \quad \text{s.t.} \quad X = L + E, \tag{3.64}$$

where $\|L\|_* = \sum_i \sigma_i(L)$ is the nuclear norm of L, i.e., the sum of its singular values, and $\|E\|_1 = \sum_{i,j} |e_{ij}|$ is the ℓ_1 norm of E viewed as a vector. This convex program is known as *principal component pursuit* (PCP).

The following theorem gives precise conditions on the rank of the matrix and the percentage of outliers under which the optimal solution of the above convex program is exactly (L_0, E_0) with overwhelming probability.

Theorem 3.11 (Robust PCA by Principal Component Pursuit (Candès et al. 2011)). *Let $X = L_0 + E_0$. Assume that $L_0 = U\Sigma V^\top$ is ν-incoherent with respect to the set of sparse matrices according to Definition 3.1. Assume also that the support of E_0 is uniformly distributed among all the sets of cardinality $D \times N$. If*

$$\operatorname{rank}(L_0) \leq \frac{\rho_d \min\{D, N\}}{\mu^2 \log^2\big(\max\{D, N\}\big)} \quad \textit{and} \quad \|E_0\|_0 \leq \rho_s ND \tag{3.65}$$

for some constant $\rho_d, \rho_s > 0$, then there is a constant c such that with probability at least $1 - c\max\{N, D\}^{-10}$, the solution (L^*, E^*) to (3.64) with $\lambda = \frac{1}{\sqrt{\max\{N,D\}}}$ is exact, i.e.,

$$L^* = L_0 \quad \textit{and} \quad E^* = E_0. \tag{3.66}$$

A complete proof and explanation for this theorem is beyond the scope of this book; interested readers are referred to (Candès et al. 2011). But this does not prevent us from understanding its implications and using it to develop practical solutions for real problems. The theorem essentially says that as long as the low-rank matrix is incoherent and its rank is bounded almost linearly from its dimension, the PCP program can correctly recover the low-rank matrix even if a constant fraction of its entries are corrupted. Other results show that under some additional benign conditions, say the signs of the entries of E_0 are random, the convex optimization can correct an arbitrarily high percentage of errors if the matrix is sufficiently large (Ganesh et al. 2010).

Alternating Direction Method of Multipliers for Principal Component Pursuit
Assuming that the conditions of Theorem 3.66 are satisfied, the next question is how to find the global minimum of the convex optimization problem in (3.64). Although in principle, many convex optimization solvers can be used, we introduce here an algorithm based on the augmented Lagrange multiplier (ALM) method suggested by (Candès et al. 2011; Lin et al. 2011).

The ALM method operates on the *augmented Lagrangian*

$$\mathscr{L}(L,E,\Lambda) = \|L\|_* + \lambda\|E\|_1 + \langle \Lambda, X - L - E\rangle + \frac{\beta}{2}\|X - L - E\|_F^2. \tag{3.67}$$

A generic Lagrange multiplier algorithm (Bertsekas 1999) would solve PCP by repeatedly setting $(L_k, E_k) = \arg\min_{L,E} \mathscr{L}(L,E,\Lambda_k)$ and then updating the Lagrange multiplier matrix by $\Lambda_{k+1} = \Lambda_k + \beta(X - L_k - E_k)$. This is also known as the *exact ALM method*.

For our low-rank and sparse decomposition problem, we can avoid having to solve a sequence of convex programs by recognizing that $\min_L \mathscr{L}(L,E,\Lambda)$ and $\min_E \mathscr{L}(L,E,\Lambda)$ both have very simple and efficient solutions. In particular, it is easy to show that

$$\arg\min_E \mathscr{L}(L,E,\Lambda) = \mathcal{S}_{\lambda\beta^{-1}}(X - L + \beta^{-1}\Lambda), \tag{3.68}$$

where $\mathcal{S}_\tau(X)$ is the soft-thresholding operator defined in (2.96) applied to each entry x of the matrix X as $\mathcal{S}_\tau(x) = \operatorname{sign}(x)\max(|x| - \tau, 0)$. Similarly, it is not difficult to show that (see Exercise 2.16)

$$\arg\min_L \mathscr{L}(L,E,\Lambda) = \mathcal{D}_{\beta^{-1}}(X - E + \beta^{-1}\Lambda), \tag{3.69}$$

where $\mathcal{D}_\tau(X)$ is the singular value thresholding operator defined in (2.95) as $\mathcal{D}_\tau(X) = U\mathcal{S}_\tau(\Sigma)V^*$, where $U\Sigma V^*$ is any singular value decomposition of X.

Thus, a more practical strategy is first to minimize $\mathscr{L}$ with respect to L (fixing E), then minimize $\mathscr{L}$ with respect to E (fixing L), and then finally update the Lagrange multiplier matrix Λ based on the residual $X - L - E$, a strategy that is summarized as Algorithm 3.8 below.

Algorithm 3.8 is a special case of a general class of algorithms known as *alternating direction method of multipliers* (ADMM), described in Appendix A. The convergence of these algorithms has been well studied and established (see e.g., (Lions and Mercier 1979; Kontogiorgis and Meyer 1989) and the many references therein, as well as discussion in (Lin et al. 2011; Yuan and Yang 2009)). Algorithm 3.8 performs excellently on a wide range of problems: relatively small numbers of iterations suffice to achieve good relative accuracy. The dominant cost of each iteration is computing L_{k+1} by singular value thresholding. This requires us to compute the singular vectors of $X - E_k + \beta^{-1}\Lambda_k$ whose corresponding singular values exceed the threshold β^{-1}. Empirically, the number of such large singular values is often bounded by $\operatorname{rank}(L_0)$, allowing the next iterate to be computed efficiently by a partial SVD.[6] The most important implementation details for this

[6]Further performance gains might be possible by replacing this partial SVD with an approximate SVD, as suggested in (Goldfarb and Ma 2009) for nuclear norm minimization.

Algorithm 3.8 (Principal Component Pursuit by ADMM (Lin et al. 2011))

1: **initialize:** $E_0 = \Lambda_0 = 0, \beta > 0$.
2: **while** not converged **do**
3: compute $L_{k+1} = \mathcal{D}_{\beta^{-1}}(X - E_k + \beta^{-1}\Lambda_k)$.
4: compute $E_{k+1} = \mathcal{S}_{\lambda\beta^{-1}}(X - L_{k+1} + \beta^{-1}\Lambda_k)$.
5: compute $\Lambda_{k+1} = \Lambda_k + \beta(X - L_{k+1} - E_{k+1})$.
6: **end while**
7: **output:** L, E.

algorithm are the choice of β and the stopping criterion. In this work, we simply choose $\beta = ND/4\|X\|_1$, as suggested in (Yuan and Yang 2009).

Some Extensions to PCP
In most practical applications, there is also small dense noise in the data. So a more realistic model for robust PCA can be $X = L + E + Z$, where Z is a Gaussian matrix that models small Gaussian noise in the given data. In this case, we can no longer expect to recover the exact solution to the low-rank matrix (which is impossible even if there are no outliers). Nevertheless, one can show that the natural convex extension

$$\min_{L,E} \quad \|L\|_* + \lambda\|E\|_1 \quad \text{s.t.} \quad \|X - L - E\|_2^2 \leq \varepsilon^2, \tag{3.70}$$

where ε is the known noise variance, gives a stable estimate to the low-rank and sparse components L and E, subject to a small residual proportional to the noise variance (Zhou et al. 2010b).

Another extension is to recover a low-rank matrix from both corrupted and compressive measurements. In other words, we try to recover the low-rank and sparse components (L, E) of $X = L + E$ from only some of its linear measurements: $\mathcal{P}_Q(X)$, where $\mathcal{P}_Q(\cdot)$ could be a general linear operator. The special case in which the operator represents a subset of the entries has been covered in the original work of principal component pursuit (Candès et al. 2011). It has been shown that under similar conditions as in Theorem 3.66, one can correctly recover the low-rank and sparse components by the following optimization:

$$\min_{L,E} \quad \|L\|_* + \lambda\|E\|_1 \quad \text{s.t.} \quad \mathcal{P}_\Omega(X) = \mathcal{P}_\Omega(L + E), \tag{3.71}$$

where as in matrix completion, $\mathcal{P}_\Omega(\cdot)$ represents projection onto the observed entries.

The case of a more general linear operator $\mathcal{P}_Q(\cdot)$ for projecting onto an arbitrary subspace Q has also been studied in (Wright et al. 2013) and is known as *compressive principal component pursuit* (CPCP). It has been shown that under fairly broad conditions (so that Q is in some sense "incoherent" to L and E), the low-rank and sparse components can be correctly recovered by the following convex program:

$$\min_{L,E} \quad \|L\|_* + \lambda \|E\|_1 \quad \text{s.t.} \quad \mathcal{P}_Q(X) = \mathcal{P}_Q(L + E). \tag{3.72}$$

We leave as an exercise for the reader (see Exercise 3.8) to derive an algorithm for solving the above problems using ideas from Lagrangian methods and alternating direction minimization methods (please refer to Appendix A).

Example 3.12 (Face Shadow Removal by PCP) As we have seen in Example 3.7, robust PCA can be used to remove shadows and specularities in face images that are typically sparse in the image domain. In this example, we apply the PCP method to the same face images in Example 3.7, which correspond to frontal face images of subject 20 under 64 different illuminations (see Figure 3.5). As before, each image is down-sampled to size 96×84. We solve the PCP problem using both the exact ALM method and the inexact method via ADMM (Algorithm 3.8). We set the parameter λ according to Theorem 3.66. The exact and the inexact ALM methods give almost identical results, but the latter is much faster than the former: 2.68 seconds for inexact ALM versus 42.0 seconds for exact ALM in MATLAB on a typical desktop computer. As a comparison, the IRLS method in Example 3.7 takes 2.68 seconds on average. Comparing with the results in Figure 3.4 obtained by the IRLS method, the results given by PCP are qualitatively better in the sense that the recovered errors are indeed sparse and correspond better to true corruptions in the face images due to shadows and specularities. In particular, we can appreciate a significant improvement in the third image. This technique is potentially useful for preprocessing training images in face recognition systems to remove such deviations from the linear model. We leave the implementation of the algorithms as a programming exercise to the reader (see Exercise 3.10).

3.3 PCA with Robustness to Outliers

Another issue that we often encounter in practice is that a small portion of the data points does not fit the subspace as well as the rest of the data. Such points are called *outliers* or *outlying samples*, and their presence can lead to a completely wrong estimate of the underlying subspace. Therefore, it is very important to develop methods for detecting and eliminating outliers from the given data.

The true nature of outliers can be very elusive. In fact, there is really no unanimous definition for what an outlier is.[7] Outliers could be atypical samples that have an unusually *large influence* on the estimated model parameters. Outliers could

[7] For a more thorough exposition of outliers in statistics, we recommend the books of (Barnett and Lewis 1983; Huber 1981).

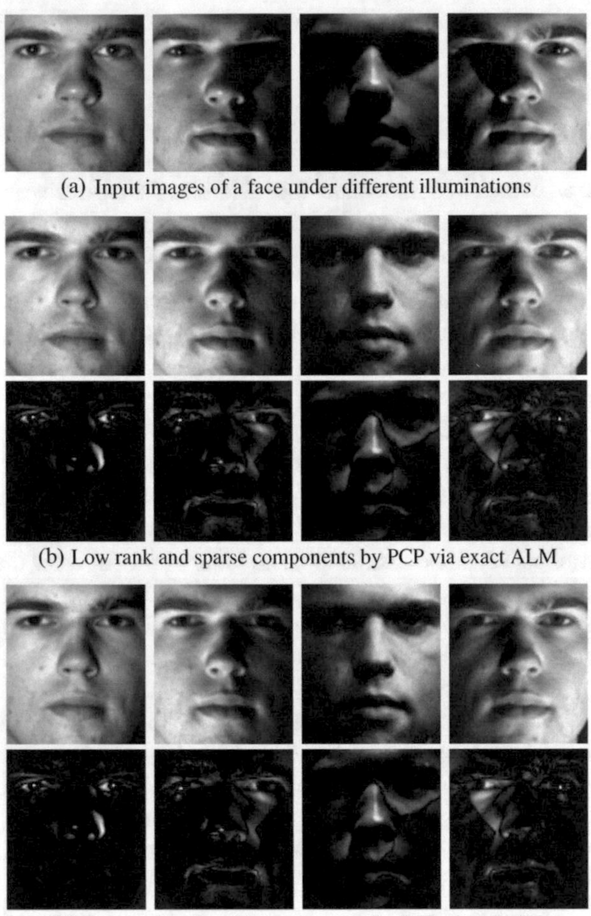

(a) Input images of a face under different illuminations

(b) Low rank and sparse components by PCP via exact ALM

(c) Low rank and sparse components by PCP via ADMM

Fig. 3.5 Removing shadows and specularities from face images by principal component pursuit. We apply Algorithm 3.7 to 64 frontal face images of subject 20 from the extended Yale B database. Each image is of size 96×84. (a) Four out of 64 representative input face images. (b)-(c) Recovered images from the low-rank component L (first row) and sparse errors E (second row).

also be perfectly valid samples from the same distribution as the rest of the data that happen to be *small-probability* instances. Alternatively, outliers could be samples drawn from a different model, and therefore they will likely *not be consistent* with the model derived from the rest of the data. In principle, however, there is no way to tell which is the case for a particular "outlying" sample point.

In this section, we will discuss two families of methods for dealing with outliers in the context of PCA. The first family will include classical methods based on the robust statistics literature described in Appendix B. The second family will include modern convex optimization techniques similar to those we have described in the previous two sections for incomplete PCA and robust PCA.

3.3.1 Outlier Detection by Robust Statistics

We begin by discussing three classical approaches from robust statistics for dealing with outliers in the context of PCA. The first method, called an influence-based method, detects outliers as points that have a large influence in the estimated subspace. The second method detects outliers as points whose probability of belonging to the subspace is very low or whose distance to the subspace is very high. Interestingly, this latter method leads to an IRLS approach to detecting outliers. The third method detects outliers by random sample consensus techniques.

Influence-Based Outlier Detection

This approach relies on the assumption that an outlier is an *atypical* sample that has an unusually large influence on the estimated subspace. This leads to an outlier detection scheme whereby the influence of a sample is determined by comparing the subspace $\hat{S} = (\hat{\boldsymbol{\mu}}, \hat{U})$ estimated with all the samples, and the subspace $\hat{S}_{(-j)} = (\hat{\boldsymbol{\mu}}_{(-j)}, \hat{U}_{(-j)})$ estimated without the jth sample. For instance, one may use a *sample influence function* based on some distance between $\hat{S}$ and $\hat{S}_{(-j)}$ such as

$$\operatorname{dist}(\hat{S}, \hat{S}_{(-j)}) = \angle\big(\operatorname{span}(\hat{U}), \operatorname{span}(\hat{U}_{(-j)})\big) \quad \text{or} \tag{3.73}$$

$$\operatorname{dist}(\hat{S}, \hat{S}_{(-j)}) = \|(I - \hat{U}\hat{U}^\top)\boldsymbol{\mu}_{(-j)}\| + \|(I - \hat{U}_{(-j)}\hat{U}_{(-j)}^\top)\boldsymbol{\mu}\|. \tag{3.74}$$

The first quantity is the largest subspace angle (see Exercise 2.8) between the linear subspace spanned by $\hat{U}$ and the linear subspace spanned by $\hat{U}_{(-j)}$. Such a distance measures the influence based on comparing only the linear part of the subspaces, which is appropriate only when the subspaces are linear but may fail otherwise. On the other hand, the second quantity is based on the orthogonal distance from point $\boldsymbol{\mu}$ in $\hat{S}$ to the subspace $\hat{S}_{(-j)}$, plus the orthogonal distance from point $\boldsymbol{\mu}_{(-j)}$ in $\hat{S}_{(-j)}$ to the subspace $\hat{S}$. This distance is more appropriate for comparing the affine part of the subspaces and can be combined with the distance between the linear parts to form a distance between affine subspaces. Given any such distance, the larger the value of the distance, the larger the influence of $\boldsymbol{x}_j$ on the estimate, and the more likely it is that $\boldsymbol{x}_j$ is an outlier. Thus, we may detect sample $\boldsymbol{x}_j$ as an outlier if its influence is above some threshold $\tau > 0$, i.e.,

$$\operatorname{dist}(\hat{S}, \hat{S}_{(-j)}) \geq \tau. \tag{3.75}$$

However, this method does not come without extra cost. We need to compute the principal components (and hence perform SVD) $N + 1$ times: once with all the samples together and another N times with one sample eliminated. There have been many studies that aim to give a formula that can accurately approximate the sample influence without performing SVD $N + 1$ times. Such a formula is called a *theoretical influence* function (see Appendix B). For a more detailed discussion about influence-based outlier rejection for PCA, we refer the interested reader to (Jolliffe 2002).

Probability-Based Outlier Detection: Multivariate Trimming, M-Estimators, and Iteratively Weighted Recursive Least Squares
In this approach, a subspace is fit to *all* sample points, including potential outliers. Outliers are then detected as the points that correspond to small-probability events or that have large fitting errors with respect to the identified subspace. A new subspace is then estimated with the detected outliers removed or down-weighted. This process is then repeated until the estimated subspace stabilizes.

More specifically, recall that in PCA, the goal is to find a low-dimensional subspace that best fits a given set of data points $\mathcal{X} \doteq \{\boldsymbol{x}_j \in \mathbb{R}^D\}_{j=1}^N$ by minimizing the least-squares error

$$\sum_{j=1}^{N} \|\boldsymbol{x}_j - \boldsymbol{\mu} - U\boldsymbol{y}_j\|^2, \tag{3.76}$$

between each point $\boldsymbol{x}_j$ and its projection onto the subspace $\boldsymbol{\mu} + U\boldsymbol{y}_j$, where $\boldsymbol{\mu} \in \mathbb{R}^D$ is any point in the subspace, $U \in \mathbb{R}^{D\times d}$ is a basis for the subspace, and $\boldsymbol{y}_j \in \mathbb{R}^d$ are the coordinates of the point in the subspace. If there are no outliers, an optimal solution to PCA can be obtained as described in Section 2.1.2, i.e.,

$$\hat{\boldsymbol{\mu}}_N = \frac{1}{N}\sum_{j=1}^{N} \boldsymbol{x}_j \quad \text{and} \quad \hat{\boldsymbol{y}}_j = \hat{U}^\top (\boldsymbol{x}_j - \hat{\boldsymbol{\mu}}_N), \tag{3.77}$$

where $\hat{U}$ is a $D \times d$ matrix whose columns are the top d eigenvectors of

$$\hat{\Sigma}_N = \frac{1}{N}\sum_{j=1}^{N} (\boldsymbol{x}_j - \hat{\boldsymbol{\mu}}_N)(\boldsymbol{x}_j - \hat{\boldsymbol{\mu}}_N)^\top. \tag{3.78}$$

If we adopt the guideline that outliers are samples that do not fit the model well or have a small probability with respect to the estimated model, then the outliers are exactly those samples that have a relatively large residual

$$\|\boldsymbol{x}_j - \hat{\boldsymbol{\mu}}_N - \hat{U}\hat{\boldsymbol{y}}_j\|^2 \quad \text{or} \quad \varepsilon_j^2 = (\boldsymbol{x}_j^\top - \boldsymbol{\mu}_N)^\top \Sigma_N^{-1} (\boldsymbol{x}_j^\top - \boldsymbol{\mu}_N), \quad j = 1, 2, \ldots, N. \tag{3.79}$$

The first error is simply the distance to the subspace, while the second error is the *Mahalanobis distance*,[8] which is obtained when we approximate the probability that a sample $\boldsymbol{x}_j$ comes from this model by a multivariate Gaussian

$$p(\boldsymbol{x}_j; \boldsymbol{\mu}_N, \hat{\Sigma}_N) = \frac{1}{(2\pi)^{D/2}\det(\hat{\Sigma}_N)^{1/2}} \exp\big(-\frac{1}{2}(\boldsymbol{x}_j^\top - \boldsymbol{\mu}_N)^\top \Sigma_N^{-1}(\boldsymbol{x}_j^\top - \boldsymbol{\mu}_N)\big). \tag{3.80}$$

In principle, we could use $p(\boldsymbol{x}_j, \boldsymbol{\mu}_N, \hat{\Sigma}_N)$ or either residual ε_j to determine whether $\boldsymbol{x}_j$ is an outlier. However, the above estimate of the subspace is obtained using all the samples, including the outliers themselves. Therefore, the estimated subspace could be completely wrong, and hence the outliers could be incorrectly detected. In order to improve the estimate of the subspace, one can recompute the model parameters after discarding or down-weighting samples that have large residuals. More specifically, let $w_j \in [0, 1]$ be a weight assigned to the jth point such that $w_j \approx 1$ if $\boldsymbol{x}_j$ is an inlier and $w_j \approx 0$ if $\boldsymbol{x}_j$ is an outlier. Then, similarly to (2.23), a new estimate of the subspace can be obtained by minimizing a reweighted least-squares error:

$$\min_{\boldsymbol{\mu},U,Y} \sum_{j=1}^N w_j\|\boldsymbol{x}_j - \boldsymbol{\mu} - U\boldsymbol{y}_j\|^2 \text{ s.t. } U^\top U = I_d \text{ and } \sum_{j=1}^N w_j\boldsymbol{y}_j = \boldsymbol{0}. \tag{3.81}$$

It can be shown (see Exercise 3.12) that the optimal solution to this problem is of the form

$$\hat{\boldsymbol{\mu}}_N = \frac{\sum_{j=1}^N w_j\boldsymbol{x}_j}{\sum_{j=1}^N w_j} \quad \text{and} \quad \hat{\boldsymbol{y}}_j = \hat{U}^\top(\boldsymbol{x}_j - \hat{\boldsymbol{\mu}}_N) \;\; \forall j \text{ s.t. } w_j > 0, \tag{3.82}$$

where $\hat{U}$ is a $D \times d$ matrix whose columns are the top d eigenvectors of

$$\hat{\Sigma}_N = \frac{\sum_{j=1}^N w_j(\boldsymbol{x}_j - \hat{\boldsymbol{\mu}}_N)(\boldsymbol{x}_j - \hat{\boldsymbol{\mu}}_N)^\top}{\sum_{j=1}^N w_j}. \tag{3.83}$$

As a consequence, under the reweighted least-squares criterion, finding a robust solution to PCA reduces to finding a robust estimate of the sample mean and the sample covariance of the data by properly setting the weights.

In what follows, we discuss two main approaches for estimating the weights.

---

[8]In fact, it can be shown that (Ferguson 1961), if the outliers have a Gaussian distribution of a different covariance matrix $a\Sigma$, then ε_i is a sufficient statistic for the test that maximizes the probability of correct decision about the outlier (in the class of tests that are invariant under linear transformations). Interested readers may want to find out how this distance is equivalent (or related) to the sample influence $\hat{\Sigma}_N^{(i)} - \hat{\Sigma}_N$ or the approximate sample influence given in (B.91).

1. *Multivariate Trimming* (MVT) is a popular robust method for estimating the sample mean and covariance of a set of points. This method assumes discrete weights

$$w_j = \begin{cases} 1 & \text{if } \boldsymbol{x}_j \text{ is an inlier,} \\ 0 & \text{if } \boldsymbol{x}_j \text{ is an outlier,} \end{cases} \tag{3.84}$$

and chooses the outliers as a certain percentage of the samples (say 10%) that have relatively large residual. This can be done by simply sorting the residuals $\{\varepsilon_j\}$ from the lowest to the highest and then choosing as outliers the desired percentage of samples with the highest residuals. Once the outliers are trimmed out, one can use the remaining samples to reestimate the subspace as in (3.82)–(3.83). Each time we have a new estimate of the subspace, we can recalculate the residual of every sample and reselect samples that need to be trimmed. We can repeat the above process until a stable estimate of the subspace is obtained. When the percentage of outliers is somewhat known, it usually takes only a few iterations for MTV to converge, and the resulting estimate is in general more robust. However, if the percentage is wrongfully specified, MVT may not converge, or it may converge to a wrong estimate of the subspace. In general, the "breakdown point" of MTV, i.e., the proportion of outliers that it can tolerate before giving a completely wrong estimate, depends only on the chosen trimming percentage.
2. *Maximum-Likelihood-Type Estimators* (M-Estimators) is another popular robust method for estimating the sample mean and covariance of a set of points. As we saw in the case of PCA with corrupted entries, this method assumes continuous weights

$$w_j = \rho(\varepsilon_j)/\varepsilon_j^2 \tag{3.85}$$

for some robust loss function $\rho(\cdot)$. The objective function then becomes

$$\sum_{j=1}^{N} \rho(\varepsilon_j). \tag{3.86}$$

Many loss functions $\rho(\cdot)$ have been proposed in the statistics literature (Huber 1981; Barnett and Lewis 1983). When $\rho(\varepsilon) = \varepsilon^2$, all weights are equal to 1, and we obtain the standard least-squares solution, which is not robust. Other robust loss functions include

(a) L_1 loss: $\rho(\varepsilon) = |\varepsilon|$;
(b) Cauchy loss: $\rho(\varepsilon) = \varepsilon_0^2 \log(1 + \varepsilon^2/\varepsilon_0^2)$;
(c) Huber loss (Huber 1981): $\rho(\varepsilon) = \begin{cases} \varepsilon^2 & \text{if } \ |\varepsilon| < \varepsilon_0, \\ 2\varepsilon_0|\varepsilon| - \varepsilon_0^2 & \text{otherwise;} \end{cases}$
(d) Geman–McClure loss (Geman and McClure 1987): $\rho(\varepsilon) = \frac{\varepsilon^2}{\varepsilon^2+\varepsilon_0^2}$,

Algorithm 3.9 (Iteratively Reweighted Least Squares for PCA with Outliers)

Input: Data matrix X, dimension d, and parameter $\varepsilon_0 > 0$.

```
initialize [μ, U, Y] = PCA(X) using PCA from Chapter 2.
repeat
    ε_j ← ‖x_j − μ − U y_j‖_2,   w_j ← ε_0^2 / (ε_j^2 + ε_0^2).
    μ ← Σ_{j=1}^N w_j (x_j − U y_j) / Σ_{j=1}^N w_j,   Σ ← Σ_{j=1}^N w_j (x_j − μ̂_N)(x_j − μ̂_N)^⊤ / Σ_{j=1}^N w_j.
    U ← top d eigenvectors of Σ.
    Y ← U^⊤ (X − μ 1^⊤).
until convergence of μ 1^⊤ + UY.

L ← UY and E ← X − L − μ 1^⊤.
```

Output: $\boldsymbol{\mu}$, U, Y, L and E.

where $\varepsilon_0 > 0$ is a parameter. Given any choice for the weights, one way of minimizing (3.86) with respect to the subspace parameters is to initialize all the weights to $w_j = 1, j = 1, \ldots, N$. This will give an initial estimate for the subspace that is the same as that given by PCA. Given this initial estimate of the subspace, one may compute the weights as $w_j = \rho(\varepsilon_j)/\varepsilon_j^2$ using any of the aforementioned robust cost functions. Given these weights, one can reestimate the subspace from (3.82)–(3.83). One can then iterate between computing the weights given the subspace and computing the subspace given the weights. This iterative process is called iteratively reweighted least squares (IRLS), as in the case of PCA with corrupted entries, and is summarized in Algorithm 3.9 for the Geman-McClure loss function. An alternative method for minimizing (3.86) is simply to do gradient descent. This method may be preferable for loss functions ρ that are differentiable, e.g., the Geman–McClure loss function. One drawback of M-estimators is that their breakdown point is inversely proportional to the dimension of the space. Thus, M-estimators become much less robust when the dimension is high.

Consensus-Based Outlier Detection

This approach assumes that the outliers are not drawn from the same subspace as the rest of the data. Hence it makes sense to try to avoid the outliers when we infer the subspace in the first place. However, without knowing which points are outliers beforehand, how can we avoid them?

One idea is to fit a subspace to a *subset* of the data instead of to all the data points. This is possible when the number of data points required to fit a subspace ($k = d$ for linear subspace or $k = d + 1$ for affine subspaces) is *much* smaller than the size N of the given data set. Of course, we should *not* expect that a randomly chosen subset will have no outliers and always lead to a good estimate of the subspace. Thus, we should try *many different subsets*:

$$\mathcal{X}_1, \mathcal{X}_2, \ldots, \mathcal{X}_m \subset \mathcal{X}, \tag{3.87}$$

where each subset $\mathcal{X}_i$ is independently drawn and contains $k \ll N$ samples.

If the number of subsets is large enough, one of the trials should contain few or no outliers and hence give a "good" estimate of the subspace. Indeed, if p is the fraction of valid samples (the "inliers"), one can show that (see Exercise B.8) with probability $q = 1 - (1 - p^k)^m$, one of the above subsets will contain only valid samples. In other words, if q is the probability that one of the selected subsets contains only valid samples, we need to randomly sample at least

$$m \geq \frac{\log(1 - q)}{\log(1 - p^k)} \tag{3.88}$$

subsets of k samples.

Now, given multiple subspaces estimated from multiple subsets, the next question is how to select a "good" subspace among them. Let $\hat{S}_i$ be the subspace fit to the set of points in $\mathcal{X}_i$. If the set $\mathcal{X}_i$ is contaminated by outliers, then $\hat{S}_i$ should be a "bad" estimate of the true subspace S, and hence few points in $\mathcal{X}$ should be well fit by $\hat{S}_i$. Conversely, if the set $\mathcal{X}_i$ contains only inliers, then $\hat{S}_i$ should be a "good" estimate of the true subspace S, and many points should be well fit by $\hat{S}_i$. Thus, to determine whether $\hat{S}_i$ is a good estimate of S, we need some criterion to determine when a point is well fit by $\hat{S}_i$ and another criterion to determine when the number of points that are well fit by $\hat{S}_i$ is sufficiently large. We declare that the subset $\mathcal{X}_i$ gives a "good" estimate $\hat{S}_i$ of the subspace S if

$$\#\{\boldsymbol{x} \in \mathcal{X} : \operatorname{dist}(\boldsymbol{x}, \hat{S}_i) \leq \tau\} \geq N_{\min}, \tag{3.89}$$

where # is the cardinality of the set, $\tau > 0$ is the threshold on the distance from any point $\boldsymbol{x} \in \mathcal{X}$ to the estimated subspace $\hat{S}$ used to determine whether a point is an inlier to $\hat{S}$, and $N_{\min}$ is a threshold on the minimum number of inliers needed to declare that the estimated subspace is "good." If the number of inliers to the subspace estimated from a given subset of the data points is too small, then the process is repeated for another sample of points until a good subspace is found or the maximum number of iterations has been exhausted. Upon termination, PCA is reapplied to all inliers in order to improve the robustness of the estimated subspace to noise. This approach to PCA with outliers is called *random sample consensus* (RANSAC) (Fischler and Bolles 1981) and is summarized in Algorithm 3.10.

One of the main advantages of RANSAC is that in theory, it can tolerate more than 50% outliers; hence it is extremely popular for practitioners who handle grossly contaminated data sets. Nevertheless, the computational cost of this scheme is proportional to the number of candidate subsets needed to ensure that the probability of choosing an outlier-free subset is large enough. This number typically grows exponentially with the subspace dimension and the number of samples. Hence, RANSAC is used mostly in situations in which the subspace dimension is low; in most of the cases we have seen, the subspace dimension does not exceed 10. Another challenge is that in order to design a successful RANSAC algorithm, one

Algorithm 3.10 (Random Sample Consensus for PCA with Outliers)

Input: Data points $\mathcal{X}$, subspace dimension d, maximum number of iterations k, threshold on fitting error τ, threshold on minimum number of inliers $N_{\min}$.

initialization $i = 0$.
while $i < k$ **do**
  $\mathcal{X}_i \leftarrow d + 1$ randomly chosen data points from $\mathcal{X}$.
  $\hat{S}_i \leftarrow \text{PCA}(\mathcal{X}_i)$.
  $\mathcal{X}_{inliers} \leftarrow \{\boldsymbol{x} \in \mathcal{X} : \text{dist}(\boldsymbol{x}, \hat{S}_i) \leq \tau\}$.
  if $|\mathcal{X}_{inliers}| \geq N_{\min}$ **then**
    $i \leftarrow k$.
  else
    $i \leftarrow i + 1$.
  end if
end while

Output: Estimated subspace $\hat{S} \leftarrow \text{PCA}(\mathcal{X}_{inliers})$ and set of inliers $\mathcal{X}_{inliers}$.

needs to choose a few key parameters carefully, such as the size of every subset (or the subspace dimension), the distance dist and the parameter τ to determine whether a point is an inlier or outlier, and the threshold $N_{\min}$ on the minimum number of inliers to the estimated subspace.

There is a vast amount of literature on RANSAC-type algorithms, especially in computer vision (Steward 1999). For more details on RANSAC and other related random sampling techniques, the reader is referred to Appendix B.

3.3.2 *Outlier Detection by Convex Optimization*

So far, we have presented classical techniques from the robust statistics literature and shown how they can be used for dealing with outliers in the context of PCA. The techniques presented so far are generally simple and intuitive. However, they do not provide clear conditions under which they can guarantee the correctness or global optimality of their solutions. To address this issue, in what follows we will present alternative approaches based on convex optimization for dealing with outliers in the context of PCA. As we will see, when the dimension of the subspace is small enough and the percentage of outliers is small enough, it is possible to perfectly recover which data points are inliers and which ones are outliers.

Outlier Detection by ℓ_1 Minimization

Let $\mathcal{X} = \{\boldsymbol{x}_j\}_{j=1}^N$ be a collection of points in $\mathbb{R}^D$. Assume that $N_{in} \leq N$ points are drawn from a linear subspace $S \subset \mathbb{R}^D$ of dimension $d \ll D$ and that the remaining $N_{out} = N - N_{in}$ data points do not belong to S. We thus have $N = N_{in} + N_{out}$ data points, where N_{in} points are inliers and N_{out} points are outliers. Assume also that there are d linearly independent data points among the inliers. Then every point

$\boldsymbol{x} \in S$ can be written as a linear combination of the inliers. In fact, every point $\boldsymbol{x} \in S$ can be written as a linear combination of at most d inliers. More generally, we can write $\boldsymbol{x} \in S$ as a linear combination of all N data points as

$$\boldsymbol{x} = \sum_{j=1}^{N} \boldsymbol{x}_j c_j = X\boldsymbol{c} \quad \text{where} \quad X = \left[\boldsymbol{x}_1, \boldsymbol{x}_2, \ldots, \boldsymbol{x}_N\right] \in \mathbb{R}^{D \times N}, \tag{3.90}$$

and set $c_j = 0$ whenever $\boldsymbol{x}_j$ is an outlier. Hence, there exists a solution $\boldsymbol{c}$ of $\boldsymbol{x} = X\boldsymbol{c}$ with at most d nonzero entries, which correspond to any d inliers that span S. Therefore, an optimal solution $\boldsymbol{c}^*$ to the following optimization problem

$$\min_{\boldsymbol{c}} \|\boldsymbol{c}\|_0 \quad \text{s.t.} \quad \boldsymbol{x} = X\boldsymbol{c} \tag{3.91}$$

should be d-sparse, i.e., it should have at most d nonzero entries, i.e., $\|\boldsymbol{c}^*\|_0 \leq d$.

Assume now that there are D linearly independent data points among both the inliers and outliers. Assume also that $\boldsymbol{x}$ does not belong to the subspace S. Then we can still express $\boldsymbol{x}$ as a linear combination of all data points as $\boldsymbol{x} = X\boldsymbol{c}$. However, when $\boldsymbol{x}$ is an arbitrary point in $\mathbb{R}^D$, we no longer expect $\boldsymbol{c}$ to be d-sparse. In fact, in general, we expect at least D entries of $\boldsymbol{c}$ to be nonzero, i.e., $\|\boldsymbol{c}\|_0 \geq D$. Of course, in some rare circumstances it could be the case that $\boldsymbol{x}$ is a linear combination of two outliers in the data, in which case we can choose $\boldsymbol{c}$ such that $\|\boldsymbol{c}\|_0 = 2$. However, such cases occur with extremely low probability.

The above discussion suggests a simple procedure to determine whether a point $\boldsymbol{x}$ is an inlier: we try to express $\boldsymbol{x}$ as a linear combination of the data points in $\mathcal{X}$ with the sparsest possible coefficients $\boldsymbol{c}$, as in (3.91). If the optimal solution $\boldsymbol{c}^*$ is d-sparse, then $\boldsymbol{x}$ is an inlier; otherwise, $\boldsymbol{x}$ is an outlier. In practice, however, we face a couple of challenges that prevent us from implementing this simple strategy.

1. The optimization problem in (3.91) is NP-hard (Amaldi and Kann 1998). Intuitively this is because there are numerous choices of d out of N nonzero entries in $\boldsymbol{c}$, and for each such choice, we need to check whether a linear system has a solution or not.
2. While in general we expect that $\|\boldsymbol{c}\|_0 \gg d$ when $\boldsymbol{x}$ is an outlier, this may not always be the case. Thus, we may be interested in characterizing whether for some distribution of the outliers we can guarantee that $\|\boldsymbol{c}\|_0 \gg d$ with high probability. Moreover, since the subspace dimension d may not be known a priori, we may want to declare $\boldsymbol{x}$ an outlier if $\|\boldsymbol{c}\|_0 > \lambda D$ for some $\lambda < 1$. This may require some mechanism for determining λ.
3. In practice, we are not trying to determine whether a generic data point $\boldsymbol{x}$ is an inlier or an outlier, but rather whether one of the given data points, say $\boldsymbol{x}_j$, is an inlier or an outlier. Trivially, $\boldsymbol{x}_j$ has a 1-sparse representation with respect to X, i.e., $\boldsymbol{x}_j = \boldsymbol{x}_j$. Thus, we need a mechanism to prevent this trivial solution.

To address the first issue, as we have learned from the brief survey of compressive sensing in Section 3.2.2, an effective technique to obtain a sparse solution is to replace the ℓ_0-minimization problem in (3.91) by the ℓ_1-minimization problem

$$\min_{\boldsymbol{c}} \|\boldsymbol{c}\|_1 \quad \text{s.t.} \quad \boldsymbol{x} = X\boldsymbol{c}. \tag{3.92}$$

In particular, it follows from Theorem 3.10 that if $X = \begin{bmatrix}\boldsymbol{x}_1, \boldsymbol{x}_2, \ldots, \boldsymbol{x}_N\end{bmatrix} \in \mathbb{R}^{D\times N}$ is an arbitrary matrix whose columns are of unit norm, i.e., $\|\boldsymbol{x}_j\|_2 = 1$ for all $j = 1, \ldots, N$ and $\boldsymbol{c}_0 \in \mathbb{R}^N$ is a d-sparse vector, then given $\boldsymbol{x} = X\boldsymbol{c}_0$, we can recover $\boldsymbol{c}_0$ by solving the optimization problem in (3.92) when the matrix X is incoherent or satisfies the RIP. In other words, the sparsest solution to the linear system $X\boldsymbol{c} = \boldsymbol{x}$ can be obtained by solving the convex ℓ_1-minimization problem in (3.92) as opposed to the NP-hard ℓ_0-minimization problem in (3.91).

The fundamental question is whether the conditions under which the solution to (3.92) coincides with that of (3.91) are satisfied by a data matrix X with N_{in} data points in a linear subspace of dimension d and N_{out} points not in the subspace. Unfortunately, this is not the case: the matrix of inliers X_{in} cannot be incoherent according to Definition 3.8, because it is not of full column rank. For instance, if $\text{rank}(X_{in}) = 1$, then X has maximum coherence $\nu(X_{in}) = 1$.

Does this mean that we cannot use ℓ_1-minimization? As it turns out, we can still use ℓ_1 minimization to recover a sparse representation of a point $\boldsymbol{x} \in S$. The reason is that the conditions in Theorem 3.10 aim to guarantee that we can recover a *unique* sparse solution, while here the solution for $\boldsymbol{c}$ is not always unique, and thus we cannot hope for the ℓ_1-minimization problem to give us a unique sparse solution to begin with. Indeed, if $\boldsymbol{x} \in S$ and $N_{in} > d$, then there may be many ways in which we may express a point in S as a linear combination of d inliers. Therefore, our goal is not to find a unique representation of $\boldsymbol{x}$ in terms of d inliers, but rather to find any representation of $\boldsymbol{x}$ in terms of any d inliers. As a consequence, we do not need the matrix of inliers to be incoherent. All we need is for the set of inliers to be incoherent with the set of outliers. More precisely, if $\mathcal{I}_{in}$ is the set of inliers and $\mathcal{I}_{out}$ is the set of outliers, all we need is that

$$\max_{j\in\mathcal{I}_{in}} \max_{k\in\mathcal{I}_{out}} |\boldsymbol{x}_j^\top \boldsymbol{x}_k| < \frac{1}{2d-1}. \tag{3.93}$$

This is in contrast to the classical condition on the mutual coherence of X in Definition 3.8, which is given by $\max_{j\neq k} |\boldsymbol{x}_j^\top \boldsymbol{x}_k| < \frac{1}{2d-1}$.

To address the second issue, we assume from now on that all N points are of unit norm, i.e., they lie in the $D-1$ dimensional sphere $\mathbb{S}^{D-1}$. We assume also that the outliers are drawn uniformly at random from $\mathbb{S}^{D-1}$. Moreover, since we will be solving an ℓ_1 minimization problem, we may want to use the ℓ_1 norm of $\boldsymbol{c}$ to determine whether $\boldsymbol{x}$ is an inlier or outlier. More specifically, when $\boldsymbol{x}$ is an inlier, we expect $\|\boldsymbol{c}\|_0 = d$; hence we expect $\|\boldsymbol{c}\|_1 = \sqrt{d}$. Likewise, when $\boldsymbol{x}$ is an outlier, we expect $\|\boldsymbol{c}\|_0 = D$, and then we expect $\|\boldsymbol{c}\|_1 = \sqrt{D}$. Therefore, we may want to declare $\boldsymbol{x}$ an outlier if $\|\boldsymbol{c}\|_1 > \lambda\sqrt{D}$ for some λ.

The third issue is relatively easy to address. When we find the sparse solution for the point $\boldsymbol{x}_j$ with respect to X, we need to enforce only that the jth entry of $\boldsymbol{c}$ is zero, so that $\boldsymbol{x}_j$ is not represented by itself. This leads to the following convex optimization problem:

$$\min_{\boldsymbol{c}} \|\boldsymbol{c}\|_1 \quad \text{s.t.} \quad \boldsymbol{x}_j = X\boldsymbol{c} \quad \text{and} \quad c_j = 0, \tag{3.94}$$

which can be solved easily using existing ℓ_1-minimization techniques.

The following result, which follows as a direct corollary of (Soltanolkotabi and Candès 2013, Theorem 1.3) (see also Theorem 8.27), shows how the optimal solution to (3.94) can be used to distinguish inliers from outliers.

Theorem 3.13. *Let S be a randomly chosen subspace of $\mathbb{R}^D$ of dimension d. Suppose there are $N_{in} = \rho d + 1$ inlier points chosen independently and uniformly at random in $S \cap \mathbb{S}^{D-1}$, where $\rho > 1$. Suppose there are N_{out} points chosen independently and uniformly at random in $\mathbb{S}^{D-1}$. Let $\mathbf{x}_j \in \mathbb{S}^{D-1}$ be the jth data point and let $\mathbf{c} \in \mathbb{R}^N$ be the solution to the ℓ_1-minimization problem in* (3.94)*. Declare $\mathbf{x}_j$ to be an outlier if $\|\mathbf{c}\|_1 > \lambda(\gamma)\sqrt{D}$, where $\gamma = \frac{N-1}{D}$, $N = N_{in} + N_{out}$, and*

$$\lambda(\gamma) = \begin{cases} \sqrt{\frac{2}{\pi}} \frac{1}{\sqrt{\gamma}}, & 1 \leq \gamma \leq e \\ \sqrt{\frac{2}{\pi e}} \frac{1}{\sqrt{\log \gamma}}, & \gamma \geq e. \end{cases} \tag{3.95}$$

If the number of outliers is such that

$$N_{out} < \frac{1}{D} \exp(c_1\sqrt{D}) - N_{in} \tag{3.96}$$

for some constant $c_1 > 0$, then the method above detects all the outliers with probability at least $1 - N_{out} \exp(-c_2 D / \log(N_{in} + N_{out}))$ for some constant $c_2 > 0$. Moreover, if the number of outliers is such that

$$N_{out} < D\rho^{c_3 \frac{D}{d}} - N_{in} \tag{3.97}$$

for some constant $c_3 > 0$, then the method above does not detect any point in S as an outlier with probability at least $1 - N_{out} \exp(-c_4 D / \log(N_{in} + N_{out})) - N_{in} \exp(-\sqrt{\rho d})$ for some constant $c_4 > 0$.

Outlier Detection by $\ell_{2,1}$ Minimization

An alternative approach to outlier detection in PCA is based on the observation that the data matrix X can be seen as a low-rank matrix with sparsely corrupted columns that correspond to the outliers. More specifically, the matrix X can be decomposed as

$$X = L_0 + E_0. \tag{3.98}$$

The jth column of L_0 is equal to $\boldsymbol{x}_j$ if it is an inlier to the subspace and is equal to $\boldsymbol{0}$ otherwise. Therefore, L_0 is of rank d and spans the same subspace as the inliers. Conversely, the jth column of E_0 is equal to $\boldsymbol{x}_j$ if it is an outlier to the subspace and is equal to $\boldsymbol{0}$ otherwise. Therefore, the nonzero columns of E_0 contain the outliers. If we assume that the fraction γ of outliers is small, then the matrix E_0 is *column sparse*.

Obviously, such a decomposition is ill posed (at least ambiguous) if the matrix X or L_0 is also column sparse. Therefore, in order for the decomposition to be unique, the matrix L_0 cannot be column sparse on the $(1-\gamma)N$ columns on which it can be nonzero. To ensure that this is the case, we need to introduce a *column incoherence* condition:

Definition 3.14 (Matrix Incoherence with Respect to Column Sparse Matrices). *A rank-d matrix $L \in \mathbb{R}^{D\times N}$ with compact SVD $L = U\Sigma V^\top$ and $(1-\gamma)N$ nonzero columns is said to be* ν-incoherent with respect to the set of column sparse matrices *if*

$$\max_j \|\boldsymbol{v}_j\|^2 \leq \frac{\nu d}{(1-\gamma)N}, \tag{3.99}$$

where $\boldsymbol{v}_j$ is the jth row of V.

Following the discussion after Definition 3.1, notice that since $V \in \mathbb{R}^{N\times d}$ is orthonormal, the largest absolute value of the entries of V is equal to 1, which happens when a column of V is 1-sparse. On the other hand, if all columns of V are so dense that all their $(1-\gamma)N$ nonzero entries are equal to each other up to sign, then each entry is equal to $\pm 1/\sqrt{(1-\gamma)N}$, and the norm of each row is $\sqrt{d/(1-\gamma)N}$. Therefore, when $\nu < 1$, the condition above controls the level of sparsity of V. As argued before, from a probabilistic perspective, this condition is rather mild in the sense that it holds for almost all generic matrices: a random (say Gaussian) matrix satisfies this condition with high probability when the dimension of the matrix is large enough. As we will see, incoherence with respect to column sparse matrices is a very useful technical condition to ensure that outlier detection is a meaningful problem.

Now, even though the incoherence condition may ensure that the above low-rank plus column-sparse decomposition problem is well posed, there is no guarantee that one can find the correct decomposition efficiently. As before, we may formulate the problem of recovering L_0 and E_0 as a rank minimization problem:

$$\min_{L,E} \quad \operatorname{rank}(L) + \lambda \|E\|_{2,0} \quad \text{s.t.} \quad X = L + E, \tag{3.100}$$

where $\|E\|_{2,0} = \sum_{j=1}^{N} 1(\|\boldsymbol{e}_j\|_2 \neq 0)$ is the number of nonzero columns in the matrix of outliers $E = [\boldsymbol{e}_1, \ldots, \boldsymbol{e}_N]$. However, since this problem is NP-hard, we need to resort to a proper relaxation. For this purpose, we can use a norm that promotes columnwise sparsity, such as the $\ell_{2,1}$ norm of E:

$$\|E\|_{2,1} = \sum_{j=1}^{N} \|\boldsymbol{e}_j\|_2, \tag{3.101}$$

which is the sum of the ℓ_2 norms of all the columns of E. Notice that if we collect all the ℓ_2 norms of the columns of E as a vector $\boldsymbol{e} = [\|\boldsymbol{e}_1\|_2, \ldots, \|\boldsymbol{e}_N\|_2]^\top$, then the above norm is essentially the ℓ_1 norm of the vector, $\|\boldsymbol{e}\|_1$; hence it measures how sparse the columns are. Notice also that $\|E\|_{2,0} = \|\boldsymbol{e}\|_0$.

Similar to the PCP optimization problem in (3.64) for PCA with robustness to intrasample outliers, we can use the convex optimization

$$\min_{L,E} \quad \|L\|_* + \lambda\|E\|_{2,1} \quad \text{s.t.} \quad X = L + E \tag{3.102}$$

to decompose sparse column outliers in the data matrix X from the low-rank component. This convex program is called *outlier pursuit.*

One can rigorously show that under certain benign conditions, the outlier pursuit program can correctly identify the set of sparse (column) outliers.

Theorem 3.15 (Robust PCA by Outlier Pursuit (Xu et al. 2010)). *Let $X = L_0 + E_0$ be a given $D \times N$ matrix. Assume that L_0 is ν-incoherent with respect to the set of column-sparse matrices according to Definition 3.14. Assume also that E_0 is supported on at most γN columns. If*

$$\operatorname{rank}(L_0) \leq \frac{c_1(1-\gamma)}{\gamma\nu}, \tag{3.103}$$

where $c_1 = \frac{9}{121}$, then the solution (L^*, E^*) to the outlier pursuit program (3.102) *with λ set to be $\frac{3}{7\sqrt{\gamma N}}$ recovers the low-dimensional column space of L_0 exactly and identifies exactly the indices of columns corresponding to outliers not lying in the column space.*

If the data also contain small noise $X = L_0 + E_0 + Z$, where Z is a random Gaussian matrix that models small noise in the data, then we can modify the outlier pursuit program as

$$\min_{L,E} \quad \|L\|_* + \lambda\|E\|_{2,1} \quad \text{s.t.} \quad \|X - L - E\|_2^2 \leq \varepsilon^2, \tag{3.104}$$

where ε is the noise variance. It can be shown that under conditions similar to those in the above theorem, this program gives a stable estimate of the correct solution. For more details, we refer the reader to (Xu et al. 2010).

Using optimization techniques introduced in Appendix A, one can easily develop ALM- or ADMM-based algorithms to solve the above convex optimization problems. We leave that to the reader as an exercise (see Exercise 3.8).

Fig. 3.6 Example images taken from the Caltech 101 data set. These images are then resized to 96×84 and used as outliers for the experiments below.

Example 3.16 (Outlier Detection among Face Images) Sometimes a face image data set can be contaminated by images of irrelevant objects, like many imperfectly sorted data sets in the Internet. In this case, it would be desirable to detect and remove such irrelevant outliers from the data set. In this example, we illustrate how to do this with the outlier detection methods introduced in this section.

As in previous experiments, we take as inliers the frontal face images of subject 20 under 64 different illumination conditions in the extended Yale B data set. For outlier images, we randomly select some pictures from the Caltech 101 data set (Fei-Fei et al. 2004) and merge them into the face image data set. Some typical examples of such pictures are shown in Figure 3.6. All the inlier and outlier images are normalized to size 96×84.

We use the outlier pursuit method, which is based on solving (3.102), to decompose the data matrix into a low-rank part L and a sparse-column term E. In this experiment, we set the parameter of the method according to Theorem 3.15 with a multiplication factor of 3, i.e., we set $\lambda = 3 \times \lambda_0$ where $\lambda_0 = \frac{3}{7\sqrt{\gamma N}}$.

Ideally, columns of E with large magnitude correspond to outliers. To show how the method performs, we apply it to data sets with increasing percentages of outliers. We compute for each column of E its ℓ_2 norm to measure whether it is an outlier. True outliers are marked in red. The results for varying percentages of outliers are shown in Figure 3.7. As we can see from the results, up to nearly 50% outliers, the outliers have significantly larger norm than the inliers.

3.4 Bibliographic Notes

PCA with Robustness to Missing Entries

The problem of completing a low-rank matrix with missing entries has a very long and rich history. Starting with the original work of (Wiberg 1976), one can refer to (Johnson 1990) for a survey on some of the early developments on this topic.

Since then, this problem has drawn tremendous interest, particularly in computer vision and pattern recognition, where researchers needed to complete data with missing entries due to occlusions. For instance, many algorithms were proposed

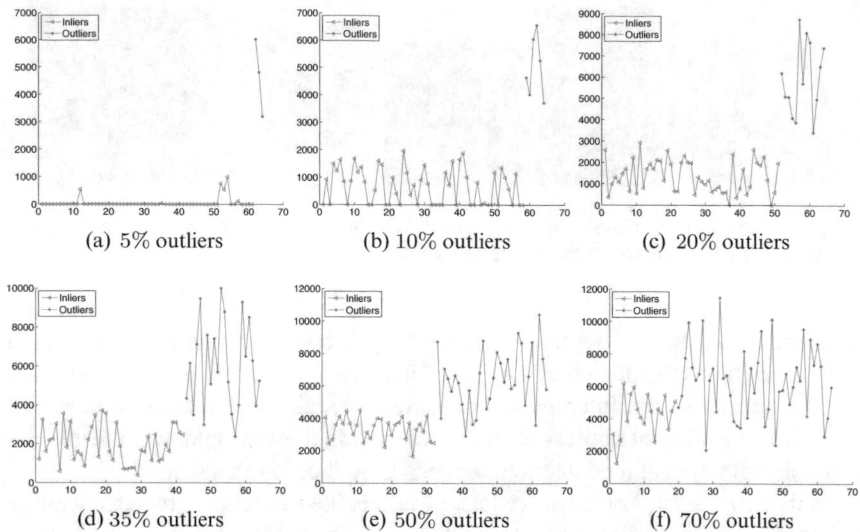

(a) 5% outliers (b) 10% outliers (c) 20% outliers

(d) 35% outliers (e) 50% outliers (f) 70% outliers

Fig. 3.7 Outlier detection among face images. In each experiment, we use 64 images, in which a certain percentage of images are selected randomly from the Caltech 101 data set as outliers, and the rest are taken randomly from the 64 illuminations of frontal face images of subject 20 in extended Yale B. We plot the column ℓ_2 norm of the matrix E given by the convex optimization method, with ground truth outliers marked as red.

to solve matrix completion problems in the late 1990s and early 2000s, including (Shum et al. 1995; Jacobs 2001; H.Aanaes et al. 2002; Brandt 2002) for the purpose of reconstructing a 3D scene from a collection of images. The power factorization method featured in this chapter was proposed in (Hartley and Schaffalitzky 2003) for the same purpose, while a variant of the EM algorithm we described appeared in (Gruber and Weiss 2004). Also, the work of (Ke and Kanade 2005) proposed the use of the ℓ_1 norm for matrix completion and recovery, which extends the original Wiberg method(Wiberg 1976) from the ℓ_2 to the ℓ_1 norm. A survey and evaluation of state-of-the-art methods for solving the matrix completion problem can be found in (Buchanan and Fitzgibbon 2005).

However, all of the work described so far has focused primarily on developing algorithms for completing a matrix, without any guarantees of correctly recovering the original low-rank matrix. The seminal work of (Recht et al. 2010; Candès and Recht 2009) has shown that under broad conditions, one can correctly recover a low-rank matrix with a significant percentage of missing entries using convex optimization (i.e., minimizing the nuclear norm of the matrix). This has inspired a host of work on developing ever stronger conditions and more efficient algorithms for low-rank matrix completion (Cai et al. 2008; Candès and Tao 2010; Keshavan et al. 2010b; Gross 2011; Keshavan et al. 2010a; Zhou et al. 2010a), including work that extends to the case of noisy data (Candès and Plan 2010).

PCA with Robustness to Corrupted Entries and Outliers
Regarding the robust recovery of a low-rank matrix, it was first proposed by (Wright et al. 2009a; Chandrasekaran et al. 2009) to use the convex relaxation (3.64) to solve the robust PCA problem. This formulation was soon followed by a rather strong theoretical justification (Candès et al. 2011) and efficient algorithms (Lin et al. 2011). This has made convex relaxation a very successful and popular technique for robust low-rank matrix recovery or outlier rejection, leading to extensions to many different settings and more scalable convex optimization algorithms.

Revival of the Factorization Approach
Due to the advent of large data sets and large-scale problems, there has been a revival of factorization (alternating minimization) approaches with theoretical guarantees of correctness for low-rank matrix completion and recovery, including the very interesting work of (Jain et al. 2012; Keshavan 2012; Hardt 2014; Jain and Netrapalli 2014). The more recent work of (Udell et al. 2015) further generalizes the factorization framework to situations in which the factors are allowed to have additional structures; and (Haeffele and Vidal 2015) combines factorization with certain nonlinear mappings typically used in a deep learning framework.

3.5 Exercises

Exercise 3.1 (Data Completion with the Subspace Known). Show that the solution to the problem (3.5) is given by the formula in (3.6).

Exercise 3.2. For the PPCA model with missing data discussed in Section 3.1.2, show that the conditional distribution of $\boldsymbol{x}_U$ given $\boldsymbol{x}_O$ is Gaussian with the following mean vector and covariance matrix:

$$\boldsymbol{\mu}_{U|O} = \boldsymbol{\mu}_U + \Sigma_{UO}\Sigma_{OO}^{-1}(\boldsymbol{x}_O - \boldsymbol{\mu}_O) \quad \text{and} \quad \Sigma_{U|O} = \Sigma_{UU} - \Sigma_{UO}\Sigma_{OO}^{-1}\Sigma_{OU}.$$

Exercise 3.3 (Orthogonal Power Iteration Method). Let $A \in \mathbb{R}^{N\times N}$ be a symmetric positive semidefinite matrix with eigenvectors $\{\boldsymbol{u}_i\}_{i=1}^N$ and eigenvalues $\{\lambda_i\}_{i=1}^N$ sorted in descending order. Assume that $\lambda_1 > \lambda_2$ and let $\boldsymbol{u}^0$ be an arbitrary vector not orthogonal to $\boldsymbol{u}_1$, i.e., $\boldsymbol{u}_1^\top \boldsymbol{u}^0 \neq 0$. Consider the sequence of vectors

$$\boldsymbol{u}_{k+1} = \frac{A\boldsymbol{u}_k}{\|A\boldsymbol{u}_k\|}. \tag{3.105}$$

1. Show that there exist $\{\alpha_i\}_{i=1}^N$ with $\alpha_1 \neq 0$ such that

$$\boldsymbol{u}^k = A^k\boldsymbol{u}^0 = \sum_{i=1}^N \alpha_i\lambda_i^k\boldsymbol{u}_i. \tag{3.106}$$

2. Use this expression to show that $\boldsymbol{u}^k$ converges to $\frac{\alpha_1}{|\alpha_1|}\boldsymbol{u}_1$ with rate $\frac{\lambda_2}{\lambda_1}$. That is, show that there exists a constant $C > 0$ such that for all $k \geq 0$,

$$\left\| \boldsymbol{u}^k - \frac{\alpha_1}{|\alpha_1|}\boldsymbol{u}_1 \right\| \leq C\left(\frac{\lambda_2}{\lambda_1}\right)^k. \tag{3.107}$$

3. Assume that $\lambda_d > \lambda_{d+1}$ and let $U^0 \in \mathbb{R}^{N\times d}$ be an arbitrary matrix whose column space is not orthogonal to the subspace $\{\boldsymbol{u}_i\}_{i=1}^d$ spanned by the top d eigenvectors of A. Consider the sequence of matrices

$$U^{k+1} = AU^k(R^k)^{-1}, \tag{3.108}$$

where $Q^kR^k = AU^k$ is the QR decomposition of AU^k. Show that U^k converges to a matrix U whose columns are the top d eigenvectors of A. Moreover, show that the rate of convergence is $\frac{\lambda_{d+1}}{\lambda_d}$.

Exercise 3.4 (Convergence of Orthogonal Power Iteration). Prove Theorem 3.4.

Exercise 3.5 (Properties of the ℓ_1 Norm). Let X be a matrix.

1. Show that the ℓ_1 norm $f(X) = \|X\|_1 = \sum_{ij} |X_{ij}|$ of X is a convex function of X.
2. Show that the subgradient of the ℓ_1 norm is given by

$$\partial\|X\|_1 = \operatorname{sign}(X) + W, \tag{3.109}$$

where W is a matrix such that $\max_{ij} |W_{ij}| \leq 1$.
3. Show that the optimal solution of

$$\min_A \quad \frac{1}{2}\|X - A\|_F^2 + \tau\|A\|_1 \tag{3.110}$$

is given by $A = \mathcal{S}_\tau(X)$, where $\mathcal{S}_\tau(x) = \operatorname{sign}(x)\max(|x| - \tau, 0)$ is the soft-thresholding operator applied entrywise to X.

Exercise 3.6 (Properties of the Weighted Nuclear Norm). Consider the following optimization problem:

$$\min_A \quad \frac{1}{2}\|X - A\|_F^2 + \tau\phi(A), \tag{3.111}$$

where $\phi(A) = \sum_{i=1}^r w_i\sigma_i(A)$ is the weighted sum of singular values of A with $w_i \geq 0$. Show that

1. $\phi(A)$ is convex when w_i is monotonically decreasing. Please derive the optimal solution under this condition.
2. Is $\phi(A)$ still convex if w_i is an increasing sequence of weights? Why?

Exercise 3.7 (Properties of the $\ell_{2,1}$ Norm).

1. Let $\boldsymbol{x}$ be a vector. Show that the subgradient of the ℓ_2 norm is given by

$$\partial\|\boldsymbol{x}\|_2 = \begin{cases} \frac{\boldsymbol{x}}{\|\boldsymbol{x}\|_2} & \text{if } \boldsymbol{x} \neq \boldsymbol{0}, \\ \boldsymbol{w} : \|\boldsymbol{w}\|_2 \leq 1\} & \text{if } \boldsymbol{x} = \boldsymbol{0}. \end{cases} \tag{3.112}$$

2. Let X be a matrix. Show that the $\ell_{2,1}$ norm $f(X) = \|X\|_{2,1} = \sum_j \|X_{\cdot j}\|_2 = \sum_j \sqrt{\sum_i X_{ij}^2}$ of X is a convex function of X.
3. Show that the subgradient of the $\ell_{2,1}$ norm is given by

$$(\partial\|X\|_{2,1})_{ij} = \begin{cases} \frac{X_{ij}}{\|X_{\cdot j}\|_2} & X_{\cdot j} \neq \boldsymbol{0} \\ W_{ij} : \|W_{\cdot j}\|_2 \leq 1 & X_{\cdot j} = \boldsymbol{0}. \end{cases} \tag{3.113}$$

4. Show that the optimal solution of

$$\min_A \quad \frac{1}{2}\|X - A\|_F^2 + \tau\|A\|_{2,1} \tag{3.114}$$

is given by $A = X\mathcal{S}_\tau(\text{diag}(\boldsymbol{x}))\text{diag}(\boldsymbol{x})^{-1}$, where $\boldsymbol{x}$ is a vector whose jth entry is given by $x_j = \|X_{\cdot j}\|_2$, and $\text{diag}(\boldsymbol{x})$ is a diagonal matrix with the entries of $\boldsymbol{x}$ along its diagonal. By convention, if $x_j = 0$, then the jth entry of $\text{diag}(\boldsymbol{x})^{-1}$ is also zero.

Exercise 3.8. Let $X = L_0 + E_0$ be a matrix formed as the sum of a low-rank matrix L_0 and a matrix of corruptions E_0, where the corruptions can be either outlying entries (gross errors) or outlying data points (outliers).

1. **(PCA with robustness to outliers).** Assuming that the matrix X is fully observed and that the matrix E_0 is a matrix of outliers, propose an algorithm for solving the outlier pursuit problem (3.102):

$$\min_{L,E} \quad \|L\|_* + \lambda\|E\|_{2,1} \quad \text{s.t.} \quad X = L + E. \tag{3.115}$$

2. **(PCA with robustness to missing entries and gross errors).** Assuming that you observe only a fraction of the entries of X as indicated by a set Ω and that the matrix E_0 is a matrix of gross errors, propose an algorithm for solving the following optimization problem:

$$\min_{L,E} \quad \|L\|_* + \lambda\|E\|_1 \quad \text{s.t.} \quad \mathcal{P}_\Omega(X) = \mathcal{P}_\Omega(L + E). \tag{3.116}$$

Exercise 3.9 (Implementation of Power Factorization (PF), Expectation Maximization (EM), and Low-Rank Matrix Completion (LRMC)). Implement the functions below using as few lines of MATLAB code as possible. Compare the performance of these methods: which method works better and which regime is best (e.g., depending on the percentage of missing entries, subspace dimension d/D)?

Function `[mu,U,Y]=pf(X,d,W)`

Parameters

`X` $D \times N$ data matrix.
`d` Number of principal components.
`W` $D \times N$ binary matrix denoting known (1) or missing (0) entries

Returned values

`mu` Mean of the data.
`U` Orthonormal basis for the subspace.
`Y` Low-dimensional representation (or principal components).

Description

Finds the d principal components of a set of points from the data X with incomplete entries as specified in W using the power factorization algorithm.

Function `[mu,U,sigma]=emppca(X,d,W)`

Parameters

`X` $D \times N$ data matrix.
`d` Number of principal components.
`W` $D \times N$ binary matrix denoting known (1) or missing (0) entries

Returned values

`mu` Mean of the data.
`U` Basis for the subspace (does not need to be orthonormal).
`sigma` Standard deviation of the noise.

Description

Finds the parameters of the PPCA model μ and $\Sigma = UU^\top + \sigma^2 I$ from the data X with incomplete entries as specified in W using the expectation maximization algorithm.

Function `A=lrmc(X,tau,W)`

Parameters

`X` $D \times N$ data matrix.
τ Parameter of the augmented Lagrangian.
`W` $D \times N$ binary matrix denoting known (1) or missing (0) entries

Returned values

`A` Low-rank completion of the matrix X.

Description

Finds the low-rank approximation of a matrix X with incomplete entries as specified in W using the low-rank matrix completion algorithm based on the augmented Lagrangian method.

Exercise 3.10 (Implementation of IRLS and ADMM Methods for Robust PCA). Implement Algorithms 3.7 and 3.8 for the functions below using as few lines of MATLAB code as possible. Compare the performance of these methods: which method works better and which regime is best (e.g., depending on percentage of corrupted entries (or corrupted data points), subspace dimension d/D)?

Function `[mu,U,Y]=rpca_irls(X,d,sigma)`

Parameters

`X`	$D \times N$ data matrix.
`d`	Number of principal components.

Returned values

`mu`	Mean of the data.
`U`	Basis for the subspace.

Description

Finds the parameters of the PCA model μ and U and the low-dimensional representation using reweighted least squares with weights $w(e) = \frac{\sigma^2}{e^2+\sigma^2}$.

Function `[L,E]=rpca_admm(X,tau,'method')`

Parameters

`X`	$D \times N$ data matrix.
τ	Parameter of the augmented Lagrangian.
`method`	'L1' for gross errors or 'L21' for outliers

Returned values

`L`	Low-rank completion of the matrix X.
`E`	Matrix of errors.

Description

Solves the optimization problem $\min_{L,E} \|L\|_* + \lambda\|E\|_\ell$ subject to $X = L + E$ where $\ell = \ell_1$ or $\ell = \ell_{2,1}$ using the ADMM algorithm.

Exercise 3.11 (Robust Face Recognition with Varying Illumination). In this exercise, you will use a small subset of the Yale B data set[9] that contains photos of ten individuals under various illumination conditions. Specifically, you will use only images from the first three individuals under ten different illumination conditions. Divide these images into two sets: *Training Set* (images 1–5 from individuals 1 to 3) and *Test Set* (images 6–10 from individuals 1–3). Notice also that there are five nonface images (accessible as images 1–5 from individual 4). We will refer to these as the *Outlier Set*. Download the file YaleB-Dataset.zip. This file contains the images along with the MATLAB function `loadimage.m`. Decompress the file and type `help loadimage` at the MATLAB prompt to see how to use this function. The function operates as follows.

[9]http://cvc.yale.edu/projects/yalefacesB/yalefacesB.html.

Function `img=loadimage(individual,condition)`

Parameters

`individual`	Number of the individual.
`condition`	Number of the image for that individual.

Returned values

`img`	The pixel image loaded from the database.

Description

Read and resize an image from the data set. The database (directory `images`) must be in the same directory as this file.

---

1. **Face completion.** Remove uniformly at random 0%, 10%, 20%, 30%, and 40% of the entries of all images of individual 1. Apply the low-rank matrix completion (LRMC) algorithm in Exercise 3.9 to these images to compute the mean face and the eigenfaces as well as to fill in the missing entries. Note that LRMC does not compute the mean face, so you will need to modify the algorithm slightly. Plot the mean face and the top three eigenfaces and compare them to what you obtained with PCA in Chapter 2. Plot also the completed faces and comment on the quality of completion as a function of the percentage of missing entries by visually comparing the original images (before removing the missing entries) to the completed ones. Plot also the error (Frobenius norm) between the original images and the completed ones as a function of the percentage of missing entries and comment on your results. Repeat for individuals 2 and 3.
2. **Face recognition with missing entries.** Remove uniformly at random 0%, 10%, 20%, 30%, and 40% of the entries of all images in the *Training Set* and *Test Set*. Apply the low-rank matrix completion (LRMC) algorithm that you implemented in part (a) to the images in the *Training Set*. Plot the projected training images $\boldsymbol{y} \in \mathbb{R}^d$ for $d = 2$ and $d = 3$ using different colors for the different classes. Do faces of different individuals naturally cluster in different regions of the low-dimensional space? Classify the faces in the *Test Set* using 1-nearest-neighbor. That is, label an image $\boldsymbol{x}$ as corresponding to individual i if its projected image $\boldsymbol{y}$ is closest to a projected image $\boldsymbol{y}_j$ of individual i. Notice that you will need to develop new code to project an image with missing entries $\boldsymbol{x}$ onto the face subspace you already estimated from the *Training Set*, which you can do as described in Section 3.1 of this book. Report the percentage of correctly classified face images for $d = 1, \ldots, 10$ and the percentage of missing entries $\{0, 10, 20, 30, 40\}\%$.
3. **Face correction.** Remove uniformly at random 0%, 10%, 20%, 30%, and 40% of the entries of all images of individual 1 and replace them by arbitrary values chosen uniformly at random from $[0, 255]$. Apply the PCP algorithm, Algorithm 3.8, for corrupted entries that you implemented in Exercise 3.10 to these images to compute the mean face and the eigenfaces as well as correct the corrupted entries. Note that RPCA does not compute the mean face, so you will

need to modify the algorithm accordingly. Plot the mean face and the top three eigenfaces and compare them to what you obtained with PCA from Chapter 2. Plot also the corrected faces and comment on the quality of correction as a function of the percentage of corrupted entries by visually comparing the original images (before removing the missing entries) to the completed ones. Plot also the error (Frobenius norm) between the original images and the corrected ones as a function of the percentage of corrupted entries and comment on your results. Repeat for individuals 2 and 3.

4. **Face recognition with corrupted entries.** Remove uniformly at random 0%, 10%, 20%, 30%, and 40% of the entries of all images of individual 1 and replace them by arbitrary values chosen uniformly at random from $[0, 255]$. Apply the RPCA algorithm for corrupted entries that you implemented in part (a) to the images in the *Training Set*. Plot the projected training images $\boldsymbol{y} \in \mathbb{R}^d$ for $d = 2$ or $d = 3$ using different colors for the different classes. Do faces of different individuals naturally cluster in different regions of the low-dimensional space? Classify the faces in the *Test Set* using 1-nearest-neighbor. That is, label an image $\boldsymbol{x}$ as corresponding to individual i if its projected image $\boldsymbol{y}$ is closest to a projected image $\boldsymbol{y}_j$ of individual i. Notice that you will need to develop new code to project an image with corrupted entries $\boldsymbol{x}$ onto the face subspace you already estimated from the *Training Set*. Report the percentage of correctly classified face images for $d = 1, \ldots, 10$ and the percentage of missing entries $\{0, 10, 20, 30, 40\}\%$.
5. **Outlier detection.** Augment the images of individual 1 with those from an *Outlier Set*. Apply the RPCA algorithm for data corrupted by outliers that you implemented in Exercise 3.10 to these images to compute the mean face and the eigenfaces as well as detect the outliers. Note that RPCA does not compute the mean face, so you will need to modify your code accordingly. Plot the mean face and the top three eigenfaces and compare them to what you obtained with PCA. Report the percentage of correctly detected outliers.
6. **Face recognition with corrupted entries.** Apply the RPCA algorithm for data corrupted by outliers that you implemented in part (e) to the images in *Training Set* ∪ *Outlier Set*. Plot the projected training images $\boldsymbol{y} \in \mathbb{R}^d$ for $d = 2$ or $d = 3$ using different colors for the different classes. Do faces of different individuals naturally cluster in different regions of the low-dimensional space? Classify the faces in the *Test Set* using 1-nearest-neighbor. That is, label an image $\boldsymbol{x}$ as corresponding to individual i if its projected image $\boldsymbol{y}$ is closest to a projected image $\boldsymbol{y}_j$ of individual i. Report the percentage of correctly detected outliers and the percentage of correctly classified face images for $d = 1, \ldots, 10$ and compare your results to those using PCA in Chapter 2.

Exercise 3.12 Show that the optimal solution to the PCA problem with robustness to outliers

$$\min_{\boldsymbol{\mu}, U, Y} \sum_{j=1}^{N} w_j \|\boldsymbol{x}_j - \boldsymbol{\mu} - U\boldsymbol{y}_j\|^2 \quad \text{s.t.} \quad U^\top U = I_d \quad \text{and} \quad \sum_{j=1}^{N} w_j \boldsymbol{y}_j = \boldsymbol{0}, \tag{3.117}$$

where $w_j \in [0, 1]$ is large when point $\boldsymbol{x}_j$ is an inlier and small otherwise, is given by

$$\hat{\boldsymbol{\mu}}_N = \frac{\sum_{j=1}^N w_j \boldsymbol{x}_j}{\sum_{j=1}^N w_j} \quad \text{and} \quad \hat{\boldsymbol{y}}_j = \hat{U}^\top (\boldsymbol{x}_j - \hat{\boldsymbol{\mu}}_N) \ \ \forall j \ \text{s.t.} \ w_j > 0, \tag{3.118}$$

where $\hat{U}$ is a $D \times d$ matrix whose columns are the top d eigenvectors of

$$\hat{\Sigma}_N = \frac{\sum_{j=1}^N w_j (\boldsymbol{x}_j - \hat{\boldsymbol{\mu}}_N)(\boldsymbol{x}_j - \hat{\boldsymbol{\mu}}_N)^\top}{\sum_{j=1}^N w_j}. \tag{3.119}$$

Chapter 4
Nonlinear and Nonparametric Extensions

One geometry cannot be more true than another; it can only be more convenient.

—Henri Poincaré

In the previous chapters, we studied the problem of fitting a low-dimensional linear or affine subspace to a collection of points. In practical applications, however, a linear or affine subspace may not be able to capture nonlinear structures in the data. For instance, consider the set of all images of a face obtained by rotating it about its main axis of symmetry. While all such images live in a high-dimensional space whose dimension is the number of pixels, there is only one degree of freedom in the data, namely the angle of rotation. In fact, the space of all such images is a one-dimensional circle embedded in a high-dimensional space, whose structure is not well captured by a one-dimensional line. More generally, a collection of face images observed from different viewpoints is not well approximated by a single linear or affine subspace, as illustrated in the following example.

Example 4.1 (PCA for Embedding Face Images under Varying Pose). To visualize the limitations of PCA on face images with pose variations, we apply PCA to a subset of the images in the extended Yale B data set. This data set consists of face images from 28 human subjects under 9 poses and 64 illumination conditions. Figures 4.1(a)–4.1(i) show the face images for subject 20 for one illumination condition and each of the nine poses, and Figure 4.1(j) shows an illustration of the geometric relationships among these nine poses. In this example, we apply PCA to the images of subject 20 from poses 5, 6, 7, and 8, and all 64 illumination conditions; thus our data set consists of $N = 256$ points. The size of each image is 192×168; thus the dimension of the data is $D = 32{,}256$. Figure 4.1(k) shows the mean face, and Figures 4.1(l)–4.1(m) show the first two eigenfaces computed by PCA. Visually, all three images are rather blurry due to the misalignment caused by pose variations.

In Figure 4.2, we represent the two principal components of each face image as a point in $\mathbb{R}^2$. The points are painted with four different colors corresponding to each

R. Vidal et al., *Generalized Principal Component Analysis*, Interdisciplinary Applied Mathematics 40, DOI 10.1007/978-0-387-87811-9_4

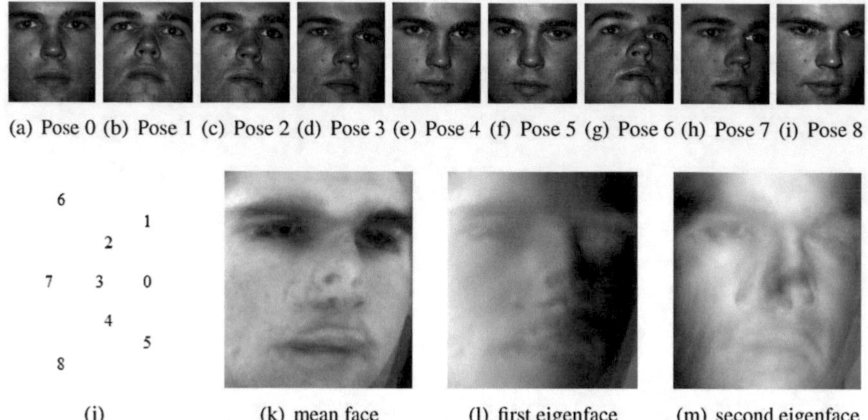

(a) Pose 0 (b) Pose 1 (c) Pose 2 (d) Pose 3 (e) Pose 4 (f) Pose 5 (g) Pose 6 (h) Pose 7 (i) Pose 8

(j) (k) mean face (l) first eigenface (m) second eigenface

Fig. 4.1 Applying PCA to face images under varying pose. (a)-(i) Sample images from nine different poses, with pose 0 representing the frontal faces that we have used in all previous experiments in the preceding two chapters. (j) Illustration of the different poses in the extended Yale B data set, with pose 0 representing the frontal faces that we have used in all previous experiments in the preceding two chapters, and poses 1 to 8 representing nonfrontal poses relative to pose 0. (k), (l), and (m) are, respectively, the mean face and the first two eigenfaces obtained by applying PCA to face images under varying pose.

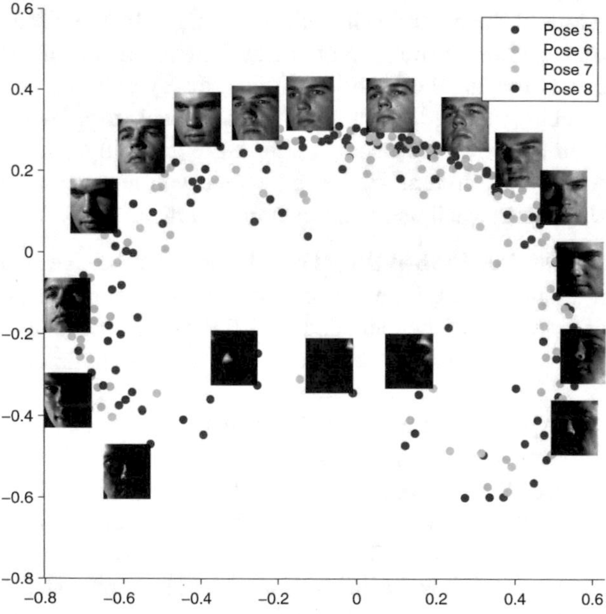

Fig. 4.2 Two-dimensional embedding obtained by applying PCA to a subset of the extended Yale B data set consisting of face images of subject 20 under 4 poses and 64 illumination conditions. Points of the same color represent images associated with the same pose but different illumination. Some images are shown next to some of the points.

one of the four poses. Images associated with some of the points are also shown in the figure. Similar to the results of previous experiments of PCA on face images in Chapter 2, we observe that the first two principal components mainly capture variations in illumination of the face images. Specifically, points on the left are face images lit from the left, points on the right are images lit from the right, points on the top are images with mainly frontal illuminations, while points on the bottom are images taken with extremely low illumination. Notice also that the variability in poses is not captured at all by the first two principal components, and images from different poses are all mixed together. This is in sharp contrast to the methods to be studied in this chapter, as we will see.

When a single low-dimensional subspace fails to describe data that have obvious nonlinear structures, as in Example 4.1, we need to go beyond linear models and consider a broader class of nonlinear models. In this chapter, we consider the problem of fitting a low-dimensional manifold to a collection of points. Specifically, let $\mathcal{X} = \{_j \in \mathbb{R}^D\}_{j=1}^N$ be a set of N points drawn from a d-dimensional manifold $\mathcal{M}$ embedded in $\mathbb{R}^D$, where $d < D$ (see e.g., Figure 4.3). The goal is to find a set of N points $\mathcal{Y} = \{y_j \in \mathbb{R}^d\}_{j=1}^N$ whose geometry resembles that of $\mathcal{X}$. To address this problem, in Section 4.1 we will present an extension of PCA, called nonlinear PCA, which is based on embedding the data into a high-dimensional space via a nonlinear mapping and then fitting a linear or affine space to the embedded data. As we will see, under certain conditions it is possible to compute the low-dimensional embedding $\mathcal{Y}$ without explicitly computing the embedded data via the so-called *kernel trick*. This will lead to a method called kernel PCA. In Section 4.2, we will present other extensions of PCA, generically called *manifold learning*, which aim to approximate the local geometry of the manifold and build a low-dimensional

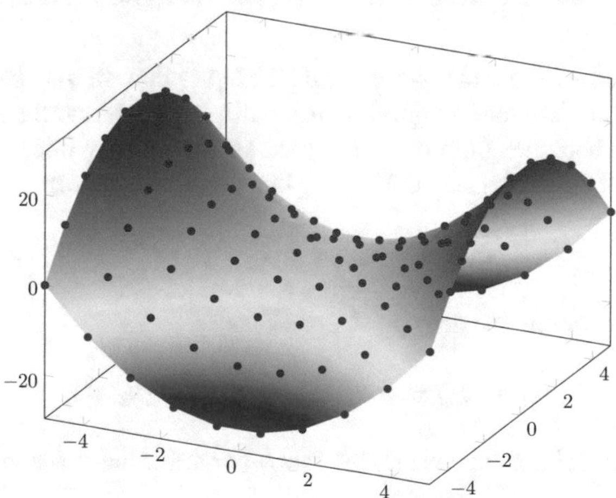

Fig. 4.3 A set of points drawn from the two-dimensional surface in $\mathbb{R}^3$, $x_3 = x_1^2 - x_2^2$. The goal is to find a two-dimensional embedding of this manifold.

nonparametric embedding of the data directly from these local approximations. Such extensions are useful for applications in which we are interested not so much in a parametric model of the manifold as in the low-dimensional points $\mathcal{Y}$ themselves. Finally, in Section 4.3, we show that manifold learning methods can also be applied to data drawn from one manifold with multiple connected components. In this case, the low-dimensional embedding $\mathcal{Y}$ can be used to cluster the data into multiple groups. Data sets that cannot be modeled by a single subspace or manifold and are instead clustered into multiple subspaces will be studied in Part II.

4.1 Nonlinear and Kernel PCA

In this section, we present a nonlinear extension of PCA called nonlinear PCA (NLPCA). The key idea behind NLPCA is that while the given data may not lie in a linear or affine subspace of $\mathbb{R}^D$, there exists a nonlinear embedding $\phi : \mathbb{R}^D \to \mathcal{H}$ into a higher-dimensional space $\mathcal{H}$ such that the embedded data lie (approximately) in a linear or affine subspace of $\mathcal{H}$. Therefore, instead of applying PCA to the given data, we apply it to the embedded data. In practice, however, the dimension of $\mathcal{H}$ may be too high to be able to compute the *nonlinear principal components* from the eigenvectors of the embedded covariance matrix. To address this issue, we also present a method called kernel PCA (KPCA). This method computes the nonlinear principal components from the eigenvectors of the so-called *kernel matrix*, which can be computed directly from the given data.

4.1.1 Nonlinear Principal Component Analysis (NLPCA)

As discussed before, the main idea behind NLPCA is that we may be able to find an embedding of the data into a high-dimensional space such that the structure of the embedded data becomes (approximately) linear. To see why this may be possible, consider a set of points $(x_1, x_2) \in \mathbb{R}^2$ lying in a conic of the form

$$c_1 x_1^2 + c_2 x_1 x_2 + c_3 x_2^2 + c_4 = 0. \tag{4.1}$$

Notice that if we define the map $\phi : \mathbb{R}^2 \to \mathbb{R}^3$ as

$$(z_1, z_2, z_3) = \phi(x_1, x_2) = (x_1^2, \sqrt{2} x_1 x_2, x_2^2), \tag{4.2}$$

then the conic in $\mathbb{R}^2$ transforms into the following affine subspace in $\mathbb{R}^3$:

$$c_1 z_1 + \frac{c_2}{\sqrt{2}} z_2 + c_3 z_3 + c_4 = 0. \tag{4.3}$$

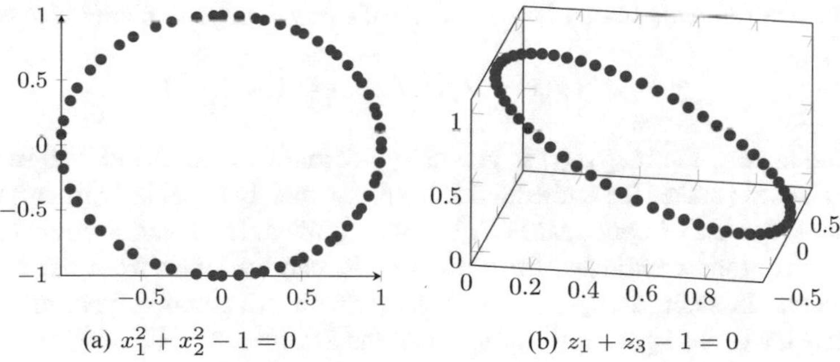

(a) $x_1^2 + x_2^2 - 1 = 0$ (b) $z_1 + z_3 - 1 = 0$

Fig. 4.4 A circle in $\mathbb{R}^2$ is embedded into a plane in $\mathbb{R}^3$ by the mapping in (4.2).

Therefore, instead of learning a nonlinear manifold in $\mathbb{R}^2$, we can simply learn an affine manifold in $\mathbb{R}^3$. This example is illustrated in Figure 4.4.

More generally, we seek a nonlinear transformation (usually an embedding)

$$\phi(\cdot):\ \mathbb{R}^D \to \mathbb{R}^M, \tag{4.4}$$

$$\boldsymbol{x} \mapsto \phi(\boldsymbol{x}), \tag{4.5}$$

such that the structure of the embedded data $\{\phi(\boldsymbol{x}_j)\}_{j=1}^N$ becomes approximately linear. In machine learning, $\phi(\boldsymbol{x}) \in \mathbb{R}^M$ is called the *feature* of the data point $\boldsymbol{x} \in \mathbb{R}^D$, and the space $\mathbb{R}^M$ is called the *feature space*.

Let $\bar{\boldsymbol{\phi}} = \frac{1}{N}\sum_{j=1}^N \phi(\boldsymbol{x}_j)$ be the sample mean in the feature space and define the mean-subtracted (centered) *embedded data matrix* as

$$\Phi \doteq [\phi(\boldsymbol{x}_1) - \bar{\boldsymbol{\phi}}, \phi(\boldsymbol{x}_2) - \bar{\boldsymbol{\phi}}, \ldots, \phi(\boldsymbol{x}_N) - \bar{\boldsymbol{\phi}}] \in \mathbb{R}^{M\times N}. \tag{4.6}$$

It follows from the results in Chapter 2 that the principal components in the feature space can be obtained from the eigenvectors of the sample covariance matrix[1]

$$\Sigma_{\phi(x)} \doteq \frac{1}{N}\sum_{j=1}^N (\phi(\boldsymbol{x}_j) - \bar{\boldsymbol{\phi}})(\phi(\boldsymbol{x}_j) - \bar{\boldsymbol{\phi}})^\top = \frac{1}{N}\Phi\Phi^\top \in \mathbb{R}^{M\times M}. \tag{4.7}$$

Specifically, let $\boldsymbol{u}_i \in \mathbb{R}^M$, $i = 1, \ldots, M$, be the M eigenvectors of $\Sigma_{\phi(x)}$, i.e.,

$$\Sigma_{\phi(x)}\boldsymbol{u}_i = \lambda_i \boldsymbol{u}_i, \quad i = 1, 2, \ldots, M. \tag{4.8}$$

[1]In principle, we should use the notation $\hat{\Sigma}_{\phi(x)}$ to indicate that it is the estimate of the actual covariance matrix. But for simplicity, we will drop the hat in the sequel and simply use $\Sigma_{\phi(x)}$. The same goes for the eigenvectors and the principal components.

Then the *d nonlinear principal components* of every data point $\boldsymbol{x}$ are given by

$$y_i \doteq \boldsymbol{u}_i^\top(\phi(\boldsymbol{x}) - \bar{\boldsymbol{\phi}}) \in \mathbb{R}, \quad i = 1, 2, \ldots, d. \tag{4.9}$$

Unfortunately, the map $\phi(\cdot)$ is generally not known beforehand, and searching for the map that makes the embedded data approximately linear is a difficult task. In such cases, the use of nonlinear PCA becomes limited. However, in some practical applications, good candidates for the map $\phi(\cdot)$ can be found from the nature of the problem. In such cases, the map, together with PCA, can be very effective in extracting the overall geometric structure of the data.

Example 4.2 (Veronese Map for an Arrangement of Subspaces). As we will see later in this book, if the data points belong to a union of multiple subspaces, then a natural choice of the transformation $\phi(\cdot)$ is the Veronese map (also known as the polynomial embedding)

$$\nu_n(\cdot): \quad \boldsymbol{x} \mapsto \nu_n(\boldsymbol{x}), \tag{4.10}$$

$$(x_1, \ldots, x_D) \mapsto (x_1^n, x_1^{n-1}x_2, \ldots, x_D^n), \tag{4.11}$$

where the monomials are ordered in degree-lexicographic order. Under such a mapping, the multiple low-dimensional subspaces are mapped into a single subspace in the feature space, which can then be identified via PCA. We will discuss this embedding in detail in Chapter 5.

4.1.2 NLPCA in a High-dimensional Feature Space

A potential difficulty associated with NLPCA is that the dimension M of the feature space can be very high. Thus, computing the principal components in the feature space may become computationally prohibitive. For instance, if we use a Veronese map of degree n, the dimension of the feature space is $M = \binom{n+D-1}{n}$, which grows exponentially fast. When M exceeds N, the eigenvalue decomposition of $\Phi\Phi^\top \in \mathbb{R}^{M\times M}$ becomes more costly than that of $\Phi^\top\Phi \in \mathbb{R}^{N\times N}$, although the two matrices have the same eigenvalues.

This motivates us to examine whether the computation of PCA in the feature space can be reduced to a computation with the lower-dimensional matrix $\Phi^\top\Phi$. The answer is actually yes. The key is to notice that despite the dimension of the feature space, every eigenvector $\boldsymbol{u} \in \mathbb{R}^M$ of $\Phi\Phi^\top$ associated with a nonzero eigenvalue is always in the span of the matrix Φ.[2] That is,

$$\Phi\Phi^\top\boldsymbol{u} = \lambda\boldsymbol{u} \quad \iff \quad \boldsymbol{u} = \Phi(\lambda^{-1}\Phi^\top\boldsymbol{u}) \in \text{range}(\Phi). \tag{4.12}$$

[2]The remaining $M - N$ eigenvectors of $\Phi\Phi^\top$ are associated with the eigenvalue zero.

Thus, if we let $\boldsymbol{w} \doteq \lambda^{-1}\Phi^\top \boldsymbol{u} \in \mathbb{R}^N$, we have $\|\boldsymbol{w}\|^2 = \lambda^{-2}\boldsymbol{u}^\top \Phi\Phi^\top \boldsymbol{u} = \lambda^{-1}$. Moreover, since $\Phi^\top \Phi \boldsymbol{w} = \lambda^{-1}\Phi^\top \Phi\Phi^\top \boldsymbol{u} = \Phi^\top \boldsymbol{u} = \lambda \boldsymbol{w}$, the vector $\boldsymbol{w}$ is an eigenvector of $\Phi^\top \Phi$ with the same eigenvalue λ. Once such a $\boldsymbol{w}$ has been computed from $\Phi^\top \Phi$, we can recover the corresponding $\boldsymbol{u}$ in the feature space as

$$\boldsymbol{u} = \Phi \boldsymbol{w}, \tag{4.13}$$

and compute the d nonlinear principal components of $\boldsymbol{x}$ under the map $\phi(\cdot)$ as

$$y_i \doteq \boldsymbol{u}_i^\top(\phi(\boldsymbol{x}) - \bar{\boldsymbol{\phi}}) = \boldsymbol{w}_i^\top \Phi^\top(\phi(\boldsymbol{x}) - \bar{\boldsymbol{\phi}}) \in \mathbb{R}, \quad i = 1, \dots, d, \tag{4.14}$$

where $\boldsymbol{w}_i \in \mathbb{R}^N$ is the ith leading eigenvector of $\Phi^\top \Phi \in \mathbb{R}^{N\times N}$.

4.1.3 Kernel PCA (KPCA)

A very interesting property of the above NLPCA method is that the computation of the nonlinear principal components involves only inner products of the features. More specifically, in order to compute the nonlinear principal components y_i, we simply need to compute the entries of the matrix $\Phi^\top \Phi$ and the entries of the vectors $\Phi^\top \phi(\boldsymbol{x})$ and $\Phi^\top \bar{\boldsymbol{\phi}} = \frac{1}{N}\sum_{j=1}^N \Phi^\top \phi(\boldsymbol{x}_j)$. In what follows, we show that all of these quantities can be obtained from inner products of the form $\phi(\boldsymbol{x})^\top \phi(\boldsymbol{y})$.

Before proceeding further, we will first give some definitions.

Definition 4.3. *The space of all square integrable functions is defined as*

$$L^2(\mathbb{R}^D) = \{f : \mathbb{R}^D \to \mathbb{R} \text{ such that } \int f(\boldsymbol{x})^2 d\boldsymbol{x} < \infty\}. \tag{4.15}$$

Definition 4.4. *A function $\kappa : \mathbb{R}^D \times \mathbb{R}^D \to \mathbb{R}$ is symmetric if for all $\boldsymbol{x}, \boldsymbol{y} \in \mathbb{R}^D$, we have $\kappa(\boldsymbol{x}, \boldsymbol{y}) = \kappa(\boldsymbol{y}, \boldsymbol{x})$. A symmetric function κ is positive semidefinite if for all $f \in L^2(\mathbb{R}^D)$, we have*

$$\iint_{\mathbb{R}^D\times\mathbb{R}^D} f(\boldsymbol{x})\kappa(\boldsymbol{x}, \boldsymbol{y})f(\boldsymbol{y})\, d\boldsymbol{x}d\boldsymbol{y} \geq 0. \tag{4.16}$$

Definition 4.5. *Let $\phi : \mathbb{R}^D \to \mathbb{R}^M$ be an embedding function. The kernel function $\kappa : \mathbb{R}^D \times \mathbb{R}^D \to \mathbb{R}$ of two vectors $\boldsymbol{x}, \boldsymbol{y} \in \mathbb{R}^D$ is defined to be the inner product of their features*

$$\kappa(\boldsymbol{x}, \boldsymbol{y}) \doteq \phi(\boldsymbol{x})^\top \phi(\boldsymbol{y}) \in \mathbb{R}. \tag{4.17}$$

One can show that κ is a symmetric positive semidefinite function in $\boldsymbol{x}$ and $\boldsymbol{y}$ (see Exercise 4.1). Let us also define the *centered kernel*[3] as

$$\tilde{\kappa}(\boldsymbol{x},\boldsymbol{y}) \doteq (\phi(\boldsymbol{x}) - \bar{\boldsymbol{\phi}})^\top(\phi(\boldsymbol{y}) - \bar{\boldsymbol{\phi}}) \in \mathbb{R}, \tag{4.18}$$

where $\bar{\boldsymbol{\phi}} = \frac{1}{N}\sum_{j=1}^{N}\phi(\boldsymbol{x}_j)$ is the *mean feature*. We may compute $\tilde{\kappa}$ from κ as

$$\tilde{\kappa}(\boldsymbol{x},\boldsymbol{y}) = \kappa(\boldsymbol{x},\boldsymbol{y}) - \frac{1}{N}\sum_{j=1}^{N}\kappa(\boldsymbol{x},\boldsymbol{x}_j) - \frac{1}{N}\sum_{i=1}^{N}\kappa(\boldsymbol{x}_i,\boldsymbol{y}) + \frac{1}{N^2}\sum_{i=1}^{N}\sum_{j=1}^{N}\kappa(\boldsymbol{x}_i,\boldsymbol{y}_j). \tag{4.19}$$

We can use these functions to compute the nonlinear principal components as follows. Define a *kernel matrix* $\mathcal{K} = [\kappa_{ij}] \in \mathbb{R}^{N\times N}$ as $\kappa_{ij} = \kappa(\boldsymbol{x}_i,\boldsymbol{x}_j)$. The *centered kernel matrix* $\tilde{\mathcal{K}} = \Phi^\top\Phi$ can be computed from $\mathcal{K}$ as

$$\tilde{\mathcal{K}} = \mathcal{K} - \frac{1}{N}\mathcal{K}\mathbf{1}\mathbf{1}^\top - \frac{1}{N}\mathbf{1}\mathbf{1}^\top\mathcal{K} + \frac{\mathbf{1}^\top\mathcal{K}\mathbf{1}}{N^2}\mathbf{1}\mathbf{1}^\top \tag{4.20}$$

$$= (I - \frac{1}{N}\mathbf{1}\mathbf{1}^\top)\mathcal{K}(I - \frac{1}{N}\mathbf{1}\mathbf{1}^\top) = J\mathcal{K}J, \tag{4.21}$$

where $J = I - \frac{1}{N}\mathbf{1}\mathbf{1}^\top$ is called the *centering matrix*.[4] Let us also define the vector $\tilde{\kappa}_{\boldsymbol{x}} = \Phi^\top(\phi(\boldsymbol{x}) - \bar{\boldsymbol{\phi}}) = [\tilde{\kappa}(\boldsymbol{x}_1,\boldsymbol{x}), \tilde{\kappa}(\boldsymbol{x}_2,\boldsymbol{x}), \ldots, \tilde{\kappa}(\boldsymbol{x}_N,\boldsymbol{x})]^\top \in \mathbb{R}^N$. This vector can be computed from $\kappa_{\boldsymbol{x}} = [\kappa(\boldsymbol{x}_1,\boldsymbol{x}), \kappa(\boldsymbol{x}_2,\boldsymbol{x}), \ldots, \kappa(\boldsymbol{x}_N,\boldsymbol{x})]^\top \in \mathbb{R}^N$ as

$$\tilde{\kappa}_{\boldsymbol{x}} = \kappa_{\boldsymbol{x}} - \frac{1}{N}\mathcal{K}\mathbf{1} - \frac{1}{N}\mathbf{1}\mathbf{1}^\top\kappa_{\boldsymbol{x}} + \frac{\mathbf{1}^\top\mathcal{K}\mathbf{1}}{N^2}\mathbf{1}. \tag{4.22}$$

With this notation, we may compute the nonlinear principal components of $\boldsymbol{x}$ as

$$y_i = \boldsymbol{w}_i^\top\Phi^\top(\phi(\boldsymbol{x}) - \bar{\boldsymbol{\phi}}) = \boldsymbol{w}_i^\top\tilde{\kappa}_{\boldsymbol{x}}, \quad i = 1,\ldots,d, \tag{4.23}$$

where $\boldsymbol{w}_i$ is the eigenvector of $\tilde{\mathcal{K}}$ associated with its ith-largest eigenvalue λ_i, and normalized so that $\|\boldsymbol{w}_i\| = \lambda_i^{-2}$. That is, $[\boldsymbol{w}_1,\ldots,\boldsymbol{w}_N] = V_d\Lambda_d^{-1/2}$, where V_d and Λ_d are obtained from the top d eigenvectors and eigenvalues in the EVD of $\tilde{\mathcal{K}} = V_{\tilde{\mathcal{K}}}\Lambda_{\tilde{\mathcal{K}}}V_{\tilde{\mathcal{K}}}^\top$. Since $\tilde{\mathcal{K}} = [\tilde{\kappa}_{\boldsymbol{x}_1},\ldots,\tilde{\kappa}_{\boldsymbol{x}_N}]$, it follows that we can compute the low-dimensional coordinates of the entire data set as

$$Y = \Lambda_d^{-1/2}V_d^\top\tilde{\mathcal{K}} = \Lambda_d^{-1/2}V_d^\top V_{\mathcal{K}}\Lambda_{\mathcal{K}}V_{\mathcal{K}}^\top = \Lambda_d^{1/2}V_d^\top. \tag{4.24}$$

[3]In PCA, we center the data by subtracting its mean. Here, we first subtract the mean of the embedded data and then compute the kernel, whence the name *centered kernel*.

[4]In PCA, if X is the data matrix, then XJ is the centered (mean-subtracted) data matrix.

In other words, the low-dimensional coordinates can be obtained from the top d eigenvectors and eigenvalues of the centered kernel matrix.

Example 4.6 (PCA as a particular case of KPCA) For the linear kernel $\kappa(\boldsymbol{x},\boldsymbol{y}) = \boldsymbol{x}^\top \boldsymbol{y}$, we have $\phi(\boldsymbol{x}) = \boldsymbol{x}$; hence KPCA reduces to PCA.

Example 4.7 For the polynomial embedding of degree 2 in (4.2), we have

$$\kappa(\boldsymbol{x},\boldsymbol{y}) = [x_1^2, \sqrt{2}x_1x_2, x_2^2][y_1^2, \sqrt{2}y_1y_2, y_2^2]^\top = (x_1y_1 + x_2y_2)^2 = (\boldsymbol{x}^\top \boldsymbol{y})^2, \quad (4.25)$$

which can be computed directly in $\mathbb{R}^2$ without the necessity of computing the embedding into $\mathbb{R}^3$.

In summary, we have shown that the nonlinear principal components can be computed directly from the kernel function $\kappa(\boldsymbol{x},\boldsymbol{y}) = \phi(\boldsymbol{x})^\top \phi(\boldsymbol{y})$ without the necessity of computing $\phi(\boldsymbol{x})$. Nonetheless, given any (positive definite) kernel function, according to a fundamental result in functional analysis, one can in principle decompose the kernel and recover the associated map $\phi(\cdot)$ if one wishes to.

Theorem 4.8 (Mercer's Theorem). *Suppose $\kappa : \mathbb{R}^D \times \mathbb{R}^D \to \mathbb{R}$ is a symmetric real-valued function such that for some $C > 0$ and almost every $(\boldsymbol{x},\boldsymbol{y})$[5] we have $|\kappa(\boldsymbol{x},\boldsymbol{y})| \leq C$. Suppose that the linear operator $\mathcal{L} : L^2(\mathbb{R}^D) \to L^2(\mathbb{R}^D)$, where*

$$\mathcal{L}(f)(\boldsymbol{x}) \doteq \int_{\mathbb{R}^D} \kappa(\boldsymbol{x},\boldsymbol{y}) f(\boldsymbol{y}) d\boldsymbol{y}, \quad (4.26)$$

is positive semidefinite. Let ψ_i be the normalized orthogonal eigenfunctions of $\mathcal{L}$ associated with the eigenvalues $\lambda_i > 0$, sorted in nonincreasing order, and let M be the number of nonzero eigenvalues. Then:

- *The sequence of eigenvalues is absolutely convergent, i.e., $\sum_{i=1}^M |\lambda_i| < \infty$.*
- *The kernel κ can be expanded as $\kappa(\boldsymbol{x},\boldsymbol{y}) = \sum_{i=1}^M \lambda_i \psi_i(\boldsymbol{x}) \psi_i(\boldsymbol{y})$ for almost all $(\boldsymbol{x},\boldsymbol{y})$. If $M = \infty$, the series is absolutely and uniformly convergent for almost all $(\boldsymbol{x},\boldsymbol{y})$.*

The interested reader may refer to (Mercer 1909) for a proof of the theorem. It follows from the theorem that given a positive semidefinite kernel κ, we can always associate with it an embedding function ϕ as

$$\phi_i(\boldsymbol{x}) = \sqrt{\lambda_i}\psi_i(\boldsymbol{x}) \quad i = 1, \ldots, M. \quad (4.27)$$

Notice that the dimension M of the embedding could be rather large, sometimes even infinite. Nevertheless, an important reason for computing the principal components with the kernel function is that we do not need to compute the embedding function or the features. Instead, we simply evaluate the dot products $\kappa(\boldsymbol{x},\boldsymbol{y})$ in the original space $\mathbb{R}^D$.

[5] "Almost every" means except for a set of measure zero.

Algorithm 4.1 (Nonlinear Kernel PCA)

Input: A set of points $\mathcal{X} = \{\boldsymbol{x}_1, \boldsymbol{x}_2, \ldots, \boldsymbol{x}_N\} \subset \mathbb{R}^D$, and a map $\phi : \mathbb{R}^D \to \mathbb{R}^M$ or a symmetric positive definite kernel function $\kappa : \mathbb{R}^D \times \mathbb{R}^D \to \mathbb{R}$.

1: Compute $\bar{\boldsymbol{\phi}} = \frac{1}{N}\sum \phi(\boldsymbol{x}_j)$ and the centered embedded data matrix Φ as in (4.6) or the centered kernel $\tilde{\kappa}$ as in (4.19).

2: Compute the centered kernel matrix

$$\tilde{\mathcal{K}} = \Phi^\top \Phi \;\text{ or }\; \big(\tilde{\kappa}(\boldsymbol{x}_i, \boldsymbol{x}_j)\big) \;\in \mathbb{R}^{N\times N}. \tag{4.28}$$

3: Compute the eigenvectors $\boldsymbol{w}_i \in \mathbb{R}^N$ of $\tilde{\mathcal{K}}$:

$$\tilde{\mathcal{K}}\boldsymbol{w}_i = \lambda_i \boldsymbol{w}_i, \tag{4.29}$$

and normalize so that $\|\boldsymbol{w}_i\|^2 = \lambda_i^{-1}$.

4: For every data point $\boldsymbol{x}$, its ith nonlinear principal component is given by

$$y_i \;=\; \boldsymbol{w}_i^\top \Phi^\top (\phi(\boldsymbol{x}) - \bar{\boldsymbol{\phi}}) \;\text{ or }\; \boldsymbol{w}_i^\top [\tilde{\kappa}(\boldsymbol{x}_1, \boldsymbol{x}), \ldots, \tilde{\kappa}(\boldsymbol{x}_N, \boldsymbol{x})]^\top, \tag{4.30}$$

for $i = 1, 2, \ldots, d$.

Output: A set of points $\{\boldsymbol{y}_j\}_{j=1}^N$ lying in $\mathbb{R}^d$, where y_{ij} is the ith nonlinear principal component of $\boldsymbol{x}_j$ for $i = 1, \ldots, d$ and $j = 1, \ldots, N$.

We summarize our discussion in this section with Algorithm 4.1.

Example 4.9 (Examples of Kernels). There are several popular choices for the nonlinear kernel function, such as the polynomial kernel and the Gaussian kernel, respectively

$$\kappa_P(\boldsymbol{x}, \boldsymbol{y}) = (\boldsymbol{x}^\top \boldsymbol{y})^n \quad\text{and}\quad \kappa_G(\boldsymbol{x}, \boldsymbol{y}) = \exp\Big(-\frac{\|\boldsymbol{x} - \boldsymbol{y}\|^2}{\sigma^2}\Big). \tag{4.31}$$

Evaluation of such functions involves only the inner product or the distance between two vectors in the original space $\mathbb{R}^D$. This is much more efficient than evaluating the inner product in the associated feature space, whose dimension grows exponentially with n and D for the first kernel and is infinite for the second kernel.

Example 4.10 (KPCA for Embedding Face Images under Varying Pose). In this example, we use the KPCA algorithm, Algorithm 4.1, to find a two-dimensional embedding of the same subset of the extended Yale B data set that we used in Example 4.1. Figure 4.5 shows the results using a Gaussian kernel with $\sigma = 0.1$. Notice that the embedding given by KPCA clearly improves on that given by PCA, which is shown in Figure 4.2. In particular, images associated with the four poses are mapped into four clearly separated (roughly straight) lines. The only part of the embedding where images from different poses are intermingled is near the origin, as shown in (b).

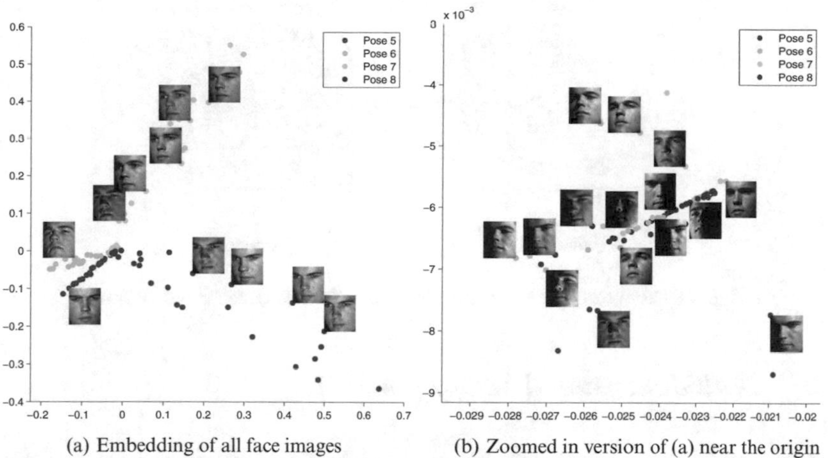

(a) Embedding of all face images

(b) Zoomed in version of (a) near the origin

Fig. 4.5 Two-dimensional embedding obtained by applying KPCA to a subset of the extended Yale B data set consisting of face images of subject 20 under 4 poses and 64 illumination conditions. Points of the same color represent images associated with the same pose but different illumination. Some images are shown next to some of the points.

4.2 Nonparametric Manifold Learning

In the previous section, we described NLPCA, a nonlinear extension of PCA based on embedding a set of data points $\mathcal{X} = \{\boldsymbol{x}_j \in \mathbb{R}^D\}_{j=1}^N$ into a high-dimensional space $\mathcal{H}$ and applying PCA in the embedded space to obtain the low-dimensional representation $\mathcal{Y} = \{\boldsymbol{y}_j \in \mathbb{R}^d\}_{j=1}^N$ with $d < D$. In this section, we present a family of manifold learning methods that search directly for the low-dimensional representation $\mathcal{Y}$ without first embedding the data into a high-dimensional space. Such methods are based on approximating the geometry of the manifold (pairwise distances, local neighborhoods, local linear relationships, etc.) and using these approximations to find a global low-dimensional embedding. For instance, Figure 4.6 shows two typical examples of submanifolds in $\mathbb{R}^3$ that can clearly be embedded in $\mathbb{R}^2$ while preserving their intrinsic geometry. Different methods differ on how certain geometric properties of $\mathcal{X}$ are intended to be preserved or approximated. In what follows, we discuss three representative popular manifold learning methods, namely multidimensional scaling (MDS), locally linear embedding (LLE), and Laplacian eigenmaps (LE). For a more comprehensive review of other manifold learning methods, we refer the reader to (Burges 2005; Lee and Verleysen 2007; Burges 2010).

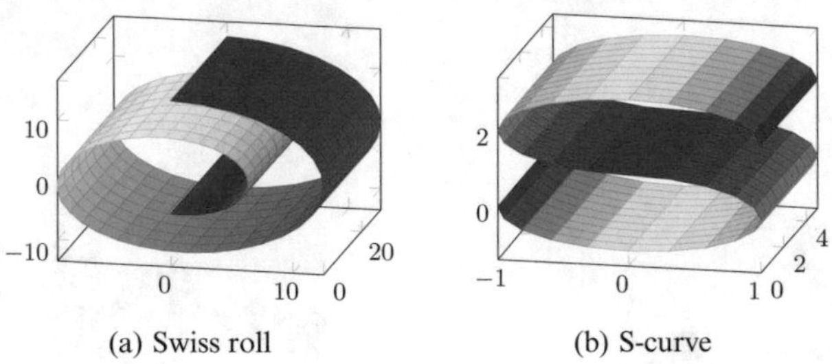

(a) Swiss roll (b) S-curve

Fig. 4.6 Two examples of manifolds typically used in manifold learning.

4.2.1 Multidimensional Scaling (MDS)

One of the oldest manifold learning methods is multidimensional scaling (MDS) (Torgerson 1958; Kruskal 1964; Gower 1966; Cox and Cox 1994). This approach aims to capture the geometry of the manifold by finding a representation $\mathcal{Y}$ whose pairwise distances approximate the pairwise distances in $\mathcal{X}$ as well as possible. To capture nonlinear structures in the data, the distance δ_{ij} between points $\boldsymbol{x}_i$ and $\boldsymbol{x}_j$ in $\mathcal{X}$ is chosen to be any distance other than the Euclidean distance $\delta_{ij} = \|\boldsymbol{x}_i - \boldsymbol{x}_j\|$, e.g., a geodesic distance. More generally, δ_{ij} can be any dissimilarity measure between pairs of points. Given a matrix of dissimilarities $\Delta = [\delta_{ij}] \in \mathbb{R}^{N \times N}$, the goal of MDS is to find a matrix $Y = \big[\boldsymbol{y}_1, \ldots, \boldsymbol{y}_N\big] \in \mathbb{R}^{d \times N}$ that minimizes the following objective:

$$\min_{Y} \sum_{i=1}^{N} \sum_{j=1}^{N} (\|\boldsymbol{y}_i - \boldsymbol{y}_j\| - \delta_{ij})^2. \tag{4.32}$$

Notice that unlike PCA, MDS operates directly on the dissimilarities; hence it does not require us to have the matrix of data points $X = \big[\boldsymbol{x}_1, \ldots, \boldsymbol{x}_N\big] \in \mathbb{R}^{D \times N}$. However, in general, the minimization over Y cannot be carried out in closed form, and gradient descent methods (see Appendix A.1.4) are typically used.[6]

Now, if instead of trying to approximate dissimilarities, we try to approximate similarities or affinities obtained from a dot product, then the solution to MDS can be obtained in closed form from the SVD of the affinity matrix. More specifically, let $A = [a_{ij}] \in \mathbb{R}^{N \times N}$ be a symmetric positive semidefinite matrix of pairwise affinities. For example, A can be defined as

$$a_{ij} = (\boldsymbol{x}_i - \boldsymbol{\mu})^\top (\boldsymbol{x}_j - \boldsymbol{\mu}),$$

[6]See (Davison 1983) for alternative optimization methods for minimizing the objective in (4.32).

where $\boldsymbol{\mu} = \frac{1}{N}X\mathbf{1}$ is the mean of the data.[7] More generally, A can be obtained after embedding the data into a high-dimensional space, as in NLPCA. In fact, we can think of A as a centered kernel matrix, which can be obtained as in (4.20).

Given A, our goal is to find a low-dimensional representation Y such that the dot products between pairs of points best approximate the given affinities, i.e., we wish to minimize

$$\sum_{i=1}^{N}\sum_{j=1}^{N}(\boldsymbol{y}_i^\top \boldsymbol{y}_j - a_{ij})^2 = \|Y^\top Y - A\|_F^2. \tag{4.33}$$

Letting $Z = Y^\top Y$ and noticing that $\text{rank}(Z) = \text{rank}(Y) = d$, we arrive at the following optimization problem:

$$\min_Z \|Z - A\|_F^2 \quad \text{such that} \quad \text{rank}(Z) = d, \quad Z = Z^\top, \quad Z \succeq 0. \tag{4.34}$$

Notice that except for the symmetric positive semidefiniteness constraint on Z, this problem is identical to the low-rank matrix approximation problem in (2.35). However, since A is symmetric positive semidefinite, this additional constraint is unnecessary. To see this, notice that if we use Theorem 2.6 to find the optimal solution, we obtain the optimal Z from the SVD of $A = U_A \Sigma_A U_A^\top$ as $Z = U \Sigma U^\top$, where U consists of the top d columns of U_A and Σ consists of the top $d \times d$ subblock of Σ_A. Notice that this solution automatically satisfies the symmetric positive semidefiniteness constraint. Given $Z = U\Sigma U^\top$, we can obtain Y as $Y = R\Sigma^{1/2}U^\top$ for any orthogonal matrix $R \in O(d)$.

In summary, when the affinity matrix A is a centered kernel matrix, MDS gives the same low-dimensional representation as KPCA, up to an arbitrary orthogonal transformation R. For further connections between MDS, PCA, and KPCA, we refer the reader to (Williams 2002).

4.2.2 *Locally Linear Embedding (LLE)*

Another popular manifold learning approach is locally linear embedding (LLE) (Roweis and Saul 2000, 2003). This approach aims to capture the geometry of a manifold $\mathcal{M}$ by exploiting the fact that the local neighborhood of a point $\boldsymbol{x} \in \mathcal{M}$ can be well approximated by the affine subspace spanned by $\boldsymbol{x}$ and its K nearest neighbors (K-NNs). These *locally linear* approximations are then used to find a low-dimensional embedding that gives a small reconstruction error with respect to such approximations.

[7] Notice that $A = JX^\top XJ$, where $J = I - \frac{1}{N}\mathbf{1}\mathbf{1}^\top$ is the centering matrix.

The first step of LLE is to approximate each data point $\boldsymbol{x}_j$ as an affine combination of its K-NNs. Intuitively, this step is analogous to approximating the tangent space of the manifold at the point $\boldsymbol{x}_j$ by the affine subspace spanned by $\boldsymbol{x}_j$ and its K-NNs. For a manifold of dimension d, the tangent space at each point is a d-dimensional affine subspace. Therefore, we need at least d NNs to reconstruct this subspace, i.e., we need to choose $K \geq d$. On the other hand, K cannot be chosen to be too large. Otherwise, each data point would be written as an affine combination of too many points, and the affine coefficients would not be unique. Since different solutions for the affine coefficients correspond to different approximations of the tangent space, the estimated local affine subspace may be inaccurate. Therefore, LLE typically requires a good knowledge of d in order to define K. Given K, the K-NNs $\{\boldsymbol{x}_{j_k}\}_{k=1}^{K}$ of each data point $\boldsymbol{x}_j$ are typically found using the Euclidean distance. However, other distances can be used as well.

To approximate each data point $\boldsymbol{x}_j$ as an affine combination of its K-NNs, we search for a matrix $C = [c_{ij}] \in \mathbb{R}^{N \times N}$ that minimizes the reconstruction error

$$E(C) = \frac{1}{2}\sum_{j=1}^{N}\Big\|\boldsymbol{x}_j - \sum_{i=1}^{N} c_{ij}\boldsymbol{x}_i\Big\|^2, \tag{4.35}$$

subject to (i) $c_{ij} = 0$ if $\boldsymbol{x}_i$ is not a K-NN of $\boldsymbol{x}_j$ and (ii) $\sum_{i=1}^{N} c_{ij} = 1$. The first constraint expresses point $\boldsymbol{x}_j$ as an affine combination of only its K-NNs, while the second constraint ensures that the combination of the K-NNs is affine.

Let $j_1, \ldots, j_K$ denote the indices of the K-NN of $\boldsymbol{x}_j$. Since $c_{ij} = 0$ when $\boldsymbol{x}_i$ is not a K-NN of $\boldsymbol{x}_j$, we need to keep track of only K affine coefficients for each point $\boldsymbol{x}_j$. Let $\boldsymbol{c}_j = [c_{j_1 j}, \ldots, c_{j_K j}]^\top \in \mathbb{R}^K$ be the vector of such coefficients and let $G_j = [g_{il}^j] \in \mathbb{R}^{K \times K}$ be the local Gram matrix at $\boldsymbol{x}_j$, which is defined as $g_{il}^j = (\boldsymbol{x}_i - \boldsymbol{x}_j)^\top(\boldsymbol{x}_l - \boldsymbol{x}_j)$ for i, l such that $\boldsymbol{x}_i$ and $\boldsymbol{x}_l$ are K-NNs of $\boldsymbol{x}_j$. With this notation, the jth term of (4.35) can be written as

$$\begin{aligned}\Big\|\boldsymbol{x}_j - \sum_{i=1}^{N} c_{ij}\boldsymbol{x}_i\Big\|^2 &= \Big\|\sum_{i=1}^{N} c_{ij}(\boldsymbol{x}_i - \boldsymbol{x}_j)\Big\|^2 = \sum_{il} c_{ij}c_{lj}(\boldsymbol{x}_i - \boldsymbol{x}_j)^\top(\boldsymbol{x}_l - \boldsymbol{x}_j) \\ &= \sum_{il} c_{ij}c_{lj}g_{il}^j = \boldsymbol{c}_j^\top G_j \boldsymbol{c}_j. \end{aligned} \tag{4.36}$$

Therefore, the optimization problem in (4.35) is equivalent to

$$\min_{\{\boldsymbol{c}_j\}} \ \frac{1}{2}\sum_{j=1}^{N}\boldsymbol{c}_j^\top G_j \boldsymbol{c}_j \quad \text{s.t.} \quad \mathbf{1}^\top \boldsymbol{c}_j = 1. \tag{4.37}$$

The Lagrangian function for (4.37) is $\mathscr{L} = \frac{1}{2}\sum_{j=1}^{N}\boldsymbol{c}_j^\top G_j \boldsymbol{c}_j + \lambda_j(1 - \mathbf{1}^\top \boldsymbol{c}_j)$. Thus, the first-order conditions for optimality are $G_j\boldsymbol{c}_j = \lambda_j \mathbf{1}$ and $\mathbf{1}^\top \boldsymbol{c}_j = 1$. Therefore, if G_j is of full rank K, then $\boldsymbol{c}_j = \lambda_j G_j^{-1}\mathbf{1}$ and $\lambda_j^{-1} = \mathbf{1}^\top G_j^{-1}\mathbf{1}$, so that

$$c_j = \frac{G_j^{-1}\mathbf{1}}{\mathbf{1}^\top G_j^{-1}\mathbf{1}} \in \mathbb{R}^K. \tag{4.38}$$

Notice that the affine coefficients c_{ij} are invariant with respect to rotations, translations, and scalings of all the data points. The invariance with respect to rotations and translations follows from the invariance properties of the Gram matrix G_j. Specifically, notice that if each $\boldsymbol{x}_j$ is transformed to $R\boldsymbol{x}_j + \boldsymbol{t}$, where $R \in SO(3)$ and $\boldsymbol{t} \in \mathbb{R}^3$, then $\boldsymbol{x}_j - \boldsymbol{x}_i$ is transformed to $R(\boldsymbol{x}_j - \boldsymbol{x}_i)$, and so G_j is not affected. The invariance with respect to scalings follows from the fact that the Gram matrix appears in both the numerator and denominator of (4.38). Therefore, the affine coefficients characterize the intrinsic geometric properties of each local neighborhood of the data in $\mathbb{R}^D$.

The second step of LLE is to find a representation $Y = [\boldsymbol{y}_1, \dots, \boldsymbol{y}_N] \in \mathbb{R}^{d\times N}$ that minimizes

$$\phi(Y) = \sum_{j=1}^{N} \Big\| \boldsymbol{y}_j - \sum_{i=1}^{N} c_{ij}\boldsymbol{y}_i \Big\|^2. \tag{4.39}$$

Notice that the objective in (4.39) is the same as the reconstruction error in (4.35), but obtained with respect to the low-dimensional representation Y rather than with respect to the original data X. Notice also that the global minimum is obtained when $Y = 0$; thus we need to impose additional constraints on the low-dimensional representation in order to avoid trivial solutions. LLE requires the low-dimensional representation Y to satisfy the following constraints:

$$\sum_{j=1}^{N} \boldsymbol{y}_j = \mathbf{0} \quad \text{and} \quad \frac{1}{N}\sum_{j=1}^{N} \boldsymbol{y}_j\boldsymbol{y}_j^\top = I. \tag{4.40}$$

The first constraint requires the low-dimensional representation to be centered at the origin, as in the case of PCA (see Chapter 2). The second constraint forces the low-dimensional representation to have unit covariance and is an arbitrary constraint to ensure that $\operatorname{rank}(Y) = d$.

To find the optimal Y, notice that the optimization problem can be written as

$$\min_{Y} \ \|Y - YC\|_F^2 \quad \text{s.t.} \quad Y\mathbf{1} = \mathbf{0} \quad \text{and} \quad \frac{1}{N}YY^\top = I. \tag{4.41}$$

Proposition 4.11 (Locally Linear Embedding). *The solution to the optimization problem* (4.41) *is given by the matrix Y whose rows are the d eigenvectors of the matrix $\mathcal{L} = (I - C)(I - C)^\top$ associated with its second- to $(d + 1)$st-smallest eigenvalues.*

Proof. Notice that $\|Y - YC\|_F^2 = \operatorname{trace}(Y(I - C)(I - C)^\top Y^\top)$. Therefore, the optimization problem (4.41) is a special case of a more general problem in (4.48),

---

Algorithm 4.2 (Locally Linear Embedding)

Input: A set of points $\{\boldsymbol{x}_j\}_{j=1}^N$ lying in a manifold $\mathcal{M}$ and integers K and d.
1: Find the K-nearest neighbors (K-NN) of each data point $\boldsymbol{x}_j, j = 1, \dots, N$, according to some distance in $\mathcal{M}$.
2: Approximate each point $\boldsymbol{x}_j \approx \sum c_{ij}\boldsymbol{x}_i$ as an affine combination of its K-NN with coefficients the c_{ij} obtained as in (4.38).
3: Let the rows of the matrix $Y = [\boldsymbol{y}_1, \dots, \boldsymbol{y}_N] \in \mathbb{R}^{d\times N}$ be the d eigenvectors of the matrix $\mathcal{L} = (I - C)(I - C)^\top$ associated with its second- to $(d + 1)$st-smallest eigenvalues.
Output: A set of points $\{\boldsymbol{y}_j\}_{j=1}^N$ lying in $\mathbb{R}^d$.

---

with $\mathcal{L}$ replaced by $(I - C)(I - C)^\top$ and $\mathcal{D}$ replaced by $\frac{I}{N}$. Therefore, the result follows by direct application of Proposition 4.14, which we will prove later.

□

In summary, LLE is a manifold learning algorithm that uses the data matrix X to construct a matrix of affine coefficients C that captures the local geometry of the manifold. The low-dimensional representation is then obtained from the eigenvectors of the matrix $\mathcal{L} = (I-C)(I-C)^\top$ associated to its second- to $(d+1)$st-smallest eigenvalues. The LLE algorithm is summarized in Algorithm 4.2.

Example 4.12 (LLE for Embedding Face Images under Varying Pose). In this example, we use the LLE algorithm, Algorithm 4.2, to find a two-dimensional embedding of the same face image data set we used in Example 4.1. Figure 4.7 shows the results using $K = 13$ nearest neighbors. We see that the embedding given by LLE also improves on that given by PCA, which is shown in Figure 4.2, but in a different way from that given by KPCA, which is shown in Figure 4.5: one direction (the x-axis) mainly captures the variations in illumination, whereas the other direction (the y-axis) spreads out the four different poses.

4.2.3 Laplacian Eigenmaps (LE)

Another popular manifold learning algorithm is Laplacian eigenmaps (LE) (Belkin and Niyogi 2002). This approach aims to capture the geometry of the manifold by finding a low-dimensional representation such that nearby points in the manifold are mapped to nearby points in the low-dimensional embedding.

More specifically, if $X = \big[\boldsymbol{x}_1, \dots, \boldsymbol{x}_N\big]$ is the data matrix, LE finds a low-dimensional embedding $Y = \big[\boldsymbol{y}_1, \dots, \boldsymbol{y}_N\big] \in \mathbb{R}^{d\times N}$ such that if $\boldsymbol{x}_i$ and $\boldsymbol{x}_j$ are close to each other, then so are $\boldsymbol{y}_i$ and $\boldsymbol{y}_j$. This is done by minimizing the objective

$$\phi(Y) = \sum_{i=1}^{N}\sum_{j=1}^{N} w_{ij}\|\boldsymbol{y}_i - \boldsymbol{y}_j\|^2, \tag{4.42}$$

subject to appropriate constraints on Y that prevent the trivial solution $Y = \mathbf{0}$.

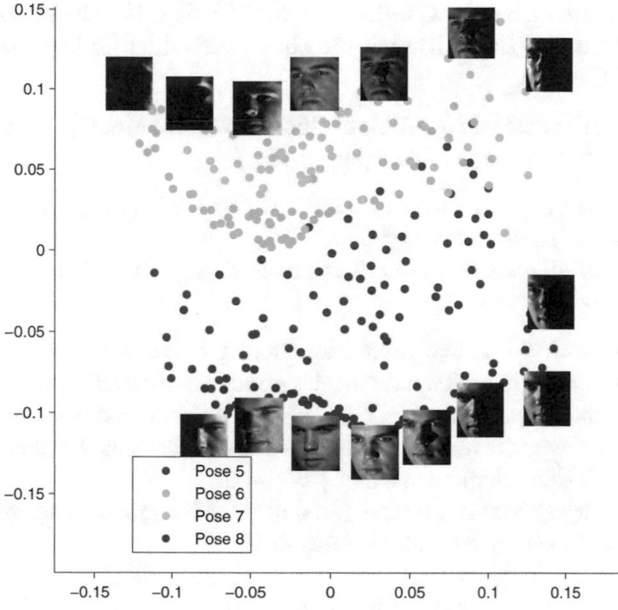

Fig. 4.7 Two-dimensional embedding obtained by applying LLE to a subset of the extended Yale B data set consisting of face images of subject 20 under 4 poses and 64 illumination conditions. Points of the same color represent images associated with the same pose but different illumination. Some images are shown next to some of the points.

The weights $w_{ij} \geq 0$ are designed so that a small penalty is paid when $\boldsymbol{x}_i$ and $\boldsymbol{x}_j$ are far, so that $\boldsymbol{y}_i$ and $\boldsymbol{y}_j$ are allowed to be far, and a large penalty is paid when $\boldsymbol{x}_i$ and $\boldsymbol{x}_j$ are close, but $\boldsymbol{y}_i$ and $\boldsymbol{y}_j$ are far. For this purpose, a local neighborhood of each point $\boldsymbol{x}_j$ is defined using the K-NN rule with some distance dist on $\mathcal{M}$, and the weights are chosen as

$$w_{ij} = \begin{cases} e^{-\frac{\text{dist}(\boldsymbol{x}_i,\boldsymbol{x}_j)^2}{2\sigma^2}} & \text{if } \boldsymbol{x}_i \text{ is a } K\text{-NN of } \boldsymbol{x}_j \text{ or vice versa,} \\ 0 & \text{else,} \end{cases} \tag{4.43}$$

where $\sigma > 0$ is a parameter.

Letting $\mathcal{D} \in \mathbb{R}^{N \times N}$ be a diagonal matrix with diagonal entries $d_{jj} = \sum_i w_{ij}$, and $W \in \mathbb{R}^{N \times N}$ the matrix of weights, we may rewrite the objective function as

$$\phi(Y) = \sum_{ij} w_{ij}\big(\|\boldsymbol{y}_i\|^2 + \|\boldsymbol{y}_j\|^2 - 2\boldsymbol{y}_i^\top \boldsymbol{y}_j\big) \tag{4.44}$$

$$= 2\sum_j d_{jj}\boldsymbol{y}_j^\top \boldsymbol{y}_j - 2\sum_{ij} w_{ij}\boldsymbol{y}_i^\top \boldsymbol{y}_j \tag{4.45}$$

$$= 2\,\text{trace}(Y\mathcal{D}Y^\top) - 2\,\text{trace}(YWY^\top) = 2\,\text{trace}(Y\mathcal{L}Y^\top), \tag{4.46}$$

where the symmetric matrix $\mathcal{L} = \mathcal{D} - W \in \mathbb{R}^{N \times N}$ is called the Laplacian matrix. The definition of the Laplacian and the above derivation lead to some important properties of $\mathcal{L}$ below.

Proposition 4.13 (Basic Properties of the Laplacian Matrix). *The Laplacian matrix $\mathcal{L} \in \mathbb{R}^{N \times N}$ has the following properties:*

- *For all $\mathbf{y} = [y_1, \ldots, y_N]^\top \in \mathbb{R}^N$, we have $\mathbf{y}^\top \mathcal{L} \mathbf{y} = \frac{1}{2} \sum_{i,j} w_{ij}(y_i - y_j)^2 \geq 0$. Hence the matrix $\mathcal{L}$ is positive semidefinite.*
- *The vector of all ones is in the null space of $\mathcal{L}$, i.e., $\mathcal{L}\mathbf{1} = \mathbf{0}$; hence the smallest eigenvalue of $\mathcal{L}$ is zero.*

It follows from the above discussion that a trivial solution to the problem $\min_Y \phi(Y)$ is $Y = \mathbf{0}$, in which case all the points are mapped to the origin. It also follows from the proposition that $Y = \mathbf{y}\mathbf{1}^\top$ is another trivial solution, in which case all data points $\boldsymbol{x}_j$ are mapped to the *same* low-dimensional embedding $\boldsymbol{y} \in \mathbb{R}^d$. Notice that both solutions are such that $\phi(Y) = 0$.

To prevent these trivial solutions, LE requires the low-dimensional representation Y to satisfy the following additional constraints:

$$Y\mathcal{D}\mathbf{1} = \mathbf{0} \quad \text{and} \quad Y\mathcal{D}Y^\top = I. \tag{4.47}$$

The first constraint requires the scaled low-dimensional representation[8] $Y\mathcal{D}$ to be orthogonal to the constant vector $\mathbf{1}$ so as to avoid the constant embedding $Y = \mathbf{y}\mathbf{1}^\top$. The second constraint ensures that $\text{rank}(Y) = d$ and helps remove an arbitrary scaling factor in the embedding. In fact, the above two constraints result from properly discretizing the solution to a continuous Laplacian embedding of a continuous manifold. Not to disturb the flow, we leave a brief introduction of the continuous Laplacian embedding to Appendix 4.A of this chapter. As one would see, the two constraints in equation (4.47) are discretized versions of their corresponding continuous counterparts in equation (4.95).

Therefore, LE finds the low-dimensional representation by solving the following minimization problem:

$$\min_Y \ \text{trace}(Y\mathcal{L}Y^\top) \quad \text{s.t.} \quad Y\mathcal{D}\mathbf{1} = \mathbf{0} \quad \text{and} \quad Y\mathcal{D}Y^\top = I. \tag{4.48}$$

The solution to this optimization problem is given by the next result.

Proposition 4.14 (Laplacian Eigenmaps). *The solution to the optimization problem* (4.48) *is given by the matrix Y whose rows are the d generalized eigenvectors of the pair $(\mathcal{L}, \mathcal{D})$ associated with its second- to $(d+1)$st-smallest generalized eigenvalues.*

[8] By scaled low-dimensional representation we mean replacing $\boldsymbol{y}_j$ by $d_{jj}\boldsymbol{y}_j$.

Proof. Notice that the Lagrangian function for this problem can be written as

$$\mathcal{L}(Y, \boldsymbol{\lambda}, \Lambda) = \text{trace}(Y\mathcal{L}Y^\top) + \boldsymbol{\lambda}^\top Y\mathcal{D}\mathbf{1} + \text{trace}(\Lambda(I - Y\mathcal{D}Y^\top)), \tag{4.49}$$

where $\boldsymbol{\lambda} \in \mathbb{R}^d$ and $\Lambda = \Lambda^\top \in \mathbb{R}^{d\times d}$ are, respectively, a vector and matrix of Lagrange multipliers. Computing the derivative of $\mathcal{L}$ with respect to Y and setting it to zero yields $2Y\mathcal{L} + \boldsymbol{\lambda}\mathbf{1}^\top\mathcal{D} - 2\Lambda Y\mathcal{D} = \mathbf{0}$. Multiplying on the right by $\mathbf{1}$ and using the constraints $\mathcal{L}\mathbf{1} = \mathbf{0}$ and $Y\mathcal{D}\mathbf{1} = \mathbf{0}$, we obtain $\boldsymbol{\lambda} = \mathbf{0}$. As a consequence,

$$Y\mathcal{L} = \Lambda Y\mathcal{D} \implies \mathcal{L}Y^\top = \mathcal{D}Y^\top\Lambda. \tag{4.50}$$

Following the same argument as in the proof of Theorem 2.3, one can show that Λ is diagonal. Therefore, the rows of Y are generalized eigenvectors of $(\mathcal{L}, \mathcal{D})$ with generalized eigenvalues in the diagonal entries of Λ. Moreover, $Y\mathcal{L}Y^\top = \Lambda Y\mathcal{D}Y^\top = \Lambda$, and so the objective value is $\text{trace}(Y\mathcal{L}Y^\top) = \text{trace}(\Lambda)$. Therefore, we must choose the smallest generalized eigenvalues of $(\mathcal{L}, \mathcal{D})$. Since $\mathbf{1}$ is an eigenvector of $\mathcal{L}$ with zero eigenvalue, and the eigenvectors of $\mathcal{L}$ must be orthogonal to $\mathcal{D}\mathbf{1}$ (because $Y\mathcal{D}\mathbf{1} = \mathbf{0}$), the rows of the optimal Y are the d generalized eigenvectors of $(\mathcal{L}, \mathcal{D})$ associated with its second- to $(d+1)$st-smallest eigenvalues, as claimed. □

The LE algorithm is summarized in Algorithm 4.3.

The reader has probably noticed that the low-dimensional embeddings given by LLE and LE are, at a high level, rather similar in several aspects:

1. They both map the original data points $\boldsymbol{x}_j \in \mathcal{M} \subset \mathbb{R}^D$ to a new set of data points $\boldsymbol{y}_j$ in $\mathbb{R}^d$ with the goal of preserving local geometric properties of the original data, rather than providing any parametric representation of $\boldsymbol{x}_j$ in its original space (as done by PCA).

Algorithm 4.3 (Laplacian Eigenmaps)

Input: A set of points $\{\boldsymbol{x}_j\}_{j=1}^N$ in a manifold $\mathcal{M}$, integers K and d, and $\sigma > 0$.

1: Find the K nearest neighbors (K-NN) of each data point $\boldsymbol{x}_j, j = 1, \ldots, N$, according to some distance dist in $\mathcal{M}$.

2: Define a matrix of weights $W \in \mathbb{R}^{N\times N}$ whose entries w_{ij} measure the affinity between two points $\boldsymbol{x}_i$ and $\boldsymbol{x}_j$ and are computed as

$$w_{ij} = \begin{cases} e^{-\frac{\text{dist}(\boldsymbol{x}_i, \boldsymbol{x}_j)^2}{2\sigma^2}} & \text{if } \boldsymbol{x}_i \text{ is a } K\text{-NN of } \boldsymbol{x}_j \text{ or vice versa,} \\ 0 & \text{else.} \end{cases} \tag{4.51}$$

3: Let $\mathcal{D}$ be a diagonal matrix with entries $d_{jj} = \sum_i w_{ij}$, and let $\mathcal{L} = \mathcal{D} - W$. Find a matrix $Y = [\boldsymbol{y}_1, \ldots, \boldsymbol{y}_N] \in \mathbb{R}^{d\times N}$ whose rows are the d generalized eigenvectors of the pair $(\mathcal{L}, \mathcal{D})$ associated with its second- to $(d+1)$st-smallest generalized eigenvalues. That is, solve for Y from $Y\mathcal{L} = \Lambda Y\mathcal{D}$, where Λ is a diagonal matrix with the generalized eigenvalues along its diagonal.

Output: A set of points $\{\boldsymbol{y}_j\}_{j=1}^N$ lying in $\mathbb{R}^d$.

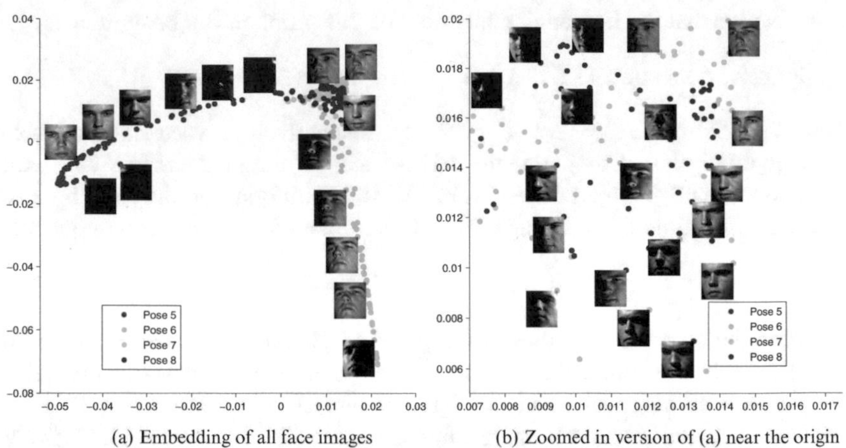

(a) Embedding of all face images

(b) Zoomed in version of (a) near the origin

Fig. 4.8 Two-dimensional embedding obtained by applying LE to a subset of the extended Yale B data set consisting of face images of subject 20 under 4 poses and 64 illumination conditions. Points of the same color represent images associated with the same pose but different illumination. Some images are shown next to some of the points.

2. They both start by computing a weight w_{ij} between any pair of points that reflects the desired geometric properties to be preserved. A weight with larger magnitude indicates that the two points are "similar" with respect to such properties.

Example 4.15 (LE for Embedding Face Images under Varying Pose). In this example, we use the LE algorithm, Algorithm 4.3, to find a two-dimensional embedding of the same face image data set we used in Example 4.1. Figure 4.8 shows the results using $K = 5$ nearest neighbors and $\sigma = 5$. We see that the embedding given by LE also improves on that given by PCA, which is shown in Figure 4.2, but differently from that given by KPCA, which is shown in Figure 4.5, and from that given by LLE, which is shown in Figure 4.7: images from two of the poses are clearly separated by LE, but the other two poses remain clustered together.

Example 4.16 (PCA, KPCA, LE, and LLE for Embedding Face Images of Two Different Subjects). In this example, we apply various linear and nonlinear embedding methods to a data set that contains frontal face images of two subjects: subject 20 and subject 21 in the extended Yale B data set. The resulting embeddings by PCA, KPCA, LE, and LLE are shown in Figure 4.9. In each figure, the two colors correspond to images associated with the two different subjects. For KPCA, we use a Gaussian kernel with $\sigma = 0.1$. For LLE, we use $K = 6$ neighbors. Finally, LE uses $K = 5$ neighbors and $\sigma = 5$. Observe that except for PCA, all the nonlinear embedding methods are able to clearly separate the images from the two individuals. Observe also that in the case of KPCA, images from the two individuals are each mapped to roughly two lines. Overall, this experiment illustrates how nonlinear manifold learning techniques are better suited for data sets whose underlying low-dimensional representation is nonlinear.

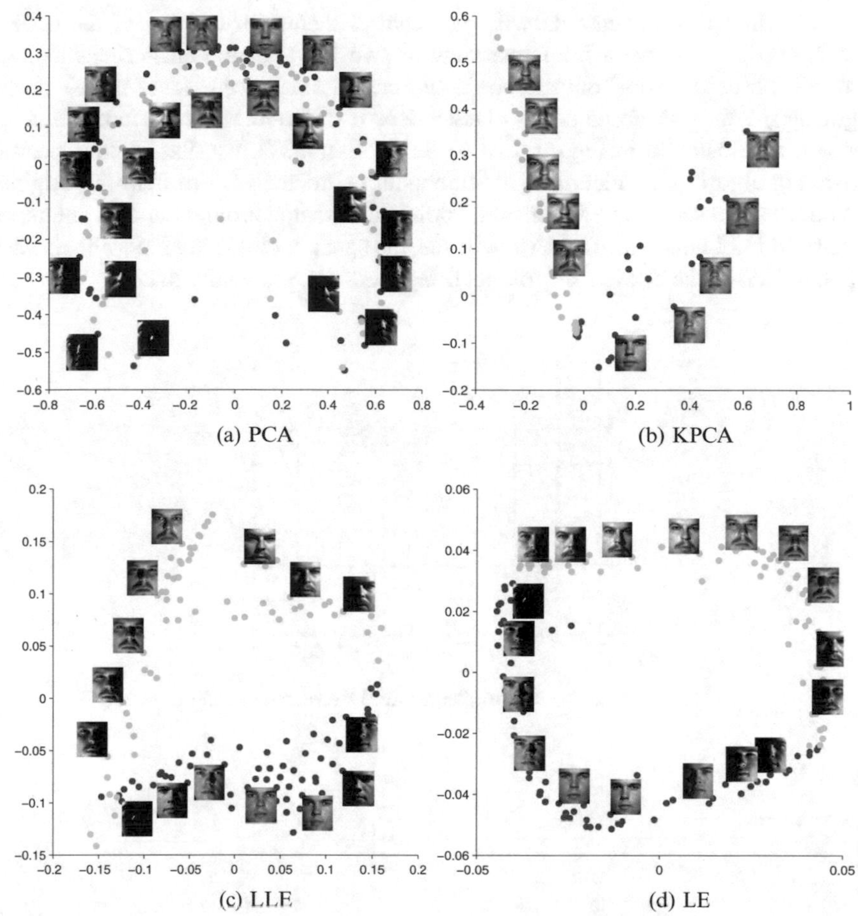

Fig. 4.9 Two-dimensional embedding obtained by PCA, KPCA, LLE, and LE for face images (frontal pose, all 64 illuminations) of two individuals (subjects 20 and 21) in the extended Yale B data set. Points of the same color correspond to frontal face images of the same subject under different illumination conditions. Sample images are shown beside some of the points.

4.3 K-Means and Spectral Clustering

As we have seen in the previous experiments with face images, when the data set contains images mixed from multiple subjects with multiple poses, it might no longer be possible to model the whole data set by a single subspace or a single nonlinear manifold. Nonetheless, after suitable nonlinear mappings, images from the same pose or from the same subject tend to form a cluster in space that is separated from those for other poses or subjects. This suggests that these nonlinear mappings may not only be useful for finding a low-dimensional representation of the data, but also simplify clustering the data if the data are mixed.

Since clustering mixed data will be a central theme for the rest of the book,[9] in this section we give a brief overview of two fundamental clustering methods that will be used throughout the book. In Section 4.3.1, we discuss the K-means algorithm, which is designed to cluster data distributed around a collection of centers, as illustrated in Figure 4.10(a). In Section 4.3.2, we discuss the spectral clustering algorithm, which uses an embedding similar to LE to map the original mixed data to a set of low-dimensional points distributed around cluster centers, as illustrated in Figure 4.10(b). As we will see, the spectral clustering algorithm is very much related to the above manifold learning methods, especially to LE.

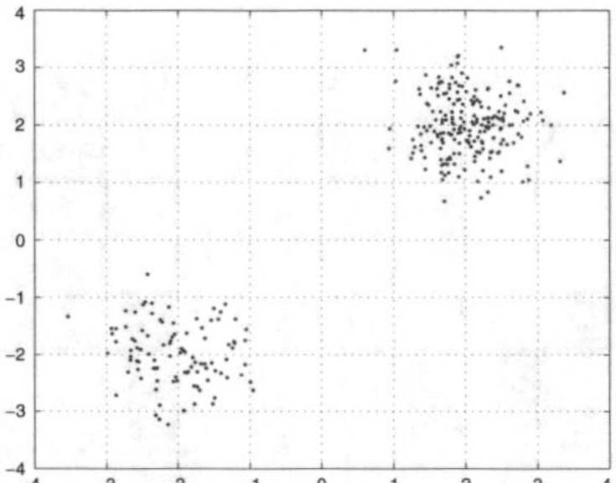

(a) 2-D data sampled around two cluster centers

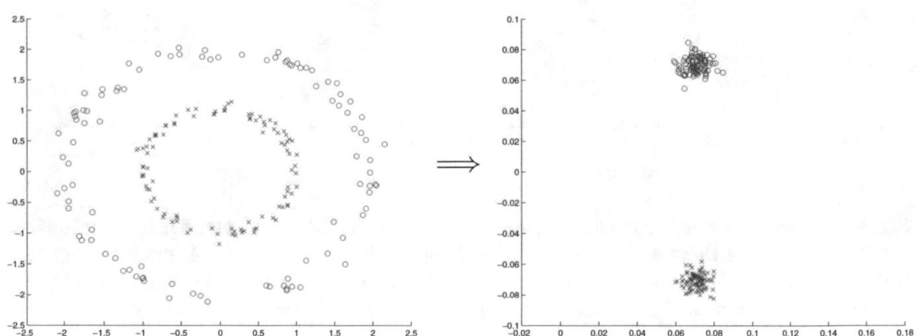

(b) 2-D data sampled around two circles is mapped to 2-D data sampled around two cluster centers

Fig. 4.10 Clustering 2D data distributed around cluster centers or around two circles. In (b), a suitable (spectral) embedding maps the original data to an embedding where the data points are clustered around two centers.

[9] As we will see in Chapters 7 and 8, spectral clustering methods will play a crucial role in many approaches to subspace clustering.

4.3.1 K-Means Clustering

The K-means algorithm is arguably one of the simplest and most widely used clustering methods. It is based on the assumption that the data points $\{x_j\}_{j=1}^N$ are distributed around a collection of n cluster centers $\{\boldsymbol{\mu}_i\}_{i=1}^n$. Assuming that n is known, the K-means algorithm aims to estimate the cluster centers by minimizing the sum of squared distances from the data points to their closest cluster centers, i.e.,

$$\min_{\boldsymbol{\mu}_1,\dots,\boldsymbol{\mu}_n} \sum_{j=1}^{N} \min_{i=1,\dots,n} \|\boldsymbol{x}_j - \boldsymbol{\mu}_i\|_2^2. \tag{4.52}$$

An important challenge in solving the above minimization problem is that the objective function[10] is not differentiable for all $\boldsymbol{\mu}_i$. Nonetheless, one can derive a simple alternating minimization algorithm by introducing a set of auxiliary variables that denote the assignments of points to cluster centers. More specifically, let $w_{ij} \in \{0,1\}$ be such that $w_{ij} = 1$ if point j is assigned to cluster i and $w_{ij} = 0$ otherwise. Then the optimization problem in (4.52) can be rewritten as

$$\begin{aligned} \min_{\{\boldsymbol{\mu}_i\},\{w_{ij}\}} \quad & \sum_{i=1}^{n}\sum_{j=1}^{N} w_{ij}\|\boldsymbol{x}_j - \boldsymbol{\mu}_i\|^2 \\ \text{s.t.} \quad & w_{ij} \in \{0,1\} \text{ and } \sum_{i=1}^{n} w_{ij} = 1, \quad j = 1,\dots,N. \end{aligned} \tag{4.53}$$

The application of the alternating minimization algorithm in Appendix A.1.5 to this problem allows us to estimate the cluster centers $\{\boldsymbol{\mu}_i\}_{i=1}^n$ and the segmentation of the data $\{w_{ij}\}_{i=1,\dots,n}^{j=1,\dots,N}$ in a straightforward manner by alternating between solving for the segmentation given the cluster centers and vice versa. Specifically:

1. If the cluster centers are known, so are the distances $\|\boldsymbol{x}_j - \boldsymbol{\mu}_i\|$. Thus, the optimization problem over $\{w_{ij}\}$ with $\{\boldsymbol{\mu}_i\}$ held fixed involves minimizing a weighted sum of the w_{ij}, subject to the constraint that for each j, there is only one i such that $w_{ij} = 1$. Therefore, to minimize the objective, for each j we must set $w_{ij} = 1$ for the i that gives the smallest distance $\|\boldsymbol{x}_j - \boldsymbol{\mu}_i\|$. This is equivalent to assigning $\boldsymbol{x}_j$ to its closest cluster center.
2. If the segmentation is known, the constraints in (4.53) are redundant, and the optimal solution for $\boldsymbol{\mu}_i$ can be obtained from

[10]Notice that the above objective is very much related to the MAP-EM algorithm for a mixture of isotropic Gaussians discussed in Appendix B.3.2.

$$-\sum_{j=1}^{N} 2w_{ij}(\boldsymbol{x}_j - \boldsymbol{\mu}_i) = \boldsymbol{0} \implies \boldsymbol{\mu}_i = \frac{\sum_{j=1}^{N} w_{ij}\boldsymbol{x}_j}{\sum_{j=1}^{N} w_{ij}}. \tag{4.54}$$

We observe that $N_i = \sum_{j=1}^{N} w_{ij}$ is the number of points assigned to the ith cluster center. Thus, we can see that the above expression for $\boldsymbol{\mu}_i$ is simply the average of the points assigned to the ith cluster center.

In other words, the K-means algorithm alternates between computing the cluster centers given the segmentation and computing the segmentation given the cluster centers, as detailed in Algorithm 4.4. Notice that at each iteration of the algorithm, the objective function in (4.52) either decreases or stays the same (see Exercise 4.10 for a proof). Notice also that since the variables w_{ij} are binary, the number of possible segmentations is finite. Since for each segmentation there is a unique solution for the cluster centers, the number of possible solutions for $\{\boldsymbol{\mu}_i\}$ is also finite. Therefore, after a finite number of iterations, the value of the objective function in (4.52) will stop decreasing, and the algorithm will converge.

Notice also that except for the case in which two or more points are at equal distance from two or more distinct cluster centers, a small perturbation of the cluster centers $\boldsymbol{\mu}_i$ does not change the assignment of points to cluster centers. Therefore, the K-means algorithm converges to a local minimum of the objective function in (4.52) in a finite number of steps.

We refer the reader to (Bottou and Bengio 1995) for a similar convergence argument, and to (Selim and Ismail 1984) for a more rigorous analysis of convergence of a generalized K-means algorithm, including cases in which the generalized method fails to converge to a local minimum. Now, even if the algorithm converges to a local minimum, in general it will not converge to a global minimum. Therefore, initialization is critical in order to obtain a good solution. A common strategy is to initialize the algorithm with n randomly chosen data points as candidate cluster centers, repeat the algorithm for multiple random initializations, and then choose the one that gives the best objective value.

Example 4.17 (K-means Clustering of Face Images under Varying Pose). In this example, we apply the K-means algorithm, Algorithm 4.4, to a subset of the AT&T face data set[11] (previously known as ORL). This data set contains photos of 40 individuals, with 10 poses for each individual. In this experiment, we use only the images for individuals 1 to 20. We first apply PCA to the images to reduce the dimension of the data from 92×112 to $D = 50$. We then cluster the PCA coefficients using K-means with 20 cluster centers. We reconstruct the 20 centroids found by K-means back to the face space by averaging the images associated with each cluster center, as shown in Figure 4.11(a). Notice that most of the centroids correspond to blurry face images. This is due to the fact that face images under varying pose

[11] AT&T Laboratories, Cambridge,
http://www.cl.cam.ac.uk/Research/DTG/attarchive/facedatabase.html.

Algorithm 4.4 (K-means)

Input: A set of points $\{\boldsymbol{x}_j\}_{j=1}^N$ and the number of groups n.

1: **Initialization:** Select n distinct data points as initial cluster centers $\boldsymbol{\mu}_1^{(0)}, \ldots, \boldsymbol{\mu}_n^{(0)}$.

2: **while** (the clusters and their centers do not converge) **do**

3: Assign each data point $\boldsymbol{x}_j$ to its closest cluster center $\boldsymbol{\mu}_i^{(k)}$, i.e.,

$$w_{ij}^{(k+1)} \leftarrow \begin{cases} 1 & \text{if } i = \arg\min_{\ell=1,\ldots,n} \|\boldsymbol{x}_j - \boldsymbol{\mu}_\ell^{(k)}\|_2^2, \\ 0 & \text{else.} \end{cases} \tag{4.55}$$

4: Update the cluster centers $\boldsymbol{\mu}_i^{(k+1)}$ to be the mean of all points $\boldsymbol{x}_j$ that belong to cluster i,

$$\boldsymbol{\mu}_i^{(k+1)} \leftarrow \frac{\sum_{j=1}^N w_{ij}^{(k+1)} \boldsymbol{x}_j}{\sum_{j=1}^N w_{ij}^{(k+1)}}. \tag{4.56}$$

If more than one cluster achieves the minimum, assign the point to one of them.

5: **end while**

Output: The n cluster centers $\{\boldsymbol{\mu}_i\}$ and the segmentation $\{w_{ij}\}$.

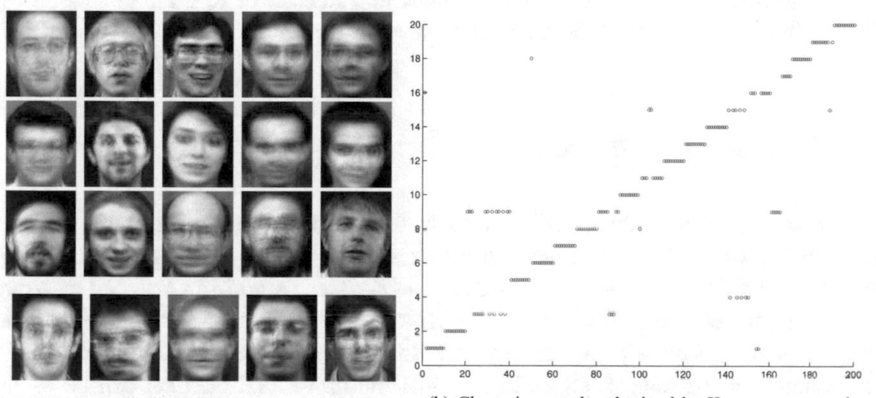

(a) 20 centroids found by K-means for the 100 training face images of the 20 individuals

(b) Clustering results obtained by K-means versus the index of each face image (10 images per class)

Fig. 4.11 Face clustering results given by the K-means algorithm on AT&T data set.

get averaged by K-means. Out of the 20 centroids, 19 correspond (approximately) to 19 different individuals, while one centroid is repeated twice. This is due to the fact that K-means, as a greedy method, has failed to converge to the globally optimal solution. As a result, the clustering error is 20%. The clustering results versus the index of the face images (which are sorted by individual with 10 images per class) are plotted in Figure 4.11(b). Overall, we can see that K-means works reasonably well for clustering face images of different individuals under varying pose, but similar illumination conditions.

Example 4.18 (K-means Clustering of Face Images under Varying Illumination) In this example, we apply the K-means algorithm, Algorithm 4.4, to a subset of the extended Yale B face data set. Specifically, we take images of two subjects (20 and 21, or 37 and 38) each under 64 different illumination conditions. We first apply PCA to the images to reduce the dimension of data to $D = 20$ and then normalize the PCA coefficients to unit ℓ_2. We then apply K-means with $n = 2$ clusters to the normalized data. The clustering errors are 50% for subjects 20 and 21, and 47.7% for subjects 37 and 38. These errors are very high, which suggests that K-means is not suitable for clustering face images under varying illumination. This is because K-means uses the Euclidean distance between two face images as a measure of similarity. Such a distance is not suitable for capturing large variations of illumination, because two face images of different individuals under the same lighting could be closer to each other than two face images of the same individual under very different lightings. This can also be explained by looking at Figure 4.9(a), which shows a two-dimensional PCA embedding for individuals 20 and 21. While we are using a 20-dimensional embedding for K-means, Figure 4.9(a) already suggests that it is very difficult to cluster the data into two groups. At the same time, Figures 4.9(b)-(d) suggest that K-means might be able to do a little better if we apply it to a KPCA, LLE, or LE embedding instead of to a PCA embedding. This will be the subject of the next subsection.

4.3.2 Spectral Clustering

The face clustering example discussed in the previous section suggests that the distribution of a mixed data set can be more complicated than simply clustering around a few cluster centers. In this case, the K-means algorithm may fail to group the data correctly. Nonetheless, as we have suggested in the example shown in Figure 4.10(b), one way to remedy the situation is to seek a suitable nonlinear embedding of the data, such as LE, so that the embedded data set can be easily clustered by K-means. But why is the phenomenon illustrated in Figure 4.10(b) possible? In other words, why is it the case that after a suitable embedding, the data points cluster around a collection of cluster centers?

In this section, we aim to answer this question by introducing a graph-theoretic approach to data clustering. In this approach, data points become vertices of a graph, and similar data points are connected by edges in the graph. Ideally, different clusters correspond to different connected components of the graph, which can be found by analyzing the null space of the *graph Laplacian*. In practice, however, there may not be a one-to-one correspondence between clusters and connected components of the graph. In this case, we will use results from matrix perturbation theory to show that some eigenvectors of the graph Laplacian provide an embedding for the data from which the clustering can be obtained more easily, thereby establishing an intimate connection between clustering and manifold learning. Finally, we will formulate the clustering problem directly as a graph partitioning problem, and show that a continuous relaxation of the associated discrete optimization problem leads to the same type of generalized eigenvalue problem solved in manifold learning.

Ideal Case Given a set of data points $\boldsymbol{x}_1, \ldots, \boldsymbol{x}_N \in \mathbb{R}^D$, we associate with it a weighted undirected graph $\mathcal{G} = (\mathcal{V}, \mathcal{E}, W)$, where $\mathcal{V} = \{1, \ldots, N\}$ is a set of N vertices and $\mathcal{E} \subset \mathcal{V} \times \mathcal{V}$ is a set of edges, which captures the "affinity" between pairs of points. For example, in a K-NN graph, each data point is connected to its K-NNs, and in an ε-neighborhood graph, each data point is connected to other points at a distance less than or equal to $\varepsilon > 0$. The (i,j)th entry $w_{ij} = w_{ji} \geq 0$ of the matrix $W \in \mathbb{R}^{N \times N}$ is a weight associated with the edge $(i,j) \in \mathcal{E}$. If $w_{ij} = 0$, the two vertices are not connected. Otherwise, the weight w_{ij} is used to describe the affinity between points $\boldsymbol{x}_i$ and $\boldsymbol{x}_j$ in terms of their properties in the original space $\mathbb{R}^D$. For instance, w_{ij} can be chosen as in LE as $w_{ij} \propto \exp(-\operatorname{dist}(\boldsymbol{x}_i, \boldsymbol{x}_j)^2/2\sigma^2)$, so that $w_{ij} \approx 1$ when points $\boldsymbol{x}_i$ and $\boldsymbol{x}_j$ are close to each other, and $w_{ij} \approx 0$ when points $\boldsymbol{x}_i$ and $\boldsymbol{x}_j$ are far from each other. Alternatively, we can set $w_{ij} = 1$ if and only if $\boldsymbol{x}_i$ is connected to $\boldsymbol{x}_j$. Since the graph is undirected, the matrix W is symmetric, i.e., $W = W^\top$. We further define the *degree* of vertex i as $d_{ii} = \sum_j w_{ij}$. Note that when $w_{ij} \in \{0, 1\}$, d_{ii} is the number of points connected to $\boldsymbol{x}_i$. Thus, the degree of vertex i is a measure of how connected point $\boldsymbol{x}_i$ is to other points. We define the *degree matrix* $\mathcal{D} \in \mathbb{R}^{N \times N}$ as a diagonal matrix whose diagonal entries are the degrees d_{ii} of the vertices. Finally, we define the *Laplacian* of the graph $\mathcal{G}$ as the matrix $\mathcal{L} \doteq \mathcal{D} - W$.

Recall from Proposition 4.13 that the matrix $\mathcal{L}$ is symmetric positive semidefinite and that the vector of all ones is in its nullspace, i.e., $\mathcal{L}\mathbf{1} = \mathbf{0}$. Recall also from Proposition 4.14 that the LE algorithm obtains a low-dimensional embedding $Y \in \mathbb{R}^{d \times N}$ from the generalized eigenvectors of $(\mathcal{L}, \mathcal{D})$ corresponding to the second- to $(d+1)$st-smallest generalized eigenvalues of $(\mathcal{L}, \mathcal{D})$. Hence, the LE method can be viewed as a special case of a very general data mapping method known as *spectral embedding*, which uses the spectrum of graph Laplacians to provide new representations for data. Now, an important assumption that we did not highlight when we introduced the LE algorithm is that we need the second-smallest generalized eigenvalue of $(\mathcal{L}, \mathcal{D})$ to be nonzero. Under this assumption, there is only one eigenvector of $\mathcal{L}$ associated with the zero eigenvalue, namely the vector $\mathbf{1}$. Moreover, one can show that the second eigenvalue of $\mathcal{L}$ is nonzero if and only if the graph is connected. Therefore, the reason for LE to require that the second eigenvalue of $\mathcal{L}$ be nonzero is precisely so that the graph $\mathcal{G}$ is connected.[12]

But what if the graph is not connected? As we will see, when $\mathcal{G}$ is not connected, there are multiple eigenvectors associated with the zero eigenvalue, each one corresponding to one of the connected components of the graph $\mathcal{G}$. This seemingly simple fact has significant implications in using spectral embedding to extract important topological properties of a graph such as whether the graph is connected, or equivalently, whether the data set has a single cluster or multiple

[12]A graph is connected when there is a path between every pair of vertices.

clusters. Moreover, the eigenvectors of $\mathcal{L}$ corresponding to the zero eigenvalues can be used to cluster the data into the connected component of the graph.

To motivate why this is the case, let us consider a simple example.

Example 4.19 Suppose the graph $\mathcal{G}$ has n connected subgraphs $\mathcal{G} = \mathcal{G}_1 \cup \mathcal{G}_2 \cup \cdots \cup \mathcal{G}_n$ with $\mathcal{G}_i \cap \mathcal{G}_j = \emptyset$. Let the number of vertices in each subgraph be $N_1, N_2, \ldots, N_n$, respectively. Consider a special weighted graph $\mathcal{G}$ where the weights are such that $w_{ij} = 1$ if and only if the two vertices i and j belong to the same connected subgraph, and otherwise $w_{ij} = 0$. Let J_m be an $m \times m$ matrix full of 1's, and I_m the $m \times m$ identity matrix. Then, if the data points are sorted according to their membership in the n connected components, the weight matrix W is a block-diagonal matrix with n diagonal submatrices J_{N_i}, and the degree matrix $\mathcal{D}$ is a block-diagonal matrix with diagonal submatrices $N_i \cdot I_{N_i}$:

$$W = \begin{bmatrix} J_{N_1} & 0 & 0 \\ 0 & \ddots & 0 \\ 0 & 0 & J_{N_n} \end{bmatrix}, \quad \mathcal{D} = \begin{bmatrix} N_1 \cdot I_{N_1} & 0 & 0 \\ 0 & \ddots & 0 \\ 0 & 0 & N_n \cdot I_{N_n} \end{bmatrix}. \tag{4.57}$$

Let $\mathbf{1}_{\mathcal{G}_i} \in \mathbb{R}^N$ denote the indicator vector for subgraph $\mathcal{G}_i$. That is, its entries are 1 for vertices in $\mathcal{G}_i$ and 0 otherwise. Then it is easy to verify that such indicator vectors are eigenvectors of the weight matrix W with the number of vertices in each subgraph N_i as the eigenvalues:

$$W\mathbf{1}_{\mathcal{G}_i} = N_i \mathbf{1}_{\mathcal{G}_i}. \tag{4.58}$$

By the definition of the degree matrix $\mathcal{D}$, these indicator vectors must be in the null space of the Laplacian of the graph $\mathcal{L} = \mathcal{D} - W$. That is,

$$\mathcal{L}\mathbf{1}_{\mathcal{G}_i} = \mathbf{0}, \ \forall i = 1, \ldots, n. \tag{4.59}$$

This simple example illustrates that the null space of the Laplacian of a specific graph encodes the membership of its vertices in different connected subgraphs. The following result shows that this property of the null space of the Laplacian is also true for more general graphs.

Proposition 4.20 (Number of Connected Subgraphs). *Given an undirected graph $\mathcal{G}$ with N vertices and $n \leq N$ connected subgraphs, i.e., $\mathcal{G} = \mathcal{G}_1 \cup \mathcal{G}_2 \cup \cdots \cup \mathcal{G}_n$ with $\mathcal{G}_i \cap \mathcal{G}_j = \emptyset$ for $i \neq j$, and a nonnegative weight matrix W, the number of zero eigenvalues of its Laplacian matrix $\mathcal{L}$ is equal to the number of connected components of the graph. Moreover, the null space null($\mathcal{L}$) is exactly spanned by the indicator vectors of these disconnected subgraphs:*

$$null(\mathcal{L}) = span\{\mathbf{1}_{\mathcal{G}_1}, \mathbf{1}_{\mathcal{G}_2}, \ldots, \mathbf{1}_{\mathcal{G}_n}\}. \tag{4.60}$$

Proof. Suppose $\boldsymbol{u} \in \mathbb{R}^N$ is an eigenvector of $\mathcal{L}$ associated with a zero eigenvalue. Then we have

$$\boldsymbol{u}^\top \mathcal{L} \boldsymbol{u} = \frac{1}{2} \sum_{i,j} w_{ij}(u_i - u_j)^2 = 0. \tag{4.61}$$

Since $w_{ij} \geq 0$, for the above equality to hold, we must have $u_i = u_j$ whenever $w_{ij} > 0$. Therefore, if two vertices i and j belong to the same connected component, the corresponding values of the eigenvector must be equal. It follows that every vector $\boldsymbol{u}$ in the null space of $\mathcal{L}$ can be written as a linear combination of the vectors $\{\mathbf{1}_{\mathcal{G}_i}\}_{i=1}^n$, i.e., $\text{null}(\mathcal{L}) = \text{span}\{\mathbf{1}_{\mathcal{G}_1}, \mathbf{1}_{\mathcal{G}_2}, \ldots, \mathbf{1}_{\mathcal{G}_n}\}$. Since these n vectors are linearly independent for $N \geq n$, we conclude that the dimension of the null space of $\mathcal{L}$ is equal to the number of connected components. □

The above property of the Laplacian matrix implies that the null space of the Laplacian matrix encodes precise information about the membership of the vertices in the n connected components. However, we cannot yet directly use this information for clustering, because we can identify the indicator vectors only up to a change of basis.

Proposition 4.21 (Null Space of Laplacian). *Every n linearly independent vectors $\boldsymbol{u}_1, \ldots \boldsymbol{u}_n \in \mathbb{R}^N$ in the null space of $\mathcal{L}$ can be written as*

$$\left[\boldsymbol{u}_1, \ldots, \boldsymbol{u}_n\right] = \left[\mathbf{1}_{\mathcal{G}_1}, \mathbf{1}_{\mathcal{G}_2}, \ldots, \mathbf{1}_{\mathcal{G}_n}\right] A \; \in \mathbb{R}^{N \times n} \tag{4.62}$$

for some nonsingular matrix $A \in \mathbb{R}^{n \times n}$.

We leave the proof of this fact as an exercise for the reader. Now, if we view the columns of the matrix $Y \doteq [\boldsymbol{u}_1, \ldots, \boldsymbol{u}_n]^\top = [\boldsymbol{y}_1, \ldots, \boldsymbol{y}_N] \in \mathbb{R}^{n \times N}$ as a new embedding of the points in $\mathbb{R}^n$, then Y has a very simple but important property: $\boldsymbol{y}_i = \boldsymbol{y}_j$ if and only if the two vertices i and j belong to the same connected component. That is, there are n distinct columns in Y, which means that all N points $\{\boldsymbol{x}_j\}$ are mapped exactly to n points in $\mathbb{R}^n$. Hence, if we are interested only in the topology of the graph, we care only about the eigenvectors associated with the zero eigenvalue, whereas LE in Section 4.2.3 uses other eigenvectors associated with nonzero eigenvalues to find a low-dimensional embedding.

General Case

So far, we have assumed that the graph has n connected components and that there is an edge connecting every two components. In practice, the data may be corrupted, and the affinity matrix W may not be such that $w_{ij} = 0$ when points i and j are in different connected components. In this case, even if the ideal graph has n connected components, the graph obtained from a corrupted matrix W may be fully connected. As result, the eigensubspace associated with the n smallest eigenvalues of $\mathcal{L}$ will be a perturbed version of the ideal eigensubspace associated with the zero eigenvalue, which is spanned by the n indicator vectors. Therefore, the embedded points $\{\boldsymbol{y}_j\}_{j=1}^N$

will no longer coincide with n points in $\mathbb{R}^n$. Ideally, we would like these points to cluster around n cluster centers in $\mathbb{R}^n$, as predicted in Figure 4.10(b). The following result, which follows from a perturbation theorem in (Davis and Cahan 1970), shows that if the perturbation to W is small enough, the N points $\{\boldsymbol{y}_j\}_{j=1}^N$ do cluster around n cluster centers in $\mathbb{R}^n$.

Theorem 4.22 (Stability of the Null Space of Laplacian). *Let $\tilde{\mathcal{L}}$ be the Laplacian of a graph with n connected components and let $0 = \tilde{\lambda}_1 = \tilde{\lambda}_2 = \cdots = \tilde{\lambda}_n < \tilde{\lambda}_{n+1} \leq \tilde{\lambda}_{n+2} \leq \cdots \leq \tilde{\lambda}_N$ be its N eigenvalues. Let $\mathcal{L}$ be a perturbed graph Laplacian and let $0 = \lambda_1 \leq \lambda_2 \leq \cdots \leq \lambda_N$ be its eigenvalues. Let $\tilde{Y} \in \mathbb{R}^{n\times N}$ be a matrix whose orthonormal rows are the indicator vectors (with length normalized to one) and let $Y \in \mathbb{R}^{n\times N}$ be a matrix whose orthonormal rows are the n eigenvectors of $\mathcal{L}$ corresponding to its n smallest eigenvalues. Then*

$$\min_{R\in O(n)} \|\tilde{Y} - RY\|_2 \leq \frac{\sqrt{2}}{\lambda_{n+1}}\|\tilde{\mathcal{L}} - \mathcal{L}\|_2. \tag{4.63}$$

In other words, if the perturbation is small enough, as measured by $\|\tilde{\mathcal{L}} - \mathcal{L}\|_2$, and λ_{n+1} is large enough, then the perturbed low-dimensional embedding Y is close to the ideal low-dimensional embedding $\tilde{Y}$ (up to an orthogonal matrix $R \in \mathbb{R}^{n\times n}$). One can show that $\lambda_{n+1} \geq \tilde{\lambda}_{n+1} - \|\tilde{\mathcal{L}} - \mathcal{L}\|_2$ (see Exercise 4.12) . Therefore, the requirement that λ_{n+1} be large enough can be translated into requiring that $\tilde{\lambda}_{n+1}$ be large enough relative to the size of the perturbation. Since $\tilde{\lambda}_{n+1}$ would have been zero had the number of clusters been $n+1$ rather than n, the requirement that $\tilde{\lambda}_{n+1}$ be large enough can be interpreted as requiring that each one of the n clusters be sufficiently well connected.

Theorem 4.22 establishes that under certain conditions, the columns of the matrix $Y = [\boldsymbol{y}_1, \ldots, \boldsymbol{y}_N] = [\boldsymbol{u}_1, \ldots, \boldsymbol{u}_n]^\top$ obtained from the n eigenvectors $\{\boldsymbol{u}_i\}_{i=1}^n$ of $\mathcal{L}$ corresponding to its n smallest eigenvalues will cluster around n "ideal" cluster centers given by the n distinct columns of $\tilde{Y}$. However, notice that the ideal cluster centers are not obtained from the true indicator vectors, as in Proposition 4.62, but from a normalized version of these indicator vectors. This is because in Theorem 4.22, the rows of the ideal embedding $\tilde{Y}$ are the indicator vectors normalized to be of unit norm. Specifically, the ith row of $\tilde{Y}$ is given by

$$\boldsymbol{h}_i \doteq \mathbf{1}_{\mathcal{G}_i}/\sqrt{|\mathcal{G}_i|}, \quad i = 1, \ldots, n, \tag{4.64}$$

where $|\mathcal{G}_i|$ is the number of vertices in the ith subgraph. This introduces an interesting "normalization" to the ideal low-dimensional embedding $\tilde{Y}$ in which the distance from the ith ideal cluster center to the origin is scaled down by $1/\sqrt{|\mathcal{G}_i|}$. Thus, the larger the subgraph $\mathcal{G}_i$, the closer its ideal cluster center is to the origin.

Given the low-dimensional embedding $Y = [\boldsymbol{u}_1, \ldots, \boldsymbol{u}_n]^\top$, we can cluster the original data $\{\boldsymbol{x}_j\}_{j=1}^N$ by clustering the low-dimensional points $\{\boldsymbol{y}_j\}_{j=1}^N$ using, for example, the K-means algorithm, Algorithm 4.4. This leads to a popular data clustering algorithm, known as *spectral clustering*, which we summarize in Algorithm 4.5.

Algorithm 4.5 (Spectral Clustering)

Input: Number of clusters n and affinity matrix $W \in \mathbb{R}^{N \times N}$ for points $\{x_j\}_{j=1}^N$.
1: Construct an affinity graph $\mathcal{G}$ with weight matrix W.
2: Compute the degree matrix $\mathcal{D} = \operatorname{diag}(W\mathbf{1})$ and the Laplacian $\mathcal{L} = \mathcal{D} - W$.
3: Compute the n eigenvectors of $\mathcal{L}$ associated with its n smallest eigenvalues.
4: Let $y_1, \dots, y_N$ be the columns of $Y \doteq [u_1, \dots, u_n]^\top \in \mathbb{R}^{n \times N}$, where $\{u_i\}_{i=1}^n$ are the eigenvectors in step 3 normalized to unit Euclidean norm.
5: Cluster the points $\{y_j\}_{j=1}^N$ into n groups using the K-means algorithm, Algorithm 4.4.
Output: The segmentation of the data into n groups.

Observe that Algorithm 4.5 is based on computing a low-dimensional embedding of the data from the eigenvectors of $\mathcal{L}$ and then clustering this low-dimensional embedding using K-means, as we suggested in Figure 4.10(b). However, the low-dimensional embedding used in Algorithm 4.5 does not coincide with any of the low-dimensional embeddings discussed in Section 4.2. The most similar low-dimensional embedding is that of LE, which uses the generalized eigenvectors of $(\mathcal{L}, \mathcal{D})$ in lieu of the eigenvectors of $\mathcal{L}$. Next, we discuss some variants to the basic spectral clustering algorithm that are very close to LE. As we shall see, such variants provide some form of normalization of the low-dimensional embedding that is beneficial from the clustering point of view.

Connections between Spectral Clustering, Mincut and Ratiocut
So far, we have assumed that the graph either has n connected components or can be approximated by a graph with n connected components via a small perturbation of the affinity. In practice, we may want to find a clustering of the data even when the perturbation from the ideal case is large. In this case, it makes sense to formulate the clustering problem as a graph partitioning problem in which we aim to divide the graph into multiple subgraphs by "cutting the weakest links." Interestingly, this approach leads to a discrete optimization problem whose continuous relaxation results in a generalized eigenvalue problem.

More specifically, let $\mathcal{A}$ and $\mathcal{B}$ be two subgraphs of $\mathcal{G}$. We define the quantity

$$w(\mathcal{A}, \mathcal{B}) \doteq \sum_{i \in \mathcal{A}, j \in \mathcal{B}} w_{ij} \tag{4.65}$$

as the sum of the weights of all edges connecting the two subgraphs. Then if we cut the graph $\mathcal{G}$ into n disjoint subgraphs, i.e., $\{\mathcal{G}_i\}_{i=1}^n$ such that $\mathcal{G} = \cup_{i=1}^n \mathcal{G}_i$ and $\mathcal{G}_i \cup \mathcal{G}_j = \emptyset$, we may measure the cost of such a cut as the sum of the weights of all the edges connecting one group to all other groups, i.e.,

$$\operatorname{Cut}(\mathcal{G}_1, \mathcal{G}_2, \dots, \mathcal{G}_n) \doteq \frac{1}{2} \sum_{i=1}^n w(\mathcal{G}_i, \mathcal{G}_i^c), \tag{4.66}$$

where $\mathcal{G}_i^c$ is the complement of $\mathcal{G}_i$. One can then formulate the clustering problem as the problem of finding the cut that minimizes the above cost. This problem is known in the literature as the mincut problem. Notice that if the graph has n connected components, there are no edges across different subgraphs, and the optimal value of the mincut is zero.

To minimize the cut, let $U = [\boldsymbol{u}_1, \ldots, \boldsymbol{u}_n]$ be the matrix whose columns are the indicator vectors for each one of the n groups as defined in (4.62). Since $u_{ji} = 1$ when point $\boldsymbol{x}_j$ belongs to subgraph $\mathcal{G}_i$ and $u_{ji} = 0$ otherwise, we have

$$\begin{aligned} \frac{1}{2}\sum_{i=1}^{n} w(\mathcal{G}_i, \mathcal{G}_i^c) &= \frac{1}{2}\sum_{i=1}^{n} \sum_{j\in\mathcal{G}_i, k\in\mathcal{G}_i^c} w_{jk} = \frac{1}{2}\sum_{i=1}^{n}\sum_{j=1}^{N}\sum_{k=1}^{N}(u_{ji} - u_{ki})^2 w_{jk} \\ &= \sum_{i=1}^{n} \boldsymbol{u}_i^\top \mathcal{L} \boldsymbol{u}_i = \operatorname{trace}(U^\top \mathcal{L} U). \end{aligned} \tag{4.67}$$

Therefore, we can formulate the mincut problem as

$$\min_{U\in\{0,1\}^{N\times n}} \operatorname{trace}(U^\top \mathcal{L} U) \quad \text{s.t.} \quad U\mathbf{1} = \mathbf{1}, \tag{4.68}$$

where the constraint enforces that each data point is assigned to one cluster.[13]

However, directly minimizing the cut often results in clusters that consist of a single vertex that has no or few connections with the rest of the graph. To avoid such trivial small clusters, we can instead minimize the so-called "ratiocut" cost

$$\operatorname{ratiocut}(\mathcal{G}_1, \mathcal{G}_2, \ldots, \mathcal{G}_n) \doteq \frac{1}{2}\sum_{i=1}^{n} \frac{w(\mathcal{G}_i, \mathcal{G}_i^c)}{|\mathcal{G}_i|} = \sum_{i=1}^{n} \frac{\operatorname{cut}(\mathcal{G}_i, \mathcal{G}_i^c)}{|\mathcal{G}_i|}, \tag{4.69}$$

which discounts groups with a small number of vertices. We leave it as an exercise to the reader (see Exercise 4.13) to check that if we normalize the indicator vector of each group by its group size as $\boldsymbol{h}_i = \mathbf{1}_{\mathcal{G}_i}/\sqrt{|\mathcal{G}_i|}$, $i = 1, \ldots, n$, then we have

$$\operatorname{ratiocut}(\mathcal{G}_1, \mathcal{G}_2, \ldots, \mathcal{G}_n) = \sum_{i=1}^{n} \boldsymbol{h}_i^\top \mathcal{L} \boldsymbol{h}_i = \operatorname{trace}(H^\top \mathcal{L} H), \tag{4.70}$$

where $H = [\boldsymbol{h}_1, \boldsymbol{h}_2, \ldots, \boldsymbol{h}_n]$. Now notice that the constraint $U\mathbf{1} = \mathbf{1}$ used in (4.68) to prevent the trivial solution $U = \mathbf{0}$ becomes $H\mathbf{1}_{\mathcal{G}_i} = \mathbf{1}_{\mathcal{G}_i}/\sqrt{|\mathcal{G}_i|}$. However, this constraint is hard to enforce, because we do not know $|\mathcal{G}_i|$. Instead, to prevent the trivial solution $H = \mathbf{0}$, we enforce the constraint $H^\top H = I$, which can be easily verified. Therefore, the objective of ratiocut can be rewritten as

$$\min_{H\in\mathcal{H}} \operatorname{trace}(H^\top \mathcal{L} H) \quad \text{s.t.} \quad H^\top H = I, \tag{4.71}$$

[13]This constraint is needed to prevent the trivial solution $U = \mathbf{0}$. Alternatively, we could enforce $U^\top U = \operatorname{diag}(|\mathcal{G}_1|, |\mathcal{G}_2|, \ldots, |\mathcal{G}_n|)$. However, this is impossible, because we do not know $|\mathcal{G}_i|$.

where $\mathcal{H}$ is the space of $N \times n$ matrices whose entries are either 0 or $1/\sqrt{|\mathcal{G}_i|}$. However, optimizing over the space $\mathcal{H}$ is also impossible, because we do not know $|\mathcal{G}_i|$. Thus, a commonly used approximation is to relax the requirement that the columns of H be normalized indicator vectors and instead allow H to be any orthogonal matrix. The optimization problem then becomes

$$\min_{H \in \mathbb{R}^{N \times n}} \operatorname{trace}(H^\top \mathcal{L} H) \quad \text{s.t.} \quad H^\top H = I. \tag{4.72}$$

We leave it as an exercise to the reader to show that the columns of the optimal H^* are exactly the n eigenvectors of $\mathcal{L}$ associated with the n smallest eigenvalues. Once H^* is known, one can further apply the K-means algorithm to cluster the row vectors of H^* to find the n clusters. This leads exactly to the spectral clustering algorithm 4.5 with $Y = H^{*\top}$.

Normalized Cut and Normalized Spectral Clustering

Instead of normalizing an indicator vector of a subgraph $\mathcal{A}$ by its size $|\mathcal{A}|$, we may also normalize it by its volume, which is defined to be

$$\operatorname{Vol}(\mathcal{A}) \doteq \sum_{i \in \mathcal{A}} d_{ii}, \tag{4.73}$$

where d_{ii} is the ith diagonal entry of the degree matrix $\mathcal{D}$. Similar to ratiocut, we may seek a partition of the graph into n components by minimizing the cost:

$$\operatorname{Ncut}(\mathcal{G}_1, \mathcal{G}_2, \ldots, \mathcal{G}_n) \doteq \frac{1}{2} \sum_{i=1}^{n} \frac{w(\mathcal{G}_i, \mathcal{G}_i^c)}{\operatorname{Vol}(\mathcal{G}_i)} = \sum_{i=1}^{n} \frac{\operatorname{cut}(\mathcal{G}_i, \mathcal{G}_i^c)}{\operatorname{Vol}(\mathcal{G}_i)}, \tag{4.74}$$

which discounts groups that have small volume. This objective function is also known as the "normalized cut."

Directly solving the Ncut problem is highly combinatorial. To simplify the expression and the problem, we may scale the indicator vectors by the volume of each subgraph and define

$$\boldsymbol{f}_i = \mathbf{1}_{\mathcal{G}_i} / \sqrt{\operatorname{Vol}(\mathcal{G}_i)}, \quad i = 1, \ldots, n. \tag{4.75}$$

Let $F = \left[\boldsymbol{f}_1, \boldsymbol{f}_2, \ldots \boldsymbol{f}_n\right]$. We leave it to the reader (see Exercise 4.13) to show that

$$\operatorname{Ncut}(\mathcal{G}_1, \mathcal{G}_2, \ldots, \mathcal{G}_n) = \operatorname{trace}(F^\top \mathcal{L} F) \quad \text{and} \quad F^\top \mathcal{D} F = I. \tag{4.76}$$

Therefore, the objective of Ncut can be rewritten as

$$\min_{F \in \mathcal{F}} \operatorname{trace}(F^\top \mathcal{L} F) \quad \text{s.t.} \quad F^\top \mathcal{D} F = I, \tag{4.77}$$

where $\mathcal{F}$ is the space of $N \times n$ matrices whose entries are either 0 or $1/\sqrt{\operatorname{Vol}(\mathcal{G}_i)}$.

Notice that optimizing over the space $\mathcal{F}$ is impossible, because we do not know $\text{Vol}(\mathcal{G}_i)$. However, if we relax the requirement that F consist of scaled indicator vectors and instead allow it to be any real matrix, then we can approximate the solution to the Ncut problem by solving the following optimization problem:

$$\min_{F \in \mathbb{R}^{N \times n}} \text{trace}(F^\top \mathcal{L} F) \quad \text{s.t.} \quad F^\top \mathcal{D} F = I. \tag{4.78}$$

Notice that this is almost the same as the optimization problem (4.48) that we have solved for LE,. The only difference is that in LE, we have the additional constraint $F^\top \mathcal{D}\mathbf{1} = \mathbf{0}$. Recall that the optimal solution for LE is given by the generalized eigenvectors of $(\mathcal{L}, \mathcal{D})$ corresponding to the second- to $(n+1)$st-smallest eigenvalues:

$$\mathcal{L} f = \lambda \mathcal{D} f. \tag{4.79}$$

We leave it as an exercise to the reader to show that the optimal solution F^* to (4.78) consists of the first n generalized eigenvectors of $(\mathcal{L}, \mathcal{D})$, or equivalently, the first n eigenvectors of the matrix $\mathcal{D}^{-1}\mathcal{L}$, since $\mathcal{D}^{-1}\mathcal{L} f = \lambda f$. Alternatively, if we define $T = \mathcal{D}^{1/2} F$, the optimization problem in (4.78) can be rewritten as

$$\min_{T \in \mathbb{R}^{N \times n}} \text{trace}(T^\top \mathcal{D}^{-1/2} \mathcal{L} \mathcal{D}^{-1/2} T) \quad \text{s.t.} \quad T^\top T = I. \tag{4.80}$$

This is almost exactly the same optimization problem as we see in ratiocut (4.72), except that we need to replace $\mathcal{L}$ with $\mathcal{D}^{-1/2}\mathcal{L}\mathcal{D}^{-1/2}$. Unlike $\mathcal{D}^{-1}\mathcal{L}$, the matrix $\mathcal{D}^{-1/2}\mathcal{L}\mathcal{D}^{-1/2}$ is a symmetrically normalized version of the Laplacian. The optimal solution T^* to the above program obviously consists of the first n eigenvectors of $\mathcal{D}^{-1/2}\mathcal{L}\mathcal{D}^{-1/2}$. Then we have $F^* = \mathcal{D}^{-1/2} T^*$.

Given the low-dimensional embedding $Y = F^{*\top}$, we can cluster the data into n groups by applying K-means to the columns of Y. This algorithm is known as the *normalized cut (Ncut)* method in the literature (Shi and Malik 2000), and is summarized in Algorithm 4.6. Notice that the only difference between this algorithm and Algorithm 4.5 is in step 3, where generalized eigenvectors are used instead of eigenvectors.

Algorithm 4.6 (Normalized Cut)

Input: Number of clusters n and affinity matrix $W \in \mathbb{R}^{N \times N}$ for points $\{\boldsymbol{x}_j\}_{j=1}^N$.

1: Construct an affinity graph $\mathcal{G}$ with weight matrix W.
2: Compute the degree matrix $\mathcal{D} = \text{diag}(W\mathbf{1})$ and the Laplacian $\mathcal{L} = \mathcal{D} - W$.
3: Compute the n generalized eigenvectors of $(\mathcal{L}, \mathcal{D})$ associated with its n smallest generalized eigenvalues.
4: Let $\boldsymbol{y}_1, \dots, \boldsymbol{y}_N$ be the columns of $Y \doteq [\boldsymbol{u}_1, \dots, \boldsymbol{u}_n]^\top \in \mathbb{R}^{n \times N}$, where $\{\boldsymbol{u}_i\}_{i=1}^n$ are the eigenvectors in step 3 normalized to unit Euclidean norm.
5: Cluster the points $\{\boldsymbol{y}_j\}_{j=1}^N$ into n groups using the K-means algorithm, Algorithm 4.4.

Output: The segmentation of the data into n groups.

Algorithm 4.7 (Normalized Spectral Clustering)

Input: Number of clusters n and affinity matrix $W \in \mathbb{R}^{N \times N}$ for points $\{x_j\}_{j=1}^N$.
1: Construct an affinity graph $\mathcal{G}$ with weight matrix W.
2: Compute the degree matrix $\mathcal{D} = \text{diag}(W\mathbf{1})$ and the Laplacian $\mathcal{L} = \mathcal{D} - W$.
3: Compute the n eigenvectors of $\mathcal{D}^{-1/2}\mathcal{L}\mathcal{D}^{-1/2}$ associated with its n smallest eigenvalues and normalize so that each row of $[u_1, \dots, u_n]$ has unit norm.
4: Let $y_1, \dots, y_N$ be the columns of $Y \doteq [u_1, \dots, u_n]^\top \in \mathbb{R}^{n \times N}$, where $\{u_i\}_{i=1}^n$ are the eigenvectors in step 3 normalized to unit Euclidean norm.
5: Cluster the points $\{y_j\}_{j=1}^N$ into n groups using the K-means algorithm, Algorithm 4.4.
Output: The segmentation of the data into n groups.

Sometimes Algorithm 4.5 is referred to as *unnormalized spectral clustering*, while Algorithm 4.6 is referred to as *normalized spectral clustering* to emphasize the fact that the former uses the unnormalized Laplacian $\mathcal{L}$, while the latter uses the normalized Laplacian $\mathcal{D}^{-1/2}\mathcal{L}\mathcal{D}^{-1/2}$ (see, e.g., (von Luxburg 2007)). In this book, we will reserve the name *normalized spectral clustering* to refer to yet another form of normalization that has been proposed in the spectral clustering literature. Observe from the relationship $T^* = \mathcal{D}^{1/2}F^*$ that each row of T^* is that of F^* scaled by the square root of the degree of the corresponding vertex. As a result, the rows associated with vertices in the same group do not necessarily have the same scale. It has been proposed in the literature (Ng et al. 2001) that one may normalize the rows of T^* to be of unit length and then cluster the normalized rows by K-means to find the n subgraphs. One of the benefits of such a normalization is to make the cluster centers well separated on the unit sphere: the cluster centers are all mutually orthogonal to each other (as shown in (Ng et al. 2001)). We summarize this method in Algorithm 4.7. This algorithm will be the default spectral clustering algorithm used in most examples in this book. Readers who are interested in a more thorough exposition and comparison of different variants of spectral clustering and want to know more about their relationships are referred the survey paper (von Luxburg 2007).

In summary, the role of the Laplacian $\mathcal{L}$ is to map, through its null space, the original data points $\{x_j\} \subset \mathbb{R}^D$ into a new set of points $\{y_j\} \subset \mathbb{R}^n$ embedded in a low-dimensional space. The original data may have complex mixed structures and deny a simple clustering solution; but the structures of the low-dimensional embedded data become much simpler, clustered around a few cluster centers. Hence they can be grouped by a simple clustering method. Of course, there is no free lunch. The difficulty in clustering the original data needs to be alleviated through the design of a good affinity measure $W = [w_{ij}]$. As can be expected, the performance of the spectral clustering method highly depends on the design of W, as we will see in the example below. In general, there is no theory that characterizes precisely how the choice of the affinity measure influences the resulting clusters. Nevertheless, in Chapter 8 we will see that when the clusters correspond to different low-dimensional subspaces, one can design affinity measures in a principled manner and with good theoretical guarantees.

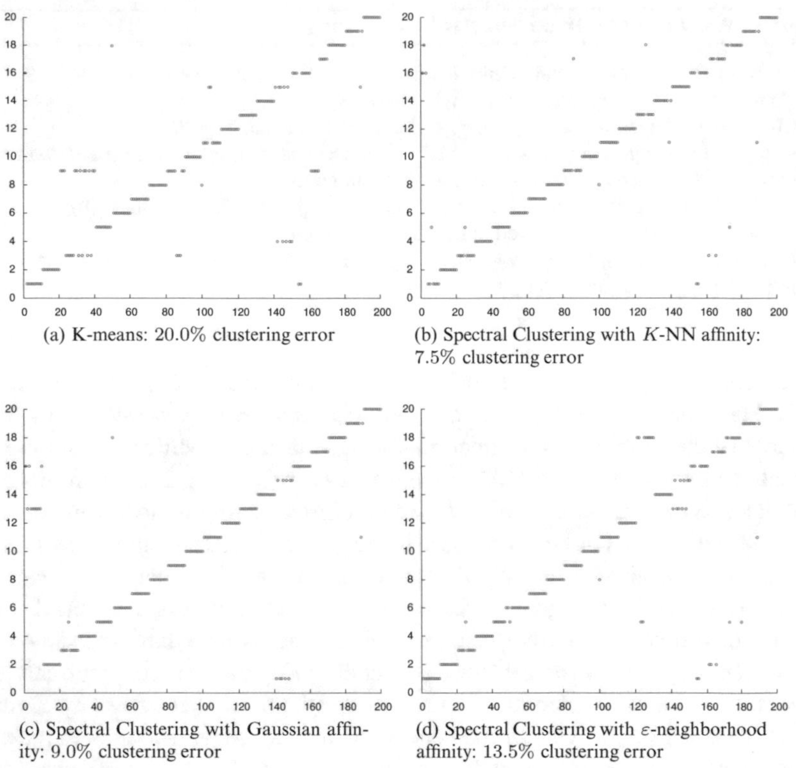

(a) K-means: 20.0% clustering error

(b) Spectral Clustering with K-NN affinity: 7.5% clustering error

(c) Spectral Clustering with Gaussian affinity: 9.0% clustering error

(d) Spectral Clustering with ε-neighborhood affinity: 13.5% clustering error

Fig. 4.12 Face clustering results given by K-means and spectral clustering with different affinity graphs. The data contains 200 face images of 20 individuals from the AT&T data set, and is projected to dimension 50 using PCA. The clustering results given by the algorithms versus the index of the face images (10 images per class) are plotted, and the clustering errors are reported in the captions.

Example 4.23 (Spectral Clustering of Face Images under Varying Pose). In this example, we apply the normalized spectral clustering algorithm, Algorithm 4.7, to the same subset of the AT&T face data set we used in Example 4.17, which consisted of face images of 20 individuals, with 10 poses for each individual. We first apply PCA to the images to reduce the dimension of the data to $D = 50$. We then apply spectral clustering to the PCA coefficients. We test three popular ways of building the affinity matrix for spectral clustering. The first one uses a K-NN affinity in which $w_{ij} = 1$ if point $\boldsymbol{x}_i$ is a K-NN of point $\boldsymbol{x}_j$ and $w_{ij} = 0$ otherwise. In the experiments, we set $K = 8$. The second one uses a Gaussian affinity $w_{ij} = \exp(-\|\boldsymbol{x}_i - \boldsymbol{x}_j\|_2^2/2\sigma^2)$. In order for the parameter σ to be invariant to the scale of the data, data points $\boldsymbol{x}_i$ and $\boldsymbol{x}_j$ are normalized to be of unit ℓ_2 norm. In the experiments, we set $\sigma = 1.0$. For the third one, we use an ε-neighborhood affinity in which $w_{ij} = 1$ if $\|\boldsymbol{x}_i - \boldsymbol{x}_j\|_2 \leq \varepsilon$ and $w_{ij} = 0$ otherwise. Again, we use the normalized data, and choose $\varepsilon = 1.0$. The clustering results of different clustering methods are given in Figure 4.12. For ease of comparison, we also include the results for K-means from Figure 4.11(b). Observe how all variants of the spectral clustering algorithm

Table 4.1 Clustering errors obtained by applying K-means and spectral clustering to subjects (20 and 21, or 37 and 38) from the extended Yale B data set.

Methods	Subjects 20, 21	Subjects 37, 38
K-means	50.0%	47.7%
Spectral clustering with kNN graph	10.2%	42.2%
Spectral clustering with Gaussian graph	48.4%	49.2%
Spectral clustering with neighborhood graph	47.7%	48.4%

improve over the K-means algorithm. In particular, spectral clustering with the K-NN affinity is able to reduce the error from 20.0% to 7.5%. As argued before, this is expected, since spectral clustering is able to discover nonlinear structures in the data and use them to produce a low-dimensional embedding where the clusters can be found more easily.

Example 4.24 (Spectral Clustering of Face Images under Varying Illumination) In this example, we apply the normalized spectral clustering algorithm, Algorithm 4.7, to the same subset of the extended Yale B data set we used in Example 4.18, which consisted of face images of two subjects (20 and 21, or 37 and 38), with 64 different illumination conditions per subject. We first apply PCA to the images to reduce the dimension of data to $D = 20$ and normalize the PCA coefficients to have unit ℓ_2 norm. We then apply spectral clustering to the normalized coefficients. We test the three popular ways of building the affinity matrix for spectral clustering. The first one uses a K-NN affinity in which $w_{ij} = 1$ if point $\boldsymbol{x}_i$ is a K-NN of point $\boldsymbol{x}_j$ and $w_{ij} = 0$ otherwise. In the experiments, we set $K = 4$. The second one uses a Gaussian affinity $w_{ij} = \exp(-\|\boldsymbol{x}_i - \boldsymbol{x}_j\|_2^2/2\sigma^2)$. In the experiments, we set $\sigma = 0.3$. For the third one, we use an ε-neighborhood affinity in which $w_{ij} = 1$ if $\|\boldsymbol{x}_i - \boldsymbol{x}_j\|_2 \leq \varepsilon$ and $w_{ij} = 0$ otherwise. We choose $\varepsilon = 1.2$.

The clustering errors are shown in Table 4.1. For ease of comparison, we also include the results for K-means from Example 4.18. We can see that spectral clustering improves with respect to K-means, since the spectral embedding is able to better capture the geometry of the data. However, the results are still unsatisfactory. This is because face images of one individual under varying illumination live in a subspace. While local Euclidean distances are better at capturing the structure of the data than Euclidean distances, local distances can still fail near the intersections of the subspaces. For example, two points near the intersection could be very close to each other, but be in different subspaces. This happens typically for underlit face images, which are all dark for all individuals and hence close to each other. This suggests the need for better methods to cluster data in multiple subspaces. This will be the subject of Part II of this book.

4.4 Bibliographic Notes

Nonlinear dimensionality reduction (NLDR) refers to the problem of finding a low-dimensional representation for a set of points lying in a nonlinear manifold embedded in a high-dimensional space. This question of how to detect and represent low-dimensional structure in high-dimensional data is fundamental to many disciplines, and several attempts have been made in different areas to address this question. For example, the number of pixels in an image can be rather large, yet most computer vision models use only a few parameters to describe the geometry, photometry, and dynamics of the scene. Since most data sets often have fewer degrees of freedom than the dimension of the ambient space, NLDR is fundamental to many problems in computer vision, machine learning, and pattern recognition.

When the data live in a low-dimensional linear subspace of a high-dimensional space, simple linear methods such as principal component analysis (PCA) (Hotelling 1933) and metric multidimensional scaling (MDS) (Cox and Cox 1994) can be used to learn the subspace and its dimension. However, when the data lie in a low-dimensional submanifold, their structure may be highly nonlinear; hence linear dimensionality reduction methods are likely to fail. This has motivated extensive efforts toward developing NLDR algorithms for computing low-dimensional embeddings. One of the first generalizations of PCA to nonlinear manifolds is the work of (Hastie 1984) and (Hastie and Stuetzle 1989) on principal curves and surfaces. The principal curve of a data set, which generalizes the notion of a principal component, is a curve that passes through the *middle* of the data points and minimizes the sum of squared distances from the data points to the curve. A more general approach, however, is to find a nonlinear embedding map, or equivalently a kernel function, such that the embedded data lie on a linear subspace. Such methods are referred to as nonlinear kernel PCA (Schölkopf et al. 1998; Schölkopf and Smola 2002). A huge family of such algorithms computes a low-dimensional representation from the eigenvectors of a matrix constructed from the local geometry of the manifold. Such algorithms include ISOMAP (Tenenbaum et al. 2000), locally linear embedding (LLE) (Roweis and Saul 2000, 2003), and its variants such as Laplacian eigenmaps (LE) (Belkin and Niyogi 2002), Hessian LLE (Donoho and Grimes 2003), local tangent space alignment (LTSA) (Zhang and Zha 2005), maximum variance unfolding (Weinberger and Saul 2004), and conformal eigenmaps (Sha and Saul 2005). For a survey of many of these algorithms, we refer the reader to (Burges 2005; Lee and Verleysen 2007; Burges 2010).

When the data points are not drawn from a single subspace or submanifold, but from a mixture of multiple low-dimensional structures, the aforementioned methods may fail. The K-means algorithm, which goes back to (Lloyd 1957; Forgy 1965; Jancey 1966; MacQueen 1967), addresses this problem by assuming arguably the simplest model for each cluster: data points in each cluster are distributed around a central point. The K-means algorithm then treats the estimation of multiple models as a "chicken-and-egg" problem, which is solved iteratively by alternating between assigning points to clusters and estimating a center for each cluster.

If each cluster has a more sophisticated structure, more advanced methods are needed. In particular, there exists a very long history and rich literature about the connections between data clustering and graph partitioning methods. The relationships between connectivity of a graph and its Laplacian has been well known as *spectral graph theory*. A standard reference on this topic is (Chung 1997). There are also several normalized spectral clustering methods that aim to give more stable clusters for real data, such as Shi and Malik (Shi and Malik 2000) and Ng, Weiss, and Jordan (Ng et al. 2001). The survey paper of von Luxburg (von Luxburg 2007) gives a more thorough review and comparison of all these methods. A more rigorous statistical justification of the spectral clustering method is given by (Shi et al. 2008), where a more thorough analysis reveals which eigenvectors of the affinity matrix should be used and under what conditions the clustering information can be fully recovered from the data. In Chapter 7 and Chapter 8, we will systematically study how to introduce good affinity matrices so that the spectral method can correctly cluster data that belong to multiple subspaces.

4.5 Exercises

Exercise 4.1 Show that the following functions are positive semidefinite kernels:

1. $\kappa(\boldsymbol{x},\boldsymbol{y}) = \phi(\boldsymbol{x})^\top \phi(\boldsymbol{y})$ for some embedding function $\phi : \mathbb{R}^D \to \mathbb{R}^M$.
2. $\kappa_P(\boldsymbol{x},\boldsymbol{y}) = (\boldsymbol{x}^\top \boldsymbol{y})^n$ for fixed $n \in \mathbb{N}$.
3. $\kappa_G(\boldsymbol{x},\boldsymbol{y}) = \exp\big(-\frac{\|x-y\|^2}{2\sigma^2}\big)$ for fixed $\sigma > 0$.

Exercise 4.2 Consider the polynomial kernel in $[-1,1]^2 \times [-1,1]^2$ defined as $\kappa(\boldsymbol{x},\boldsymbol{y}) = (\boldsymbol{x}^\top \boldsymbol{y})^2 = (x_1y_1 + x_2y_2)^2$. Define the operator

$$\mathcal{L}(f)(\boldsymbol{x}) = \int \kappa(\boldsymbol{x},\boldsymbol{y}) f(\boldsymbol{y}) d\boldsymbol{y}. \tag{4.81}$$

Show that the eigenfunctions of $\mathcal{L}$ corresponding to nonzero eigenvalues are of the form $\psi(\boldsymbol{x}) = c_1x_1^2 + c_2x_1x_2 + c_3x_2^2$. Show that there are three such eigenfunctions, where (c_1, c_2, c_3) and λ are obtained from

$$\begin{bmatrix} 4/5 & 0 & 4/9 \\ 0 & 8/9 & 0 \\ 4/9 & 0 & 4/5 \end{bmatrix} \begin{bmatrix} c_1 \\ c_2 \\ c_3 \end{bmatrix} = \lambda \begin{bmatrix} c_1 \\ c_2 \\ c_3 \end{bmatrix}. \tag{4.82}$$

Exercise 4.3 (Karhunen–Loève Transform) The Karhunen–Loève transform (KLT) can be thought of as a generalization of PCA from a (finite-dimensional) random vector $\boldsymbol{x} \in \mathbb{R}^D$ to an (infinite-dimensional) random process $x(t), t \in \mathbb{R}$, where $x(t)$ is a square-integrable, zero-mean, second-order stationary random process whose auto correlation function is defined as $\kappa(t,\tau) \doteq \mathbb{E}[x(t)x(\tau)]$ for all $t,\tau \in \mathbb{R}$. Assume that for some $C > 0$ and almost every (t,τ), we have $|\kappa(t,\tau)| \leq C$.

1. Show that $\kappa(t,\tau)$ is a positive definite kernel, i.e., show that for all $y(t)$, $\iint y(t)\kappa(t,\tau)y(\tau)dtd\tau \geq 0$.
2. Show that $\kappa(t,\tau)$ has a family of orthonormal eigenfunctions $\{\phi_i(t)\}_{i=1}^{\infty}$ that are defined as

$$\int \kappa(t,\tau)\phi_i(\tau)\, d\tau = \lambda_i\phi_i(t), \quad i = 1, 2, \ldots. \tag{4.83}$$

3. Show that with respect to the eigenfunctions, the original random process can be decomposed as

$$x(t) = \sum_{i=1}^{\infty} x_i\phi_i(t), \tag{4.84}$$

where $\{x_i\}_{i=1}^{\infty}$ is a set of uncorrelated random variables.

Exercise 4.4 (Full Rank of Gaussian RBF Gram Matrices) Suppose that you are given N distinct points $\{\boldsymbol{x}_j\}_{j=1}^N$. If $\sigma \neq 0$, then the matrix $\mathcal{K} \in \mathbb{R}^{N\times N}$ given by

$$\mathcal{K}_{ij} = \exp\left(-\frac{\|\boldsymbol{x}_i - \boldsymbol{x}_j\|^2}{2\sigma^2}\right) \tag{4.85}$$

has full rank.

Exercise 4.5 Let $\{\boldsymbol{x}_j \in \mathbb{R}^D\}_{j=1}^N$ be a set of points that you believe live in a manifold of dimension d. Imagine you have applied PCA, KPCA with kernel κ, and LLE with K-NN to the data. Assume now that you are given a new point $\boldsymbol{x} \in \mathbb{R}^D$ and you wish to find its corresponding point $\boldsymbol{y} \in \mathbb{R}^d$ according to each of the three methods. How would you compute $\boldsymbol{y} \in \mathbb{R}^d$ without applying PCA, KPCA, or LLE from scratch to the $N+1$ points? Under what conditions is the solution you propose equivalent to applying PCA, KPCA, or LLE to the $N+1$ points?

Exercise 4.6 Implement the KPCA algorithm, Algorithm 4.1, for an arbitrary kernel function `kernel.m`. The format of your function should be as follows.

Function	`[y]=kpca(x,d,kernel,params)`
Parameters	
`x`	$D \times N$ matrix whose columns are the data points
`d`	dimension of the low-dimensional embedding
`kernel`	name of the MATLAB function that computes the kernel `k = kernel(x1,x2,params)`
`params`	parameters needed by the kernel function, such as the degree in the polynomial kernel or the standard deviation in the Gaussian kernel
Returned values	
`y`	$d \times N$ matrix containing the projected coordinates
Description	
Computes the kernel principal components of a set of points	

Also implement the functions `k = poly_kernel(x1,x2,n)` for the polynomial kernel $\kappa(\boldsymbol{x}_1, \boldsymbol{x}_2) = (\boldsymbol{x}_1^\top \boldsymbol{x}_2)^n$ and `k = gauss_kernel(x1,x2,sigma)` for the Gaussian kernel $\kappa(\boldsymbol{x}_1, \boldsymbol{x}_2) = \exp(-\|\boldsymbol{x}_1 - \boldsymbol{x}_2\|^2/\sigma^2)$, where `k` $\in \mathbb{R}^{N \times N}$ and `x1, x2` $\in \mathbb{R}^{D \times N}$. Apply your function to the synthetic data generated using the code available at http://www.kernel-machines.org/code/kpca_toy.m.

Exercise 4.7 Implement the LLE algorithm, Algorithm 4.2. The format of your function should be as follows.

Function	`[y]=lle(x,d,K)`
Parameters	
`x`	$D \times N$ matrix whose columns are the data points
`d`	dimension of the low-dimensional embedding
`K`	number of nearest neighbors
Returned values	
`y`	$d \times N$ matrix containing the projected coordinates
Description	
Computes the LLE embedding of a set of points	

Apply your function to the S-curve and Swiss roll data sets generated using the code available at http://www.cs.nyu.edu/~roweis/lle/code/scurve.m and http://www.cs.nyu.edu/~roweis/lle/code/swissroll.m. Compare your results to those obtained using the authors' code of the LLE algorithm, which is available at http://www.cs.nyu.edu/~roweis/lle/code/lle.m.

Exercise 4.8 Implement the LE algorithm, Algorithm 4.3. The format of your function should be as follows.

Function	`[y]=le(x,d,K,sigma)`
Parameters	
`x`	$D \times N$ matrix whose columns are the data points
`d`	dimension of the low-dimensional embedding
`K`	number of nearest neighbors
`sigma`	standard deviation of the Gaussian kernel
Returned values	
`y`	$d \times N$ matrix containing the projected coordinates
Description	
Computes the LE embedding of a set of points	

Apply your function to the S-curve and Swiss roll data sets generated using the code available at http://www.cs.nyu.edu/~roweis/lle/code/scurve.m and http://www.cs.nyu.edu/~roweis/lle/code/swissroll.m

Exercise 4.9 Apply PCA, KPCA, LE, and LLE to the frontal face images of subjects 20 and 21 in the extended Yale B data set to obtain a two-dimensional embedding such as the one shown in Figure 4.9. In each figure, use two different colors to distinguish the images associated with the two different subjects. For KPCA, vary the parameter σ and comment on the effect of this parameter on the resulting embedding. For LLE, vary the parameter K and comment on the effect of this parameter on the resulting embedding. For LE, vary both K and σ and comment on the effect of this parameter on the resulting embedding.

Exercise 4.10 Let $f(\boldsymbol{\mu}_1, \ldots, \boldsymbol{\mu}_n)$ be the objective function in (4.52). Show that the iterations of Algorithm 4.4 are such that $f(\boldsymbol{\mu}_1^{(k+1)}, \ldots, \boldsymbol{\mu}_n^{(k+1)}) \leq f(\boldsymbol{\mu}_1^{(k)}, \ldots, \boldsymbol{\mu}_n^{(k)})$.

Exercise 4.11 [K-Means for Image Segmentation] Apply the K-means algorithm to the segmentation of color (RGB) images. Play with the number of segments and the choice of the window size (i.e., instead of using only the RGB values at the pixel, use also the RGB values in a window of surrounding pixels concatenated as a feature vector).

Exercise 4.12 Let $\tilde{\mathcal{L}}$ be the Laplacian of a graph with n connected components. Let $\mathcal{L}$ be a perturbed graph Laplacian. Show that

$$\lambda_{n+1} \geq \tilde{\lambda}_{n+1} - \|\mathcal{L} - \tilde{\mathcal{L}}\|_2, \tag{4.86}$$

where $\tilde{\lambda}_{n+1}$ and λ_{n+1} are the $(n+1)$st eigenvalues of $\tilde{\mathcal{L}}$ and $\mathcal{L}$, respectively.

Exercise 4.13 Let $\mathcal{G}$ be an undirected weighted graph with vertex set $\mathcal{V} = \{1, 2, \ldots, N\}$, edge set $\mathcal{E} = \{(i,j) : i,j \in \mathcal{V}\}$, and weights $w_{ij} = w_{ji} \geq 0$. Let $W \in \mathbb{R}^{N \times N}$ be the weighted adjacency matrix, $\mathcal{D} \in \mathbb{R}^{N \times N}$ the (diagonal) degree matrix with entries $d_i = \sum_{j=1}^{N} w_{ij}$, and $\mathcal{L} = \mathcal{D} - W \in \mathbb{R}^{N \times N}$ the Laplacian matrix. Let $\mathcal{G}_1, \ldots, \mathcal{G}_n$ be a partition of $\mathcal{G}$, that is $\mathcal{G} = \mathcal{G}_1 \cup \cdots \cup \mathcal{G}_n$, and $\mathcal{G}_i \cap \mathcal{G}_j = \emptyset$, $\forall i \neq j = 1, \ldots, n$.

1. Let $H \in \mathbb{R}^{N\times n}$ be defined as in (4.64). Show that $\boldsymbol{h}_i^\top \mathcal{L}\boldsymbol{h}_i = \frac{\text{Cut}(\mathcal{G}_i, \mathcal{G}_i^c)}{|\mathcal{G}_i|}$ and

$$\text{ratiocut}(\mathcal{G}_1, \cdots, \mathcal{G}_n) \doteq \sum_{i=1}^{n} \frac{\text{Cut}(\mathcal{G}_i, \mathcal{G}_i^c)}{|\mathcal{G}_i|} = \text{trace}(H^\top \mathcal{L} H). \tag{4.87}$$

2. Show that the optimal solution to the relaxed ratiocut problem (4.72) consists of the first n eigenvectors of $\mathcal{L}$.
3. Let $F \in \mathbb{R}^{N\times n}$ be defined as in (4.75). Show that $F^\top \mathcal{D} F = I$ and

$$\text{Ncut}(\mathcal{G}_1, \cdots, \mathcal{G}_n) \doteq \sum_{i=1}^{n} \frac{\text{Cut}(\mathcal{G}_i, \mathcal{G}_i^c)}{\text{Vol}(\mathcal{G}_i)} = \text{trace}(F^\top \mathcal{L} F). \tag{4.88}$$

4. Show that the optimal solution to the relaxed Ncut problem (4.78) consists of the first n eigenvectors of $\mathcal{D}^{-1}\mathcal{L}$.
5. Let $\boldsymbol{y} = [y_1, \ldots, y_N]^\top \in \mathbb{R}^N$. Show that

$$\boldsymbol{y}^\top \mathcal{D}^{-1/2} \mathcal{L} \mathcal{D}^{-1/2} \boldsymbol{y} = \frac{1}{2} \sum_{i=1}^{N} \sum_{j=1}^{N} w_{ij} \Big(\frac{y_i}{\sqrt{d_{ii}}} - \frac{y_j}{\sqrt{d_{jj}}} \Big)^2. \tag{4.89}$$

6. Show that $\mathcal{D}^{-1/2}\mathcal{L}\mathcal{D}^{-1/2}$ is symmetric positive semidefinite.

Exercise 4.14 Implement the spectral clustering, normalized cut, and normalized spectral clustering algorithms, Algorithms 4.5, 4.6, 4.7. The format of your function should be as follows.

Function `[segment]=spectral_clustering(x,n,K,sigma,method)`

Parameters

`x`	$D \times N$ matrix whose columns are the data points
`n`	number of groups
`K`	number of nearest neighbors
`sigma`	standard deviation of the Gaussian kernel
`method`	"unnormalized," "Ncut," or "normalized"

Returned values

`segment`	$1 \times N$ vector containing the group number associated with each data point

Description

Clusters the data using the spectral clustering algorithm

Compare the three variants of spectral clustering on the two circles data set in Figure 4.10(b) for different choices of the parameter. Comment on the effect of K and σ.

Exercise 4.15 Apply the K-means, spectral clustering, normalized cut, and normalized spectral clustering algorithms to the images for individuals 1 and 20 in the

AT&T face data set used in Examples 4.17 and 4.23. Apply PCA to the images to reduce the dimension of the data to $d = 50$ and then apply each algorithm to the PCA coefficients. For each spectral clustering algorithm, use three affinity graphs: K-NN affinity, ε-neighborhood affinity, and Gaussian affinity. Vary the parameters of each method K, ε, and σ and comment on the effect of these parameters on the quality of the clustering results.

4.A Laplacian Eigenmaps: Continuous Formulation

Laplacian eigenmaps (LE) (Belkin and Niyogi 2002) is a popular dimensionality reduction method that aims to capture the geometry of a manifold by finding a low-dimensional representation such that nearby points in the manifold are mapped to nearby points in the low-dimensional embedding. In this chapter, we have seen how such a goal can be achieved for a collection of sample points drawn from the manifold. Nevertheless, the original derivation of LE draws inspiration from a similar goal for embedding a continuous manifold into a (low-dimensional) Euclidean space. To complement the discrete LE method described in this chapter, we describe LE in the continuous setting in this appendix.

In the continuous setting, the goal of LE is to find d functions from a compact manifold $\mathcal{M}$ to the real line $\mathbb{R}$ that preserve *locality*, i.e., functions that map nearby points in the manifold to nearby points on the real line. When $\mathcal{M} = \mathbb{R}^D$, we have that

$$\begin{aligned}|f(\boldsymbol{x}) - f(\boldsymbol{y})| &= |\langle \nabla f(\boldsymbol{x}), (\boldsymbol{x}-\boldsymbol{y})\rangle + o(\|\boldsymbol{x}-\boldsymbol{y}\|)| \\ &\leq \|\nabla f(\boldsymbol{x})\| \, \|\boldsymbol{x}-\boldsymbol{y}\| + o(\|\boldsymbol{x}-\boldsymbol{y}\|).\end{aligned}$$

Therefore, the function f preserves locality when $\|\nabla f(\boldsymbol{x})\|$ is small for all $\boldsymbol{x}$. This suggests choosing $\int \|\nabla f(\boldsymbol{x})\|^2 d\boldsymbol{x}$ as a measure of whether locality is preserved on average.

More generally, let $f : \mathcal{M} \to \mathbb{R}$ be a map from a compact manifold $\mathcal{M}$ to the real line and assume that it is twice differentiable, i.e., $f \in C^2(\mathcal{M})$. We can find a function that maps nearby points in the manifold to nearby points on the real line by solving the following optimization problem:

$$\min_{f \in C^2(\mathcal{M})} \int_{\mathcal{M}} \|\nabla f(\boldsymbol{x})\|^2 \, d\boldsymbol{x} \quad \text{s.t.} \quad \|f\|^2 = \int_{\mathcal{M}} f(\boldsymbol{x})^2 d\boldsymbol{x} = 1, \tag{4.90}$$

where $\nabla f \in T_{\boldsymbol{x}}\mathcal{M}$ is the gradient of f, and the constraint $\|f\| = 1$ is added to prevent the trivial solution $f \equiv 0$. We can solve the above optimization problem using the method of Lagrange multipliers. The Lagrangian function is given by

$$\mathscr{L}(f,\lambda) = \int_{\mathcal{M}} \big(\|\nabla f(\boldsymbol{x})\|^2 + \lambda(f^2(\boldsymbol{x}) - 1)\big) d\boldsymbol{x}. \tag{4.91}$$

Using calculus of variations, we can compute the gradient of $\mathscr{L}$ with respect to f as

$$\nabla_f \mathscr{L} = -2\Delta f + 2\lambda f, \tag{4.92}$$

where Δ is the Laplace–Beltrami operator on $\mathcal{M}$, which can be expressed in tangent coordinates z_i as $\Delta f = \sum_i \frac{\partial^2 f}{\partial z_i^2}$. Setting the gradient to zero, we obtain

$$\Delta f = \lambda f; \tag{4.93}$$

hence f is an eigenfunction of the linear operator Δ with associated eigenvalue λ. Notice that the optimal value of the problem in (4.90) is the associated eigenvalue:

$$\int_{\mathcal{M}} \|\nabla f(\boldsymbol{x})\|^2 \, d\boldsymbol{x} = \int_{\mathcal{M}} \Delta f(\boldsymbol{x}) f(\boldsymbol{x}) \, d\boldsymbol{x} = \lambda \int_{\mathcal{M}} f^2(\boldsymbol{x}) \, d\boldsymbol{x} = \lambda. \tag{4.94}$$

Therefore, we conclude that the function f that solves the optimization problem in (4.90) is the eigenfunction of Δ associated with its smallest eigenvalue. It is easy to see that such an eigenfunction is the constant function $f(\boldsymbol{x}) \equiv c$, which is associated with the zero eigenvalue. This function maps all points in the manifold $\mathcal{M}$ to a single point on the real line $\mathbb{R}$, which is a trivial embedding.

To find a nontrivial embedding, we need to find eigenfunctions associated with nonzero eigenvalues. These eigenfunctions must be orthogonal to the constant function, i.e., their integral must be zero. Therefore, such eigenfunctions must satisfy

$$\Delta f = \lambda f, \quad \int_{\mathcal{M}} f(\boldsymbol{x}) \, d\boldsymbol{x} = 0, \quad \int_{\mathcal{M}} f(\boldsymbol{x})^2 d\boldsymbol{x} = 1. \tag{4.95}$$

Following ideas similar to those of the proof of the PCA theorem in Chapter 2, but adapted to functional spaces, one can show that the optimal d-dimensional embedding is given by the d eigenfunctions of the Laplace–Beltrami operator Δ corresponding to the second- to $(d+1)$st-smallest eigenvalues.

Notice that the Laplacian embedding that we introduced in this chapter is essentially a discrete version of the above continuous Laplacian embedding. In particular, if $\{\boldsymbol{x}_j\}_{j=1}^N$ is a set of points sampled from the manifold $\mathcal{M}$ and $y_j = f(\boldsymbol{x}_j)$ for $j = 1, \ldots, N$, then for appropriately chosen weights w_{ij}, the objective function and the constraints can be discretized as

$$\frac{1}{2} \int_{\mathcal{M}} \|\nabla f(\boldsymbol{x})\|^2 \, d\boldsymbol{x} \approx \frac{1}{2} \sum_{i=1}^{N} \sum_{j=1}^{N} w_{ij}(y_i - y_j)^2 = \boldsymbol{y}^\top \mathcal{L} \boldsymbol{y} \tag{4.96}$$

$$\int_{\mathcal{M}} f(\boldsymbol{x})d\boldsymbol{x} \approx \sum_{j=1}^{N} y_j d_{jj} = \boldsymbol{y}^\top \mathcal{D}\mathbf{1} \tag{4.97}$$

$$\int_{\mathcal{M}} f(\boldsymbol{x})^2 d\boldsymbol{x} \approx \sum_{j=1}^{N} y_j^2 d_{jj} = \boldsymbol{y}^\top \mathcal{D}\boldsymbol{y}, \tag{4.98}$$

where $\mathcal{L} = \mathcal{D} - W$ is the discrete graph Laplacian and $\mathcal{D} = \operatorname{diag}\{d_{jj}\}$ is the diagonal weight matrix $\mathcal{D}$ with $d_{jj} = \sum_i w_{ij}$. Thus, one can see that the constraints we introduced in equation (4.47) are essentially discretized versions of the last two constraints in (4.95), and that the diagonal weight matrix $\mathcal{D}$ in (4.47) precisely corresponds to the density of samples on the manifold according to the measure $d\boldsymbol{x}$.

Part II
Modeling Data with Multiple Subspaces

Chapter 5
Algebraic-Geometric Methods

As long as algebra and geometry have been separated, their progress has been slow and their uses limited; but when these two sciences have been united, they have lent each mutual forces, and have marched together towards perfection.

—Joseph Louis Lagrange

In this chapter, we consider a generalization of PCA in which the given sample points are drawn from an unknown arrangement of subspaces of unknown and possibly different dimensions. We first present a series of simple examples that demonstrate that the subspace clustering problem can be solved noniteratively via certain algebraic methods. These solutions lead to a general-purpose algebrogeometric algorithm for subspace clustering. We conveniently refer to the algorithm as *algebraic subspace clustering* (ASC). To better isolate the difficulties in the general problem, we will develop the algorithm in two steps. The first step is to develop a basic algebraic clustering algorithm by assuming a known number of subspaces; and in the second step, we deal with an unknown number of subspaces and develop a recursive version of the algebraic subspace clustering algorithm. The algorithms in this chapter will be derived under ideal noise-free conditions and assume no probabilistic model. Nevertheless, the algebraic techniques involved are numerically well conditioned, and the algorithms are designed to tolerate moderate amounts of noise. Dealing with large amounts of noise or even outliers will be the subject of Chapter 6 and Chapter 8.

In order to make the material accessible to a larger audience, in this chapter we focus primarily on the development of a (conceptual) algorithm. We leave a more formal study of subspace arrangements and rigorous derivation of all their algebraic properties that support the algorithms of this chapter to Appendix C.

R. Vidal et al., *Generalized Principal Component Analysis*, Interdisciplinary Applied Mathematics 40, DOI 10.1007/978-0-387-87811-9_5

5.1 Problem Formulation of Subspace Clustering

In mathematics (especially in algebraic geometry), a collection of subspaces is formally known as a subspace arrangement:

Definition 5.1 (Subspace Arrangement). *A* subspace arrangement *is defined as a finite collection of n linear subspaces in* $\mathbb{R}^D$*:* $\mathcal{A} \doteq \{S_1, \ldots, S_n\}$*. The union of the subspaces is denoted by* $Z_{\mathcal{A}} \doteq S_1 \cup S_2 \cup \cdots \cup S_n$*.*

For simplicity, we will use the term "subspace arrangement" to refer to both $\mathcal{A}$ and $Z_{\mathcal{A}}$.

Imagine that we are given a set of sample points drawn from an arrangement of an unknown number of subspaces that have unknown and possibly different dimensions. Our goal is to simultaneously estimate these subspaces and cluster the points into their corresponding subspaces. Versions of this problem are known in the literature as *subspace clustering*, *multiple eigenspaces* (Leonardis et al. 2002), and *mixtures of principal component analyzers* (Tipping and Bishop 1999a), among others. To be precise, we will first state the problem that we will study in this chapter, which we refer to as "multiple-subspace clustering," or simply as "subspace clustering," to be suggestive of the problem of fitting multiple subspaces to the data.

Notice that in the foregoing problem statement, we have not yet specified the objective for the optimality of the solution. We will leave the interpretation of that open for now and will delay the definition until the context is more specific. Although the problem seems to be stated in a purely geometric fashion, it is easy to reformulate it in a statistical fashion. For instance, we have assumed here that the subspaces do not have to be orthogonal to each other. In a statistical setting, this is essentially equivalent to assuming that these subspaces are not necessarily uncorrelated. Within each subspace, one can also relate all the geometric and statistical notions associated with "principal components" in the classical PCA: the orthonormal basis chosen for each subspace usually corresponds to a decomposition of the random variable into uncorrelated principal components *conditioned on the subspace*.

5.1.1 Projectivization of Affine Subspaces

Note that a linear subspace always passes through the origin, but an affine subspace does not. So, would the above problem statement lose any generality by restricting it to linear subspaces? The answer to this question is no. In fact, every proper affine subspace in $\mathbb{R}^D$ can be converted to a proper linear subspace in $\mathbb{R}^{D+1}$ by lifting every point of it through the so-called homogeneous coordinates:

Definition 5.2 (Homogeneous Coordinates). *The homogeneous coordinates of a point* $\boldsymbol{x} = [x_1, x_2, \ldots, x_D]^\top \in \mathbb{R}^D$ *are defined as* $[x_1, x_2, \ldots, x_D, 1]^\top$*.*

Given a set of points in an affine subspace, it is easy to prove that their homogeneous coordinates span a linear subspace. More precisely:

Fact 5.3 (Homogeneous Representation of Affine Subspaces) *The homogeneous coordinates of points on a d-dimensional affine subspace in $\mathbb{R}^D$ span a $(d + 1)$-dimensional linear subspace in $\mathbb{R}^{D+1}$. This representation is one-to-one.*

Figure 5.1 shows an example of the homogeneous representation of three lines in $\mathbb{R}^2$. The points on these lines span three linear planes in $\mathbb{R}^3$ that pass through the origin.

Definition 5.4 (Central Subspace Arrangements). *We say that an arrangement of subspaces is* central *if every subspace passes through the origin, i.e., every subspace is a linear subspace.*

According to this definition, the homogeneous representation of an (affine) subspace arrangement in $\mathbb{R}^D$ gives a central subspace arrangement in $\mathbb{R}^{D+1}$. Therefore, Problem 5.1 does not lose any generality. From now on, we may assume that our data set is drawn from a central subspace arrangement, in which all subspaces are linear, not affine, subspaces, unless otherwise stated. In a statistical setting, this is equivalent to assuming that each subset of samples has zero mean.

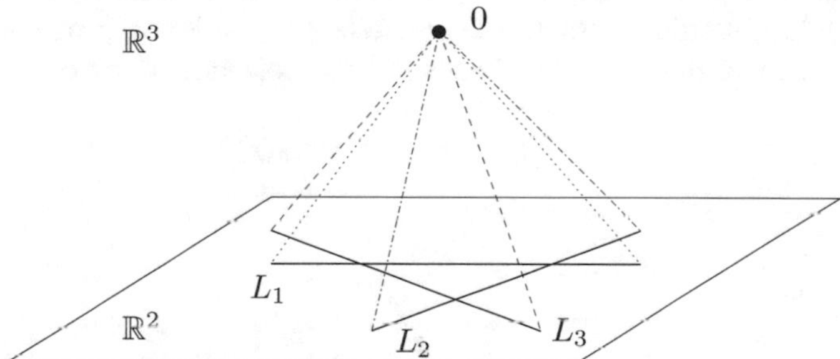

Fig. 5.1 Lifting of three (affine) lines in $\mathbb{R}^2$ to three linear subspaces in $\mathbb{R}^3$ via the homogeneous representation.

Problem 5.1 (Multiple-Subspace Clustering).

Given a set of sample points $\mathcal{X} = \{\boldsymbol{x}_j \in \mathbb{R}^D\}_{j=1}^N$ drawn from $n \geq 1$ distinct linear subspaces $S_i \subset \mathbb{R}^D$ of dimensions $d_i < D$, $i = 1, 2, \ldots, n$, identify each subspace S_i without knowing which sample points belong to which subspace. More specifically, by identifying the subspaces, we mean the following:

1. Identifying the number of subspaces n and their dimensions $d_i = \dim(S_i)$;
2. Identifying an orthonormal basis for each subspace S_i (or equivalently a basis for its orthogonal complement $S_i^{\perp}$);
3. Clustering the N points into the subspaces to which they belong.

5.1.2 Subspace Projection and Minimum Representation

The are many cases in which the given data points live in a space of very high dimension. For instance, in many computer vision problems, the dimension of the ambient space D is the number of pixels in an image, which is normally in the range 10^6. In such cases, the complexity of any subspace clustering solution becomes computationally prohibitive. It is therefore important for us to seek situations in which the dimension of the ambient space can be significantly reduced.

Fortunately, in most practical applications, we are interested in modeling the data by subspaces of relatively small dimensions ($d \ll D$). Thus one can avoid dealing with high-dimensional data sets by first projecting them onto a lower-dimensional (sub)space. An example is shown in Figure 5.2, where two lines L_1 and L_2 in $\mathbb{R}^3$ are projected onto a plane P. In this case, clustering the two lines in the three-dimensional space $\mathbb{R}^3$ is equivalent to clustering the two projected lines in the two-dimensional plane P.

In general, we will distinguish between two different kinds of "projections." The first kind corresponds to the case in which the span of all the subspaces is a proper subspace of the ambient space, i.e., $\operatorname{span}(\cup_{i=1}^n S_i) \subset \mathbb{R}^D$. In this case, one may simply apply PCA (Chapter 2) to eliminate the redundant dimensions. The second kind corresponds to the case in which the largest dimension of the subspaces, denoted by $d_{\max}$, is strictly less than $D-1$. When $d_{\max}$ is known,[1] one may choose a $(d_{\max}+1)$-dimensional subspace P such that by projecting $\mathbb{R}^D$ onto this subspace,

$$\pi_P : \boldsymbol{x} \in \mathbb{R}^D \quad \mapsto \quad \boldsymbol{x}' = \pi_P(\boldsymbol{x}) \in P, \tag{5.1}$$

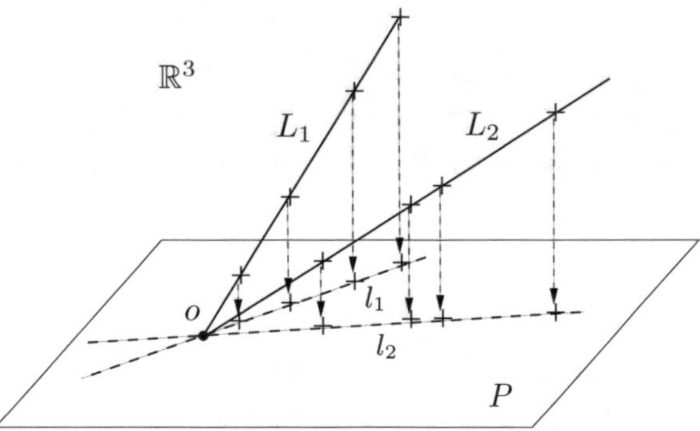

Fig. 5.2 Samples on two 1-dimensional subspaces L_1, L_2 in $\mathbb{R}^3$ projected onto a 2-dimensional plane P. The number and separation of the lines is preserved by the projection.

[1] For example, in 3D motion segmentation from affine cameras, it is known that the subspaces have dimension at most four (Costeira and Kanade 1998; Kanatani 2001; Vidal and Hartley 2004).

the dimension of each original subspace S_i is preserved,[2] and there is a one-to-one correspondence between S_i and its projection—no reduction in the number of subspaces n,[3] as stated in the following theorem.

Theorem 5.5 (Segmentation-Preserving Projections). *If a set of vectors $\{\boldsymbol{x}_j\}$ all lie in n linear subspaces of dimensions $\{d_i\}_{i=1}^n$ in $\mathbb{R}^D$, and if π_P represents a linear projection onto a subspace P of dimension D', then the points $\{\pi_P(\boldsymbol{x}_j)\}$ lie in at most n linear subspaces of P of dimensions $\{d_i'(\leq d_i)\}_{i=1}^n$. Furthermore, if $D > D' > d_{\max}$, then there is an open and dense set of projections that preserve the number and dimensions of the subspaces.*

Thanks to Theorem 5.5, if we are given a data set X drawn from an arrangement of low-dimensional subspaces in a high-dimensional space, we can first project X onto a generic subspace of dimension $D' = d_{\max} + 1$ and then model the data with a subspace arrangement in the projected subspace, as illustrated by the following sequence of steps:

$$X \subset \mathbb{R}^D \xrightarrow{\ \pi_P\ } X' \subset P \longrightarrow \cup_{i=1}^n \pi_P(S_i) \xrightarrow{\ \pi_P^{-1}\ } \cup_{i=1}^n S_i. \tag{5.2}$$

However, even though the set of $(d_{\max}+1)$-dimensional subspaces $P \subset \mathbb{R}^D$ that preserve the separation and dimension of the subspaces is an open and dense set, it remains unclear as to what a "good" choice for P is, especially when there is noise in the data. For simplicity, one may randomly select a few projections and choose the one that results in the smallest fitting error. Another alternative is to apply PCA regardless and project the data onto the $(d_{\max}+1)$-dimensional principal subspace.

One solution for choosing P is attributed to (Broomhead and Kirby 2000). The technique was originally designed for dimension reduction of differential manifolds.[4] We here adopt it for subspace arrangements. Instead of directly using the original data matrix $X = [\boldsymbol{x}_1, \boldsymbol{x}_2, \ldots, \boldsymbol{x}_N]$, we gather the vectors (also called "secants") defined by every pair of points $\boldsymbol{x}_i, \boldsymbol{x}_j \in X$,

$$\boldsymbol{y}_{ij} \doteq \boldsymbol{x}_i - \boldsymbol{x}_j \quad \in \mathbb{R}^D, \tag{5.3}$$

and construct a matrix consisting of $\boldsymbol{y}_{ij}$ as columns:

$$Y \doteq [\boldsymbol{y}_{12}, \boldsymbol{y}_{13}, \ldots, \boldsymbol{y}_{(N-1)N}] \quad \in \mathbb{R}^{D \times M}, \tag{5.4}$$

---

[2]This requires that P be transversal to each $S_i^{\perp}$, i.e., $\text{span}\{P, S_i^{\perp}\} = \mathbb{R}^D$ for every $i = 1, 2, \ldots, n$. Since n is finite, this transversality condition can be easily satisfied. Furthermore, the set of positions for P that violate the transversality condition is only a zero-measure closed set (Hirsch 1976).

[3]This requires that all $\pi_P(S_i)$ be transversal to each other in P, which is guaranteed if we require P to be transversal to $S_i^{\perp} \cap S_{i'}^{\perp}$ for $i, i' = 1, 2, \ldots, n$. All P's that violate this condition form again only a zero-measure set.

[4]It is essentially based on Whitney's classical proof of the fact that every differential manifold can be embedded in a Euclidean space.

where $M = (N-1)N/2$. Then the principal components of Y span the subspace in which the distance (and hence the separateness) between the projected points is preserved the most. Therefore, the optimal subspace that maximizes the separateness of the projected points is given by the $d_{\max}+1$ principal components of Y. More precisely, if $Y = U\Sigma V^{\top}$ is the SVD of Y, then the optimal subspace P is given by the first $d_{\max}+1$ columns of U.

5.2 Introductory Cases of Subspace Clustering

Notice that to apply the K-subspaces and EM algorithms, we need to know three things in advance: the number of subspaces, their dimensions, and initial estimates of the bases of the subspaces. In practice, this may not be the situation, and many difficulties may arise. The optimizing process in both algorithms is essentially a local iterative descent scheme. If the initial estimates of the bases of the subspaces are far off from the global optimum, the process is likely to converge to a local minimum. More seriously, if the number of subspaces and their dimensions were wrong, the process might never converge or might converge to meaningless solutions. Furthermore, when the number and dimensions of the subspaces are unknown and the samples are noisy (or contaminated by outliers), model selection becomes a much more elusive problem, as we have alluded to earlier in the introductory chapter.

In this and the next few chapters, we will systematically address these difficulties and aim to arrive at global noniterative solutions to subspace clustering that require less or none of the above initial information. Before we delve into the most general case, we first examine, in this section, a few important special cases. The reason is twofold: firstly, many practical problems fall into these cases already and the simplified solutions can be directly applied; and secondly, the analysis of these special cases offers some insights into a solution to the general case.

5.2.1 Clustering Points on a Line

Let us begin with an extremely simple clustering problem: clustering a collection of points $\{x_1, x_2, \ldots, x_N\}$ on the real line $\mathbb{R}$ around a collection of cluster centers $\{\mu_1, \mu_2, \ldots, \mu_n\}$. In spite of its simplicity, this problem shows up in various clustering problems. For instance, in intensity-based image segmentation, one wants to separate the pixels of an image into different regions, with each region corresponding to a significantly different level of intensity (a one-dimensional quantity). More generally, the point clustering problem is very much at the heart of popular clustering techniques such as K-means and spectral clustering for clustering data in spaces of any dimension (which we have discussed at the end of Chapter 4). Furthermore, as we will see throughout this chapter (and the book), the same basic ideas introduced through this simple example can also be applied to clustering points from arrangements of more complex structures such as lines, hyperplanes, subspaces, and even surfaces.

In the sequel, we introduce a not so conventional solution to the point clustering problem. The new formulation on which the solution is based is neither geometric (like K-subspaces) nor statistical (like EM). Instead, the solution is purely *algebraic*.

Let $x \in \mathbb{R}$ be any of the data points. In an ideal situation in which each data point perfectly matches one of the cluster centers, we know that there exists a constant μ_i such that $x = \mu_i$. This means that

$$(x = \mu_1) \vee (x = \mu_2) \vee \cdots \vee (x = \mu_n). \tag{5.5}$$

The $\vee$ in the preceding equation stands for the logical connective "or." This is equivalent to that x satisfies the following polynomial equation of degree n in x:

$$p_n(x) \doteq (x - \mu_1)(x - \mu_2)\cdots(x - \mu_n) = \sum_{k=0}^{n} c_k x^{n-k} = 0. \tag{5.6}$$

Since the polynomial equation $p_n(x) = 0$ must be satisfied by every data point, we have that

$$\boldsymbol{V}_n\, \boldsymbol{c}_n \doteq \begin{bmatrix} x_1^n & x_1^{n-1} & \cdots & x_1 & 1 \\ x_2^n & x_2^{n-1} & \cdots & x_2 & 1 \\ \vdots & \vdots & & \vdots & \vdots \\ x_N^n & x_N^{n-1} & \cdots & x_N & 1 \end{bmatrix} \begin{bmatrix} 1 \\ c_1 \\ \vdots \\ c_n \end{bmatrix} = 0, \tag{5.7}$$

where $\boldsymbol{V}_n \in \mathbb{R}^{N\times(n+1)}$ is a matrix of embedded data points, and $\boldsymbol{c}_n \in \mathbb{R}^{n+1}$ is the vector of coefficients of $p_n(x)$.

In order to determine the number of groups n and then the vector of coefficients $\boldsymbol{c}_n$ from (5.7), notice that for n groups, there is a unique polynomial of degree n whose roots are the n cluster centers. Since the coefficients of this polynomial must satisfy equation (5.7), in order to have a unique solution we must have that $\operatorname{rank}(\boldsymbol{V}_n) = n$. This rank constraint on $\boldsymbol{V}_n \in \mathbb{R}^{N\times(n+1)}$ enables us to determine the number of groups n as[5]

$$n \doteq \min\{i : \operatorname{rank}(\boldsymbol{V}_i) = i\}. \tag{5.8}$$

Example 5.6 (Two Clusters of Points). The intuition behind this formula is as follows. Consider, for simplicity, the case of $n = 2$ groups, so that

[5]Notice that the minimum number of points needed is $N \geq n$, which is *linear* in the number of groups. We will see in future chapters that this is no longer the case for more general clustering problems.

Algorithm 5.1 (Algebraic Point Clustering Algorithm).

Let $\mathcal{X} = \{x_j\}_{j=1}^N \subset \mathbb{R}$ be a given collection of $N \geq n$ points clustering around an unknown number n of cluster centers $\{\mu_i\}_{i=1}^n$. The number of clusters, the cluster centers, and the clustering of the data can be determined as follows:

1. **Number of Clusters**. Let $V_i \in \mathbb{R}^{N \times (i+1)}$ be a matrix containing the last $i+1$ columns of V_n. Determine the number of clusters as
$$n \doteq \min\{i : \operatorname{rank}(V_i) = i\}.$$
2. **Cluster Centers.** Solve for c_n from $V_n c_n = 0$. Set $p_n(x) = \sum_{k=0}^n c_k x^{n-k}$. Find the cluster centers μ_i as the n roots of $p_n(x)$.
3. **Clustering.** Assign point x_j to cluster $i = \arg\min_{l=1,\dots,n}(x_j - \mu_l)^2$.

$p_n(x) = p_2(x) = (x - \mu_1)(x - \mu_2)$, with $\mu_1 \neq \mu_2$. Then it is clear that there is no polynomial equation $p_1(x) = x - \mu$ of degree one that is satisfied by *all* the points. Similarly, there are infinitely many polynomial equations of degree 3 or more that are satisfied by all the points, namely any multiple of $p_2(x)$. Thus the degree $n = 2$ is the only one for which there is a unique polynomial that fits all the points.

Once the minimum polynomial $p_n(x)$ that fits all the data points is found, we can solve the equation $p_n(x) = 0$ for its n roots. These roots, by definition, are the centers of the clusters. We summarize the overall solution as Algorithm 5.1.

Notice that the above algorithm is described in a purely algebraic fashion and is more of a conceptual than practical algorithm. It does not minimize any geometric errors or maximize any probabilistic likelihood functions. In the presence of noise in the data, one has to implement each step of the algorithm in a numerically more stable and statistically more robust way. For example, with noisy data, the matrix V_n will most likely be of full rank. In this case, the vector of coefficients c_n should be solved in a least-squares sense as the singular vector of V_n associated with the smallest singular value. It is also possible that the $p_n(x)$ obtained from c_n may have some complex roots, because the constraint that the polynomial must have real roots is never enforced in solving for the coefficients in the least-squares sense.[6] In practice, for well-separated clusters with moderate noise, the roots normally give decent estimates of the cluster centers.

Although clustering points on a line may seem a rather simple problem, it can be easily generalized to the problem of clustering points in a plane (see Exercise 5.1). Furthermore, it is also a key step of a very popular data clustering algorithm: *spectral clustering*. See Exercise 5.2.

[6]However, in some special cases, one can show that this will never occur. For example, when $n = 2$, the least-squares solution for c_n is $c_2 = \operatorname{Var}[x]$, $c_1 = E[x^2]E[x] - E[x^3]$ and $c_0 = E[x^3]E[x] - E[x^2]^2 \leq 0$; hence $c_1^2 - 4c_0c_2 \geq 0$, and the two roots of the polynomial $c_0x^2 + c_1x + c_2$ are always real.

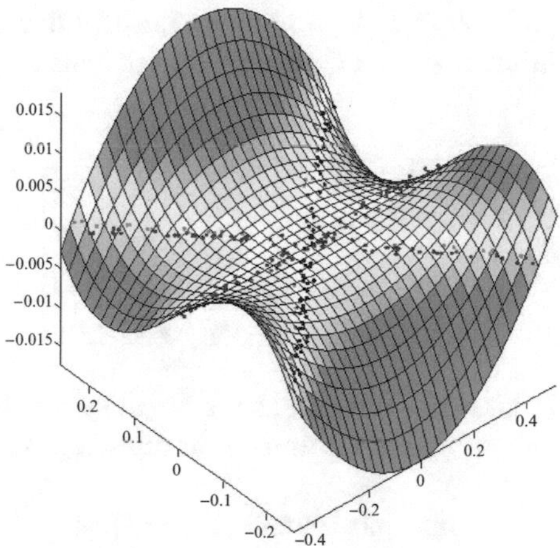

Fig. 5.3 A polynomial in two variables whose zero set is three lines in $\mathbb{R}^2$.

5.2.2 Clustering Lines in a Plane

Let us now consider the case of clustering data points to a collection of n lines in $\mathbb{R}^2$ passing through the origin, as illustrated in Figure 5.3. Each of the lines can be represented as

$$L_i \doteq \{\boldsymbol{x} = [x, y]^\top : b_{i1}x + b_{i2}y = 0\}, \quad i = 1, 2, \ldots, n. \tag{5.9}$$

Given a point $\boldsymbol{x} = [x, y]^\top$ on one of the lines, we must have that

$$(b_{11}x + b_{12}y = 0) \vee \cdots \vee (b_{n1}x + b_{n2}y = 0). \tag{5.10}$$

Therefore, even though each individual line is described with one polynomial equation of degree one (a linear equation), an arrangement of n lines can be described with a polynomial of degree n, namely

$$p_n(\boldsymbol{x}) = (b_{11}x + b_{12}y) \cdots (b_{n1}x + b_{n2}y) = \sum_{k=0}^{n} c_k x^{n-k} y^k = 0. \tag{5.11}$$

An example is shown in Figure 5.3.

The polynomial $p_n(\boldsymbol{x})$ allows us to algebraically eliminate the clustering of the data at the beginning of the model estimation, because the equation $p_n(\boldsymbol{x}) = 0$ is satisfied by every data point regardless of whether it belongs to L_1, L_2, ..., or L_n. Furthermore, even though $p_n(\boldsymbol{x})$ is nonlinear in each data point $\boldsymbol{x} = [x, y]^\top$, $p_n(\boldsymbol{x})$ is actually linear in the vector of coefficients $\boldsymbol{c} = [c_0, c_1, \ldots, c_n]^\top$. Therefore, given enough data points $\{\boldsymbol{x}_j = [x_j, y_j]^\top\}_{j=1}^N$, one can linearly fit this polynomial to the data. Indeed, if n is known, we can obtain the coefficients of $p_n(\boldsymbol{x})$ by solving the equation

$$\boldsymbol{V}_n\boldsymbol{c}_n = \begin{bmatrix} x_1^n & x_1^{n-1}y_1 & \cdots & x_1y_1^{n-1} & y_1^n \\ x_2^n & x_2^{n-1}y_2 & \cdots & x_2y_2^{n-1} & y_2^n \\ \vdots & \vdots & & \vdots & \vdots \\ x_N^n & x_N^{n-1}y_N & \cdots & x_Ny_N^{n-1} & y_N^n \end{bmatrix} \begin{bmatrix} c_0 \\ c_1 \\ \vdots \\ c_n \end{bmatrix} = 0. \tag{5.12}$$

Similar to the case of points in a line, the above linear system has a unique solution if and only if $\mathrm{rank}(\boldsymbol{V}_n) = n$; hence the number of lines is given by

$$n \doteq \min\{i : \mathrm{rank}(\boldsymbol{V}_i) = i\}. \tag{5.13}$$

Given the vector of coefficients $\boldsymbol{c}_n$, we are now interested in estimating the equations of each line from the associated polynomial $p_n(\boldsymbol{x})$. We know that each line is determined by its normal vector $\boldsymbol{b}_i = [b_{i1}, b_{i2}]^\top$, $i = 1, 2, \ldots, n$. For the sake of simplicity, let us consider the case $n = 2$. A simple calculation shows that the derivative of $p_2(\boldsymbol{x})$ is given by

$$\nabla p_2(\boldsymbol{x}) = (b_{21}x + b_{22}y)\boldsymbol{b}_1 + (b_{11}x + b_{12}y)\boldsymbol{b}_2. \tag{5.14}$$

Therefore, if the point $\boldsymbol{x}$ belongs to L_1, then $(b_{11}x + b_{12}y) = 0$, and hence $\nabla p_2(\boldsymbol{x}) \sim \boldsymbol{b}_1$. Similarly, if $\boldsymbol{x}$ belongs to L_2, then $\nabla p_2(\boldsymbol{x}) \sim \boldsymbol{b}_2$. This means that given any point $\boldsymbol{x}$, without knowing which line contains the point, we can obtain the equation of the line passing through the point by simply evaluating the derivative of $p_2(\boldsymbol{x})$ at $\boldsymbol{x}$. This fact should come as no surprise and is valid for any number of lines n. Therefore, if we are given one point $\{\boldsymbol{y}_i \in L_i\}$ on each line,[7] we can determine the normal vectors as $\boldsymbol{b}_i \sim \nabla p_n(\boldsymbol{y}_i)$. We summarize the overall solution for clustering points to multiple lines as Algorithm 5.2.

The reader may have realized that the problem of clustering points on a line is very much related to the problem of clustering lines in the plane. In point clustering, for each data point x there exists a cluster center μ_j such that $x - \mu_i = 0$. By working in homogeneous coordinates, one can convert it into a line clustering problem: for each data point $\boldsymbol{x} = [x, 1]^\top$, there is a line $\boldsymbol{b}_i = [1, -\mu_i]^\top$ passing through the point. Figure 5.4 shows an example of how three cluster centers are converted into three lines via homogeneous coordinates. Indeed, notice that if we let $y = 1$ in the matrix $\boldsymbol{V}_n$ in (5.12), we obtain exactly the matrix $\boldsymbol{V}_n$ in (5.7). Therefore, the vector of coefficients $\boldsymbol{c}_n$ is the same for both algorithms, and the two polynomials are related by $p_n(x, y) = y^n p_n(x/y)$. Therefore, the point clustering problem can be solved either by polynomial factorization (Algorithm 5.1) or by polynomial differentiation (Algorithm 5.2).

[7] We will discuss in the next subsection how to automatically obtain one point per subspace from the data when we generalize this problem to clustering points on hyperplanes.

Algorithm 5.2 (Algebraic Line Clustering Algorithm).

Let $\mathcal{X} = \{x_j\}_{j=1}^N$ be a collection of $N \geq n$ points in $\mathbb{R}^2$ clustering around an unknown number n of lines whose normal vectors are $\{b_i\}_{i=1}^n$. The number of lines, the normal vectors, and the clustering of the data can be determined as follows:

1. **Number of Lines**. Let V_i be defined as in (5.12). Determine the number of clusters as
$$n \doteq \min\{i : \operatorname{rank}(V_i) = i\}.$$
2. **Normal Vectors.** Solve for c_n from $V_n c_n = 0$ and set $p_n(x, y) = \sum_{k=0}^n c_k x^{n-k} y^k$. Determine the normal vectors as
$$b_i = \frac{\nabla p_n(y_i)}{\|\nabla p_n(y_i)\|} \in \mathbb{R}^2, \qquad i = 1, 2, \ldots, n,$$
where y_i is a point in the ith line.
3. **Clustering.** Assign point x_j to line $i = \arg\min_{\ell=1,\ldots,n} (b_\ell^\top x_j)^2$.

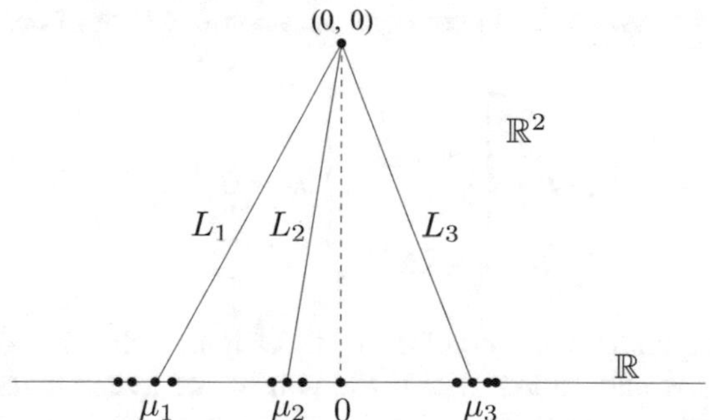

Fig. 5.4 Using homogeneous coordinates to convert the point clustering problem into the line clustering problem.

5.2.3 Clustering Hyperplanes

In this section, we consider another particular case of Problem 5.1 in which all the subspaces are hyperplanes of equal dimension $d_1 = \cdots = d_n = d = D - 1$. This case shows up in a wide variety of clustering and segmentation problems in computer vision, including vanishing point detection and motion segmentation. We will discuss these applications in greater detail in later chapters.

We start by noticing that every $(D-1)$-dimensional subspace $S_i \subset \mathbb{R}^D$ can be defined in terms of a nonzero *normal* vector $b_i \in \mathbb{R}^D$ as follows:[8]

$$S_i \doteq \left\{x \in \mathbb{R}^D : b_i^\top x \doteq b_{i1}x_1 + b_{i2}x_2 + \cdots + b_{iD}x_D = 0\right\}. \tag{5.15}$$

[8] Since the subspaces S_i are all different from each other, we assume that the normal vectors $\{b_i\}_{i=1}^n$ are pairwise linearly independent.

Therefore, a point $\boldsymbol{x} \in \mathbb{R}^D$ lying in one of the hyperplanes S_i must satisfy the formula

$$(\boldsymbol{b}_1^\top \boldsymbol{x} = 0) \vee (\boldsymbol{b}_2^\top \boldsymbol{x} = 0) \vee \cdots \vee (\boldsymbol{b}_n^\top \boldsymbol{x} = 0), \tag{5.16}$$

which is equivalent to the following homogeneous polynomial of degree n in $\boldsymbol{x}$ with real coefficients:

$$p_n(\boldsymbol{x}) = \prod_{i=1}^{n} (\boldsymbol{b}_i^\top \boldsymbol{x}) = \sum c_{n_1,n_2,\ldots,n_D} x_1^{n_1} x_2^{n_2} \cdots x_D^{n_D} = \nu_n(\boldsymbol{x})^\top \boldsymbol{c}_n = 0, \tag{5.17}$$

where $c_{n_1,\ldots,n_D} \in \mathbb{R}$ represents the coefficient of the monomial $x_1^{n_1} x_2^{n_2} \cdots x_D^{n_D}$, $\boldsymbol{c}_n$ is the vector of all coefficients, and $\nu_n(\boldsymbol{x})$ is the stack of all possible monomials, known as the Veronese map of $\boldsymbol{x}$ (see Appendix C for a more formal introduction). The number of linearly independent monomials is $M_n \doteq \binom{D+n-1}{n}$; hence $\boldsymbol{c}_n$ and $\nu_n(\boldsymbol{x})$ are vectors in $\mathbb{R}^{M_n}$.

After applying (5.17) to the given collection of N sample points $\{\boldsymbol{x}_j\}_{j=1}^N$, we obtain the following system of linear equations on the vector of coefficients: $\boldsymbol{c}_n$

$$\boldsymbol{V}_n \, \boldsymbol{c}_n \doteq \begin{bmatrix} \nu_n(\boldsymbol{x}_1)^\top \\ \nu_n(\boldsymbol{x}_2)^\top \\ \vdots \\ \nu_n(\boldsymbol{x}_N)^\top \end{bmatrix} \boldsymbol{c}_n = 0 \quad \in \mathbb{R}^N. \tag{5.18}$$

We now study under what conditions we can solve for n and $\boldsymbol{c}_n$ from equation (5.18). To this end, notice that if the number of hyperplanes n was known, we could immediately recover $\boldsymbol{c}_n$ as the eigenvector of $\boldsymbol{V}_n^\top \boldsymbol{V}_n$ associated with its smallest eigenvalue. However, since the above linear system (5.18) depends explicitly on the number of hyperplanes n, we cannot estimate $\boldsymbol{c}_n$ directly without knowing n in advance. Recall from Example C.30 that the vanishing ideal I of a hyperplane arrangement is always principal, i.e., generated by a single polynomial of degree n. The number of hyperplanes n then coincides with the degree of the first nontrivial homogeneous component I_n of the vanishing ideal. This leads to the following theorem.

Theorem 5.7 (Number of Hyperplanes). *Assume that a collection of $N \geq M_n - 1$ sample points $\{\boldsymbol{x}_j\}_{j=1}^N$ on n different $(D-1)$-dimensional subspaces of $\mathbb{R}^D$ is given. Let $\boldsymbol{V}_i \in \mathbb{R}^{N \times M_i}$ be the matrix defined as in (5.18), but computed with polynomials of degree i. If the sample points are in general position and at least $D-1$ points correspond to each hyperplane, then*

$$\operatorname{rank}(\boldsymbol{V}_i) \begin{cases} = M_i & i < n, \\ = M_i - 1 & i = n, \\ < M_i - 1 & i > n. \end{cases} \tag{5.19}$$

Therefore, the number n of hyperplanes is given by

$$n = \min\{i : \operatorname{rank}(V_i) = M_i - 1\}. \tag{5.20}$$

In the presence of noise, one cannot directly estimate n from (5.20), because the matrix V_i is always of full rank. In this case, one can use the model selection criteria introduced in Chapter 2 to determine the rank.

Theorem 5.7 and the linear system in equation (5.18) allow us to determine the number of hyperplanes n and the vector of coefficients c_n, respectively, from sample points $\{x_j\}_{j=1}^N$. The rest of the problem now becomes how to recover the normal vectors $\{b_i\}_{i=1}^n$ from c_n. Imagine, for the time being, that we were given a set of n points $\{y_i\}_{i=1}^n$, each one lying in only one of the n hyperplanes, that is, $y_i \in S_i$ for $i = 1, 2, \ldots, n$. Now let us consider the derivative of $p_n(x)$ evaluated at each y_i. We have

$$\nabla p_n(x) = \frac{\partial p_n(x)}{\partial x} = \frac{\partial}{\partial x} \prod_{i=1}^n (b_i^\top x) = \sum_{i=1}^n (b_i) \prod_{\ell \neq i} (b_\ell^\top x). \tag{5.21}$$

Because $\prod_{\ell \neq m} (b_\ell^\top y_i) = 0$ for $i \neq m$, one can obtain each of the normal vectors as

$$b_i = \frac{\nabla p_n(y_i)}{\|\nabla p_n(y_i)\|}, \quad i = 1, 2, \ldots, n. \tag{5.22}$$

Therefore, if we know one point in each one of the hyperplanes, the hyperplane clustering problem can be solved analytically by simply evaluating the partial derivatives of $p_n(x)$ at each of the points with known labels.

Consider now the case in which we do not know the membership of any of the data points. We now show that one can obtain one point per hyperplane by intersecting a random line with each of the hyperplanes. To this end, consider a random line $L \doteq \{tv + x_0,\ t \in \mathbb{R}\}$ with direction v and base point x_0. We can obtain one point in each hyperplane by intersecting L with the union of all the hyperplanes.[9] Since at the intersection points we must have $p_n(tv + x_0) = 0$, the n points $\{y_i\}_{i=1}^n$ can be obtained as

$$y_i = t_i v + x_0, \quad i = 1, 2, \ldots, n, \tag{5.23}$$

where $\{t_i\}_{i=1}^n$ are the roots of the followin univariate polynomial of degree n:

$$q_n(t) = p_n(tv + x_0) = \prod_{i=1}^n \left(t b_i^\top v + b_i^\top x_0\right) = 0. \tag{5.24}$$

We summarize our discussion so far as Algorithm 5.3 for clustering hyperplanes.

[9] Except when the chosen line is parallel to one of the hyperplanes, which corresponds to a zero-measure set of lines.

Algorithm 5.3 (Algebraic Hyperplane Clustering Algorithm).

Let $\mathcal{X} = \{x_j\}_{j=1}^N \subset \mathbb{R}^D$ be a given collection of points clustered around an unknown number n of hyperplanes with normals $\{b_i\}_{i=1}^n$. The number of planes, the normal vectors, and the clustering of the data can be determined as follows:

1. **Number of Hyperplanes**. Let V_i be defined as in (5.18). Determine the number of clusters as

$$n \doteq \min\{i : \operatorname{rank}(V_i) = M_i - 1\}.$$

2. **Normal Vectors.** Solve for c_n from $V_n c_n = 0$ and set $p_n(x) = c_n^\top \nu_n(x)$. Choose x_0 and v at random and compute the n roots $t_1, t_2, \dots, t_n \in \mathbb{R}$ of the univariate polynomial $q_n(t) = p_n(tv + x_0)$. Determine the normal vectors as

$$b_i = \frac{\nabla p_n(y_i)}{\|\nabla p_n(y_i)\|}, \qquad i = 1, 2, \dots, n,$$

where $y_i = x_0 + t_i v$ is a point in the ith hyperplane.
3. **Clustering.** Assign point x_j to hyperplane $i = \arg\min_{l=1,\dots,n} (b_l^\top x_j)^2$.

5.3 Subspace Clustering Knowing the Number of Subspaces

In this section, we derive a general solution to the subspace clustering problem (Problem 5.1) in the case in which the number of subspaces n is *known*. However, in contrast to the special cases we saw in the previous section, the dimensions of the subspaces can be different from one another. In Section 5.3.1, we illustrate the basic ideas of dealing with subspaces of different dimensions via a simple example. Through Sections 5.3.2–5.3.4, we give a detailed derivation and proof for the general case. The final algorithm is summarized in Section 5.3.5.

5.3.1 An Introductory Example

To motivate and highlight the key ideas, in this section we study a simple example of clustering data points lying in subspaces of different dimensions in $\mathbb{R}^3$: a line $S_1 = \{x : x_1 = x_2 = 0\}$ and a plane $S_2 = \{x : x_3 = 0\}$, as shown in Figure 5.5.

We can describe the union of these two subspaces as

$$S_1 \cup S_2 = \{x : (x_1 = x_2 = 0) \vee (x_3 = 0)\} = \{x : (x_1 x_3 = 0) \wedge (x_2 x_3 = 0)\}.$$

Therefore, even though each individual subspace is described with polynomials of degree one (linear equations), the union of two subspaces is described with two polynomials of degree two, namely $p_{21}(x) = x_1 x_3$ and $p_{22}(x) = x_2 x_3$. In general, we can represent any two subspaces of $\mathbb{R}^3$ as the set of points satisfying a set of homogeneous polynomials of the form

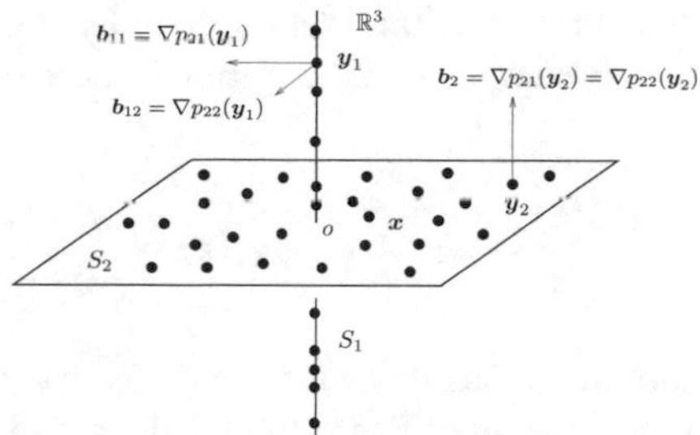

Fig. 5.5 Data samples drawn from a union of one plane and one line (through the origin o) in $\mathbb{R}^3$. The derivatives of the two vanishing polynomials $p_{21}(\boldsymbol{x}) = x_1x_2$ and $p_{22}(\boldsymbol{x}) = x_1x_3$ evaluated at a point $\boldsymbol{y}_1$ on the line give two normal vectors to the line. Similarly, the derivatives at a point $\boldsymbol{y}_2$ in the plane give the normal vector to the plane.

$$c_1x_1^2 + c_2x_1x_2 + c_3x_1x_3 + c_4x_2^2 + c_5x_2x_3 + c_6x_3^2 = 0. \tag{5.25}$$

Although these polynomials are nonlinear in each data point $[x_1, x_2, x_3]^\top$, they are actually linear in the vector of coefficients $\boldsymbol{c} = [c_1, c_2, \ldots, c_6]^\top$. Therefore, given enough data points, one can linearly *fit* these *polynomials* to the *data*.

Given the collection of polynomials that vanish on the data points, we are now interested in estimating a basis for each subspace. In our example, let $P_2(\boldsymbol{x}) = [p_{21}(\boldsymbol{x}),\ p_{22}(\boldsymbol{x})]$ and consider the derivatives of $P_2(\boldsymbol{x})$ at two representative points of the two subspaces $\boldsymbol{y}_1 = [0, 0, 1]^\top \in S_1$ and $\boldsymbol{y}_2 = [1, 1, 0]^\top \in S_2$:

$$\nabla P_2(\boldsymbol{x}) = \begin{bmatrix} x_3 & 0 \\ 0 & x_3 \\ x_1 & x_2 \end{bmatrix} \Longrightarrow \nabla P_2(\boldsymbol{y}_1) = \begin{bmatrix} 1 & 0 \\ 0 & 1 \\ 0 & 0 \end{bmatrix} \text{ and } \nabla P_2(\boldsymbol{y}_2) = \begin{bmatrix} 0 & 0 \\ 0 & 0 \\ 1 & 1 \end{bmatrix}. \tag{5.26}$$

Then the columns of $\nabla P_2(\boldsymbol{y}_1)$ span the orthogonal complement to the first subspace $S_1^\perp$, and the columns of $\nabla P_2(\boldsymbol{y}_2)$ span the orthogonal complement to the second subspace $S_2^\perp$ (see Figure 5.5). Thus the dimension of the line is given by $d_1 = 3 - \operatorname{rank}(\nabla P_2(\boldsymbol{y}_1)) = 1$, and the dimension of the plane is given by $d_2 = 3 - \operatorname{rank}(\nabla P_2(\boldsymbol{y}_2)) = 2$. Therefore, if we are given one point in each subspace, we can obtain the *subspace bases* and their *dimensions* from the *derivatives of the polynomials* at the given points.

The final question is how to choose one representative point per subspace. With perfect data, we may choose a first point as any of the points in the data set. With noisy data, we may first define a distance from each point in $\mathbb{R}^3$ to the union of the subspaces,[10] and then choose a point in the data set that minimizes this distance. Say we pick $\boldsymbol{y}_2 \in S_2$ as such a point. We can then compute the normal vector $\boldsymbol{b}_2 = [0, 0, 1]^\top$ to S_2 from $\nabla P(\boldsymbol{y}_2)$ as above. How do we now pick a second point

[10]For example, the squared algebraic distance to $S_1 \cup S_2$ is $p_{21}(\boldsymbol{x})^2 + p_{22}(\boldsymbol{x})^2 = (x_1^2 + x_2^2)x_3^2$.

in S_1 but not in S_2? As it turns out, this can be done by *polynomial division*. We can divide the original polynomials by $b_2^\top x$ to obtain new polynomials of degree $n-1=1$:

$$p_{11}(x) = \frac{p_{21}(x)}{b_2^\top x} = x_1 \quad \text{and} \quad p_{12}(x) = \frac{p_{22}(x)}{b_2^\top x} = x_2.$$

Since these new polynomials vanish on S_1 but not on S_2, we can use them to define a new distance to S_1 only,[11] and then find a point y_1 in S_1 but not in S_2 as the point in the data set that minimizes this distance.

The next sections show how this simple example can be systematically generalized to multiple subspaces of unknown and possibly different dimensions by *polynomial fitting* (Section 5.3.2), *differentiation* (Section 5.3.3), and *division* (Section 5.3.4).

5.3.2 *Fitting Polynomials to Subspaces*

Now consider a subspace arrangement $\mathcal{A} = \{S_1, S_2, \ldots, S_n\}$ with $\dim(S_i) = d_i, i = 1, 2, \ldots, n$. Let $\mathcal{X} = \{x_1, x_2, \ldots, x_N\}$ be a sufficiently large number of sample points in general position drawn from $Z_{\mathcal{A}} = S_1 \cup S_2 \cup \cdots \cup S_n$. As we may know from Appendix C, the vanishing ideal $I(Z_{\mathcal{A}})$, i.e., the set of all polynomials that vanish on $Z_{\mathcal{A}}$, is much more complicated than those in the special cases we studied earlier in this chapter.

Nevertheless, since we assume that we know the number of subspaces n, we have only to consider the set of polynomials of degree n that vanish on $Z_{\mathcal{A}}$, i.e., the homogeneous component I_n of $I(Z_{\mathcal{A}})$. As we know from Appendix C, these polynomials uniquely determine $Z_{\mathcal{A}}$. Furthermore, as the result of Corollary C.38, we know that if the subspace arrangement is transversal, then I_n is generated by the products of n linear forms that vanish on the n subspaces, respectively. More precisely, suppose the subspace S_i is of dimension d_i and let $k_i = D - d_i$. Let

$$B_i \doteq [b_1, b_2, \ldots, b_{k_i}] \quad \in \mathbb{R}^{D\times(k_i)}$$

be a set of base vectors for the orthogonal complement $S_i^\perp$ of S_i. The vanishing ideal $I(S_i)$ of S_i is generated by the set of linear forms

$$\{l(x) \doteq b^\top x, \ b \in B_i\}.$$

[11] For example, the squared algebraic distance to S_1 is $p_{11}(x)^2 + p_{12}(x)^2 = x_1^2 + x_2^2$.

Then every polynomial $p_n(\boldsymbol{x}) \in I_n$ can be written as a summation of products of the linear forms

$$p_n(\boldsymbol{x}) = \sum l_1(\boldsymbol{x}) l_2(\boldsymbol{x}) \cdots l_n(\boldsymbol{x}),$$

where $l_i \in I(S_i)$.

Using the Veronese map (defined by C.1 in Appendix C), each polynomial in I_n can also be written as

$$p_n(\boldsymbol{x}) = \boldsymbol{c}_n^\top \nu_n(\boldsymbol{x}) = \sum c_{n_1, n_2, \ldots, n_D} x_1^{n_1} x_2^{n_2} \cdots x_D^{n_D} = 0, \tag{5.27}$$

where $c_{n_1, n_2, \ldots, n_D} \in \mathbb{R}$ represents the coefficient of the monomial $\boldsymbol{x}^{\boldsymbol{n}} = x_1^{n_1} x_2^{n_2} \cdots x_D^{n_D}$. Although the polynomial equation is nonlinear in each data point $\boldsymbol{x}$, it is *linear* in the vector of coefficients $\boldsymbol{c}_n$. Indeed, since each polynomial $p_n(\boldsymbol{x}) = \boldsymbol{c}_n^\top \nu_n(\boldsymbol{x})$ must be satisfied by every data point, we have $\boldsymbol{c}_n^\top \nu_n(\boldsymbol{x}_j) = 0$ for all $j = 1, 2, \ldots, N$. Therefore, the vector of coefficients $\boldsymbol{c}_n$ must satisfy the system of linear equations

$$\boldsymbol{V}_n(D)\, \boldsymbol{c}_n \doteq \begin{bmatrix} \nu_n(\boldsymbol{x}_1)^\top \\ \nu_n(\boldsymbol{x}_2)^\top \\ \vdots \\ \nu_n(\boldsymbol{x}_N)^\top \end{bmatrix} \boldsymbol{c}_n = 0 \ \in \mathbb{R}^N, \tag{5.28}$$

where $\boldsymbol{V}_n(D) \in \mathbb{R}^{N \times M_n(D)}$ is called the *embedded data matrix*.

Clearly, the coefficient vector of every polynomial in I_n is in the null space of the data matrix $\boldsymbol{V}_n(D)$. For every polynomial obtained from the null space of $\boldsymbol{V}_n(D)$ to be in I_n, we need to have

$$\dim(\text{Null}(\boldsymbol{V}_n(D))) = \dim(I_n) \doteq h_I(n),$$

where $h_I(n)$ is the Hilbert function of the ideal $I(Z_{\mathcal{A}})$ (see Appendix C). Or equivalently, the rank of the data matrix $\boldsymbol{V}_n(D)$ needs to satisfy

$$\text{rank}(\boldsymbol{V}_n(D)) = M_n(D) - h_I(n) \tag{5.29}$$

in order that I_n can be exactly recovered from the null space of $\boldsymbol{V}_n(D)$. As a result of the algebraic sampling theory in Appendix C, the above rank condition is typically satisfied with $N \geq (M_n(D) - 1)$ data points in general position.[12] A basis of I_n,

$$I_n = \text{span}\{p_{n\ell}(\boldsymbol{x}),\ \ell = 1, 2, \ldots, h_I(n)\}, \tag{5.30}$$

[12]In particular, it requires at least d_i points from each subspace S_i.

can be computed from the set of $h_I(n)$ singular vectors of $V_n(D)$ associated with its $h_I(n)$ zero singular values. In the presence of moderate noise, we can still estimate the coefficients of the polynomials in a least-squares sense from the singular vectors associated with the $h_I(n)$ smallest singular values.

As discussed in Sections 4.1.1 and 4.1.3, the basic modeling assumption in NLPCA and KPCA is that there exists an embedding of the data into a higher-dimensional feature space F such that the features live in a linear subspace of F. However, there is no general methodology for finding the correct embedding for an arbitrary problem. Equation (5.28) shows that the commonly used polynomial embedding $\nu_n(\cdot)$ is the right one to use when the data live in an arrangement of subspaces, because the embedded data points $\{\nu_n(\boldsymbol{x}_j)\}_{j=1}^N$ indeed live in a subspace of $\mathbb{R}^{M_n(D)}$. Notice that each vector $\boldsymbol{c}_n$ is simply a normal vector to the embedded subspace, as illustrated in Figure 5.6.

5.3.3 Subspaces from Polynomial Differentiation

Given a basis for the set of polynomials representing an arrangement of subspaces, we are now interested in determining a basis and the dimension of each subspace. In this section, we show that one can estimate the dimensions and the bases by differentiating all the polynomials $\{p_{n\ell}\}$ obtained from the null space of the embedded data matrix $V_n(D)$.

Let $p_n(\boldsymbol{x})$ be any polynomial in I_n. Since $p_n \in I(Z_{\mathcal{A}}) \subset I(S_i)$, where $I(S_i)$ is generated by linear forms $l(\boldsymbol{x}) = \boldsymbol{b}^\top \boldsymbol{x}$ with $\boldsymbol{b} \in S_i^\perp$, p_n is of the form

$$p_n = l_1 g_1 + l_2 g_2 + \cdots + l_{k_i} g_{k_i} \tag{5.31}$$

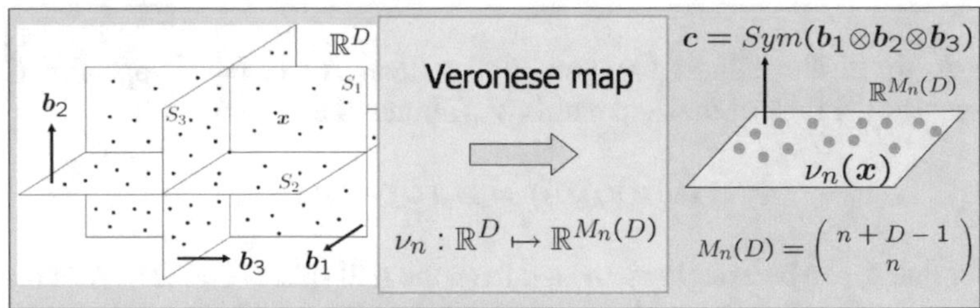

Fig. 5.6 The polynomial embedding maps a union of subspaces of $\mathbb{R}^D$ into a single subspace of $\mathbb{R}^{M_n(D)}$ whose normal vectors $\{\boldsymbol{c}_n\}$ are the coefficients of the polynomials $\{p_n\}$ defining the subspaces. The normal vectors to the embedded subspace $\{\boldsymbol{c}_n\}$ are related to the normal vectors to the original subspaces $\{\boldsymbol{b}_i\}$ via the symmetric tensor product.

for $l_1, l_2, \ldots, l_{k_i} \in I(S_i)$ and some polynomials $g_1, g_2, \ldots, g_{k_i}$.[13] The derivative of p_n is

$$\nabla p_n = \sum_{j=1}^{k_i} (g_j \nabla l_j + l_j \nabla g_j) = \sum_{j=1}^{k_i} (g_j \boldsymbol{b}_j + l_j \nabla g_j). \tag{5.32}$$

Let $\boldsymbol{y}_i$ be a point in the subspace S_i but not in any other subspaces in the arrangement $Z_{\mathcal{A}}$. Then $l_j(\boldsymbol{y}_i) = 0, j = 1, 2, \ldots, k_i$. Thus, the derivative of p_n evaluated at $\boldsymbol{y}_i$ is a superposition of the vectors $\boldsymbol{b}_j$:

$$\nabla p_n(\boldsymbol{y}_i) = \sum_{j=1}^{k_i} g_j(\boldsymbol{y}_i) \boldsymbol{b}_j \quad \in S_i^{\perp}. \tag{5.33}$$

This fact should come as no surprise. The zero set of each polynomial p_n is just a surface in $\mathbb{R}^D$. Therefore, its derivative $\nabla p_n(\boldsymbol{y}_i)$ at a nonsingular point $\boldsymbol{y}_i \in S_i$ gives a vector orthogonal to the surface. Since an arrangement of subspaces is locally flat, i.e., in a neighborhood of $\boldsymbol{y}_i$ the surface is merely the subspace S_i, it follows that the derivative at $\boldsymbol{y}_i$ lives in the orthogonal complement $S_i^{\perp}$ of S_i. By evaluating the derivatives of *all* the polynomials in I_n at the same point $\boldsymbol{y}_i$, we obtain a set of normal vectors that span the orthogonal complement of S_i. We summarize the above facts as Theorem 5.8. Figure 5.5 illustrates the theorem for the case of a plane and a line described in Section 5.3.1.

Theorem 5.8 (Subspace Bases and Dimensions by Polynomial Differentiation). *If the data set $\mathcal{X}$ is such that* $\dim(\mathit{Null}(\boldsymbol{V}_n(D))) = \dim(I_n) = h_I(n)$ *and one generic point $\boldsymbol{y}_i$ is given for each subspace S_i, then we have*

$$S_i^{\perp} = \mathit{span}\left\{ \frac{\partial}{\partial \boldsymbol{x}} \boldsymbol{c}_n^{\top} \nu_n(\boldsymbol{x}) \Big|_{\boldsymbol{x}=\boldsymbol{y}_i}, \ \forall \boldsymbol{c}_n \in \mathit{Null}(\boldsymbol{V}_n(D)) \right\}. \tag{5.34}$$

Therefore, the dimensions of the subspaces are given by

$$d_i = D - \mathit{rank}\big(\nabla P_n(\boldsymbol{y}_i)\big) \quad \mathit{for} \quad i = 1, 2, \ldots, n, \tag{5.35}$$

where $P_n(\boldsymbol{x}) \doteq [p_{n1}(\boldsymbol{x}), \ldots, p_{nh_I(n)}(\boldsymbol{x})] \in \mathbb{R}^{1 \times h_I(n)}$ is a row of linearly independent polynomials in I_n, and $\nabla P_n(\boldsymbol{x}) \doteq \big[\nabla p_{n1}(\boldsymbol{x}), \ldots, \nabla p_{nh_I(n)}(\boldsymbol{x})\big] \in \mathbb{R}^{D \times h_I(n)}$.

Proof. (Sketch only). The fact that the derivatives span the entire normal space is the consequence of the general dimension theory for algebraic varieties (Bochnak et al. 1998; Harris 1992; Eisenbud 1996). For a (transversal) subspace arrangement, one can also prove the theorem using the fact that polynomials in I_n are generated by the products of n linear forms that vanish on the n subspaces, respectively. □

[13] In fact, from discussions in the preceding subsection, we know that the polynomials $g_j, j = 1, \ldots, k_i$ are products of linear forms that vanish on the remaining $n - 1$ subspaces.

Given c_n, the computation of the derivative of $p_n(x) = c_n^\top \nu_n(x)$ can be done algebraically:

$$\nabla p_n(x) = c_n^\top \nabla \nu_n(x) = c_n^\top E_n \nu_{n-1}(x),$$

where $E_n \in \mathbb{R}^{M_n(D)\times M_{n-1}(D)}$ is a constant matrix containing only the exponents of the Veronese map $\nu_n(x)$. Thus, the computation does *not* involve taking derivatives of the (possibly noisy) data.

5.3.4 Point Selection via Polynomial Division

Theorem 5.8 suggests that one can obtain a basis for each $S_i^\perp$ directly from the derivatives of the polynomials representing the union of the subspaces. However, in order to proceed, we need to have one point per subspace, i.e., we need to know the vectors $\{y_1, y_2, \ldots, y_n\}$. In this section, we show how to select these n points in the *unsupervised learning scenario* in which we do not know the label for any of the data points.

In Section 5.2.3, we showed that in the case of hyperplanes, one can obtain one point per hyperplane by intersecting a random line L with the union of all hyperplanes.[14] This solution, however, does not generalize to subspaces of arbitrary dimensions. For instance, in the case of data lying on a line and in a plane shown in Figure 5.5, a randomly chosen line L may not intersect the line. Furthermore, because polynomials in the null space of $V_n(D)$ are no longer factorizable, their zero set is no longer a union of hyperplanes; hence the points of intersection with L may not lie in any of the subspaces.

In this section, we propose an alternative algorithm for choosing one point per subspace. The idea is that we can always choose a point y_n lying in one of the subspaces, say S_n, by checking that $P_n(y_n) = 0$. Since we are given a set of data points $\mathcal{X} = \{x_1, x_2, \ldots, x_N\}$ lying in the subspaces, in principle we can choose y_n to be any of the data points. However, in the presence of noise and outliers, a random choice of y_n may be far from the true subspaces. One may be tempted to choose a point in the data set $\mathcal{X}$ that minimizes $\|P_n(x)\|$, as we did in our introductory example in Section 5.3.1. However, such a choice has the following problems:

1. The value $\|P_n(x)\|$ is merely an *algebraic* error, i.e., it does not really represent the *geometric* distance from x to its closest subspace. In principle, finding the geometric distance from x to its closest subspace is a hard problem, because we do not know the normal bases $\{B_1, B_2, \ldots, B_n\}$.

[14] This can always be done, except when the chosen line is parallel to one of the subspaces, which corresponds to a zero-measure set of lines.

2. Points $\boldsymbol{x}$ lying close to the intersection of two or more subspaces are more likely to be chosen, because two or more factors in $p_n(\boldsymbol{x}) = (\boldsymbol{b}_1^\top \boldsymbol{x})(\boldsymbol{b}_2^\top \boldsymbol{x}) \cdots (\boldsymbol{b}_n^\top \boldsymbol{x})$ are approximately zero, which yields a smaller value for $|p_n(\boldsymbol{x})|$. In fact, we can see from (5.33) that for an arbitrary $\boldsymbol{x}$ in the intersection, the vector $\nabla p_n(\boldsymbol{x})$ needs to be a common normal vector to two or more subspaces. If the subspaces have no common normal vector, then $\|\nabla p_n(\boldsymbol{x})\| = 0$. Thus, one should avoid choosing points close to the intersection, because they typically give very noisy estimates of the normal vectors.

We could avoid these two problems if we could compute the distance from each point to the subspace passing through it. However, we cannot compute such a distance yet, because we do not know the subspaces' bases. The following lemma shows that we can compute a first-order approximation to such a distance from P_n and its derivatives.

Lemma 5.9. *Let $\hat{\boldsymbol{x}}$ be the projection of $\boldsymbol{x} \in \mathbb{R}^D$ onto its closest subspace. The Euclidean distance from $\boldsymbol{x}$ to $\hat{\boldsymbol{x}}$ is given by*

$$\|\boldsymbol{x} - \hat{\boldsymbol{x}}\| = n\sqrt{P_n(\boldsymbol{x})\big(\nabla P_n(\boldsymbol{x})^\top \nabla P_n(\boldsymbol{x})\big)^\dagger P_n(\boldsymbol{x})^\top} + O\big(\|\boldsymbol{x} - \hat{\boldsymbol{x}}\|^2\big),$$

where $P_n(\boldsymbol{x}) = [p_{n1}(\boldsymbol{x}), \ldots, p_{nh_I(n)}(\boldsymbol{x})] \in \mathbb{R}^{1 \times h_I(n)}$ is a row vector with all the polynomials, $\nabla P_n(\boldsymbol{x}) = \big[\nabla p_{n1}(\boldsymbol{x}), \ldots, \nabla p_{nh_I(n)}(\boldsymbol{x})\big] \in \mathbb{R}^{D \times h_I(n)}$, and $A^\dagger$ is the Moore–Penrose inverse of A.

Proof. The projection $\hat{\boldsymbol{x}}$ of a point $\boldsymbol{x}$ onto the zero set of the polynomials $\{p_{n\ell}\}_{\ell=1}^{h_I(n)}$ can be obtained as the solution to the following constrained optimization problem:

$$\min \|\hat{\boldsymbol{x}} - \boldsymbol{x}\|^2, \quad \text{s.t.} \quad p_{n\ell}(\hat{\boldsymbol{x}}) = 0, \quad \ell = 1, 2, \ldots, h_I(n). \tag{5.36}$$

Using Lagrange multipliers $\lambda \in \mathbb{R}^{h_I(n)}$, we can convert this problem into the unconstrained optimization problem

$$\min_{\hat{\boldsymbol{x}}, \lambda} \|\hat{\boldsymbol{x}} - \boldsymbol{x}\|^2 + P_n(\hat{\boldsymbol{x}})\lambda. \tag{5.37}$$

From the first-order conditions with respect to $\hat{\boldsymbol{x}}$, we have

$$2(\hat{\boldsymbol{x}} - \boldsymbol{x}) + \nabla P_n(\hat{\boldsymbol{x}})\lambda = 0. \tag{5.38}$$

After multiplying on the left by $(\nabla P_n(\hat{\boldsymbol{x}}))^\top$ and $(\hat{\boldsymbol{x}} - \boldsymbol{x})^\top$, respectively, we obtain

$$\lambda = 2\big(\nabla P_n(\hat{\boldsymbol{x}})^\top \nabla P_n(\hat{\boldsymbol{x}})\big)^\dagger \nabla P_n(\hat{\boldsymbol{x}})^\top \boldsymbol{x}, \quad \|\hat{\boldsymbol{x}} - \boldsymbol{x}\|^2 = \frac{1}{2}\boldsymbol{x}^\top \nabla P_n(\hat{\boldsymbol{x}})\lambda, \tag{5.39}$$

where we have used the fact that $(\nabla P_n(\hat{\boldsymbol{x}}))^\top \hat{\boldsymbol{x}} = 0$. After substituting the first equation into the second, we obtain that the squared distance from $\boldsymbol{x}$ to its closest subspace can be expressed as

$$\|\hat{\boldsymbol{x}} - \boldsymbol{x}\|^2 = \boldsymbol{x}^\top \nabla P_n(\hat{\boldsymbol{x}}) \big(\nabla P_n(\hat{\boldsymbol{x}})^\top \nabla P_n(\hat{\boldsymbol{x}})\big)^\dagger \nabla P_n(\hat{\boldsymbol{x}})^\top \boldsymbol{x}. \tag{5.40}$$

After expanding in Taylor series about $\boldsymbol{x}$ and noticing that $\nabla P_n(\boldsymbol{x})^\top \boldsymbol{x} = nP_n(\boldsymbol{x})^\top$, we obtain

$$\|\hat{\boldsymbol{x}} - \boldsymbol{x}\|^2 \approx n^2 P_n(\boldsymbol{x}) \big(\nabla P_n(\boldsymbol{x})^\top \nabla P_n(\boldsymbol{x})\big)^\dagger P_n(\boldsymbol{x})^\top, \tag{5.41}$$

which completes the proof. □

Thanks to Lemma 5.9, we can immediately choose a candidate $\boldsymbol{y}_n$ lying in (close to) one of the subspaces and not in the intersection as

$$\boldsymbol{y}_n = \operatorname*{argmin}_{\boldsymbol{x} \in \mathcal{X}: \nabla P_n(\boldsymbol{x}) \neq 0} P_n(\boldsymbol{x}) \big(\nabla P_n(\boldsymbol{x})^\top \nabla P_n(\boldsymbol{x})\big)^\dagger P_n(\boldsymbol{x})^\top \tag{5.42}$$

and compute a basis $B_n \in \mathbb{R}^{D \times (D-d_n)}$ for $S_n^\perp$ by applying PCA to $\nabla P_n(\boldsymbol{y}_n)$.

In order to find a point $\boldsymbol{y}_{n-1}$ lying in (close to) one of the remaining $(n-1)$ subspaces but not in (far from) S_n, we could in principle choose $\boldsymbol{y}_{n-1}$ as in (5.42) after removing the points in S_n from the data set $\mathcal{X}$. With noisy data, however, this depends on a threshold and is not very robust. Alternatively, we can find a new set of polynomials $\{p_{(n-1)\ell}(\boldsymbol{x})\}$ defining the algebraic set $\cup_{i=1}^{n-1} S_i$. In the case of hyperplanes, there is only one such polynomial, namely

$$p_{n-1}(\boldsymbol{x}) \doteq (\boldsymbol{b}_1 \boldsymbol{x})(\boldsymbol{b}_2 \boldsymbol{x}) \cdots (\boldsymbol{b}_{n-1}^\top \boldsymbol{x}) = \frac{p_n(\boldsymbol{x})}{\boldsymbol{b}_n^\top \boldsymbol{x}} = \boldsymbol{c}_{n-1}^\top \nu_{n-1}(\boldsymbol{x}).$$

Therefore, we can obtain $p_{n-1}(\boldsymbol{x})$ by *polynomial division*. Notice that dividing $p_n(\boldsymbol{x})$ by $\boldsymbol{b}_n^\top \boldsymbol{x}$ is a *linear problem* of the form

$$R_n(\boldsymbol{b}_n) \boldsymbol{c}_{n-1} = \boldsymbol{c}_n, \tag{5.43}$$

where $R_n(\boldsymbol{b}_n) \in \mathbb{R}^{M_n(D) \times M_{n-1}(D)}$. This is because solving for the coefficients of $p_{n-1}(\boldsymbol{x})$ is equivalent to solving the equations $(\boldsymbol{b}_n^\top \boldsymbol{x})(\boldsymbol{c}_{n-1}^\top \nu_{n-1}(\boldsymbol{x})) = \boldsymbol{c}_n^\top \nu_n(\boldsymbol{x})$ for all $\boldsymbol{x} \in \mathbb{R}^D$. These equations are obtained by equating the coefficients, and they are linear in $\boldsymbol{c}_{n-1}$, because $\boldsymbol{b}_n$ and $\boldsymbol{c}_n$ are already known.

Example 5.10 If $n = 2$ and $\boldsymbol{b}_2 = [b_1, b_2, b_3]^\top$, then the matrix $R_2(\boldsymbol{b}_2)$ is given by

$$R_2(\boldsymbol{b}_2) = \begin{bmatrix} b_1 & b_2 & b_3 & 0 & 0 & 0 \\ 0 & b_1 & 0 & b_2 & b_3 & 0 \\ 0 & 0 & b_1 & 0 & b_2 & b_3 \end{bmatrix}^\top \in \mathbb{R}^{6 \times 3}.$$

In the case of subspaces of arbitrary dimensions, we cannot directly divide the entries of the polynomial vector $P_n(\boldsymbol{x})$ by $\boldsymbol{b}_n^\top \boldsymbol{x}$ for any column $\boldsymbol{b}_n$ of B_n, because the polynomials $\{p_{n\ell}(\boldsymbol{x})\}$ may not be factorizable. Furthermore, they do not necessarily have the common factor $\boldsymbol{b}_n^\top \boldsymbol{x}$. The following theorem resolves this difficulty by showing how to compute the polynomials associated with the remaining subspaces $\cup_{i=1}^{n-1} S_i$.

Theorem 5.11 (Choosing One Point per Subspace by Polynomial Division). *If the data set $\mathcal{X}$ is such that* $\dim(null(\boldsymbol{V}_n(D))) = \dim(I_n)$, *then the set of homogeneous polynomials of degree $(n-1)$ associated with the algebraic set $\cup_{i=1}^{n-1} S_i$ is given by $\{\boldsymbol{c}_{n-1}^\top \nu_{n-1}(\boldsymbol{x})\}$, where the vectors of coefficients $\boldsymbol{c}_{n-1} \in \mathbb{R}^{M_{n-1}(D)}$ must satisfy*

$$\boldsymbol{V}_n(D) R_n(\boldsymbol{b}_n) \boldsymbol{c}_{n-1} = 0, \quad \forall\, \boldsymbol{b}_n \in S_n^\perp. \tag{5.44}$$

Proof. We first prove the necessity. That is, every polynomial $\boldsymbol{c}_{n-1}^\top \nu_{n-1}(\boldsymbol{x})$ of degree $n-1$, that vanishes on $\cup_{i=1}^{n-1} S_i$ satisfies the above equation. Since a point $\boldsymbol{x}$ in the original algebraic set $\cup_{i=1}^{n} S_i$ belongs to either $\cup_{i=1}^{n-1} S_i$ or S_n, we have $\boldsymbol{c}_{n-1}^\top \nu_{n-1}(\boldsymbol{x}) = 0$ or $\boldsymbol{b}_n^\top \boldsymbol{x} = 0$ for all $\boldsymbol{b}_n \in S_n^\perp$. Hence $p_n(\boldsymbol{x}) \doteq (\boldsymbol{c}_{n-1}^\top \nu_{n-1}(\boldsymbol{x}))(\boldsymbol{b}_n^\top \boldsymbol{x}) = 0$, and $p_n(\boldsymbol{x})$ must be a linear combination of polynomials in $P_n(\boldsymbol{x})$. If we denote $p_n(\boldsymbol{x})$ by $\boldsymbol{c}_n^\top \nu_n(\boldsymbol{x})$, then the vector of coefficients $\boldsymbol{c}_n$ must be in the null space of $\boldsymbol{V}_n(D)$. From $\boldsymbol{c}_n^\top \nu_n(\boldsymbol{x}) = (\boldsymbol{c}_{n-1}^\top \nu_{n-1}(\boldsymbol{x}))(\boldsymbol{b}_n^\top \boldsymbol{x})$, the relationship between $\boldsymbol{c}_n$ and $\boldsymbol{c}_{n-1}$ can be written as $R_n(\boldsymbol{b}_n)\boldsymbol{c}_{n-1} = \boldsymbol{c}_n$. Since $\boldsymbol{V}_n(D)\boldsymbol{c}_n = 0$, $\boldsymbol{c}_{n-1}$ needs to satisfy the linear system of equations $\boldsymbol{V}_n(D) R_n(\boldsymbol{b}_n)\boldsymbol{c}_{n-1} = 0$.

We now prove the sufficiency. That is, if $\boldsymbol{c}_{n-1}$ is a solution to (5.44), then $\boldsymbol{c}_{n-1}^\top \nu_{n-1}(\boldsymbol{x})$ is a homogeneous polynomial of degree $(n-1)$ that vanishes on $\cup_{i=1}^{n-1} S_i$. Since $\boldsymbol{c}_{n-1}$ is a solution to (5.44), then for all $\boldsymbol{b}_n \in S_n^\perp$, we have that $\boldsymbol{c}_n = R_n(\boldsymbol{b}_n)\boldsymbol{c}_{n-1}$ is in the null space of $\boldsymbol{V}_n(D)$. Now, from the construction of $R_n(\boldsymbol{b}_n)$, we also have that $\boldsymbol{c}_n^\top \nu_n(\boldsymbol{x}) = (\boldsymbol{c}_{n-1}^\top \nu_{n-1}(\boldsymbol{x}))(\boldsymbol{b}_n^\top \boldsymbol{x})$. Hence, for every $\boldsymbol{x} \in \cup_{i=1}^{n-1} S_i$ but not in S_n, we have $\boldsymbol{c}_{n-1}^\top \nu_{n-1}(\boldsymbol{x}) = 0$, because there is a $\boldsymbol{b}_n$ such that $\boldsymbol{b}_n^\top \boldsymbol{x} \neq 0$. Therefore, $\boldsymbol{c}_{n-1}^\top \nu_{n-1}(\boldsymbol{x})$ is a homogeneous polynomial of degree $(n-1)$ that vanishes on $\cup_{i=1}^{n-1} S_i$. □

Thanks to Theorem 5.11, we can obtain a basis $\{p_{(n-1)\ell}(\boldsymbol{x}), \ell = 1, 2, \ldots, h_I(n-1)\}$ for the polynomials vanishing on $\cup_{i=1}^{n-1} S_i$ from the intersection of the null spaces of $\boldsymbol{V}_n(D) R_n(\boldsymbol{b}_n) \in \mathbb{R}^{N \times M_{n-1}(D)}$ for all $\boldsymbol{b}_n \in S_i^\perp$. By evaluating the derivatives of the polynomials $p_{(n-1)\ell}$, we can obtain normal vectors to S_{n-1} and so on. By repeating this process, we can find a basis for each of the remaining subspaces. The overall subspace clustering and estimation process involves polynomial fitting, differentiation, and division.

5.3.5 The Basic Algebraic Subspace Clustering Algorithm

In practice, we may avoid computing P_i for $i < n$ by using a heuristic distance function to choose the points $\{\boldsymbol{y}_1, \boldsymbol{y}_2, \ldots, \boldsymbol{y}_n\}$ as follows. Since a point in $\cup_{\ell=i}^{n} S_\ell$ must satisfy $\|B_i^\top \boldsymbol{x}\| \|B_{i+1}^\top \boldsymbol{x}\| \cdots \|B_n^\top \boldsymbol{x}\| = 0$, we can choose a point $\boldsymbol{y}_{i-1}$ on $\cup_{\ell=1}^{i-1} S_\ell$ as

Algorithm 5.4 (ASC: Algebraic Subspace Clustering).

Given a set of samples $\mathcal{X} = \{\boldsymbol{x}_1, \boldsymbol{x}_2, \dots, \boldsymbol{x}_N\}$ in $\mathbb{R}^D$, fit n linear subspaces with dimensions $d_1, d_2, \dots, d_n$:

1: Set $\boldsymbol{V}_n(D) \doteq [\nu_n(\boldsymbol{x}_1), \nu_n(\boldsymbol{x}_2), \dots, \nu_n(\boldsymbol{x}_N)]^\top \in \mathbb{R}^{N \times M_n(D)}$.
2: **for all** $i = n : 1$ **do**
3: Solve $\boldsymbol{V}_i(D)\boldsymbol{c} = 0$ to obtain a basis $\{\boldsymbol{c}_{i\ell}\}_{\ell=1}^{h_I(i)}$ of null$(\boldsymbol{V}_i(D))$, where the number of polynomials $h_I(i)$ is obtained as in Appendix C.
4: Set $P_i(\boldsymbol{x}) = [p_{i1}(\boldsymbol{x}), p_{i2}(\boldsymbol{x}), \dots, p_{ih_I(i)}(\boldsymbol{x})] \in \mathbb{R}^{1 \times h_I(i)}$, where $p_{i\ell}(\boldsymbol{x}) = \boldsymbol{c}_{i\ell}^\top \nu_i(\boldsymbol{x})$ for $\ell = 1, 2, \dots, h_I(i)$.
5: Compute

$$\boldsymbol{y}_i = \underset{\boldsymbol{x} \in \mathcal{X}: \nabla P_i(\boldsymbol{x}) \neq 0}{\arg\min} P_i(\boldsymbol{x}) \big(\nabla P_i(\boldsymbol{x})^\top \nabla P_i(\boldsymbol{x})\big)^\dagger P_i(\boldsymbol{x})^\top,$$

$$B_i \doteq [\boldsymbol{b}_{i1}, \boldsymbol{b}_{j2}, \dots, \boldsymbol{b}_{i(D-d_i)}] = \mathrm{PCA}\big(\nabla P_i(\boldsymbol{y}_i)\big),$$

$$\boldsymbol{V}_{i-1}(D) = \boldsymbol{V}_i(D) \left[R_i^\top(\boldsymbol{b}_{i1}), R_i^\top(\boldsymbol{b}_{i2}), \dots, R_i^\top(\boldsymbol{b}_{i(D-d_i)})\right]^\top.$$

6: **end for**
7: **for all** $j = 1 : N$ **do**
8: Assign point $\boldsymbol{x}_j$ to subspace S_i if $i = \arg\min_{\ell=1,2,\dots,n} \|B_\ell^\top \boldsymbol{x}_j\|^2$.
9: **end for**

$$\boldsymbol{y}_{i-1} = \underset{\boldsymbol{x} \in X: \nabla P_n(\boldsymbol{x}) \neq 0}{\arg\min} \frac{\sqrt{P_n(\boldsymbol{x})(\nabla P_n(\boldsymbol{x})^\top \nabla P_n(\boldsymbol{x}))^\dagger P_n(\boldsymbol{x})^\top} + \delta}{\|B_i^\top \boldsymbol{x}\| \|B_{i+1}^\top \boldsymbol{x}\| \cdots \|B_n^\top \boldsymbol{x}\| + \delta}, \tag{5.45}$$

where $\delta > 0$ is a small number chosen to avoid cases in which both the numerator and the denominator are zero (e.g., with perfect data).

We summarize the results of this section as the following algebraic subspace clustering (ASC) algorithm, Algorithm 5.4, for clustering a known number of subspaces of unknown and possibly different dimensions from sample data points $\mathcal{X} = \{\boldsymbol{x}_1, \boldsymbol{x}_2, \dots, \boldsymbol{x}_N\}$.

Example 5.12 (Algebraic Subspace Clustering on Synthetic Data) In this experiment, we evaluate on synthetic data the performance of two variants of the algebraic subspace clustering algorithm: one is to use the heuristic distance measure introduced above (5.45) for selecting a point closest to each subspace, referred to as the minimum distance method; the other is to fit a hyperplane to each subspace as a superset and select a point on each hyperplane by intersecting it with a random line as done in Algorithm 5.3, referred to as the line-intersection method. In this experiment, we set $\delta = 10^{-5}$ in (5.45).

We randomly generate three subspaces S_1, S_2, S_3 in $\mathbb{R}^4$ of dimensions d_1, d_2, d_3. From each subspace we randomly sample 100 points, corrupted by zero-mean Gaussian noise of standard deviation σ in the orthogonal complement of the subspaces. Figures 5.7(a)–5.7(d) show the clustering error rate of these algebraic algorithms for various levels of noise and for various values of d_1, d_2, d_3, averaged over 300 independent trials.

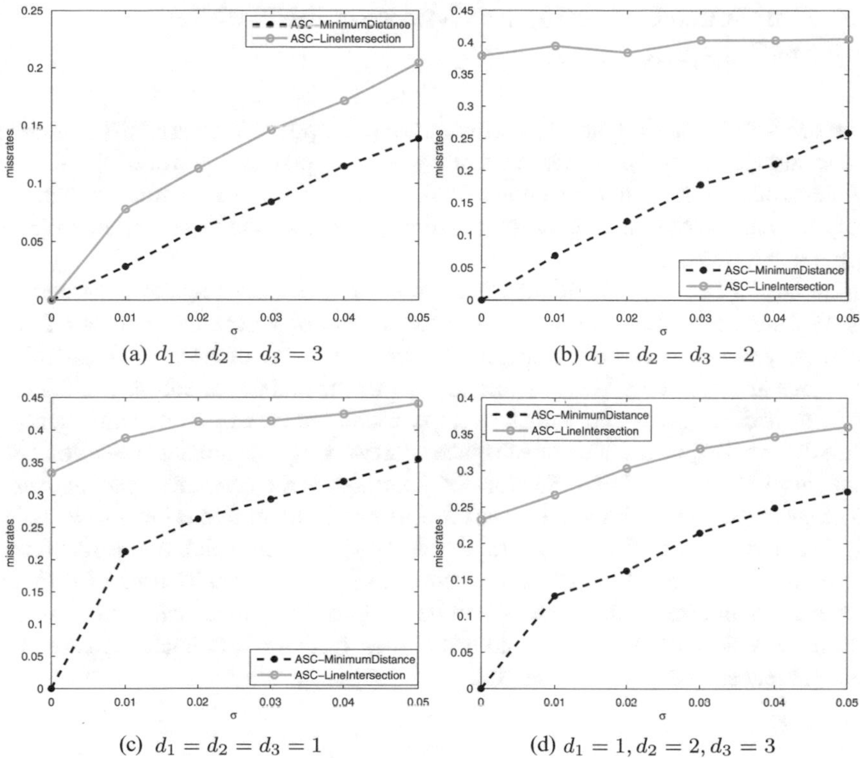

Fig. 5.7 Misclustering rates versus noise level for the two algebraic hyperplane/subspace clustering algorithms for three subspaces of various dimensions in $\mathbb{R}^4$.

In particular, in Figure 5.7(a) we have the case of three hyperplanes in $\mathbb{R}^4$. As expected, both algorithms work correctly when there is no noise, and their performance degrades as the noise increases. The line-intersection method is obviously more sensitive to noise. However, for subspaces in $\mathbb{R}^4$ with dimension lower than 2 (Figures 5.7(b)–5.7(d)), we see that the line-intersection method produces errors even in the noiseless case. The reason is evident: a random line almost surely does not intersect a 2-dimensional linear subspace in $\mathbb{R}^4$. Hence the points found almost surely do not lie on any of the subspaces. In contrast, the minimum-distance method performs well for subspaces of lower dimensions in the noiseless case. Both methods are very sensitive to noise. This suggests that there is need to further improve their clustering accuracy in the presence of noise, which we will do through other techniques in later chapters of the book.

5.4 Subspace Clustering not Knowing the Number of Subspaces

The solution to the subspace clustering problem proposed in Section 5.3.5 assumes prior knowledge of the number of subspaces n. In practice, however, the number of subspaces n may not be known beforehand; hence we cannot estimate the polynomials representing the subspaces directly, because the linear system in (5.28) depends explicitly on n.

Earlier, in Section 5.2, we presented some special cases (e.g., arrangements of hyperplanes) for which one can recover the number of subspaces from data. In this section, we show that by exploiting the algebraic structure of the vanishing ideals of subspace arrangements, it is possible to recover the number of subspaces, together with their dimensions and their bases. As usual, we first examine some subtlety with determining the number of subspaces via two simple examples in Section 5.4.1 and illustrate the key ideas. Section 5.4.2 considers the case of *perfect subspace arrangements* in which all subspaces are of equal dimension $d = d_1 = \cdots = d_n$. We derive a set of rank constraints on the data from which one can estimate n and d. Section 5.4.3 considers the most general case of subspaces of different dimensions and shows that n and d can be computed in a recursive fashion by first fitting subspaces of larger dimensions and then further clustering these subspaces into subspaces of smaller dimensions.

5.4.1 Introductory Examples

Imagine that we are given a set of points $\mathcal{X} = \{\boldsymbol{x}_1, \boldsymbol{x}_2, \ldots, \boldsymbol{x}_N\}$ lying on two lines in $\mathbb{R}^3$, say

$$S_1 = \{\boldsymbol{x} : x_2 = x_3 = 0\} \quad \text{and} \quad S_2 = \{\boldsymbol{x} : x_1 = x_3 = 0\}. \tag{5.46}$$

If we form the matrix of embedded data points $\boldsymbol{V}_n(D)$ for $n = 1$ and $n = 2$, respectively,

$$\boldsymbol{V}_1(3) = \begin{bmatrix} \vdots & & \vdots \\ x_1 & x_2 & x_3 \\ \vdots & & \vdots \end{bmatrix} \text{ and } \boldsymbol{V}_2(3) = \begin{bmatrix} \vdots & & & & & \vdots \\ x_1^2 & x_1x_2 & x_1x_3 & x_2^2 & x_2x_3 & x_3^2 \\ \vdots & & & & & \vdots \end{bmatrix},$$

we obtain $\text{rank}(\boldsymbol{V}_1(3)) = 2 < 3$ and $\text{rank}(\boldsymbol{V}_2(3)) = 2 < 6$.[15] Therefore, we cannot determine the number of subspaces as the degree n such that the matrix $\boldsymbol{V}_n(D)$ drops rank (as we did in Section 5.2.3 for the case of hyperplanes), because we would obtain $n = 1$, which is not the correct number of subspaces.

How do we determine the correct number of subspaces in this case? As discussed in Section 5.1.2, a linear projection onto a low-dimensional subspace preserves the number and dimensions of the subspaces. In our example, if we project the data onto the plane $P = \{\boldsymbol{x} : x_1 + x_2 + x_3 = 0\}$ and then embed the projected data, we obtain

$$\boldsymbol{V}_1(2) = \begin{bmatrix} \vdots & \vdots \\ x_1 & x_2 \\ \vdots & \vdots \end{bmatrix} \quad \text{and} \quad \boldsymbol{V}_2(2) = \begin{bmatrix} \vdots & & \vdots \\ x_1^2 & x_1x_2 & x_2^2 \\ \vdots & & \vdots \end{bmatrix}.$$

In this case, $\text{rank}(\boldsymbol{V}_1(2)) = 2 \not< 2$, but $\text{rank}(\boldsymbol{V}_2(2)) = 2 < 3$. Therefore, the first time the matrix $\boldsymbol{V}_n(d+1)$ drops rank is when $n = 2$ and $d = 1$. This suggests, as we will formally show in Section 5.4.2, that when the subspaces are of equal dimension, one can determine d and n as the minimum values for which there is a projection onto a $(d+1)$-dimensional subspace such that the matrix $\boldsymbol{V}_n(d+1)$ drops rank.

Unfortunately, the situation is not so simple for subspaces of different dimensions. Imagine now that in addition to the two lines S_1 and S_2, we are also given data points in a plane $S_3 = \{\boldsymbol{x} : x_1 + x_2 = 0\}$ (so that the overall configuration is similar to that shown in Figure 1.2). In this case, we have $\text{rank}(\boldsymbol{V}_1(3)) = 3 \not< 3$, $\text{rank}(\boldsymbol{V}_2(3)) = 5 < 6$, and $\text{rank}(\boldsymbol{V}_3(3)) = 6 < 10$. Therefore, if we try to determine the number of subspaces as the degree of the embedding for which the embedded data matrix drops rank, we will obtain $n = 2$, which is incorrect again. The reason for this is clear: we can either fit the data with one polynomial of degree $n = 2$, which corresponds to the plane S_3 and the plane P spanned by the two lines, or we can fit the data with four polynomials of degree $n = 3$, which vanish precisely on the two lines S_1, S_2, and the plane S_3.

In cases like this, one needs to resort to a more sophisticated algebraic process to identify the correct number of subspaces. As in the previous example, we can first search for the minimum degree n and dimension d such that $\boldsymbol{V}_n(d+1)$ drops rank. In our example, we obtain $n = 2$ and $d = 2$. By applying the algebraic subspace clustering algorithm to this data set, we will partition it into two planes P and S_3. Once the two planes have been estimated, we can reapply the same process to each plane. The plane P will be separated into two lines S_1 and S_2, as described in the previous example, while the plane S_3 will remain unchanged. This recursive process stops when every subspace obtained can no longer be separated into lower-dimensional subspaces. We will give a more detailed description of this in Section 5.4.3.

[15]The reader is encouraged to verify these facts numerically and do the same for the examples in the rest of this section.

5.4.2 Clustering Subspaces of Equal Dimension

In this section, we derive explicit formulas for the number of subspaces n and their dimensions $\{d_i\}$ in the case of subspaces of equal dimension $d = d_1 = d_2 = \cdots = d_n$. Notice that this is a generalized version of the two-lines example that we discussed in the previous section. In the literature, arrangements of subspaces of equal dimensions are called *pure arrangements*. This type of arrangement is important for a wide range of applications in computer vision (Costeira and Kanade 1998; Kanatani 2002; Vidal and Ma 2004), pattern recognition (Belhumeur et al. 1997; Vasilescu and Terzopoulos 2002), as well as identification of hybrid linear systems (Overschee and Moor 1993; Ma and Vidal 2005).

Theorem 5.13 (Subspaces of Equal Dimension). *Let $\{\boldsymbol{x}_j\}_{j=1}^N$ be a given collection of $N \geq M_n(d+1) - 1$ sample points lying in n different d-dimensional subspaces of $\mathbb{R}^D$. Let $\boldsymbol{V}_i(\ell+1) \in \mathbb{R}^{N \times M_i(\ell+1)}$ be the embedded data matrix defined in (5.28), but computed with the Veronese map ν_i of degree i applied to the data projected onto a generic $(\ell+1)$-dimensional subspace of $\mathbb{R}^D$. If the sample points are in general position and at least d points are drawn from each subspace, then the dimension of the subspaces is given by*

$$d = \min\{\ell : \exists\, i \geq 1 \text{ such that } \operatorname{rank}(\boldsymbol{V}_i(\ell+1)) < M_i(\ell+1)\}, \tag{5.47}$$

and the number of subspaces can be obtained as

$$n = \min\{i : \operatorname{rank}(\boldsymbol{V}_i(d+1)) = M_i(d+1) - 1\}. \tag{5.48}$$

Proof. For simplicity, we divide the proof into the following three cases:

Case 1: d known
Imagine for a moment that d was known, and that we wanted to compute n only. Since d is known, following our analysis in Section 5.1.2, we can first project the data onto a $(d+1)$-dimensional space $P \subset \mathbb{R}^D$ so that they become n d-dimensional hyperplanes in P (see Theorem 5.5). Now compute the matrix $\boldsymbol{V}_i(d+1)$ as in (5.28) by applying the Veronese map of degree $i = 1, 2, \ldots$ to the projected data. From our analysis in Section 5.2.3, there is a unique polynomial of degree n representing the union of the projected subspaces, and the coefficients of this polynomial must lie in the null space of $\boldsymbol{V}_n(d+1)$. Thus, given $N \geq M_n(d+1) - 1$ points in general position, with at least d points in each subspace, we have that $\operatorname{rank}(\boldsymbol{V}_n(d+1)) = M_n(d+1) - 1$. Furthermore, there cannot be a polynomial of degree less than n that is satisfied by all the data,[16] whence $\operatorname{rank}(\boldsymbol{V}_i(d+1)) = M_i(d+1)$ for $i < n$.

[16]This is guaranteed by the algebraic sampling theorem in Appendix C.

Consequently, if d is known, we can compute n by first projecting the data onto a $(d+1)$-dimensional space and then obtaining

$$n = \min\{i : \text{rank}(\boldsymbol{V}_i(d+1)) = M_i(d+1) - 1\}. \tag{5.49}$$

Case 2: n known
Consider now the opposite case in which n is known, but d is unknown. Let $\boldsymbol{V}_n(\ell+1)$ be defined as in (5.28), but computed from the data projected onto a generic $(\ell+1)$-dimensional subspace of $\mathbb{R}^D$. When $\ell < d$, we have a collection of $(\ell+1)$-dimensional subspaces in an $(\ell+1)$-dimensional space, which implies that $\boldsymbol{V}_n(\ell+1)$ must be of full rank. If $\ell = d$, then from equation (5.49), we have that $\text{rank}(\boldsymbol{V}_n(\ell+1)) = M_n(\ell+1)-1$. When $\ell > d$, then equation (5.28) has more than one solution, and thus $\text{rank}(\boldsymbol{V}_n(\ell+1)) < M_n(\ell+1) - 1$. Therefore, if n is known, we can compute d as

$$d = \min\{\ell : \text{rank}(\boldsymbol{V}_n(\ell+1)) = M_n(\ell+1) - 1\}. \tag{5.50}$$

Case 3: n and d unknown
We are left with the case in which both n and d are unknown. As before, if $\ell < d$, then $\boldsymbol{V}_i(\ell+1)$ is of full rank for all i. When $\ell = d$, $\boldsymbol{V}_i(\ell+1)$ is of full rank for $i < n$, drops rank by one if $i = n$, and drops rank by more than one if $i > n$. Thus one can set d to be the smallest integer ℓ for which there exists an i such that $\boldsymbol{V}_i(\ell+1)$ drops rank, that is,

$$d = \min\{\ell : \exists i \geq 1 \text{ such that } \text{rank}(\boldsymbol{V}_i(\ell+1)) < M_i(\ell+1)\}.$$

Given d, one can compute n as in equation (5.49). □

Therefore, in principle, both n and d can be retrieved if sufficient data points are drawn from the subspaces. The subspace clustering problem can be subsequently solved by first projecting the data onto a $(d+1)$-dimensional subspace and then applying the algebraic subspace clustering algorithm (Algorithm 5.4) to the projected data points.

In the presence of noise, one may not be able to estimate d and n from equations (5.47) and (5.48), respectively, because the matrix $\boldsymbol{V}_i(\ell+1)$ may be of full rank for all i and ℓ. As before, we can use the criteria introduced in Section 2.3 of Chapter 2 to determine the rank of $\boldsymbol{V}_i(\ell+1)$. However, in practice this requires a search for up to possibly $(D-1)$ values for d and $\lceil N/(D-1) \rceil$ values for n. In our experience, the rank conditions work well when either d or n is known. There are still many open issues in the problem of finding a good search strategy and model selection criterion for n and k when both of them are unknown. Some of these issues will be addressed by other methods to be introduced in Chapter 6 and Chapter 8.

5.4.3 Clustering Subspaces of Different Dimensions

In this section, we consider the problem of clustering an unknown number of subspaces of unknown and possibly different dimensions from sample points.

First of all, we notice that the simultaneous recovery of the number and dimensions of the subspaces may be an ill-conditioned problem if we are not clear about what we are looking for. For example, in the extreme cases, one may interpret the sample set $\mathcal{X}$ as N 1-dimensional subspaces, with each subspace spanned by each of the sample points $\boldsymbol{x} \in \mathcal{X}$; or one may view the whole $\mathcal{X}$ as belonging to one D-dimensional subspace, i.e., $\mathbb{R}^D$ itself.

Although the above two trivial solutions can be easily rejected by imposing some conditions on the solutions,[17] other more difficult ambiguities may also arise in cases such as that of Figure 1.2 in which two lines and a plane can also be interpreted as two planes. More generally, when the subspaces are of different dimensions, one may not be able to determine the number of subspaces directly from the degree of the polynomials fitting the data, because the degree of the polynomial of minimum degree that fits a collection of subspaces is always less than or equal to the number of subspaces.

To resolve the difficulty in determining the number and dimension of subspaces, notice that the *algebraic set* $Z_{\mathcal{A}} = \cup_{i=1}^n S_i$ can be decomposed into irreducible subsets S_i; an irreducible algebraic set is also called a *variety*. The decomposition of $Z_{\mathcal{A}}$ into $\{S_1, S_2, \ldots, S_n\}$ is always unique. Therefore, as long as we are able to correctly determine from the given sample points the underlying algebraic set $Z_{\mathcal{A}}$ or the associated (radical) ideal $I(Z_{\mathcal{A}})$, in principle the number of subspaces n and their dimensions $\{d_1, d_2, \ldots, d_n\}$ can always be uniquely determined in a purely algebraic fashion. In Figure 1.2, for instance, the first interpretation (two lines and one plane) would be the right one, and the second one (two planes) would be incorrect, because the two lines, which span one of the planes, do not form an irreducible algebraic set.

Having established that the problem of subspace clustering is equivalent to decomposing the algebraic ideal associated with the subspaces, we are left with deriving a computable scheme to achieve the goal.

From every homogeneous component I_k of

$$I(Z_{\mathcal{A}}) = I_m \oplus I_{m+1} \oplus \cdots \oplus I_n \oplus \cdots,$$

we may compute a subspace arrangement Z_k such that $Z_{\mathcal{A}} \subseteq Z_k$ is a subspace embedding (see Section C.3). For each $k \geq m$, we can evaluate the derivatives of polynomials in I_k on subspace S_i and denote the collection of derivatives by

$$D_{k,i} \doteq \cup_{\boldsymbol{x} \in S_i} \{\nabla f\,|_{\boldsymbol{x}},\ \forall f \in I_k\}, \quad i = 1, 2, \ldots, n. \tag{5.51}$$

[17]To reject the N-lines solution, one can put a cap on the maximum number of groups n_{max}; and to reject $\mathbb{R}^D$ as the solution, one can simply require that the maximum dimension of every subspace be strictly less than D.

Obviously, we have the following relationship:

$$D_{k,i} \subseteq D_{k+1,i} \subseteq S_i^{\perp}, \quad \forall k \geq m. \tag{5.52}$$

Then for each I_k, we can define a new subspace arrangement as

$$Z_k \doteq D_{k,1}^{\perp} \cup D_{k,2}^{\perp} \cup \cdots \cup D_{k,n}^{\perp}. \tag{5.53}$$

Notice that it is possible that $D_{k,i} = D_{k,i'}$ for different i and i', and Z_k contains fewer than n subspaces. We summarize the above derivation as the following theorem.

Theorem 5.14 (A Filtration of Subspace Arrangements). *Let $I(Z_{\mathcal{A}}) = I_m \oplus I_{m+1} \oplus \cdots \oplus I_n \oplus \cdots$ be the ideal of a subspace arrangement $Z_{\mathcal{A}}$. Let Z_k be the subspace arrangement defined by the derivatives of $I_k, k \geq m$ as above. Then we obtain a filtration of subspace arrangements*

$$Z_m \supseteq Z_{m+1} \supseteq \cdots \supseteq Z_n = Z_{\mathcal{A}},$$

and each subspace of $Z_{\mathcal{A}}$ is embedded in one of the subspaces of Z_k.

The above theorem naturally leads to a recursive scheme that allows us to determine the correct number and dimensions of the subspaces in $Z_{\mathcal{A}}$. Specifically, we start with $n = 1$ and increase n until there is at least one polynomial of degree n fitting all the data, i.e., until the matrix $V_n(D)$ drops rank for the first time. For such an n, we can use Algorithm 5.4 to separate the data into n subspaces. Then we can further separate each one of these n groups of points using the same procedure. The stopping criterion for the recursion is when all the groups cannot be further separated or the number of groups n reaches some n_{max}.[18]

5.5 Model Selection for Multiple Subspaces

However, if the data points in the sample set $\mathcal{X}$ are corrupted by random noise, the above recursive scheme may fail to return a meaningful solution. In fact, up till now, we have been purposely avoiding a fundamental difficulty in our problem: it is inherently *ambiguous* in fitting multiple subspaces for any given data set, especially if the number of subspaces and their dimensions are not given a priori. When the sample points in $\mathcal{X}$ are noisy or are in fact drawn from a nonlinear manifold, any multisubspace model will unlikely fit the data perfectly except for the pathological cases: 1. All points are viewed as in one D-dimensional subspace, the ambient space. 2. Every point is viewed as lying in an individual one-dimensional subspace.

[18]For example, the inequality $M_n(D) \leq N$ imposes a constraint on the maximum possible number of groups n_{max}.

In general, the greater the number of planes we use, the higher accuracy we may achieve in fitting any given data set. Thus, a fundamental question we like to address in this section is the following:

> *Among a class of subspace arrangements, what is the "optimal" model that fits a given data set?*

From a practical viewpoint, we also need to know under what conditions the optimal model exists and is unique, and more importantly, how to compute it efficiently.

In Appendix B, we have seen that in general, any model selection criterion aims to strike a balance between the complexity of the resulting model and the fidelity of the model to the given data. However, its exact form often depends on the class of models of interest as well as how much information is given about the model in advance. If we are to apply any of the model-selection criteria (or their concepts) to subspace arrangements, at least two issues need to be addressed:

1. We need to know how to measure the model complexity of arrangements of subspaces (possibly of different dimensions).
2. Since the choice of a subspace arrangement involves both continuous parameters (the subspace bases) and discrete parameters (the number of subspaces and their dimensions), we need to know how to properly balance the model complexity and the modeling error for subspace arrangements.

In the rest of this section, we provide a specific model selection criterion for subspace arrangements. The most fundamental idea behind the proposed criterion is that the optimal model should lead to the most compact or sparse representation for the data set.

5.5.1 Effective Dimension of Samples of Multiple Subspaces

Definition 5.15 (Effective Dimension). *Given an arrangement of n subspaces $Z_{\mathcal{A}} \doteq \cup_{i=1}^n S_i$ in $\mathbb{R}^D$ of dimension $d_i < D$, and N_i sample points $\mathcal{X}_i$ drawn from each subspace S_i, the* effective dimension *of the entire set $\mathcal{X} = \cup_{i=1}^n \mathcal{X}_i$ of $N = \sum_{i=1}^n N_i$ sample points is defined to be*

$$\mathrm{ED}(\mathcal{X}, Z_{\mathcal{A}}) \doteq \frac{1}{N}\Big(\sum_{i=1}^{n} d_i(D-d_i) + \sum_{i=1}^{n} N_i d_i\Big). \tag{5.54}$$

We contend that $\mathrm{ED}(\mathcal{X}, Z_{\mathcal{A}})$ is the "average" number of (unquantized) real numbers that one needs to assign to $\mathcal{X}$ per sample point in order to specify the configurations of the n subspaces and the relative locations of the sample points in the subspaces. In the first term of equation (5.54), $d_i(D-d_i)$ is the total number of real numbers (known as the Grassmannian coordinates)[19] needed to specify a d_i-dimensional subspace S_i in $\mathbb{R}^D$; in the second term of (5.54), $N_i d_i$ is the total number of real numbers needed to specify the d_i coordinates of the N_i sample points in the subspace S_i. In general, if there is more than one subspaces in $Z_{\mathcal{A}}$, $\mathrm{ED}(\mathcal{X}, thenZ_{\mathcal{A}})$ can be a rational number instead of an integer for the conventional dimension.

Notice that we choose here real numbers as the basic "units" for measuring complexity of the model in a manner similar to that in the theory of sparse representation. Indeed, if the set of basis vectors of the subspaces is given, the second term of the effective dimension is essentially the sum of the ℓ^0 norm of the data points each represented as a linear combination of the bases. In general, the existence of a sparse linear representation always relies on the fact that the underlying model is an arrangement of a large number of subspaces. Of course, the compactness of the model can potentially be measured by more accurate units than real numbers. Binary numbers, or "bits," have traditionally been used in information theory for measuring the complexity of a data set. We will thoroughly examine that direction in the next chapter and will subsequently reveal the relationships among different measures such as the ℓ^0 norm, ℓ^1 norm, and (binary) coding length.

In the above definition, the effective dimension of $\mathcal{X}$ depends on the subspace arrangement $Z_{\mathcal{A}}$. This is because in general, there could be many subspace structures that can fit $\mathcal{X}$. For example, we could interpret the whole data set as lying in one D-dimensional subspace, and we would obtain an effective dimension D. On the other hand, we could interpret every point in $\mathcal{X}$ as lying in a one-dimensional subspace spanned by itself. Then there will be N such one-dimensional subspaces in total, and the effective dimension, according to the above formula, will also be D. In general, such interpretations are obviously somewhat redundant. Therefore, we define the *effective dimension* of a given sample set $\mathcal{X}$ to be the minimum among all possible models that can fit the data set:[20]

$$\mathrm{ED}(\mathcal{X}) \doteq \min_{Z_{\mathcal{A}}:\mathcal{X}\subset Z_{\mathcal{A}}} \mathrm{ED}(\mathcal{X}, Z_{\mathcal{A}}). \tag{5.55}$$

Example 5.16 (Effective Dimension of One Plane and Two Lines). Figure 1.2 shows data points drawn from one plane and two lines in $\mathbb{R}^3$. Obviously, the points in the two lines can also be viewed as lying in the plane that is spanned by the two lines. However, that interpretation would result in an increase of the effective dimension, since one would need two coordinates to specify a point in a plane, as opposed to one on a line. For instance, suppose there are fifteen points on each line and thirty points

[19]Notice that to represent a d-dimensional subspace in a D-dimensional space, we need only specify a basis of d linearly independent vectors for the subspace. We may stack these vectors as rows of a $d \times D$ matrix. Every nonsingular linear transformation of these vectors spans the same subspace. Thus, without loss of generality, we may assume that the matrix is of the normal form $[I_{d\times d}, G]$ where G is a $d \times (D-d)$ matrix consisting of the so-called Grassmannian coordinates.

[20]The space of subspace arrangements is topologically compact and closed; hence the minimum effective dimension is always achievable and hence well defined.

in the plane. When we use two planes to represent the data, the effective dimension is $\frac{1}{60}(2\times2\times3-2\times2^2+60\times2) = 2.07$; when we use one plane and two lines, the effective dimension is reduced to $\frac{1}{60}(2\times2\times3-2^2-2\times1+30\times1+30\times2) = 1.6$. In general, if the number of points N is arbitrarily large (say approaching infinity), then depending on the distributions of points on the lines or the plane, the effective dimension can be anything between 1 and 2, the true dimensions of the subspaces.

As suggested by the above example, the arrangement of subspaces that leads to the minimum effective dimension normally corresponds to a "natural" and hence compact representation of the data in the sense that it achieves the best compression (or dimension reduction) among all possible multiple-subspace models.

5.5.2 Minimum Effective Dimension of Noisy Samples

In practice, real data are corrupted by noise; hence we do not expect that the optimal model will fit the data perfectly. The conventional wisdom is to strike a good balance between the complexity of the chosen model and the data fidelity (to the model). See Appendix B.4 for a more detailed discussion about numerous model selection criteria. To measure the data fidelity, let us denote the projection of each data point $\boldsymbol{x}_j \in \mathcal{X}$ to the closest subspace by $\hat{\boldsymbol{x}}_j$ and let $\hat{\mathcal{X}} = \{\hat{\boldsymbol{x}}_j\}_{j=1}^N$. Then the total error residual can be measured by

$$\|\mathcal{X} - \hat{\mathcal{X}}\|^2 = \sum_{j=1}^{N} \|\boldsymbol{x}_j - \hat{\boldsymbol{x}}_j\|^2. \tag{5.56}$$

Since all model selection criteria exercise the same rationale as above, we here adopt the geometric-AIC (G-AIC) criterion (2.87),[21] which leads to the following objective for selecting the optimal multiple-subspace model:

$$Z_{\mathcal{A}}^* = \arg\min_{Z_{\mathcal{A}}:\hat{\mathcal{X}}\subset Z_{\mathcal{A}}} \frac{1}{N}\|\mathcal{X} - \hat{\mathcal{X}}\|^2 + 2\sigma^2 \mathrm{ED}(\hat{\mathcal{X}}, Z_{\mathcal{A}}), \tag{5.57}$$

where σ^2 is the noise variance of the data. However, this optimization problem can be very difficult to solve: The variance σ^2 might not be known a priori, and we need to search for the global minimum in the configuration space of all subspace arrangements, which is not a smooth manifold and has very complicated topological and geometric structures. The required computation is typically prohibitive.

[21] We here adopt the G-AIC criterion only to illustrate the basic ideas. In practice, depending on the problem and application, it is possible that other model selection criteria may be more appropriate.

To alleviate some of the difficulty, we may in practice instead minimize the effective dimension subject to a maximum allowable error tolerance. That is, among all the multiple-subspace models that fit the data within a given error bound, we choose the one with the smallest effective dimension. To this end, we define the minimum effective dimension *subject to an error tolerance* τ as

$$\mathrm{MED}(\mathcal{X},\tau) \doteq \min_{Z_{\mathcal{A}}} \mathrm{ED}(\hat{\mathcal{X}}, Z_{\mathcal{A}}) \quad \text{s.t.} \quad \|\mathcal{X}-\hat{\mathcal{X}}\|_{\infty} \leq \tau, \tag{5.58}$$

where $\hat{\mathcal{X}}$ is the projection of $\mathcal{X}$ onto the subspaces in $Z_{\mathcal{A}}$, and the error norm $\|\cdot\|_{\infty}$ indicates the maximum norm: $\|\mathcal{X}-\hat{\mathcal{X}}\|_{\infty} = \max_{1\leq j\leq N}\|\boldsymbol{x}_j-\hat{\boldsymbol{x}}_j\|$. Based on the above definition, the effective dimension of a data set is then a notion that depends on the error tolerance. In the extreme, if the error tolerance is arbitrarily large, the "optimal" subspace model for any data set can simply be the (zero-dimensional) origin; if the error tolerance is zero instead, for data with random noise, each sample point needs to be treated as a one-dimensional subspace in $\mathbb{R}^D$ of its own, and that brings the effective dimension up close to D.

In many applications, the notion of maximum allowable error tolerance is particularly relevant. For instance, in image representation and compression, the task is often to find a linear or hybrid linear model to fit the imagery data subject to a given peak signal-to-noise ratio (PSNR).[22] The resulting effective dimension directly corresponds to the number of coefficients needed to store the resulting representation. The smaller the effective dimension, the more compact or compressed the final representation. In Chapter 9, we will see exactly how the minimum effective dimension principle is applied to image representation. The same principle can be applied to any situation in which one tries to fit a piecewise linear model to a data set whose structure is nonlinear or unknown.

5.5.3 *Recursive Algebraic Subspace Clustering*

Unlike the geometric AIC (5.57), the MED objective (5.58) is relatively easy to achieve. For instance, the recursive ASC scheme that we discussed earlier at the end of Section 5.4.3 can be easily modified to minimize the effective dimension subject to an error tolerance: we allow the recursion to proceed only if the effective dimension would decrease while the resulting subspaces still fits the data with the given error bound.

---

[22]In this context, the noise is the difference between the original image and the approximate image (the signal).

To summarize the above discussions, in principle we can use the following algorithm to recursively identify subspaces in an arrangement $Z_{\mathcal{A}}$ from a set of noisy samples $\mathcal{X} = \{\boldsymbol{x}_1, \boldsymbol{x}_2, \ldots, \boldsymbol{x}_N\}$.

Be aware that when the data are noisy, it sometimes can be very difficult to determine the correct dimension of the null space of the matrix $\boldsymbol{V}_n(D)$ from its singular-value spectrum. If the dimension is determined incorrectly, it may result in either underestimating or overestimating the number of fitting polynomials. In general, if the number of polynomials were underestimated, the resulting subspaces would overfit the data;[23] and if the number of polynomials were overestimated, the resulting subspaces would underfit the data.

Obviously, both overfitting and underfitting result in incorrect estimates of the subspaces. However, do they necessarily result in equally bad clustering of the data? The answer is *no*. Between overfitting and underfitting, we actually would favor overfitting. The reason is that though overfitting results in subspaces that are larger than the original subspaces, it is a zero-measure event that an overestimated subspace contains simultaneously more than one original subspace. Thus, the grouping of the data points may still be correct. For instance, consider the extreme case that we choose only one polynomial that fits the data. Then the derivatives of the polynomial, evaluated at one point per subspace, lead to n hyperplanes. Nevertheless, these overfitting hyperplanes will in general result in a correct grouping of the data points. One can verify this with the introductory example we discussed in Section 5.3.1. Either of the two polynomials $p_{21}(\boldsymbol{x}) = x_1 x_3$ and $p_{22}(\boldsymbol{x}) = x_2 x_3$ leads to two hyperplanes that cluster the line and the plane correctly.

Example 5.17 (Recursive ASC on Synthetic Data) Figure 5.8 demonstrates an example of applying the recursive ASC algorithm, Algorithm 5.5, to cluster synthetic data points drawn from two lines (100 points each) and one plane (400 points) in $\mathbb{R}^3$ corrupted by 5% uniform noise (Figure 5.8 topleft). Given a reasonable error tolerance, the algorithm stops after two levels of recursion (Figure 5.8 top right). Note that the pink line (top right) or group 4 (bottom left) is a "ghost" line at the intersection of the original plane and the plane spanned by the two lines.[24] Figure 5.8 bottom right is the plot of MED of the same data set subject to different levels of error tolerance. As we see, the effective dimension decreases monotonically with the increase in error tolerance.

---

[23]That is, the dimensions of some of the subspaces estimated could be larger than the true ones.

[24]This is exactly what we would have expected, since the recursive ASC first clusters the data into two planes. Points on the intersection of the two planes get assigned to either plane depending on the random noise. If needed, the points on the ghost line can be merged with the plane by some simple postprocessing.

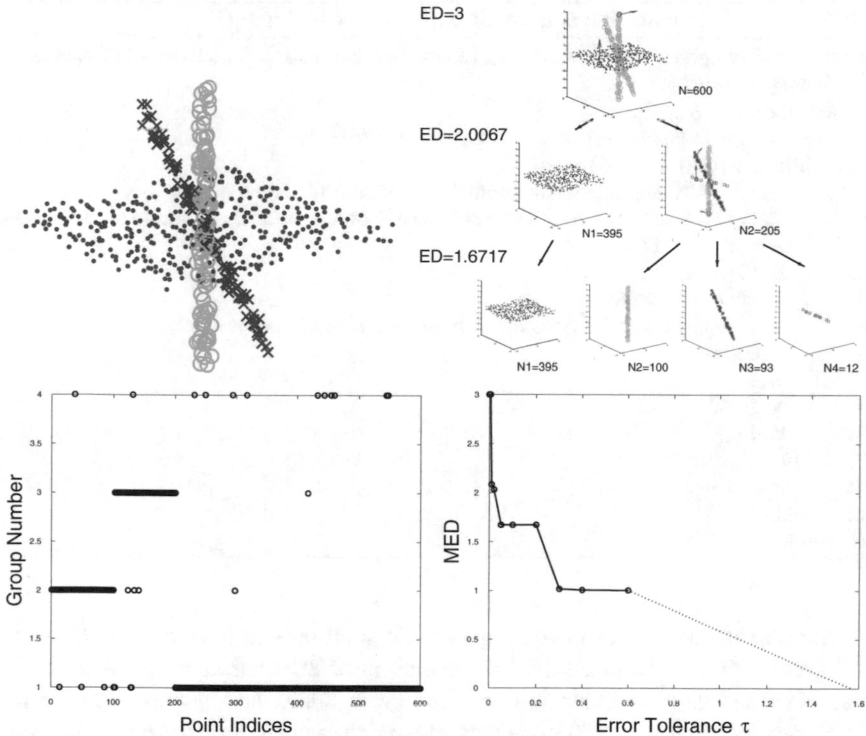

Fig. 5.8 Simulation results. Top left: sample points drawn from two lines and a plane in $\mathbb{R}^3$ with 5% uniform noise; top right: the process of recursive clustering by the recursive ASC algorithm, Algorithm 5.5, with the error tolerance $\tau = 0.05$; bottom left: group assignment for the points; bottom right: plot of MED versus error tolerance.

5.6 Bibliographic Notes

Algebraic Clustering Algorithms and Extensions

The difficulty with initialization for the iterative clustering algorithms that we have presented in the previous chapter has motivated the recent development of algebrogeometric approaches to subspace clustering that do *not* require initialization. (Kanatani 2001; Boult and Brown 1991; Costeira and Kanade 1998) demonstrated that when the subspaces are orthogonal, of equal dimensions, and with trivial intersection, one can use the SVD of the data to define a similarity matrix from which the clustering of the data can be obtained using spectral clustering techniques. Unfortunately, this method is sensitive to noise in the data, as pointed out in (Kanatani 2001; Wu et al. 2001), where various improvements are proposed. When the intersection of the subspaces is nontrivial, the clustering of the data is usually obtained in an ad hoc fashion, again using clustering algorithms such as K-means. A basis for each subspace is then obtained by applying PCA to each group. For the

Algorithm 5.5 (Recursive Algebraic Subspace Clustering).

Given a set of samples $\mathcal{X} = \{\boldsymbol{x}_1, \boldsymbol{x}_2, \ldots, \boldsymbol{x}_N\}$ in the ambient space $\mathbb{R}^D$, find a set of subspaces that fit $\mathcal{X}$ subject to an error $\tau > 0$:

for all $k = 1 : n_{max}$ **do**
  Set $\boldsymbol{V}_k(D) \doteq [\nu_k(\boldsymbol{x}_1), \nu_k(\boldsymbol{x}_2), \ldots, \nu_k(\boldsymbol{x}_N)]^\top \in \mathbb{R}^{M_k(D) \times N}$.
  if $\text{rank}(\boldsymbol{V}_k(D)) < M_k(D)$ **then**
    Use the ASC algorithm, Algorithm 5.4, to partition $\mathcal{X}$ into k subsets $\mathcal{X}_1, \ldots, \mathcal{X}_k$.
    Apply PCA and fit each $\mathcal{X}_i$ with a subspace S_i of dimension d_i, subject to the error τ. Let $Z = S_1 \cup \cdots \cup S_k$.
    if $\text{ED}(\mathcal{X}, Z) < D$ **then**
      for $i = 1 : k$ **do**
        Apply **recursive ASC** for $\mathcal{X}_i$ (with S_i as the ambient space).
      end for
    else
      Break.
    end if
  else
    $k \leftarrow k + 1$.
  end if
end for

special case of two planes in $\mathbb{R}^3$, a geometric solution was developed in (Shizawa and Mase 1991) in the context of segmentation of 2D transparent motions. In the case of subspaces of codimension one, i.e., *hyperplanes*, an algebraic solution was developed in (Vidal et al. 2003b), where the hyperplane clustering problem is shown to be equivalent to homogeneous polynomial factorization.

The algebraic subspace clustering algorithm for the most general case[25] was later developed in (Vidal et al. 2004); and the decomposition of the polynomial(s) was based on differentiation, a numerically better conditioned operation. The algebraic clustering algorithm was successfully applied to solve the motion segmentation problem in computer vision (Vidal and Ma 2004). The generalization to arrangements of both linear and quadratic surfaces was first studied in (Rao et al. 2005).

Algebraic Properties of Subspace Arrangements

The importance of using subspace arrangements to model real-world high-dimensional data and the early success of the basic ASC algorithms had motivated mathematicians to provide a more thorough characterization of subspace arrangements in terms of their vanishing ideals. A complete characterization of the Hilbert functions of the ideals for subspace arrangements was given in (Derksen 2007), which serves as the theoretical foundation for this chapter. In Appendix C, we have sketched the basic algebraic concepts, results, and additional references about subspace arrangements. One may also refer to (Ma et al. 2008) for a comprehensive review of recent developments of this topic.

[25]That is, an arbitrary number of subspaces of arbitrary dimensions.

Effective Dimension, Sparsity, and Compression

The notion of minimum effective dimension was first introduced in the context of the recursive algebraic subspace clustering method studied in (Huang et al. 2004). We now understand that effective dimension is essentially a sparsity measure in terms of the ℓ^0-norm. In future chapters, we will examine other surrogate measures of model compactness for subspace arrangements, including coding length in Chapter 6 and the convex ℓ^1-norm in Chapter 8.

Robustness and Outlier Rejection

There has been much work on the estimation of polynomials that best fit a given set of noisy samples. In Exercise 5.10, we will study one such approach that works well in the context of algebraic subspace clustering. The approach essentially follows that of (Taubin 1991).

If there are also outliers in the given sample set, the problem becomes a more difficult robust model estimation problem. There is vast body of literature on robust statistics; see Appendix B.5 for a brief review. Sample influence is always believed to be an important index for detecting outliers. Certain first-order approximations of the influence value were developed at roughly the same time as the sample influence function was proposed (Campbell 1978; Critchley 1985), when computational resources were scarcer than they are today. In the literature, formulas that approximate an influence function are referred to as *theoretical influence functions*. Usually, the percentage of outliers can be determined by the influence of the candidate outliers on the model estimated (Hampel et al. 1986).

In the basic ASC algorithm, Algorithm 5.4, we see that the key is to be able to robustly estimate the covariance of the samples in the lifted space, i.e. the matrix $V_n(D)^\top V_n(D)$. Among the class of robust covariance estimators (see Appendix B.5), the multivariate trimming (MVT) method (Gnanadesikan and Kettenring 1972) has always been one of the most popular for practitioners, probably because of its computational efficiency for high-dimensional data as well as its tolerance of a large percentage of outliers. Its application to ASC is posed as Exercise 5.12.

Random sampling techniques such as the least median estimate (LME) (Hampel 1974; Rousseeuw 1984) and random sample consensus (RANSAC) (Fischler and Bolles 1981) have been widely used in many engineering areas, especially in pattern recognition and computer vision (Steward 1999). They are very effective when the model is relatively simple. For instance, RANSAC is known to be very effective in making the classic PCA robust, i.e., in estimating a single subspace in the presence of outliers. However, if there are multiple subspaces, RANSAC is known to work well when the dimensions of all the subspaces are the same (Torr 1998). If the subspaces have different dimensions, a Monte Carlo scheme can be used to estimate one subspace at a time (Torr and Davidson 2003; Schindler and Suter 2005). However, the performance degrades very quickly with the increase in the number of subspaces and the percentage of outliers. This has been observed in the careful experimental comparison done in (Yang et al. 2006). ASC combined with MVT has been shown to perform generally better on most of the simulated data sets.

In the next chapter, we are going to see an entirely new approach to clustering data from multiple subspaces. Rather than fitting a global model to the arrangement or one model for each subspace, the new method forms subspace-like clusters by merging one sample point at a time. As we will see, one distinctive feature of such an agglomerative approach is its striking ability to handle a high percentage of outliers, far more robust than the methods we have discussed or exercised so far.

5.7 Exercises

Exercise 5.1 (Clustering Points in a Plane). Describe how Algorithm 5.1 can also be applied to a set of points in the plane $\{\boldsymbol{x}_j \in \mathbb{R}^2\}_{j=1}^N$ that are distributed around a collection of cluster centers $\{\boldsymbol{\mu}_i \in \mathbb{R}^2\}_{i=1}^n$ by interpreting the data points as complex numbers: $\{z \doteq x + y\sqrt{-1} \in \mathbb{C}\}$. In particular, discuss what happens to the coefficients and roots of the fitting polynomial $p_n(z)$.

Exercise 5.2 (Connection of Algebraic Clustering with Spectral Clustering). Spectral clustering is a very popular data clustering method. In spectral clustering, one is given a set of N data points (usually in a multidimensional space) and an $N \times N$ pairwise similarity matrix $S = (s_{ij})$. The entries s_{ij} of S measure the likelihood of two points belonging to the same cluster: $s_{ij} \to 1$ when points i and j likely belong to the same group and $s_{ij} \to 0$ when points i and j likely belong to different groups.

1. First examine the special case in which the N data points have two clusters and the similarity matrix S is *ideal*; that is, $s_{ij} = 1$ if and only if points i and j belong to the same cluster and $s_{ij} = 0$ otherwise. What do the eigenvectors of S look like, especially the one(s) that correspond to nonzero eigenvalue(s)? Argue how the entries of the eigenvectors encode information about the membership of the points.
2. Show how Algorithm 5.1 can be used to cluster the points based on the eigenvector of the similarity matrix. Based on Exercise 5.1, show how to cluster the points by using two eigenvectors simultaneously.
3. Generalize your analysis and conclusions to the case of n clusters.

Since many popular image segmentation algorithms are based on spectral clustering (on certain similarity measure between pixels), you may use the above algorithm to improve the segmentation results.

Exercise 5.3 (Level Sets and Normal Vectors). Let $f(\boldsymbol{x}) : \mathbb{R}^D \to \mathbb{R}$ be a smooth function. For a constant $c \in \mathbb{R}$, the set $S_c \doteq \{\boldsymbol{x} \in \mathbb{R}^D | f(\boldsymbol{x}) = c\}$ is called a level set of the function f; S_c is in general a $(D-1)$-dimensional submanifold. Show that if $\|\nabla f(\boldsymbol{x})\|$ is nonzero at a point $\boldsymbol{x}_0 \in S_c$, then the gradient $\nabla f(\boldsymbol{x}_0) \in \mathbb{R}^D$ at $\boldsymbol{x}_0$ is orthogonal to all tangent vectors of the level set S_c.

Exercise 5.4 (Hyperplane Embedding from a Single Polynomial). Consider a subspace arrangement $Z_{\mathcal{A}} = S_1 \cup S_2 \cup \cdots \cup S_n \subset \mathbb{R}^D$; $f(\boldsymbol{x})$ is a polynomial that vanishes on $Z_{\mathcal{A}}$. Show that if we differentiate $f(\boldsymbol{x})$ at points on $Z_{\mathcal{A}}$, we always obtain an arrangement of hyperplanes that contain $Z_{\mathcal{A}}$.

Exercise 5.5 (Multiple Algebraic Subspace Clustering). For each $f = 1, 2, \ldots, F$, let $\{\boldsymbol{x}_{fj} \in \mathbb{R}^D\}_{j=1}^N$ be a collection of N points lying in n hyperplanes with normal vectors $\{\boldsymbol{b}_{fi}\}_{i=1}^n$. For each $j = 1, 2, \ldots, N$, the sequence $\{\boldsymbol{x}_{fj} \in \mathbb{R}^D\}_{f=1}^F$ represents the trajectory of the jth data point. Assume that each sequence of data points $\{\boldsymbol{x}_{fj} \in \mathbb{R}^D\}_{f=1}^F$ is associated with one sequence of hyperplanes with normal vectors $\{\boldsymbol{b}_{fi}\}_{f=1}^F$. That is, for each $j = 1, 2, \ldots, N$, there is an $i = 1, 2, \ldots, n$ such that for all $f = 1, 2, \ldots, F$, we have $\boldsymbol{b}_{fi}^\top \boldsymbol{x}_{1j} = 0$. Propose an extension of the ASC algorithm that computes the normal vectors in such a way that $\boldsymbol{b}_{1i}, \boldsymbol{b}_{2i}, \ldots \boldsymbol{b}_{Fi}$ correspond to each other.

Hint: If $p_{fn}(\boldsymbol{x}) = \boldsymbol{c}_f^\top \nu_n(\boldsymbol{x}) = (\boldsymbol{b}_{f1}^\top \boldsymbol{x})(\boldsymbol{b}_{f2}^\top \boldsymbol{x}) \cdots (\boldsymbol{b}_{fn}^\top \boldsymbol{x})$ *and the jth set of points* $\boldsymbol{x}_{1j}, \boldsymbol{x}_{2j}, \ldots, \boldsymbol{x}_{Fj}$ *corresponds to the ith group of hyperplanes, then* $\boldsymbol{b}_{fi} \sim \nabla p_{fn}(\boldsymbol{x}_{fj})$.

Exercise 5.6 (Properties of the Veronese map). Consider the Veronese map $\nu_n : [x_1, \ldots, x_D]^\top \mapsto [\ldots, \boldsymbol{x}^n, \ldots]^\top$, where $\boldsymbol{x}^n = x_1^{n_1} x_2^{n_2} \ldots x_D^{n_D}$ ranges over all monomials of degree $n = \sum_{i=1}^{D} n_i$ in the variables $x_1, x_2, \ldots, x_D$, sorted in degree-lexicographic order, and let $\boldsymbol{x}, \boldsymbol{y} \in \mathbb{R}^D$.

1. **Inner product invariance:** Show that the polynomial kernel $k(\boldsymbol{x}, \boldsymbol{y}) = (\boldsymbol{y}^\top \boldsymbol{x})^n$ can be written in terms of the Veronese map as $k(\boldsymbol{x}, \boldsymbol{y}) = \nu_n(\boldsymbol{y})^\top M \nu_n(\boldsymbol{x})$, where $M \in \mathbb{R}^{M_n(D) \times M_n(D)}$ is a diagonal matrix, and its $(n_1, n_2, \ldots, n_D)$th entry is $\frac{n!}{n_1! n_2! \ldots n_D!}$ with $\sum_{i=1}^D n_i = n$.
   Hint: Use the multinomial theorem.
2. **Linear invariance:**
   (a) Show that $\nu_n(\alpha \boldsymbol{x} + \boldsymbol{y}) = \sum_{i=0}^n \alpha^i f_i(\boldsymbol{x}, \boldsymbol{y})$, where $f_i(\boldsymbol{x}, \boldsymbol{y}) \in \mathbb{R}^{M_n(D)}$ is a bihomogeneous polynomial of degree i in $\boldsymbol{x}$ and $(n - i)$ in $\boldsymbol{y}$ for $i = 0, \ldots, n$.
   (b) Let S_n be the space of homogeneous polynomials of degree n in D variables. Define the transformation $T : S_n \to S_n$ such that $T(p_n(\boldsymbol{x})) = p_n(A\boldsymbol{x})$, where $A \in \mathbb{R}^{D \times D}$. Show that the transformation T is linear.
   (c) Show that for all $A \in \mathbb{R}^{D \times D}$, there exists an $\tilde{A} \in \mathbb{R}^{M_n(D) \times M_n(D)}$ such that for all $\boldsymbol{x}$, $\nu_n(A\boldsymbol{x}) = \tilde{A} \nu_n(\boldsymbol{x})$.
3. **Rotation invariance:** Show that for $D = 3$ and all $R \in SO(3)$, there exists $\tilde{R} \in SO(M_n(D))$ such that for all $\boldsymbol{x}$, $\nu_n(R\boldsymbol{x}) = \tilde{R} \nu_n(\boldsymbol{x})$.

Exercise 5.7 (Two Subspaces in General Position). Consider two linear subspaces of dimension d_1 and d_2 respectively in $\mathbb{R}^D$. We say that they are in general position if an arbitrarily small perturbation of the position of the subspaces does not change the dimension of their intersection. Show that two subspaces are in general position if and only if

$$\dim(S_1 \cap S_2) = \min\{d_1 + d_2 - D; 0\}. \tag{5.59}$$

Exercise 5.8. Implement the basic algebraic subspace clustering algorithm, Algorithm 5.4, and test the algorithm for different subspace arrangements with different levels of noise.

Exercise 5.9. Consider a collection of points $\{x_j \in \mathbb{R}^3\}_{j=1}^P$ lying in three subspaces of $\mathbb{R}^3$,

$$S_1 = \{x : x_3 = 0\}, \tag{5.60}$$

$$S_2 = \{x : x_1 = 0 \wedge x_2 + x_3 = 0\}, \tag{5.61}$$

$$S_3 = \{x : x_1 = 0 \wedge x_2 - x_3 = 0\}. \tag{5.62}$$

1. Show that the data can be fit with a set of m homogeneous polynomials of degree $n = 2$ in three variables. Determine the value of m. Write down the m polynomials explicitly. What is the minimum number of points P, and how should such points be distributed in S_1, S_2, and S_3 so that the m polynomials can be uniquely determined? Show how to determine m and the polynomials from data. Compute the gradient of each of the m polynomials at a data point $y_1 \in S_1$, $y_2 \in S_2$ and $y_3 \in S_3$. Is it possible to segment the data into the three subspaces using these gradients? If so, say how. If not, say what segmentation can be obtained from the gradients.
2. Answer questions in part 1 with $n = 3$.
3. Answer all questions in part 1 with $n = 4$. If your answer to the last question in part 1 is yes, then explain why the data can be segmented correctly into the three subspaces, even though the degree of the polynomials is greater than the number of subspaces.

Exercise 5.10 (Estimating Vanishing Polynomials). In the next two exercises, we study two ways of estimating the vanishing polynomials of a subspace arrangement from noisy samples. Since the data are noisy, a sample point x is only close to the zero set of the fitting polynomials $P(x) = [p_1(x), p_2(x), \ldots, p_m(x)]^\top$. Let $\hat{x}$ be the closest point to x on the zero set of $P(x)$.

1. Show that the approximate square distance from x to $\hat{x}$ is given by

$$\|x - \hat{x}\|^2 \approx P(x)^\top \left(\nabla P(x) \nabla P(x)^\top\right)^\dagger P(x). \tag{5.63}$$

This distance is known as the *Sampson distance*. From this, conclude that given a set of sample points $\mathcal{X} = \{x_1, \ldots, x_N\}$, in order to minimize the mean square fitting error $\frac{1}{N}\sum_{j=1}^N \|x_j - \hat{x}_j\|_2^2$, we can approximately minimize the average Sampson distance

$$\frac{1}{N}\sum_{j=1}^{N} P(x_j)^\top \left(\nabla P(x_j) \nabla P(x_j)^\top\right)^\dagger P(x_j). \tag{5.64}$$

2. However, for every nonsingular matrix $M \in \mathbb{R}^{m \times m}$, the values $\tilde{P}(\boldsymbol{x}) = MP(\boldsymbol{x})$ define the same zero set. Show that in order to reduce this redundancy, we can normalize the following matrix to an identity:

$$\frac{1}{N}\sum_{j=1}^{N} \nabla P(\boldsymbol{x}_j)\nabla P(\boldsymbol{x}_j)^\top = I_{m\times m}. \tag{5.65}$$

Thus, the problem of minimizing the average Sampson distance now becomes a constrained optimization problem:

$$\begin{aligned} P^* = {} & \arg\min_P \tfrac{1}{N}\textstyle\sum_{j=1}^N P(\boldsymbol{x}_j)^\top \left(\nabla P(\boldsymbol{x}_j)\nabla P(\boldsymbol{x}_j)^\top\right)^\dagger P(\boldsymbol{x}_j), \\ & \text{subject to} \quad \tfrac{1}{N}\textstyle\sum_{j=1}^N \nabla P(\boldsymbol{x}_j)\nabla P(\boldsymbol{x}_j)^\top = I_{m\times m}. \end{aligned} \tag{5.66}$$

3. Since the average of $\nabla P(\boldsymbol{x}_j)\nabla P(\boldsymbol{x}_j)^\top$ is an identity, we can approximate each by an identity too. Then the above problem becomes

$$\begin{aligned} P^* = {} & \arg\min_P \tfrac{1}{N}\textstyle\sum_{j=1}^N \|P(\boldsymbol{x}_j)\|^2, \\ & \text{subject to} \quad \tfrac{1}{N}\textstyle\sum_{j=1}^N \nabla P(\boldsymbol{x}_j)\nabla P(\boldsymbol{x}_j)^\top = I_{m\times m}. \end{aligned} \tag{5.67}$$

Now show that the vector of coefficients of each polynomial in P^* is a generalized eigenvector for a properly defined pair of matrices W and B. That is, they are solutions $\boldsymbol{c}_i^*$ to the following equation:

$$W\boldsymbol{c}_i^* = \lambda_i B\boldsymbol{c}_i^*, \quad i = 1, 2, \ldots, m. \tag{5.68}$$

Exercise 5.11 (Fisher Discriminant Analysis for Subspaces). We now illustrate how concepts from discriminant analysis can be adopted to estimate better-fitting polynomials. We use an arrangement of hyperplanes to demonstrate the basic ideas. In this case, the fitting polynomial has the form

$$p(\boldsymbol{x}) = \prod_{i=1}^{n} (\boldsymbol{b}_i^\top \boldsymbol{x}) = \boldsymbol{c}^\top \nu_n(\boldsymbol{x}) = 0 \tag{5.69}$$

with n the number of (different) hyperplanes and $\boldsymbol{b}_i$ the normal vector to the ith plane. In this case, it is very easy to find the coefficient vector $\boldsymbol{c}$, since the kernel of the data matrix $\boldsymbol{V}_n(D)$ is only one-dimensional.

1. In the presence of noise, it is likely that $p(\boldsymbol{x}) \neq 0$, but we would like to find the coefficient vector $\boldsymbol{c}$ that minimizes the following average least-squares fitting error $\frac{1}{N}\sum_{j=1}^N |p(\boldsymbol{x}_j)|^2$. Show that the solution $\boldsymbol{c}^*$ is the eigenvector associated with the smallest eigenvalue of the matrix:

$$W \doteq \left(\frac{1}{N} V_n(D)^\top V_n(D)\right). \tag{5.70}$$

In the spirit of discriminant analysis, the matrix W will be called the *within-subspace scatter matrix.*

2. Let us examine the derivative of the polynomial at each of the data samples. Let $\boldsymbol{x}_1 \in S_1$. Show that the norm of the derivative $\nabla p(\boldsymbol{x}_1)$ is

$$\left\|\nabla p(\boldsymbol{x}_1)\right\|^2 = \left|\left(\prod_{i=2}^{n} \boldsymbol{b}_i^\top \boldsymbol{x}_1\right)\right|^2. \tag{5.71}$$

Thus, the average of the quantity $\|\nabla p(\boldsymbol{x}_1)\|^2$ over all $\boldsymbol{x}_1$ in S_1 gives a good measure of "distance" from S_1 to $\bigcup_{i=2}^{n} S_i$, the union of the other subspaces. For the purpose of clustering, we would like to find the coefficient vector $\boldsymbol{c}$ that maximizes the following quantity:

$$\max \frac{1}{N} \sum_{j=1}^{N} \left\|\nabla p(\boldsymbol{x}_j)\right\|^2 = \boldsymbol{c}^\top \left(\frac{1}{N} \sum_{j=1}^{N} \nabla \nu_n(\boldsymbol{x}_j) \nabla \nu_n(\boldsymbol{x}_j)^\top\right) \boldsymbol{c} \doteq \boldsymbol{c}^\top B \boldsymbol{c}. \tag{5.72}$$

In the spirit of discriminant analysis, we will call B the *between-subspace scatter matrix.*

3. Therefore, we would like to seek a fitting polynomial that simultaneously minimizes the polynomial evaluated at each of the samples while maximizing the norm of the derivative at each point. This can be achieved by minimizing the ratio of these two metrics:

$$\boldsymbol{c}^* = \arg\min_{\boldsymbol{c}} \frac{\boldsymbol{c}^\top W \boldsymbol{c}}{\boldsymbol{c}^\top B \boldsymbol{c}}. \tag{5.73}$$

Show that the solution to this problem is given by the generalized eigenvector $\boldsymbol{c}$ that is associated with the smallest generalized eigenvalue λ of (W, B):

$$W\boldsymbol{c} = \lambda B \boldsymbol{c}. \tag{5.74}$$

When B is nonsingular, $\boldsymbol{c}$ is simply the eigenvector of $B^{-1}W$ associated with the smallest eigenvalue.

Exercise 5.12 (Robust Estimation of Fitting Polynomials). We know that samples from an arrangement of n subspaces, their Veronese lifting, all lie on a single subspace span$(\boldsymbol{V}_n(D))$. The coefficients of the fitting polynomials are simply the null space of $\boldsymbol{V}_n(D)$. If there is noise, the lifted samples approximately span a subspace, and the coefficients of the fitting polynomials are eigenvectors associated with the small eigenvalues of $\boldsymbol{V}_n(D)^\top \boldsymbol{V}_n(D)$. However, if there are outliers, the lifted samples together no longer span a subspace. Notice that this is the same situation that robust statistical techniques such as multivariate trimming (MVT) are designed to deal with. See Appendix B.5 for more details. In this exercise, show how to combine MVT with ASC so that the resulting algorithm will be robust to outliers. Implement your scheme and find out the highest percentage of outliers that the algorithm can handle (for various subspace arrangements).

Chapter 6
Statistical Methods

Statistics in the hands of an engineer are like a lamppost to a drunk—they're used more for support than illumination.

—A.E. Housman

The algebraic-geometric approach to subspace clustering described in the previous chapter provides a fairly complete characterization of the algebra and geometry of multiple subspaces, which leads to simple and elegant subspace clustering algorithms. However, while these methods can handle some noise in the data, they do not make explicit assumptions about the distribution of the noise or the data inside the subspaces. Therefore, the estimated subspaces need not be optimal from a statistical perspective, e.g., in a maximum likelihood (ML) sense.

In this chapter, we study the subspace clustering problem from a statistical estimation perspective. We assume that the data points are noisy (or contaminated) samples drawn from a model based on the union of low-dimensional subspaces and develop algorithms for estimating that model from data. As we have seen in the study of (probabilistic) PCA in Chapter 2, noisy data points that belong to a single subspace can be modeled as samples from a (nearly degenerate) Gaussian. Therefore, a natural approach is to model data from multiple subspaces as samples from a mixture of (nearly degenerate) Gaussians. However, the associated model estimation problem is very challenging due to the following chicken-and-egg problem: To estimate the models for the subspaces (say by PPCA), we need to know which data points belong to which cluster; and to cluster the data points according to their respective subspaces, we need to know the model for each subspace.

In the first part of this chapter, we will present an extremely simple and intuitive approach to subspace clustering called *K-subspaces*, which alternates between assigning each data point to its closest subspace and estimating a subspace for each group of points using PCA. The K-subspaces algorithm can be seen as a

R. Vidal et al., *Generalized Principal Component Analysis*, Interdisciplinary Applied Mathematics 40, DOI 10.1007/978-0-387-87811-9_6

generalization of the K-means algorithm described in Section 4.3.1 from clustering data around cluster centers (0-dimensional subspaces) to clustering data in a union of subspaces. Thus, similar to K-means, one of the disadvantages of K-subspaces is that its performance depends on having a good initialization for either the subspaces or the segmentation. The K-subspaces algorithm can also be seen as a generalization of the geometric PCA algorithm described in Section 2.1.2 from one to multiple subspaces. As such, another disadvantage of K-subspaces is that it does not provide a proper generative model for the data.

In the second part of this chapter, we will approach the subspace clustering problem using conventional statistical methods for the estimation of mixture models. We will first introduce a generative model for data in a union of subspaces called *mixtures of probabilistic principal component analysis* (MPPCA), which generalizes the PPCA model described in Section 2.2 from one to multiple subspaces. We will show that the parameters of the MPPCA model can be estimated using the EM and MAP-EM methods described in Appendix B. However, since the distribution associated with a subspace is nearly degenerate, these general methods need to be properly customized to work for subspaces. As we will see, both the EM and MAP-EM algorithms rely on a very simple and intuitive iterative optimization procedure that alternates between estimating the subspaces (model estimation) and assigning data points to their subspaces (data clustering). As such, they can be seen as probabilistic versions of the K-subspaces algorithm. As before, one disadvantage of these methods is that their performance depends strongly on having a good initialization for either the subspaces or the segmentation of the data. Moreover, the model class (number of subspaces and their dimensions) needs to be known beforehand.

In the third part of the chapter, we will approach the subspace clustering and model selection problems using techniques from data compression. We will show that the problem of finding a model that maximizes the likelihood function, as done by the EM method, is essentially equivalent to minimizing the coding length of the data. We will also show that it is possible to accurately estimate the coding length for any subspace-like subset and that we can minimize the overall coding length (hence maximize the likelihood) by clustering the data in an agglomerative fashion. Unlike the EM method, the coding approach does not require any prior knowledge about the model class, nor does it conduct any intermediate model estimation during its optimization process. Moreover, the agglomerative process naturally separates outliers, since encoding them together would significantly increase the coding length. Since agglomerative methods have been widely used in the knowledge discovery literature, the framework presented in this chapter will show how such agglomerative methods can be naturally derived and justified from a data compression perspective. If properly designed, they are equivalent to principled statistical estimation methods for mixture models and yet have many computational and practical advantages.

6.1 K-Subspaces

A very simple way of improving the performance of algebraic algorithms in the case of noisy data is to use iterative refinement. Intuitively, given an initial segmentation, we can fit a subspace to each group using geometric PCA. Then, given a PCA model for each subspace, we can assign each data point to its closest subspace. By iterating these two steps, we can obtain a refined estimate of the subspaces and of the segmentation. This is the basic idea behind the K-planes (Bradley and Mangasarian 2000) and K-subspaces (Tseng 2000; Agarwal and Mustafa 2004) algorithms, which generalize the K-means algorithm (Lloyd 1957; Forgy 1965; Jancey 1966; MacQueen 1967) from data distributed around multiple cluster centers to data drawn, respectively, from multiple hyperplanes or from multiple affine subspaces of any dimensions.

6.1.1 K-Subspaces Model

Let $\mathcal{X} = \{\boldsymbol{x}_1, \boldsymbol{x}_2, \ldots, \boldsymbol{x}_N\}$ be a set of points drawn approximately from n affine subspaces of $\mathbb{R}^D$,

$$S_i = \{\boldsymbol{x} : \boldsymbol{x} = U_i \boldsymbol{y} + \boldsymbol{\mu}_i\}, \quad i = 1, \ldots, n, \tag{6.1}$$

where $U_i \in \mathbb{R}^{D \times d_i}$ is an orthonormal basis for the subspace, $\boldsymbol{y} \in \mathbb{R}^{d_i}$ is a low-dimensional representation of point $\boldsymbol{x}$ in subspace S_i, and $\boldsymbol{\mu}_i \in \mathbb{R}^D$ is an arbitrary point in the subspace. Let $w_{ij} \in \{0, 1\}$ be such that $w_{ij} = 1$ if point j belongs to subspace i and $w_{ij} = 0$ otherwise. Assuming that the number of subspaces n and their dimensions $\{d_i\}_{i=1}^n$ are known, K-subspaces aims to estimate the points $\{\boldsymbol{\mu}_i\}_{i=1}^n$, the subspace bases $\{U_i\}_{i=1}^n$, and the segmentation of the data $\{w_{ij}\}_{i=1,\ldots,n}^{j=1,\ldots,N}$ by solving the following optimization problem:

$$\begin{aligned} \min_{\{\boldsymbol{\mu}_i\},\{U_i\},\{w_{ij}\}} \quad & \sum_{i=1}^{n}\sum_{j=1}^{N} w_{ij}\, \mathrm{dist}(\boldsymbol{x}_j, S_i)^2 \\ \text{s.t.} \quad & U_i^\top U_i = I, \\ & w_{ij} \in \{0, 1\} \text{ and } \sum_{i=1}^{n} w_{ij} = 1, \quad j = 1, \ldots, N. \end{aligned} \tag{6.2}$$

Here $\mathrm{dist}(\boldsymbol{x}_j, S_i) = \min_{\boldsymbol{y} \in \mathbb{R}^{d_i}} \|\boldsymbol{x}_j - \boldsymbol{\mu}_i - U_i \boldsymbol{y}\| = \|(I - U_i U_i^\top)(\boldsymbol{x}_j - \boldsymbol{\mu}_i)\|$ is the distance from point $\boldsymbol{x}_j$ to subspace S_i.

Since $w_{ij} = 1$ when point $\boldsymbol{x}_j$ belongs to subspace S_i and $w_{ij} = 0$ otherwise, the objective function in (6.2) is simply the sum of the squared distances from each data point to its own subspace. The first constraint in (6.2) requires that the subspace basis be orthonormal. This constraint is also enforced by geometric PCA in (2.23). The last two constraints in (6.2) ensure that each point is assigned to only one subspace. These are the same constraints enforced by K-means in (4.53). Moreover, notice that when $n = 1$, we have $w_{ij} = 1$, and so the optimization problem solved by K-subspaces in (6.2) is equivalent to that solved by geometric PCA in (2.23). Also, when $d = 0$, all terms involving U_i and $\boldsymbol{y}_j$ disappear, and so the optimization problem solved by K-subspaces in (6.2) reduces to that solved by K-means in (4.53). Therefore, K-subspaces is a natural generalization of both geometric PCA and K-means.

6.1.2 K-Subspaces Algorithm

The K-subspaces algorithm solves the optimization problem in (6.2) using an alternating minimization approach. Given an assignment of data points to subspaces, i.e., given $\{w_{ij}\}$, the optimization problem in (6.2) decouples into n optimization problems of the form

$$\min_{\boldsymbol{\mu}_i, U_i} \sum_{j=1}^{N} w_{ij} \|(I - U_i U_i^\top)(\boldsymbol{x}_j - \boldsymbol{\mu}_i)\|^2 \text{ s.t. } U_i^\top U_i = I, \ i = 1, \ldots, n. \tag{6.3}$$

Since $w_{ij} \in \{0, 1\}$, each problem is equivalent to the geometric PCA problem in (2.23) but restricted to the data points $\boldsymbol{x}_j$ assigned to the ith subspace, i.e., the points such that $w_{ij} = 1$. Therefore, the optimal solution for the subspace parameters is given by

$$\boldsymbol{\mu}_i = \frac{\sum_{j=1}^{N} w_{ij} \boldsymbol{x}_j}{\sum_{j=1}^{N} w_{ij}}, \quad i = 1, \ldots, n, \tag{6.4}$$

$$U_i = \text{top } d_i \text{ eigenvectors of } \Big(\sum_{j=1}^{N} w_{ij} (\boldsymbol{x}_j - \boldsymbol{\mu}_i)(\boldsymbol{x}_j - \boldsymbol{\mu}_i)^\top \Big), \ i = 1, \ldots, n, \tag{6.5}$$

$$\boldsymbol{y}_j = U_i^\top (\boldsymbol{x}_j - \boldsymbol{\mu}_i) \quad \text{for all } j \text{ such that } w_{ij} = 1. \tag{6.6}$$

Given $\{\boldsymbol{\mu}_i\}$, $\{U_i\}$, $\{\boldsymbol{y}_j\}$, the optimal value for w_{ij} is obtained by assigning each data point to its closest subspace, i.e.,

$$w_{ij} = \begin{cases} 1 & \text{if } i = \arg\min_{\ell=1,\ldots,n} \|\boldsymbol{x}_j - \boldsymbol{\mu}_\ell - U_\ell \boldsymbol{y}_j\|^2, \\ 0 & \text{otherwise.} \end{cases} \tag{6.7}$$

Algorithm 6.1 (K-Subspaces).

Input: Data $\{\boldsymbol{x}_j\}_{j=1}^N$, number of subspaces n, and subspace dimensions $\{d_i\}_{i=1}^n$.

1: Initialize subspace parameters with a set of mean vectors $\boldsymbol{\mu}_i \in \mathbb{R}^D$ and orthogonal matrices $U_i \in \mathbb{R}^{D\times d_i}$.

2: **while** not converged **do**

3: **Segmentation:** For each $j = 1, \ldots, N$, assign point $\boldsymbol{x}_j$ to subspace S_i as:

$$w_{ij} = \begin{cases} 1 & \text{if } i = \arg\min_{\ell=1,\ldots,n} \|(I - U_\ell U_\ell^\top)(\boldsymbol{x}_j - \boldsymbol{\mu}_\ell)\|^2, \\ 0 & \text{else.} \end{cases} \tag{6.8}$$

If more than one subspace achieves the minimum, assign the point arbitrarily to one of the subspaces that achieves the minimum.

4: **Estimation:** For each $i = 1, \ldots, n$, estimate the parameters of S_i by applying PCA to the set of points $\boldsymbol{x}_j$ such that $w_{ij} = 1$, which gives

$$\boldsymbol{\mu}_i = \frac{\sum_{j=1}^N w_{ij}\boldsymbol{x}_j}{\sum_{j=1}^N w_{ij}}, \tag{6.9}$$

$$U_i = \text{top } d_i \text{ eigenvectors of } \Big(\sum_{j=1}^N w_{ij}(\boldsymbol{x}_j - \boldsymbol{\mu}_i)(\boldsymbol{x}_j - \boldsymbol{\mu}_i)^\top\Big), \tag{6.10}$$

$$\boldsymbol{y}_j = U_i^\top(\boldsymbol{x}_j - \boldsymbol{\mu}_i) \quad \text{for all } j \text{ such that } w_{ij} = 1. \tag{6.11}$$

5: **end while**

Output: Subspace parameters $\{U_i, \boldsymbol{\mu}_i\}_{i=1}^n$, low-dimensional representations $\{\boldsymbol{y}_j\}_{j=1}^N$, and segmentation of the data $\{w_{ij}\}_{i=1,\ldots,n}^{j=1,\ldots,N}$.

The K-subspaces algorithm then proceeds by alternating between applying PCA to each group of points and assigning each data point to its closest subspace, as detailed in Algorithm 6.1.

6.1.3 Convergence of the K-Subspaces Algorithm

Observe that at each step of the K-subspaces algorithm, the objective function either decreases or stays the same (see Exercise 6.2). Also, the objective function is nonnegative, and hence bounded below. Moreover, since the number of possible assignments of points to subspaces is finite, there is an open neighborhood of subspace parameters that produce the same assignments. Therefore, after a finite number of iterations, the objective function will stop decreasing, the assignments will not change, and the subspace estimates will not change. Notice also that except for the case that two or more points are at an equal distance from two or more distinct subspaces, a small perturbation of the subspaces S_i does not change the assignment of points to subspaces. Therefore, the K-subspaces algorithm is, in general, guaranteed to converge to a local minimum of the objective function in a finite number of iterations. We refer the reader to (Selim and Ismail 1984) for a more rigorous analysis of convergence of a generalized K-means algorithm (which includes K-subspaces as a particular case), including cases in which the generalized method fails to converge to a local minimum.

Now, even if the algorithm converges to a local minimum, in general it will not converge to a global minimum. Therefore, initialization is critical in order to obtain a good solution. One can initialize K-subspaces with an initial assignment of points to subspaces. Alternatively, one can initialize the K-subspaces algorithm with a set of subspace parameters. A common strategy is to start the algorithm from multiple random initializations, and then choose the one that gives the best objective value.

6.1.4 Advantages and Disadvantages of K-Subspaces

The main advantage of K-subspaces is its simplicity, since it alternates between assigning points to subspaces and estimating the subspaces via PCA. Another advantage is that it can handle both linear and affine subspaces explicitly. The third advantage is that it converges in a finite number of iterations. However, K-subspaces suffers from a number of drawbacks. First, its convergence to the global optimum depends on a good initialization. If a random initialization is used, several restarts are often needed to find the global optimum. Alternatively, we may use the ASC algorithm described in Chapter 5 for initialization. We refer the reader to (Aldroubi and Zaringhalam 2009; Zhang et al. 2010) for two additional initialization methods. Second, K-subspaces is sensitive to outliers, partly due to the use of the ℓ_2-norm. This issue can be addressed by using a robust norm, such as the ℓ_1-norm, as done by the median K-flats algorithm (Zhang et al. 2009). However, this results in a more complex algorithm, which requires solving a robust PCA problem at each iteration. Alternatively, one can resort to nonlinear minimization techniques, which are guaranteed to converge only to a local minimum. Third, K-subspaces requires n and $\{d_i\}_{i=1}^n$ to be known beforehand. One possible avenue to be explored is to use the model selection criteria for mixtures of subspaces proposed in (Huang et al. 2004). We refer the reader to (Aldroubi and Zaringhalam 2009; Aldroubi et al. 2008) for a more detailed analysis of some of the aforementioned issues.

6.2 Mixture of Probabilistic PCA (MPPCA)

As discussed before, the geometric models do not explicitly model noise from a probabilistic perspective. In this section, we present a generalization of the PPCA model introduced in Section 2.2 from one to multiple subspaces and show how the parameters of the mixture model can be estimated using the EM algorithm.

6.2.1 MPPCA Model

Let $\mathcal{X} = \{x_1, x_2, \ldots, x_N\} \subset \mathbb{R}^D$ be a set of points that are drawn independently from a "noisy union" of n subspaces $\{S_i\}_{i=1}^n$ of dimensions $\{d_i\}_{i=1}^n$ according to the following probabilistic generative model. First, we draw a random variable $z \in \{1, 2, \ldots, n\}$ from a multinomial distribution with parameters $\{\pi_i\}_{i=1}^n$, where $\pi_i = P(z = i) \in (0, 1)$ is such that $\sum_{i=1}^n \pi_i = 1$ and denotes the prior probability of selecting subspace S_i. Second, given that $z = i$, we generate a point $x \in \mathbb{R}^D$ according to the PPCA model introduced in Section 2.2. That is,

$$x = \mu_i + B_i y + \varepsilon_i, \tag{6.12}$$

where $\mu_i \in \mathbb{R}^D$ and $B_i \in \mathbb{R}^{D \times d_i}$ are model parameters that represent, respectively, a point in subspace S_i and a (not necessarily orthonormal) basis for subspace S_i. On the other hand, y and ε_i are independent zero-mean random (Gaussian) variables with covariances $\Sigma_y = I_{d_i}$ and $\Sigma_{\varepsilon_i} = \sigma_i^2 I_D$ that represent, respectively, a low-dimensional representation of x in S_i and a noise vector. Therefore, the distribution of x given that it belongs to subspace S_i is Gaussian with mean μ_i and covariance $\Sigma_i = B_i B_i^\top + \sigma_i^2 I_D$. That is,

$$p_{\theta_i}(x \mid z = i) = \frac{1}{(2\pi)^{D/2} \det(\Sigma_i)^{1/2}} \exp\left(- \frac{(x-\mu_i)^\top \Sigma_i^{-1} (x-\mu_i)}{2} \right), \tag{6.13}$$

where $\theta_i = (\mu_i, B_i, \sigma_i)$ denotes the parameters of the ith PPCA model. The distribution of a point drawn from a mixture of n PPCA models is then given by

$$p_\theta(x) = \sum_{i=1}^n p_{\theta_i}(x \mid z = i) P(z = i) = \sum_{i=1}^n p_{\theta_i}(x \mid z = i) \pi_i, \tag{6.14}$$

where $\theta = (\theta_1, \ldots, \theta_n, \pi_1, \ldots, \pi_n)$ denotes the MPPCA model parameters.

6.2.2 Maximum Likelihood Estimation for MPPCA

Our goal is twofold: (1) We would like to cluster (or group) all the sample points into their respective subspaces; (2) we would like to obtain a good estimate of the parameters of each subspace. As we have discussed in Appendix B, the "goodness" of a statistical estimate can be measured in different ways. For mixture models, a conventional approach is the ML estimate, reviewed in Appendix B.2. The ML estimator aims to find the parameters θ that maximize the likelihood function for the given data. That is,

$$\max_{\theta \in \Theta} \prod_{j=1}^N p_\theta(x_j) \quad \text{or equivalently} \quad \min_{\theta \in \Theta} \sum_{j=1}^N - \log p_\theta(x_j). \tag{6.15}$$

In Appendix B.2.1, we derived the EM algorithm for solving this problem. Starting from an initial guess θ^0 for the model parameters, the EM algorithm alternates between the following two steps until convergence (see Algorithm B.1):

E-step: For fixed $\theta = \theta^k = (\theta_1^k, \ldots, \theta_n^k, \pi_1^k, \ldots, \pi_n^k)$, compute

$$w_{ij}^k \doteq p_{\theta^k}(z = i \mid \boldsymbol{x}_j) = \frac{p_{\theta_i^k}(\boldsymbol{x}_j \mid z = i)\pi_i^k}{\sum_{i=1}^n p_{\theta_i^k}(\boldsymbol{x}_j \mid z = i)\pi_i^k}. \tag{6.16}$$

M-step: For fixed w_{ij}^k, solve for θ by maximizing the expected log-likelihood

$$\begin{aligned} \theta^{k+1} &= \arg\max_{\theta \in \Theta} \sum_{j=1}^N \sum_{i=1}^n w_{ij}^k \log p_\theta(\boldsymbol{x}_j, z = i) \\ &= \arg\max_{\theta \in \Theta} \sum_{j=1}^N \sum_{i=1}^n w_{ij}^k \log \big(p_{\theta_i}(\boldsymbol{x}_j \mid z = i)\pi_i\big). \end{aligned} \tag{6.17}$$

Observe that the E-step can be computed in closed form by direct substitution of (6.13) in (6.16). Thus, the main question is how to solve for θ in the M-step.

A Naive EM Algorithm for MPPCA
In Appendix B.3.1, we derived a closed-form solution for the optimal update of θ for a mixture of Gaussians. Since the MPPCA model is a mixture of Gaussians, it seems rather tempting to directly apply the results in B.3.1 to the above model. In particular, if we maximize the expected log-likelihood with respect to $(\pi_i, \boldsymbol{\mu}_i, \Sigma_i)$, we obtain

$$\pi_i^{k+1} = \frac{\sum_{j=1}^N w_{ij}^k}{\sum_{j=1}^N \sum_{i=1}^n w_{ij}^k}, \quad \boldsymbol{\mu}_i^{k+1} = \frac{\sum_{j=1}^N w_{ij}^k \boldsymbol{x}_j}{\sum_{j=1}^N w_{ij}^k}, \tag{6.18}$$

$$\Sigma_i^{k+1} = \frac{\sum_{j=1}^N w_{ij}^k (\boldsymbol{x}_j - \boldsymbol{\mu}_i^{k+1})(\boldsymbol{x}_j - \boldsymbol{\mu}_i^{k+1})^\top}{\sum_{j=1}^N w_{ij}^k}. \tag{6.19}$$

The naive EM algorithm then alternates between the E- and M-steps, both of which can be computed in closed form.

Let $\hat{\Sigma}_i$ be the covariance matrix to which the naive EM algorithm converges. We may find a basis B_i for subspace S_i and the standard deviation σ_i of the noise ε_i by applying Theorem 2.9 to $\hat{\Sigma}_i$. Specifically,

$$\widehat{B}_i = U(\Lambda - \hat{\sigma}_i^2 I)^{1/2} R, \text{ and } \hat{\sigma}_i^2 = \frac{1}{D - d_i} \sum_{j=d_i+1}^D \lambda_j, \tag{6.20}$$

where $U \in \mathbb{R}^{D\times d_i}$ is a matrix whose columns are the top d_i eigenvectors of $\hat{\Sigma}_i$, $\Lambda \in \mathbb{R}^{d_i\times d_i}$ is a diagonal matrix whose diagonal entries are the corresponding top d_i eigenvalues, and λ_j is the jth eigenvalue of $\hat{\Sigma}_i$.

Finally, if one needs to get a partition of the data into their respective subspaces, each point $\boldsymbol{x}_j$ is assigned to cluster $\mathcal{X}_{c_j}$ with

$$c_j = \arg\max_i \hat{w}_{ij}, \tag{6.21}$$

where $\hat{w}_{ij}$ is the converged a posteriori distribution of z given $\boldsymbol{x}_j$.

EM Algorithm for MPPCA

The naive EM algorithm estimates the mixture of subspaces as a mixture of generic Gaussian distributions. While this algorithm is extremely simple,[1] its derivation is not correct. The reason is that by treating the PPCA components as generic Gaussians, we are treating all entries of the covariance matrix Σ_i as independent parameters.[2] In reality, the covariance matrix for each component of the MPPCA model has a special form $\Sigma_i = B_iB_i^\top + \sigma_i^2 I$. Thus, the true number of free parameters in (B_i, σ_i) can be much smaller than the number of parameters in Σ_i, especially when the dimensions of the subspaces are much lower than the dimension of the ambient space, which is often the case for many of the problems we are concerned with in this book. Therefore, in the M-step, we should not treat the entries of Σ_i as free parameters. Instead, we should optimize over (B_i, σ_i).

Specifically, after substituting (6.13) into (6.17), we can see that the M-step reduces to

$$(\theta_i^{k+1}, \pi_i^{k+1}) = \arg\min_{\theta_i,\pi_i} \sum_{j=1}^{N} w_{ij}^k \Big(\frac{(\boldsymbol{x}_j - \boldsymbol{\mu}_i)^\top \Sigma_i^{-1}(\boldsymbol{x}_j - \boldsymbol{\mu}_i) + \log\det(\Sigma_i)}{2} - \log \pi_i \Big). \tag{6.22}$$

The optimal solution for π_i, subject to $\sum_{i=1}^{n} \pi_i = 1$, is the same as that in (6.18). To obtain the optimal solution for $\theta_i = (\boldsymbol{\mu}_i, B_i, \sigma_i)$, notice that since $\Sigma_i = B_iB_i^\top + \sigma_i^2 I$, the problem above is almost identical the PPCA problem we studied in Section 2.2.2. In particular, the above objective function is the same as the likelihood function (2.56), except for the weights w_{ij}^k. Hence a similar derivation to that in Theorem 2.9 gives us the following optimal solutions for θ_i (we leave the detailed derivation as an exercise to the reader):

[1] The reader is encouraged to implement it and test it with simulated data.

[2] If the dimension of the ambient space D is high, the degrees of freedom in the model are very large and the extrema of the likelihood function may not be so salient when the number of samples is not significantly larger than the dimension. Therefore, the optimization algorithm may converge very slowly to an extremum, since so many parameters are not properly regularized. There has been a rich literature in statistics about how to properly regularize the estimates of high-dimensional Gaussians, especially when the number of samples is limited.

$$\boldsymbol{\mu}_i^{k+1} = \frac{\sum_{j=1}^{N} w_{ij}^k \boldsymbol{x}_j}{\sum_{j=1}^{N} w_{ij}^k}, \quad B_i^{k+1} = U(\Lambda - (\sigma_i^{k+1})^2 I)^{1/2} R, \quad (\sigma_i^{k+1})^2 = \frac{\sum_{j=d_i+1}^{D} \lambda_j}{D - d_i}, \tag{6.23}$$

where U is the matrix with the top d_i eigenvectors of Σ_i^{k+1} given in (6.19), Λ_1 is the matrix with the corresponding top d_i eigenvalues, $R \in \mathbb{R}^{d_i \times d_i}$ is an arbitrary orthogonal matrix, and λ_j is the jth-largest eigenvalue of Σ_i^{k+1}. Hence, as discussed in the proof of Theorem 2.9, the correct optimal estimate for the covariance Σ_i that maximizes the likelihood function in the M-step is

$$\Sigma_i^{k+1} = B_i^{k+1}(B_i^{k+1})^\top + (\sigma_i^{k+1})^2 I. \tag{6.24}$$

Notice that this estimate (by the EM algorithm for MPPCA) does not coincide with the one in (6.19) (by the EM algorithm for mixtures of Gaussians). The key difference is that the estimate in (6.24) essentially "regularizes" the covariance estimate in (6.19) to be a low-rank matrix. This is because the update in (6.23) essentially conducts a soft thresholding of the singular values of the "unregularized" empirical estimate of the covariance matrix given by (6.19). As we saw in Chapters 2 and 3, this kind of soft (or hard) thresholding arises very commonly in situations in which we try to enforce a matrix estimate to be of low rank.

Finally, if one needs to get a partition of the data into their respective subspaces, each point $\boldsymbol{x}_j$ is assigned to cluster $\mathcal{X}_{c_j}$ with

$$c_j = \arg\max_i \hat{w}_{ij}, \tag{6.25}$$

where $\hat{w}_{ij}$ is the converged a posteriori distribution of z given $\boldsymbol{x}_j$.

For completeness, we have summarized the EM algorithm for a Mmixture of subspaces (or MPPCAs) as Algorithm 6.2.

6.2.3 *Maximum a Posteriori (MAP) Estimation for MPPCA*

The EM algorithm for MPPCA is based on alternating between computing the expected log-likelihood (E-step), which involves taking the expectation with respect to the latent variables, and maximizing the expected log-likelihood (M-step). An alternative approach to taking the expectation is to directly maximize over the latent variables. As we will see in this section, this results in an approximate EM algorithm in which, in the E-step, each data point is assigned to the PPCA model that maximizes the posterior of the latent variables, whence the name MAP-EM.

More specifically, let $z_j \in \{1, 2, \dots, n\}$ be the latent variable denoting the PPCA model that generated $\boldsymbol{x}_j$. The MAP-EM algorithm aims to find the model parameters and latent variables that maximize the complete log likelihood, i.e.,

Algorithm 6.2 (EM for MPPCA).

Input: Sample points $\mathcal{X} = \{x_j\}_{j=1}^N$, number of subspaces n, their dimensions $\{d_i\}_{i=1}^n$, and initial estimates for the prior probabilities π_i^0 (often to be $1/n$), the subspace bases B_i^0, the mean vectors μ_i^0, and noise variance $(\sigma_i^0)^2$.

Initialization: $k \leftarrow 0$ and $\Sigma_i^0 \leftarrow B_i^0 B_i^{0\top} + (\sigma_i^0)^2 I$ for $i = 1, \dots, n$.

while not converged **do**

E-step: Update the a posteriori distribution of z given x_j as

$$w_{ij}^k \leftarrow \frac{p_{\theta_i^k}(x_j \mid z = i)\pi_i^k}{\sum_{i=1}^n p_{\theta_i^k}(x_j \mid z = i)\pi_i^k}, \tag{6.26}$$

where $p_{\theta_i^k}(x_j \mid z = i)$ is a Gaussian with mean μ_i^k and covariance Σ_i^k.

M-step: Update the estimates for $\{\pi_i, \mu_i, \Sigma_i\}_{i=1}^n$ as

$$\pi_i^{k+1} \leftarrow \frac{\sum_{j=1}^N w_{ij}^k}{\sum_{j=1}^N \sum_{i=1}^n w_{ij}^k}, \quad \mu_i^{k+1} \leftarrow \frac{\sum_{j=1}^N w_{ij}^k x_j}{\sum_{j=1}^N w_{ij}^k}, \tag{6.27}$$

$$\Sigma_i^{k+1} \leftarrow \frac{\sum_{j=1}^N w_{ij}^k (x_j - \mu_i^{k+1})(x_j - \mu_i^{k+1})^\top}{\sum_{j=1}^N w_{ij}^k}. \tag{6.28}$$

Let the columns of U and the diagonal entries of Λ be, respectively, the top d_i eigenvectors and eigenvalues of Σ_i^{k+1}. Update the subspace basis B_i and the noise variance σ_i^2, and reset the data covariance Σ_i as

$$\begin{aligned} B_i^{k+1} &\leftarrow U(\Lambda - (\sigma_i^{k+1})^2 I)^{1/2} R, \quad (\sigma_i^{k+1})^2 = \frac{1}{D - d_i} \sum_{i=d_i+1}^{D} \lambda_i, \\ \Sigma_i^{k+1} &\leftarrow B_l^{k+1} (B_i^{k+1})^\top + (\sigma_i^{k+1})^? I. \end{aligned} \tag{6.29}$$

Set $k \leftarrow k + 1$.

end while

Assign data point x_j to the class $c_j = \arg\max_i \hat{w}_{ij}$.

Output: MPPCA parameters $\{\pi_i, \mu_i, B_i, \sigma_i^2, \Sigma_i\}_{i=1}^n$ and segmentation c.

$$\max_{\theta \in \Theta} \max_{\{z_j\}} \prod_{j=1}^N p_\theta(x_j, z_j) \text{ or equivalently } \min_{\theta \in \Theta} \min_{\{z_j\}} \sum_{j=1}^N -\log p_\theta(x_j, z_j). \tag{6.30}$$

This problem is equivalent to

$$\begin{aligned} \min_{\theta \in \Theta} \sum_{j=1}^N \min_{z_j} \big(-\log p_\theta(x_j, z_j) \big) &\equiv \min_{\theta \in \Theta} \sum_{j=1}^N \min_{i=1,\dots,n} \big(-\log(p_{\theta_i}(x_j | z = i)\pi_i) \big) \\ &\equiv \min_{\theta \in \Theta} \min_{\{w_{ij}\}} \sum_{j=1}^N \sum_{i=1}^n w_{ij} \operatorname{dist}(x_j, S_i)^2, \end{aligned} \tag{6.31}$$

where

$$\begin{aligned}\operatorname{dist}(\boldsymbol{x}_j, S_i)^2 &= -\log\left(p_{\theta_i}(\boldsymbol{x}_j \mid z = i)\pi_i\right) \\ &= \left(\frac{1}{2}((\boldsymbol{x}_j - \boldsymbol{\mu}_i)^\top \Sigma_i^{-1}(\boldsymbol{x}_j - \boldsymbol{\mu}_i) + \log((2\pi)^D \det(\Sigma_i))) - \log \pi_i\right)\end{aligned} \tag{6.32}$$

is a "probabilistic distance"[3] from point $\boldsymbol{x}_j$ to subspace S_i, which is parameterized by (π_i, θ_i), and $w_{ij} \in \{0, 1\}$ is defined as

$$w_{ij} = \begin{cases} 1 & \text{if } i = \operatorname*{arg\,min}_{\ell=1,\dots,n} \operatorname{dist}(\boldsymbol{x}_j, S_\ell)^2 = \operatorname*{arg\,max}_{\ell=1,\dots,n} p_{\theta_\ell}(\boldsymbol{x}_j \mid z = \ell)\pi_\ell \\ 0 & \text{otherwise.} \end{cases} \tag{6.33}$$

Observe the striking similarly between the objective of MAP-EM in (6.31) and that of K-subspaces in (6.2). Thus, following the optimization strategy for K-subspaces, we can solve the problem in (6.31) using alternating minimization:

E-step: Given θ, solve for w_{ij} such that $\sum_{i=1}^n w_{ij} = 1$. The optimal solution is given by (6.33). Therefore, when $\{(\theta_i, \pi_i)\}$ are known, w_{ij} assigns point $\boldsymbol{x}_j$ to the PPCA model that maximizes the posterior probability of z given $\boldsymbol{x}$, whence the name MAP-EM.

M-step: Given w_{ij}, solve for $\theta \in \Theta$. This problem is identical to the M-step in (6.22), whose solution is given by (6.18) and (6.23).

For completeness, we have summarized the MAP-EM algorithm for a mixture of subspaces (or MPPCAs) as Algorithm 6.3.

6.2.4 Relationship between K-Subspaces and MPPCA

Notice that the MAP-EM algorithm is essentially the same as the EM algorithm for MPPCA, except for the computation of w_{ij}: in EM, $w_{ij} \in [0, 1]$ denotes the probability that point $\boldsymbol{x}_j$ belongs to subspace S_i, while in MAP-EM, $w_{ij} \in \{0, 1\}$ indicates whether $\boldsymbol{x}_j$ is assigned to S_i. This is the reason why sometimes the former method is known as *soft* EM, while the latter is known as *hard* EM.

Notice also that because $w_{ij} \in \{0, 1\}$, the M-step of the MAP-EM algorithm does not require using all the data points to solve for θ_i. Indeed, if we look at the solutions in (6.23), we see that we can solve for θ_i using only the data points that are assigned to subspace S_i. Thus, the MAP-EM algorithm can be restated as follows:

[3]While the quantity is not an actual distance, notice that it is nonnegative because it is minus the logarithm of a probability.

Algorithm 6.3 (MAP-EM for MPPCA).

Input: Sample points $\mathcal{X} = \{\boldsymbol{x}_j\}_{j=1}^N$, number of subspaces n, their dimensions $\{d_i\}_{i=1}^n$, and initial estimates for the prior probabilities π_i^0 (often to be $1/n$), the subspace bases B_i^0, the mean vectors $\boldsymbol{\mu}_i^0$, and noise variance $(\sigma_i^0)^2$.

Initialization: $k \leftarrow 0$ and $\Sigma_i^0 \leftarrow B_i^0 B_i^{0\top} + (\sigma_i^0)^2 I$ for $i = 1, \ldots, n$.

while not converged **do**

E-step: Assign data points to PPCAs as

$$w_{ij}^{k+1} = \begin{cases} 1 & \text{if } i = \arg\max_{\ell=1,\ldots,n} p_{\theta_\ell^k}(\boldsymbol{x}_j \mid z = \ell)\pi_\ell \\ 0 & \text{otherwise,} \end{cases} \tag{6.34}$$

where $p_{\theta_i^k}(\boldsymbol{x}_j \mid z = i)$ is a Gaussian with mean $\boldsymbol{\mu}_i^k$ and covariance Σ_i^k.

M-step: Update the estimates for $\{\pi_i, \boldsymbol{\mu}_i, \Sigma_i\}_{i=1}^n$ as

$$\pi_i^{k+1} \leftarrow \frac{\sum_{j=1}^N w_{ij}^k}{\sum_{j=1}^N \sum_{i=1}^n w_{ij}^k}, \quad \boldsymbol{\mu}_i^{k+1} \leftarrow \frac{\sum_{j=1}^N w_{ij}^k \boldsymbol{x}_j}{\sum_{j=1}^N w_{ij}^k}, \tag{6.35}$$

$$\Sigma_i^{k+1} \leftarrow \frac{\sum_{j=1}^N w_{ij}^k (\boldsymbol{x}_j - \boldsymbol{\mu}_i^{k+1})(\boldsymbol{x}_j - \boldsymbol{\mu}_i^{k+1})^\top}{\sum_{j=1}^N w_{ij}^k}. \tag{6.36}$$

Let the columns of U and the diagonal entries of Λ be, respectively, the top d_i eigenvectors and eigenvalues of Σ_i^{k+1}. Update the subspace basis B_i and the noise variance σ_i^2, and reset the data covariance Σ_i as

$$\begin{aligned} B_i^{k+1} &\leftarrow U(\Lambda - (\sigma_i^{k+1})^2 I)^{1/2} R, \quad (\sigma_i^{k+1})^2 = \frac{1}{D - d_i} \sum_{i=d_i+1}^{D} \lambda_i, \\ \Sigma_i^{k+1} &\leftarrow B_i^{k+1}(B_i^{k+1})^\top + (\sigma_i^{k+1})^2 I. \end{aligned} \tag{6.37}$$

Set $k \leftarrow k + 1$.

end while

Assign data point $\boldsymbol{x}_j$ to the class $c_j = \arg\max_i \hat{w}_{ij}$.

Output: MPPCA parameters $\{\pi_i, \boldsymbol{\mu}_i, B_i, \sigma_i^2, \Sigma_i\}_{i=1}^n$ and segmentation $\boldsymbol{c}$.

E-step: Given the subspace parameters (θ_i, π_i), assign each data point to its closest subspace as in (6.33).

M-step: Given the assignments of points to subspaces, w_{ij}, solve for θ_i by applying PPCA to all the data points assigned to subspace S_i. In addition, solve also for π_i by counting the proportion of data points assigned to S_i.

Therefore, the MAP-EM algorithm for MPPCA is identical to K-subspaces, except for two key differences: the distance used for assigning points to subspaces is different (K-subspaces uses a geometric distance, while MAP-EM uses a "probabilistic distance"), and the method for fitting the subspace to the data in each group is different (K-subspaces uses PCA, while MAP-EM uses PPCA). It is hence natural to ask whether there exist some conditions under which MAP-EM for MPPCA reduces to K-subspaces, in the same way as MAP-EM for mixtures of Gaussians reduces to K-means.

A sufficient condition for the MAP-EM algorithm for PPCA to reduce to K-subspaces is that the "probabilistic distance" reduce to the Euclidean distance, i.e.,

$$\Big(\frac{1}{2}((\boldsymbol{x}_j - \boldsymbol{\mu}_i)^\top \Sigma_i^{-1}(\boldsymbol{x}_j - \boldsymbol{\mu}_i) + \log(\det(\Sigma_i))) - \log \pi_i\Big) \propto \|(I - U_i U_i^\top)(\boldsymbol{x}_j - \boldsymbol{\mu}_i)\|_2^2, \tag{6.38}$$

where $\propto$ means equality up to scale or up to a constant term. If we assume that the prior probabilities of choosing any subspace are all equal, i.e., $\pi_i = \frac{1}{n}$, then the third term of the "probabilistic distance" becomes a constant. Moreover, let $U_i\Lambda_iU_i^\top$ be the compact SVD of $B_iB_i^\top$ and recall that $\Sigma_i = B_iB_i^\top + \sigma_i^2 I_D$. It follows from (2.70) that $\det(\Sigma_i) = \det(\Lambda_i + \sigma_i^2 I_{d_i})\sigma_i^{2(D-d_i)}$. Therefore, if we assume that all subspaces are of equal dimensions $d_1 = \cdots = d_n = d$, the variances of the noises are the same for all subspaces, i.e., $\sigma_1 = \cdots = \sigma_n = \sigma$, and the distribution of the data inside the subspaces are the same for all subspaces,[4] i.e., $\Lambda_1 = \cdots = \Lambda_n = \Lambda$, then the second term of the "probabilistic distance" is a constant.

Now notice that $\Sigma_i = (B_iB_i^\top + \sigma^2 I_D) = (U_i(\Lambda + \sigma^2 I_d)U_i^\top + \sigma^2(I - U_iU_i^\top))$. Thus

$$\Sigma_i^{-1} = U_i(\Lambda + \sigma^2 I_d)^{-1}U_i^\top + \sigma^{-2}(I - U_iU_i^\top). \tag{6.39}$$

Therefore, if we assume that σ^2 is much smaller than each of the diagonal entries of Λ, i.e., if we assume that the signal-to-noise ratio (SNR) is high, then $\Sigma_i^{-1} \approx \sigma^{-2}(I - U_iU_i^\top)$, and so

$$(\boldsymbol{x}_j - \boldsymbol{\mu}_i)^\top \Sigma_i^{-1}(\boldsymbol{x}_j - \boldsymbol{\mu}_i) \approx \frac{1}{\sigma^2}\|(I - U_iU_i^\top)(\boldsymbol{x}_j - \boldsymbol{\mu}_i)\|_2^2. \tag{6.40}$$

In summary, we have shown that when the prior probabilities of selecting a subspace are the same, the dimensions of the subspaces are the same, the distribution of the data inside the subspaces is the same across all subspaces, and the variance of the noise is the same across all subspaces and is small compared to the variance of the data, then the MAP-EM algorithm produces approximately the same assignments as the K-subspaces algorithm.

[4]Notice that this means that while the subspaces are allowed to have different orientations (different U_i), the distribution of the data points inside the subspaces is the same.

6.3 Compression-Based Subspace Clustering

So far, we have introduced several methods for estimating and clustering multiple subspaces, utilizing algebraic, iterative, or statistical techniques. A common assumption behind all these approaches is that a good estimate of the underlying subspace model (parameters) is necessary for clustering the data. Such parametric methods often require knowing the number of subspaces and their dimensions in advance. In practice, however, it is very desirable to have a method that can automatically determine such information from a given data set itself.

As discussed in Section 5.5, the estimation of the number of subspaces and their dimensions is a very difficult problem that is generally ill posed. In particular, we noted that one can fit N data points drawn from a union of multiple subspaces in $\mathbb{R}^D$ with a single subspace of dimension D, or with N subspaces of dimension 1. However, neither of these two pathological cases might correspond to the true union of subspaces from which the data were sampled. To address this problem, in Section 5.5 we studied the effective dimension (ED) of a data set as a means to measure the "compactness" of a representation of the data. The ED measures the "average" number of (unquantized) real numbers needed to encode each of the data points and is a function of the number of subspaces, their dimensions, and the number of data points in each subspace. We showed how the ED can be used in conjunction with classical model selection criteria discussed in Appendix B.4 to estimate the number of subspaces and their dimensions. We also showed that a "compact" representation of the data can be obtained by minimizing the ED while approximating the given data to within a certain error tolerance.

In this section, we present an alternative approach to finding the "most compact" representation of data drawn from a union of subspaces. In this approach, the compactness of a representation is measured in terms of the number of bits needed to encode each of the data points, i.e., its *coding length*. A compact representation is then found as the one that minimizes the coding length while approximating the given data to within a certain error tolerance. By varying the tolerance, we obtain models of different complexity, among which we can choose the one that remains the most stable for large ranges of the tolerance parameter.

6.3.1 Model Estimation and Data Compression

When the number of subspaces and their dimensions are unknown, one can view the subspace modeling and clustering process as one of seeking a model that gives the "most compact" representation for the data. This simple idea is actually the most fundamental principle behind almost all model selection methods (Rissanen 1978; Hansen and Yu 2001). In principle, by a compact model we mean a model with the smallest possible number of subspaces, each of the smallest possible dimension. In practice, however, the aforementioned pathological cases tell us that we cannot simply minimize one quantity, say the number of subspaces, while leaving the other one free to vary. Therefore, defining an appropriate measure of "compactness" is critical.

Dimensionality is an abstract measure for compactness that is in many ways rather rough: two data sets that span the same dimension could have drastically different variances and volumes. In practice, a much more accurate measure for data compactness is its physical storage: the number of binary bits needed to represent the data set. Interestingly, this measure is closely related to the ML estimate that we have studied in previous methods. To see this, consider a set of data points $\mathcal{X} = \{\boldsymbol{x}_1, \boldsymbol{x}_2, \dots, \boldsymbol{x}_N\}$ drawn from a mixture of distributions: $p_\theta(\boldsymbol{x}) \doteq \sum_{i=1}^{n} \pi_i p_{\theta_i}(\boldsymbol{x})$, where $\theta = (\theta_1, \dots, \theta_n, \pi_1, \dots, \pi_n)$. The ML estimate of θ is given by

$$\hat{\theta} = \arg\max_{\theta \in \Theta} \sum_{j=1}^{N} \log p_\theta(\boldsymbol{x}_j). \tag{6.41}$$

The ML criterion is equivalent to minimizing the negative log-likelihood: $\sum_{j=1}^{N} -\log p_\theta(\boldsymbol{x}_j)$. At the optimal estimate $\hat{\theta}$, the value of the negative log-probability $-\log p_{\hat{\theta}}(\boldsymbol{x}_j)$ is the coding length required to store the data point $\boldsymbol{x}_j$ using the optimal Shannon coding scheme for the distribution $p_\theta(\boldsymbol{x})$ (Cover and Thomas 1991). Therefore, the expected coding length for the entire data set $\mathcal{X}$ is

$$l_N(\mathcal{X}) = \frac{1}{N} \sum_{j=1}^{N} -\log p_{\hat{\theta}}(\boldsymbol{x}_j). \tag{6.42}$$

Since the samples $\boldsymbol{x}_j$ are i.i.d. samples from $p_\theta(\boldsymbol{x})$, by the law of large numbers, as $N \to \infty$, the coding length $l_N(\mathcal{X})$ converges to the quantity

$$\hat{H}(\boldsymbol{x}) = \int -p_\theta(\boldsymbol{x}) \log p_{\hat{\theta}}(\boldsymbol{x})\, d\boldsymbol{x}. \tag{6.43}$$

Notice that $H(\boldsymbol{x}) = \int -p_\theta(\boldsymbol{x}) \log p_\theta(\boldsymbol{x})\, d\boldsymbol{x}$ is the entropy of the random variable $\boldsymbol{x}$. The difference $\hat{H}(\boldsymbol{x}) - H(\boldsymbol{x})$ is the so-called *KL divergence* in (B.30):

$$KL\big(p_\theta(\boldsymbol{x}) \,||\, p_{\hat{\theta}}(\boldsymbol{x})\big) = \int p_\theta(\boldsymbol{x}) \log\left(\frac{p_\theta(\boldsymbol{x})}{p_{\hat{\theta}}(\boldsymbol{x})}\right) d\boldsymbol{x}. \tag{6.44}$$

This quantity is always nonnegative and is zero only when $p_\theta(\boldsymbol{x}) = p_{\hat{\theta}}(\boldsymbol{x})$. In other words, $H(\boldsymbol{x})$ is the lowest possible expected coding length for data drawn from the distribution $p_\theta(\boldsymbol{x})$, and every other incorrect estimate of the distribution will lead to a higher coding length.

However, achieving the minimum coding length, or entropy, requires precise knowledge of the distribution $p_\theta(\boldsymbol{x})$, while in reality, only a finite number of samples of the distribution are given. As we have seen in the case of K-subspaces and EM, inferring the (mixture) distribution parameters from the samples can be a rather difficult nonlinear or even nonsmooth optimization problem itself. Thus, the goal of this chapter is to seek a good "surrogate" for the coding length for the given data set. By minimizing such a surrogate, we can indirectly approximate the entropy of the data set.

This leads to the question of what constitutes a good surrogate. Clearly, a good surrogate should be easily and directly computable from the given sample data. The coding length given by the surrogate should be a close approximation (say a tight upper bound) to the minimum coding length required for the given sample data and should converge asymptotically to the optimal coding length as the number of samples goes to infinity. Many nonparametric approximations of the distribution from the samples can be used to construct such a surrogate. Since in our case, we are dealing with a mixture of subspace-like Gaussians, not a generic distribution, we can do much better. As we will see, in our case we can obtain a closed-form formula that gives an accurate upper bound for the coding length.

Once we are able to accurately evaluate the coding length of a data set (against a hypothesized model), we can decide whether a particular clustering of the data set leads to a shorter coding length. The optimal clustering of the data is naturally the one that minimizes the overall coding length. In this way, we establish an equivalence between data clustering and data compression. As we will see in this section, one distinguished character of such a compression-based clustering method is that the principle of model selection is naturally implemented in the process of optimization. Hence the method does not require any prior knowledge about the number and dimensions of the subspaces for modeling the data. In addition, outliers will be handled in an easy and unified way: they will simply be assigned to special clusters that are less compressible than the rest of data, which can be well fit by low-dimensional subspaces.

In the remainder of this section, we will show how the overall coding length can be minimized in a simple agglomerative fashion, which leads to an extremely efficient and robust algorithm for clustering data drawn from a mixture of linear subspaces or Gaussians. Although we will conduct extensive simulations and give some illustrative examples on real data to validate this method, we will wait till Chapter 10 to see how well the compression-based clustering method works on extensive real-world imagery data. In the data mining or information retrieval literature, *agglomerative clustering* methods are very popular nonparametric methods for clustering mixed data (Kamvar et al. 2002). The analysis and results in this chapter help provide a solid statistical justification for such methods, at least for a mixture of Gaussian distributions.

6.3.2 Minimium Coding Length via Agglomerative Clustering

In this section, we give a self-contained summary of the main ideas behind the compression-based approach to subspace clustering and the agglomerative algorithm. We leave more detailed mathematical analysis and justification to

Sections 6.3.3 and 6.3.4. Readers who are interested only in the algorithm and experiments may skip those sections and go directly to Section 6.4 without much loss of continuity.

Lossy Coding of Multivariate Gaussian Data
Given a set of data vectors $\mathcal{X} = \{\boldsymbol{x}_1, \boldsymbol{x}_2, \ldots, \boldsymbol{x}_N\} \subset \mathbb{R}^D$, we use X to denote the matrix whose columns are the data points, i.e., $X = [\boldsymbol{x}_1, \boldsymbol{x}_2, \ldots, \boldsymbol{x}_N] \in \mathbb{R}^{D \times N}$. An encoding scheme maps each $\boldsymbol{x}_j \in \mathcal{X}$ to a sequence of binary bits, from which the original vector can be recovered as $\hat{\boldsymbol{x}}_j$ using a decoding scheme. A coding scheme is said to be lossless if $\hat{\boldsymbol{x}}_j = \boldsymbol{x}_j$ for all data. Here we consider a lossy coding scheme. That is, the original vector is recovered up to (on average) an allowable distortion: $\mathbb{E}[\|\boldsymbol{x}_j - \hat{\boldsymbol{x}}_j\|^2] \le \varepsilon^2$.

The total length of the encoded sequences of all the vectors in $\mathcal{X}$ is denoted by $L(X) \in \mathbb{R}_+$. To achieve good compression, we want to choose a coding scheme and an associated coding length function $L(\cdot)$ that are optimal for the family of distributions of interest. In information theory (Cover and Thomas 1991), the optimal number of binary bits needed to encode a random vector $\boldsymbol{x}$ of distribution $p(\boldsymbol{x})$ subject to the distortion ε^2 is given by the so-called rate-distortion function of the distribution $p(\boldsymbol{x})$. When the data points $\mathcal{X}$ are i.i.d. samples from a zero-mean[5] multivariate Gaussian distribution $\mathcal{N}(0, \Sigma)$, it is known that the function $R = \frac{1}{2}\log_2 \det(I + \frac{D}{\varepsilon^2}\Sigma)$ gives a good approximation to the optimal rate-distortion function (Cover and Thomas 1991).[6] Since $\hat{\Sigma} = \frac{1}{N}XX^\top$ is an estimate of the covariance Σ, the average number of bits needed per vector in $\mathcal{X}$ is

$$R(X) \doteq \frac{1}{2}\log_2 \det\big(I + \frac{D}{\varepsilon^2 N}XX^\top\big). \tag{6.45}$$

For readers who are less familiar with information theory, we will give an intuitive derivation of this formula in Section 6.3.3. Furthermore, we will give a constructive proof in Appendix 6.A that the same formula gives a (tight) upper bound on the average coding length for any finite number of samples from the Gaussian distribution.

To represent the N data vectors in $\mathcal{X}$, we will need $N \cdot R(X)$ bits. In addition, since the optimal codes are associated with the distribution of data in $\mathbb{R}^D$ (and the decoder will need that information to recover the data), to represent this distribution

[5]For simplicity, in the main text, we will derive and present the main results with the zero-mean assumption. However, all the formulas, results, and algorithms can be readily extended to the nonzero-mean case, as shown in Appendix 6.A.

[6]Strictly speaking, the rate-distortion function for the Gaussian distribution $\mathcal{N}(0, \Sigma)$ is $R = \frac{1}{2}\log_2 \det\big(\frac{D}{\varepsilon^2}\Sigma\big)$ when $\frac{\varepsilon^2}{D}$ is smaller than the smallest eigenvalue of Σ. Thus the above formula is a good approximation when the distortion ε is relatively small. However, when $\frac{\varepsilon^2}{D}$ is larger than some eigenvalues of Σ, the rate-distortion function becomes more complicated (Cover and Thomas 1991). Nevertheless, the approximate formula $R = \frac{1}{2}\log_2 \det(I + \frac{D}{\varepsilon^2}\Sigma)$ can be viewed as the rate-distortion function of the "regularized" distribution that works for all ranges of ε.

we will also need $D \cdot R(X)$ additional bits,[7] yielding an overall coding length for the data set $\mathcal{X}$ of

$$L(X) \doteq (N+D) \cdot R(X) = \frac{N+D}{2} \log_2 \det\left(I + \frac{D}{\varepsilon^2 N} XX^\top\right). \tag{6.46}$$

A more detailed derivation and explanation of the expression for $L(X)$ is given in Section 6.3.3, in which we will also study various good properties of this function. For the purpose of data clustering, though, it suffices to note that in addition to being (approximately) optimal for data drawn from a Gaussian distribution, the same formula $L(X)$ also provides a good upper bound on the total number of bits needed to code a finite number of sample vectors drawn from a Gaussian distribution (see Appendix 6.A for a proof).

Clustering via Data Compression
Given a set of samples $\mathcal{X} = \{\boldsymbol{x}_j \in \mathbb{R}^D\}_{j=1}^N$, one can always view them as drawn from a single Gaussian distribution and encode the data matrix X subject to distortion ε^2 using $L(X)$ bits. However, if the samples are drawn from a mixture of Gaussian distributions or subspaces, it may be more efficient to encode $\mathcal{X}$ as the union of multiple (disjoint) clusters: $\mathcal{X} = \mathcal{X}_1 \cup \mathcal{X}_2 \cup \cdots \cup \mathcal{X}_n$. If each cluster is coded separately, the total number of bits needed is

$$L^s(X_1, X_2, \ldots, X_n) \doteq \sum_{i=1}^{n} L(X_i) + |X_i|\big(-\log_2(|X_i|/N)\big), \tag{6.47}$$

where $|X_i|$ indicates the number of column vectors (i.e., the cardinality of the cluster $\mathcal{X}_i$). In the above expression, the term $\sum_{i=1}^n |X_i|\big(-\log_2(|X_i|/N)\big)$ is the number of bits needed to code (losslessly) the membership of the N samples in the n clusters (e.g., using Huffman coding (Cover and Thomas 1991)).[8]

Given a fixed coding scheme with its associated coding length function $L(\cdot)$, the optimal clustering is the one that minimizes the mixed coding length $L^s(\cdot)$ over all possible partitions of $\mathcal{X}$. Due to the properties of the rate-distortion function (6.45) for Gaussian data, we will see that softening the objective function (6.47) by allowing a probabilistic (or soft) membership does *not* further reduce the (expected) overall coding length (see Theorem 6.5 in Section 6.3.4). Notice also that the objective in (6.47) is a function of the distortion ε. In principle, one may add a "penalty" term, such as $ND \cdot \log \varepsilon$, to the overall coding length[9] L^s so as to determine

[7] This can be viewed as the cost of coding the D principal axes of the data covariance $\frac{1}{N} XX^\top$.

[8] Here we assume that the ordering of the samples is random and entropy coding is the best we can do to code the membership. However, if the samples are ordered such that nearby samples are more likely to belong to the same cluster (e.g., in segmenting pixels of an image), the second term can and should be replaced by a tighter estimate.

[9] This particular penalty term is justified by noticing that $ND \cdot \log \varepsilon$ is (within an additive constant) the number of bits required to code the residual $\boldsymbol{x} - \hat{\boldsymbol{x}}$ up to (very small) distortion $\delta \ll \varepsilon$.

Algorithm 6.4 (Agglomerative Lossy Compression for Subspace Clustering).

Input: Data points $\mathcal{X} = \{\boldsymbol{x}_j \in \mathbb{R}^D\}$ and distortion parameter $\varepsilon^2 > 0$.

Initialize $\mathcal{S} = \{\mathcal{X}_j = \{\boldsymbol{x}_j\} : \boldsymbol{x}_j \in \mathcal{X}\}$.
while $|\mathcal{S}| > 1$ **do**
  choose two distinct sets $\mathcal{X}_i, \mathcal{X}_j \in \mathcal{S}$ such that $L([X_i, X_j]) - L^s(X_i, X_j)$ is minimal.
  if $L([X_i, X_j]) - L^s(X_i, X_j) \geq 0$ **then**
    break
  else
    $\mathcal{S} \leftarrow (\mathcal{S} \setminus \{\mathcal{X}_i, \mathcal{X}_j\}) \cup \{\mathcal{X}_i \cup \mathcal{X}_j\}$.
  end if
end while

Output: $\mathcal{S}$.

the optimal distortion ε^*. The resulting objective $\min_\varepsilon L^s + ND \cdot \log \varepsilon$ will then correspond to an optimal coding length that depends only on the data. Nevertheless, very often we leave ε as a free parameter to be set by the user. In practice, this allows the user to potentially obtain hierarchical clustering of the data at different scales of quantization. We will thoroughly examine how the value of ε affects the final clustering through experiments in Section 6.4.

Minimizing the Coding Length

Finding the global minimum of the overall coding length L^s over all partitions of the data set is a daunting combinatorial optimization problem, which is intractable for large data sets. Nevertheless, the coding length can be effectively minimized in a steepest descent fashion, as outlined in Algorithm 6.4. The minimization proceeds in a "bottom-up" fashion: initially, every sample is treated as its own cluster. At each iteration, two clusters $\mathcal{X}_i$ and $\mathcal{X}_j$ are chosen so that merging them results in the greatest decrease in the coding length. We use $[X_i, X_j]$ to denote the concatenation of the two data matrices X_i and X_j. The algorithm terminates when the coding length cannot be further reduced by merging any pair of clusters. This method is known as the agglomerative lossy compression (ALC) method in the literature (Ma et al. 2007).

As we will see later, in Section 6.4, extensive simulations and experiments demonstrate that the ALC algorithm is very effective in clustering data drawn from a mixture of Gaussians or subspaces. ALC tolerates significant numbers of outliers, and automatically determines the optimal number of clusters for any given distortion: the smaller the value of ε, the larger the number of clusters. However, as a greedy descent scheme, the algorithm is not guaranteed always to find the globally optimal clustering for any given $(\mathcal{X}, \varepsilon)$.[10] In our experience, we have found that the main factor affecting the global convergence of the algorithm seems to be the

[10]However, it may be possible to improve the convergence using more complicated split-and-merge strategies (Ueda et al. 2000). In addition, due to Theorem 6.5 of Section 6.3.4, the globally (asymptotically) optimal clustering can also be computed via concave optimization (Benson 1994), at the cost of potentially exponential computation time.

density of the samples relative to the distortion ε^2. In Section 6.4, we will give strong empirical evidence for the convergence of the algorithm over a wide range of ε.

A naive implementation of Algorithm 6.4 maintains a table containing $L([X_i, X_j])$ for all i, j and requires $O(N^3 + N^2D^3)$ time, where N is the number of samples and D the dimension of the space. However, there are many possible ways to further improve the efficiency or convergence of the ALC algorithm. For instance, one may adopt more advanced split-and-merge strategies (such as those in (Ueda et al. 2000)) or random techniques (such as (Fischler and Bolles 1981)) to improve the speed and effectiveness of the algorithm. Nonetheless, it is possible that in the future, one could even develop more efficient and effective algorithms to minimize the coding length function that are entirely different from the agglomerative approach presented here.

Notice also that the greedy merging process in Algorithm 6.4 is similar in spirit to classical agglomerative clustering methods, especially the so-called Ward's method (Ward 1963). However, whereas Ward's method assumes isotropic Gaussians, the above compression-based approach is capable of clustering Gaussians with arbitrary covariance, including nearly degenerate distributions. Classical agglomerative approaches have been shown to be inappropriate for such situations (Kamvar et al. 2002). In this sense, the change in coding length provides a principled means of measuring similarity between arbitrary Gaussians. This approach also demonstrates significant robustness to uniform outliers, another situation in which linkage algorithms (Hastie et al. 2001) fail.

Remark 6.1. *As discussed in (Wright et al. 2009b), the good performance of ALC can be partly justified from a classification perspective. More specifically, suppose that at some step of Algorithm 6.4 we have already determined n clusters* $\mathcal{X}_1, \ldots, \mathcal{X}_n$. *Suppose also that* $\boldsymbol{x}$ *is a data point that has not yet been merged with any of the n clusters. Then, if we restrict line 3 of Algorithm 6.4 to comparing only the sets* $\{\boldsymbol{x}\}$ *and* $\mathcal{X}_i$ *for all* $i = 1, \ldots, n$, *then we can understand the optimal choice in line 3 as trying to classify data point* $\boldsymbol{x}$ *to one of the current n clusters. The way in which point* $\boldsymbol{x}$ *is classified is by comparing how many additional bits are needed to encode it with each one of the clusters. This is measured as*

$$\delta L(\boldsymbol{x}, i) = L([X_i, \boldsymbol{x}]) - L(X_i) + L(i), \quad i = 1, \ldots, n, \tag{6.48}$$

where the last term is the cost of losslessly coding the label y for $\boldsymbol{x}$ *as* $y = i$. *Thus, step 3 of the ALC algorithm simply assigns* $\boldsymbol{x}$ *to the cluster that minimizes the number of additional bits needed to code* $(\boldsymbol{x}, \hat{y})$. *This is exactly the minimum incremental coding length (MICL) classification criterion introduced in (Wright et al. 2009b), which can be written as*

$$\hat{y}(\boldsymbol{x}) \doteq \arg\min_{y=1,\ldots,n} \delta L(\boldsymbol{x}, y). \tag{6.49}$$

Somewhat surprisingly, this seemingly naive rule of deciding which cluster to merge for $\boldsymbol{x}$ *is in fact nearly optimal when the size of the clusters is large enough. Or more precisely, classification based on the MICL criterion is, asymptotically, equivalent to a regularized version of the optimal maximum a posteriori (MAP) classifier. The interested reader is referred to (Wright et al. 2009b) for a rigorous justification.*

6.3.3 Lossy Coding of Multivariate Data

In this section, we give a more detailed justification of the coding rate and length functions introduced in the previous section. In the next section, we provide a more thorough analysis of the compression-based approach to subspace clustering. Readers who are less concerned with technical details may skip these two sections without much loss of continuity.

The optimal coding scheme and the optimal coding rate of a random vector $\boldsymbol{x}$ with known probability distribution $p(\boldsymbol{x})$ have been well studied in information theory (see (Cover and Thomas 1991) and references therein). However, here we are dealing with a finite set of i.i.d. samples $\mathcal{X} = \{\boldsymbol{x}_j \in \mathbb{R}^D\}_{j=1}^N$ of $\boldsymbol{x}$. Such a data set can be viewed as a nonparametric distribution itself—each vector $\boldsymbol{x}_j$ in $\mathcal{X}$ occurs with equal probability $1/N$. In this case, the optimal coding scheme for the distribution $p(\boldsymbol{x})$ is no longer optimal for $\mathcal{X}$, and the formula for the coding length no longer accurate. Nevertheless, some of the basic ideas for deriving the optimal coding rate can still be extended to the nonparametric setting.

In this section, borrowing ideas from information theory, we derive a tight bound on the coding length or rate for the given data $\mathcal{X}$. In Appendix 6.A, we give an alternative derivation of the bound. Although both approaches essentially arrive at the same estimate, they together reveal that the derived coding rate and length functions hold under different conditions:

1. The derivation in this section shows that for small ε, the formula for $R(X)$ gives a good approximation to the (asymptotically) optimal rate-distortion function of a Gaussian distribution.
2. The derivation in Appendix 6.A shows that the same coding rate and length formulas work for every finite set of vectors $\mathcal{X}$ that span a subspace.

The Rate-Distortion Function
For simplicity, we assume here that the given data have zero mean, i.e., $\sum_j \boldsymbol{x}_j = \boldsymbol{0}$. The reader may refer to Appendix 6.A for the case in which the mean is not zero. Let ε^2 be the squared error allowable for encoding every vector $\boldsymbol{x}_j$. That is, if $\hat{\boldsymbol{x}}_j$ is an approximation of $\boldsymbol{x}_j$, we allow $\mathbb{E}[\|\boldsymbol{x}_j - \hat{\boldsymbol{x}}_j\|^2] \leq \varepsilon^2$. In other words, on average, the allowable squared error for each entry of $\boldsymbol{x}_j$ is ε^2/D.

The solution to coding the vectors in $\mathcal{X}$, subject to the mean squared error ε^2, can be explained by *sphere packing*, which is normally adopted in information theory (Cover and Thomas 1991). Here we are allowed to perturb each vector $\boldsymbol{x}_j \in \mathcal{X}$ within a sphere of radius ε in $\mathbb{R}^D$. In other words, we are allowed to distort each entry of $\boldsymbol{x}_j$ with an (independent) random variable of variance ε^2/D. Without loss of generality, we may model the error as an independent additive Gaussian noise:

$$\hat{\boldsymbol{x}}_j = \boldsymbol{x}_j + \boldsymbol{z}_j, \quad \text{with} \quad \boldsymbol{x}_j \sim \mathcal{N}(\boldsymbol{0}, \Sigma) \quad \text{and} \quad \boldsymbol{z}_j \sim \mathcal{N}\Big(\boldsymbol{0}, \frac{\varepsilon^2}{D} I\Big). \tag{6.50}$$

Then the covariance matrix of the vectors $\{\hat{\boldsymbol{x}}_j\}$ can be approximated as

$$\mathbb{E}\Big[\frac{1}{N}\sum_{j=1}^{N} \hat{\boldsymbol{x}}_j \hat{\boldsymbol{x}}_j^\top\Big] = \frac{\varepsilon^2}{D} I + \Sigma \approx \frac{\varepsilon^2}{D} I + \frac{1}{N} XX^\top \in \mathbb{R}^{D\times D}, \tag{6.51}$$

where $\frac{1}{N}XX^\top$ is the ML estimate of Σ. As a consequence, the volume of the region spanned by the column vectors of $\hat{X} = [\hat{\boldsymbol{x}}_1, \ldots, \hat{\boldsymbol{x}}_N]$ is proportional to (the square root of the determinant of the covariance matrix)

$$\operatorname{vol}(\hat{X}) \propto \sqrt{\det\Big(\frac{\varepsilon^2}{D} I + \frac{1}{N} XX^\top\Big)}. \tag{6.52}$$

Similarly, the volume spanned by each random vector $\boldsymbol{z}_j$ is proportional to

$$\operatorname{vol}(\boldsymbol{z}) \propto \sqrt{\det\Big(\frac{\varepsilon^2}{D} I\Big)}. \tag{6.53}$$

In order to encode each vector, we can partition the region spanned by all the vectors into nonoverlapping spheres of radius ε. When the volume $\operatorname{vol}(\hat{X})$ of the region is significantly larger than the volume of the sphere, the total number of spheres that we can pack into the region is approximately equal to

$$\#\text{ of spheres} = \operatorname{vol}(\hat{X})/\operatorname{vol}(\boldsymbol{z}). \tag{6.54}$$

Thus, to know each vector $\boldsymbol{x}_j$ with an accuracy up to ε^2, we need only specify the sphere containing the vector $\boldsymbol{x}_j$ (see Figure 6.1). If we use binary numbers to label all the spheres in the region of interest, the number of bits needed is

$$\begin{aligned} R(X) &\doteq \log_2(\#\text{of spheres}) = \log_2\big(\operatorname{vol}(\hat{X})/\operatorname{vol}(\boldsymbol{z})\big) \\ &= \frac{1}{2}\log_2\det\big(I + \frac{D}{N\varepsilon^2} XX^\top\big), \end{aligned} \tag{6.55}$$

where the last equality uses the fact $\det(A)/\det(B) = \det(B^{-1}A)$.

Remark 6.2 (Relationships to the rate-distortion function of a Gaussian). *If the samples $\boldsymbol{x}_j$ are drawn from a Gaussian distribution $\mathcal{N}(\boldsymbol{0}, \Sigma)$, then $\frac{1}{N}XX^\top$ converges*

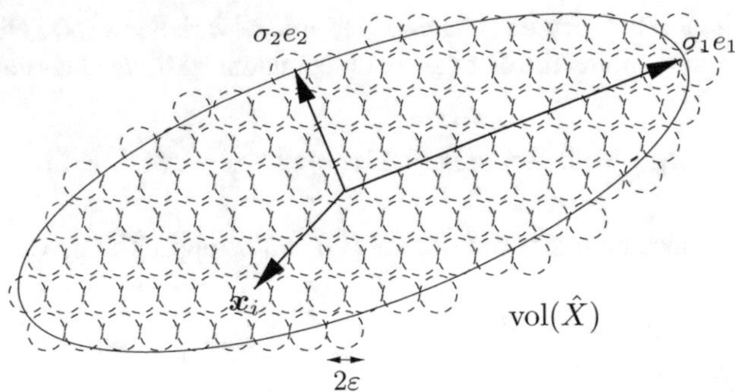

Fig. 6.1 Encoding of a set of vectors in a region in $\mathbb{R}^D$ with an accuracy up to ε^2. To know the vector $\boldsymbol{x}_j$, we need only know the label of the corresponding sphere. Here $e_1, e_2 \in \mathbb{R}^D$ represent the singular vectors of the matrix $\hat{X}$, and $\sigma_1, \sigma_2 \in \mathbb{R}$ the singular values.

to the covariance matrix Σ. Thus, we have $R(X) \to \frac{1}{2}\log_2\det\left(I+\frac{D}{\varepsilon^2}\Sigma\right)$ as $N \to \infty$. When $\frac{\varepsilon^2}{D} \leq \lambda_{\min}(\Sigma)$, the optimal rate distortion for i.i.d. samples from $\mathcal{N}(0,\Sigma)$ is $\frac{1}{2}\log_2\det\left(\frac{D}{\varepsilon^2}\Sigma\right)$, to which (6.55) *provides a good approximation. In general, the optimal rate distortion is a complicated formula given by reverse-waterfilling on the eigenvalues of Σ (see Theorem 13.3.3 of (Cover and Thomas 1991)). The approximation* (6.55) *provides an upper bound that holds for all ε and is tight when ε is small relative to the eigenvalues of the covariance.*

The formula for $R(X)$ can also be viewed as the rate-distortion function of the data $\mathcal{X}$ regularized by a noise of variance $\frac{\varepsilon^2}{D}$ as in equation (6.50). *The covariance $\hat{\Sigma}$ of the perturbed vectors $\hat{\boldsymbol{x}}_j$ always satisfies $\frac{\varepsilon^2}{D} \leq \lambda_{\min}(\hat{\Sigma})$, allowing for a simple analytic expression for the rate distortion for all ε. This regularized rate distortion has the further advantage of agreeing with the bound for the coding length of finitely many vectors that span a subspace, derived in Appendix 6.A.*

Notice that the formula for $R(X)$ is accurate only in the asymptotic sense, i.e., when we are dealing with a large number of samples and the error ε is small (relative to the magnitude of the data $\mathcal{X}$). We want to emphasize that the above derivation of the coding rate does not give an actual coding scheme. The construction of efficient coding schemes that achieve the optimal rate-distortion bound is itself a difficult problem (see, for example, (Hamkins and Zeger 2002) and references therein). However, for the purpose of measuring the quality of clustering and compression, all that matters is that *in principle*, a scheme attaining the optimal rate $R(X)$ exists.

The Coding Length Function

Given the coding rate $R(X)$, the total number of bits needed to encode the N vectors in $\mathcal{X}$ is

$$N \cdot R(X) = \frac{N}{2}\log_2\det\left(I + \frac{D}{N\varepsilon^2}XX^\top\right). \tag{6.56}$$

From a data communication point of view, $N \cdot R(X)$ bits are already sufficient, since both the transmitter and the receiver share the same code book, that is, they both know the region spanned by $\mathcal{X}$ in $\mathbb{R}^D$. However, from the data representation or compression point of view, we need more bits to represent the code book itself. This is equivalent to specifying all the principal axes of the region spanned by the data, i.e., the singular values/vectors of X; see Figure 6.1. Since the number of principal axes is D, we need $D \cdot R(X)$ additional bits to encode them. Therefore, the total number of bits needed to encode the N vectors in $\mathcal{X} \subset \mathbb{R}^D$ subject to the squared error ε^2 is[11]

$$L(X) \doteq (N+D)R(X) = \frac{N+D}{2}\log_2 \det\big(I + \frac{D}{N\varepsilon^2}XX^\top\big). \tag{6.57}$$

Appendix 6.A provides an alternative derivation of the same coding length function $L(X)$ as an upper bound for a finite number of samples. If the data $\mathcal{X}$ have nonzero mean, we need more bits to encode the mean, too. See in Appendix 6.A how the coding length function should be properly modified in that case.

Properties of the Coding Length Function

Commutative Property. Since $XX^\top \in \mathbb{R}^{D\times D}$ and $X^\top X \in \mathbb{R}^{N\times N}$ have the same nonzero eigenvalues, the coding length function can also be expressed as

$$L(X) = \frac{N+D}{2}\log_2 \det\big(I + \frac{D}{N\varepsilon^2}XX^\top\big) \tag{6.58}$$

$$= \frac{N+D}{2}\log_2 \det\big(I + \frac{D}{N\varepsilon^2}X^\top X\big). \tag{6.59}$$

Thus if $D \ll N$, the second expression will be less costly for computing the coding length. As we saw in Section 4.1.3, the matrix $X^\top X$, which depends only on the inner products between pairs of data vectors, is known in the statistical learning literature as the *kernel matrix*. This property suggests that the ideas and the algorithm presented in Section 6.3.2 can be readily extended to cluster data sets that have *nonlinear* structures, by choosing a proper kernel function.

Invariant Property. Notice that in the zero-mean case, the coding length function $L(X)$ is invariant under an orthogonal transformation of the data $\mathcal{X}$. That is, for every orthogonal matrix $U \in O(D)$ or $V \in O(N)$, we have

$$L(UX) = L(X) = L(XV). \tag{6.60}$$

In other words, the length function depends only on the singular values of X (or eigenvalues of $XX^\top$). This equality suggests that one may choose any

[11]Compared to the MDL criterion (2.85), if the term $N \cdot R(X)$ corresponds to the coding length for the data, the term $D \cdot R(X)$ then corresponds to the coding length for the model parameter θ.

orthonormal basis (e.g., Fourier, wavelets) to represent and encode the data, and the number of bits needed should always be the same. This agrees with the fact that the chosen coding length (or rate) is optimal for a Gaussian distribution. However, if the data are non-Gaussian or nonlinear, a proper transformation can still be useful for compressing the data.[12] Here we are essentially seeking a partition, rather than a transformation, of the non-Gaussian (or nonlinear) data set such that each subset is sufficiently Gaussian (or subspace-like) and hence cannot be compressed any further, either by (orthogonal) transformation or clustering.

6.3.4 Coding Length of Mixed Gaussian Data

Now suppose we have partitioned the set of N vectors $\mathcal{X} = \{\boldsymbol{x}_1, \boldsymbol{x}_2, \ldots, \boldsymbol{x}_N\}$ into n nonoverlapping clusters $\mathcal{X} = \mathcal{X}_1 \cup \mathcal{X}_2 \cup \cdots \cup \mathcal{X}_n$. The corresponding data matrix has the form $X = [X_1, X_2, \ldots, X_n]$, where X_i is a submatrix associated with $\mathcal{X}_i$. Then the total number of bits needed to encode the clustered data is $L^s(X_1, X_2, \ldots, X_n) = \sum_{i=1}^n L(X_i) + |X_i|\big(-\log_2(|X_i|/N)\big)$. Here the superscript s is used to indicate the coding length after clustering.

Clustering and Compression

To better understand under what conditions a data set should or should not be partitioned in order to reduce the overall coding rate or length, in what follows we provide two representative examples. To simplify the analysis, we assume that the data set can be partitioned into two subsets of an equal number of vectors, i.e., $X_1, X_2 \in \mathbb{R}^{D\times N}$. We also assume that $N \gg D$, so that we can ignore the asymptotically insignificant terms in the coding rate and length functions.

Example 6.3 (Uncorrelated Subsets). Notice that in general, we have

$$\begin{aligned} L(X_1) + L(X_2) &= \frac{N}{2}\log_2\det\big(I + \frac{D}{N\varepsilon^2}X_1X_1^\top\big) + \frac{N}{2}\log_2\det\big(I + \frac{D}{N\varepsilon^2}X_2X_2^\top\big) \\ &\leq \frac{2N}{2}\log_2\det\big(I + \frac{D}{2N\varepsilon^2}(X_1X_1^\top + X_2X_2^\top)\big) = L([X_1, X_2]), \end{aligned}$$

where the inequality follows from the concavity of the function $\log_2\det(\cdot)$ (see Theorem 7.6.7 of (Horn and Johnson 1985)). Thus, if the difference $L([X_1, X_2]) - \big(L(X_1) + L(X_2)\big)$ is large, the overhead needed to encode the membership of the clustered data (here one bit per vector) becomes insignificant. If we further assume that X_2 is a rotated version of X_1, i.e., $X_2 = UX_1$ for some $U \in O(D)$, one can show that the difference $L([X_1, X_2]) - \big(L(X_1) + L(X_2)\big)$ is (approximately) maximized when X_2 becomes orthogonal to X_1. We call two clusters X_1, X_2 *uncorrelated* if

[12]For a more thorough discussion on why some transformations (such as wavelets) are useful for data compression, the reader may refer to (Donoho et al. 1998).

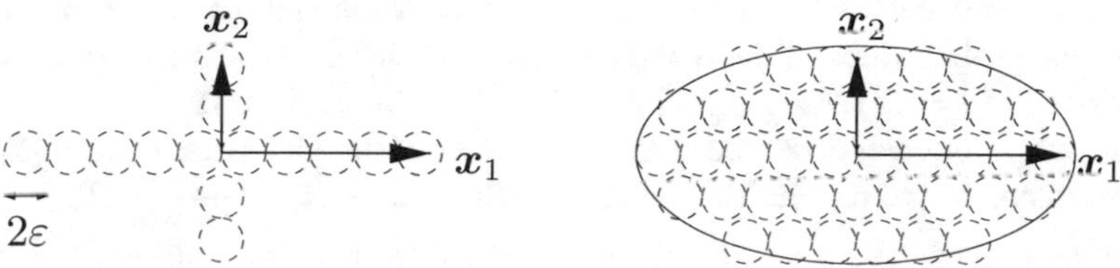

Fig. 6.2 The number of spheres (code words) of two different schemes for coding two orthogonal vectors. Left: encoding the two vectors separately. Right: encoding the two vectors together.

$X_1^\top X_2 = 0$. Thus, segmenting the data into uncorrelated clusters typically reduces the overall coding length. From the viewpoint of sphere packing, Figure 6.2 explains the reason.

Example 6.4 (Strongly Correlated Subsets). We say that two clusters X_1, X_2 are strongly correlated if they span the same subspace in $\mathbb{R}^D$. Or somewhat equivalently, we may assume that X_1 and X_2 have approximately the same sample covariance, that is, $X_2X_2^\top \approx X_1X_1^\top$. Under this assumption, we have

$$\begin{aligned} L(X_1) + L(X_2) &= \frac{N}{2}\log_2\det\big(I + \frac{D}{N\varepsilon^2}X_1X_1^\top\big) + \frac{N}{2}\log_2\det\big(I + \frac{D}{N\varepsilon^2}X_2X_2^\top\big) \\ &\approx \frac{2N}{2}\log_2\det\big(I + \frac{D}{2N\varepsilon^2}(X_1X_1^\top + X_2X_2^\top)\big) = L([X_1, X_2]). \end{aligned}$$

Since $L^s(X_1, X_2) = L(X_1) + L(X_2) + H(|X_1|, |X_2|)$, the overhead needed to encode the membership becomes significant, and the segmented data require more bits than the unsegmented data. In other words, when two clusters have approximately the same covariance, it is better to encode them together.

Optimality of Deterministic Clustering

So far, we have considered only partitioning the data $\mathcal{X}$ into n nonoverlapping clusters. That is, each vector is assigned to a cluster with probability either 0 or 1. We call such a clustering "deterministic." In this section, we examine an important question: *is there a probabilistic partitioning of the data that can achieve an even lower coding rate?* By a probabilistic partitioning, we mean a more general class of partitions in which we assign each vector $\boldsymbol{x}_j$ to the cluster i according to a probability $\pi_{ij} \in [0, 1]$, with $\sum_{i=1}^n \pi_{ij} = 1$ for all $j = 1, 2, \ldots, N$.

To facilitate counting the coding length of such (probabilistically) partitioned data, for each point $\boldsymbol{x}_j$ we introduce a random variable $y_{ij} \in \{0, 1\}$ with multinomial distribution $p(y_{ij} = 1) = \pi_{ij}$. For convenience, we define two matrices

$$Y_i \doteq \begin{bmatrix} y_{i1} & 0 & \cdots & 0 \\ 0 & y_{i2} & \ddots & \vdots \\ \vdots & \ddots & \ddots & 0 \\ 0 & \cdots & 0 & y_{iN} \end{bmatrix}, \quad \Pi_i \doteq \begin{bmatrix} \pi_{i1} & 0 & \cdots & 0 \\ 0 & \pi_{i2} & \ddots & \vdots \\ \vdots & \ddots & \ddots & 0 \\ 0 & \cdots & 0 & \pi_{iN} \end{bmatrix} \in \mathbb{R}^{N\times N}, \tag{6.61}$$

where Y_i collects the observed memberships of the N data points in cluster i, and Π_i collects the probabilities of the N data points in cluster i. These matrices satisfy the conditions $\mathbb{E}[Y_i] = \Pi_i$ and $\sum_{i=1}^{n} Y_i = \sum_{i=1}^{n} \Pi_i = I_{N\times N}$, $\Pi_i \succeq 0$.

Obviously, for each observed instance, the size of the ith cluster is trace (Y_i), and the covariance of the ith cluster Σ_i can be estimated as $\hat{\Sigma}_i = \frac{1}{\operatorname{trace}(Y_i)} XY_iX^\top$. Thus, the coding rate of the ith cluster (if viewed as a Gaussian distribution $\mathcal{N}(0, \hat{\Sigma}_i)$) is given by

$$R(X_i) \doteq \frac{1}{2}\log_2 \det\big(I + \frac{D}{\operatorname{trace}(Y_i)\varepsilon^2} XY_iX^\top\big). \tag{6.62}$$

Substituting the above covariance and cluster size into the segmented coding length function (6.47) yields that the total number of bits required to encode the data X clustered according to the (random) assignment $Y = \{Y_i\}_{i=1}^n$ is

$$\begin{aligned} L^s(X, Y) \doteq \sum_{i=1}^{n} & \frac{\operatorname{trace}(Y_i) + D}{2} \log_2 \det\big(I + \frac{D}{\operatorname{trace}(Y_i)\varepsilon^2} XY_iX^\top\big) \\ & + \operatorname{trace}(Y_i)\Big(-\log_2 \frac{\operatorname{trace}(Y_i)}{N}\Big). \end{aligned} \tag{6.63}$$

As we will see in Theorem 6.5, $L^s(X, Y)$ is a concave function of the assignment Y. We have $L^s(X, \mathbb{E}[Y]) \geq \mathbb{E}[L^s(X, Y)]$ (using that $f(\mathbb{E}[x]) \geq \mathbb{E}[f(x)]$ for concave functions). Since $\mathbb{E}[Y] = \Pi$, we have that $L^s(X, \Pi)$, defined as

$$\begin{aligned} L^s(X, \Pi) \doteq \sum_{i=1}^{n} & \frac{\operatorname{trace}(\Pi_i) + D}{2} \log_2 \det\big(I + \frac{D}{\operatorname{trace}(\Pi_i)\varepsilon^2} X\Pi_iX^\top\big) \\ & + \operatorname{trace}(\Pi_i)\Big(-\log_2 \frac{\operatorname{trace}(\Pi_i)}{N}\Big), \end{aligned} \tag{6.64}$$

gives an upper bound for the expected coding length for such probabilistic clustering. Similarly, the expected number of bits needed to encode each vector is bounded by

$$\begin{aligned} R^s(X, \Pi) &\doteq \frac{1}{N} L^s(X, \Pi) \\ &= \sum_{i=1}^{n} \frac{\operatorname{trace}(\Pi_i)}{N}\Big(R(X_i) - \log_2 \frac{\operatorname{trace}(\Pi_i)}{N}\Big) + \frac{D}{N} R(X_i). \end{aligned} \tag{6.65}$$

Thus, one may consider that the optimal partition Π^* is the global minimum of the expected overall coding length $L^s(X, \Pi)$, or equivalently the average coding rate $R^s(X, \Pi)$. To some extent, one can view the minimum value of $R^s(X, \Pi)$ as a good approximation to the actual entropy of the given data set $\mathcal{X}$.[13]

---

[13]Especially when the data $\mathcal{X}$ indeed consist of a mixture of subsets and each cluster is a typical set of samples from an (almost degenerate) Gaussian distribution.

Notice that the second term, $\frac{D}{N}R(X_i)$, in the expression of $R^s(X, \Pi)$ is insignificant when the number of samples is large, $N \gg D$. It roughly corresponds to how many extra bits are needed to encode the basis of the subspace, and it becomes significant only when the number of samples is small. The first term in the expression of $R^s(X, \Pi)$ is the only part that matters asymptotically (i.e., as the number of vectors in each cluster goes to infinity) and we denote it by

$$\begin{aligned} R^{s,\infty}(X, \Pi) \doteq \sum_{i=1}^{n} \frac{\text{trace}\,(\Pi_i)}{2N} \log_2 \det\big(I + \frac{D}{\varepsilon^2\,\text{trace}\,(\Pi_i)} X\Pi_i X^\top\big) \\ - \frac{\text{trace}\,(\Pi_i)}{N} \log_2 \Big(\frac{\text{trace}\,(\Pi_i)}{N}\Big). \end{aligned} \tag{6.66}$$

Thus, the global minimum of $R^{s,\infty}(X, \Pi)$ determines the optimal clustering when the sample size is large.

Theorem 6.5 (Concavity of Asymptotic Coding Length). *The asymptotic part $R^{s,\infty}(X, \Pi)$ of the rate-distortion function $R^s(X, \Pi)$ is a concave function of Π in the convex domain $\Omega \doteq \{\Pi : \sum_{i=1}^n \Pi_i = I, \Pi_i \succeq 0\}$.*

Proof. Let $\mathcal{S}$ be the set of all $N \times N$ nonnegative definite symmetric matrices. We will show that $R^{s,\infty}(X, \Pi)$ is concave as a function from $\mathcal{S}^n \to \mathbb{R}$, and also when restricted to the domain of interest, $\Omega \subset \mathcal{S}^n$.

First consider the second term of $R^{s,\infty}(X, \Pi)$. Notice that $\sum_{i=1}^n \text{trace}\,(\Pi_i) = N$ is a constant. So we must show the concavity of the function $g(P) \doteq -\text{trace}\,(P) \log_2 \text{trace}\,(P)$ only for $P \in \mathcal{S}$. To see this, notice that the function $f(x) = -x \log_2 x$ is concave and that $g(P) = f(\text{trace}\,(P))$ is the composition of f with the trace (a linear function of P). Thus, $g(P)$ is concave in P.

Now consider the first term of $R^{s,\infty}(X, \Pi)$. Let

$$h(\Pi_i) \doteq \text{trace}\,(\Pi_i) \log_2 \det\big(I + \frac{D}{\varepsilon^2\,\text{trace}\,(\Pi_i)} X\Pi_i X^\top\big). \tag{6.67}$$

It is well known in information theory that the function $q(P) \doteq \log_2 \det(P)$ is concave for $P \in \mathcal{S}$ and $P \succ 0$ (see Theorem 7.6.7 of (Horn and Johnson 1985)). Now define $r : \mathcal{S} \to \mathbb{R}$ to be

$$r(\Pi_i) \doteq \log_2 \det(I + \alpha X\Pi_i X^\top) = q(I + \alpha X\Pi_i X^\top). \tag{6.68}$$

Since r is just the concave function q composed with an affine transformation $\Pi_i \mapsto I + \alpha X\Pi_i X^\top$, r is concave (see Section 3.2.3 of (Boyd and Vandenberghe 2004)). Let $\psi : \mathcal{S} \times \mathbb{R}_+ \to \mathbb{R}$ be defined as

$$\psi(\Pi_i, t) \doteq t \cdot \log_2 \det\big(I + \frac{N}{\varepsilon^2 t} X\Pi_i X^\top\big) = t \cdot r\big(\frac{1}{t}\Pi_i\big). \tag{6.69}$$

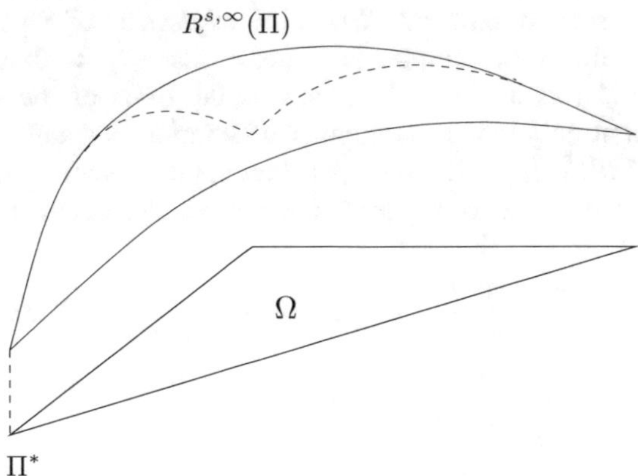

Fig. 6.3 The function $R^{s,\infty}(X, \Pi)$ is a concave function of Π over a convex domain Ω, which is in fact a polytope in the space $\mathbb{R}^{nN}$. The minimal coding length is achieved at a vertex Π^* of the polytope.

According to Theorem 3.2.6 of (Boyd and Vandenberghe 2004), ψ is concave. Notice that $H \doteq \{(\Pi_i, t) : t = \text{trace}(\Pi_i)\}$ is a linear subspace in the product space of $\mathbb{R}$ and the space of all symmetric matrices. So $H \cap (\mathcal{S} \times \mathbb{R}_+)$ is a convex set, and the desired function, $h(\Pi_i) = \psi(\Pi_i, \text{trace}(\Pi_i))$, is just the restriction of ψ to this convex set. Thus, h is concave.

Since $R^{s,\infty}(X, \Pi)$ is a sum of concave functions in Π_i, it is concave as a function from $\mathcal{S}^n$ to $\mathbb{R}$, and so is its restriction to the convex set Ω in $\mathcal{S}^n$. □

Since $R^{s,\infty}(X, \Pi)$ is concave, its global minimum Π^* is always reached at the boundary, or more likely, at a vertex of the convex domain Ω, as shown in Figure 6.3. At the vertex of Ω, the entries π_{ij} of Π^* are either 0's or 1's. This means that even if we allow soft assignment of each point to the n clusters according to some probabilistic distribution, the optimal solution with the minimal coding length can always be approximately achieved by assigning each point to one of the clusters with probability one. This is why Algorithm 6.4 does not consider any probabilistic clustering.

Another implication of the above theorem is that the problem of minimizing the coding length is essentially a concave optimization problem. Many effective concave optimization algorithms can be adopted to find the globally optimal clustering, such as the simplex algorithm (Benson 1994). However, such generic concave optimization algorithms typically have high (potentially exponential) complexity. In the next section, we will show with extensive simulations and experiments that the greedy algorithm proposed in Section 6.3.2 (Algorithm 6.4) is already effective in minimizing the coding length.

Remark 6.6. *Interestingly, in multiple-channel communications, the goal is instead to maximize the channel capacity, which has very much the same formula as the coding rate function (Tse and Viswanath 2005). The above theorem suggests* *that a higher channel capacity may be achieved inside the convex domain Ω, i.e., by probabilistically assigning the transmitters to a certain number of clusters. Since the coding rate function is concave, the maximal channel capacity can be very easily computed via convex optimization (Boyd and Vandenberghe 2004).*

6.4 Simulations and Applications

In this section, we will conduct simulations on synthetic data to examine the effectiveness of the coding length function described in Section 6.3 and evaluate the performance of the ALC algorithm, Algorithm 6.4. We will also demonstrate the ALC algorithm on a few real examples of clustering gene expression data, segmenting natural images, and clustering face images. In the face clustering example, we will also compare the ALC algorithm with the K-subspaces and MPPCA algorithms discussed in Sections 6.1 and 6.2, respectively. The reader is referred to Chapter 10 and Chapter 11 for more extensive applications of these methods to real data.

6.4.1 Statistical Methods on Synthetic Data

Clustering Linear Subspaces of Different Dimensions

We first demonstrate the ability of the ALC algorithm, Algorithm 6.4, to cluster noisy samples drawn from a mixture of linear subspaces of different dimensions $d \in \{1, 2, \dots, 7\}$ in $\mathbb{R}^D$ with $D \in \{3, 5, 7, 8\}$. For every d-dimensional subspace, $100d$ samples are drawn uniformly from a ball of diameter 1 lying in the subspace. Each sample is corrupted by independent Gaussian noise of standard deviation $\varepsilon_0 = 0.04$.[14]

We compare the results of the ALC algorithm with parameter $\varepsilon = \varepsilon_0$ to those of the EM algorithm of (Ghahramani and Hinton 1996) for a mixture of factor analyzers (FA).[15] Since the EM-FA algorithm gives a probabilistic assignment, we use an ML classification step[16] to obtain hard assignments. We modified the EM-FA algorithm of (Ghahramani and Hinton 1996) slightly to allow it to work

---

[14] Notice that the data are not drawn from a mixture of Gaussians. As we have mentioned before, although we have derived the coding length function largely based on the mixture of Gaussians model, the coding length function actually gives a good estimate of the coding length for any subspace-like data; see Appendix 6.A for more details. Hence this experiment tests how well the coding length works when we deviate from the basic mixture of Gaussians model.

[15] Factor analysis is a probabilistic model for a subspace that generalizes the PPCA model discussed in Section 2.2: in FA, the noise covariance is diagonal, while in PPCA, it is a scaled identity matrix.

[16] That is, each sample is assigned to its cluster according to the maximum likelihood rule among all subspaces estimated by the EM algorithm.

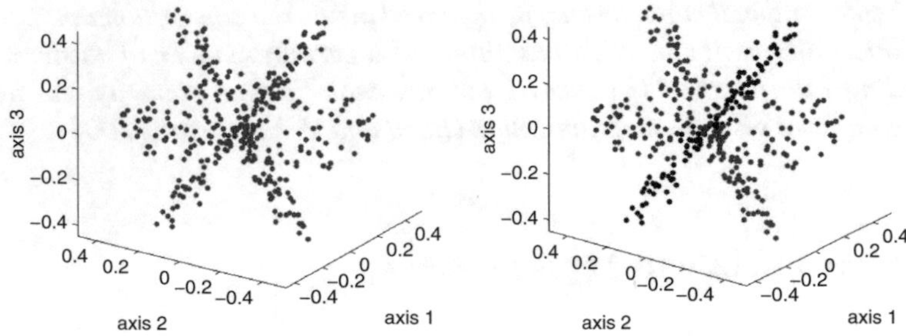

Fig. 6.4 Result of applying the ALC algorithm, Algorithm 6.4, to data drawn approximately from a union of three subspaces of dimensions $(2, 1, 1)$ in $\mathbb{R}^3$. Left: noisy input data. Right: output clusters.

Table 6.1 Simulation results for data drawn from mixtures of noisy linear subspaces. Clustering error percentages are averaged over 25 trials. The ALC algorithm correctly identifies the number and dimension of the subspaces in all 25 trials, for all configurations. Far right column: results using EM for a mixture of subspaces with different dimensions (Ghahramani and Hinton 1996) with random initialization.

True subspace dimensions	Identified dimensions	Clustering(%) error (ALC)	Clustering (%) error (EM-FA)
$(2, 1, 1)$ in $\mathbb{R}^3$	2, 1, 1	3.38	60.67
$(2, 2, 1)$ in $\mathbb{R}^3$	2, 2, 1	10.00	31.02
$(4, 2, 2, 1)$ in $\mathbb{R}^5$	4, 2, 2, 1	1.47	56.64
$(6, 3, 1)$ in $\mathbb{R}^7$	6, 3, 1	0.23	33.84
$(7, 5, 2, 1, 1)$ in $\mathbb{R}^8$	7, 5, 2, 1, 1	1.96	57.71

for a mixture of subspaces with different dimensions. To avoid the model selection issue, we provided the EM-FA algorithm with the correct number of subspaces and their dimensions. However, there is no need to provide such information for the ALC algorithm.

Figure 6.4 shows one representative result of the ALC algorithm, while Table 6.1 summarizes the average clustering accuracy for several configurations tested. In each case, the algorithm stops at the correct number of clusters, and the dimensions of the clusters $\mathcal{X}_i$ match those of the generating subspaces.[17] For all five configurations, the average percentage of samples assigned to the correct cluster was at least 90.0%. The main cause of clustering error is points that lie near the intersection of multiple subspaces. Due to noise, it may actually be more efficient to code such points to one of the other subspaces. Notice also that in all cases, ALC dramatically outperforms EM-FA, despite requiring no knowledge of the subspace dimensions.

[17]The dimension of each cluster $\mathcal{X}_i$ is identified using principal component analysis (PCA) by thresholding the singular values of the data matrix X_i with respect to ε.

Table 6.2 Size of the range of $\log \varepsilon$ for which the ALC algorithm converges to the correct number and dimension of clusters, for each of the arrangements considered in Figure 6.4.

Subspace dimensions	$(2, 1, 1)$ in $\mathbb{R}^3$	$(2, 2, 1)$ in $\mathbb{R}^3$	$(4, 2, 2, 1)$ in $\mathbb{R}^5$	$(6, 3, 1)$ in $\mathbb{R}^7$	$(7, 5, 2, 1, 1)$ in $\mathbb{R}^8$
$\log_{10} \frac{\varepsilon_{max}}{\varepsilon_{min}}$	2.5	1.75	2.0	2.0	.75

Since in practice, ε_0 is unknown, it is important to investigate the sensitivity of the results to the choice of ε. Table 6.2 gives, for each of the examples in Table 6.1, the range of ε for which ALC converges to the true number of subspaces and true subspace dimensions. Notice that for each of the configurations considered, there exists a significant range of ε for which the greedy algorithm converges.

Global Convergence

Empirically, we find that the ALC algorithm does not suffer from many of the difficulties with local minima that plague iterative clustering algorithms such as K-subspaces, or statistical parameter estimation algorithms such as EM. The convergence appears to depend mostly on the density of the samples relative to the distortion ε. For example, if the number of samples is fixed at $N = 1200$, and the data are drawn from three $\lceil \frac{D}{2} \rceil$-dimensional subspaces in $\mathbb{R}^D$, the algorithm converges to the correct solution for $D = 2$ up to $D = 56$. Here, we choose $\varepsilon = \varepsilon_0 = 0.008$. Beyond $D = 56$, the algorithm fails to converge to the three true subspaces, because the samples have become too sparse. For $D > 56$, the computed clustering gives a higher coding length than the ground truth clustering, suggesting that the algorithm converged to a local minimum.

The same observation occurs for subspaces with different dimensions. For example, we randomly draw 800 samples from four subspaces of dimensions 20, 15, 15, and 10 in $\mathbb{R}^{40}$, and added noise with $\varepsilon_0 = 0.14$. The results of ALC on these data for different values of the distortion parameter ε are shown in Figure 6.5. It can be observed that when the distortion ε is very small, ALC does not necessarily converge to the optimal coding length. Nevertheless, the number of clusters, 4, is still identified correctly by the algorithm when ε becomes relatively large.

Recall also that as described in Section 6.3.2, ε can potentially be chosen automatically by minimizing $L^s + ND \cdot \log \varepsilon$, where the second term approximates (up to a constant) the number of bits needed to code the residual. The green curve in Figure 6.5 shows the value of this penalized coding length. Notice that its minimum falls very near the true value of $\log \varepsilon$. We observe similar results for other simulated examples: the penalty term is generally effective in selecting a relevant ε.

Robustness to Outliers

We test the robustness of the ALC algorithm to outliers on the example of two lines and a plane in $\mathbb{R}^3$, illustrated in Figure 6.6. In this example, 158 samples are drawn uniformly from a 2D disk of diameter 1, and 100 samples are drawn uniformly from each of the two line segments of length 1. The additive noise level is $\varepsilon_0 = 0.03$.

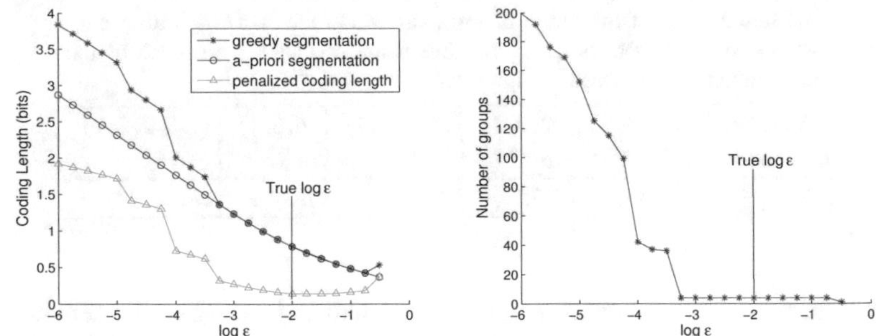

Fig. 6.5 Left: the coding length found by the greedy algorithm (the red curve) compared to the ground truth (the blue curve labeled "a priori segmentation") for data drawn from four linear subspaces of dimensions 20, 15, 15, 10 in $\mathbb{R}^{40}$. The green curve shows the penalized coding length $L^s + ND \cdot \log \varepsilon$. Right: the number of clusters found by the greedy algorithm; it converges to the correct number, 4, when the distortion is relatively large.

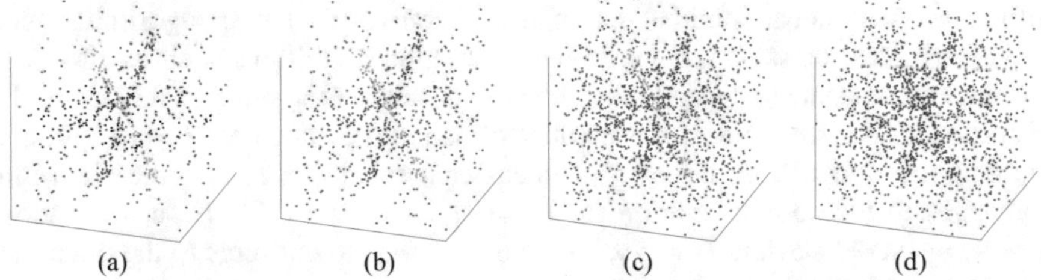

Fig. 6.6 Clustering results for data drawn from three linear subspaces, corrupted by N_o outliers. (a) $N_o = 300$ (45.6% outliers). (b) $N_o = 400$ (52.8% outliers). (c) $N_o = 1100$ (75.4% outliers). (d) $N_o = 1200$ (77.0% outliers).

The data set is contaminated by N_o outliers, whose three coordinates are uniformly distributed on $[-0.5, 0.5]$.

As the number of outliers increases, the clustering results of ALC exhibit several distinct phases. For $N_o \leq 300$ (45.6% outliers), ALC always finds the correct clustering, and the outliers are merged into a single (three-dimensional) cluster. From $N_o = 400$ (52.8% outliers) up to $N_o = 1100$ (75.4% outliers), the two lines are correctly identified, but samples in the plane are merged with the outliers. For $N_o = 1200$ (77.4% outliers) and higher, all of the data samples are merged into one cluster, as the distribution of the data becomes essentially random in the ambient space. Figure 6.6 shows the results for $N_o = 300, 400, 1100, 1200$. Notice that the effect of adding the outliers resembles the effect of ice (the lines and the plane) being melted away by warm water. This suggests a similarity between the artificial process of data clustering and the physical process of phase transition.

Number of Clusters versus Distortion Level

Figure 6.7 shows how the number of clusters changes as ε varies. In this experiment, $N = 358$ points are drawn from two lines and a plane and then perturbed by noise

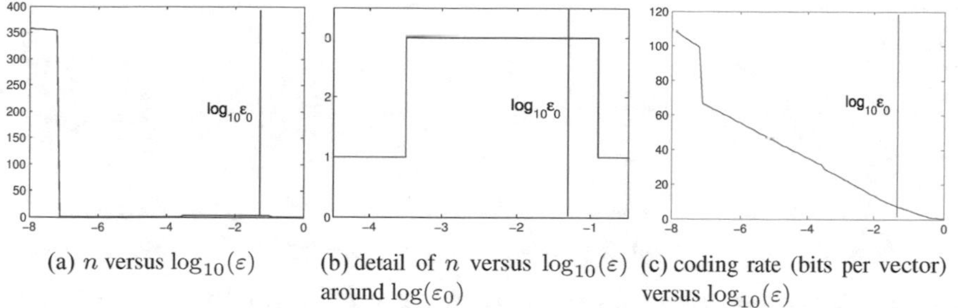

(a) n versus $\log_{10}(\varepsilon)$ (b) detail of n versus $\log_{10}(\varepsilon)$ around $\log(\varepsilon_0)$ (c) coding rate (bits per vector) versus $\log_{10}(\varepsilon)$

Fig. 6.7 Estimated number of clusters n as a function of the distortion level ε when the noise level is $\varepsilon_0 = 0.05$.

with standard deviation $\varepsilon_0 = 0.05$. Notice that the number of clusters experiences distinct phases, with abrupt transitions around several critical values of ε. For sufficiently small ε, each data point forms its own cluster. However, as ε increases, the cost of coding the cluster membership begins to dominate, and all the points are clustered together in a single three-dimensional subspace (the ambient space). Around the true noise level, ε_0, there is another stable phase, corresponding to the three true subspaces. Finally, as ε becomes large, the number of clusters reverts to 1, since it becomes most efficient to represent the points using a single zero-dimensional subspace (the origin).

This behavior contrasts with the phase transition discussed in (Rose 1998). There, the number of clusters increases monotonically throughout the simulated annealing process. Because the formulation in this chapter allows the dimension of the clusters to vary, the number of clusters does not decrease monotonically with ε. Notice, however, that the phase corresponding to the correct clustering is stable over several orders of magnitude of the parameter ε. This is important, since in practice, the true noise level ε_0 is usually unknown.

Another interesting thing to notice is that the coding rate $R^s(X)$ in many regions is mostly a linear function of $-\log_{10}\varepsilon$: $R^s(X) \approx -\beta\log_{10}\varepsilon + \alpha$, for some constants $\alpha, \beta > 0$, which is a typical characteristic of the rate-distortion function of Gaussians.

Clustering of Affine Subspaces

Appendix 6.A shows how the coding length function should be properly modified when the data are not of zero mean. Here, we show how the modified ALC algorithm works for affine subspaces. In this example, $N = 358$ samples are drawn from three linear subspaces in $\mathbb{R}^3$, and their centers are translated to $[2.1, 2.2, 2]^\top$, $[2.4, 1.9, 2.1]^\top$, $[1.9, 2.5, 1.9]^\top$. The data are then corrupted by noise with $\varepsilon_0 = 0.01, 0.03, 0.05, 0.08$. Figure 6.8 shows the clustering results for different noise levels. Although not shown in the figure, we found that experimentally when $10^{-7} < \varepsilon < 0.1$, the algorithm always identifies the correct number of subspaces with $\varepsilon = \varepsilon_0$. However, when $\varepsilon \leq 10^{-7}$, the density of the samples within the subspace becomes more important than the distortion orthogonal to the subspace,

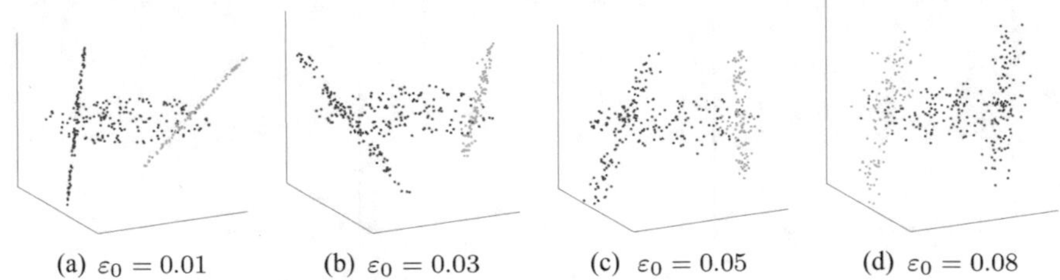

Fig. 6.8 Clustering results for data drawn from three affine subspaces for different levels of noise $\varepsilon_0 = 0.01, 0.03, 0.05, 0.08$. The parameter ε in the algorithm is chosen to be $\varepsilon = \varepsilon_0$.

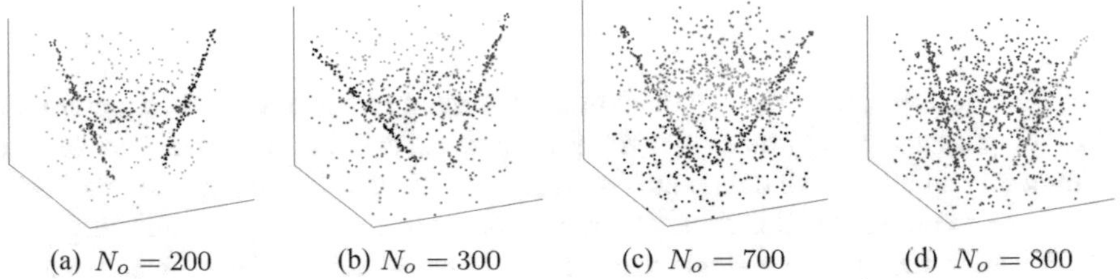

Fig. 6.9 Clustering results for data drawn from three affine subspaces with N_o outliers. The parameter ε in the algorithm is $\varepsilon = \varepsilon_0 = 0.02$. (a) $N_o = 200$ (35.8% outliers), (b) $N_o = 300$ (45.6% outliers), (c) $N_o = 700$ (66.2% outliers), (d) $N_o = 800$ (69.1% outliers).

and the algorithm no longer converges. However, for such small distortion, there always exists a large stable phase (with respect to changing ε) that gives rise to the correct number of subspaces, $n = 3$. When $\varepsilon_0 > 0.1$, the algorithm starts to fail and merges the data samples into one or two clusters.

We now fix the level of noise at $\varepsilon_0 = 0.02$ and add N_o outliers whose three coordinates are uniformly distributed in the range of $[1.5, 2.5]$, which is the same as the range of the inliers. When the number of outliers is $N_o = 200$ (35.8% outliers), the ALC algorithm finds the correct clustering, and all the outliers are clustered into one cluster. From $N_o = 300$ (45.6% outliers) to $N_o = 700$ (66.2% outliers), the ALC algorithm still identifies the two lines and one plane. However, the outliers above and below the plane are clustered into two separate clusters. For more than $N_o = 800$ (69.1% outliers), the algorithm identifies the two lines, but samples from the plane are merged with the outliers into one cluster. Figure 6.9 shows the clustering results for $N_o = 200, 300, 700, 800$.

Model Selection for Affine Subspaces and Nonzero-Mean Gaussians

We compare the ALC algorithm to the algorithms of (Figueiredo and Jain 2002) and (Ghahramani and Beal 2000) on three examples of mixed data drawn from affine subspaces and nonzero-mean Gaussians. The first example consists of outlier-free data samples drawn from three affine subspaces (two lines and one plane) and corrupted by noise with $\varepsilon_0 = 0.01$. Samples are drawn as in the previous examples. The means of the three clusters are fixed (as in the previous examples), but the

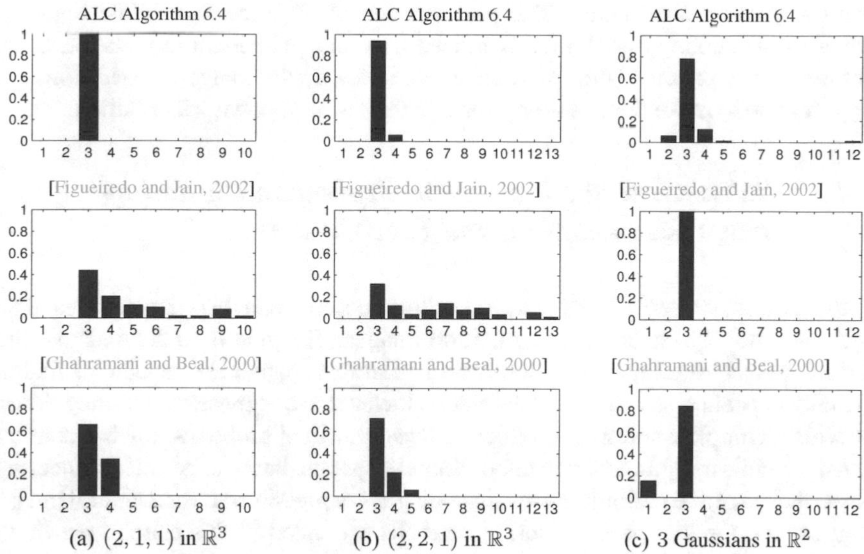

(a) $(2, 1, 1)$ in $\mathbb{R}^3$ (b) $(2, 2, 1)$ in $\mathbb{R}^3$ (c) 3 Gaussians in $\mathbb{R}^2$

Fig. 6.10 Frequency of estimated number of subspaces n in 50 trials. The left and center columns show results for randomly generated arrangements of affine subspaces. The right column shows results for data sets generated from three full-rank Gaussians, as in (Figueiredo and Jain 2002). For all cases, the correct number of clusters is $n = 3$.

orientations of the two lines are chosen randomly. The second example uses data drawn from three affine subspaces (two planes and one line), with 158 points drawn from each plane and 100 from the line, again with $\varepsilon_0 = 0.01$. The orientations of onc plane and of the line are chosen randomly. The final distribution tested is a mixture of $n = 3$ full-rank Gaussians in $\mathbb{R}^2$, with means $[2, 0]$, $[0, 0]$, $[0, 2]$ and covariance $\mathrm{diag}(2, 0.2)$ (this is Figure 3 of (Figueiredo and Jain 2002)). Here $N = 900$ points are sampled (with equal probability) from the three Gaussians.

For the first two examples, wc run the ALC algorithm with $\varepsilon = \varepsilon_0 = 0.01$. For the third example, we set $\varepsilon = 0.2$. We repeat each trial 50 times. Figure 6.10 shows a histogram of the number of clusters arrived at by the three algorithms. For all algorithms, all clustering results with $n = 3$ are essentially correct (clustering error $< 4\%$). However, for degenerate, or subspace-like, data (Figure 6.10(a) and Figure 6.10(b)), the ALC algorithm was the most likely to converge to the correct cluster number. For full-rank Gaussians (Figure 6.10(c)), the ALC algorithm performs quite well, but is outperformed by (Figueiredo and Jain 2002), which finds the correct clustering in all 50 trials. The failures of ALC occur because the greedy descent converges to a local minimum of the coding length rather than the global minimum.

Please note that (Figueiredo and Jain 2002) was not explicitly designed for degenerate distributions, whereas (Ghahramani and Beal 2000) was not designed for full-rank distributions. Also note that the samples in this experiment were drawn from a uniform distribution. The performance of all three algorithms improves when the generating distribution is indeed Gaussian. The main implication of the comparison is therefore that ALC succeeds under a wide range of conditions and requires one to make fewer assumptions on the underlying data distribution.

6.4.2 Statistical Methods on Gene Expression Clustering, Image Segmentation, and Face Clustering

In this section, we apply some of the algorithms described in this chapter to real data, such as gene expression data and natural images. The goal is to demonstrate that the subspace clustering algorithms described in this chapter are capable of finding visually appealing structures in real data. However, we emphasize that they do not provide a complete solution to either of these practical problems. Such a solution usually entails a significant amount of domain-specific knowledge and engineering. Nevertheless, these preliminary results with gene expression data and natural images suggest that the subspace clustering methods presented in this chapter provide a generic solution for clustering mixed data that is simple and effective enough to be easily customized for a broad range of practical problems.

Clustering Gene Expression Data

In this example, we show the results of applying Algorithm 6.4 to gene expression data. The data set[18] consists of 13,872 vectors in $\mathbb{R}^{19}$, each of which describes the expression level of a single gene at different time points during an experiment on anthrax sporulation. A random subset of 600 vectors is visualized in Figure 6.11

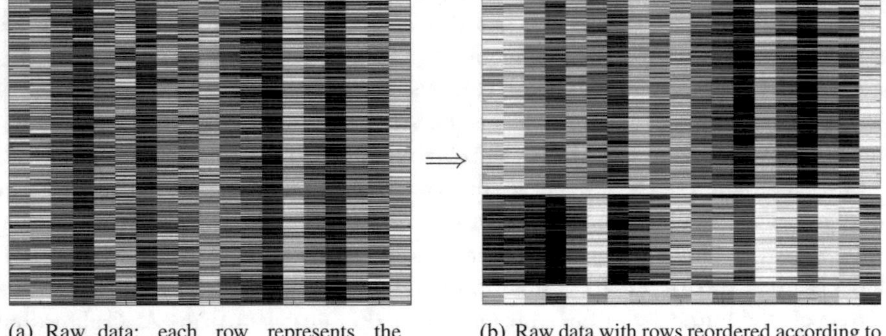

(a) Raw data: each row represents the expression level of one gene versus time

(b) Raw data with rows reordered according to three distinct clusters found by Algorithm 6.4

Fig. 6.11 Results of applying the compression-based subspace clustering algorithm to clustering of microarray data set GDS930.

[18]GDS930, available at http://www.ncbi.nlm.nih.gov/projects/geo.

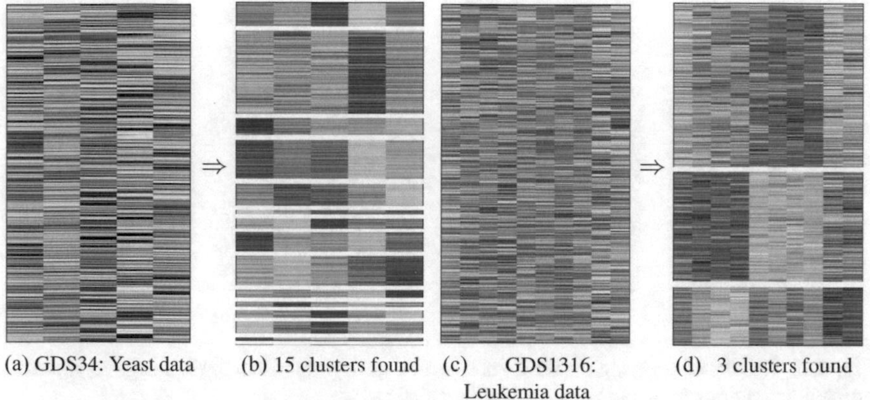

(a) GDS34: Yeast data (b) 15 clusters found (c) GDS1316: Leukemia data (d) 3 clusters found

Fig. 6.12 Results of applying the compression-based subspace clustering algorithm to clustering two microarray data sets: a yeast data set (a)–(b) and a leukemia data set (c)–(d).

(left). Here, rows correspond to genes and columns to time points. We cluster these vectors without any preprocessing, using Algorithm 6.4 with $\varepsilon = 1$. The algorithm finds three distinct clusters, which are displayed in Figure 6.11 (right) by reordering the rows.

Figure 6.12 shows clustering results on two additional gene expression data sets. The first data set[19] consists of 8448 vectors in $\mathbb{R}^5$, describing the expression levels of yeast genes at five different time points during a heat shock experiment. Figure 6.12 (a) shows expression levels for a randomly selected subset of 1200 genes. We cluster these vectors using the ALC algorithm, Algorithm 6.4, with $\varepsilon = 0.1$. ALC discovers a number of visually coherent clusters, shown in Figure 6.12 (b). The second data set[20] consists of 45,101 vectors in $\mathbb{R}^{10}$, each of which corresponds to the expression level of a single gene under varying experimental conditions (this experiment investigated Down syndrome-related leukemias). We run Algorithm 6.4 with $\varepsilon = 1$ on a subset of 800 of these vectors (shown in Figure 6.12 (c)). Three large, distinct clusters emerge, visualized in Figure 6.12 (d) by reordering the rows of the data.

Segmentation of Natural Images

In this example, we consider the problem of segmenting an image into multiple regions corresponding to different intensity, color, texture, or appearance patterns. We assume that we are able to extract an appearance descriptor at each pixel in the image and that all the descriptors associated with one region can be well approximated by a low-dimensional subspace of the space of descriptors. Under these assumptions, we may cast the image segmentation problem as a subspace

[19]GDS34, available at http://www.ncbi.nlm.nih.gov/projects/geo.

[20]GDS1316, available at http://www.ncbi.nlm.nih.gov/projects/geo.

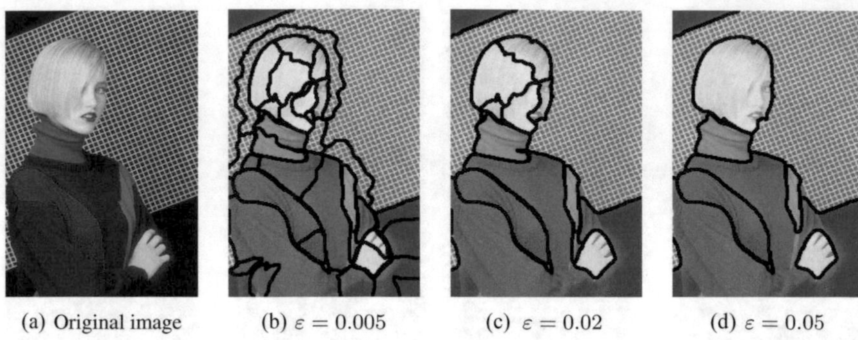

(a) Original image (b) $\varepsilon = 0.005$ (c) $\varepsilon = 0.02$ (d) $\varepsilon = 0.05$

Fig. 6.13 Results of applying the ALC subspace clustering algorithm to the image segmentation problem for various values of the distortion parameter ε.

clustering problem in which each subspace corresponds to one region in the image. This approach to image segmentation will be discussed in great detail in Chapter 10. As we will see, the ALC algorithm obtains good (unsupervised) image segmentation results even using features as simple as a (Gaussian) window of raw pixel intensities as an appearance descriptor for each pixel. In Figure 6.13, we show one example in which we use the ALC algorithm with different levels of distortion ε to segment the set of all 5×5 windows around all pixels. As we can see from the results, when ε is too small, the image is oversegmented, and as ε increases, so does the size of each region. Therefore, the larger the distortion allowed, the larger the granularity of the resulting segments. Readers who are interested in the subject of image segmentation please see Chapter 10 for a more careful treatment of this problem and more extensive experimental evaluation and justification.

Clustering Face Images under Varying Illumination
In this example, we illustrate the performance of the K-subspaces, MPPCA, and ALC algorithms on a subset of the extended Yale B data set consisting of frontal face images of two subjects (20 and 21, or 37 and 38) viewed under 64 different illumination conditions. The cropped and aligned images are of size 192×168 pixels. To reduce the computational complexity and memory requirements, these images are down-sampled to 48×42 pixels, and then PCA is further applied to reduce the dimension of the data to $D = 18$.

First, we apply the K-subspaces algorithm. The dimension of each subspace is set to $d = 5$. For the initialization of the mean vectors $\boldsymbol{\mu}_i \in \mathbb{R}^D$ and subspace bases $U_i \in \mathbb{R}^{D \times d}$ for each of the two subspaces $i = 1, 2$, we take all 128 images, compute their mean face and the first $d = 5$ eigenfaces, and then add random perturbations. The K-subspaces algorithm is then run with 10 random initializations, and the result that gives the smallest value for the K-subspaces objective function is chosen as the final output. Figure 6.14 provides an example of the convergence procedure for

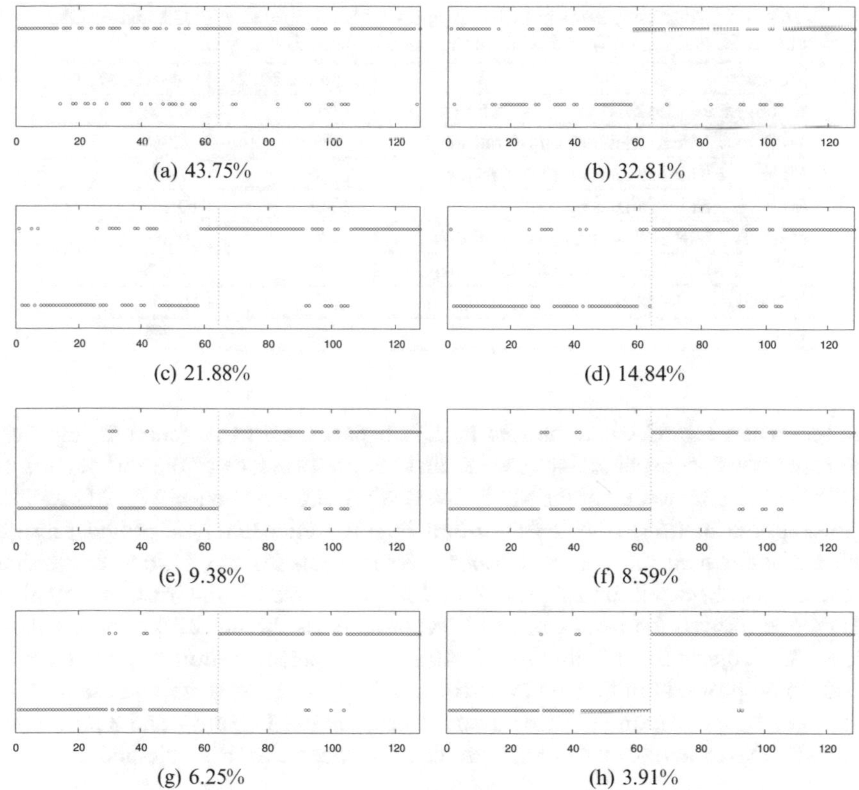

Fig. 6.14 Clustering faces under varying illumination using the K-subspaces algorithm. Figure (a) illustrates the segmentation at the initialization, with red points on each sides of the blue dotted line representing face images from different subjects. The following figures show the clustering results given by the successive iterations. Figure (h) is the final clustering output. Clustering errors are reported in the captions.

clustering images of subjects 20 and 21. Observe that as the iterations proceed, the clustering error reduces from 43.75% to 3.91%.

Next, we test the MPPCA algorithms (EM and MAP-EM) on the same data. As before, the dimension of the subspaces is set to $d = 5$, and the mean vectors $\boldsymbol{\mu}_i$ and subspace bases U_i are initialized as random perturbations of the mean and principal basis of the entire data set. The MPPCA algorithms are run with 10 random initializations, and the result that gives the smallest objective is chosen as the final output. We also test the ALC algorithm with $\varepsilon = 400$ and use the ALC algorithm to initialize MPPCA.

Table 6.3 shows the clustering errors of all methods. Observe that K-subspaces with random initialization does better than either version of MPPCA with random initialization. This is because the MPPCA model has more parameters than the K-subspaces model, and hence it is more difficult to initialize. Indeed, we can

Table 6.3 Clustering errors obtained by applying K-subspaces, MPPCA and ALC to subjects 20 and 21, or 37 and 38 from the extended Yale B data set.

Methods	Subjects 20, 21	Subjects 37, 38
K-subspaces with random initialization	3.9%	8.6%
MPPCA-EM with random initialization	15.6%	21.9%
MPPCA-EM initialized by K-subspaces	3.1%	7.8%
MPPCA-EM initialized by ALC	2.3%	10.2%
MPPCA-MAP-EM with random initialization	13.3%	20.3%
MPPCA-MAP-EM initialized by K-subspaces	3.9%	7.8%
MPPCA-MAP-EM initialized by ALC	2.3%	10.2%
ALC	3.1%	10.2%

see that when MPPCA is initialized by K-subspaces, MPPCA generally improves the results of K-subspaces. Notice also that the performance of the EM version of MPPCA is similar to that of the MAP-EM version, showing that MAP-EM provides a good approximation to the EM algorithm. Finally, notice that ALC provides results that are better than those of K-subspaces for subjects 20 and 21, and worse than those of K-subspaces for subject 37 and 38. Moreover, using ALC to initialize MPPCA improves the results of MPPCA for subjects 20 and 21, but makes them worse for subjects 37 and 38. Overall, all methods perform similarly, but MPPCA needs to be properly initialized to perform well. Readers who are interested in the subject of face clustering segmentation are referred to Chapters 7 and 8 for a more extensive experimental evaluation with other subspace clustering methods.

6.5 Bibliographic Notes

Iterative Methods

As discussed in Chapter 4, the K-means algorithm (Lloyd 1957; Forgy 1965; Jancey 1966; MacQueen 1967) was originally developed to cluster data distributed around multiple cluster centers. This is done by alternating between computing a cluster center for each group of points and assigning points to their closest cluster centers. The K-means algorithm was generalized to the case of hyperplanes in (Bradley and Mangasarian 2000), which proposed the K-planes algorithm in which one alternates between fitting a hyperplane to each group of points and assigning points to their closest hyperplane. The K-subspaces algorithm (Tseng 2000; Agarwal and Mustafa 2004) featured in this chapter further generalized K-planes from multiple hyperplanes to multiple affine subspaces of any dimensions. The K-subspaces algorithm alternates between fitting a subspace to each group of points using geometric PCA and assigning each point to its closest subspace. The original K-means algorithm is hence a particular case of K-subspaces in which each affine subspace has dimension zero.

Extensions of K-subspaces that we have not covered in this book include utilizing norms other than the squared Euclidean distance to fit multiple subspaces to the data. For example, the median K-flats algorithm (Zhang et al. 2009) uses the ℓ_1-norm. Also, the work of (Aldroubi et al. 2008; Aldroubi and Zaringhalam 2009) allows for various distances between points and subspaces, and considers not only finite-dimensional subspaces, but also subspaces of a Hilbert space. More generally, all these algorithms can be viewed as particular cases of a family of generalized K-means algorithms, which had been studied earlier by (Selim and Ismail 1984). In generalized K-means, the aim is to fit multiple geometric models to data by minimizing the sum of squared distances from data points to models. As such, in principle, any distance and any model can be used. The K-means algorithm is simply a particular case in which the distance is the Euclidean distance and the model is a cluster center. Likewise, K-subspaces is a particular case in which the model is an affine subspace and the distance is the distance from the point to the subspace. An important difference among different generalized K-means methods is whether a model can be easily estimated given the segmentation: this is the case for K-means, K-planes, and K-subspaces. The other main difference is whether the algorithm still converges. As we have seen in this chapter, this is the case for K-means, K-planes, and K-subspaces, but this need not be the case in general, as shown in (Selim and Ismail 1984).

Regarding applications, the K-subspaces algorithm was used in solving the face clustering problem (Ho et al. 2003), where face images associated with each subject are assumed to lie on a single subspace. From then on, the K-subspaces algorithm has been used as a baseline algorithm for segmenting videos with multiple moving objects. We will discuss this problem in great detail in Chapter 11.

Mixtures of Subspaces, Mixtures of Probabilistic Principal Components Analyzers, and Mixtures of Factor Analyzers

As discussed in Appendix B, the EM algorithm was originally developed in (Dempster et al. 1977) to find the ML parameters of a general class of statistical models. When the statistical model is a mixture model, the EM algorithm alternates between soft assignment of points to models and estimating the parameters for each model. While the application EM to find the parameters of a mixture of Gaussians had been known for a while, the PPCA model described in Section 2.2 is a particular case of a Gaussian model in which the covariance matrix has a "subspace-like" structure. The work of (Tipping and Bishop 1999a) was the first to show how the EM algorithm can be applied to this special mixture of PPCA model, while the work of (Ghahramani and Beal 2000) showed how the EM algorithm can be applied to a mixture of factor analyzers (FA), which are a generalization of PPCA. As we have seen in this chapter, the resulting algorithm alternates between soft assignment of points to subspace models (PPCA or FA) and estimating the parameters of each (PPCA or FA) model. We refer the reader to Appendix B for a more general review of the EM algorithm and to (Wu 1983) for a rigorous analysis of its convergence. For a more thorough and complete exposition of EM, one may refer to (Neal and Hinton 1998) or the book (McLanchlan and Krishnan 1997).

Regarding applications, the EM algorithm for mixtures of FA was used in (Frey et al. 1998) for face classification and in (Yang et al. 2000) for face detection. The EM algorithm for mixtures of PPCA has also been used as a baseline algorithm for segmenting videos with multiple moving objects, and we will discuss this problem in great detail in Chapter 11.

Compression-Based Clustering
Unfortunately, as we have alluded to above, iterative methods such as K-subspaces and EM are sensitive to initialization; hence they may not converge to the global optimum. In addition, they often assume that the number of the subspaces and their dimensions are known in advance, and the data are assumed to be drawn from such models without any outliers. These issues have severely limited the performance and generality of such methods in solving practical problems in computer vision and image processing (Shi and Malik 1998; Torr et al. 2001). That is why we sought alternative methods for clustering data from a mixture of subspaces, including the compression-based method introduced in this chapter and many other methods to be introduced in the next few chapters.

The analysis in this chapter has revealed strong connections between data clustering and data compression: the correct clustering is associated with the actual entropy of the data. Compression as a principle has been proposed for data clustering before, e.g., (Cilibrasi and Vitányi 2005), introducing an interpoint normalized compression distance that works for various data types. However, here the coding length gives a measure of distance between different *subsets* of the data. The agglomerative approach to minimizing the lossy coding length of mixed Gaussians was introduced in the work of (Ma et al. 2007). The method was known as the agglomerative lossy compression (ALC) method.

The simulations and experiments have suggested potential connections with certain phase transition phenomena that often appear in statistical physics. From a theoretical standpoint, it would be highly desirable to obtain analytical conditions on the critical values of the distortion and the outlier density that can explain and predict the phase transition behaviors. So far, only the case with zero-dimensional subspaces, i.e., vector quantization (VQ), has been well characterized (Rose 1998).

Relations to Other Metrics
It has been shown that computing the lossy ML estimate is approximately (up to first order, asymptotically) equivalent to minimizing the coding rate of the data subject to a distortion ε (Madiman et al. 2004):

$$\hat{\theta}_{LML} = \arg\min_{\theta,\pi} R(\hat{p}(X), \theta, \varepsilon), \tag{6.70}$$

where $\hat{p}(X)$ is the empirical estimate of the probabilistic distribution from a set of sample data $\mathcal{X} = \{\boldsymbol{x}_j\}_{j=1}^N$. From the dimension-reduction perspective, one would attempt to directly minimize the dimension, or $\text{rank}(X_i)$, of each subset $\mathcal{X}_i$. It is well known that the ε-regularized $\log\det(\cdot)$ is a good continuous surrogate for the discrete-valued rank function (Fazel et al. 2003). Techniques from compressed sensing have recently shown that when the rank is sufficiently low, the minimum of such surrogates coincides with the minimum-rank solution (Recht et al. 2010). In Chapter 8, we will explore other convex surrogates for measuring compactness of high-dimensional data. Although such convex surrogates may seem to be a grossly relaxed measure of data dimension, the surprise is that they lead to a provably correct solution to the subspace clustering problem under fairly general conditions. From the data-compression viewpoint, however, one might be more interested in an accurate estimate of the total volume of the data set. Notice that $\det(X_iX_i^\top + \varepsilon^2 I)$ is an ε-regularized volume of the subset $\mathcal{X}_i$, which is well defined even if the points in $\mathcal{X}_i$ lie on a proper subspace. Thus, minimizing the lossy coding length function L^s indeed unifies and generalizes other statistical or geometric metrics popular for data compression and clustering.

6.6 Exercises

Exercise 6.1 (K-subspaces and K-means). Let $\{\boldsymbol{x}_j \in \mathbb{R}^D\}_{j=1}^N$ be a collection of points lying in n affine subspaces

$$S_i = \{\boldsymbol{x} : \boldsymbol{x} = \boldsymbol{\mu}_i + U_i \boldsymbol{y}\} \qquad i = 1, \ldots, n \tag{6.71}$$

of dimensions d_i, where $\boldsymbol{\mu}_i \in \mathbb{R}^D$, $U_i \in \mathbb{R}^{D \times d_i}$ has orthonormal columns, and $\boldsymbol{y} \in \mathbb{R}^{d_i}$. Assume that within each subspace S_i the data are distributed around m_i cluster centers $\{\boldsymbol{\mu}_{ik} \in \mathbb{R}^D\}_{i=1\ldots n}^{k=1\ldots m_i}$.

1. Assume that n, d_i, and m_i are known and propose a clustering algorithm similar to K-means and K-subspaces to estimate the model parameters $\boldsymbol{\mu}_i$, U_i, $\boldsymbol{y}_j^i$, and $\boldsymbol{\mu}_{ik}$, and the segmentation of the data according to the $\sum_{i=1}^n m_i$ groups. More specifically, write down the cost function to be minimized and the constraints among the model parameters (if any), and use the method of Lagrange multipliers to find the optimal model parameters given the segmentation.
2. Assume that n, d_i, and m_i are unknown. How would you modify the cost function in item 1?

Exercise 6.2. Consider the objective function of the K-subspaces algorithm:

$$f(\{\boldsymbol{\mu}_i\}_{i=1}^n, \{U_i\}_{i=1}^n) = \min_{\substack{\{\boldsymbol{\mu}_i\}_{i=1}^n \\ \{U_i : U_i^\top U_i = I\}_{i=1}^n}} \sum_{j=1}^N \min_{i=1,\ldots,n} \|(I - U_i U_i^\top)(\boldsymbol{x}_j - \boldsymbol{\mu}_i)\|^2. \tag{6.72}$$

Show that the iterations of Algorithm 6.1 are such that

$$f(\{\boldsymbol{\mu}_i^{(k+1)}\}_{i=1}^n, \{U_i^{(k+1)}\}_{i=1}^n) \leq f(\{\boldsymbol{\mu}_i^{(k)}\}_{i=1}^n, \{U_i^{(k)}\}_{i=1}^n). \tag{6.73}$$

Exercise 6.3 (EM for Isotropic Gaussians). Based on Section B.2.1 or Section B.3.1 in Appendix B, derive a simplified EM algorithm for a mixture of isotropic Gaussians. That is, all the component Gaussians are of the form $\mathcal{N}(\mu_i, \sigma_i^2 I)$.

Exercise 6.4 (EM for Mixture of PPCAs). Based on Theorem 2.9 in Chapter 2, show that in the maximization step of the EM algorithm for mixture of subspaces, the optimal estimates are updated through the equations given in Algorithm 6.2. In particular, the maximization against the subspace parameters (U_i, σ_i) are given by the equations (6.23) and (6.24).

Exercise 6.5 (Clustering Three Planes in $\mathbb{R}^3$). Implement (in MATLAB) the K-subspaces algorithm, Algorithm 6.1, and the EM algorithm, Algorithm 6.2. Randomly generate three planes in $\mathbb{R}^3$ and draw a number of sample points in the planes with small noise. Use the algorithms to cluster the samples. Play with the level of noise (added to the samples) and the number of random initializations of the algorithm. Report the average clustering error as a function of noise.

Exercise 6.6 (RANSAC for Multiple Subspaces). For the same data sets generated for the above exercise, add a fixed percentage of outliers (say 5%–20%) uniformly drawn around the region that the subspaces occupy. Apply the robust statistical methods, in particular RANSAC, described in Appendix B.5.3 to detect the planes. Try two different strategies:

1. Try to use RANSAC to extract one plane at a time (meaning points on the two other planes also become outliers.)
2. Try to use RANSAC to extract points close to all three planes together (Hint: find a parametric representation for all three planes).

Compare the performance of RANSAC, K-subspaces, and EM on the same data sets. Discuss what you need to do when the dimension of the (sub)spaces and the number of subspaces becomes large, and assess the computational complexity of the RANSAC method.

Exercise 6.7 (Compression-Based Agglomerative Clustering for Multiple Subspaces). Implement the ALC algorithm, Algorithm 6.4, and test it on the same data sets you have generated for the previous exercises. Compare the performance with RANSAC for data sets in different dimensions, of different numbers of subspaces, and with different levels of outliers.

Exercise 6.8 (Large Quantization Error). Show that when $\varepsilon \to \infty$, the coding length function L^s reaches a minimum when all sample points are merged into one cluster.

Exercise 6.9 (Small Quantization Error: Zero-Mean Case). Now we characterize what happens when $\varepsilon \to 0$. Here we assume that all clusters have zero mean and consider the coding length function (6.46).

1. Show that for two sample points $\boldsymbol{x}_1, \boldsymbol{x}_2$ that are linearly independent, as $\varepsilon \to 0$, $L([\boldsymbol{x}_1, \boldsymbol{x}_2]) > L^s([\boldsymbol{x}_1], [\boldsymbol{x}_2])$.
2. Now for an arbitrary set $\mathcal{X} = \{\boldsymbol{x}_1, \ldots, \boldsymbol{x}_N\}$, suppose $N > D^2$. Then as $\varepsilon \to 0$, clustering all points together will result in a shorter coding length than leaving each point as its own cluster.
3. Show that under the same conditions as in the previous question, if the data are drawn from some nonsingular distribution, then for sufficiently small ε, with probability 1, the minimum coding length is achieved by merging all samples into one cluster.

The first two facts show that for extremely small ε, the agglomerative Algorithm 6.4 will mostly likely get stuck in a local minimum, since it does not merge any pair of points at all. This is what we have seen in the simulations. The third fact shows that for a generic distribution, with the sample size fixed, for an extremely small ε, assigning all samples into one cluster is actually the optimal solution. Notice that none of this invalidates the proposed algorithm, since it is expected to work in the regime whereby the sample density is comparable to ε. But these facts do suggest that when the data are rather undersampled, one should modify (6.46) to better impose the global subspace structures.

Exercise 6.10 (Small Quantization Error: Affine Case). Show that if we use the coding length for the nonzero mean (affine) case (6.90), given in Appendix 6.A, then as $\varepsilon \to 0$, keeping each point separate gives a smaller coding length than clustering all points together. This is opposite to the zero-mean case.

6.A Lossy Coding Length for Subspace-like Data

In Section 6.3.3, we have shown that in principle, one can construct a coding scheme for a given set of data vectors $X = [\boldsymbol{x}_1, \ldots, \boldsymbol{x}_N] \in \mathbb{R}^{D \times N}$ such that the average number of bits needed to encode each vector is bounded by

$$R(X) = \frac{1}{2} \log_2 \det\left(I + \frac{D}{N\varepsilon^2} X X^\top\right), \tag{6.74}$$

as if X were drawn from a zero-mean multivariate Gaussian distribution with estimated covariance $\Sigma = \frac{1}{N} X X^\top$. However, in the nonparametric setting (i.e., with a finite number of samples), we do not know whether the above coding length is still any good.

In this appendix, we provide a constructive proof that $L(X) = (N + D)R(X)$ indeed gives a tight upper bound on the number of bits needed to encode X. One interesting feature of the construction is that the coding scheme apparently relies on coding the subspace spanned by the vectors (i.e., the singular vectors) and the coordinates of the vectors with respect to the subspace. Thus geometrically, minimizing the coding length (via clustering) is essential to reducing the dimension of each subset of the data and the variance of each subset within each subspace.

The Zero-Mean Case: Linear Subspaces

For simplicity, we first assume that the given vectors $X = [\boldsymbol{x}_1, \dots, \boldsymbol{x}_N]$ have zero mean, and we will study the non-zero-mean case later. Consider the singular value decomposition (SVD) of the data matrix $X = U\Sigma V^\top$. Let $B = [b_{ij}] = \Sigma V^\top$. The column vectors of $U = [u_{ij}]$ form a basis for the subspace spanned by vectors in X, and the column vectors of B are the coordinates of the vectors with respect to this basis.

For coding purposes, we store the approximated matrices $U + \delta U$ and $B + \delta B$. The matrix X can be recovered as

$$X + \delta X \doteq (U + \delta U)(B + \delta B) = UB + \delta UB + U\delta B + \delta U \delta B. \tag{6.75}$$

Then $\delta X \approx \delta UB + U\delta B$, since entries of $\delta U \delta B$ are negligible when ε is small (relative to the data X). The squared error introduced to the entries of X are

$$\begin{aligned}\sum_{i,j} \delta x_{ij}^2 &= \operatorname{trace}\left(\delta X \delta X^\top\right) \\ &\approx \operatorname{trace}\left(U\delta B \delta B^\top U^\top + \delta U B B^\top \delta U^\top + \delta U B \delta B^\top U^\top + U \delta B B^\top \delta U^\top\right).\end{aligned}$$

We may further assume that the coding errors δU and δB are zero-mean independent random variables. Using the fact that trace(AB) = trace(BA), the expected squared error becomes

$$\mathbb{E}\big(\operatorname{trace}(\delta X \delta X^\top)\big) = \mathbb{E}\big(\operatorname{trace}(\delta B \delta B^\top)\big) + \mathbb{E}\big(\operatorname{trace}(\Sigma^2 \delta U^\top \delta U)\big). \tag{6.76}$$

Now let us encode each entry b_{ij} with a precision $\varepsilon' = \frac{\varepsilon}{\sqrt{D}}$ and u_{ij} with a precision $\varepsilon_j'' = \frac{\varepsilon\sqrt{N}}{\sqrt{\lambda_j D}}$, where λ_j is the jth eigenvalue of $XX^\top$.[21] This is equivalent to assuming that the error δb_{ij} is uniformly distributed in the interval $\big[-\frac{\varepsilon}{\sqrt{D}}, \frac{\varepsilon}{\sqrt{D}}\big]$ and δu_{ij} is uniformly distributed in the interval $\big[-\frac{\varepsilon\sqrt{N}}{\sqrt{\lambda_j D}}, \frac{\varepsilon\sqrt{N}}{\sqrt{\lambda_j D}}\big]$. Under such a coding precision, it is easy to verify that

$$\mathbb{E}\big(\operatorname{trace}(\delta X \delta X^\top)\big) \le \frac{2\varepsilon^2 N}{3} < \varepsilon^2 N. \tag{6.77}$$

Then the mean squared error per vector in X is

$$\frac{1}{N}\mathbb{E}\big(\operatorname{trace}(\delta X \delta X^\top)\big) < \varepsilon^2. \tag{6.78}$$

[21]Notice that ε_j'' normally does not increase with the number of vectors N, because λ_j increases proportionally to N.

The number of bits needed to store the coordinates b_{ij} with precision $\varepsilon' = \frac{\varepsilon}{\sqrt{D}}$ is

$$\sum_{i=1}^{D}\sum_{j=1}^{N}\frac{1}{2}\log_2\left(1+\left(\frac{b_{ij}}{\varepsilon'}\right)^2\right)=\frac{1}{2}\sum_{i=1}^{D}\sum_{j=1}^{N}\log_2\left(1+\frac{b_{ij}^2 D}{\varepsilon^2}\right) \tag{6.79}$$

$$\leq \frac{N}{2}\sum_{i=1}^{D}\log_2\left(1+\frac{D\sum_{j=1}^{N}b_{ij}^2}{N\varepsilon^2}\right)=\frac{N}{2}\sum_{i=1}^{D}\log_2\left(1+\frac{D\lambda_i}{N\varepsilon^2}\right). \tag{6.80}$$

In the above inequality, we have used the concavity of the log function:

$$\frac{\log(1+a_1)+\cdots+\log(1+a_n)}{n}\leq \log\left(1+\frac{a_1+\cdots+a_n}{n}\right) \tag{6.81}$$

for nonnegative real numbers $a_1, a_2, \ldots, a_n \geq 0$.

Similarly, the number of bits needed to store the entries of the singular vectors u_{ij} with precision $\varepsilon'' = \frac{\varepsilon\sqrt{N}}{\sqrt{\lambda_i}D}$ is

$$\sum_{i=1}^{D}\sum_{j=1}^{D}\frac{1}{2}\log_2\left(1+\left(\frac{u_{ij}}{\varepsilon''}\right)^2\right)=\frac{1}{2}\sum_{i=1}^{D}\sum_{j=1}^{D}\log_2\left(1+\frac{u_{ij}^2 D^2\lambda_j}{N\varepsilon^2}\right) \tag{6.82}$$

$$\leq \frac{D}{2}\sum_{j=1}^{D}\log_2\left(1+\frac{D^2\lambda_j\sum_{i=1}^{D}u_{ij}^2}{N\varepsilon^2}\right)=\frac{D}{2}\sum_{j=1}^{D}\log_2\left(1+\frac{D\lambda_j}{N\varepsilon^2}\right). \tag{6.83}$$

Thus for U and B together, we need a total of

$$L(X)=\frac{N+D}{2}\sum_{i=1}^{D}\log_2\left(1+\frac{D\lambda_i}{N\varepsilon^2}\right)=\frac{N+D}{2}\log_2\det\left(I+\frac{D}{N\varepsilon^2}XX^\top\right). \tag{6.84}$$

We thus have proved the statement given at the beginning of this section: $L(X) = (N+D)\cdot R(X)$ gives a good upper bound on the number of bits needed to encode X.

The Non-Zero-Mean Case: Affine Subspaces

In the above analysis, we have assumed that the data in the matrix $X = [\boldsymbol{x}_1, \ldots, \boldsymbol{x}_N]$ have zero mean. In general, these vectors may have a nonzero mean. In other words, the points represented by these vectors may lie in an affine subspace instead of a linear subspace.

If the data are not of zero mean, let $\boldsymbol{\mu} \doteq \frac{1}{N}\sum_{j=1}^{N}\boldsymbol{x}_j \in \mathbb{R}^D$ and define the matrix

$$V \doteq \boldsymbol{\mu}\cdot 1_{1\times N} = [\boldsymbol{\mu}, \boldsymbol{\mu}, \ldots, \boldsymbol{\mu}] \in \mathbb{R}^{D\times N}. \tag{6.85}$$

Then $\bar{X} \doteq X - V$ is a matrix whose column vectors have zero mean. We may apply the same coding scheme in the previous section to $\bar{X}$.

Let $\bar{X} = U\Sigma V^\top \doteq UB$ be the singular value decomposition of $\bar{X}$. Let $\delta U, \delta B, \delta\boldsymbol{\mu}$ be the error in coding $U, B, \boldsymbol{\mu}$, respectively. Then the error induced on the matrix X is

$$\delta X = \delta\boldsymbol{\mu} \cdot \mathbf{1}_{1\times N} + U\delta B + \delta U B. \tag{6.86}$$

Assuming that $\delta U, \delta B, \delta\boldsymbol{\mu}$ are zero-mean independent random variables, the expected total squared error is

$$\mathbb{E}\,\mathrm{trace}(\delta X \delta X^\top)) = N\mathbb{E}(\delta\boldsymbol{\mu}^\top\delta\boldsymbol{\mu}) + \mathbb{E}(\mathrm{trace}(\delta B\delta B^\top)) + \mathbb{E}(\mathrm{trace}(\Sigma\delta U^\top\delta U)).$$

We encode entries of B and U with the same precision as before. We encode each entry $\mu_i, i = 1, \dots, D$ of the mean vector $\boldsymbol{\mu}$ with the precision $\varepsilon' = \frac{\varepsilon}{\sqrt{D}}$ and assume that the error $\delta\mu_i$ is a uniform distribution in the interval $\big[-\frac{\varepsilon}{\sqrt{D}}, \frac{\varepsilon}{\sqrt{D}}\big]$. Then we have $N\mathbb{E}(\delta\boldsymbol{\mu}^\top\delta\boldsymbol{\mu}) = \frac{N\varepsilon^2}{3}$. Using equation (6.77) for the zero-mean case, the total squared error satisfies

$$\mathbb{E}\big(\mathrm{trace}(\delta X\delta X^\top)\big) \le \frac{N\varepsilon^2}{3} + \frac{2N\varepsilon^2}{3} = N\varepsilon^2. \tag{6.87}$$

Then the mean squared error per vector in X is still bounded by ε^2:

$$\frac{1}{N}\mathbb{E}\big(\mathrm{trace}(\delta X\delta X^\top)\big) \le \varepsilon^2. \tag{6.88}$$

Now, in addition to the $L(\bar{X})$ bits needed to encode U and B, the number of bits needed to encode the mean vector $\boldsymbol{\mu}$ with precision $\varepsilon' = \frac{\varepsilon}{\sqrt{D}}$ is

$$\sum_{i=1}^{D}\frac{1}{2}\log_2\Big(1+\big(\frac{\mu_i}{\varepsilon'}\big)^2\Big) = \frac{1}{2}\sum_{i=1}^{D}\log_2\Big(1+\frac{D\mu_i^2}{\varepsilon^2}\Big) \le \frac{D}{2}\log_2\Big(1+\frac{\boldsymbol{\mu}^\top\boldsymbol{\mu}}{\varepsilon^2}\Big), \tag{6.89}$$

where the last inequality follows from the inequality (6.81).

Thus, the total number bits needed to store X is

$$L(X) = \frac{N+D}{2}\log_2\det\big(I + \frac{D}{N\varepsilon^2}\bar{X}\bar{X}^\top\big) + \frac{D}{2}\log_2\Big(1+\frac{\boldsymbol{\mu}^\top\boldsymbol{\mu}}{\varepsilon^2}\Big). \tag{6.90}$$

Notice that if X is actually of zero mean, we have $\boldsymbol{\mu} = 0$, $\bar{X} = X$, and the above expression for $L(X)$ is exactly the same as before.

Chapter 7
Spectral Methods

The art of doing mathematics consists in finding that special case which contains all the germs of generality.

—David Hilbert

The preceding two chapters studied the subspace clustering problem using algebraic-geometric and statistical techniques, respectively. Under the assumption that the data are not corrupted, we saw in Chapter 5 that algebraic-geometric methods are able to solve the subspace clustering problem in full generality, allowing for an arbitrary union of different subspaces of any dimensions and in any orientations, as long as sufficiently many data points in general configuration are drawn from the union of subspaces. However, while algebraic-geometric methods are able to deal with moderate amounts of noise, they are unable to deal with outliers. Moreover, even in the noise-free setting, the computational complexity of linear-algebraic methods for fitting polynomials grows exponentially with the number of subspaces and their dimensions. As a consequence, algebraic-geometric methods are most effective for low-dimensional problems with moderate amounts of noise.

On the other hand, the statistical methods described in Chapter 6 are able to deal with noise and to a certain extent with outliers. However, they generally require the number of subspaces and their dimensions to be known beforehand. Moreover, they are based on solving nonconvex optimization problems, and so one cannot guarantee that an optimal solution has been obtained. Thus, statistical methods are most effective for problems for which we have a good knowledge of the number and dimensions of the subspaces and the distribution of noise and outliers.

Despite their differences, algebraic-geometric and statistical approaches to subspace clustering both aim to directly identify a *parametric model* for each of the subspaces in the original space where the data are distributed. As such, these methods can be seen as generalizations of the geometric and statistical interpretations of PCA discussed in Chapter 2. In many practical applications,

R. Vidal et al., *Generalized Principal Component Analysis*, Interdisciplinary Applied Mathematics 40, DOI 10.1007/978-0-387-87811-9_7

however, it is not always necessary to identify a parametric model for the subspaces. In face clustering, for example, it often suffices to find a partition of the data into multiple groups. For such purposes, any *nonparametric* representation of the data that preserves the topological connectivity of the original data can be used. This was the main motivation behind many of the manifold learning methods discussed in Chapter 4, which aim to find a low-dimensional representation of the data that best preserves some *affinity* among the original data points. In particular, the spectral clustering method discussed in Section 4.3.2 is based on constructing an affinity matrix that captures whether two points belong to the same group. This affinity is then used to map the data to a low-dimensional space in which certain clustering algorithms such as K-means can be used to cluster the data.

In this chapter, we will study nonparametric approaches to subspace clustering based on spectral clustering. As we will see, the main difficulty in applying spectral clustering to the subspace clustering problem lies in constructing a good affinity matrix, which captures whether two points belong to the same subspace. Therefore, this chapter will concentrate on the problem of building an affinity matrix for subspace clustering. We will present various approaches to building the affinity matrix. Some methods are guaranteed to provide the correct affinity when the data are perfect or the dimensions of the subspaces are known, but they are computationally expensive. Other methods are computationally more efficient, but they do not provide an affinity that is guaranteed to be correct.

7.1 Spectral Subspace Clustering

Before proceeding any further, let us briefly review the spectral clustering method introduced in Section 4.3.2 for clustering N data points $\{\boldsymbol{x}_j\}_{j=1}^N$ into n groups. Spectral clustering is based on constructing a weighted graph $\mathcal{G} = (\mathcal{V}, \mathcal{E}, W)$, where $\mathcal{V} = \{1, \ldots, N\}$ is the set of nodes, $\mathcal{E} \subset \mathcal{V} \times \mathcal{V}$ is the set of edges, and $W \in \mathbb{R}^{N \times N}$ is a symmetric nonnegative *affinity matrix* whose (j, k)th entry, w_{jk}, measures the affinity between points $\boldsymbol{x}_j$ and $\boldsymbol{x}_k$. Ideally, $w_{jk} = 1$ if points j and k are in the same group and $w_{jk} = 0$ if points j and k are in different groups. In practice, a typical affinity is

$$w_{jk} = \exp\Big(-\frac{1}{2\sigma^2}\operatorname{dist}(\boldsymbol{x}_j, \boldsymbol{x}_k)^2\Big), \tag{7.1}$$

where $\operatorname{dist}(\boldsymbol{x}_j, \boldsymbol{x}_k)$ is some distance between points j and k and $\sigma > 0$ is a parameter. Let $\mathcal{D} = \operatorname{diag}(W\mathbf{1})$ be a diagonal matrix whose jth diagonal entry gives the degree $d_{jj} = \sum_k w_{jk}$ of node j, and let $\mathcal{L} = \mathcal{D} - W \in \mathbb{R}^{N \times N}$ be the graph's Laplacian matrix. Spectral clustering obtains a clustering of the data by applying the K-means algorithm to the columns of the matrix $Y = [\boldsymbol{u}_1, \boldsymbol{u}_2, \ldots, \boldsymbol{u}_n]^\top \in \mathbb{R}^{n \times N}$, where $\{\boldsymbol{u}_i\}_{i=1}^n$ are the eigenvectors of $\mathcal{L}$ associated with its n smallest eigenvalues. Figure 7.1 gives an example of applying spectral clustering to a 2D data set. Notice that the original data

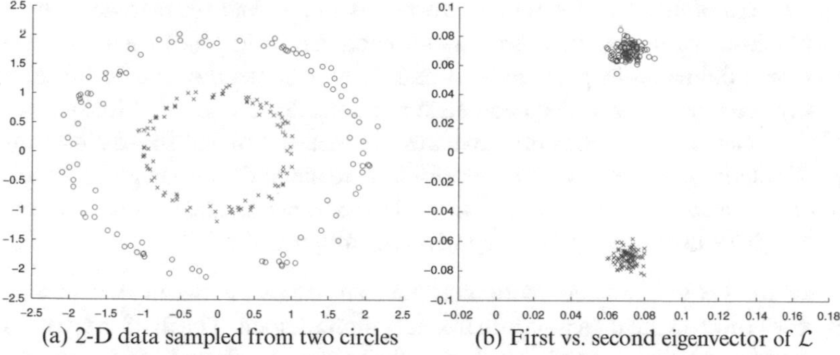

(a) 2-D data sampled from two circles (b) First vs. second eigenvector of $\mathcal{L}$

Fig. 7.1 Applying spectral clustering to data drawn from two concentric circles in $\mathbb{R}^2$. The affinity is computed as $w_{jk} = \exp\left(\frac{-\|x_j - x_k\|_2^2}{2\sigma^2}\right)$, where $\sigma = 0.1$. The figure on the right plots the second eigenvector of the Laplacian $\mathcal{L}$ (y-axis) versus the first (x-axis).

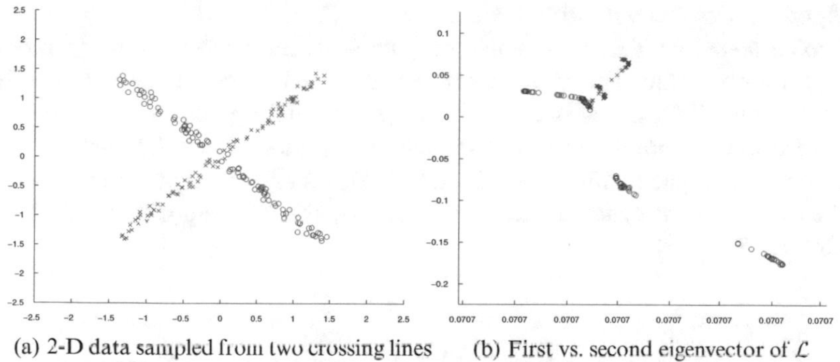

(a) 2-D data sampled from two crossing lines (b) First vs. second eigenvector of $\mathcal{L}$

Fig. 7.2 Applying spectral clustering to data drawn from two lines. The affinity is computed as $w_{jk} = \exp(\frac{-\text{dist}(x_j, x_k)^2}{2\sigma^2})$, where $\sigma = 0.1$. The figure on the right plots the second eigenvector of the Laplacian $\mathcal{L}$ (y-axis) versus the first (x-axis).

are not linearly separable and that central clustering algorithms such as K-means would fail on this data set. Spectral clustering, on the other hand, transforms the original problem into a central clustering problem that is easy to solve by K-means.

One of the main challenges in directly applying spectral clustering to the subspace clustering problem is defining a good affinity matrix. As illustrated in Figure 7.2, two points could be very close to each other but lie in different subspaces (e.g., near the intersection of two subspaces). Conversely, two points could be far from each other but belong to the same subspace. Therefore, the distance-based affinity in (7.1) is not appropriate for subspace clustering. Moreover, notice that one cannot define an affinity measure for subspace clustering by looking at pairs of points alone, as done in the case of the distance-based affinity in (7.1). Indeed, notice that two points in the plane always lie on a line containing the two points.

Thus, we cannot determine whether two points in the data set belong to the same line by looking at the coordinates of the two points in isolation. Instead, we need to look at three points in the plane at least and check whether they are collinear. More generally, we need to consider geometric relationships among multiple (if not all) points in order to construct an effective affinity measure for subspace clustering.

In this chapter, we review a few representative methods for designing an affinity matrix for subspace clustering and discuss their advantages and limitations. Broadly speaking, these methods can be categorized as follows:

1. *Local methods:* These methods compute an affinity between two points that depends only on the data points in a local neighborhood of each of the two points. Examples include the local subspace affinity and local best fit flat methods (see Section 7.2), which compute an affinity by comparing the subspaces estimated from a neighborhood of each data point, and the locally linear manifold clustering method (see Section 7.3), which computes an affinity from the locally linear relationships among a point and its nearest neighbors obtained using the LLE algorithm described in Section 4.2.2.
2. *Global methods:* These methods compute an affinity between two points that depends on all the data points. Examples include the spectral curvature affinity (see Section 7.4), which is computed from the affinities between a point and all subsets of $O(d)$ points in the data set, where d is the subspace dimension, and the algebraic subspace affinity (see Section 7.5), which is computed by comparing the subspaces estimated at each data point by the ASC algorithm described in Chapter 5.

7.2 Local Subspace Affinity (LSA) and Spectral Local Best-Fit Flats (SLBF)

As we alluded to earlier, we cannot define an affinity for subspace clustering using information from pairs of points alone. For a subspace of dimension d, we need at least $d+1$ points to determine whether they lie in the same linear subspace and $d+2$ points to determine whether they lie in the same affine subspace. Now, if for each data point $\boldsymbol{x}_j$ we knew d or $d+1$ other data points that are in the same subspace as $\boldsymbol{x}_j$, then we could estimate a local subspace S_j from $\boldsymbol{x}_j$ and the other d or $d+1$ points in the same global subspace as $\boldsymbol{x}_j$. Then, if two points $\boldsymbol{x}_j$ and $\boldsymbol{x}_k$ are in the same global subspace, their local subspaces S_j and S_k should be the same, while if they are in different global subspaces, their local subspaces should be different. Therefore, we could define an affinity between two points by defining an affinity $w_{jk} = \mathrm{aff}(S_j, S_k)$ between their local subspaces. The challenge is that we do not know a priori which points belong to which subspace; hence we do not know which points are most likely to belong to the same subspace.

Local methods for building a subspace clustering affinity, such as *local subspace affinity* (LSA) (Yan and Pollefeys 2006) and *spectral local best-fit flats* (SLBF)

(Zhang et al. 2010), overcome this difficulty by relying on the observation that a point and its nearest neighbors typically belong to the same subspace. Under this assumption, we can use any of the techniques introduced in the first part of the book (e.g., PCA) to fit a affine subspace to each point and its nearest neighbors. If $\boldsymbol{x}_j$ belongs to a subspace of dimension d_j, we need at least $d_j + 1$ points in general position to fit a d_j-dimensional affine subspace (also known as a flat). In practice, we can choose $K > d_j$ neighbors; hence d_j does not need to be known exactly: we need only an upper bound. Given K, we can fit a subspace $S_j = (\boldsymbol{\mu}_j, U_j)$ to each point $\boldsymbol{x}_j$ and its K-NN, $\boldsymbol{x}_{j_1}, \dots, \boldsymbol{x}_{j_K}$, where $\boldsymbol{\mu}_j \in \mathbb{R}^D$ is a point in the subspace and $U_j \in \mathbb{R}^{D \times d_j}$ is a basis for the subspace. In practice, since the subspace should pass through $\boldsymbol{x}_j$, we can let $\boldsymbol{\mu}_j = \boldsymbol{x}_j$; hence U_j is a basis for the linear subspace spanned by the vectors $\{\boldsymbol{x}_{j_1} - \boldsymbol{x}_j, \dots, \boldsymbol{x}_{j_K} - \boldsymbol{x}_j\}$. When d_j is unknown, we can estimate it using the model selection techniques described in Section 2.3.

Once the local subspaces $\{S_j\}_{j=1}^N$ have been estimated, we can use them to define a subspace clustering affinity $w_{jk} = \text{aff}(S_j, S_k)$ between $\boldsymbol{x}_j$ and $\boldsymbol{x}_k$. In what follows, we describe a few methods for building such an affinity.

Affinities Based on Principal Angles

A simple approach to defining a distance between two linear subspaces is to use the notion of *principal angles* introduced in Exercise 2.8. In the case of data lying in a union of hyperplanes, the idea is to exploit the fact that two points in the same hyperplane share the same normal vector to the hyperplane. Therefore, if $\boldsymbol{b}_j$ is the normal to the hyperplane passing through point $\boldsymbol{x}_j$, and similarly for $\boldsymbol{x}_k$, then the angle $\theta_{jk} = \angle(\boldsymbol{b}_j, \boldsymbol{b}_k)$ between $\boldsymbol{b}_j$ and $\boldsymbol{b}_k$ is zero when points $\boldsymbol{x}_j$ and $\boldsymbol{x}_k$ are in the same hyperplane, and nonzero otherwise. Hence, we can define the affinity as

$$w_{jk} = \cos^q(\theta_{jk}) \quad \text{or} \quad w_{jk} = \exp(-\sin^q(\theta_{jk})), \tag{7.2}$$

where $q > 0$ is a user-defined parameter. Note that if $\boldsymbol{x}_j$ and $\boldsymbol{x}_k$ are in the same hyperplane, then $w_{jk} = 1$; otherwise, $w_{jk} < 1$.

We can extend the above affinity to linear subspaces of any dimensions using the notion of principal angles between subspaces.

Definition 7.1 (Principal Angles). *Let S_j and S_k be two linear subspaces of $\mathbb{R}^D$ of dimensions d_j and d_k, respectively. The principal angles between subspaces S_j and S_k, $\Theta(S_j, S_k) = \{\theta_{jk}^m\}_{m=1}^{\min\{d_j, d_k\}}$, where $\theta_{jk}^m \in [0, \pi/2]$, are defined recursively for $m = 1, \dots, \min\{d_j, d_k\}$, as*

$$\cos(\theta_{jk}^1) = \max_{\boldsymbol{u}_j \in S_j} \max_{\boldsymbol{u}_k \in S_k} \left\{ \langle \boldsymbol{u}_j, \boldsymbol{u}_k \rangle : \|\boldsymbol{u}_j\| = \|\boldsymbol{u}_k\| = 1 \right\} = \angle(\boldsymbol{u}_j^1, \boldsymbol{u}_k^1) \tag{7.3}$$

$$\begin{aligned}\cos(\theta_{jk}^m) = \max_{\boldsymbol{u}_j \in S_j} \max_{\boldsymbol{u}_k \in S_k} \Big\{ \langle \boldsymbol{u}_j, \boldsymbol{u}_k \rangle : \|\boldsymbol{u}_j\| = \|\boldsymbol{u}_k\| = 1, \boldsymbol{u}_j \perp \boldsymbol{u}_j^i, \boldsymbol{u}_k \perp \boldsymbol{u}_k^i, \\ \forall i \in \{1, \dots, m-1\} \Big\} = \angle(\boldsymbol{u}_j^m, \boldsymbol{u}_k^m).\end{aligned} \tag{7.4}$$

Algorithm 7.1 (Local Subspace Affinity)

Input: A set of points $\{\boldsymbol{x}_j \in \mathbb{R}^D\}_{j=1}^N$ lying in a union of n subspaces $\{S_i\}_{i=1}^n$, and the integers K and n.

1: Find the K-NN $\boldsymbol{x}_{j_1}, \ldots, \boldsymbol{x}_{j_K}$ of each data point $\boldsymbol{x}_j, j = 1, \ldots, N$, according to some distance dist in $\mathbb{R}^D$.

2: Fit a local subspace S_j to $\boldsymbol{x}_j, \boldsymbol{x}_{j_1}, \ldots, \boldsymbol{x}_{j_K}$ and estimate its dimension d_j.

3: Compute an affinity matrix W as defined on the right-hand side of (7.5).

4: Cluster the data into n groups by applying Algorithm 4.7 to W.

Output: The segmentation of the data into n groups.

Remark 7.2. *As shown in Exercise 2.8, if U_j is an orthogonal basis for S_j and similarly for S_k, then $\cos(\theta_{jk}^m)$ is equal to the mth-largest singular value of $U_j^\top U_k$. This gives an efficient way of computing the principal angles directly from the subspace bases.*

Now let S_j be the subspace passing through $\boldsymbol{x}_j$, let $d_j = \dim(S_j)$, and similarly for $\boldsymbol{x}_k$. Observe that when $\boldsymbol{x}_j$ and $\boldsymbol{x}_k$ belong to the same subspace, $S_j \subset S_k$, or $S_k \subset S_j$, we have $\theta_{jk}^m = 0$ for all $m = 1, \ldots, d_{jk}$. Otherwise, there is an m such that $\theta_{jk}^m \neq 0$. Therefore, we can use the principal angles to define an affinity matrix as

$$w_{jk} = \prod_{m=1}^{\min\{d_j,d_k\}} \cos^2(\theta_{jk}^m) \quad \text{or} \quad w_{jk} = \exp\Big(- \sum_{m=1}^{\min\{d_j,d_k\}} \sin^2(\theta_{jk}^m)\Big). \tag{7.5}$$

The former was introduced in (Wolf and Shashua 2003) as a kernel for subspaces, while the latter was introduced in (Yan and Pollefeys 2006) and is known as *local subspace affinity* (LSA). In either case, notice that if points $\boldsymbol{x}_j$ and $\boldsymbol{x}_k$ belong to the same subspace, $S_j \subset S_k$ or $S_k \subset S_j$, we have $w_{jk} = 1$; otherwise, $w_{jk} < 1$.

Algorithm 7.1 summarizes the LSA algorithm of (Yan and Pollefeys 2006). Notice that the actual algorithm in (Yan and Pollefeys 2006) includes an additional preprocessing step, whereby the data are projected onto a lower-dimensional space using PCA and then normalized to have unit norm. Since this additional step is not essential, we have omitted it from the algorithm description.

Affinities Based on Geometric Distances

One disadvantage of the affinities in (7.5) is that they are applicable only to linear subspaces, because the principal angles depend only on the subspace basis. To see this, notice, for example, that when two affine subspaces are parallel to each other, all their principal angles are equal to zero; hence w_{jk} is equal to 1 not only for points j and k in the same subspace, but also for points j and k in two different subspaces. Therefore, in the case of data drawn from affine subspaces, w_{jk} needs to be modified to incorporate an appropriate distance between points $\boldsymbol{x}_j$ and $\boldsymbol{x}_k$.

One approach to building an affinity for data lying in a union of subspaces that is applicable to both linear and affine subspaces is to incorporate the distance between points $\boldsymbol{x}_j$ and $\boldsymbol{x}_j$ to the affinity as

$$w_{jk} = \exp(-\|\boldsymbol{x}_j - \boldsymbol{x}_k\|_2^2) \exp\Big(- \sum_{m=1}^{\min\{d_j,d_k\}} \sin^2(\theta_{jk}^m)\Big). \tag{7.6}$$

Another approach is to use some distance from $\boldsymbol{x}_j$ to the subspace passing through $\boldsymbol{x}_k$, e.g., the Euclidean distance $\operatorname{dist}(\boldsymbol{x}_j, S_k) = \|(I - U_k U_k^\top)\boldsymbol{x}_j\|$. Since this quantity is not symmetric, i.e., $d(\boldsymbol{x}_j, S_k) \neq d(\boldsymbol{x}_k, S_j)$, the *spectral local best-fit flats* (SLBF) algorithm (Zhang et al. 2010) considers their geometric mean

$$d_{jk} = \sqrt{\operatorname{dist}(\boldsymbol{x}_j, S_k)\operatorname{dist}(\boldsymbol{x}_k, S_j)}, \tag{7.7}$$

and defines the affinity as

$$w_{jk} = \exp\Big(- \frac{d_{jk}}{2\sigma_j^2}\Big) + \exp\Big(- \frac{d_{jk}}{2\sigma_k^2}\Big), \tag{7.8}$$

where σ_j measures how well point j and its neighbors are fit by S_j, and similarly for σ_k. We refer the reader to (Zhang et al. 2010) for details on how to compute σ_j.

Algorithm 7.2 summarizes the SLBF algorithm of (Zhang et al. 2010). Notice that the actual algorithm in (Zhang et al. 2010) does not require one to specify the parameter K. Instead, multiple neighborhoods are considered around each data point, and the neighborhood that gives the best local fit is selected. For the sake of simplicity, we have omitted this additional step and assumed that K is given.

The main advantage of local algorithms for building a subspace clustering affinity, such as LSA and SLBF, is that they are conceptually simple. Indeed, they are natural generalizations of classical spectral clustering methods, where instead of comparing two data points, one compares two data points and their local neighborhoods. Therefore, from a computational perspective, the additional costs relative to standard spectral clustering are to (1) find the nearest neighbors for each data point and (2) fit a localsubspace to each data point. Another advantage of local

Algorithm 7.2 (Spectral Local Best-Fit Flats)

Input: A set of points $\{\boldsymbol{x}_j \in \mathbb{R}^D\}_{j=1}^N$ lying in a union of n subspaces $\{S_i\}_{i=1}^n$, and the integers K and n.

1: Find the K-NN $\boldsymbol{x}_{j_1}, \dots, \boldsymbol{x}_{j_K}$ of each data point $\boldsymbol{x}_j, j = 1, \dots, N$, according to some distance dist in $\mathbb{R}^D$.
2: Fit a local subspace S_j to $\boldsymbol{x}_j, \boldsymbol{x}_{j_1}, \dots, \boldsymbol{x}_{j_K}$ and estimate the dimension d_j.
3: Compute an affinity matrix W as defined in (7.8).
4: Cluster the data into n groups by applying Algorithm 4.7 to W.

Output: The segmentation of the data into n groups.

methods is that outliers are likely to be rejected, because they are far from all the points, and so they are not considered neighbors of the inliers.

On the other hand, local methods have two main drawbacks. First, the neighbors of a point could contain points in different subspaces, especially near the intersection of two subspaces. Points with contaminated neighborhoods will lead to erroneous estimates of the local subspaces, which can cause errors in the affinities, hence in the clustering. This problem could be partially ameliorated by reducing the number of nearest neighbors. However, this could lead to unreliable estimates of the local subspaces in the presence of noise, or worse, the selected neighbors may not span the underlying subspace. Therefore, a fundamental challenge with local methods is how to select the proper size of the neighborhood: the neighborhood should be small enough that only points in the same subspace are chosen and large enough that the neighbors span the local subspace. SLBF tries to remedy some of these issues by choosing the size of the neighborhood automatically. However, this increases computational complexity, because one needs to perform local PCA multiple times, once for each neighborhood size, in order to select the best neighborhood size for each data point. This issue is one of the motivations for the sparse and low-rank methods to be described in Chapter 8.

7.3 Locally Linear Manifold Clustering (LLMC)

Local methods, such as LSA and LSBF, perform subspace clustering by applying the spectral clustering algorithm to an affinity matrix obtained by comparing local subspaces estimated from each data point and its nearest neighbors. Therefore, these algorithms can be seen as generalizations of the Laplacian eigenmaps (LE) algorithm (see Section 4.2.3, Algorithm 4.3) from manifold embedding to subspace clustering.

In this section, we describe a variant of the *locally linear manifold clustering* (LLMC) method proposed in (Goh and Vidal 2007), which performs subspace clustering by applying spectral clustering to an affinity obtained from the locally linear representations computed by the LLE algorithm (see Section 4.2.2, Algorithm 4.2). Therefore, the LLMC algorithm can be seen as a generalization of the LLE algorithm from manifold embedding to subspace clustering.

Let us first recall that the first step of LLE is to represent each point $\boldsymbol{x}_j$ as an affine combination of its K-NN. The coefficients c_{jk} of these linear combinations are found by minimizing the sum of the reconstruction errors for all the data points

$$\sum_{j=1}^{N} \|\boldsymbol{x}_j - \sum_{k \neq j} c_{kj} \boldsymbol{x}_k\|^2 = \|X - XC\|_F^2, \tag{7.9}$$

subject to $\sum_{k \neq j} c_{kj} = 1$ and $c_{kj} = 0$ if $\boldsymbol{x}_k$ is not a K-NN of $\boldsymbol{x}_j$. The optimal coefficients can be found in closed form, as shown in (4.38).

Let us now recall that the second step of LLE is to find a low-dimensional representation $Y = [\boldsymbol{y}_1, \ldots, \boldsymbol{y}_N] \in \mathbb{R}^{d \times N}$ that solves the following problem:

$$\min_Y \sum_{j=1}^N \left\| \boldsymbol{y}_j - \sum_{k=1}^N c_{kj} \boldsymbol{y}_k \right\|^2 \text{ s.t. } \sum_{j=1}^N \boldsymbol{y}_j = \boldsymbol{0} \text{ and } \frac{1}{N} \sum_{j=1}^N \boldsymbol{y}_j \boldsymbol{y}_j^\top = I, \tag{7.10}$$

which is equivalent to

$$\min_Y \ \operatorname{trace}(Y \mathcal{L} Y^\top) \quad \text{s.t.} \quad Y\boldsymbol{1} = \boldsymbol{0} \quad \text{and} \quad \frac{1}{N} YY^\top = I, \tag{7.11}$$

where

$$\mathcal{L} = (I - C)(I - C)^\top. \tag{7.12}$$

As shown in Proposition 4.11, the optimization problem in (7.11) is given by the eigenvectors of $\mathcal{L}$ corresponding to the second- to the $(d+1)$th-smallest eigenvalues.

Now observe that the optimization problem in (7.11) is very similar to that solved by the spectral clustering algorithm, which can be written as

$$\min_Y \ \operatorname{trace}(Y \mathcal{L} Y^\top) \quad \text{s.t.} \quad YY^\top = I. \tag{7.13}$$

The main difference is that the centering constraint $Y\boldsymbol{1} = \boldsymbol{0}$ is dropped, since our goal is clustering rather than embedding. Therefore, we may be tempted to think that (7.13) is a spectral clustering problem with affinity matrix

$$W = C + C^\top - CC^\top, \tag{7.14}$$

and hence the eigenvectors of $\mathcal{L}$ give the segmentation of the data into a union of subspaces as in Proposition 4.14. However, this need not be the case, because the matrix W in (7.14) need not be nonnegative; hence the matrix $\mathcal{L}$ need not be a graph's Laplacian (see Exercise 7.1). Therefore, while the solution of (7.13) continues to be the eigenvectors of $\mathcal{L}$ associated with the smallest eigenvalues, we can no longer use Proposition 4.14 to assert that the null space of $\mathcal{L}$ gives the segmentation of the data.

It is shown in (Goh and Vidal 2007) that when every point and its K-NNs are always in the same subspace, then there are n vectors $\{\boldsymbol{v}_i\}_{i=1}^n$ in the null space of $\mathcal{L}$ that give the segmentation of the data, i.e., $v_{ij} = 1$ if point j belongs to subspace i, and $v_{ij} = 0$ otherwise (see Exercise 7.2). Therefore, the null space of $\mathcal{L}$ still gives the segmentation of the data in this case. However, it is also shown in (Goh and Vidal 2007) that these vectors are not the only vectors in the null space of $\mathcal{L}$ (see Exercise 7.3). Therefore, a procedure for selecting the *segmentation eigenvectors* $\{\boldsymbol{v}_i\}_{i=1}^n$ from the null space of $\mathcal{L}$ is needed before spectral clustering can be applied.

Algorithm 7.3 (Locally Linear Manifold Clustering)

Input: A set of points $\{x_j \in \mathbb{R}^D\}_{j=1}^N$ lying in a union of n subspaces $\{S_i\}_{i=1}^n$, and the integers K and n.

1: Find the K-NN $x_{j_1}, \dots, x_{j_K}$ of each data point $x_j, j = 1, \dots, N$, according to some distance dist in $\mathbb{R}^D$.

2: Approximate each point $x_j \approx \sum c_{ij} x_i$ as an affine combination of its K-NN with coefficients the c_{ij} obtained as in (4.38).

3: Compute an affinity matrix W as defined in (7.15).

4: Cluster the data into n groups by applying Algorithm 4.7 to W.

Output: The segmentation of the data into n groups.

Nonetheless, we can avoid the aforementioned difficulties by replacing the second step of the LLE algorithm by a spectral clustering step. In particular, if every point and its K-NNs are always in the same subspace, by construction the matrix C is such that $c_{kj} = 0$ if points x_j and x_k are in different subspaces. Therefore, we can use the matrix C directly to build an affinity. Since the matrix C is not necessarily symmetric or nonnegative, we define the following affinity matrix:

$$W = |C| + |C^\top|. \tag{7.15}$$

Algorithm 7.3 summarizes the LLMC algorithm of (Goh and Vidal 2007), modified to use the affinity in (7.15) rather than the one in (7.14).

A first advantage of LLMC is its robustness to outliers. This is because, as in the case of LSA and SLBF, outliers are often far from the inliers; hence it is unlikely that they are chosen as neighbors of the inliers. Another important advantage of LLMC is that it is also applicable to nonlinear subspaces, while all the other methods discussed so far are applicable only to linear (or affine) subspaces. This is because the LLMC algorithm is effectively looking for locally linear relationships among neighboring points, without enforcing a global subspace structure. However, LLMC suffers from the same disadvantage of LSA, namely that it has problems with points near the intersections, because it is not always the case that a point and its K-NNs are in the same subspace. Also, properly choosing the number of nearest neighbors is a challenge. Nonetheless, these issues could be resolved by choosing the neighborhood automatically, as done by SLBF.

7.4 Spectral Curvature Clustering (SCC)

The local methods discussed so far, such as LSA, SLBF, LLMC, construct an affinity between two data points by looking only at the nearest neighbors of each data point. However, if the neighborhood of a point contains points from different subspaces, these local methods could fail. To remedy this issue, this and the next sections describe global methods for building a subspace clustering affinity.

The approach we describe in this section is based on multiway clustering (Agarwal et al. 2005; Govindu 2005; Chen and Lerman 2009b), which is a generalization of spectral clustering from pairwise affinities to multiway affinities. As the name suggests, an M-way affinity is an affinity among M points that tries to capture whether these points belong to the same group. In the case of data lying in a union of subspaces of the same dimension d, the idea is to consider an arbitrary neighborhood of *any* $d+1$ points for each data point, measure how likely these $d+2$ points are to belong to the same subspace, and use this $(d+2)$-way affinity to construct a pairwise affinity between two points, so that spectral clustering can be applied to it.

For the sake of simplicity, let us first consider the case of data lying in a union of lines in the plane. In this case, we can select any three points and check whether they are collinear. If the coordinates of the points are $\boldsymbol{x}_1 = (x_1, y_1)$, $\boldsymbol{x}_2 = (x_2, y_2)$, $\boldsymbol{x}_3 = (x_3, y_3)$, we can determine whether they are collinear by checking whether the area of the triangle formed by the three points is zero. We can compute this area as

$$\text{area}(\boldsymbol{x}_1, \boldsymbol{x}_2, \boldsymbol{x}_3) = \begin{vmatrix} x_1 & y_1 & 1 \\ x_2 & y_2 & 1 \\ x_3 & y_3 & 1 \end{vmatrix}. \tag{7.16}$$

Alternatively, we can fit a circle to the three points and use the radius of this circle to compute the Menger curvature, which is defined as the reciprocal of the radius. Then, if the points are collinear, the Menger curvature is zero. Interestingly, the Menger curvature is a function of the area and of the sides of the triangle formed by the three points and can be computed as

$$\kappa_M(\boldsymbol{x}_1, \boldsymbol{x}_2, \boldsymbol{x}_3) = \frac{4\text{area}(\boldsymbol{x}_1, \boldsymbol{x}_2, \boldsymbol{x}_3)}{\|\boldsymbol{x}_1 - \boldsymbol{x}_2\| \|\boldsymbol{x}_2 - \boldsymbol{x}_3\| \|\boldsymbol{x}_3 - \boldsymbol{x}_1\|}. \tag{7.17}$$

Yet another alternative is to use the polar curvature, which is simply the Menger curvature multiplied by half the square of the diameter of the three points, $\text{diam}(\boldsymbol{x}_1, \boldsymbol{x}_2, \boldsymbol{x}_3)^2 = \|\boldsymbol{x}_1 - \boldsymbol{x}_2\|^2 + \|\boldsymbol{x}_2 - \boldsymbol{x}_3\|^2 + \|\boldsymbol{x}_3 - \boldsymbol{x}_1\|^2$, which gives

$$\kappa_p(\boldsymbol{x}_1, \boldsymbol{x}_2, \boldsymbol{x}_3) = \frac{1}{2}\text{diam}(\boldsymbol{x}_1, \boldsymbol{x}_2, \boldsymbol{x}_3)^2 \kappa_M(\boldsymbol{x}_1, \boldsymbol{x}_2, \boldsymbol{x}_3) \tag{7.18}$$

$$= 2\text{diam}(\boldsymbol{x}_1, \boldsymbol{x}_2, \boldsymbol{x}_3) \sqrt{\sum_{j=1}^{3} \frac{\text{area}(\boldsymbol{x}_1, \boldsymbol{x}_2, \boldsymbol{x}_3)^2}{\prod_{i \neq j} \|\boldsymbol{x}_i - \boldsymbol{x}_j\|^2}}. \tag{7.19}$$

More generally, let $X_{d+2} = \{\boldsymbol{x}_j\}_{j=1}^{d+2}$ be $d+2$ points in $\mathbb{R}^D$, and let $\text{vol}(X_{d+2})$ be the volume of the $(d+1)$-simplex formed by these points. This volume should be zero if the $d+2$ points are in the same affine subspace of dimension d. We can also compute the polar curvature of X_{d+2}, which is defined as

$$\kappa_p(X_{d+2}) = \operatorname{diam}(X_{d+2}) \sum_{j=1}^{d+2} \sqrt{\frac{(d+1)!^2 \operatorname{vol}^2(X_{d+2})}{\prod_{\substack{1\leq i \leq d+2 \\ i\neq j}} \|\boldsymbol{x}_i - \boldsymbol{x}_j\|^2}}. \tag{7.20}$$

As before, if the $d+2$ points lie in the same subspace, we have $\kappa_p = 0$.

The spectral curvature clustering (SCC) method introduced in (Chen and Lerman 2009b) constructs an affinity for subspace clustering based on the concept of polar curvature described above. In particular, given N data points $\{\boldsymbol{x}_j\}_{j=1}^N$, let $X_{d+2} = \{\boldsymbol{x}_{j_\ell}\}_{\ell=1}^{d+2}$ be any random choice of $d+2$ data points. For each such choice, the SCC method defines a multiway affinity as

$$\mathcal{A}_{j_1 j_2,\ldots,j_{d+2}} = \exp\left(-\frac{1}{2\sigma^2} \operatorname{diam}^2(X_{d+2}) \sum_{\ell=1}^{d+2} \frac{(d+1)!^2 \operatorname{vol}^2(X_{d+2})}{\prod_{\substack{1\leq m \leq d+2 \\ m\neq \ell}} \|\boldsymbol{x}_{j_m} - \boldsymbol{x}_{j_\ell}\|^2}\right) \tag{7.21}$$

if $j_1, j_2, \ldots, j_{d+2}$ are distinct, and zero otherwise, where $\sigma > 0$ is a parameter. Given this multiway affinity, we can define an affinity between any pair of points as

$$w_{jk} = \sum_{j_2,\ldots,j_{d+1} \in \{1,\ldots,N\}} \mathcal{A}_{j,j_2,\ldots,j_{d+2}} \mathcal{A}_{k,j_2,\ldots,j_{d+2}}. \tag{7.22}$$

Algorithm 7.4 summarizes the SCC algorithm of (Chen and Lerman 2009b). Notice that the actual algorithm in (Chen and Lerman 2009b) uses a procedure for initializing K-means within the spectral clustering step. For the sake of simplicity, we have omitted this additional step.

The main advantage of SCC over LSA, SLBF, and LLMC is that it uses points from the entire data set to define the affinity between two points, while LSA, SLBF, and LLMC restrict themselves to the nearest neighbors of each point. This ultimately results in a better similarity measure. Also, SCC is better justified theoretically. In particular, when the data points are sampled from a mixture of distributions concentrated around multiple affine subspaces, SCC can perform better than the above local methods, as shown in (Chen and Lerman 2009a). In addition, SCC can be extended to nonlinear manifolds using kernel methods, as shown in (Chen et al. 2009).

Algorithm 7.4 (Spectral Curvature Clustering)

Input: A set of points $\{\boldsymbol{x}_j \in \mathbb{R}^D\}_{j=1}^N$ lying in a union of n subspaces $\{S_i\}_{i=1}^n$, and the integers d and n.

1: For each choice of $d+2$ data points, compute the multiway affinity in (7.21).
2: Compute the pairwise affinity matrix W as defined in (7.22).
3: Cluster the data into n groups by applying Algorithm 4.7 to W.

Output: The segmentation of the data into n groups.

On the other hand, one of the main disadvantages of the SCC algorithm is that it requires computing $O(N^{d+2})$ entries of $\mathcal{A}$ and summing over $O(N^{d+1})$ elements of $\mathcal{A}$. Therefore, the computational complexity of SCC grows exponentially with the dimension of the subspaces. A practical implementation of SCC uses a fixed number c of $(d+1)$-tuples ($c \ll N^{d+1}$) for each point to build the similarity W. A choice of $c \approx c_0 n^{d+2}$ is suggested in (Chen and Lerman 2009b), which is much smaller, but still exponential in d. In practice, the method appears to be not too sensitive to the choice of c but more to how the $d+1$ points are chosen. In (Chen and Lerman 2009b), it is argued that a uniform sampling strategy does not perform well, because many samples could contain subspaces of different dimensions. To avoid this, two stages of sampling are performed. The first stage is to obtain an initial clustering of the data. In the second stage, the initial clusters are used to guide the sampling and thus obtain a better affinity.

Another drawback of SCC is that it requires the subspaces to be of known and equal dimension d. In practice, the algorithm can still be applied to subspaces of different dimensions by choosing $d = d_{\max}$, but it becomes unclear how the notion of "volume" should be properly modified in the spectral curvature formula.

7.5 Spectral Algebraic Subspace Clustering (SASC)

In this subsection, we present another global approach to building a subspace clustering affinity based on the ASC algorithm described in Chapter 5. Recall that ASC is based on fitting a set of polynomials to the data and computing the normals to each subspace from the gradients of these polynomials at n data points, one per subspace. As it turns out, we can use the normal vectors computed by ASC to define a subspace clustering affinity. The key idea is that instead of computing the normal vectors at n points only, we can compute them at each of the N data points. In this way, we assign to each data point a set of normal vectors. Then, we can define an affinity between two points using any affinity between the two subspaces spanned by the two sets of normal vectors.

Let us begin with the simple case of data lying in a union of hyperplanes. As discussed in Chapter 5, in this case there is a single polynomial $p(\boldsymbol{x})$ that vanishes in the union of hyperplanes. Moreover, we can use $p(\boldsymbol{x})$ to estimate the normal to the hyperplane passing through $\boldsymbol{x}_j$ as (see Algorithm 5.3)

$$\boldsymbol{b}_j = \frac{\nabla p(\boldsymbol{x}_j)}{\|\nabla p(\boldsymbol{x}_j)\|}, \quad \forall j = 1, \ldots, N. \tag{7.23}$$

Given the normal vectors $\boldsymbol{b}_j$ and $\boldsymbol{b}_k$ to the hyperplanes passing through $\boldsymbol{x}_j$ and $\boldsymbol{x}_k$, respectively, we can use them to compute the angle $\theta_{jk} = \angle(\boldsymbol{b}_j, \boldsymbol{b}_k)$ between the hyperplanes. This angle should be zero if the two points are in the same hyperplane. Therefore, the we can define an affinity for data points in a union of hyperplanes as

$$w_{jk} = \cos^q(\theta_{jk}) = \frac{|\langle \nabla p(\boldsymbol{x}_j), \nabla p(\boldsymbol{x}_k)\rangle|^q}{\|\nabla p(\boldsymbol{x}_j)\|^q \|\nabla p(\boldsymbol{x}_k)\|^q}, \tag{7.24}$$

where $q > 0$ is a user-defined parameter. Notice that this is the same affinity used by LSA and SLBF in (7.2). The main difference is that LSA and LSBF compute this affinity from local estimates of the normal vectors, while in (7.24), this affinity is computed from global estimates of these normals obtained from the polynomial $p(\boldsymbol{x})$, which is fit globally to all the data points.

A second approach to defining an affinity from the normals to the hyperplane is to use the distance from a point to a hyperplane. In particular, note that if point $\boldsymbol{x}_j$ belongs to the hyperplane passing through $\boldsymbol{x}_k$, we must have $\boldsymbol{b}_k^\top \boldsymbol{x}_j = 0$ or $\langle \nabla p(\boldsymbol{x}_k), \boldsymbol{x}_j\rangle = 0$. In fact, notice that if $\boldsymbol{b}_k$ is a unit vector, then $|\boldsymbol{b}_k^\top \boldsymbol{x}_j|$ is simply the distance from point $\boldsymbol{x}_j$ to the hyperplane passing through $\boldsymbol{x}_k$. Therefore, we can define an affinity for points in a union of hyperplanes as

$$w_{jk} = 1 - \frac{|\langle \boldsymbol{x}_j, \nabla p(\boldsymbol{x}_k)\rangle|^q}{\|\boldsymbol{x}_j\| \|\nabla p(\boldsymbol{x}_k)\|^q}, \tag{7.25}$$

where $q > 0$ is a user-defined parameter, or alternatively as its symmetric version

$$w_{jk} = 1 - \frac{1}{2}\frac{|\langle \boldsymbol{x}_j, \nabla p(\boldsymbol{x}_k)\rangle|^q}{\|\boldsymbol{x}_j\| \|\nabla p(\boldsymbol{x}_k)\|^q} - \frac{1}{2}\frac{|\langle \boldsymbol{x}_k, \nabla p(\boldsymbol{x}_j)\rangle|^q}{\|\boldsymbol{x}_k\| \|\nabla p(\boldsymbol{x}_j)\|^q}. \tag{7.26}$$

Let us now consider the case of arbitrary subspaces of any dimensions. Let S_j be the subspace passing through $\boldsymbol{x}_j$, let $S_j^\perp$ be its orthogonal complement, and let $d_j = \dim(S_j)$. As described in Theorem 5.8, we can estimate $S_j^\perp$ as the span of the derivatives of all the polynomials $p_{n\ell}$ that vanish on the union of subspaces, i.e.,

$$S_j^\perp = \operatorname{span}\Big\{\nabla p_{n\ell}(\boldsymbol{x}_j)\Big\}. \tag{7.27}$$

Then the principal angles $\{\theta_{jk}^m\}_{m=1}^{\min\{D-d_j, D-d_k\}}$ between $S_j^\perp$ and $S_k^\perp$ can be used to define an affinity matrix as

$$w_{jk} = \prod_{m=1}^{\min\{D-d_j, D-d_k\}} \cos^2(\theta_{jk}^m). \tag{7.28}$$

Notice that the affinity in (7.28) is very similar to that on the left-hand side of (7.5). However, there are two key differences. First, the affinity in (7.5) is computed from the principal angles between the subspaces S_j and S_k, while the affinity in (7.28) is computed from the principal angles between their orthogonal complements, $S_j^\perp$ and $S_k^\perp$, respectively. Second, the affinity in (7.5) is computed from the principal angles between locally estimated subspaces, while the affinity in (7.28) is computed from the principal angles between globally estimated subspaces.

7.6 Simulations and Applications

7.6.1 Spectral Methods on Synthetic Data

In this subsection we illustrate the performance of various spectral subspace clustering methods using synthetically generated data. We randomly generate $n = 3$ subspaces in $\mathbb{R}^D$ of varying dimensions $\{d_i\}_{i=1}^n$. We consider two different settings. In the first one, the ambient dimension is low ($D = 4$), and the subspace dimensions are high relative to the ambient dimension ($d_i \in \{1, 2, 3\}$). In the second one, the ambient dimension is high ($D = 100$), and the subspace dimensions are low relative to the ambient dimensions ($d_i \in \{5, 7, 9\}$). For each setting, we randomly sample $N_i = 100$ or $N_i = 150$ points from each subspace, respectively, and add zero-mean Gaussian noise with covariance $\sigma^2(I - U_i U_i^\top)$, where U_i is an orthonormal basis for the subspace and $\sigma \in [0, 0.05]$. Subsequently, the noisy points are normalized to have unit Euclidean norm. Given these N data points, the task is to cluster them into their respective subspaces using spectral clustering. For this purpose, we use many of the methods described in this chapter for constructing the subspace clustering affinity. In addition, since many of the methods described in this chapter are related to the single-manifold learning methods discussed in Chapter 4, we also use some manifold learning methods to generate the affinity matrices. In particular, we compare the following methods:

1. KPCA-linear, i.e., we use the polynomial kernel of degree one in (4.31) as an affinity.
2. KPCA-Gaussian, i.e., we use the Gaussian kernel of spread $\sigma = 0.1$ in (4.31) as an affinity.
3. LE, i.e., we use the weight matrix W in (4.43) as an affinity. The number of nearest neighbors is set to $K = 12$, and the spread parameter is set to $\sigma = 0.5$ in (4.43).
4. LSA, i.e., we use the affinity on the right-hand side of (7.5). When $D = 4$, the number of nearest neighbors is set to $K = 3$ and the subspace dimension to $d = 3$. When $D = 100$, the number of nearest neighbors is set to $K = 12$ and the subspace dimension to $d = 9$.
5. LSA-GT is the same as LSA, except that the subspace dimension is set to the ground truth dimension in the case of equidimensional subspaces, and to the maximum dimension in the case of subspaces of different dimensions.
6. LLMC, i.e., we use the matrix W in (7.15) obtained from the coefficient matrix C in (4.38) as an affinity. The number of nearest neighbors is set to $K = 12$.
7. SCC, i.e., we use the affinity in (7.22) with the subspace dimensions set to $d = 3$ if $D = 4$, and $d = 9$ if $D = 100$.
8. SCC-GT is the same as SCC, except that the subspace dimension is set to the ground truth dimension in the case of equidimensional subspaces, and to the maximum dimension in the case of subspaces of different dimensions.
9. SASC-angle, i.e., we use the affinity in (7.24) with $q = 1$.
10. SASC-distance, i.e., we use the affinity in (7.26) with $q = 1$.

Each of these methods produces an $N \times N$ matrix that if needed, we convert to a valid affinity matrix by taking the absolute value of its entries and symmetrizing it. This affinity is then given as input to the spectral clustering method in Algorithm 4.7 to produce $n = 3$ clusters from the generalized eigenvectors of $(\mathcal{L}, \mathcal{D})$, where $\mathcal{L}$ is the Laplacian and $\mathcal{D}$ is the degree matrix associated with the affinity.

High-Dimensional Subspaces of a Low-Dimensional Ambient Space
Figure 7.3 shows the clustering error rates of different methods as a function of the noise level for data drawn from three subspaces of $\mathbb{R}^4$ of dimensions $d_i \in \{1, 2, 3\}$. By looking at the results, we can draw the following conclusions.

1. The affinities obtained by manifold learning methods such as KPCA and LE fail in all cases even for zero noise. This is expected, because manifold learning methods are designed for a single manifold, not for multiple subspaces. The only manifold learning method that shows good performance is KPCA-linear for $d_1 = d_2 = d_3 = 1$. This is because in this case, the kernel is simply the cosine of

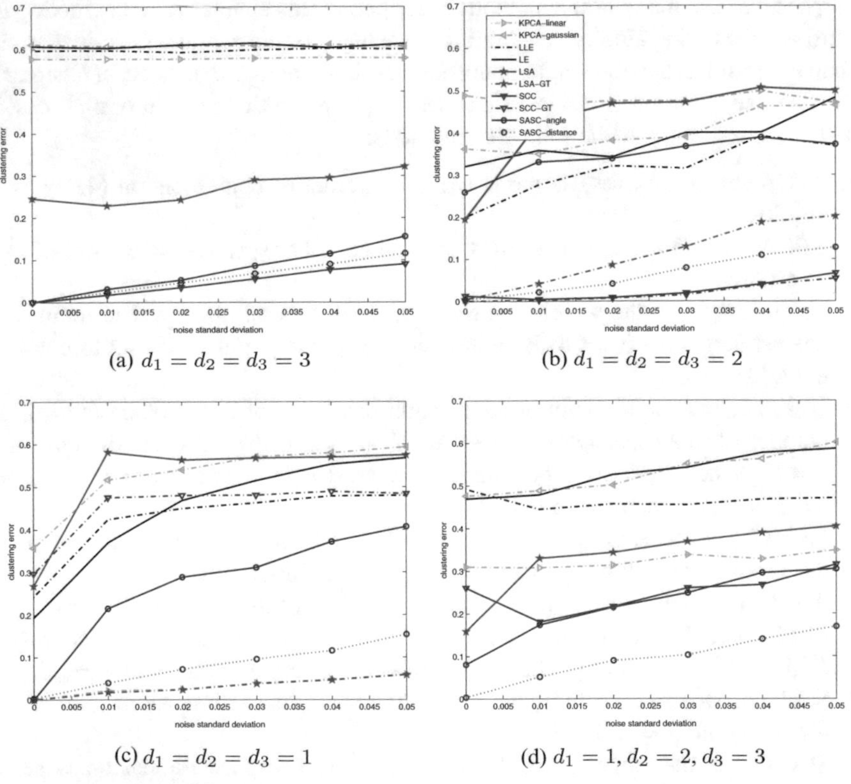

(a) $d_1 = d_2 = d_3 = 3$

(b) $d_1 = d_2 = d_3 = 2$

(c) $d_1 = d_2 = d_3 = 1$

(d) $d_1 = 1, d_2 = 2, d_3 = 3$

Fig. 7.3 Clustering errors for spectral subspace clustering algorithms applied to three randomly generated subspaces in $\mathbb{R}^4$ of dimensions $1 \leq d_i \leq 3$. The errors are averages over 100 independent experiments.

the angle between data points, which coincides with the angle between the lines and hence encodes information about the underlying segmentation into multiple subspaces.

2. The performance of LSA and LLMC is comparable to that of manifold learning methods, except for LSA when $d_1 = d_2 = d_3 = 3$. This shows that LSA is sensitive to having data points near the intersection of the subspaces, which affects the selection of the nearest neighbors, as well as to proper knowledge of the dimensions of the subspaces. The latter can also be seen by noticing that LSA-GT, which uses the true subspaces dimensions, consistently outperforms LSA, which uses only an upper bound on the subspace dimensions. Moreover, notice that the performance of LSA-GT improves as the dimension of the subspaces is reduced. This is arguably because there are fewer data points near the intersection of the subspaces.
3. SCC seems to be less sensitive to its dimension parameter. For example, in the case of equidimensional subspaces of dimension 2, using the ground truth dimension $d = 2$ or the upper bound $d_{\max} = 3$ does not affect the results. However, for the case of lines, SCC-GT behaves poorly, while SCC with the upper bound $d_{\max} = 3$ becomes numerically unstable.
4. Finally, both algebraic methods SASC-angle and SASC-distance perform similarly for the case of hyperplanes. However, for lower-dimensional spaces, the performance of SASC-angle degrades significantly, while SASC-distance remains robust. Interestingly, SASC-distance is the only method that gives zero error in the absence of noise for the more intricate case of subspaces of different dimensions.

Low-Dimensional Subspaces of a High-Dimensional Ambient Space

Here, we perform a similar experiment, except that the ambient dimension is $D = 100$ and the subspace dimensions are $d_i \in \{5, 7, 9\}$. In this case, neither SASC-angle nor SASC-distance can be applied directly, because the Veronese map of degree 3 embeds $\mathbb{R}^{100}$ into $\mathbb{R}^{171700}$, which implies that we would need at least 171,700 data points to compute a vanishing polynomial. To alleviate this difficulty, we project the 100-dimensional (noisy) points onto their first 10 principal components and subsequently normalize them to have unit norm. We use these latter points to run SASC-angle and SASC-distance. Figure 7.4 shows the clustering error rates (in logarithmic scale) of different methods as a function of the noise level. By looking at the results, we can draw the following conclusions.

1. First, in sharp contrast with the case of high-dimensional subspaces of a low-dimensional space, KPCA-linear, LLMC, and LSA perform perfectly across all dimension configurations and all considered levels of noise. For LLMC and LSA, the explanation is straightforward: the low dimensionality of the subspaces leads to correct choices for nearest neighbors. For KPCA-linear, we notice that the low dimensionality of the subspaces leads to high values of the entries of the intracluster submatrices of the kernel matrix (recall that these entries are just cosines of angles between points).
2. Second, notice that the method that performs the worst is SASC-angle. This is expected in all dimension configurations except $(9, 9, 9)$, in which the underlying

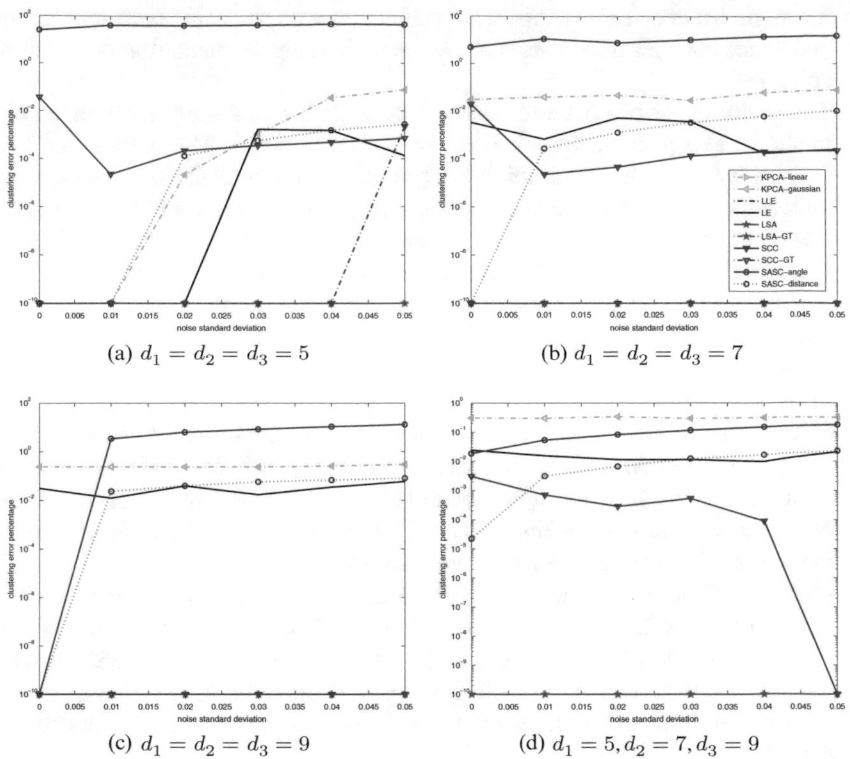

(a) $d_1 = d_2 = d_3 = 5$

(b) $d_1 = d_2 = d_3 = 7$

(c) $d_1 = d_2 = d_3 = 9$

(d) $d_1 = 5, d_2 = 7, d_3 = 9$

Fig. 7.4 Clustering errors for spectral-based algorithms applied to three randomly generated subspaces of dimensions $d_i \in \{5, 7, 9\}$ in $\mathbb{R}^{100}$. The errors are averages over 100 independent experiments.

subspaces are effectively hyperplanes of $\mathbb{R}^{10}$. Indeed, in this case, SASC-angle gives zero error for zero noise. However, as soon as the points become noisy, its performance degrades. This is consistent with its behavior in the experiments for $D = 4$, except that here the degradation in the performance is much more rapid. This is because the effect of the projection is to reduce the angle between the orthogonal complements of the subspaces.[1] In this case, the orthogonal complements after the projection are just the lines defined by the normals to the 9-dimensional hyperplanes.

[1]As a simple illustration of this phenomenon, consider two lines in $\mathbb{R}^3$ and their projection onto some plane of $\mathbb{R}^3$ that is not orthogonal to any of the lines: the angle between the lines is reduced after the projection. As the angles between the normals become smaller, the ability of SASC-angle to distinguish between the different hyperplanes is reduced.

3. Third, SASC-distance performs satisfactorily, indicating that it is much more robust to variations of the subspace dimensions than SASC-angle. As expected, KPCA-Gaussian is significantly less accurate. This is because the Gaussian kernel promotes similarities between points that lie close to each other in the Euclidean sense, which need not be the case for points lying in the same subspace. Interestingly, LE, which also uses the Gaussian kernel, performs much better than KPCA-Gaussian. The reason is that LE builds the affinity by applying the Gaussian kernel only among nearest neighbors, which by the low dimensionality are expected to be more accurate. Finally, SCC performs quite well even when the subspace dimensions are different, indicating that SCC becomes more robust to subspace dimension variations in the low-rank regime.

7.6.2 Spectral Methods on Face Clustering

In this subsection, we consider the problem of clustering face images of multiple individuals acquired with a fixed pose and varying illumination. Figure 7.5 gives an example of face images of multiple individuals that are clustered into three groups. As discussed in Chapter 2, it has been shown that under the Lambertian reflectance assumption, images of a subject with a fixed pose and varying illumination lie close to a linear subspace of dimension 9 (Basri and Jacobs 2003). Therefore, a collection of face images of multiple subjects lie close to a union of 9-dimensional subspaces, and we may use any of the methods described in this chapter to cluster a given set of face images.

We illustrate the performance of various spectral subspace clustering methods on the problem of clustering frontal face images of two subjects (20 and 21, or 37 and 38) from the extended Yale B data set under 64 varying illumination conditions. Each image is cropped to 192×168 pixels, which cover the face of the individual only. To reduce the computational cost and the memory requirements of all algorithms, we down-sample the images to 48×42 pixels and treat each 2,016-dimensional vectorized image as a data point. In Chapter 4, we used a similar data set to illustrate the performance of different manifold learning techniques for building low-dimensional representations of data in a union of subspaces. In this experiment, we use the data set mainly to illustrate the different affinities obtained

Fig. 7.5 Face clustering: given face images of multiple subjects (top), the goal is to find images that belong to the same subject (bottom).

Algorithm 7.5 (Spectral Algebraic Subspace Clustering)

Input: A set of points $\{\boldsymbol{x}_j \in \mathbb{R}^D\}_{j=1}^N$ lying in a union of n subspaces $\{S_i\}_{i=1}^n$, and the parameter $q > 0$.

1: Fit set of polynomials $\{p_{n\ell}\}$ of degree n to the data.
2: Find a basis for the subspace spanned by the vectors $\{\nabla p_{n\ell}(\boldsymbol{x}_j)\}$, i.e., a basis for the orthogonal complement of the subspace S_j passing through $\boldsymbol{x}_j$.
3: For each pair of points, compute the angle-based affinity in (7.24) or (7.28), or the distance-based affinity in (7.26).
4: Cluster the data into n groups by applying Algorithm 4.7 to W.

Output: The segmentation of the data into n groups.

by the different methods and see how effective they are in terms of clustering the two subjects using spectral clustering. For comparison, we also show the affinities obtained by manifold learning methods such as KPCA, LLMC, and LE, as described in the previous subsection. The parameters of the different methods are set as follows.

1. KPCA, i.e., we use the polynomial kernel of degree one (KPCA-linear) or a Gaussian kernel with spread $\sigma = 0.15$ (KPCA-Gaussian) in (4.31) as an affinity.
2. LE, i.e., we use the weight matrix W in (4.43) as an affinity with the number of nearest neighbors chosen as $K = 6$ and the spread chosen as $\sigma = 0.15$.
3. LSA, i.e., Algorithm 7.1 with the number of nearest neighbors chosen as $K = 9$ and the dimension of the subspaces chosen as $d = 9$.
4. LLMC, i.e., Algorithm 7.3 with the number of nearest neighbors chosen as $K = 6$.
5. SCC, i.e., Algorithm 7.4 with the subspace dimension chosen as $d = 9$.
6. SASC, i.e., Algorithm 7.5 with the angle-based affinity with parameter $q = 10$ (SASC-angle), or with the distance-based affinity with parameter $q = 1$ (SASC-distance).

Figure 7.6 shows the affinity matrices of different methods as well as the clustering error percentage for face images from subjects 20 and 21 (Figures 7.6 (a)–7.6 (h)) and subjects 37 and 38 (Figures 7.6 (i)–7.6 (p)). By looking at the figures, we can draw the following conclusions.

1. KPCA-linear does not capture the two-subspace structure in the data. This is to be expected, since KPCA with the polynomial kernel of degree one is effectively equivalent to applying classical PCA to the data and then computing the dot products between data points in the projected space.
2. The nonlinear manifold learning methods produce a sparse affinity where the image of one individual is connected to very few other images of the same individual and occasionally also to images of the other individual. This is because KPCA-Gaussian, LLMC, and LE are all based on pairwise distances, either for constructing the affinity or for selecting nearest neighbors. Since for each individual, the images are ordered in a fixed illumination pattern where, e.g., the

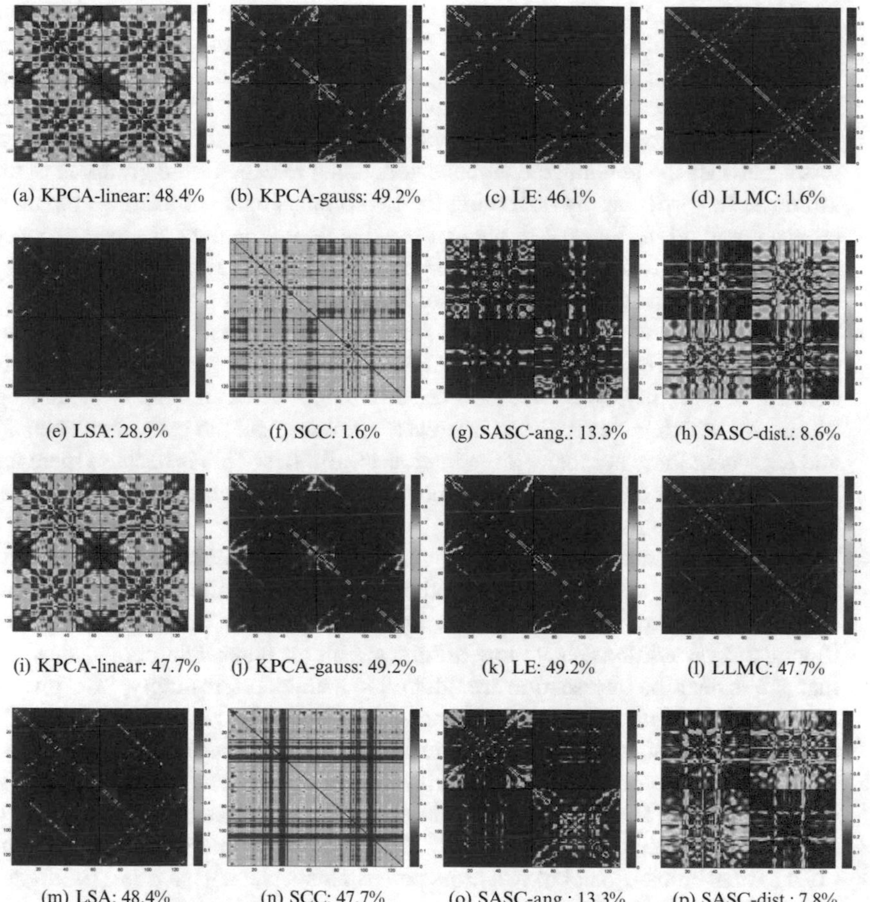

(a) KPCA-linear: 48.4% (b) KPCA-gauss: 49.2% (c) LE: 46.1% (d) LLMC: 1.6%

(e) LSA: 28.9% (f) SCC: 1.6% (g) SASC-ang.: 13.3% (h) SASC-dist.: 8.6%

(i) KPCA-linear: 47.7% (j) KPCA-gauss: 49.2% (k) LE: 49.2% (l) LLMC: 47.7%

(m) LSA: 48.4% (n) SCC: 47.7% (o) SASC-ang.: 13.3% (p) SASC-dist.: 7.8%

Fig. 7.6 Affinity matrices of face images produced by different methods. We take frontal face images of subjects 20 and 21 (Figures 7.6 (a)–7.6 (h)) and subjects 37 and 38 (Figures 7.6 (i)–7.6 (p)) under 64 different illumination conditions from extended Yale B. For spectral ASC, the images are projected to dimension 9 using PCA. For all other methods, the images are downsampled to 48×42. The affinity is obtained by taking absolute value, symmetrizing, normalizing to the range $(0, 1)$, then setting the diagonal to zero. The ground truth clustering is given as the horizontal and vertical black lines in each figure, so that points 1–64 correspond to the first individual, and points 65–128 correspond to the second individual. Clustering error percentages for each method are reported in the captions.

first few (1–13) and the last few (58–64) are bright frontal lighting patterns, a cross pattern is formed in the two diagonal blocks of the affinity matrix, because images of the same individual under similar lighting directions are close to each other. A similar, albeit less pronounced, effect is visible in the affinities between two individuals viewed from similar lighting directions (see cross pattens in off-diagonal blocks, especially for individuals 37 and 38). This means that

KPCA-Gaussian, LLMC, and LE are expected to produce a good affinity when the pairwise distances do capture the two-subspace structure (as is the case for individuals 20 and 21), but they are expected to produce a bad affinity otherwise. Interestingly, spectral clustering produces very bad results for KPCA-Gaussian, LE, and LLMC, except for LLMC with subjects 20 and 21. We observed that this is because the generalized eigenvectors failed to capture the grouping in the affinities. In particular, by observing the two-dimensional embeddings obtained by each method in Figure 4.9, we can see that while the data are well grouped according to the two subjects, it is very hard to obtain the correct clustering by applying K-means to the low-dimensional embeddings.

3. The LSA affinity is similar to that obtained by manifold learning methods. This can be explained by the fact that LSA also uses a nearest neighbor approach to construct the affinity. However, there are differences between the case of subjects 20 and 21, which have few large interclass affinities, and the case of subjects 37 and 28, where there are many large interclass affinities. This is because there are many points near the intersection of both subspaces, which correspond to faces with low illumination.
4. The SCC affinity does not seem to capture the two-subspace structure. Most data points have high affinities with all other data points, while a few data points (image faces near faces 40 and 104 that correspond to faces with very low illumination conditions) have low affinities with all other data points, showing that points near the intersection are difficult to handle. Interestingly, SCC gives a nearly perfect result for subjects 20 and 21, and a very bad result for subjects 37 and 38, showing again the lack of robustness of spectral clustering.
5. SASC produces very different affinities depending on whether angles or distances are used. The angle-based affinity is relatively sparse and captures the two-subspace structure relatively well. Indeed, observe that the affinities in the off-diagonal blocks do not have a cross pattern. Instead, they have an axis aligned pattern. This reduces the clustering error, because a point in one subspace is now connected to many other points in the same subspace, and very few in the other subspace. This is reflected in the clustering results, which are now more stable between one pair of subjects and the other. The distance-based affinity, on the other hand, gives high pairwise affinities across the board. But still, the intraclass affinities are generally higher than the interclass affinities, and this method gives the smallest clustering error.

Overall, while some of the methods described in this chapter seem to work well for low-dimensional data, none of these methods seems to work well on the face clustering problem. This suggests the need for developing better affinities for high-dimensional data sets, which we will do in the next chapter.

7.7 Exercises

Exercise 7.1. Consider the affinity matrix used by the LLMC algorithm as defined in (7.14):

$$W = C + C^\top - CC^\top. \tag{7.29}$$

Show that this matrix need not be nonnegative; hence it is not a valid affinity for subspace clustering. Can you derive conditions on C under which W is guaranteed to be nonnegative?

Exercise 7.2. Let $\mathcal{L} = (I - C)(I - C)^\top$ be the matrix in (7.12), which is obtained by applying the LLE algorithm to a data matrix X whose columns are drawn from a union of n subspaces. Assume that for all $j = 1, \ldots, N$, the K-NN of $\boldsymbol{x}_j$ lie in the same subspace as $\boldsymbol{x}_j$. Show that there are n vectors $\{\boldsymbol{v}_i\}_{i=1}^n$ in the null space of $\mathcal{L}$ that give the segmentation of the data, i.e., for all $j = 1, \ldots, N$, $v_{ij} = 1$ if point j belongs to subspace i, and $v_{ij} = 0$ otherwise.

Exercise 7.3. Let $\mathcal{L} = (I - C)(I - C)^\top$ be the matrix in (7.12), which is obtained by applying the LLE algorithm to a data matrix X whose columns are drawn from a union of n subspaces. Let X_i, for $i = 1, \ldots, n$, be the submatrix of X whose columns contain the data points in the ith subspace. Show that

$$\mathcal{L}X_i = 0 \quad \forall\, i = 1, \ldots, n. \tag{7.30}$$

Chapter 8
Sparse and Low-Rank Methods

A mathematical theory is not to be considered complete until you have made it so clear that you can explain it to the first man whom you meet on the street.

—David Hilbert

The previous chapter studies a family of subspace clustering methods based on spectral clustering. In particular, we have studied both local and global methods for defining a subspace clustering affinity, and have noticed that we seem to be facing an important dilemma. On the one hand, local methods compute an affinity that depends only on the data points in a local neighborhood of each data point. Local methods can be rather efficient and somewhat robust to outliers, but they cannot deal well with intersecting subspaces. On the other hand, global methods utilize geometric information derived from the entire data set (or a large portion of it) to construct the affinity. Global methods might be immune to local mistakes, but they come with a big price: their computational complexity is often exponential in the dimension and number of subspaces. Moreover, none of the methods comes with a theoretical analysis that guarantees the correctness of clustering. Therefore, a natural question that arises is whether we can construct a subspace clustering affinity that utilizes global geometric relationships among all the data points, is computationally tractable when the dimension and number of subspaces are large, and is guaranteed to provide the correct clustering under certain conditions.

In this chapter, we will present a family of methods for constructing a subspace clustering affinity that satisfies the above requirements. These methods will capture global geometric relationships among all data points by expressing each data point as a linear combination of all other data points and then enforcing a prior on the matrix of coefficients, such as it being of low rank or sparse. Using convex relaxations of these priors, these methods will lead to computationally efficient ways of computing these global affinity matrices. We will characterize conditions

R. Vidal et al., *Generalized Principal Component Analysis*, Interdisciplinary Applied Mathematics 40, DOI 10.1007/978-0-387-87811-9_8

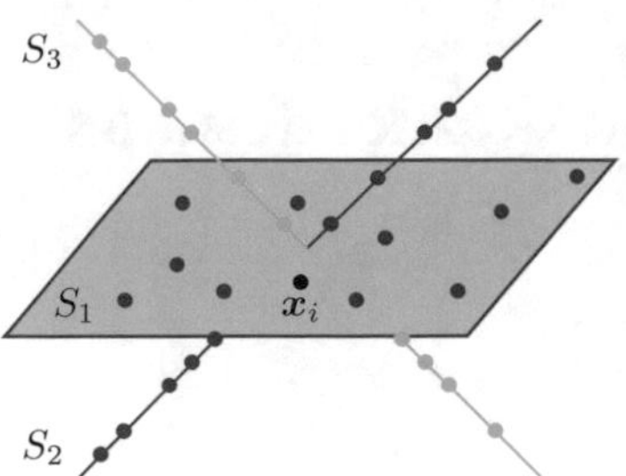

Fig. 8.1 Data drawn from three subspaces in $\mathbb{R}^3$.

under which these convex relaxation methods are guaranteed to provide the correct clustering. We will also extend these methods and their theoretical results to the case of data contaminated by noise and outliers.

Self-Expressiveness and Subspace-Preserving Representations

The approaches described in this chapter are based on two important properties of data points lying in a union of linear subspaces. The first property is called *self-expressiveness*. This property states that each data point in a union of subspaces can be expressed as a linear combination of all other data points, i.e.,

$$\boldsymbol{x}_j = \sum_{k \neq j} c_{kj} \boldsymbol{x}_k, \quad \text{or} \quad X = XC \text{ and } \operatorname{diag}(C) = \mathbf{0}, \tag{8.1}$$

where $X = \left[\boldsymbol{x}_1, \ldots, \boldsymbol{x}_N\right] \in \mathbb{R}^{D \times N}$ is the data matrix and $C = [c_{kj}] \in \mathbb{R}^{N \times N}$ is the matrix of coefficients. This property is illustrated in Figure 8.1, which shows a collection of data points drawn from a union of one plane and two lines in $\mathbb{R}^3$. Clearly, each data point can be expressed as a linear combination of other data points, e.g., any other three data points in general configuration.

The second property is called *subspace-preserving*. This property states that each data point in a union of subspaces can be expressed as a linear combination of other data points in the same subspace, i.e., point $\boldsymbol{x}_j \in S_i$ can be expressed as

$$\boldsymbol{x}_j = \sum_{k \neq j : \boldsymbol{x}_k \in S_i} c_{kj} \boldsymbol{x}_k. \tag{8.2}$$

Such a representation c_{kj} *preserves* the clustering of the data according to the multiple subspaces, whence the name *subspace-preserving*. This property is illustrated in Figure 8.1, where we can see that each point in a line is a scalar multiple of any other point in the same line, and every point in the plane is a linear combination of any other two linearly independent points in the same plane. More generally, this example suggests that some subspace-preserving representations could be *sparse*, the nonzero coefficients could correspond to other points in the same subspace, and the number of nonzero coefficients could be equal to the dimension of the subspace.

Therefore, we could find a subspace-preserving representation of each data point by enforcing a *sparsity prior* on the columns of the matrix C. While we have motivated our analysis with a sparse prior in mind, we will see that other priors on C are also possible, such as it being of low rank.

Low-Rank and Sparse Representations
The above discussion suggests that we can find a subspace-preserving matrix C by solving an optimization problem of the form

$$\min_{C} \ \|C\|_{\mathcal{C}} \quad \text{s.t.} \quad X = XC \ \text{ and } \ \operatorname{diag}(C) = \mathbf{0}, \tag{8.3}$$

where $\|\cdot\|_{\mathcal{C}}$ is a regularizer on the matrix C, such as it being of low rank or sparse. Since the above problem is NP-hard for many regularizers $\|\cdot\|_{\mathcal{C}}$, e.g., $\operatorname{rank}(C)$ or $\|C\|_0$, we will use convex relaxations of these regularizers, e.g., $\|C\|_*$ or $\|C\|_1$. An important observation is that under certain conditions, the solution of the relaxed problem is still subspace-preserving. Such conditions typically require the subspaces to be sufficiently separated and the data to be well distributed inside the subspaces. These conditions are stronger than those required by algebraic-geometric methods, which make no assumption on the relative orientation of the subspaces. However, the complexity of solving the associated convex optimization problems is polynomial in the problem dimensions, rather than exponential, and furthermore, under such conditions, the globally optimal solution for the convex program can be shown to be subspace-preserving. Thus, the additional conditions are a small price to pay.

Robustness to Noise, Corruptions, and Outliers
Another feature of the methods that will be described in this chapter is that they can be extended to handle errors (noise, corruptions, or outliers) by solving an optimization problem of the form

$$\min_{C,E} \ \|C\|_{\mathcal{C}} + \lambda \|E\|_{\mathcal{E}} \quad \text{s.t.} \quad X = XC + E \quad \text{and} \quad \operatorname{diag}(C) = \mathbf{0}, \tag{8.4}$$

where $E = [e_{ij}] \in \mathbb{R}^{D \times N}$ is the matrix of errors, and $\|\cdot\|_{\mathcal{C}}$ and $\|\cdot\|_{\mathcal{E}}$ are suitable norms or regularizers. While the regularizers can be chosen to be nonconvex, e.g., the number of corrupted data points, we will see that under certain conditions on the arrangement of the subspaces, the distribution of the data, and the distribution of the errors, it is still possible to guarantee that the matrix C is subspace-preserving.

Chapter Organization
The remainder of this chapter is organized as follows. Section 8.1 introduces the *self-expressiveness property* and the *subspace-preserving property* more formally. Section 8.2 introduces the low-rank subspace clustering (LRSC) algorithm, which solves the subspace clustering problem using a low-rank prior. Section 8.2.1 investigates theoretical conditions for its correctness in the case of uncorrupted data, and Sections 8.2.2 and 8.2.3 extend the algorithm to data corrupted by noise,

corrupted entries, and outliers. Section 8.3 introduces the sparse subspace clustering (SSC) algorithm, which solves the subspace clustering problem using a sparse prior. Section 8.3.1 investigates theoretical conditions for its correctness in the case of uncorrupted data and describes optimization algorithms for solving the associated optimization problems. Sections 8.3.2, 8.3.3, and 8.3.4 generalize the SSC algorithm to deal with outliers, noise, and corrupted entries in the data. Section 8.3.5 shows how to extend the SSC algorithm to the more general class of affine subspaces. Section 8.4 verifies the theoretical analysis through experiments on synthetic data and evaluates LRSC and SSC on the face clustering problem.

8.1 Self-Expressiveness and Subspace-Preserving Representations

This section introduces two properties that will be at the heart of the low-rank and sparse subspace clustering algorithms to be presented in the next sections. The first one is the *self-expressiveness property*, which states that a point in a union of subspaces can always be expressed as a linear combination of other points in the union of subspaces. The second one is the *subspace-preserving property*, which states that a point in a union of subspaces can always be expressed as a linear combination of other points in its own subspace. This last property will be necessary for ensuring the correctness of low-rank and sparse subspace clustering algorithms.

8.1.1 Self-Expressiveness Property

Let $\{\boldsymbol{x}_j\}_{j=1}^N$ be a set of points drawn from a union of n different linear subspaces $\{S_i\}_{i=1}^n$ of dimensions $\{d_i\}_{i=1}^n$. Denote the matrix containing all data points by

$$X \triangleq \left[\boldsymbol{x}_1, \ldots, \boldsymbol{x}_N\right] = \left[X_1, \ldots, X_n\right] \Gamma^\top, \tag{8.5}$$

where $X_i \in \mathbb{R}^{D \times N_i}$ is a matrix containing the $N_i > d_i$ points that lie in subspace S_i, and $\Gamma \in \mathbb{R}^{N \times N}$ is an unknown permutation matrix, which sorts the columns of X according to which subspace they belong to. The *expressiveness property* of the set of points is defined as follows.

Definition 8.1 (Expressiveness Property). *A set of points $\{\boldsymbol{x}_j\}_{j=1}^N$ drawn from a union of subspaces $\cup_{i=1}^n S_i$ is said to be expressive if every point $\boldsymbol{x} \in \cup_{i=1}^n S_i$ can be expressed as a linear combination of the data points $\{\boldsymbol{x}_j\}_{j=1}^N$. That is, for each $\boldsymbol{x} \in \cup_{i=1}^n S_i$, there exists a vector of coefficients $\boldsymbol{c} \in \mathbb{R}^N$ such that*

$$\boldsymbol{x} = X\boldsymbol{c}, \tag{8.6}$$

where $X = \left[\boldsymbol{x}_1, \ldots, \boldsymbol{x}_N\right] \in \mathbb{R}^{D \times N}$ is the data matrix. We call any such $\boldsymbol{c}$ a representation of $\boldsymbol{x}$ with respect to X.

Evidently, if $\text{rank}(X) = D$, the data set can be used to represent any point in $\mathbb{R}^D$. However, we are interested only in representing points in $\boldsymbol{x} \in \cup_{i=1}^n S_i$. Thus we would like to have a weaker condition that guarantees expressiveness. It is easy to see that a set of points drawn from a union of linear subspaces is expressive if for all $i = 1, \ldots, n$, we have $\text{rank}(X_i) = d_i$. For each $\boldsymbol{x} \in S_i$ can be expressed as a linear combination of d_i linearly independent columns in X_i, hence in X. Since we have assumed that this is true for all $i = 1, \ldots, n$, we conclude that each $\boldsymbol{x} \in \cup_{i=1}^n S_i$ can be expressed as a linear combination of the columns of X. Now, since the assumption that $\text{rank}(X_i) = d_i$ is easily satisfied by a set of points in general position within S_i, we will assume from now on that the given data points are in general position in $\cup_{i=1}^n S_i$, so that the data set is always *expressive*.

Let us now consider the particular case that $\boldsymbol{x}$ is one of the data points, say $\boldsymbol{x}_j$. Obviously, we can express $\boldsymbol{x}_j$ in terms of itself. However, this representation would not be very useful for clustering. A more interesting case is to express a data point in terms of other data points. This leads to the *self-expressiveness property* of data points in a union of linear subspaces, which is defined as follows.

Definition 8.2 (Self-Expressiveness Property). *A set of points $\{\boldsymbol{x}_j\}_{j=1}^N$ drawn from a union of subspaces $\cup_{i=1}^n S_i$ is said to satisfy the self-expressiveness property if each data point can be expressed as a linear combination of all the other data points. That is, for each $j = 1, \ldots, N$, there exists a vector of coefficients $\boldsymbol{c}_j \triangleq \left[c_{1j}, c_{2j}, \ldots, c_{Nj}\right]^\top \in \mathbb{R}^N$ such that*

$$\boldsymbol{x}_j = \sum_{k \neq j} c_{kj} \boldsymbol{x}_k \quad \Longleftrightarrow \quad \boldsymbol{x}_j = X\boldsymbol{c}_j \quad \text{and} \quad c_{jj} = 0, \tag{8.7}$$

where $X = \left[\boldsymbol{x}_1, \ldots, \boldsymbol{x}_N\right] \in \mathbb{R}^{D \times N}$ is the data matrix. Equivalently, the data matrix X is said to satisfy the self-expressiveness property if

$$\exists\, C \in \mathbb{R}^{N \times N} \quad \text{s.t.} \quad X = XC \quad \text{and} \quad \text{diag}(C) = \mathbf{0}, \tag{8.8}$$

where $C = [c_{kj}] \in \mathbb{R}^{N \times N}$ is the matrix of coefficients.

In order for a data set to be self-expressive, we simply need the set of all data points minus one point to be expressive. More specifically, let $X_i^{-j} \in \mathbb{R}^{D \times (N_i - 1)}$ be the data matrix X_i from subspace S_i with its jth column removed. We have the following result, whose proof we leave as an exercise (see Exercise 8.2).

Lemma 8.3. *A set of points drawn from a union of subspaces satisfies the self-expressiveness property if for all $i = 1, \ldots, n$ and $j = 1, \ldots, N_i$, we have* $\text{rank}(X_i^{-j}) = d_i$.

Intuitively, the above lemma states that data drawn from a union of subspaces satisfy the self-expressiveness property if sufficiently many data points in general configuration are drawn from each of the subspaces. Since this condition for self-expressiveness is extremely mild, we will assume from now on that it holds.

8.1.2 Subspace-Preserving Representation

An expressive data set can be seen as an overcomplete dictionary that can be used to represent every point in $\cup_{i=1}^n S_i$ as in (8.6). However, notice that the representation of a point $\boldsymbol{x} \in \cup_{i=1}^n S_i$ with respect to X need not be unique. Indeed, if $\boldsymbol{x} \in S_i$ and $N_i > d_i$, there are many choices of d_i or more columns of X that can be used to generate $\boldsymbol{x}$. More generally, we could use data points from subspaces other than S_i to generate $\boldsymbol{x} \in S_i$. Since our goal is to determine the subspace containing the point $\boldsymbol{x}$, we are interested in a representation of $\boldsymbol{x}$ that involves columns of X_i only. We call such a representation *subspace-preserving*, since it preserves the clustering of the data points.

Definition 8.4 (Subspace-Preserving Representation). *A representation $\mathbf{c} \in \mathbb{R}^N$ of a point $\boldsymbol{x} \in S_i$ in terms of a self-expressive dictionary $X = \big[\boldsymbol{x}_1, \ldots, \boldsymbol{x}_N\big]$ is called subspace-preserving if the nonzero entries of $\mathbf{c}$ correspond to points in S_i, i.e.,*

$$\forall j = 1, \ldots, N, \quad c_j \neq 0 \implies \boldsymbol{x}_j \in S_i. \tag{8.9}$$

A representation $C \in \mathbb{R}^{N \times N}$ of a self-expressive matrix X in terms of itself is called subspace-preserving if each column of C is subspace-preserving, i.e., if

$$c_{jk} = 0 \quad \textit{when points } j \textit{ and } k \textit{ are in different subspaces.} \tag{8.10}$$

It follows that a subspace-preserving representation has the following structure:

$$C = \Gamma \begin{bmatrix} C_1 & \mathbf{0} & \cdots & \mathbf{0} \\ \mathbf{0} & C_2 & \cdots & \mathbf{0} \\ \vdots & \vdots & \ddots & \vdots \\ \mathbf{0} & \mathbf{0} & \cdots & C_n \end{bmatrix} \Gamma^\top, \tag{8.11}$$

where Γ is a permutation matrix. Therefore, in principle we may use any subspace-preserving representation to directly cluster the data into multiple subspaces. In practice, however, we may not be able to construct a representation that is perfectly subspace-preserving. That is, it may be the case that $|c_{jk}|$ is small but not necessarily zero when points j and k are in different subspaces. In this case, its better to use the spectral clustering algorithm described in Chapter 4 to cluster the data. In a nutshell, we build a similarity graph whose nodes are the data points and whose edges connect two points according to an affinity measure. Ideally, each connected component of the graph corresponds to points in the same subspace. Hence, the segmentation of the data into the subspaces can be obtained by separating the components of the graph. Since the affinity matrix needs to be symmetric and nonnegative, an immediate choice of the similarity matrix is $W = |C| + |C|^\top$. In other words, each node j connects itself to a node k by an edge whose weight is equal to $|c_{jk}| + |c_{kj}|$. The reason for the symmetrization is that in general, a data point $\boldsymbol{x}_j$ can be written as a linear combination of other points including $\boldsymbol{x}_k$. However, $\boldsymbol{x}_k$ may not necessarily choose $\boldsymbol{x}_j$ in its representation. By this particular choice of the weight, we make sure that nodes j and k get connected to each other if either $\boldsymbol{x}_j$ or $\boldsymbol{x}_k$ is in the representation of the other.

8.2 Low-Rank Subspace Clustering (LRSC)

In this section, we introduce the low-rank subspace clustering (LRSC) algorithm for clustering a set of points drawn from a union of subspaces. This algorithm addresses the subspace clustering problem in two steps. In the first step, it finds a low-rank representation of the data subject to the self-expressiveness constraint. In the second step, it uses the low-rank representation to build an affinity matrix for spectral clustering. We first introduce the LRSC algorithm in the case of uncorrupted data. We show that the low-rank representation can be computed in closed form from the SVD of the data matrix, and that this representation is subspace-preserving under the assumption that the subspaces are *independent*. We then extend LRSC to data corrupted by noise. We show that the low-rank representation can still be computed in closed form from the SVD of the data, but the representation is no longer guaranteed to be subspace-preserving. We then extend LRSC to the case of data corrupted by outliers or gross errors and show that under certain conditions, one can compute the low-rank representation from a clean data matrix obtained by applying a robust PCA algorithm to the corrupted data matrix.

8.2.1 LRSC with Uncorrupted Data

The LRSC algorithm aims to find a subspace-preserving representation C of the data matrix X such that the coefficient matrix C is of low rank. In other words, it tries to solve the following low-rank minimization problem:

$$\min_{C} \ \operatorname{rank}(C) \quad \text{s.t.} \quad X = XC. \tag{8.12}$$

Notice that this problem is essentially the same as that in (8.3) with the regularizer chosen as $\|C\|_{\mathcal{C}} = \operatorname{rank}(C)$. The main difference is that we have dropped the constraint $\operatorname{diag}(C) = \mathbf{0}$, which is used to prevent a point from being expressed in terms of itself. We can drop this constraint because here we are trying to minimize the rank of C, and diagonally dominant matrices have high rank. Therefore, the optimization problem in (8.12) automatically prevents the trivial solution $C = I$.

At first sight, we may wonder why we are searching for a low-rank representation to begin with. In particular, we may ask whether a low-rank representation exists, and if it does, whether it is subspace-preserving. In what follows, we show that under certain conditions, a low-rank representation does exist and can be computed in closed form from the SVD of X. We also show that this low-rank representation is a solution to the following relaxed convex optimization problem:

$$\min_{C} \quad \|C\|_* \quad \text{s.t.} \quad X = XC, \tag{8.13}$$

where we replace the rank by the nuclear norm. We also derive conditions on the subspaces under which the low-rank representation is subspace-preserving.

Closed-Form Solution to the Low-Rank Minimization Problem
Observe that the self-expressiveness constraint $X = XC$ implies that

$$r \triangleq \operatorname{rank}(X) \leq \operatorname{rank}(C). \tag{8.14}$$

Therefore, a necessary condition for a low-rank solution C of $X = XC$ to exist is that X be of low rank, i.e., $r < \min\{D, N\}$. Notice that this need not be the case in general. For example, data points sampled from 10 subspaces of dimension 10 in $\mathbb{R}^{100}$ could span the whole ambient space. Therefore, in order for problem (8.12) to be well posed, we will assume throughout this section that $r < \min\{D, N\}$.

Now let $X = U_1 \Sigma_1 V_1^\top$ be the rank-r SVD of X and let $C^* = V_1 V_1^\top$. Then

$$\operatorname{rank}(C^*) = \operatorname{rank}(X) \quad \text{and} \quad XC^* = U_1 \Sigma_1 V_1^\top V_1 V_1^\top = U_1 \Sigma_1 V_1^\top = X. \tag{8.15}$$

In other words, C^* is a matrix of the smallest possible rank such that $X = XC$. Hence, C^* is an optimal solution to (8.12). However, it is not the unique optimal solution, since we have the following result.

Theorem 8.5. *Let $X = U \Lambda V^\top$ be the SVD of X. Let $V = [V_1 \; V_2]$, where the columns of V_1 are the singular vectors of X corresponding to its nonzero singular values. The optimal solutions to the program* (8.12) *are given by*

$$C^* = V_1 V_1^\top + V_2 B V_1^\top, \tag{8.16}$$

where $B \in \mathbb{R}^{(N-r)\times r}$. Moreover, the optimal value is exactly $r = \operatorname{rank}(X)$.

Proof. We have already shown that $V_1 V_1^\top$ is a particular solution to the linear system of equations $XC = X$. Since $V_2 \in \mathbb{R}^{N\times(N-r)}$ gives a basis for the null space of X, the general solution to $XC = X$ is given by $C = V_1 V_1^\top + V_2 A$, where $A \in \mathbb{R}^{(N-r)\times N}$ is an arbitrary matrix. We have already shown that the optimal value of (8.12) is $r = \operatorname{rank}(X)$. Therefore, the matrix A must be constrained so that $\operatorname{rank}(V_1 V_1^\top + V_2 A) = r$. Noticing that $V_1 V_1^\top + V_2 A = \begin{bmatrix} V_1 \; V_2 \end{bmatrix} \begin{bmatrix} V_1^\top \\ A \end{bmatrix}$, where $\begin{bmatrix} V_1 \; V_2 \end{bmatrix}$ is of full rank, we conclude that $\operatorname{rank}\left(\begin{bmatrix} V_1^\top \\ A \end{bmatrix} \right) = r$, and so $A = B V_1^\top$ for some $B \in \mathbb{R}^{(N-r)\times r}$. □

Among this family of solutions, we are going to be primarily interested in the particular solution $C^* = V_1V_1^\top$. This matrix was introduced in (Costeira and Kanade 1998) in the context of the motion segmentation problem in computer vision under the name *shape interaction matrix*. This matrix is symmetric; hence it is also a solution to the following optimization problem:

$$\min_C \ \operatorname{rank}(C) \quad \text{s.t.} \quad X = XC, \quad \text{and} \quad C = C^\top. \tag{8.17}$$

Since our ultimate goal is to build a symmetric affinity matrix from C, it seems natural to add the constraint $C = C^\top$ to the optimization problem in (8.12), especially sinvr this does not make the problem more difficult to solve.

Convex Relaxation of the Low-Rank Minimization Problem
As we will soon see, there is much advantage to be gained from a reformulation of (8.17) in terms of a convex program, since its variations will be computationally more tractable when we begin to incorporate additional structures into the basic self-expressive model. For example, we can relax (8.17) as

$$\min_C \ \|C\|_* \quad \text{subject to} \quad X = XC \ \text{ and } \ C = C^\top. \tag{8.18}$$

Notice that the relaxed program is convex on C, but not strictly convex. Therefore, we do not know a priori whether this problem has a unique minimum. The following theorem shows that $C^* = V_1V_1^\top$ is the unique minimum.

Theorem 8.6 (LRSC for Uncorrupted Data (Vidal and Favaro 2014)). *Let $X = U\Lambda V^\top$ be the SVD of X, where the diagonal entries of $\Lambda = \operatorname{diag}\{\lambda_i\}$ are the singular values of X in decreasing order. The optimal solution to the program* (8.18) *is given by*

$$C^* = V_1V_1^\top, \tag{8.19}$$

where $V = [V_1\ V_2]$ is partitioned according to the sets $\mathcal{I}_1 = \{i : \lambda_i > 0\}$ and $\mathcal{I}_2 = \{i : \lambda_i = 0\}$. Moreover, the optimal value is exactly $r = \operatorname{rank}(X)$.

Proof. Let $C = U_C\Delta U_C^\top$ be the eigenvalue decomposition (EVD) of C. Then $X = XC$ can be rewritten as $U_1\Lambda_1V_1^\top = U_1\Lambda_1V_1^\top U_C\Delta U_C^\top$, which reduces to

$$V_1^\top U_C = V_1^\top U_C\Delta, \tag{8.20}$$

since $U_1^\top U_1 = I$ and $U_C^\top U_C = I$. Let $W = V^\top U_C = [w_1, \dots, w_N]$. Then $w_j = w_j\delta_j$ for all $j = 1, \dots, N$. This means that $\delta_j = 1$ if $w_j \neq \mathbf{0}$ and δ_j is arbitrary otherwise. Since our goal is to minimize $\|C\|_* = \|\Delta\|_* = \sum_{j=1}^N |\delta_j|$, we need to set as many δ_j to zero as possible. Since $X = XC$ implies that $\operatorname{rank}(X) \leq \operatorname{rank}(C)$, we can set at most $N - r$ of the δ_j to zero, and the remaining r of the δ_j must be equal to one. Now, if $\delta_j = 0$, then $w_j = V_1^\top U_C e_j = \mathbf{0}$, where e_j is the jth column of the identity matrix.

This means that the columns of U_C associated to $\delta_j = 0$ must be orthogonal to the columns of V_1, and hence the columns of U_C associated with $\delta_j = 1$ must be in the range of V_1. Thus, $U_C = \begin{bmatrix} V_1R_1 \; V_2R_2 \end{bmatrix} \Pi$ for some orthogonal matrices R_1 and R_2 and permutation matrix Π. Therefore, the optimal C^* is

$$C^* = U_C \Delta U_C^\top = \begin{bmatrix} V_1R_1 \; V_2R_2 \end{bmatrix} \begin{bmatrix} I & \mathbf{0} \\ \mathbf{0} & \mathbf{0} \end{bmatrix} \begin{bmatrix} V_1R_1 \; V_2R_2 \end{bmatrix}^\top = V_1 V_1^\top, \tag{8.21}$$

as claimed. □

Interestingly, $C = V_1 V_1^\top$ is also the optimal solution to the following convex program, without explicitly enforcing the symmetry on C (Wei and Lin 2010):

$$\min \|C\|_* \quad \text{subject to} \quad X = XC. \tag{8.22}$$

We leave the derivation as an exercise to the reader (see Exercise 8.1).

Subspace-Preserving Property of LRSC for Independent Subspaces
Now we see that the matrix $C^* = V_1 V_1^\top$ arises repeatedly as the optimal solution to the above low-rank formulations. The next question is under what conditions this low-rank representation is subspace-preserving, so that we can use it to build a similarity for subspace clustering. Theorem 8.9 below shows that C^* is subspace-preserving if the subspaces are *independent*, as defined next.

Definition 8.7. *A collection of subspaces* $\{S_i\}_{i=1}^n$ *is said to be* independent *if the dimension of their sum is equal to the sum of their dimensions, i.e., if* $\dim(\bigoplus_{i=1}^n S_i) = \sum_{i=1}^n \dim(S_i)$, *where* $\oplus$ *denotes the sum operator for subspaces, which is defined as* $S_1 \oplus S_2 = \{\boldsymbol{x} = \boldsymbol{x}_1 + \boldsymbol{x}_2, \boldsymbol{x}_1 \in S_1, \boldsymbol{x}_2 \in S_2\}$.

Example 8.8. As an example, the three 1-dimensional subspaces shown in Figure 8.2 (left) are independent, since they span a 3-dimensional space and the sum of their dimensions is also 3. On the other hand, the subspaces shown in Figure 8.2 (right) are not independent, since they span a 2-dimensional space, while the sum of their dimensions is 3.

Theorem 8.9 ((Vidal et al. 2008)). *Let* $X = [X_1, \ldots, X_n]\Gamma^\top$ *be a matrix whose columns are drawn from a union of n independent subspaces, where* X_i *is a rank-*d_i *matrix containing* $N_i > d_i$ *points from subspace i of dimension* d_i*, and* Γ *is a permutation matrix. Let* $X = U_1 \Lambda V_1^\top$ *be its compact SVD and let* $C = V_1 V_1^\top$*. Then the matrix C is subspace-preserving, i.e., its entries* c_{jk} *have the following property:*

$$c_{jk} = 0 \quad \textit{if points } j \textit{ and } k \textit{ are in different subspaces.} \tag{8.23}$$

Proof. To see this, let $A_i \in \mathbb{R}^{N_i \times (N_i - d_i)}$ be a matrix whose columns form an orthonormal basis for the null space of X_i, that is, $X_i A_i = 0$ and $A_i^\top A_i = I$. Consider now the matrix

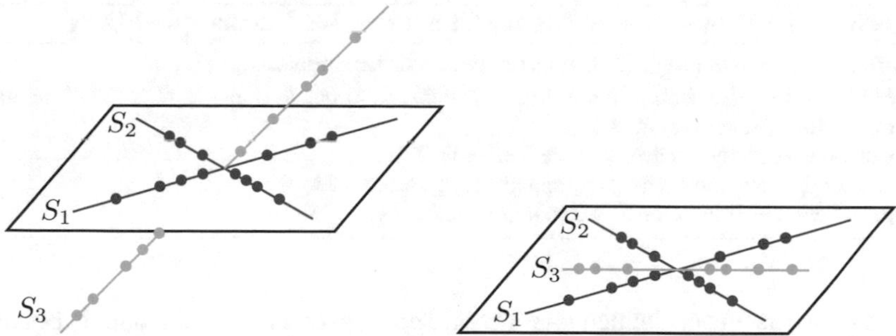

Fig. 8.2 Left: the three 1-dimensional subspaces are independent, since they span the 3-dimensional space and the sum of their dimensions is also 3. Right: the three 1-dimensional are disjoint, since any two subspaces intersect at the origin.

$$A = \Gamma^\top \begin{bmatrix} A_1 & 0 & \cdots & 0 \\ 0 & A_2 & & \vdots \\ \vdots & & \ddots & 0 \\ 0 & \cdots & 0 & A_n \end{bmatrix} \in \mathbb{R}^{N\times(N-d)}, \tag{8.24}$$

where $d = \sum_{i=1}^n d_i$. It is clear that the $N - d$ columns of A are orthonormal and lie in the null space of X. On the other hand, the columns of the matrix $V_2 \in \mathbb{R}^{N\times(N-r)}$ defined before in the proof of Theorem 8.5 form a basis for the right null space of X. Since the subspaces are independent, we have that $d = r$; hence V_2 and A have the same dimension, and so the columns of A also form a basis for the right null space of X. Thus, there exists an orthogonal matrix B such that $V_2 = AB$. Combining this with the fact that $VV^\top = I$, we have

$$VV^\top = V_1V_1^\top + V_2V_2^\top = C + AA^\top = I. \tag{8.25}$$

From this, it follows that

$$\Gamma C \Gamma^\top = I - (\Gamma A)(\Gamma A)^\top, \tag{8.26}$$

which is block-diagonal. This means that $c_{jk} = 0$ if points j and k are in different subspaces, as desired. □

Low-Rank Subspace Clustering Affinity

In conclusion, when the subspaces are independent and sufficiently many points in general configuration are drawn from each subspace, the low-rank representation matrix $C = V_1V_1^\top$ is subspace-preserving; hence it can be used to build a subspace clustering affinity. Since the matrix C is symmetric, we may use $|C|$ directly as an affinity. More generally, we can define an affinity matrix W as

$$w_{jk} = |C_{jk}|^q, \tag{8.27}$$

Algorithm 8.1 (Low-Rank Subspace Clustering for Uncorrupted Data)

Input: A set of points $\{x_j\}_{j=1}^N$ lying in a union of n linear subspaces $\{S_i\}_{i=1}^n$.

1: Solve the low-rank optimization program in (8.18) to obtain a matrix $C \in \mathbb{R}^{N \times N}$ as in the closed-form expression in (8.19).

2: Compute an affinity matrix W as defined in (8.27).

3: Cluster the data into n groups by applying Algorithm 4.7 to W.

Output: Segmentation of the data into n groups: $X_1, X_2, \ldots, X_n$.

where $q > 0$ is a user-defined parameter. The segmentation of the data is obtained by applying spectral clustering to this affinity matrix.

Summary of the LRSC Algorithm

In summary, the LRSC algorithm for data points that lie perfectly in a union of linear subspaces proceeds as shown in Algorithm 8.1.

8.2.2 LRSC with Robustness to Noise

In the previous subsection, we considered the low-rank subspace clustering problem under the assumption that the N data points are drawn perfectly from a union of low-dimensional subspaces. In practice, the data matrix may be contaminated by noise, corrupted entries, or outliers; hence the self-expressiveness constraint $X = XC$ may not hold exactly.

A simple way of modeling errors proposed in (Rao et al. 2008, 2010) and further developed in (Elhamifar and Vidal 2009, 2013) is to relax the self-expressiveness constraint to

$$X = XC + E, \tag{8.28}$$

where E is the matrix of errors. This model has the advantage of maintaining the linearity of the self-expressive constraint, which will lead to convex optimization problems on C and E, as we shall see. However, a shortcoming of this model is that it does not directly model the errors in the original data.

An alternative error model proposed in (Wei and Lin 2010) and further developed in (Favaro et al. 2011; Vidal and Favaro 2014) assumes that there is an unknown data matrix A whose columns lie perfectly in a union of subspaces. The corrupted data are then generated by adding the error matrix E directly to A subject to the self-expressiveness constraint on A, i.e.,

$$X = A + E, \qquad A = AC. \tag{8.29}$$

This model has the advantage of being more natural, in the sense of modeling the errors directly in the original data. However, the main disadvantage is that the model is nonlinear due to the product of A and C, which are both unknown.

In spite of their differences, an interesting observation is that both error models can be related. In fact, starting from the nonconvex model in (8.29), notice that we may eliminate the clean data matrix A to arrive at

$$(X-E)=(X-E)C \implies X=XC+E-EC \implies X=XC+\tilde{E}, \tag{8.30}$$

where $\tilde{E}=E-EC$. Therefore, the convex model in (8.29) can be seen as a special case of the nonconvex model in (8.29) in which the nonlinear part EC is neglected. This might be appropriate when E represents small noise, but it might be inadequate when E represents large corruptions.

In this section, we discuss both models in more detail in the case that E corresponds to noise or small errors. In spite of their differences, we show that both models lead to closed-form solutions for C, which are very much related.

Convex Error Model

To find a low-rank representation C that is robust to noise under the model (8.28), we can penalize the sum of the squared errors $\|E\|_F^2$, which leads to the following convex program:

$$\min_C \|C\|_* + \frac{\tau}{2}\|X-XC\|_F^2 \quad \text{s.t.} \quad C=C^\top, \tag{8.31}$$

where $\tau>0$ is a tradeoff parameter. Notice that this program is convex in C, but not strictly convex. Therefore, we do not know a priori whether the minimizer is unique. The following theorem shows that the minimizer is unique and can be computed in closed form from the SVD of X.

Theorem 8.10 (LRSC for Noisy Data). *Let $X=U\Lambda V^\top$ be the SVD of X, where the diagonal entries of $\Lambda=diag(\{\lambda_i\})$ are the singular values of X in decreasing order. The optimal solution to the convex program* (8.31) *is given by*

$$C=V\mathcal{P}_{\frac{1}{\sqrt{\tau}}}(\Lambda)V^\top = V_1\left(I-\frac{1}{\tau}\Lambda_1^{-2}\right)V_1^\top, \tag{8.32}$$

where the operator $\mathcal{P}_\varepsilon$ acts on the diagonal entries of Λ as

$$\mathcal{P}_\varepsilon(x) \doteq \begin{cases} 1-\frac{\varepsilon^2}{x^2} & x>\varepsilon \\ 0 & x\leq\varepsilon \end{cases} \tag{8.33}$$

and $U=[U_1\ U_2]$, $\Lambda=diag(\Lambda_1,\Lambda_2)$, and $V=[V_1\ V_2]$ are partitioned according to the sets $\mathcal{I}_1=\{i:\lambda_i>\frac{1}{\sqrt{\tau}}\}$ and $\mathcal{I}_2=\{i:\lambda_i\leq\frac{1}{\sqrt{\tau}}\}$. Moreover, the optimal value of the program (8.31) *is*

$$\Phi_\tau(X) \doteq \sum_{i\in\mathcal{I}_1}\left(1-\frac{1}{2\tau}\lambda_i^{-2}\right)+\frac{\tau}{2}\sum_{i\in\mathcal{I}_2}\lambda_i^2. \tag{8.34}$$

Proof. Let $C = U_C \Delta U_C^\top$ be the eigenvalue decomposition (EVD) of C. The cost function of the convex program (8.31) can be rewritten as

$$\|U_C \Delta U_C^\top\|_* + \frac{\tau}{2}\|U \Lambda V^\top (I - U_C \Delta U_C^\top)\|_F^2 = \tag{8.35}$$
$$\|\Delta\|_* + \frac{\tau}{2}\|\Lambda V^\top U_C (I - \Delta) U_C^\top\|_F^2 = \|\Delta\|_* + \frac{\tau}{2}\|\Lambda W (I - \Delta)\|_F^2,$$

where $W = V^\top U_C$. To minimize this cost with respect to W, we need to consider only the last term of the cost function, i.e.,

$$\|\Lambda W (I - \Delta)\|_F^2 = \operatorname{trace}\left((I - \Delta)^2 W^\top \Lambda^2 W\right). \tag{8.36}$$

Let $\sigma_i(A)$ be the ith-largest singular value of an arbitrary matrix A. Applying a von Neumann-type singular value inequality to (8.36) (see Lemma 2.5 as well as (Vidal and Favaro 2014)), we obtain that for all orthonormal matrices W,

$$\min_W \operatorname{trace}\left((I - \Delta)^2 W^\top \Lambda^2 W\right) = \sum_{i=1}^N \sigma_i\left((I - \Delta)^2\right) \sigma_{n-i+1}(\Lambda^2), \tag{8.37}$$

where the minimum is achieved by a permutation matrix $W = \Pi^\top$ such that the diagonal entries of $\Pi \Lambda^2 \Pi^\top$ are in ascending order. Let the ith-largest entry of $(I - \Delta)^2$ and Λ^2 be, respectively, $(1 - \delta_i)^2 = \sigma_i\left((I - \Delta)^2\right)$ and $\nu_{n-i+1}^2 = \lambda_i^2 = \sigma_i(\Lambda^2)$. Then the optimal value of (8.31) after minimizing W is

$$\min_W \|\Delta\|_* + \frac{\tau}{2}\|\Lambda W (I - \Delta)\|_F^2 = \sum_{i=1}^N |\delta_i| + \frac{\tau}{2}\sum_{i=1}^N \nu_i^2 (1 - \delta_i)^2. \tag{8.38}$$

To find the optimal Δ, we can solve for each δ_i independently as $\delta_i = \arg\min_\delta |\delta| + \frac{\tau}{2}\nu_i^2 (1 - \delta)^2$. The solution to this problem can be found in closed form using the soft-thresholding operator, which gives

$$\delta_i = \mathcal{S}_{\frac{1}{\tau \nu_i^2}}(1) = \begin{cases} 1 - \frac{1}{\tau \nu_i^2} & \nu_i > 1/\sqrt{\tau} \\ 0 & \nu_i \le 1/\sqrt{\tau} \end{cases}. \tag{8.39}$$

Then $\delta_i = \mathcal{P}_{\frac{1}{\sqrt{\tau}}}(\lambda_{n-i+1})$, which can be compactly written as $\Delta = \Pi \mathcal{P}_{\frac{1}{\sqrt{\tau}}}(\Lambda) \Pi^\top$. Therefore,

$$\Pi^\top \Delta \Pi = \mathcal{P}_{\frac{1}{\sqrt{\tau}}}(\Lambda) = \begin{bmatrix} I - \frac{1}{\tau}\Lambda_1^{-2} & 0 \\ 0 & 0 \end{bmatrix}, \tag{8.40}$$

where $\Lambda = \operatorname{diag}(\Lambda_1, \Lambda_2)$ is partitioned according to the sets $\mathcal{I}_1 = \{i : \lambda_i > 1/\sqrt{\tau}\}$ and $\mathcal{I}_2 = \{i : \lambda_i \leq 1/\sqrt{\tau}\}$.

To find the optimal W, notice from Lemma 2 of (Vidal and Favaro 2014) that the equality $\operatorname{trace}\big((I - \Delta)^2 W^\top \Lambda^2 W\big) = \sum_{i=1}^N (1-\delta_i)^2 \lambda_{n-i+1}^2$ is achieved if and only if there exists an orthonormal matrix U_X such that

$$(I - \Delta)^2 = U_X (I - \Delta)^2 U_X^\top \text{ and } W^\top \Lambda^2 W = U_X \Pi \Lambda^2 \Pi^\top U_X^\top. \tag{8.41}$$

Since the SVD of a matrix is unique up to the sign of the singular vectors associated with different singular values and up to a rotation and sign of the singular vectors associated with repeated singular values, we conclude that $U_X = I$ up to the aforementioned ambiguities of the SVD of $(I - \Delta)^2$. Likewise, we have that $W^\top = U_X \Pi$ up to the aforementioned ambiguities of the SVD of Λ^2. Now, if Λ^2 has repeated singular values, then $(I - \Delta)^2$ has repeated eigenvalues at the same locations. Therefore, $W^\top = U_X \Pi = \Pi$ up to a block-diagonal transformation, where each block is an orthonormal matrix that corresponds to a repeated singular value of Δ.

Nonetheless, even though W may not be unique, the optimal matrix C^* is always unique and equal to

$$\begin{aligned} C &= U_C \Delta U_C^\top = V W \Delta W^\top V^\top = V \Pi^\top \Delta \Pi V^\top \\ &= \begin{bmatrix} V_1 & V_2 \end{bmatrix} \begin{bmatrix} I - \frac{1}{\tau}\Lambda_1^{-2} & 0 \\ 0 & 0 \end{bmatrix} \begin{bmatrix} V_1 & V_2 \end{bmatrix}^\top = V_1 (I - \frac{1}{\tau}\Lambda_1^{-2}) V_1^\top. \end{aligned} \tag{8.42}$$

Finally, the optimal C^* is such that $XC = U_1(\Lambda_1 - \frac{1}{\tau}\Lambda_1^{-1})V_1^\top$ and $X - XC = U_2 \Lambda_2 V_2^\top + \frac{1}{\tau} U_1 \Lambda_1 V_1^\top$. This shows (8.34), because

$$\|C\|_* + \frac{\tau}{2}\|X - XC\|_F^2 = \sum_{i \in \mathcal{I}_1} \left(1 - \frac{1}{\tau}\lambda_i^{-2}\right) + \frac{\tau}{2}\left(\sum_{i \in \mathcal{I}_1} \frac{\lambda_i^{-2}}{\tau^2} + \sum_{i \in \mathcal{I}_2} \lambda_i^2\right),$$

as claimed. □

Notice that the optimal solution for C is obtained by applying a nonlinear thresholding $\mathcal{P}_{\frac{1}{\sqrt{\tau}}}$ to the singular values of X: the singular values smaller than $\frac{1}{\sqrt{\tau}}$ are mapped to zero, while larger singular values are mapped closer to one. Notice also that as $\tau \to \infty$, we recover the solution for uncorrupted data.

In addition, notice that the optimal value of the convex program is a decomposable function of the singular values of X, as are the Frobenius norm $\|X\|_F^2 = \sum \lambda_i^2$ and nuclear norm $\|X\|_* = \sum \lambda_i$ of X. However, unlike $\|X\|_F^2$ or $\|X\|_*$, $\Phi_\tau(X)$ is not a convex function of X, because $\Phi_\tau(\lambda)$ is quadratic near zero and saturates as λ increases, as illustrated in Figure 8.3. Interestingly, as $\tau \to \infty$, $\Phi_\tau(X)$ approaches $\operatorname{rank}(X)$. Therefore, we may view $\Phi_\tau(X)$ as a nonconvex continuous relaxation of $\operatorname{rank}(X)$.

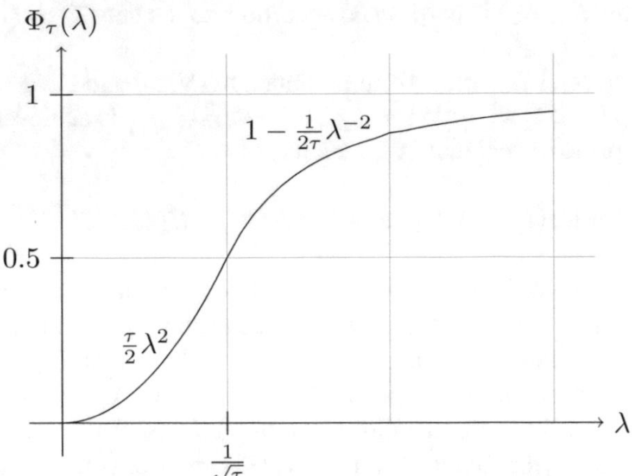

Fig. 8.3 Plot of $\Phi_\tau(\lambda)$.

Overall, the main advantage of the error model in (8.28) is that the low-rank representation C can be computed in closed form, similar to the case of uncorrupted data. However, one disadvantage is that the solution depends on the parameter τ, which needs to be tuned. Also, we are not aware of an extension of Theorem 8.9 to the case of noisy data. Therefore, we have no guarantees for the correctness of clustering. Yet another shortcoming of the model in (8.28) is that it does not directly model the noise in the original data, as discussed before.

Nonconvex Error Model

To find a low-rank representation C that is robust to noise under the model in (8.29), we can penalize the sum of the squared errors $\|E\|_F^2$, as before, which leads to the following program:

$$\min_{A,C,E} \|C\|_* + \frac{\tau}{2}\|E\|_F^2 \quad \text{s.t.} \quad X = A + E,\ A = AC,\ C^\top = C. \tag{8.43}$$

However, notice that while the objective function is convex and most of the constraints are linear, the self-expressiveness constraint $A = AC$ is no longer linear, and hence the program in (8.43) is no longer convex. Nonetheless, we can still solve for C in closed form as stated in the following theorem.

Theorem 8.11 (LRSC with Nonconvex Error Model (Vidal and Favaro 2014)). *Let $X = U\Lambda V^\top$ be the SVD of X, where the diagonal entries of $\Lambda = diag\{\lambda_i\}$ are the singular values of X in decreasing order. The optimal solution to the program* (8.43) *is given by*

$$A^* = U_1\Lambda_1 V_1^\top, \quad C^* = V_1 V_1^\top, \quad \textit{and} \quad E^* = U_2\Lambda_2 V_2^\top, \tag{8.44}$$

where $U = [U_1\ U_2]$, $\Lambda = diag(\Lambda_1, \Lambda_2)$, *and* $V = [V_1\ V_2]$ *are partitioned according to the sets* $\mathcal{I}_1 = \{i : \lambda_i > \sqrt{2/\tau}\}$ *and* $\mathcal{I}_2 = \{i : \lambda_i \leq \sqrt{2/\tau}\}$. *Moreover, the optimal value of the program* (8.31) *is*

$$\|C^*\|_* + \frac{\tau}{2}\|E^*\|_F^2 = |\mathcal{I}_1| + \frac{\tau}{2}\sum_{i \in \mathcal{I}_2} \lambda_i^2. \tag{8.45}$$

Proof. Thanks to Theorem 8.5, we can solve for C given A as $C^* = V_1 V_1^\top$, and the optimal value of the optimization problem with respect to C is rank(A), i.e.,

$$\text{rank}(A) = \min_C\{\|C\|_* : X = A + E,\ A = AC,\ C^\top = C\}. \tag{8.46}$$

Substituting this back into (8.43), we arrive at

$$\min_{A,E}\ \text{rank}(A) + \frac{\tau}{2}\|E\|_F^2 \quad \text{s.t.} \quad X = A + E. \tag{8.47}$$

This last problem is exactly the PCA problem discussed in (2.91), whose optimal solution can be obtained in closed form from the SVD of $X = U\Lambda V^\top$ as

$$A^* = U\mathcal{H}_{\sqrt{\frac{2}{\tau}}}(\Lambda)V^\top = U_1\Lambda_1 V_1^\top, \tag{8.48}$$

where $\mathcal{H}_\varepsilon$ is the hard thresholding operator defined in (2.90). Finally, since $X = U_1\Lambda_1 V_1^\top + U_2\Lambda_2 V_2^\top$, we obtain the desired expression for E^*. The optimal value of the program (8.31) follows by direct substitution. □

Notice that the optimal solution for (A, C) is effectively obtained in two steps:

1. Obtain the clean data matrix A form the noisy data matrix X via hard singular value thresholding with a threshold $\sqrt{2/\tau}$.
2. Obtain the low-rank representation C from the clean data matrix A using the result in Theorem 8.6 for uncorrupted data.

The combination of these two steps yields an optimal solution for C that is obtained by applying a nonlinear thresholding to the singular values of X: the singular values smaller than $\sqrt{2/\tau}$ are mapped to zero, while larger singular values are mapped to one. Notice also that as $\tau \to \infty$, we recover the solution for uncorrupted data.

In addition, notice the striking similarity between the two solutions for C in (8.32) and (8.44) under the two noise models in (8.28) and (8.29), respectively. Under the convex model in (8.28), we map the singular values of X above a threshold of $1/\sqrt{\tau}$ to a number below one, while under the nonconvex model in (8.29), we map the singular values of X above a threshold of $\sqrt{2/\tau}$ to one.

Finally, notice that the solution C for the nonconvex model is self-expressive, provided that the clean data matrix is equal to the true noise-free data matrix. Therefore, the question of correctness of clustering reduces to a question of how well the clean data matrix A can be recovered from the noisy data matrix X. We leave this as an open question.

8.2.3 LRSC with Robustness to Corruptions

Let us now consider the case in which the given data might contain some corrupted entries or outliers. In this case, we would like to find a low-rank self-expressive representation that is robust to such corruptions. As in the case of noisy data, we will consider two models for the corruptions.

Convex Error Model
In this case, the self-expressive constraint is written as $X = XC + E$, where C is the low-rank representation, and E is the matrix of corruptions or outliers. Therefore, we can find a low-rank representation C that is robust to corrupted entries or outliers by solving the following convex program:

$$\min_{C,E} \|C\|_* + \lambda\|E\|_{\mathcal{E}} \quad \text{s.t.} \quad X = XC + E, \quad C = C^\top, \tag{8.49}$$

where $\|\cdot\|_{\mathcal{E}}$ is a penalty on the error. In the case of sparse corrupted entries, E is a sparse matrix; hence we can penalize the error with the ℓ_1 norm $\|E\|_1$. In the case of outliers, a few columns of E are corrupted; hence we can penalize this error with the $\ell_{2,1}$ norm $\|E\|_{2,1}$. This latter model was originally proposed in (Liu et al. 2010) under the name low-rank representation (LRR). However, this name is perhaps too generic, since there are many low-rank representations that are not necessarily designed for subspace clustering purposes, e.g., those discussed in Chapter 3. For this reason, we adopt the name LRSC, which is more suggestive of the fact that a low-rank representation is being used for subspace clustering. With this in mind, a more appropriate name for the method in (Liu et al. 2010) might be LRSC with robustness to outliers, or simply robust LRSC.

Now, returning to problem (8.49), notice that this problem is convex, but not strictly convex. Hence, we do not know a priori whether it has a unique solution. In the case of noisy data, where $\|E\|_{\mathcal{E}} = \|E\|_F^2$, we were able to show that the problem admits a unique closed-form solution as in Theorem 8.10. However, in the case of the ℓ_1 or $\ell_{2,1}$ norm, we are not aware of a closed-form solution, and the issue of whether (8.49) has a unique solution remains an open question in the literature.

Now, since the problem is convex, one can find a solution by applying the ADMM algorithm in Appendix A. We leave this as an exercise to the reader.

Nonconvex Error Model
In this case, the self-expressiveness constraint is written as $X = A + E$ and $A = AC$, as in (8.29). Therefore, we can find a low-rank representation C that is robust to corrupted entries or outliers by solving the following convex program:

$$\min_{A,C,E} \|C\|_* + \lambda\|E\|_{\mathcal{E}} \quad \text{s.t.} \quad X = A + E,\ A = AC,\ C^\top = C. \tag{8.50}$$

Following the derivation in the proof of Theorem 8.11, we can use Theorem 8.5 to first minimize with respect to C given A, which gives

$$\operatorname{rank}(A) = \min_{C}\{\|C\|_* : X = A + E,\ A = AC,\ C^\top = C\}. \tag{8.51}$$

Substituting this back into (8.43), we arrive at

$$\min_{A,E} \quad \operatorname{rank}(A) + \lambda\|E\|_{\mathcal{E}} \quad \text{s.t.} \quad X = A + E. \tag{8.52}$$

Notice that this problem is very much related to the RPCA problem discussed in Chapter 3. In particular, when $\|E\|_{\mathcal{E}} = \|E\|_0$, the problem in (8.52) reduces to the problem in (3.58), and when $\|E\|_{\mathcal{E}} = \|E\|_{0,1}$, the problem in (8.52) reduces to the problem in (3.100). However, as discussed in Chapter 3, both problems are NP-hard, and under certain conditions, we can resort to the following relaxation:

$$\min_{A,E} \quad \|A\|_* + \lambda\|E\|_{\mathcal{E}} \quad \text{s.t.} \quad X = A + E, \tag{8.53}$$

where $\|\cdot\|_{\mathcal{E}}$ is either the ℓ_1 or the $\ell_{2,1}$ norm. Therefore, we may use the algorithms discussed in Chapter 3 (e.g., Algorithm 3.8) to solve this problem.

Similar to the case of noisy data, in the above approach, an optimal solution for (A, C) is obtained in two steps:

1. Obtain the clean data matrix A from the noisy data matrix X by applying an RPCA algorithm to X.
2. Obtain the low-rank representation C from the clean data matrix A using the result in Theorem 8.6 for uncorrupted data.

The combination of these two steps is very appealing, since it allows us to solve the nonconvex problem in (8.50) by solving the two convex problems in (8.53) and (8.18), respectively. This is possible under the assumption that the relaxation from (8.52) to (8.53) is valid. This requires the conditions established in Theorem 3.66 or Theorem 3.15 to hold. Intuitively, such conditions require the clean data matrix to be incoherent, the rank of the clean data matrix A to be small enough, and the percentage of corruptions or outliers to be small enough. The main problem is that those conditions were derived under the assumption that the columns of A are drawn from a single low-dimensional subspace, while here the columns of A are drawn from a union of low-dimensional subspaces. Therefore, we do not know whether these conditions are directly applicable. In particular, one case in which these conditions fail occurs when A is of high rank, which can happen even if each of the subspaces is of low rank (e.g., 10 subspaces of dimension 10 in $\mathbb{R}^{100}$ could produce a high-rank matrix A).

In summary, the main advantages of the LRSC algorithm are its computational simplicity and the fact that one can find a low-rank representation in closed form in the case of uncorrupted data or in the case of data corrupted by noise. On the other hand, the main disadvantages of the LRSC algorithm are that it requires the union of subspaces to be of low rank, that it is guaranteed to produce a subspace-preserving representation under a strong assumption of uncorrupted data drawn from independent subspaces, and that while it can be extended to corrupted data, there are very few theoretical results that justify its correctness under those circumstances. In the next section, we present an alternative algorithm that addresses many of these issues by searching for a sparse representation instead of a low-rank representation.

8.3 Sparse Subspace Clustering (SSC)

In this section, we introduce the sparse subspace clustering (SSC) algorithm for clustering a set of points drawn from a union of subspaces using sparse representation techniques. The SSC algorithm addresses the subspace clustering problem in two steps. In the first step, for each data point, we find a few other points that belong to the same subspace. To do so, we solve a global sparse optimization program whose solution encodes information about the memberships of data points to the underlying subspace of each point. In the second step, we use this information in a spectral clustering framework to infer the clustering of the data.

We first introduce SSC in the case of uncorrupted data and derive conditions under which the sparse representation is subspace-preserving. One such condition is that the subspaces be independent, which is the same condition required by LRSC. However, we will show that SSC is able to recover a subspace-preserving representation under much broader conditions, which do not require independence. These conditions require the subspaces to be sufficiently separated and the data to be well distributed inside the subspaces. Moreover, when the subspaces are drawn at random and the data points are drawn at random inside the subspaces, we show that SSC gives a subspace-preserving representation with overwhelming probability. We then extend SSC to data corrupted by noise, outliers, or gross errors, and show that similar theoretical conditions for the correctness of SSC can be derived. Overall, we will see that SSC can be applied under broader conditions than LRSC, but the drawback is that the solutions can no longer be computed in closed form.

8.3.1 SSC with Uncorrupted Data

The SSC algorithm aims to find a subspace-preserving representation of the data such that the coefficient vector $\boldsymbol{c}_j$ associated with each data point $\boldsymbol{x}_j$ is sparse. In other words, SSC tries to solve the following sparse representation problem:

$$\min_{\boldsymbol{c}_j} \ \|\boldsymbol{c}_j\|_0 \quad \text{s.t.} \quad \boldsymbol{x}_j = X\boldsymbol{c}_j, \quad c_{jj} = 0, \quad j = 1, \ldots, N. \tag{8.54}$$

Letting $X = [\boldsymbol{x}_1, \ldots, \boldsymbol{x}_N]$, $C = [\boldsymbol{c}_1, \ldots, \boldsymbol{c}_N]$, and $\|C\|_{0,1} = \sum_j \|\boldsymbol{c}_j\|_0$, we see that these N optimization problems are equivalent to the following problem:

$$\min_{C} \ \|C\|_{0,1} \quad \text{s.t.} \quad X = XC, \quad \operatorname{diag}(C) = \boldsymbol{0}, \tag{8.55}$$

which is a particular case of (8.3) with the regularizer chosen as $\|C\|_C = \|C\|_{0,1}$.

In what follows, we show that a sparse representation exists and can be computed by solving the following relaxed convex optimization problem:

$$\min_C \|C\|_1 \quad \text{s.t.} \quad X = XC, \quad \text{diag}(C) = \mathbf{0}, \tag{8.56}$$

where we replace the number of nonzero entries in C by its ℓ_1 norm. We also derive conditions on the subspaces and the data points under which the solution to the above optimization problem is subspace-preserving.

Sparse Representation in a Union of Subspaces

As discussed in the introduction to this chapter, the main motivation for solving the sparse optimization problem in (8.54) is that every point $\boldsymbol{x} \in S_i$ in a subspace of dimension d_i can be expressed as a linear combination of d_i other points in S_i, i.e., $\boldsymbol{x} = X\boldsymbol{c}$ with $\|\boldsymbol{c}\|_0 = d_i$. Since we assume that d_i is much smaller than the dimension of the ambient space D, the vector $\boldsymbol{c}$ will be sparse if $d_i \ll N_i$, where N_i is the number of data points in subspace i. Now, even under the additional requirement that the number of nonzero entries of $\boldsymbol{c}$ be equal to d_i, we notice that such a subspace-preserving representation is still not unique, because every choice of d_i linearly independent columns of X_i will generate every point in S_i. Nonetheless, since our goal is to cluster the data into multiple subspaces, it does not matter which subset of d_i columns we select as long as the chosen d_i columns belong to S_i. In other words, all that matters for clustering purposes is that the sparse representation $\boldsymbol{c}$ be subspace-preserving.

To find a subspace-preserving representation of a point $\boldsymbol{x} \in \cup_{i=1}^n S_i$, we need to restrict the set of solutions of $\boldsymbol{x} = X\boldsymbol{c}$ by enforcing the desired subspace-preserving pattern on the entries of $\boldsymbol{c}$. However, since we do not know the subspace to which $\boldsymbol{x}$ belongs, we do not know which sparsity pattern to enforce. That is why our approach to finding a subspace-preserving $\boldsymbol{c}$ is to minimize its number of nonzero entries, i.e.,

$$\min_{\boldsymbol{c}} \|\boldsymbol{c}\|_0 \quad \text{s.t.} \quad \boldsymbol{x} = X\boldsymbol{c}. \tag{8.57}$$

This approach makes intuitive sense, because a subspace-preserving solution is also sparse. However, a sparse solution need not be subspace-preserving. To see this, let us consider the data points in Figure 8.1 and suppose we wish to express the point $\boldsymbol{x} \in S_1$ as a sparse linear combination of all the other data points. Clearly, the point can be written as a linear combination of any two points in the plane. Hence, we can construct a solution with two nonzero coefficients. However, notice that we may also write $\boldsymbol{x}$ as a linear combination of one point in S_2 and one point in S_3. In this case, the sparsity level is still two, but the nonzero coefficients correspond to points in different subspaces. This motivates the following question:

> Under what conditions on the subspaces and/or the data is a solution of (8.57) guaranteed to be subspace-preserving?

Now, even if we could guarantee that a sparse solution is always subspace-preserving, the problem of finding a sparse solution to a linear system of equations is in general NP-hard (Amaldi and Kann 1998). Since we are interested in efficiently finding a subspace-preserving representation of $\boldsymbol{x}$ in the dictionary X, we consider minimizing the tightest convex relaxation of the ℓ_0 seminorm, i.e.,

$$\min_{\boldsymbol{c}} \|\boldsymbol{c}\|_1 \quad \text{s.t.} \quad \boldsymbol{x} = X\boldsymbol{c}. \tag{8.58}$$

This problem can be solved efficiently using convex programming tools (Boyd and Vandenberghe 2004; Kim et al. 2007; Boyd et al. 2010) and is known to prefer sparse solutions (Donoho 2006; Candès and Tao 2005; Tibshirani 1996). However, the standard conditions for sparse recovery[1] discussed in Chapter 3 (e.g., incoherence of X) may not apply here, due to the special structure of the dictionary X, whose atoms are drawn from a union of low-dimensional subspaces. Moreover, the standard conditions for sparse recovery assume that the solution for $\boldsymbol{c}$ is unique, which is not the case here, as discussed before. Thus, while we cannot hope to find conditions under which (8.57) and (8.58) are equivalent, we can still ask for conditions under which the ℓ_1 solution is subspace-preserving. This motivates the following question:

> Under what conditions on the subspaces and/or the data is a solution of (8.58) guaranteed to be subspace-preserving?

In the remainder of this section, we study sufficient conditions on the subspaces and/or the data points under which every solution of (8.57) or (8.58) is guaranteed to be subspace-preserving. We investigate such conditions for two classes of subspace arrangements: *independent* (Definition 8.7) and *disjoint* (Definition 8.15). We then extend the results to the case of arbitrary subspace arrangements. Finally, we extend the results to the case of random subspace arrangements, where both the subspaces and the data are drawn uniformly at random.

Subspace-Preserving Property of SSC for a Union of Independent Subspaces

Let us first consider data points that lie in a union of independent subspaces, which is the underlying model of many subspace clustering algorithms. The next theorem shows that in the case of expressive data drawn from independent subspaces, the ℓ_0-minimization program in (8.57), the ℓ_1-minimization program in (8.58), and more generally the ℓ_q-minimization in (8.59) for $q < \infty$ always recover a subspace-preserving representation of the data points.

Theorem 8.12 ((Elhamifar and Vidal 2013)). *Consider a collection of N distinct data points drawn from n independent subspaces $\{S_i\}_{i=1}^n$ of dimensions $\{d_i\}_{i=1}^n$. Let X_i denote N_i data points in S_i, where* $\operatorname{rank}(X_i) = d_i$*, and let X_{-i} denote data points in all subspaces except S_i. Then, for each $q < \infty$, for every S_i, and for every nonzero $\boldsymbol{x}$ in S_i, the ℓ_q-minimization program*

[1] That is, for the equivalence between (8.57) and (8.58).

$$\begin{bmatrix} c_+^* \\ c_-^* \end{bmatrix} = \arg\min \left\| \begin{bmatrix} c_+ \\ c_- \end{bmatrix} \right\|_q \quad \text{s.t.} \quad \boldsymbol{x} = [X_i \ X_{-i}] \begin{bmatrix} c_+ \\ c_- \end{bmatrix}, \tag{8.59}$$

recovers a subspace-preserving representation, i.e., $\boldsymbol{c}_+^* \neq \boldsymbol{0}$ *and* $\boldsymbol{c}_-^* = \boldsymbol{0}$.

Proof. It follows from (8.59) that

$$\boldsymbol{x} - X_i \boldsymbol{c}_+^* = X_{-i} \boldsymbol{c}_-^*. \tag{8.60}$$

Notice that the left-hand side of this equation corresponds to a point in the subspace S_i, while the right-hand side corresponds to a point in the sum of the other subspaces $\bigoplus_{j \neq i} S_j$. Since the subspaces $\{S_i\}_{i=1}^n$ are assumed to be independent, so are the two subspaces S_i and $\bigoplus_{j \neq i} S_j$; hence they intersect only at the origin. Therefore, $\boldsymbol{x} = X_i \boldsymbol{c}_+^*$ and $X_{-i} \boldsymbol{c}_-^* = \boldsymbol{0}$, and so $\left[\boldsymbol{c}_+^{*\top} \ \boldsymbol{0}^\top\right]^\top$ is a feasible solution of the optimization problem in (8.59). For the sake of contradiction assume that $\boldsymbol{c}_-^* \neq \boldsymbol{0}$. We have

$$\left\| \begin{bmatrix} \boldsymbol{c}_+^* \\ \boldsymbol{0} \end{bmatrix} \right\|_q < \left\| \begin{bmatrix} \boldsymbol{c}_+^* \\ \boldsymbol{c}_-^* \end{bmatrix} \right\|_q, \tag{8.61}$$

for all $q < \infty$, which contradicts the optimality of $\left[\boldsymbol{c}_+^{*\top} \ \boldsymbol{c}_-^{*\top}\right]^\top$. Therefore, we must have $\boldsymbol{c}_+^* \neq \boldsymbol{0}$ and $\boldsymbol{c}_-^* = \boldsymbol{0}$, whereby we obtain the desired result. □

Interestingly, the ℓ_q-solution is subspace-preserving for every $q < \infty$, not just for $q = 0$ or $q = 1$. Moreover, the ℓ_q-solution is subspace-preserving without any assumption on the distribution of the data points within each subspace, other than $\operatorname{rank}(X_l) = d_l$. However, this comes at the price of having a more restrictive model for the union of subspaces, which must be composed of independent subspaces. Now, while the ℓ_q-solution is subspace-preserving for every $q < \infty$, notice that not all norms will produce a solution where the number of nonzero coefficients is equal to the dimension of the subspace to which the point $\boldsymbol{x}$ belongs. For example, $q = 2$ promotes a solution that typically has many nonzero elements corresponding to a minimum-energy representation of the given vector. Overall, by decreasing the value of q from infinity toward zero, we expect the sparsity of the solution to increase, as illustrated in Figure 8.4. The extreme case of $q = 0$ corresponds to finding the sparsest representation of the given point, since the ℓ_0 seminorm counts the number of nonzero elements of the solution.

The following two lemmas show that for a generic point $\boldsymbol{x} \in \cup_{i=1}^n S_i$, the solutions of the ℓ_0 and ℓ_1 minimization problems are each such that the number of nonzero coefficients is less than or equal to the dimension of the subspace to which the point $\boldsymbol{x}$ belongs. That is, if $\boldsymbol{x} \in S_i$, then the solutions are d_i-sparse.

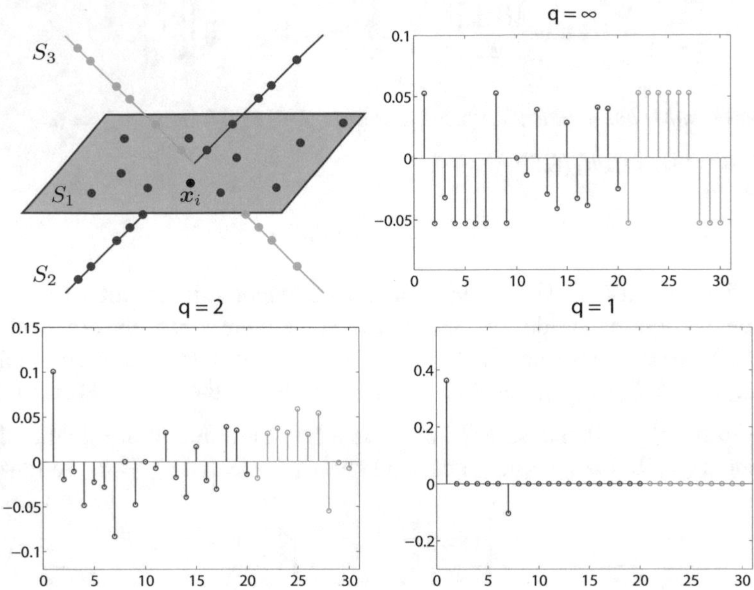

Fig. 8.4 Three subspaces in $\mathbb{R}^3$ with 10 data points in each subspace, ordered such that the fist and the last 10 points belong to S_1 and S_3, respectively. The solution of the ℓ_q-minimization program $\min_{c_i} \|c_i\|_q$ s.t. $x_i = Xc_i$ and $c_{ii} = 0$, where x_i lies in S_1 and $q = 1, 2, \infty$, is shown. Note that as the value of q decreases, the sparsity of the solution increases. For $q = 1$, the solution corresponds to choosing two other points lying in S_1.

Lemma 8.13. *Let $x \in S_i$, where $\dim(S_i) = d_i$. Under the assumptions of Theorem 8.12, every solution of the ℓ_0-minimization program in* (8.57) *is d_i-sparse.*

Proof. Since $x \in S_i$, there exist $c_i \in \mathbb{R}^{d_i}$ such that $x = X_i c_i$, where X_i is a rank-d_i matrix whose d_i columns are data points in S_i. Let the columns of X_{-i} denote the remaining data points and rewrite the ℓ_0-minimization program in (8.57) as

$$\min \left\| \begin{bmatrix} c_+ \\ c_- \end{bmatrix} \right\|_0 \quad \text{s.t.} \quad x = [X_i \ X_{-i}] \begin{bmatrix} c_+ \\ c_- \end{bmatrix}. \tag{8.62}$$

Since $\begin{bmatrix} c_i^\top & \mathbf{0}^\top \end{bmatrix}^\top$ is a feasible solution to the above optimization program with at most d_i nonzero entries, we conclude that every optimal solution to the ℓ_0-minimization program in (8.57) must be d_i-sparse. □

Lemma 8.14. *Let $x \in S_i$, where $\dim(S_i) = d_i$. Under the assumptions of Theorem 8.12, every solution of the ℓ_1-minimization program in* (8.58) *is d_i-sparse.*

Proof. Recall the relationship between the ℓ_1-norm of the optimal solution of (8.58) and the symmetrized convex hull of the columns of X (Donoho 2005):

$$\mathcal{P}(X) \triangleq \operatorname{conv}(\pm x_1, \pm x_2, \cdots, \pm x_N). \tag{8.63}$$

More precisely, the ℓ_1-norm of the optimal solution of (8.58) corresponds to the smallest $\alpha > 0$ such that the scaled polytope $\alpha\mathcal{P}$ reaches $\boldsymbol{x}$ (Donoho 2005). Since each face of this polytope corresponds to the convex hull of at most d_i data points (up to sign), the optimal $\boldsymbol{c}$ will have at most d_i nonzero entries. □

The preceding two lemmas guarantee that the solution to either the ℓ_0 or ℓ_1 optimization problem will have at most d_i nonzero entries. However, they do not guarantee that it will have exactly d_i nonzero entries. While it is possible for $\boldsymbol{x}$ to be generated by fewer than d_i data points, e.g., if $\boldsymbol{x}$ is one of the data points or lies in one of the faces of the polytope of dimension less than d_i, such cases represent a zero-measure set in $\cup_{i=1}^n S_i$. Thus, the generic situation will be that the number of nonzero coefficients will be equal to d_i. Of course, we prefer to use the convex optimization problem, which can be solved more efficiently.

In summary, we have shown that when the subspaces are independent and the data set is expressive, the solution to the ℓ_1-minimization program in (8.58) for a generic point $\boldsymbol{x} \in \cup_{i=1}^n S_i$ is such that the number of nonzero coefficients is equal to the dimension of the subspace to which $\boldsymbol{x}$ belongs, and the nonzero coefficients correspond to data points in the same subspace as $\boldsymbol{x}$. This theoretical result will be the basis for the SSC algorithm, to be discussed at the end of this subsection.

Subspace-Preserving Property of SSC for a Union of Disjoint Subspaces
We consider now the more general class of disjoint subspaces and investigate conditions under which the optimization program in (8.58) recovers a subspace-preserving representation of each data point.

Definition 8.15. A collection of subspaces $\{S_i\}_{i=1}^n$ is said to be *disjoint* if every pair of subspaces intersect only at the origin. In other words, for every pair of subspaces, we have $\dim(S_i \oplus S_j) = \dim(S_i) + \dim(S_j)$.

Example 8.16 As an example, both subspace arrangements shown in Figure 8.2 are disjoint, since each pair of subspaces intersect at the origin.

Notice that based on the above definitions, the notion of disjointness is weaker than the notion of independence, because an independent subspace model is also disjoint, but the converse is not necessarily true. As a simple example, note that no $n \geq 3$ distinct lines in $\mathbb{R}^2$ are independent because the dimension of their sum is 2, which is smaller than the sum of their dimensions n. However, they are disjoint, because every pair of distinct lines intersect only at the origin.[2]

In what follows, we show that when the smallest angle between any two subspaces is greater than a bound determined by the distribution of the data points across all the subspaces, a subspace-preserving representation can be found using ℓ_1 minimization. The smallest principal angle is defined as follows (see Definition 7.1 for the definition of all principal angles).

[2]Notice that both independent and disjoint subspace arrangements are transversal, according to the definition in Appendix C.

Definition 8.17. *The* smallest principal angle *between two subspaces S_i and S_j, denoted by θ_{ij}, is defined as*

$$\cos(\theta_{ij}) \triangleq \max_{\boldsymbol{v}\in S_i} \max_{\boldsymbol{z}\in S_j}\{\langle \boldsymbol{v}, \boldsymbol{z}\rangle : \|\boldsymbol{v}\|_2 = \|\boldsymbol{z}\|_2 = 1\}. \tag{8.64}$$

Note that two disjoint subspaces intersect only at the origin; hence their smallest principal angle is greater than zero, and $\cos(\theta_{ij}) \in [0, 1)$.

The next theorem shows that in the case of expressive data drawn from disjoint subspaces, the ℓ_1-minimization program in (8.58) always recovers a subspace-preserving representation of the data points.

Theorem 8.18 (SSC for Disjoint Subspaces (Elhamifar and Vidal 2013)). *Consider a collection of data points drawn from n disjoint subspaces $\{S_i\}_{i=1}^n$ of dimensions $\{d_i\}_{i=1}^n$. Let $\mathbb{W}_i$ be the set of all full-rank submatrices $\tilde{X}_i \in \mathbb{R}^{D\times d_i}$ of X_i, where* $\operatorname{rank}(X_i) = d_i$. *If the condition*

$$\sqrt{d_i}\max_{j\neq i} \cos(\theta_{ij}) < \frac{1}{\|X_{-i}\|_{\infty,2}} \max_{\tilde{X}_i\in\mathbb{W}_i} \sigma_{d_i}(\tilde{X}_i) \tag{8.65}$$

holds, then for every nonzero $\boldsymbol{x}$ in S_i, the ℓ_1-minimization program

$$\begin{bmatrix} \boldsymbol{c}_+^* \\ \boldsymbol{c}_-^* \end{bmatrix} = \arg\min \left\| \begin{bmatrix} \boldsymbol{c}_+ \\ \boldsymbol{c}_- \end{bmatrix} \right\|_1 \quad s.t.\ \boldsymbol{x} = \begin{bmatrix} X_i\ X_{-i} \end{bmatrix} \begin{bmatrix} \boldsymbol{c}_+ \\ \boldsymbol{c}_- \end{bmatrix}, \tag{8.66}$$

recovers a subspace-preserving solution, i.e., $\boldsymbol{c}_+^* \neq \boldsymbol{0}$ and $\boldsymbol{c}_-^* = \boldsymbol{0}$.[3]

The condition in (8.65) has a very intuitive interpretation.

- The left-hand side depends exclusively on the geometry of the subspaces and their relative arrangements and not on the data points sampled from the subspaces. One factor is the dimension of the subspaces, which is constrained to be sufficiently small. The other factor is the cosine of the smallest principal angle between S_i and any other subspace, which is also constrained to be small enough; hence the smallest angle between subspaces should be large enough.
- The right-hand side depends exclusively on the distribution of the data points sampled from the subspaces and not on the geometry of the subspaces. The numerator depends on the distribution of the data in S_i. Specifically, it depends on $\sigma_{d_i}(\tilde{X}_i)$, which is the d_ith singular value of $\tilde{X}_i$. This number is zero if the data are degenerate, i.e., $\operatorname{rank}(X_i) < d_i$. Now, since $\tilde{X}_i$ is a full-rank submatrix of X_i and the maximum over all possible choices is taken, we simply need d_i data points in S_i to be sufficiently well distributed so that $\sigma_{d_i}(\tilde{X}_i)$ is large enough. The denominator depends on the distribution of the data points in other subspaces,

[3] $\|Y_{-i}\|_{\infty,2}$ is the maximum ℓ_2-norm of the columns of Y_{-i}.

especially their ℓ_2 norm. Thus, the value of the right-hand side can be rather high when the norms of the data points are oddly distributed, e.g., when the maximum norm of data points in S_i is much smaller than the maximum norm of data points in all other subspaces. Since the segmentation of data in a union of linear subspaces does not change when data points are scaled, we can apply the ℓ_1 minimization program after normalizing the data points to have unit Euclidean norms. In this case, the sufficient condition in (8.65) reduces to

$$\sqrt{d_i} \max_{j \neq i} \cos(\theta_{ij}) < \max_{\tilde{X}_i \in \mathbb{W}_i} \sigma_{d_i}(\tilde{X}_i), \tag{8.67}$$

and the right-hand side measures how well distributed the data are inside S_i.

In summary, the condition in (8.65) requires the subspaces to be sufficiently well separated and the data to be sufficiently well distributed inside the subspaces. We illustrate this result with the following example.

Example 8.19. Consider the example of three subspaces in $\mathbb{R}^3$, the plane and two lines shown in Figure 8.5, where the data points in the subspaces are normalized to have unit Euclidean norm. Assume that $\boldsymbol{x} \in S_1$ also lives in the sum of S_2 and S_3. Note that $\boldsymbol{x}$ can be written as a linear combination of $\boldsymbol{x}_1$ and $\boldsymbol{x}_2$ as

$$\boldsymbol{x} = c_1\boldsymbol{x}_1 + c_2\boldsymbol{x}_2, \quad c_1 = \frac{\sin(\nu - \nu')}{\sin(\nu)}, \; c_2 = \frac{\sin(\nu')}{\sin(\nu)}. \tag{8.68}$$

When ν becomes small, hence $\boldsymbol{x}_1$ and $\boldsymbol{x}_2$ get close to a line in S_1, and $\boldsymbol{x}$ is along the direction orthogonal to $\boldsymbol{x}_1$ and $\boldsymbol{x}_2$, for instance, $\nu' = \pi/2 + \nu$, then we have

$$c_1 = \frac{-1}{\sin(\nu)}, \; c_2 \approx \frac{1}{\sin(\nu)} \implies |c_1| + |c_2| \approx \frac{2}{\sin(\nu)}. \tag{8.69}$$

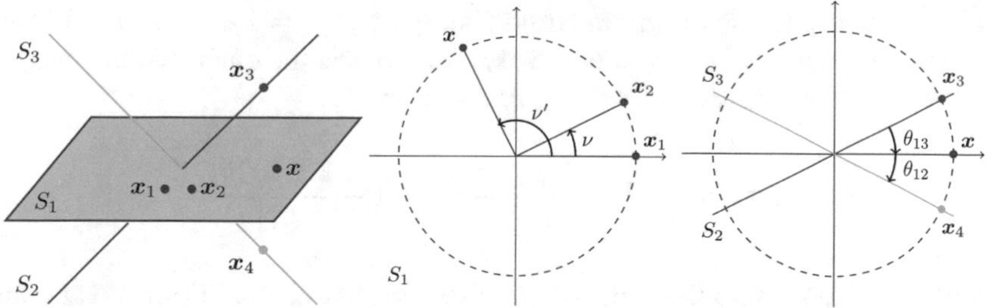

Fig. 8.5 Left: Three subspaces in $\mathbb{R}^3$. Middle: The ℓ_1 norm of reconstructing $\boldsymbol{x}$ from $\{\boldsymbol{x}_1, \boldsymbol{x}_2\} \in S_1$ increases as $\boldsymbol{x}_1$ and $\boldsymbol{x}_2$ get close together and $\boldsymbol{x}$ gets close to the orthogonal direction to $\boldsymbol{x}_1$ and $\boldsymbol{x}_2$. Right: The ℓ_1 norm of reconstructing $\boldsymbol{x}$ from $\{\boldsymbol{x}_3, \boldsymbol{x}_4\}$ decreases as the smallest principal angle between S_1 and the other two subspaces, S_2 and S_3, increases.

It is easy to verify that for $\tilde{X}_1 = [\boldsymbol{x}_1 \ \ \boldsymbol{x}_2]$, we have $\sigma_2(\tilde{X}_1) = \sin(\nu)$. Thus, when ν decreases, i.e., when the distribution of the points in S_1 becomes nearly degenerate, $\tilde{X}_1$ decreases, and the ℓ_1 norm of the reconstruction of $\boldsymbol{x}$ in terms of points from the same subspace S_1 increases. On the other hand, $\boldsymbol{x}$ can also be written as a linear combination of $\boldsymbol{x}_3$ and $\boldsymbol{x}_4$ from S_2 and S_3, respectively, as

$$\boldsymbol{x} = c_3\boldsymbol{x}_3 + c_4\boldsymbol{x}_4, \ c_3 = \frac{\sin(\theta_{13})}{\sin(\theta_{12} + \theta_{13})}, \ c_4 = \frac{\sin(\theta_{12})}{\sin(\theta_{12} + \theta_{13})}, \tag{8.70}$$

where θ_{12} and θ_{13} indicate the principal angles for (S_1, S_2) and (S_1, S_3), respectively. Assuming $\theta_{12} = \theta_{13} = \theta$, we have

$$c_3 = c_4 = \frac{1}{2\cos(\theta)} \implies |c_3| + |c_4| = \frac{1}{\cos(\theta)}, \tag{8.71}$$

which decreases to 1 as θ decreases to 0.

Example 8.20. In this example, we present simulation results that illustrate the theoretical conditions under which a sparse representation is subspace-preserving. We consider three disjoint subspaces $\{S_i\}_{i=1}^3$ of equal dimension $d = 4$ embedded in an ambient space of dimension $D = 50$. To ensure that the subspaces are not independent, we generate subspace bases $\{U_i \in \mathbb{R}^{D\times d}\}_{i=1}^3$ such that each subspace lies in the sum of the other two subspaces, i.e., $\operatorname{rank}([U_1 \ \ U_2 \ \ U_3]) = 2d$, and such that the smallest principal angles θ_{12} and θ_{23} are equal to $\theta \in [6, 60]$. Thus, we can verify the effect of the smallest principal angle in the subspace-preserving property by changing the value of θ. To investigate the effect of the data distribution on the subspace-preserving property, we randomly generate the same number of data points $N_g \in [d + 1, 32d]$ in each subspace. Typically, as the number of data points in a subspace increases, the probability of the data being close to a degenerate subspace decreases.[4] For each pair (θ, N_g), we generate a set of points on the three d-dimensional subspaces. For each point, we solve the ℓ_1-minimization program in (8.80). Let w_{jk} denote the ground truth affinity, i.e., $w_{jk} = 1$ if points i and j are in the same subspace, and $w_{jk} = 0$ otherwise. We evaluate the quality of the matrix C by measuring the *subspace-preserving property error*

$$\text{spp error} = 1 - \frac{\sum_{jk} w_{jk}|c_{jk}|}{\sum_{jk} |c_{jk}|} = 1 - \frac{\|W \circ C\|_1}{\|C\|_1}, \tag{8.72}$$

where each term inside the sum indicates the fraction of the ℓ_1-norm that comes from points in the correct subspace. The subspace-preserving property error is zero when $\boldsymbol{x}_j$ chooses points in its own subspace only, while the error is equal to one

[4] To remove the effect of different scalings of data points, i.e., to consider only the effect of the principal angle and number of points, we normalize the data points.

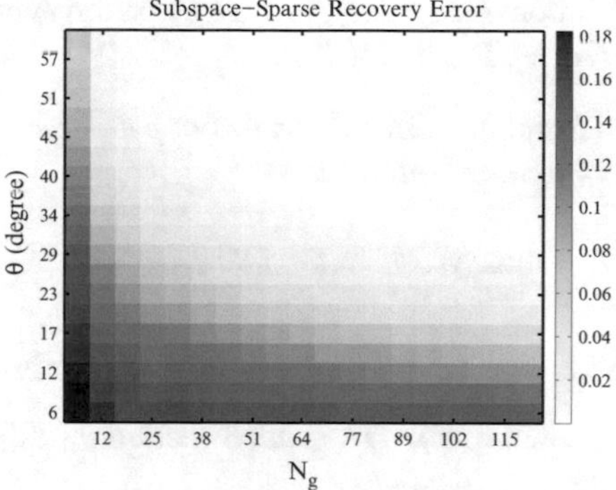

Fig. 8.6 Subspace-preserving property error for three disjoint subspaces. Increasing the number of points or smallest principal angle decreases the errors.

when $\boldsymbol{x}_j$ chooses points from other subspaces. The average results for 100 trials are shown in Figure 8.6. Note that when either θ or N_g is small, the error is large, as predicted by the theoretical analysis. On the other hand, as θ or N_g increases, the error decreases, and for (θ, N_g) sufficiently large, we obtain zero error.

Subspace-Preserving Property of SSC for a Union of Arbitrary Subspaces
We will now consider a more general class of subspaces, which need not be independent or disjoint, and investigate conditions under which the optimization program in (8.58) recovers a subspace-preserving representation of each data point. Interestingly, the conditions we will present are weaker than the one in (8.67), which guarantees that we can recover a subspace-preserving representation for *any* point in a union of *disjoint* subspaces. In practice, since our goal is to cluster a given set of data points, we need to guarantee a subspace-preserving representation only for the given data points. Therefore, rather than computing the smallest principal angle from the dot products between an arbitrary point in one subspace and an arbitrary point in other subspaces as in (8.64), we can restrict our attention to a few dot products between the so-called *dual directions* of a data point in one subspace and the data points in other subspaces. Before defining dual directions and deriving the new conditions, we will need some additional notation.

As before, let $X = [\boldsymbol{x}_1, \ldots, \boldsymbol{x}_N] = [X^1, \ldots, X^n]\Gamma^\top \in \mathbb{R}^{D\times N}$ denote the data matrix, where $X^i \in \mathbb{R}^{D\times N_i}$ denotes the data in the ith subspace,[5] and let X^i_{-j} be the matrix X^i with the jth column removed. We will assume that each data point is normalized to be of unit norm, so that $\|\boldsymbol{x}_j\|_2 = 1$ for all $j = 1, \ldots, N$. Let $X^i = U^i Y^i$ be a factorization of X^i such that $U^i \in \mathbb{R}^{D\times d_i}$ is an orthonormal basis of the subspace, while $Y^i = [\boldsymbol{y}^i_1, \ldots, \boldsymbol{y}^i_{N_i}] \in \mathbb{R}^{d_i\times N_i}$ is a matrix of coordinates whose columns are of

[5]So far, we have used subscripts to indicate both data points and subspaces. In this part, to avoid confusion between the indices for data points and the indices for subspaces, we will use subscripts to indicate data points and superscripts to indicate subspaces.

unit norm. Let $\mathcal{P}(X) = \text{conv}(\pm\boldsymbol{x}_1, \pm\boldsymbol{x}_2, \ldots, \pm\boldsymbol{x}_N)$ be the symmetrized convex hull of the columns of X, and let $\mathcal{P}^i_{-j} = \mathcal{P}(X^i_{-j})$.

Definition 8.21 (Dual Points). *Consider a vector $\boldsymbol{y} \in \mathbb{R}^d$ and a matrix $Y \in \mathbb{R}^{d\times N}$, and let Λ^* be the set of optimal solutions to*

$$\max_{\boldsymbol{\lambda}\in\mathbb{R}^d} \langle \boldsymbol{y}, \boldsymbol{\lambda}\rangle \quad s.t. \quad \|Y^\top \boldsymbol{\lambda}\|_\infty \leq 1. \tag{8.73}$$

The dual point $\boldsymbol{\lambda}(\boldsymbol{y}, Y)$ is defined as a point in Λ^* with minimum Euclidean norm.

Definition 8.22 (Dual Directions). *The dual direction $\boldsymbol{v}^i_j \in S_i$ corresponding to the dual point $\boldsymbol{\lambda}^i_j = \boldsymbol{\lambda}(\boldsymbol{y}^i_j, Y^i_{-j})$ is defined as*

$$\boldsymbol{v}^i_j = U^i \frac{\boldsymbol{\lambda}^i_j}{\|\boldsymbol{\lambda}^i_j\|_2}, \quad i = 1, \ldots, n, \quad j = 1, \ldots, N_i. \tag{8.74}$$

Notice that the dual point $\boldsymbol{\lambda}^i_j \in \mathbb{R}^d$ depends on $\boldsymbol{y}^i_j \in \mathbb{R}^d$, which in turn depends on $\boldsymbol{x}^i_j \in S_i$. Hence, the dual direction $\boldsymbol{v}^i_j \in S_i$ effectively depends on $\boldsymbol{x}^i_j$. Thus, we will also refer to $\boldsymbol{v}^i_j$ as the dual direction corresponding to the point $\boldsymbol{x}^i_j \in S_i$.

Definition 8.23 (Subspace Incoherence). *The subspace incoherence of the points in the ith subspace with respect to the points in all other subspaces is defined by*

$$\mu_i = \max_{k:\boldsymbol{x}_k\notin S_i} \|V^{i\top}\boldsymbol{x}_k\|_\infty = \max_{j:\boldsymbol{x}_j\in S_i} \max_{k:\boldsymbol{x}_k\notin S_i} |\langle \boldsymbol{v}^i_j, \boldsymbol{x}_k\rangle|, \tag{8.75}$$

where $V^i = [\boldsymbol{v}^i_1, \ldots, \boldsymbol{v}^i_{N_i}] \in \mathbb{R}^{D\times N_i}$ is the matrix of dual directions.

Observe that the optimization problem in (8.73) is the dual problem to $\min_{\boldsymbol{c}} \|\boldsymbol{c}\|_1$ s.t. $\boldsymbol{y} = Y\boldsymbol{c}$ (see Exercise 8.4). Observe also that if both $\boldsymbol{x}$ and the columns of X belong to a subspace of dimension d, then there exist $\boldsymbol{y}$ and Y such that $\boldsymbol{x} = U\boldsymbol{y}$ and $X = UY$, where U is an orthonormal basis for the subspace. Then $U\boldsymbol{\lambda}(\boldsymbol{y}, Y)$ is equal to the dual point $\boldsymbol{\lambda}(\boldsymbol{x}, X)$ associated to the primal problem $\min_{\boldsymbol{c}} \|\boldsymbol{c}\|_1$ s.t. $\boldsymbol{x} = X\boldsymbol{c}$. Therefore, the dual direction $\boldsymbol{v}^i_j$ is simply an optimal solution to the dual of the problem $\min_{\boldsymbol{c}} \|\boldsymbol{c}\|_1$ s.t. $\boldsymbol{x}^i_j = X^i_{-j}\boldsymbol{c}$, which expresses the jth point in the ith subspace as a linear combination of the remaining points in the same subspace.

Definition 8.24 (Inradius). *The inradius of a convex body $\mathcal{P}$, denoted by $r(\mathcal{P})$, is defined as the radius of the largest Euclidean ball inscribed in $\mathcal{P}$.*

Theorem 8.25 (SSC for Arbitrary Subspaces (Soltanolkotabi and Candès 2013)). *Consider a collection of data points $\{\boldsymbol{x}_j\}_{j=1}^N$ drawn from n subspaces $\{S_i\}_{i=1}^n$ of dimensions $\{d_i\}_{i=1}^n$. Assume that each data point is normalized so that $\|\boldsymbol{x}_j\|_2 = 1$. If the condition*

$$\max_{j:x_j \in S_i} \max_{k:x_k \notin S_i} |\langle \boldsymbol{v}_j^i, \boldsymbol{x}_k \rangle| < \min_{j:x_j \in S_i} r(\mathcal{P}_{-j}^i) \tag{8.76}$$

holds, then the ℓ_1*-minimization program in* (8.56) *recovers a subspace-preserving representation* C.

Notice that the condition in (8.76) is weaker than that in (8.67), because

$$\begin{aligned} \max_{j:x_j \in S_i} \max_{k:x_k \notin S_i} |\langle \boldsymbol{v}_j^i, \boldsymbol{x}_k \rangle| &\leq \max_{\boldsymbol{v} \in S_i} \max_{j \neq i} \max_{\boldsymbol{z} \in S_j} \{\langle \boldsymbol{v}, \boldsymbol{z} \rangle : \|\boldsymbol{v}\|_2 = \|\boldsymbol{z}\|_2 = 1\} \\ &= \max_{j \neq i} \cos(\theta_{ij}), \end{aligned} \tag{8.77}$$

and

$$\frac{1}{\sqrt{d_i}} \max_{\tilde{X}_i \in \mathbb{W}_i} \sigma_{d_i}(\tilde{X}_i) < \min_{j:x_j \in S_i} r(\mathcal{P}_{-j}^i), \tag{8.78}$$

where the last inequality follows from Exercise 8.6.

Subspace-Preserving Property of SSC for a Union of Random Subspaces
So far, we have considered models in which the subspaces are fixed and the data points inside the subspaces are fixed. As it turns out, if we allow the subspaces to be drawn uniformly at random and the data points to be drawn uniformly at random inside the subspaces, the ℓ_1-minimization program succeeds in finding a subspace-preserving representation as long as the dimensions of the subspaces are within at most a logarithmic factor from the ambient dimension. More specifically, we have the following result.

Theorem 8.26 (SSC for Random Subspaces (Soltanolkotabi and Candès 2013))**.** *Assume that there are n subspaces, each of dimension d, chosen independently and uniformly at random. Furthermore, suppose there are $\rho d + 1$ points chosen independently and uniformly at random on each subspace, so that the total number of points is $N = n(\rho d + 1)$. Assume further that the points are normalized to be of unit norm. Then there exists a constant $c(\rho) > 1$ such that the solution of* (8.58) *is subspace-preserving with probability at least $1 - 2/N - Ne^{-\sqrt{\rho d}}$ if*

$$d < \frac{c^2(\rho)\log(\rho)}{12\log(N)} D. \tag{8.79}$$

Solution to the Sparse Minimization Problem
So far, we have discussed theoretical conditions under which a sparse representation is subspace-preserving. Here, we concentrate on the problem of finding such a sparse representation and using it to cluster the data.

The first step of the SSC algorithm is to express each data point as a sparse linear combination of all other data points. More specifically, for each data point $\boldsymbol{x}_j$, the following ℓ_1-minimization program is solved:

$$\min_{\boldsymbol{c}_j} \|\boldsymbol{c}_j\|_1 \quad \text{s.t.} \quad \boldsymbol{x}_j = X\boldsymbol{c}_j \quad \text{and} \quad c_{jj} = 0. \tag{8.80}$$

We first notice that this program can be reduced to that in (8.58) by removing the jth column of X and the jth entry of $\boldsymbol{c}_j$ from the optimization. Thus, we can find the sparse representations of all data points by solving a standard ℓ_1-minimization program for each point independently. However, notice that such an optimization problem is convex, but not smooth, due to the ℓ_1 norm. Therefore, conventional interior-point or subgradient methods are relatively slow. This has motivated an increasing interest in efficiently solving ℓ_1 minimization programs, which has led to numerous efficient algorithms, such as homotopy continuation, proximal-gradient, gradient projection, iterative thresholding and shrinkage. We refer the reader to (Boyd and Vandenberghe 2004; Kim et al. 2007; Boyd et al. 2010) for details.

While efficient methods for solving an ℓ_1-minimization program exist, such methods may not be the most appropriate for solving the N programs in (8.80). This is because the N programs use almost the same dictionary, and hence they share some common structure. This inspired the work of (Elhamifar and Vidal 2013), which proposes to solve all N optimization programs simultaneously. More specifically, let $C \triangleq \left[\boldsymbol{c}_1, \ldots, \boldsymbol{c}_N\right] \in \mathbb{R}^{N\times N}$ be the matrix whose jth column $\boldsymbol{c}_j$ corresponds to the sparse representation of $\boldsymbol{x}_j$, and let $\operatorname{diag}(C) \in \mathbb{R}^N$ be the vector containing the diagonal elements of C. We can rewrite the optimization program in (8.80) for all data points $j = 1, \ldots, N$ in matrix form as in (8.56), i.e.,

$$\min_C \|C\|_1 \quad \text{s.t.} \quad X = XC, \;\; \operatorname{diag}(C) = \mathbf{0}, \tag{8.81}$$

where $\min \|C\|_1 = \sum_{j=1}^N \|\boldsymbol{c}_j\|_1 = \sum_{i,j=1}^N |c_{ij}|$ is the ℓ_1-norm of C. While many convex optimization algorithms can be used to solve this problem, we use the *alternating direction method of multipliers* (ADMM), described in Appendix A.

We begin by introducing an auxiliary matrix variable Z as follows:

$$\min_{C,Z} \|C\|_1 \quad \text{s.t.} \quad X = XZ, \;\; Z = C - \operatorname{diag}(C). \tag{8.82}$$

The augmented Lagrangian of the above optimization problem is given by

$$\begin{aligned} \mathscr{L}(C, Z, \Lambda_1, \Lambda_2) &= \|C\|_1 + \langle \Lambda_1, X - XZ \rangle + \langle \Lambda_2, Z - (C - \operatorname{diag}(C)) \rangle \\ &\quad + \frac{\mu_1}{2}\|X - XZ\|_F^2 + \frac{\mu_2}{2}\|Z - (C - \operatorname{diag}(C))\|_F^2 \qquad (8.83) \\ &= \|C\|_1 + \frac{\mu_1}{2}\left\|X - XZ + \frac{\Lambda_1}{\mu_1}\right\|_F^2 + \frac{\mu_2}{2}\left\|Z - (C - \operatorname{diag}(C)) + \frac{\Lambda_2}{\mu_2}\right\|_F^2 + \eta, \end{aligned}$$

where $\Lambda_1 \in \mathbb{R}^{D\times N}$ and $\Lambda_2 \in \mathbb{R}^{N\times N}$ are matrices of Lagrange multipliers, $\mu_1, \mu_2 > 0$ are parameters, and η is a term that does not depend on C or Z.

Algorithm 8.2 (Matrix ℓ_1 Minimization by ADMM)

Input: Data matrix X

initialize: $C^0 = \mathbf{0}$, $\Lambda_1^0 = \mathbf{0}$, $\Lambda_2^0 = \mathbf{0}$, $\mu_1 > 0$, $\mu_2 > 0$.
while not converged **do**
  compute $Z^{k+1} = (\mu_1 X^\top X + \mu_2 I)^{-1}\big(\mu_1 X^\top(X + \frac{\Lambda_1^k}{\mu_1}) + \mu_2(C^k - \frac{\Lambda_2^k}{\mu_2})\big)$;
  compute $C^{k+1} = \mathcal{S}_{\frac{1}{\mu_2}}\big(Z^{k+1} + \frac{\Lambda_2^k}{\mu_2}\big)$;
  compute $C^{k+1} = C^{k+1} - \operatorname{diag}(C^{k+1})$;
  compute $\Lambda_1^{k+1} = \Lambda_1^k + \mu_1(X - XZ^{k+1})$;
  compute $\Lambda_2^{k+1} = \Lambda_2^k + \mu_2(Z^{k+1} - C^{k+1})$;
end while

Output: Sparse representation C.

A generic Lagrange multiplier algorithm (Bertsekas 1999) would solve the problem $\max_{\Lambda_1,\Lambda_2}\min_{C,Z}\mathcal{L}(C,Z,\Lambda_1,\Lambda_2)$ by iterating the following two steps:

$$
\begin{aligned}
(C^k, Z^k) &= \arg\min_{C,Z} \mathcal{L}\big(C, Z, \Lambda_1^{k-1}, \Lambda_2^{k-1}\big),\\
(\Lambda_1^k, \Lambda_2^k) &= (\Lambda_1^{k-1}, \Lambda_2^{k-1}) + \big(\mu_1(X - XZ^k), \mu_2(Z^k - C^k)\big).
\end{aligned}
\tag{8.84}
$$

For our problem, we can avoid having to solve a sequence of convex programs by recognizing that $\min_C \mathcal{L}(C,Z,\Lambda_1,\Lambda_2)$ and $\min_Z \mathcal{L}(C,Z,\Lambda_1,\Lambda_2)$ both have very simple and efficient solutions. Thus, a more practical strategy is to first minimize $\mathcal{L}$ with respect to Z (fixing C), then minimize $\mathcal{L}$ with respect to C (fixing Z), and then finally update the Lagrange multiplier matrices Λ_1 and Λ_2 based on the residuals $X - XZ$ and $Z - C$, a strategy that is summarized as Algorithm 8.2 below. The optimal Z given C, Λ_1, and Λ_2 is given by

$$
Z = (\mu_1 X^\top X + \mu_2 I)^{-1}\Big(\mu_1 X^\top\big(X + \frac{\Lambda_1}{\mu_1}\big) + \mu_2\big(C - \frac{\Lambda_2}{\mu_2}\big)\Big). \tag{8.85}
$$

To compute the optimal $C = [c_{ij}]$ given $Z = [z_{ij}]$ and $\Lambda_2 = [\lambda_{ij}]$, we need to minimize

$$
\sum_i |c_{ii}| + \sum_{i \neq j} |c_{ij}| + \frac{1}{\mu_2}\big(c_{ij} - (z_{ij} + \frac{\lambda_{ij}}{\mu_2})\big)^2, \tag{8.86}
$$

which gives

$$
c_{ij} = \begin{cases} 0 & \text{if } \ i = j, \\ \mathcal{S}_{\frac{1}{\mu_2}}\big(z_{ij} + \frac{\lambda_{ij}}{\mu_2}\big) & \text{otherwise,} \end{cases} \tag{8.87}
$$

where $\mathcal{S}_\varepsilon$ is the shrinkage-thresholding operator defined in (2.96). The above solution for C can be written in matrix form as

$$C = \tilde{C} - \mathrm{diag}(\tilde{C}), \quad \text{where} \quad \tilde{C} = \mathcal{S}_{\frac{1}{\mu_2}}\Big(Z + \frac{\Lambda_2}{\mu_2}\Big), \tag{8.88}$$

and $\mathcal{S}_\varepsilon$ is extended to matrices by applying it to each element. Given the new C and Z, the update of the Lagrange multipliers is done as described before. The convergence of this kind of algorithm has been well studied and established (see, e.g., (Lions and Mercier 1979; Kontogiorgis and Meyer 1989) and the references therein, as well as discussion in (Lin et al. 2011; Yuan and Yang 2009)).

Sparse Subspace Clustering Affinity

After solving the optimization program in (8.81), we obtain a sparse representation for each data point whose nonzero elements ideally correspond to points from the same subspace. The next step of the algorithm is to infer the segmentation of the data into different subspaces using the sparse coefficients. To address this problem, we use the spectral clustering algorithm described in Chapter 4. This algorithm needs a symmetric nonnegative affinity matrix as an input. While a subspace-preserving matrix C does have the property that $c_{jk} = 0$ when points j and k are in different subspaces, the matrix C is not necessarily symmetric or nonnegative. However, we can easily build a nonnegative and symmetric affinity as follows:

$$W = |C| + |C|^\top. \tag{8.89}$$

Summary of the SSC Algorithm with Uncorrupted Data

In summary, the SSC algorithm for data points that lie perfectly in a union of linear subspaces proceeds as shown in Algorithm 8.3.

8.3.2 SSC with Robustness to Outliers

In the previous subsection, we considered the sparse subspace clustering problem under the assumption that the N data points are drawn perfectly from a union of low-dimensional subspaces. Here, we consider the case that N_d data points are drawn perfectly from a union of subspaces while N_0 data points are outliers to the union of subspaces. That is, we are given a corrupted dictionary X composed of $N = N_d + N_0$

Algorithm 8.3 (Sparse Subspace Clustering for Uncorrupted Data)

Input: A set of points $\{\boldsymbol{x}_j\}_{j=1}^N$ lying in a union of n linear subspaces $\{S_i\}_{i=1}^n$.
1: Solve the sparse optimization program (8.81) to obtain a matrix $C \in \mathbb{R}^{N\times N}$.
2: Compute an affinity matrix W as $W = |C| + |C|^\top$.
3: Cluster the data into n groups by applying Algorithm 4.7 to W.
Output: Segmentation of the data into n groups: $X_1, X_2, \dots, X_n$.

data points. As usual, for the inliers we do not know which points belong to which subspace. Moreover, we do not know which points are inliers and which points are outliers. Therefore, given a point $\boldsymbol{x}$, our first goal is to determine whether the point is an inlier or an outlier. Then, if it is an inlier, our goal is to find other points in the data set that are in the same subspace as $\boldsymbol{x}$, i.e., to find a subspace-preserving representation of $\boldsymbol{x}$ in terms of the dictionary X as before.

The work of (Soltanolkotabi and Candès 2013) proposes a very simple approach to answer these questions. The main idea is to solve the ℓ_1-minimization program in (8.58) as usual. Then, if the point $\boldsymbol{x}$ is an inlier, we expect the solution $\boldsymbol{c}$ to be sparse, while if the point is an outlier, we expect the solution to be less sparse. The point $\boldsymbol{x}$ is declared to be an outlier if the ℓ_1 norm of its representation $\boldsymbol{c}$ is above a threshold. More specifically, a point is declared to be an outlier if

$$\|\boldsymbol{c}\|_1 > \lambda(\gamma)\sqrt{D}, \tag{8.90}$$

where $\gamma = \frac{N-1}{D}$ is the density of the data points, and λ is a threshold ratio function

$$\lambda(\gamma) = \begin{cases} \sqrt{\frac{2}{\pi}}\frac{1}{\sqrt{\gamma}} & 1 \leq \gamma \leq e \\ \sqrt{\frac{2}{\pi e}}\frac{1}{\sqrt{\log \gamma}} & \gamma \geq e. \end{cases} \tag{8.91}$$

The following result shows that this very simple procedure can reliably detect all outliers without making any assumption about the orientation of the subspaces or the distribution of the points within each subspace. Furthermore, if the points on each subspace are uniformly distributed, this scheme will not wrongfully detect a subspace point as an outlier.

Theorem 8.27 (SSC with Outliers (Soltanolkotabi and Candès 2013)). *Assume that there are N_d points to be clustered together with N_0 outliers sampled uniformly at random on the $(D-1)$-dimensional unit sphere ($N = N_0 + N_d$). The scheme in* (8.90) *detects all of the outliers with high probability as long as*

$$N_0 < \frac{1}{D}e^{c\sqrt{D}} - N_d, \tag{8.92}$$

where c is a numerical constant. Furthermore, suppose the subspaces are d-dimensional and of arbitrary orientation, and that each contains $\rho d + 1$ points sampled independently and uniformly at random. Then with high probability, the scheme in (8.90) *does not detect any subspace point as an outlier, provided that*

$$N_0 < D\rho^{c_2 D/d} - N_d, \tag{8.93}$$

where $c_2 = c^2(\rho)/(2e^{2\pi})$.

Since this theorem guarantees that we can perfectly detect both the inliers and the outliers with overwhelming probability, we can perform subspace clustering

Algorithm 8.4 (Sparse Subspace Clustering with Outliers)

Input: A set of points $\{x_j\}_{j=1}^N$ lying in a union of n linear subspaces $\{S_i\}_{i=1}^n$.
1: Solve the sparse optimization program (8.81) to obtain a matrix $C \in \mathbb{R}^{N \times N}$.
2: For each $j = 1, \dots, N$ declare x_j to be an outlier iff $\|c_j\|_1 > \lambda(\gamma)\sqrt{D}$. Let $X_0 \in \mathbb{R}^{D \times N_0}$ be the matrix of outliers.
3: Solve the sparse optimization program (8.81) for the remaining N_d points to obtain a matrix $C \in \mathbb{R}^{N_d \times N_d}$.
4: Compute an affinity matrix W as $W = |C| + |C|^\top$.
5: Cluster the data into n groups $X_1, X_2, \dots, X_n$ by applying Algorithm 4.7 to W.

Output: Segmentation of the data into $n + 1$ groups: $X_0, X_1, X_2, \dots, X_n$.

by applying the SSC algorithm for uncorrupted data to the inliers. In summary, the SSC algorithm for data points contaminated by outliers proceeds as shown in Algorithm 8.4.

8.3.3 SSC with Robustness to Noise

Let us now consider the sparse subspace clustering problem under the assumption that the data points are contaminated by noise or small errors. Following the model in (8.28), we model the errors directly in the self-expressiveness constraint as

$$X = XC + E, \tag{8.94}$$

where E is the matrix of errors. Under this model, we can find a sparse representation C that is robust to noise or small errors by penalizing the sum of squared errors $\|E\|_F^2$. This leads to the following program:

$$\min_C \quad \|C\|_1 + \frac{\tau}{2}\|X - XC\|_F^2 \quad \text{s.t.} \quad \operatorname{diag}(C) = \mathbf{0}, \tag{8.95}$$

where $\tau > 0$ is a parameter. Notice that this problem is a particular case of (8.4) with $\|C\|_{\mathcal{C}} = \|C\|_1$ and $\|E\|_{\mathcal{E}} = \frac{\tau}{2}\|E\|_F^2$.

Subspace-Preserving Property of SSC for Deterministic Noise

As before, we are interested in understanding the conditions under which the solution to the above program is subspace-preserving, i.e., $c_{jk} = 0$ when points j and k are in different subspaces. However, notice that there is an important difference between solving the SSC problem for noisy data in (8.95) and solving the SSC problem for uncorrupted data in (8.56). In the uncorrupted case, the self-expressiveness constraint $X = XC$ implies that $c_j \neq \mathbf{0}$ whenever $x_j \neq \mathbf{0}$. This, together with the subspace-preserving property, implies that point x_j will be connected to another point in its own subspace. However, this need not be the case for problem (8.95), because the optimal solution could be $C = \mathbf{0}$ when τ is small enough. To guarantee that each column of C is different from zero, the parameter τ needs to exceed a certain threshold, as stated in the following lemma.

Lemma 8.28. *Consider the optimization program* (8.95). *If*

$$\tau \leq \tau_{\min,2} \triangleq \frac{1}{\min_{j=1,\ldots,N} \max_{k\neq j} |\boldsymbol{x}_j^\top \boldsymbol{x}_k|}, \tag{8.96}$$

then there exists at least one data point $\boldsymbol{x}_j$ for which the optimal solution is $\boldsymbol{c}_j = \boldsymbol{0}$.

Proof. Consider the optimization program in (8.80). If $\boldsymbol{c}_j = \boldsymbol{0}$ is an optimal solution, then the zero vector must be contained in the subgradient of the objective function $\|\boldsymbol{c}_j\|_1 + \frac{\tau}{2}\|\boldsymbol{x}_j - \sum_{k\neq j} \boldsymbol{x}_k c_{jk}\|_2^2$ at $\boldsymbol{c}_j = \boldsymbol{0}$. That is, $0 \in w_j - \tau \boldsymbol{x}_j^\top \boldsymbol{x}_k$ for all $k \neq j$, where $|w_j| \leq 1$. This implies that $\tau |\boldsymbol{x}_j^\top \boldsymbol{x}_k| \leq 1$ for all $k \neq j$. Therefore, in order for $\boldsymbol{c}_j = \boldsymbol{0}$ to be an optimal solution, we need $\tau \leq 1/\max_{k\neq j} |\boldsymbol{x}_j^\top \boldsymbol{x}_k|$. Consider now the optimization program (8.95). It follows that in order for one of the columns of C to be zero, we need $\tau \leq 1/\min_{j=1,\ldots,N} \max_{k\neq j} |\boldsymbol{x}_j^\top \boldsymbol{x}_k|$. □

It follows that the conditions for correctness of clustering must depend not only on the separation between subspaces and the distribution of the data inside the subspaces (as in the case of uncorrupted data), but also on properly choosing τ. In addition, we expect that the separation between subspaces and/or the spread of the data points inside each subspace should increase as the amount of noise increases. Therefore, we expect that sufficient conditions for the correctness of clustering are that the subspaces be sufficiently separated, the data be well distributed in the subspaces, the noise be small enough, and the parameter τ be properly chosen.

Before stating these conditions more precisely, we need to extend the notions of dual points (Definition 8.21), dual directions (Definition 8.22), and subspace incoherence (Definition 8.23) to the case in which the data are contaminated by noise. To that end, let $A = [\boldsymbol{a}_1, \ldots, \boldsymbol{a}_N] \in \mathbb{R}^{D\times N}$ be a matrix whose columns are drawn from a union of n subspaces of $\mathbb{R}^D$, $\{S_i\}_{i=1}^n$, of dimensions $\{d_i\}_{i=1}^n$. Assume that the columns of A are of unit Euclidean norm, and let $A^i \in \mathbb{R}^{D\times N_i}$ be the submatrix of A containing the N_i points in S_i, and A_{-j}^i the submatrix of A^i with its jth column removed. Let $X = [\boldsymbol{x}_1, \ldots, \boldsymbol{x}_N] \in \mathbb{R}^{D\times N}$ denote a noisy version of A, and denote the columns of X corresponding to A^i and A_{-j}^i by X^i and X_{-j}^i, respectively. We have the following definitions.

Definition 8.29 (Dual Points). *The dual point $\boldsymbol{\lambda}(\boldsymbol{x}, X, \tau) \in \mathbb{R}^D$ of a point $\boldsymbol{x} \in \mathbb{R}^D$ with respect to a matrix $X \in \mathbb{R}^{D\times N}$ is defined as the optimal solution to*

$$\max_{\boldsymbol{\lambda}\in\mathbb{R}^D} \langle \boldsymbol{x}, \boldsymbol{\lambda} \rangle - \frac{1}{2\tau}\|\boldsymbol{\lambda}\|_2^2 \quad \text{s.t.} \quad \|X^\top \boldsymbol{\lambda}\|_\infty \leq 1. \tag{8.97}$$

Definition 8.30 (Projected Dual Directions). *The dual direction corresponding to a dual point $\boldsymbol{\lambda}(\boldsymbol{x}, X, \tau)$ and projected onto a d-dimensional subspace $S \subset \mathbb{R}^D$ is defined as*

$$\boldsymbol{v}(\boldsymbol{x}, X, S, \tau) = \frac{P_S \boldsymbol{\lambda}}{\|P_S \boldsymbol{\lambda}\|_2}. \tag{8.98}$$

Definition 8.31 (Projected Subspace Incoherence). *Let $\boldsymbol{v}_j^i = \boldsymbol{v}(\boldsymbol{x}_j^i, X_{-j}^i, S_i, \tau)$ be the dual direction of the jth column $\boldsymbol{x}_j^i$ of X^i with respect to X_{-j}^i and projected onto S_i. The subspace incoherence between the (noisy) points in the ith subspace and the (clean) points in all other subspaces is defined by*

$$\mu_i = \max_{k:\boldsymbol{a}_k \notin S_i} \|V^{i\top} \boldsymbol{a}_k\|_\infty = \max_{j:\boldsymbol{x}_j \in S_i} \max_{k:\boldsymbol{a}_k \notin S_i} |\langle \boldsymbol{v}_j^i, \boldsymbol{a}_k \rangle|, \tag{8.99}$$

where $V^i = [\boldsymbol{v}_1^i, \dots, \boldsymbol{v}_{N_i}^i] \in \mathbb{R}^{D \times N_i}$ is the matrix of projected dual directions.

Recall the definition of inradius $r(\mathcal{P})$ of a convex polytope $\mathcal{P}$ (Definition 8.24). Let $\mathcal{P}_{-j}^i$ be the convex polytope formed by the columns of the clean matrix A_{-j}^i. Let $r = \min_{i=1,\dots,n} r_i$, where $r_i = \min_{j=1,\dots,N_i} r(\mathcal{P}_{-j}^i)$. We have the following result.

Theorem 8.32 (SSC for Deterministic Noise (Wang and Xu 2013)). *Let μ_i be the ith projected subspace incoherence. Let $\delta = \|X - A\|_{2,\infty}$ be an upper bound on the noise, and $\delta_1 = \max_i \|U_i^\top (X - A)\|_{2,\infty}$ an upper bound on the projection of the noise onto the subspaces, where U_i is an orthonormal basis for the ith subspace. If for all $i = 1, \dots, n$, we have that*

$$\mu_i < r_i, \quad \delta \leq \min_{i=1,\dots,n} \frac{r(r_i - \mu_i)}{7r_i + 2} \quad \textit{and} \tag{8.100}$$

$$\frac{1}{r - 2\delta - \delta^2} < \tau < \min_{i=1,\dots,n} \frac{r_i - \mu_i - 2\delta_1}{\delta(1+\delta)(2 + r_i - \delta_1)}, \tag{8.101}$$

then the solution to problem (8.95) *is subspace-preserving, and its columns are nonzero.*

Notice that when $\delta = 0$, the condition in (8.100) reduces to $\mu_i < r_i$, which is the condition of Theorem 8.25. The main difference is that in the uncorrupted case, we solve for C subject to the self-expressiveness constraint $X = XC$, which corresponds to problem (8.95) with $\tau = \infty$, while in Theorem 8.32, we solve the matrix LASSO problem with $\frac{1}{r} < \tau$. Therefore, Theorem 8.32 generalizes Theorem 8.25, since it allows us to obtain a subspace-preserving representation from a more general optimization problem. Notice also that the bound δ on the noise level requires r_i to be larger than μ_i by a certain margin that depends on δ. Therefore, as δ increases, the condition is harder to satisfy. Likewise, as δ increases, the upper bound on τ reduces. In summary, the theorem captures our intuition that the clustering problem becomes harder as the noise increases.

Solving the Sparse Optimization Problem with Noise

Notice that the problem in (8.95) is equivalent to N LASSO problems (Tibshirani 1996), one for each column of C. However, as in the case of uncorrupted data, it is more convenient to solve all N problems simultaneously. The reader can show that the optimization problem in (8.95) is equivalent to (see Exercise 8.3)

$$\min_{C,Z} \quad \|C\|_1 + \frac{\tau}{2}\|X - XZ\|_F^2 \quad \text{s.t.} \quad Z = C - \operatorname{diag}(C). \tag{8.102}$$

As before, we can solve this problem using ADMM. More specifically, the augmented Lagrangian of the above optimization problem is given by

$$\begin{aligned}\mathscr{L}(C,Z,\Lambda_2) =& \|C\|_1 + \frac{\tau}{2}\|X - XZ\|_F^2 + \\ & \langle \Lambda_2, Z - (C - \operatorname{diag}(C))\rangle + \frac{\mu_2}{2}\|Z - (C - \operatorname{diag}(C))\|_F^2 \\ =& \|C\|_1 + \frac{\tau}{2}\|X - XZ\|_F^2 + \frac{\mu_2}{2}\left\|Z - (C - \operatorname{diag}(C)) + \frac{\Lambda_2}{\mu_2}\right\|_F^2 + \eta,\end{aligned} \tag{8.103}$$

where $\Lambda_2 \in \mathbb{R}^{N\times N}$ is a matrix of Lagrange multipliers, $\mu_2 > 0$ is a parameter, and η is a term that does not depend on C or Z. It is easy to see that this augmented Lagrangian is a particular case of that in (8.83); hence the optimization problem can be solved using Algorithm 8.5, which is a particular case of Algorithm 8.2.

Summary of the SSC Algorithm with Noisy Data
In summary, the SSC algorithm for data points contaminated by noise proceeds as shown in Algorithm 8.6.

Algorithm 8.5 (Matrix LASSO Minimization by ADMM)

Input: Data matrix X
initialize: $C^0 = \mathbf{0}, \Lambda_2^0 = \mathbf{0}, \mu_2 > 0.$
while not converged **do**
  compute $Z^{k+1} = (\tau X^\top X + \mu_2 I)^{-1}\left(\tau X^\top X + \mu_2(C^k - \frac{\Lambda_2^k}{\mu_2})\right)$;
  compute $C^{k+1} = \mathcal{S}_{\frac{1}{\mu_2}}\left(Z^{k+1} + \frac{\Lambda_2^k}{\mu_2}\right)$;
  compute $C^{k+1} = C^{k+1} - \operatorname{diag}(C^{k+1})$;
  compute $\Lambda_2^{k+1} = \Lambda_2^k + \mu_2(Z^{k+1} - C^{k+1})$;
end while
Output: Sparse representation C.

Algorithm 8.6 (Sparse Subspace Clustering for Noisy Data)

Input: A set of points $\{x_j\}_{j=1}^N$ lying in a union of n linear subspaces $\{S_i\}_{i=1}^n$.
Solve the sparse optimization program (8.95) to obtain a matrix $C \in \mathbb{R}^{N\times N}$.
Compute an affinity matrix W as $W = |C| + |C|^\top$.
Cluster the data into n groups by applying Algorithm 4.7 to W.
Output: Segmentation of the data into n groups: $X_1, X_2, \ldots, X_n$.

8.3.4 SSC with Robustness to Corrupted Entries

Let us now consider the case in which the data points are sampled from a union of subspaces and corrupted by outlying entries. Following the model in (8.28), we model the errors directly in the self-expressiveness constraint as

$$X = XC + E. \tag{8.104}$$

Under this model, we can find a sparse representation C that is robust to sparse corrupted entries by penalizing the ℓ_1 norm of the errors $\|E\|_1$. This leads to the following program:

$$\min_{C,E} \ \|C\|_1 + \tau\|E\|_1, \quad \text{s.t.} \quad X = XC + E, \quad \operatorname{diag}(C) = \mathbf{0}, \tag{8.105}$$

where $\tau > 0$ is a parameter. Notice that this problem is a particular case of (8.4) with $\|C\|_{\mathcal{C}} = \|C\|_1$ and $\|E\|_{\mathcal{E}} = \tau\|E\|_1$.

As in the case of noisy data, it is possible for the optimal solution of (8.105) to be $C = \mathbf{0}$ and $X = E$. To avoid this trivial solution, we need to choose the regularization parameter above a certain threshold, as stated in the following lemma.

Lemma 8.33. *Consider the optimization program* (8.105). *If*

$$\tau \leq \tau_{\min,1} \triangleq \frac{1}{\min_{j=1,\ldots,N} \max_{k\neq j} \|\boldsymbol{x}_k\|_1}, \tag{8.106}$$

then there exists at least one data point $\boldsymbol{x}_j$ for which the optimal solution is $(\boldsymbol{c}_j, \boldsymbol{e}_j) = (\mathbf{0}, \boldsymbol{x}_j)$.

Proof. Consider the optimization program $\min \|\boldsymbol{c}_j\|_1 + \tau\|\boldsymbol{x}_j - \sum_{k\neq j} \boldsymbol{x}_k c_{jk}\|_1$. If $\boldsymbol{c}_j = \mathbf{0}$ is an optimal solution, then the zero vector must be contained in the subgradient of the objective function at $\boldsymbol{c}_j = \mathbf{0}$. That is, $0 \in w_j - \tau \boldsymbol{x}_k^\top \operatorname{sign}(\boldsymbol{x}_j)$ for all $k \neq j$, where $|w_j| \leq 1$. This implies that $\tau\|\boldsymbol{x}_k\|_1 \leq 1$ for all $k \neq j$. Therefore, in order for $\boldsymbol{c}_j = \mathbf{0}$ to be an optimal solution, we need $\tau \leq 1/\max_{k\neq j} \|\boldsymbol{x}_k\|_1$. Consider now the optimization program (8.105). It follows that in order for one of the columns of C to be zero, we need $\tau \leq 1/\min_{j=1,\ldots,N} \max_{k\neq j} \|\boldsymbol{x}_k\|_1$. □

Having established conditions to avoid a trivial solution, we will now describe an ADMM algorithm for solving the optimization problem in (8.105). We begin by introducing an auxiliary matrix variable Z as follows:

$$\min_{C,Z,E} \|C\|_1 + \tau\|E\|_1 \quad \text{s.t.} \quad X = XZ + E, \ Z = C - \operatorname{diag}(C). \tag{8.107}$$

The augmented Lagrangian of the above optimization problem is given by

$$\begin{aligned}\mathcal{L}(C,Z,E,\Lambda_1,\Lambda_2) &= \|C\|_1 + \langle\Lambda_1, X-XZ-E\rangle + \langle\Lambda_2, Z-(C-\operatorname{diag}(C))\rangle \\ &\quad + \tau\|E\|_1 + \frac{\mu_1}{2}\|X-XZ-E\|_F^2 + \frac{\mu_2}{2}\|Z-(C-\operatorname{diag}(C))\|_F^2 \\ &= \|C\|_1 + \tau\|E\|_1 + \frac{\mu_1}{2}\left\|X-XZ-E+\frac{\Lambda_1}{\mu_1}\right\|_F^2 + \frac{\mu_2}{2}\left\|Z-(C-\operatorname{diag}(C))+\frac{\Lambda_2}{\mu_2}\right\|_F^2 + \eta,\end{aligned}$$

where $\Lambda_1 \in \mathbb{R}^{D\times N}$ and $\Lambda_2 \in \mathbb{R}^{N\times N}$ are matrices of Lagrange multipliers, $\mu_1 > 0$, $\mu_2 > 0$ are parameters, and η is a term that does not depend on C, Z, or E.

The optimal Z given C, E, Λ_1, and Λ_2 is given by

$$Z = (\mu_1 X^\top X + \mu_2 I)^{-1}\Big(\mu_1 X^\top\big(X - E + \frac{\Lambda_1}{\mu_1}\big) + \mu_2\big(C - \frac{\Lambda_2}{\mu_2}\big)\Big). \tag{8.108}$$

Then, as shown in (8.88), the optimal C given Z, E, and Λ_2, is given by

$$C = \tilde{C} - \operatorname{diag}(\tilde{C}), \quad \text{where} \quad \tilde{C} = \mathcal{S}_{\frac{1}{\mu_2}}\Big(Z + \frac{\Lambda_2}{\mu_2}\Big). \tag{8.109}$$

Then, the optimal E given Z and Λ_1 is given by

$$E = \mathcal{S}_{\frac{\tau}{\mu_1}}\Big(X - XZ + \frac{\Lambda_1}{\mu_1}\Big). \tag{8.110}$$

Given the new C, Z, and E, the update of the Lagrange multipliers is done as $\Lambda 1 \leftarrow \Lambda_1 + \mu_1(X - XZ - E)$ and $\Lambda_2 \leftarrow \Lambda_2 + \mu_2(Z - C)$. This leads to Algorithm 8.7 for clustering data corrupted by outliers in a union of subspaces.

Algorithm 8.7 (Sparse Subspace Clustering with Corrupted Entries)

Input: Data matrix X

initialize: $C^0 = \mathbf{0}$, $E^0 = \mathbf{0}$, $\Lambda_1^0 = \mathbf{0}$, $\Lambda_2^0 = \mathbf{0}$, $\mu_1 > 0$, $\mu_2 > 0$.
while not converged **do**
 set $Z^{k+1} = (\mu_1 X^\top X + \mu_2 I)^{-1}\big(\mu_1 X^\top(X - E^k + \frac{\Lambda_1^k}{\mu_1}) + \mu_2(C^k - \frac{\Lambda_2^k}{\mu_2})\big)$;
 compute $C^{k+1} = \mathcal{S}_{\frac{1}{\mu_2}}\big(Z^{k+1} + \frac{\Lambda_2^k}{\mu_2}\big)$;
 compute $C^{k+1} = C^{k+1} - \operatorname{diag}(C^{k+1})$;
 compute $E^{k+1} = \mathcal{S}_{\frac{\tau}{\mu_2}}\big(X - XZ^{k+1} + \frac{\Lambda_1^k}{\mu_1}\big)$;
 compute $\Lambda_1^{k+1} = \Lambda_1^k + \mu_1(X - XZ^{k+1} - E^{k+1})$;
 compute $\Lambda_2^{k+1} = \Lambda_2^k + \mu_2(Z^{k+1} - C^{k+1})$;
end while

Output: Sparse representation C.

8.3.5 SSC for Affine Subspaces

In the previous subsections, we considered the sparse subspace clustering problem under the assumption that the data points are drawn from a union of n *linear* subspaces. In practice, however, we may need to cluster data lying in a union of *affine* rather than *linear* subspaces. A simple way to deal with affine subspaces is to use the projectivization technique discussed in Section 5.1.1, where each data point in $\boldsymbol{x} \in \mathbb{R}^D$ is embedded into $\mathbb{R}^{D+1}$ by augmenting it with a one as $(\boldsymbol{x}^\top, 1)^\top$. In this way, each d_i-dimensional affine subspace S_i is considered a subset of a (d_i+1)-dimensional linear subspace that includes S_i and the origin. Thus, a union of n affine subspaces of $\mathbb{R}^D$ of dimensions $\{d_i\}_{i=1}^n$ is considered a subset of a union of n linear subspaces of $\mathbb{R}^{D+1}$ of dimensions $\{d_i+1\}_{i=1}^n$. This suggests that any linear subspace clustering algorithm can be applied to the embedded data. However, this approach has the drawback of possibly increasing the dimension of the intersection of two subspaces, which in some cases can result in indistinguishability of subspaces from each other. For example, two different lines $x = -1$ and $x = +1$ in the x-y plane form the same 2-dimensional linear subspace after including the origin; hence they become indistinguishable.

To directly deal with affine subspaces, we use the fact that every point $\boldsymbol{x}$ in an affine subspace $S \subset \mathbb{R}^D$ of dimension $d < D$ can be written as an affine combination of $d+1$ other points, $\{\boldsymbol{x}_j\}_{j=0}^d$, also in S, i.e.,

$$\boldsymbol{x} = c_1\boldsymbol{x}_1 + c_2\boldsymbol{x}_2 + \cdots + c_d\boldsymbol{x}_d, \quad \sum_{j=0}^{d} c_j = 1, \tag{8.111}$$

where the points $\{\boldsymbol{x}_j\}_{j=0}^d$ are affinely independent.[6] Therefore, if the columns of $X = \left[\boldsymbol{x}_1, \ldots, \boldsymbol{x}_N\right]$ lie in a union of n affine subspaces of $\mathbb{R}^D$, $\{S_i\}_{i=1}^n$, then each $\boldsymbol{x}_j \in S_i$ can be written as an affine combination of d_i+1 other points in S_i. Following Definition 8.2, we assume that the data matrix is *affinely self-expressive*, i.e.,

$$\forall j \ \exists \boldsymbol{c}_j \in \mathbb{R}^N \quad \text{s.t.} \quad \boldsymbol{x}_j = X\boldsymbol{c}_j, \quad \mathbf{1}^\top \boldsymbol{c}_j = \mathbf{1}, \quad c_{jj} = 0, \tag{8.112}$$

or equivalently, in matrix form,

$$\exists C \in \mathbb{R}^{N\times N} \quad \text{s.t.} \quad X = XC, \quad \mathbf{1}^\top C = 1, \quad \operatorname{diag}(C) = \mathbf{0}, \tag{8.113}$$

where $C = \left[\boldsymbol{c}_1, \boldsymbol{c}_2, \ldots, \boldsymbol{c}_N\right] \in \mathbb{R}^{N\times N}$.

Observe that $\boldsymbol{c}_j \in \mathbb{R}^N$ has at most d_i+1 nonzero entries that correspond to d_i+1 other data points in S_i. Observe also that the property in (8.113) is nearly identical

[6] A set of points $\{\boldsymbol{x}_j\}_{j=0}^d$ is said to be affinely dependent if there exist scalars $c_0, \ldots, c_d$ not all zero such that $\sum_{j=0}^d c_j\boldsymbol{x}_j = \mathbf{0}$ and $\sum_{j=1}^d c_j = 1$.

to the self-expressiveness property in (8.8), except that we have the additional constraint that all coefficients must add up to one, i.e., $\mathbf{1}^\top C = \mathbf{1}^\top$. Therefore, the SSC algorithms discussed in the previous subsections remain identical, except that we need to add this additional constraint to the optimization problems they solve. For example, in the case of noiseless data, we need to solve

$$\min_C \|C\|_1 \quad \text{s.t.} \quad X = XC, \quad \mathbf{1}^\top C = \mathbf{1}^\top, \quad \operatorname{diag}(C) = \mathbf{0}; \tag{8.114}$$

in the case of noisy data, we need to solve

$$\min_C \|C\|_1 + \frac{\tau}{2}\|X - XC\|_F^2 \quad \text{s.t.} \quad \mathbf{1}^\top C = \mathbf{1}^\top, \quad \operatorname{diag}(C) = \mathbf{0}; \tag{8.115}$$

and in the case of data with corrupted entries, we need to solve

$$\min_C \|C\|_1 + \tau\|X - XC\|_1, \quad \text{s.t.} \quad \mathbf{1}^\top C = \mathbf{1}^\top, \quad \operatorname{diag}(C) = \mathbf{0}. \tag{8.116}$$

8.4 Simulations and Applications

8.4.1 Low-Rank and Sparse Methods on Synthetic Data

In this subsection, we compare the performance of low-rank and sparse subspace clustering methods to that of other spectral subspace clustering methods using synthetically generated data. In particular, we investigate the effect of the kind of subspaces (independent or disjoint) as well as the effect of the number and dimensions of the subspaces on the clustering performance of different methods.

We randomly generate $n \in \{3, 5\}$ subspaces in $\mathbb{R}^{30}$ of varying dimensions $2 \leq d_i \leq 5$. For each trial, we randomly sample $N_i = 10d_i$ points from each subspace, respectively, and add zero-mean Gaussian noise with covariance $\sigma^2(I - U_iU_i^\top)$, where U_i is an orthonormal basis for the subspace and $\sigma \in [0, 0.35]$. Subsequently, the noisy points are normalized to have unit Euclidean norm. Given these N data points, we cluster them into their respective subspaces using different subspace clustering methods and compute the average and median clustering errors for each algorithm over 100 random trials.

Tables 8.1 and 8.2 show the clustering errors for the noise-free and noisy data points, respectively, with $\sigma = 0.1$. From the results, we can draw the following conclusions:

1. The performance of LRSC depends on the subspace model. For independent subspaces, it yields very small clustering errors, which is expected, since this algorithm has theoretical guarantees for independent subspaces. On the other hand, for disjoint subspaces, LRSC gives large errors, suggesting that it cannot work well beyond the independent subspace model.

Table 8.1 Clustering error (%) of different algorithms on synthetic noise-free data for different subspace dimensions $d = (d_1, d_2, \ldots, d_n)$, number of subspaces n, and subspace models (independent or disjoint). For LRSC, LRSC$_{2,1}$ and LRSC$_1$ mean that either the $\ell_{2,1}$ or the ℓ_1 norm is used to penalize the errors, i.e., $\|E\|_{\mathcal{E}} = \|E\|_{2,1}$ or $\|E\|_{\mathcal{E}} = \|E\|_1$, respectively. Also, the LRSC affinity matrix is constructed as in (8.27) with $q = 1$ or $q = 4$.

Algorithm	LSA	SCC	LRSC$_{2,1}$ $q=1$	LRSC$_{2,1}$ $q=4$	LRSC$_1$ $q=1$	SSC
independent subspaces						
$d=(3,3,3)$						
Mean	2.23	0.00	0.00	0.04	0.00	0.00
Median	2.22	0.00	0.00	0.00	0.00	0.00
$d=(2,3,5)$						
Mean	2.17	0.81	0.10	0.00	0.00	0.00
Median	2.00	1.00	0.00	0.00	0.00	0.00
$d=(4,4,4,4,4)$						
Mean	0.63	0.00	0.00	0.04	0.00	0.00
Median	0.50	0.00	0.00	0.00	0.00	0.00
$d=(1,2,3,4,5)$						
Mean	5.57	3.78	0.83	0.00	0.17	0.00
Median	2.00	2.00	0.67	0.00	0.00	0.00
disjoint subspaces						
$d=(3,3,3)$						
Mean	9.77	0.00	15.09	7.38	11.41	0.97
Median	8.89	0.00	13.33	5.56	8.89	0.00
$d=(2,3,5)$						
Mean	5.98	0.88	7.13	1.21	4.64	0.11
Median	6.00	1.00	4.50	0.00	4.00	0.00
$d=(4,4,4,4,4)$						
Mean	18.98	0.00	42.73	32.03	39.46	2.46
Median	18.50	0.00	43.00	32.50	40.50	2.00
$d=(1,2,3,4,5)$						
Mean	5.23	6.97	27.15	5.84	22.49	0.95
Median	4.67	4.33	28.33	4.67	24.33	0.00

2. For independent subspaces, SSC yields very small clustering errors, which is expected, since SSC always works under the independent subspace model, as we showed in the theoretical analysis. For disjoint subspaces, the clustering errors of SSC slightly increase but are still small. This is expected, since for disjoint subspaces, SSC works under some conditions on the subspace angles and the data distribution, which can be violated for some data points or subspaces.
3. The clustering performance of SCC does not depend on the subspace models (independent or disjoint). However, it depends on whether subspaces have the

Table 8.2 Clustering error (%) of different algorithms on synthetic noisy data for different subspace dimensions $d = (d_1, d_2, \dots, d_n)$, number of subspaces n, and subspace models (independent or disjoint). For LRSC, $\text{LRSC}_{2,1}$ and LRSC_1 mean that either the $\ell_{2,1}$ or the ℓ_1 norm is used to penalize the errors, i.e., $\|E\|_{\mathcal{E}} = \|E\|_{2,1}$ or $\|E\|_{\mathcal{E}} = \|E\|_1$, respectively. Also, the LRSC affinity matrix is constructed as in (8.27) with $q = 1$ or $q = 4$.

Algorithm	LSA	SCC	$\text{LRSC}_{2,1}$ $q = 1$	$\text{LRSC}_{2,1}$ $q = 4$	LRSC_1 $q = 1$	SSC
independent subspaces						
$d = (3, 3, 3)$						
Mean	2.48	0.00	0.00	0.04	0.00	0.00
Median	2.22	0.00	0.00	0.00	0.00	0.00
$d = (2, 3, 5)$						
Mean	1.76	0.21	0.36	0.00	0.03	0.02
Median	1.00	0.00	0.00	0.00	0.00	0.00
$d = (4, 4, 4, 4, 4)$						
Mean	0.70	0.00	0.00	0.04	0.00	0.00
Median	0.50	0.00	0.00	0.00	0.00	0.00
$d = (1, 2, 3, 4, 5)$						
Mean	0.85	27.86	1.45	0.02	0.67	0.02
Median	0.67	29.00	0.67	0.00	0.00	0.00
disjoint subspaces						
$d = (3, 3, 3)$						
Mean	9.92	0.00	13.79	7.76	9.70	0.81
Median	8.89	0.00	12.22	5.56	7.78	0.00
$d = (2, 3, 5)$						
Mean	6.69	0.28	6.15	1.34	3.92	0.12
Median	5.00	0.00	4.00	0.00	2.00	0.00
$d = (4, 4, 4, 4, 4)$						
Mean	19.25	0.00	42.50	31.68	38.15	5.71
Median	19.50	0.00	43.25	31.50	38.50	3.00
$d = (1, 2, 3, 4, 5)$						
Mean	5.51	30.85	24.35	5.57	21.93	3.19
Median	5.31	33.67	24.67	4.00	22.00	0.67

same or different dimensions. More specifically, SCC obtains very low clustering errors when the subspaces have the same or very close dimensions for both independent and disjoint subspaces. On the other hand, the clustering error of SCC is large when the subspaces have very different dimensions. This comes from the fact that SCC uses the maximum dimension, $d_{\max}$, of the subspaces to compute the affinity among $d_{max} + 2$ data points. As a result, it is possible that points in different subspaces of small dimensions obtain a large affinity, i.e., that they are considered by the algorithm to be from the same subspace.

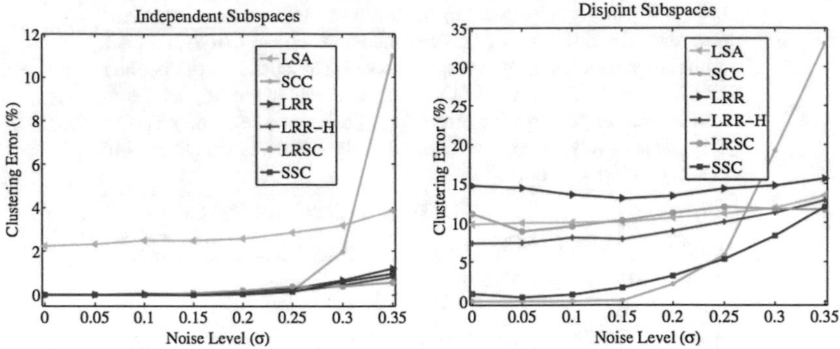

Fig. 8.7 Average clustering errors of different subspace clustering algorithms as a function of the noise level σ for independent subspace model (left) and disjoint subspace model (right).

4. Unlike other algorithms that obtain nearly zero clustering errors for independent subspaces, LSA obtains larger clustering errors for both independent and disjoint subspaces. This comes from the fact that LSA computes the affinity between pairs of points by first fitting a local subspace to each data point and its nearest neighbors. Since the neighborhood of a data point may contain points from different subspaces, the locally fitted subspace may not be close to the true underlying subspace at the point, hence degrading the performance of the algorithm. Moreover, for disjoint subspaces, the probability of having points on other subspaces that are close to the given point increases, thereby increasing the clustering error of LSA. Figure 8.7 shows the average clustering errors of different algorithms as a function of the noise level σ for independent and disjoint subspace models for $d_1 = d_2 = d_3 = 3$. As the results show, all algorithms except LSA obtain very low clustering errors for small and moderate amounts of noise for independent subspaces (the performance of SCC degrades when the noise level is above $\sigma = 0.25$). On the other hand, for disjoint subspaces, only SSC and SCC obtain low clustering errors for small and moderate amounts of noise, while other algorithms obtain large errors for all levels of noise.

8.4.2 *Low-Rank and Sparse Methods on Face Clustering*

In this subsection, we will apply low-rank and sparse subspace clustering methods to the face clustering problem. As discussed in Chapter 2, under the Lambertian reflectance model, face images of a single individual from a fixed viewpoint and varying illumination lie approximately in a 9-dimensional subspace of the ambient space whose dimension is the number of pixels in the image. In practice, as discussed in Chapter 3, a few pixels deviate from the Lambertian model due to cast shadows and specularities, which can be modeled as sparse outlying entries.

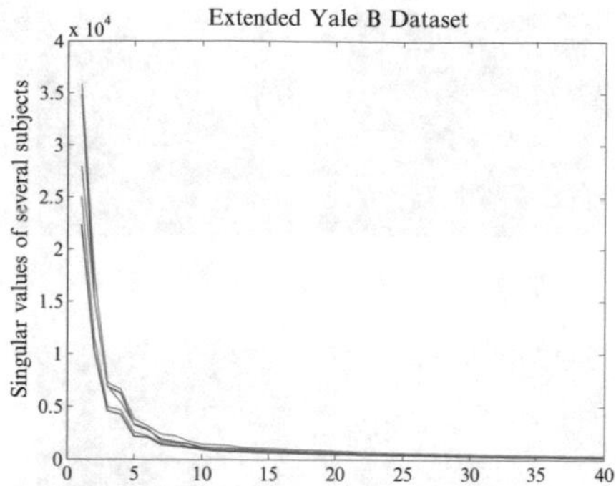

Fig. 8.8 Singular values of five data matrices corresponding to the face images of five different individuals in the extended Yale B data set. The face images of each individual span a subspace of dimension around 9.

Therefore, the face clustering problem reduces to the problem of clustering a set of images according to multiple subspaces and corrupted by sparse gross errors.

We will compare low-rank and sparse subspace clustering methods to the statistical and spectral methods described in Chapters 4, 6, and 7 on the extended Yale B data set (Lee et al. 2005). Figure 7.5 shows sample images from the database, which includes 64 frontal face images of 38 individuals acquired under 64 different lighting conditions. Each image is cropped to 192×168 pixels, which cover the face of the individual only. To reduce the computational cost and the memory requirements of all algorithms, we down-sample the images to 48×42 pixels and treat each 2016-dimensional vectorized image as a data point. Figure 8.8 shows the singular values of five data matrices corresponding to the face images of five different individuals in the data set. Note that the singular value curve has a knee around 9, corroborating the approximate 9-dimensionality of the face data in each subject. In addition, the singular values gradually decay to zero, showing that the data are corrupted by errors. Thus, the face images of n subjects can be modeled as corrupted data points lying close to a union of 9-dimensional subspaces.

Face Clustering Affinities for Two Subjects from the Extended Yale B Data Set
First, we compare the affinity matrices provided by different methods. For this purpose, we follow the same experimental setup as in Section 7.6.2, where we used several methods to compute a subspace clustering affinity among face images of two subjects (20 and 21, or 37 and 38) in the extended Yale B data set. Here, we simply add the results for LRSC and SSC on the same data. For LRSC, we use the model for noisy data in (8.31), with the parameter set to $\tau = 60$. For SSC, we use the model with sparse outlying entries in (8.105) with the parameter set to $\tau = 20\tau_{\min,1}$, where $\tau_{\min,1}$ is defined in (8.106). The results are shown in Figure 8.9, which includes some of the results in Figure 7.6 for LSA, SCC, and SASC.

Observe that LRSC and SSC both produce a cross pattern for the intraclass affinities, which is similar to that of other techniques such as LSA, LLMC, and

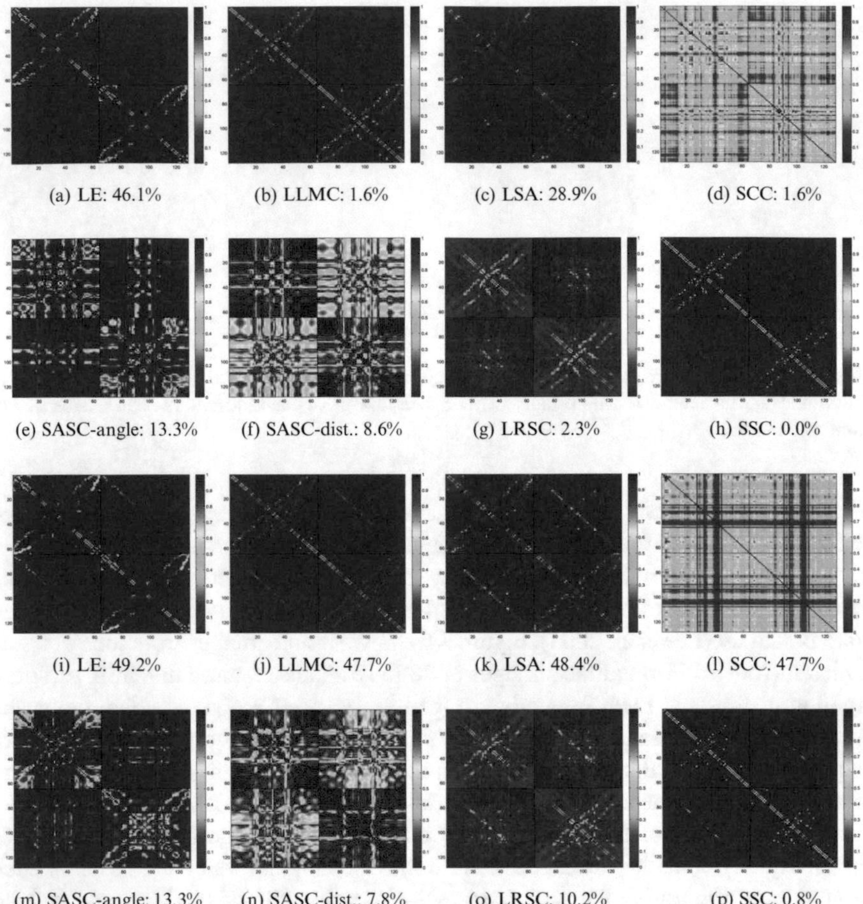

(a) LE: 46.1% (b) LLMC: 1.6% (c) LSA: 28.9% (d) SCC: 1.6%

(e) SASC-angle: 13.3% (f) SASC-dist.: 8.6% (g) LRSC: 2.3% (h) SSC: 0.0%

(i) LE: 49.2% (j) LLMC: 47.7% (k) LSA: 48.4% (l) SCC: 47.7%

(m) SASC-angle: 13.3% (n) SASC-dist.: 7.8% (o) LRSC: 10.2% (p) SSC: 0.8%

Fig. 8.9 Affinity matrices of face images produced by different methods. We take frontal face images of subjects 20 and 21 (Figures 8.9(a)–8.9(h)) and subjects 37 and 38 (Figures 8.9(i)–8.9(p)) under 64 different illumination conditions from extended Yale B. For SASC, the images are projected to dimension 9 using PCA. For all other methods, the images are down-sampled to 48×42. The affinity is obtained by taking absolute value, symmetrizing, normalizing to the range $(0, 1)$, then setting diagonal to be zero. The ground truth clustering is given as the horizontal and vertical black lines in each figure, so that points 1–64 correspond to the first individual, and points 65–128 correspond to the second individual. Clustering error percentages for each method are reported in the captions.

LE. The main difference is that the intraclass similarities of LRSC and SSC are generally higher that those of other methods. Moreover, observe that the affinity of SSC is sparser than that of LRSC, as expected. Also, notice the similarities between the affinities of LLMC and SSC. This is expected, since both methods rely on the self-expressiveness constraint, with LLMC restricting the nonzero coefficients to correspond to nearest neighbors and SSC searching for the best nonzero coefficients.

The main difference is that SSC produces higher intraclass affinities and lower interclass affinities than other methods, which ultimately results in better clustering. Overall, we observe that SASC-distance, LRSC, and SSC produce clustering results that are more accurate and more stable across individuals. Also, by comparing to the results of K-means and spectral clustering with simple distance-based affinity (see Examples 4.18 and 4.24 and Table 4.1), we can see the advantage of exploiting the multisubspace structure in the data to construct affinities that are specifically designed for the subspace clustering problem.

Face Clustering Errors on a Few Subjects from the Extended Yale B Data Set
Next, we present a more detailed evaluation of the clustering performance of different algorithms on 2–10 subjects from the extended Yale B database. Following the experimental setup of (Elhamifar and Vidal 2013), we divide the 38 subjects into 4 groups, where the first three groups correspond to subjects 1–10, 11–20 and 21–30, and the fourth group corresponds to subjects 31–38. For each of the first three groups, we consider all choices of $n \in \{2, 3, 5, 8, 10\}$ subjects, and for the last group, we consider all choices of $n \in \{2, 3, 5, 8\}$ subjects. Finally, we apply all subspace clustering algorithms for each trial, i.e., for each set of n subjects. The parameters for the different methods are set as follows.

1. LSA, i.e., Algorithm 7.1 with the number of nearest neighbors chosen as $K = 7$ and the dimension of the subspaces chosen as $d = 5$.
2. SCC, i.e., Algorithm 7.4 with the dimension of the subspaces chosen as $d = 9$.
3. LRSC. We evaluate two versions of this method. One of them, which we denote by $\text{LRSC}_{2,1}$ and is also known as LRR, is the low-rank method with robustness to outliers that is based on solving the optimization problem in (8.49) with $\|E\|_{\mathcal{E}} = \lambda\|E\|_{2,1}$ and $\lambda = 0.18$. Given the solution to (8.49), we construct the affinity as in (8.27) with $q = 1$ or $q = 4$. The other one, which we denote by LRSC_1, is the low-rank method with robustness to outlying entries, which is based on solving the optimization problem in (8.50) with $\|E\|_{\mathcal{E}} = \lambda\|E\|_1$ and $\lambda = 0.008$ using ADMM.
4. SSC. We evaluate the model with robustness to sparse outlying entries, which is based on solving the optimization problem in (8.105) with the parameter set to $\tau = 20\tau_{\min,1}$, where $\tau_{\min,1}$ is defined in (8.106).

In Table 8.3, we report the results in Table 3 of (Elhamifar and Vidal 2013), which shows the average and median subspace clustering errors of different algorithms. The results are obtained by first applying the RPCA-ADMM algorithm, Algorithm 3.8, to the face images of each subject and then applying different subspace clustering algorithms to the low-rank component of the data obtained by RPCA-ADMM. While this cannot be done in practice, because the clustering of the data is not known beforehand, the purpose of this experiment is to show that when the data are uncorrupted, LRSC and SSC correctly identify the subspaces. As we can see, LSA and SCC do not perform well, even with decorrupted data. Notice also that $\text{LRSC}_{2,1}$ with $q = 4$ does not perform well for more than eight subjects, showing that using higher-order powers on the obtained low-rank coefficient matrix

Table 8.3 Clustering error (%) of different algorithms on a few subjects from the extended Yale B database after applying RPCA-ADMM separately to the images from each subject.

Algorithm	LSA	SCC	$\text{LRSC}_{2,1}$ $q = 1$	$\text{LRSC}_{2,1}$ $q = 4$	LRSC_{1} $q = 1$	SSC
2 Subjects						
Mean	6.15	1.29	0.09	0.05	**0.00**	0.06
Median	**0.00**	**0.00**	**0.00**	**0.00**	**0.00**	**0.00**
3 Subjects						
Mean	11.67	19.33	0.12	0.10	**0.00**	0.08
Median	2.60	8.59	**0.00**	**0.00**	**0.00**	**0.00**
5 Subjects						
Mean	21.08	47.53	0.16	0.15	**0.00**	0.07
Median	19.21	47.19	**0.00**	**0.00**	**0.00**	**0.00**
8 Subjects						
Mean	30.04	64.20	4.50	11.57	**0.00**	0.06
Median	29.00	63.77	0.20	15.43	**0.00**	**0.00**
10 Subjects						
Mean	35.31	63.80	0.15	13.02	**0.00**	0.89
Median	30.16	64.84	**0.00**	13.13	**0.00**	0.31

does not always improve the result of $\text{LRSC}_{2,1}$. SSC and LRSC_1, on the other hand, perform very well, with LRSC_1 achieving perfect performance.

In Table 8.4, we report the results in Table 4 of (Elhamifar and Vidal 2013), which shows the average and median subspace clustering errors of different algorithms. The results are obtained by first applying the RPCA-ADMM algorithm, Algorithm 3.8, to all face images and then applying different subspace clustering algorithms to the low-rank component of the data obtained by RPCA-ADMM. Observe that applying RPCA-ADMM to all data points simultaneously may not be as effective as applying RPCA-ADMM to data points in each subject separately. This comes from the fact that RPCA-ADMM tends to bring the data points into a common low-rank subspace, which can result in decreasing the principal angles between subspaces and decreasing the distances between data points in different subjects. This can explain the increase in the clustering error of all clustering algorithms with respect to the results in Table 8.3. Still, notice that the clustering error for SSC is low for all different numbers of subjects. Specifically, SSC obtains 2.09% and 11.46% for clustering of data points in 2 and 10 subjects, respectively.

Table 8.5 shows the results of applying different clustering algorithms to the original data without first applying RPCA-ADMM to each group. Notice that the performance of LSA and SCC deteriorates dramatically, showing that these methods are very sensitive to gross errors. The performance of $\text{LRSC}_{2,1}$ with $q = 1$ is better, but the errors are still very high, especially as the number of subjects increases. In this case, changing q to $q = 4$ does help to significantly reduce the clustering error. SSC, on the other hand, performs very well, achieving a clustering error

Table 8.4 Clustering error (%) of different algorithms on a few subjects from the extended Yale B data set after applying RPCA-ADMM simultaneously to all the data in each trial.

Algorithm	LSA	SCC	$\text{LRSC}_{2,1}$ $q=1$	$\text{LRSC}_{2,1}$ $q=4$	LRSC_{1} $q=1$	SSC
2 Subjects						
Mean	32.53	9.29	7.27	5.72	5.67	**2.09**
Median	47.66	7.03	6.25	3.91	4.69	**0.78**
3 Subjects						
Mean	53.02	32.00	12.29	10.01	8.72	**3.77**
Median	51.04	37.50	11.98	9.38	8.33	**2.60**
5 Subjects						
Mean	58.76	53.05	19.92	15.33	10.99	**6.79**
Median	56.87	51.25	19.38	15.94	10.94	**5.31**
8 Subjects						
Mean	62.32	66.27	31.39	28.67	16.14	**10.28**
Median	62.50	64.84	33.30	31.05	14.65	**9.57**
10 Subjects						
Mean	62.40	63.07	35.89	32.55	21.82	**11.46**
Median	62.50	60.31	34.06	30.00	25.00	**11.09**

Table 8.5 Clustering error (%) of different algorithms on a few subjects from the extended Yale B data set without pre-processing the data.

Algorithm	LSA	SCC	$\text{LRSC}_{2,1}$ $q=1$	$\text{LRSC}_{2,1}$ $q=4$	LRSC_{1} $q=1$	SSC
2 Subjects						
Mean	32.80	16.62	9.52	2.54	5.32	**1.86**
Median	47.66	7.82	5.47	0.78	4.69	**0.00**
3 Subjects						
Mean	52.29	38.16	19.52	4.21	8.47	**3.10**
Median	50.00	39.06	14.58	2.60	7.81	**1.04**
5 Subjects						
Mean	58.02	58.90	34.16	6.90	12.24	**4.31**
Median	56.87	59.38	35.00	5.63	11.25	**2.50**
8 Subjects						
Mean	59.19	66.11	41.19	14.34	23.72	**5.85**
Median	58.59	64.65	43.75	10.06	28.03	**4.49**
10 Subjects						
Mean	60.42	73.02	38.85	22.92	30.36	**10.94**
Median	57.50	75.78	41.09	23.59	28.75	**5.63**

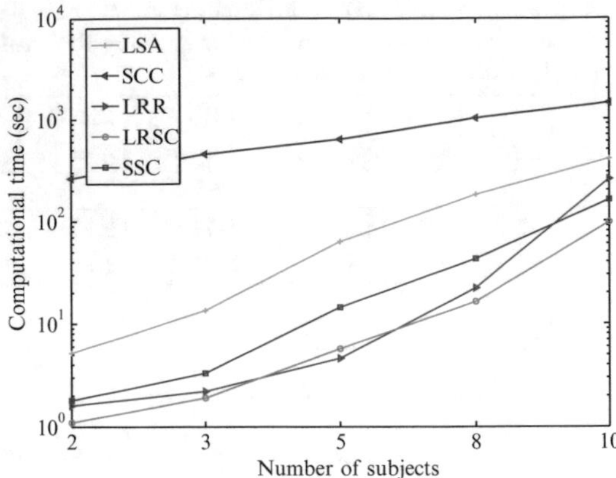

Fig. 8.10 Average computing time (seconds) of the algorithms on the extended Yale B database as a function of the number of subjects.

of about 10% for 10 subjects. Notice that this error is smaller than that obtained by applying RPCA-ADMM to all data points. This is due to the fact that SSC directly incorporates the corruption model of the data by sparse outlying entries into the sparse optimization program, giving it the ability to perform clustering on the corrupted data.

Figure 8.10 shows the average computing time of each algorithm as a function of the number of subjects (or equivalently the number of data points). Note that the computing time of SCC is drastically higher than that of other algorithms. This comes from the fact that the complexity of SCC increases exponentially in the dimension of the subspaces, which in this case is $d = 9$. On the other hand, SSC, LRR, and LRSC use fast and efficient convex optimization techniques that keep their computing time lower than that of other algorithms. Overall, LRR and LRSC are the fastest methods.

Face Clustering Errors on All Subjects from the Extended Yale B Data Set
Next, we present a more detailed evaluation of the clustering performance of different algorithms on a varying number of subjects from the extended Yale B database. In each experiment, we randomly pick n subjects out of all 38 subjects and cluster all images from these n subjects. Table 8.6 reports the average subspace clustering errors for $n \in \{2, 5, 10, 20, 30, 38\}$, where all results are averaged over 100 random choices of subjects except for the case of $n = 38$, where there is only one possible choice. The tested methods are the following:

1. K-means and spectral clustering algorithms described in Chapter 4, i.e., Algorithms 4.4 and 4.5, respectively. For spectral clustering, we evaluate the K-NN affinity with the number of neighbors set to $K = 5$, the Gaussian affinity with standard deviation $\sigma = 0.3$, and the ε-neighborhood affinity with $\varepsilon = 1.5$.

Table 8.6 Clustering error (%) of different algorithms on all subjects from the extended Yale B data set. The data set consists of face images of $n \in \{2, 10, 20, 30, 38\}$ randomly chosen individuals under 64 different illumination conditions. Images are down-sampled to size 48×42 as features. NA denotes running error in the process of the code.

No. subjects	2	5	10	20	30	38
K-means	48.7	77.1	86.8	90.5	92.1	92.6
SC: kNN	21.7	40.0	48.1	50.2	54.4	56.8
SC: Gauss	48.5	77.2	87.0	92.0	93.7	94.5
SC: neighborhood	44.0	65.3	78.6	83.9	86.4	88.9
K-subspaces	7.3	33.8	48.0	57.2	60.9	64.5
EM-MPPCA	16.1	47.4	56.5	62.2	65.0	67.1
LSA	34.0	58.9	64.8	65.2	68.6	68.7
SCC	15.1	NA	NA	NA	NA	90.9
SASC-angle	24.8	66.5	78.7	NA	NA	NA
SASC-distance	19.1	66.1	73.0	NA	NA	NA
SSC	0.8	2.0	6.8	13.5	17.9	22.8
LRSC	6.8	11.2	18.4	31.1	31.4	32.1

2. K-subspaces and MPPCA-EM algorithms described in Chapter 6, i.e., Algorithms 6.1 and 6.2, respectively. For all methods, we use random initializations for the parameters and set the subspace dimension to $d = 5$. To reduce computational complexity of these methods, we use PCA to project the original data to dimension $D = 9n$.
3. LSA, SCC, and SASC algorithms described in Chapter 7, i.e., Algorithms 7.1, 7.4, and 7.5, respectively. For LSA, we set the subspace dimension to $d = 5$ and the number of neighbors as $K = 7$. For SCC, we also set the subspace dimension to $d = 9$. For SASC, we use the angle-based affinity with $q = 10$ and the distance-based affinity with $q = 1$.
4. LRSC and SSC algorithms described in Chapter 8. For LRSC, we use the model for noisy data in (8.31), with the parameter set to $\tau = 150$. For SSC, we use the model with sparse outlying entries in (8.105) with the parameter set to $\tau = 30\tau_{\min,1}$, where $\tau_{\min,1}$ is defined in (8.106).

Table 8.6 reports the average subspace clustering errors of these algorithms. We have the following observations.

1. First, notice that the K-means and spectral clustering methods are not designed for subspace clustering, so they generally show much higher clustering errors than the subspace clustering methods.
2. Second, when comparing the three groups of subspace clustering methods, we can see that LRSC and SSC give the best performance, followed by the iterative and statistical methods, K-subspaces, and EM-MPPCA, respectively. Notice also

that in general, the spectral clustering methods do not perform well, except for the case $n = 2$.

3. Third, notice that while the K-subspaces method works fairly well for the case $n = 2$, its performance deteriorates very fast as the number of groups increases. This might be because the initialization of the algorithm is more difficult when there are many groups.
4. The LSA method is built on the observation that a point and its nearest neighbors are usually from the same subspace. Intuitively, this assumption is more realistic if subspaces are densely sampled. But for the extended Yale B database, there are only 64 samples for each subspace, which may explain why the method does not work well even for the case of two subjects.
5. As discussed in Chapter 3, a few pixels from the face images may deviate from the subspace model due to cast shadows and specularities. Since SASC does not have an explicit way of dealing with such corruptions, this may explain why the SASC-angle and SASC-distance methods do not give good results. In contrast, LRSC and SSC can regularize their objective functions to account for corruptions, so their performance is much better.
6. The theoretical conditions for correctness of LRSC require the subspaces to be independent (see Theorem 8.9). In contrast, the results for SSC show that it can recover subspace-preserving representations under much broader conditions, e.g., in Theorem 8.25 it is shown that the subspaces can even intersect, which may explain why SSC outperforms LRSC in the experiments.

8.5 Bibliographic Notes

Low-Rank Subspace Clustering

The optimization problem solved by LRSC was originally introduced in (Liu et al. 2010; Wei and Lin 2010; Liu et al. 2013), although the subspace clustering affinity matrix used by LRSC had been proposed much earlier by (Costeira and Kanade 1998) outside the context of spectral clustering, and the theoretical correctness of this affinity had already been studied in (Kanatani 2001; Vidal et al. 2008). The closed-form solution in the case of uncorrupted data is due to (Wei and Lin 2010), the closed-form solution in the case of noisy data is due to (Favaro et al. 2011; Vidal and Favaro 2014), the nonconvex methods for dealing with noise corruptions are due to (Wei and Lin 2010; Favaro et al. 2011; Vidal and Favaro 2014), while the method for dealing with outliers was proposed in the original paper (Liu et al. 2010, 2013).

Sparse Subspace Clustering

The origins of SSC can be traced back to a conversation between René Vidal and Robert Azencott at the Johns Hopkins University library in 2007. At the time, René was trying to explain the algebraic subspace clustering techniques described in Chapter 5, when Robert wondered whether techniques from sparsity could be applied to the problem. At the time, René was looking into ways of making algebraic techniques robust to noise and outliers, but it wasn't clear how techniques from sparsity could be applied. Because of that, René had been working with Alvina Goh on methods based on locality rather than sparsity, which led to the initial development of LRMC in (Goh and Vidal 2007). Here, the idea of writing a point as a linear combination of neighboring points and using the coefficients for clustering was already present. However, the idea of using sparse representation theory was not. After the presentation of this work at CVPR07 (Computer Vision and Pattern Recognition 2007), René Vidal and Yi Ma started talking about sparse representation theory and how it could be used to address the issue of outliers in motion segmentation. These interactions led to the work of (Rao et al. 2008), which uses ℓ_1-minimization together with the self-expressiveness constraint for detecting outlier entries. However, in that work the sparse coefficients are not used for segmentation: once the data have been cleaned, the ALC algorithm described in Chapter 6 was used for clustering. The idea of using the sparse coefficients for clustering was in the air at an IMA workshop on "Multi-Manifold Data Modeling and Its Applications" that Ehsan Elhamifar, Yi Ma, and René Vidal attended. However, it was not until the work of (Elhamifar and Vidal 2009) that these ideas were rigorously formalized, and their theoretical correctness was demonstrated in the case of independent subspaces. Ehsan then extended the theoretical analysis to disjoint subspaces in (Elhamifar and Vidal 2010). These ideas were also presented at a tutorial in CVPR10, which Emannuel Candès attended. This motivated Emmanuel and Mahdi Soltanolkotabi to study more general conditions for correctness that ultimately appeared in (Soltanolkotabi and Candès 2013). In particular, a major stepping-stone was the development of theoretical conditions for handling outliers. Finally, the theoretical analysis for the conditions of correctness of SSC in the presence of noise appeared in (Wang and Xu 2013; Soltanolkotabi et al. 2014).

8.6 Exercises

Exercise 8.1. Let $X = U_1 \Sigma_1 V_1^\top$ be the compact SVD of X. Show that the optimal solution to

$$\min_{C} \ \|C\|_* \quad \text{s.t.} \quad X = XC \tag{8.117}$$

is given by $C = V_1 V_1^\top$.

Exercise 8.2. Prove Lemma 8.3.

Exercise 8.3. Show that the optimization problem in (8.95) is equivalent to

$$\min_{C,Z} \quad \|C\|_1 + \frac{\tau}{2}\|X - XZ\|_F^2 \quad \text{s.t.} \quad Z = C - \operatorname{diag}(C). \tag{8.118}$$

Exercise 8.4. Show that the dual of the optimization problem

$$\min_{c} \|c\|_1 \quad \text{s.t.} \quad Xc = x \tag{8.119}$$

is given by

$$\max_{\lambda} \langle x, \lambda \rangle \quad \text{s.t.} \quad \|X^\top \lambda\|_\infty \leq 1. \tag{8.120}$$

Exercise 8.5. Show that the dual of the optimization problem

$$\min_{c} \|c\|_1 + \frac{\tau}{2}\|Xc - x\|_2^2 \tag{8.121}$$

is given by

$$\max_{\lambda \in \mathbb{R}^D} \langle x, \lambda \rangle - \frac{1}{2\tau}\|\lambda\|_2^2 \quad \text{s.t.} \quad \|X^\top \lambda\|_\infty \leq 1. \tag{8.122}$$

Exercise 8.6. Let $X = \left[x_1, \ldots, x_N\right]$, let $\mathcal{P}(X) = \text{conv}(\pm x_1, \pm x_2, \ldots, \pm x_N)$ be the symmetrized convex hull of the columns of X, and let $r(\mathcal{P}(X))$ be the inradius of $\mathcal{P}(X)$, i.e., the radius of the smallest ball contained in $\mathcal{P}(X)$. Show that

$$\min_{c}\{\|c\|_1 : Xc = x\} \leq \frac{\|x\|_2}{r(\mathcal{P}(X))}. \tag{8.123}$$

Further assume that the columns of X are drawn from a union of n subspaces $\{S_i\}_{i=1}^n$ of dimensions $\{d_i\}_{i=1}^n$. Let $X^i \in \mathbb{R}^{D \times N_i}$ be the submatrix of X containing the points in the ith subspace and assume that $\text{rank}(X^i) = d_i$. Let $X^i_{-j} \in \mathbb{R}^{D\times(N_i-1)}$ is the submatrix of X^i with the jth column removed. Show that

$$\frac{1}{\sqrt{d_i}} \max_{\tilde{X}_i \in \mathbb{W}_i} \sigma_{d_i}(\tilde{X}_i) < \min_{j: x_j \in S_i} r(\mathcal{P}^i_{-j}), \tag{8.124}$$

where $\mathbb{W}_i$ is the set of all full-rank submatrices $\tilde{X}_i \in \mathbb{R}^{D\times d_i}$ of X^i.

Part III
Applications

Chapter 9
Image Representation

Everything should be made as simple as possible, but not simpler.

—Albert Einstein

In this and the following chapters, we demonstrate why multiple subspaces can be a very useful class of models for image processing and how the subspace clustering techniques may facilitate many important image processing tasks, such as image representation, compression, image segmentation, and video segmentation.

9.1 Seeking Compact and Sparse Image Representations

Since smart phones and electronic cameras become popular, the quantity of images being captured and transmitted daily has been increasing at an explosive rate. It has become a pressing problem to find more efficient and compact representations of images for many purposes, including storage, transmitting, enhancing, and understanding. In this chapter, we will not be able to provide a complete solution to the image compression problem. Nevertheless, we will try to illustrate some of the basic concepts behind image compression and show how the multiple-subspace models that we have studied in this book should be able to shed some new light on the developing future, arguably more efficient, image compression methods.

Except for a few image representations such as fractal-based approaches (Fisher 1995), most existing image representations apply certain linear transformations to the image so that the energy of the (transformed) image will be concentrated in the coefficients of a small set of bases of the transformation, also known as a sparse representation. To be more precise, if we view an image (or patches of an image) as a signal $\boldsymbol{x} \in \mathbb{R}^n$, we seek a linear representation of $\boldsymbol{x}$ as

$$\boldsymbol{x} = \Phi\alpha,$$

R. Vidal et al., *Generalized Principal Component Analysis*, Interdisciplinary Applied Mathematics 40, DOI 10.1007/978-0-387-87811-9_9

for some $\Phi = [\phi_1, \phi_2, \ldots, \phi_m] \in \mathbb{R}^{n \times m}, \alpha \in \mathbb{R}^m$. The columns of the matrix Φ are sometimes called base vectors or atoms. In classical image compression schemes such as JPEG, it is typical to choose $m = n$ and let Φ be a full-rank matrix representing an invertible linear transform such as the Fourier transform, discrete cosine transform (JPEG), or wavelet transform (JPEG-2000). The essential goal of such a linear transform is to ensure that in the transformed domain, $\alpha \in \mathbb{R}^m$ has only very sparse nonzero entries and hence is much more compressible than the original image signal $\boldsymbol{x}$.

Finding such a sparser representation is typically the first step in subsequent (lossy) compression and transmission of images or videos.[1] The result can also be used for other purposes such as image segmentation,[2] classification, and object recognition.

Most of the popular methods for obtaining a sparse representation of images can be classified into several categories. We give here a brief overview of some representative methods.

9.1.1 Prefixed Linear Transformations

Methods in this category seek to transform all images (or patches) using a *prefixed* linear transformation. That is, we apply the same linear transform $\Phi \in \mathbb{R}^{n \times n}$ regardless of the input image or image patch $\boldsymbol{x}$:

$$\boldsymbol{x} = \Phi\alpha, \quad \text{or} \quad \alpha = \Phi^{-1}\boldsymbol{x}. \tag{9.1}$$

Each image is then represented as a superposition of a fixed set of base vectors (as columns of the transformation matrix Φ). All these methods evolved essentially from the classical Fourier transform. One variation of the (discrete) Fourier transform, the discrete cosine transform (DCT), serves as the core of the JPEG standard (Wallace 1991).

Due to the Gibbs phenomenon, DCT is poor at approximating discontinuities in the imagery signal. Wavelets (DeVore et al. 1992; Donoho et al. 1998; Mallat 1999; Shapiro 1993) were developed to remedy this problem, and they have been shown to be very effective for representing 1-dimensional piecewise smooth signals with discontinuities. JPEG-2000 adopted wavelets as its standard.

However, because wavelet transforms deal only with 1D discontinuities, they are not well-suited to represent 2D singularities along edges or contours. Anisotropic bases such as wedgelets (Donoho 1999), curvelets (Candès and Donoho 2002), countourlets (Do and Vetterli 2002), and bandlets (LePennec and Mallat 2005) have been proposed explicitly to capture different 2D discontinuities. These x-lets have

[1] Which involves further quantization and entropy coding of the so-obtained sparse signals α.

[2] A topic we will study in more detail from the compression perspective in the next section.

been shown to be (approximately) optimal for representing objects with singularities along C^2-smooth edges.[3] Nevertheless, natural images, especially images that have complex textures and patterns, do not consist solely of discontinuities along C^2-smooth edges. This is probably the reason why in practice, these edge-based methods do not seem to outperform (separable) wavelets on complex images.

More generally, one should not expect that a (fixed) "gold-standard" invertible transformation would work optimally for all images (and signals) in the world. Furthermore, conventional image (or signal) processing methods were developed primarily for grayscale images. For color images or other multiple-valued images, one has to apply them to each value separately (e.g., one color channel at a time). The strong correlation that is normally present among the multiple values or colors is unfortunately ignored by this class of methods.

9.1.2 Adaptive, Overcomplete, and Hybrid Representations

To remedy limitations of a pre-fixed transform, another category of methods aims to identify the optimal (in some sense) representation that is *adaptive* to specific statistics or structures of each image or a special class of images.[4]

For instance, suppose we are working with a set of images or image patches as columns of a data matrix $X = [\boldsymbol{x}_1, \boldsymbol{x}_2, \ldots, \boldsymbol{x}_N]$. If these images or patches are highly correlated, then these columns, viewed as vectors lying in a high-dimensional space, span a rather low-dimensional subspace. Then we can identify a basis for this subspace via the Karhunen–Loève transform (KLT) or principal component analysis (PCA) (Effros and Chou 1995):

$$X \approx U \Sigma V^{\top} = UA,$$

where (U, Σ, V) is a compact singular value decomposition of the data: $U \in \mathbb{R}^{n \times d}$, $V \in \mathbb{R}^{d \times N}$, and $\Sigma \in \mathbb{R}^{d \times d}$ is a diagonal matrix of all significant singular values. As we see, if $d \ll n$, then each n-dimensional image $\boldsymbol{x}_j$ can be represented with a d-dimensional vector α_j. Of course, for any other image $\boldsymbol{x}$ belonging to this category, we can use the learned basis $\Phi = U$ to (approximately) represent it with a signal of much lower dimension:

$$\boldsymbol{x} \approx \Phi \alpha = U \alpha, \quad \alpha \in \mathbb{R}^d. \tag{9.2}$$

[3]Here, "optimality" means that the transformation achieves the optimal asymptotic for approximating the class of functions considered (DeVore 1998).

[4]Here, in contrast to the case of pre-fixed transformations, "optimality" means the representation obtained is the optimal one within the class of models considered, in the sense that it minimizes certain discrepancies between the model and the data.

Notice that here U is a very tall matrix; hence the representation is naturally more compact than the original signal.

In theory, the basis identified via PCA provides the optimal linear representation, assuming that the imagery data satisfy a unimodal (subspace-like) distribution. However, in reality, this assumption is rarely true. Different classes of images may have rather different statistical characteristics; hence the basis learned from one class may not apply well to other types of images. In fact, even within a single image, patches typically exhibit multimodal statistics, since the image usually contains many heterogeneous regions with significantly different statistical characteristics (e.g., Figure 9.2). As we have seen in previous chapters, heterogeneous data can be better represented using a mixture of parametric models, one for each homogeneous subset. Such a mixture of models is often referred to as a *hybrid model*.

Vector quantization (VQ) (Gersho and Gray 1992) is a popular method in image processing and compression that assumes that the imagery data (say patches) are clustered around many different centers. The centers of those clusters can be found only via algorithms such as K-means. From the dimension-reduction point of view, VQ represents the imagery data with many 1-dimensional subspaces. Or more precisely, in VQ, one tries to represent an image or image patch as

$$\boldsymbol{x} = \Phi\alpha = [\mu_1, \mu_2, \ldots, \mu_m]\alpha, \quad \|\alpha\|_0 = 1, \tag{9.3}$$

where μ_i are the cluster centers and α is a vector that is required to have only one nonzero coefficient.[5] This model typically leads to an excessive number of clusters if we want to accurately represent very complicated textures.[6]

To make the above models more effective, one could relax the requirement that the vector α can have only one nonzero coefficient. An increasingly popular model allows the image (or patch) to be represented as a sparse superposition of the columns of a matrix Φ:

$$\boldsymbol{x} = \Phi\alpha = [\phi_1, \phi_2, \ldots, \phi_m]\alpha, \quad \|\alpha\|_0 \leq k, \tag{9.4}$$

where, in contrast to the VQ model (9.3), here α can have multiple nonzero entries but be sparse enough. To alleviate the limitations of the linear transformation model (9.1) and the PCA model (9.2), the matrix Φ does not need to be a square or tall matrix. In fact, it is often chosen to be overcomplete or redundant (Bruckstein et al. 2009; Elad et al. 2010), i.e., $m \geq n$. Notice that the set of all k-sparse signals with respect to Φ lies on the union of k-dimensional subspaces, each spanned by a set of k linearly independent columns of Φ. Notice that such a sparse model contains

[5] In fact, in the VQ model, the coefficients are assumed to be binary.

[6] Be aware that compared to methods in the first category, representations in this category typically need additional memory to store the information about the resulting model itself, e.g., the basis of the subspace in PCA, the cluster means in VQ.

a total of $\binom{m}{k}$ subspaces. Given an overcomplete basis, also known as a dictionary in the literature, many methods and theories have been developed in the past few years for effectively and efficiently computing a sparse representation α for a given input signal $\boldsymbol{x}$ (Donoho and Elad 2003; Candès 2006; Candès and Wakin 2008).

One opening issue with the sparse representation approach is how to determine the dictionary $\Phi \in \mathbb{R}^{n\times m}$? Also, even if we identify such a dictionary, would all the $\binom{m}{k}$ subspaces be necessary for representing the images? We know that a real image often naturally partitions into multiple regions of color and textures, say K regions $R_1, \dots, R_K$. It is reasonable to represent image patches in each region R_i by a low-dimensional subspace with a basis U_i. If we concatenate bases of all the subspaces together, then any patch $\boldsymbol{x}$ in the image can be represented in the form

$$\boldsymbol{x} = \Phi\alpha = [U_1, U_2, \dots, U_K]\alpha, \tag{9.5}$$

regardless of the subspace to which it belongs. Here α is naturally sparse if $\boldsymbol{x}$ belongs to one of the subspaces, since it will have at most $d_i = \dim(U_i)$ nonzero coefficients. However, this model is different from the above sparse model (9.4) in that we do not need arbitrary sparse combinations of the columns of Φ: α has nonzero entries associated with only one of the blocks U_i. We say that such an α is block-sparse. Therefore, the model essentially describes a union of only K subspaces. Since each component model represents a linear subspace, we often refer to such a model as a *hybrid linear model*. In some sense, such a model remedies limitations of the PCA model (9.2), the VQ model (9.3), and the sparse model (9.4). It strikes a good balance between simplicity and expressiveness for representing natural images.

The remaining issue is how to effectively identify a best hybrid linear model for any given image set and evaluate how much more effective such a hybrid linear model actually is compared to conventional image representations discussed above. Obviously, the techniques that we have introduced in earlier chapters of this book can be applied to identify these subspaces, including the number of subspaces, their dimensions, and bases. Nevertheless, in this chapter, we will see how to determine which subspace clustering and estimation algorithm is the one most suitable for a given application and how the algorithm needs to be properly customized and used in the special setting of the application.

9.1.3 Hierarchical Models for Multiscale Structures.

One important characteristic of natural images is that their statistical characteristics are typically self-similar across different spatial scales. Sometimes, this is also referred to as scale-invariance of natural image statistics. This is a very important property: it essentially suggests that if one class of models applies well to an image, one should expect that it should apply equally well to down-sampled versions of the image as well.

Many existing frequency-domain techniques harness this characteristic (Burt and Adelson 1983). For instance, wavelets, curvelets, and fractals have all demonstrated effectiveness in decomposing the original imagery signal into multiple scales (or subbands). As the result of such a *multiscale* decomposition, the low-dimensional structures of the image at different scales can be modeled in a multiresolution fashion. As we will see, this significantly reduces the dimension of the signals to which we need to apply the models and hence reduces the overall computational complexity.

In this chapter, we will show how to apply the above hybrid linear model (9.5) in a multiscale fashion so as to obtain even more compact image representations. The resulting scheme is a *multiscale hybrid linear model*, which can be simply described as follows:

> *Given an image, at each scale level of its down-sampled pyramid, fit the (residual) image by a (multiple-subspace) hybrid linear model.*

Compared to a hybrid linear model at a single scale, the multiscale scheme can reduce not only the complexity of the resulting representation but also the overall computational cost. Surprisingly, as we will demonstrate, such a simple scheme is able to generate representations for natural images that are more compact, even with the overhead needed to store the model, than most state-of-the-art representations, including DCT, PCA, and wavelets.

9.2 Image Representation with Multiscale Hybrid Linear Models

9.2.1 Linear versus Hybrid Linear Models

In this section, we introduce and examine the hybrid linear model for image representation. The relationship between hybrid linear models across different spatial scales will be discussed in Section 9.2.2.

An image $\mathcal{I}$ with width W, height H, and c color channels resides in a very high-dimensional space $\mathbb{R}^{W\times H\times c}$. We may first reduce the dimension by dividing the image into a set of nonoverlapping $b \times b$ blocks.[7] Each $b \times b$ block is then stacked into a vector $\boldsymbol{x} \in \mathbb{R}^D$, where $D = b^2c$ is the dimension of the ambient space. For example, if $c = 3$ and $b = 2$, then $D = 12$. In this way, the image $\mathcal{I}$ is converted to a set of vectors $\mathcal{X} = \{\boldsymbol{x}_j \in \mathbb{R}^D\}_{j=1}^N$, where $N = WH/b^2$ is the total number of vectors.

The most commonly adopted distance measure for image compression is the mean square error (MSE) between the original image $\mathcal{I}$ and (compressed) approximate image $\hat{\mathcal{I}}$:

[7]Therefore, b needs to be a common divisor of W and H.

$$\varepsilon_I^2 = \frac{1}{WHc}\|\hat{\mathcal{I}} - \mathcal{I}\|^2. \tag{9.6}$$

Since in the following derivation we will be approximating the (block) vectors $\{\boldsymbol{x}_j\}_{j=1}^N$ rather than individual image pixels, it is more convenient for us to define the mean square error (MSE) *per vector*, which is different from ε_I^2 by a scale factor,

$$\varepsilon^2 = \frac{1}{N}\sum_{j=1}^{N}\|\hat{\boldsymbol{x}}_j - \boldsymbol{x}_j\|^2 = \frac{b^2}{WH}\sum_{j=1}^{N}\|\hat{\boldsymbol{x}}_j - \boldsymbol{x}_j\|^2 = \frac{b^2}{WH}\|\hat{\mathcal{I}} - \mathcal{I}\|^2 = (b^2c)\varepsilon_I^2. \tag{9.7}$$

The peak signal-to-noise ratio (PSNR) of the approximate image is defined to be[8]

$$\text{PSNR} \doteq -10\log\varepsilon_I^2 = -10\log\frac{\varepsilon^2}{b^2c}. \tag{9.8}$$

Linear Models.
If we assume that the vectors $\boldsymbol{x}$ are drawn from a (nearly degenerate) Gaussian distribution or a linear subspace, the optimal model subject to a given PSNR can be inferred by principal component analysis (PCA) (Pearson 1901; Hotelling 1933; Jolliffe 2002) or equivalently the Karhunen–Loève transform (KLT) (Effros and Chou 1995). The effectiveness of such a linear model relies on the assumption that although D can be large, all the vectors $\boldsymbol{x}$ may lie on a subspace of a much lower dimension in the ambient space $\mathbb{R}^D$. Figure 9.1 illustrates this assumption.

Let $\bar{\boldsymbol{x}} = \frac{1}{N}\sum_{j=1}^N \boldsymbol{x}_i$ be the mean of the imagery data vectors, and let $X \doteq [\boldsymbol{x}_1 - \bar{\boldsymbol{x}}, \boldsymbol{x}_2 - \bar{\boldsymbol{x}}, \ \ldots\ , \boldsymbol{x}_N - \bar{\boldsymbol{x}}] = U\Sigma V^\top$ be the SVD of the mean-subtracted

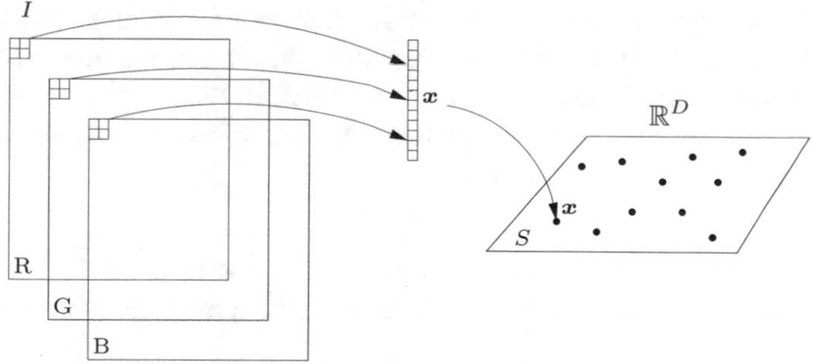

Fig. 9.1 In a linear model, the imagery data vectors $\{\boldsymbol{x}_j \in \mathbb{R}^D\}$ reside in an (affine) subspace S of dimension $d \ll D$.

[8]Here by default, the peak value of the imagery data is normalized to 1.

data matrix X. Then all the vectors $\boldsymbol{x}_j$ can be represented as a linear superposition: $\boldsymbol{x}_j = \bar{\boldsymbol{x}} + \sum_{k=1}^{D} \alpha_j^k \phi_k, j = 1, \ldots, N$, where $\{\phi_k\}_{k=1}^D$ are just the columns of the matrix U.

The matrix $\Sigma = \text{diag}(\sigma_1, \sigma_2, \ldots, \sigma_D)$ contains the ordered singular values $\sigma_1 \geq \sigma_2 \geq \cdots \geq \sigma_D$. It is well known that the optimal linear representation of $\boldsymbol{x}_j$ subject to the MSE ε^2 is obtained by keeping the first d (principal) components

$$\hat{\boldsymbol{x}}_j \doteq \bar{\boldsymbol{x}} + \sum_{k=1}^{d} \alpha_j^k \phi_k, \quad j = 1, \ldots, N, \tag{9.9}$$

where d is chosen to be

$$d = \min(k), \quad \text{s.t.} \quad \frac{1}{N} \sum_{i=k+1}^{D} \sigma_i^2 \leq \varepsilon^2. \tag{9.10}$$

The model complexity of the linear model, denoted by Ω, is the total number of coefficients needed for representing the model $\{\alpha_j^k, \phi_k, \bar{\boldsymbol{x}}\}$ and subsequently a lossy approximation $\hat{\mathcal{I}}$ of the image $\mathcal{I}$. It is given by

$$\Omega(N, d) \doteq Nd + d(D - d + 1), \tag{9.11}$$

where the first term is the number of coefficients $\{\alpha_j^k\}$ to represent $\{\hat{\boldsymbol{x}}_j - \bar{\boldsymbol{x}}\}_{j=1}^N$ with respect to the basis $\Phi = \{\phi_k\}_{k=1}^d$, and the second term is the number of Grassmannian coordinates[9] needed for representing the basis Φ and the mean vector $\bar{\boldsymbol{x}}$. The second term is often called *overhead*.[10] Notice that the original set of vectors $\{\boldsymbol{x}_j\}$ contains ND coordinate entries. If $\Omega \ll ND$, then the new representation, although lossy, is more compact. The search for such a compact representation is at the heart of every (lossy) image compression method. When the image $\mathcal{I}$ is large and the block size b is small, N will be much larger than D, so that the overhead will be much smaller than the first term. However, in order to compare this method fairly with other methods, in the subsequent discussions and experiments, we always count the total number of coefficients needed for the representation, including the overhead.

Hybrid Linear Models

The linear model is very efficient when the distribution of the data $\{\boldsymbol{x}_j\}$ is indeed unimodal. However, if the image $\mathcal{I}$ contains several heterogeneous textural regions

---

[9]Notice that to represent a d-dimensional subspace in a D-dimensional space, we need only specify a basis of d linearly independent vectors for the subspace. We may stack these vectors as rows of a $d \times D$ matrix. Any nonsingular linear transformation of these vectors span the same subspace. Thus, without loss of generality, we may assume that the matrix is of the normal form $[I_{d \times d}, G]$, where G is a $d \times (D - d)$ matrix consisting of the so-called Grassmannian coordinates.

[10]Notice that if one uses a preselected basis, such as discrete Fourier transform, discrete cosine transform (JPEG), or wavelets (JPEG-2000), there is no such overhead.

$\{\mathcal{I}_i\}_{i=1}^n$, the data vectors $\{\boldsymbol{x}_j\}$ can no longer be modeled well as samples from a single subspace. It is more natural to assume that they come from a number of subspaces, with one subspace corresponding to one homogeneous textural region. Since complexity of texture can be different in different regions, the dimensions of the subspaces can also be different. Figure 9.2 displays the first three principal components of the data vector $\boldsymbol{x}_j$ (as dots in $\mathbb{R}^3$) of an image. Note the clear multimodal characteristic in the data.

Suppose that a natural image $\mathcal{I}$ can be segmented into n disjoint regions $\mathcal{I} = \cup_{i=1}^n \mathcal{I}_i$ with $\mathcal{I}_i \cap \mathcal{I}_{i'} = \emptyset$ for $i \neq i'$. In each region $\mathcal{I}_i$, we may assume that the linear model (9.9) is valid for the subset of vectors $\{\boldsymbol{x}_{i,j}\}_{j=1}^{N_i}$ in $\mathcal{I}_i$:

$$\hat{\boldsymbol{x}}_{i,j} = \bar{\boldsymbol{x}}_i + \sum_{k=1}^{d_i} \alpha_j^k \phi_{i,k}, \quad j = 1, \ldots, N_i. \tag{9.12}$$

Intuitively, the hybrid linear model can be illustrated by Figure 9.3.

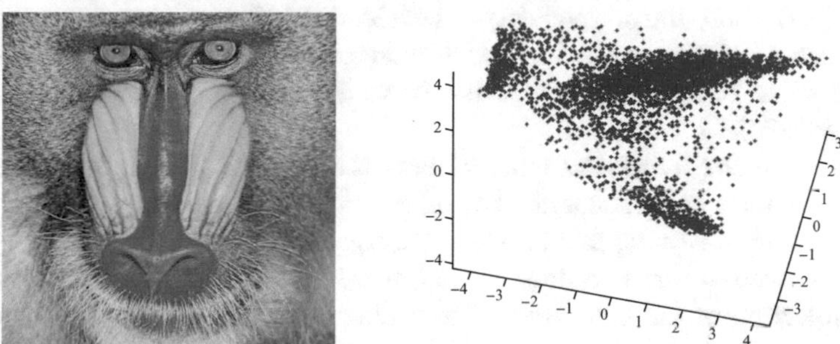

Fig. 9.2 Left: The baboon image. Right: The coordinates of each dot are the first three principal components of the patch vectors $\boldsymbol{x}_j$. There is a clear multimodal structure in the data.

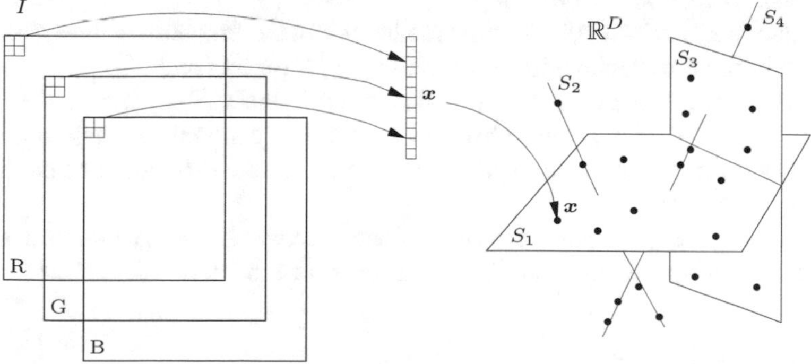

Fig. 9.3 In hybrid linear models, the imagery data vectors $\{\boldsymbol{x}_j\}$ reside in multiple (affine) subspaces, which may have different dimensions.

As in the linear model, the dimension d_j of each subspace is determined by a common desired MSE ε^2 using equation (9.10). The model complexity, i.e., the total number of coefficients needed to represent the hybrid linear model $\{\phi_{i,k}, \hat{\boldsymbol{x}}_{i,j}\}$, is[11]

$$\Omega = \Omega(N_1, d_1) + \cdots + \Omega(N_n, d_n) = \sum_{i=1}^{n} \big(N_i d_i + d_i(D - d_i + 1)\big). \quad (9.13)$$

Notice that Ω is similar to the effective dimension (ED) of a subspace arrangement model defined in Chapter 5. Thus, finding a representation that minimizes Ω is the same as minimizing the effective dimension of the imagery data set.[12]

Instead, if we model the union of all the vectors $\cup_{i=1}^{n}\{\boldsymbol{x}_{i,j}\}_{j=1}^{N_i}$ with a single subspace (subject to the same MSE), then the dimension of the subspace in general needs to be $d = \min\{d_1 + \cdots + d_n, D\}$. It is easy to verify from the definition (9.11) that under reasonable conditions (e.g., n is bounded from being too large), we have

$$\Omega(N, d) \; > \; \Omega(N_1, d_1) + \cdots + \Omega(N_n, d_n). \quad (9.14)$$

Thus, if a hybrid linear model can be identified for an image, the resulting representation will in general be much more compressed than that with a single linear or affine subspace. This will also be verified by experiments on real images in Section 9.2.3.

However, such a hybrid linear model alone is not able to generate a representation that is as compact as that generated by other competitive methods such as wavelets. There are at least two aspects in which the above model can be further improved. Firstly, we need to further reduce the negative effect of overhead by incorporating a pre-projection of the data onto a lower-dimensional space. Secondly, we need to implement the hybrid linear model in a multiscale fashion. We will discuss the former aspect in the remainder of this section and leave the issues of multiscale implementation to the next section.

Dimension Reduction via Projection

In the complexity of the hybrid linear model (9.13), the first term is always smaller than that of the linear model (9.11), because $d_i \leq d$ for all i and $\sum_{i=1}^{n} N_i = N$. The second overhead term, however, can be larger than in that of the linear model (9.11), because the bases of multiple subspaces now must be stored. We here propose a method to further reduce the overhead by separating the estimation of the hybrid model into two steps.

In the first step, we may project the data vectors $\mathcal{X} = \{\boldsymbol{x}_j\}$ onto a lower-dimensional subspace (e.g., via PCA) so as to reduce the dimension of the ambient

---

[11] We also need a very small number of binary bits to store the membership of the vectors. But those extra bits are insignificant compared to Ω and often can be ignored.

[12] In fact, the minimal Ω can also be associated with the Kolmogorov entropy or with the minimum description length (MDL) of the imagery data.

space from D to D'. The justification for such a subspace projection has been discussed earlier in Section 5.1.2. Here, the dimension D' is chosen to achieve an MSE $\frac{1}{2}\varepsilon^2$. The data vectors in the lower ambient space $\mathbb{R}^{D'}$ are denoted by $\mathcal{X}' = \{\boldsymbol{x}'_j\}$. In the second step, we identify a hybrid linear model for $\{\boldsymbol{x}'_j\}$ within the lower-dimensional ambient space $\mathbb{R}^{D'}$. In each subspace, we determine the dimension d_i subject to the MSE $\frac{1}{2}\varepsilon^2$. The two steps combined achieve an overall MSE ε^2, but they can actually reduce the total model complexity to

$$\Omega = \sum_{i=1}^{n} \big(N_i d_i + d_i(D' - d_i + 1)\big) + D(D' + 1). \tag{9.15}$$

This Ω will be smaller than the Ω in equation (9.13), because D' is smaller than D. The reduction of the ambient space will also make the identification of the hybrid linear model (say via the algebraic subspace clustering algorithm) much faster.

If the number n of subspaces is given, algorithms like the algebraic subspace clustering algorithm introduced in Chapter 5 and the statistical EM algorithm in Chapter 6 can always find a segmentation. The basis $\{\phi_{i,k}\}$ and dimension d_i of each subspace are determined by the desired MSE ε^2. As n increases, the dimension of the subspaces may decrease, but the overhead required to store the bases may increase. The optimal n^* can therefore be found recursively by minimizing Ω for different n's, as shown in Figure 9.4.

In our experience, we have found that n is typically in the range from 2 to 6 for natural images, especially in the multiscale implementation that we will introduce next.

Algorithm 9.1 describes the pseudocode for estimating the hybrid linear model of an image $\mathcal{I}$, in which the *SubspaceSegmentation*$(\cdot)$ function is implemented (for the experiments in this chapter) using the algebraic subspace clustering algorithm given in Chapter 5.

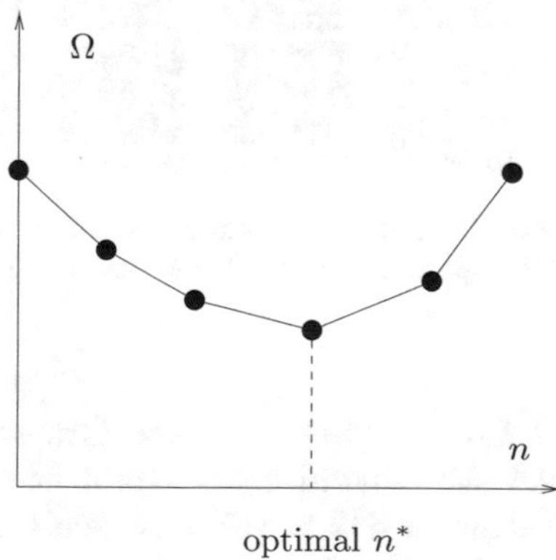

Fig. 9.4 The optimal n^* can be found by minimizing Ω with respect to n.

Algorithm 9.1 (Hybrid Linear Model Estimation).

function $\hat{\mathcal{I}} = \text{HybridLinearModel}(\mathcal{I}, \varepsilon^2)$
$\{\boldsymbol{x}_j\} = \text{StackImageIntoVectors}(\mathcal{I})$;
$\{\boldsymbol{x}'_j\}, \{\phi_k\}, \{\alpha^k_j\} = \text{PCA}(\{\boldsymbol{x}_j - \bar{\boldsymbol{x}}\}, \frac{1}{2}\varepsilon^2)$;
for each possible n **do**
  $\{\boldsymbol{x}'_{i,j}\} = \text{SubspaceSegmentation}(\{\boldsymbol{x}'_j\}, n)$;
  $\{\hat{\boldsymbol{x}'}_{i,j}\}, \{\phi_{i,k}\}, \{\alpha^k_{i,j}\} = \text{PCA}(\{\boldsymbol{x}'_{i,j} - \bar{\boldsymbol{x}}'_i\}, \frac{1}{2}\varepsilon^2)$;
  compute Ω_n;
end for
$\Omega_{opt} = \min(\Omega_n)$;
$\hat{\mathcal{I}} = \text{UnstackVectorsIntoImage}(\{\hat{\boldsymbol{x}'}_{i,j}\}$ with $\Omega_{opt})$;
output $\{\alpha^k_j\}, \{\phi_k\}, \bar{\boldsymbol{x}}, \{\alpha^k_{i,j}\}, \{\phi_{i,k}\}, \{\bar{\boldsymbol{x}}'_i\}$ with Ω_{opt};
return $\hat{\mathcal{I}}$.

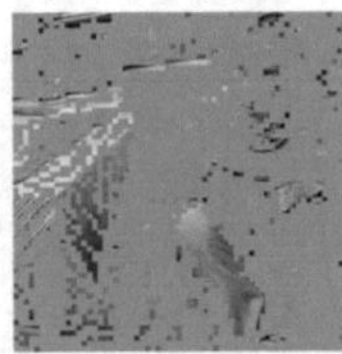
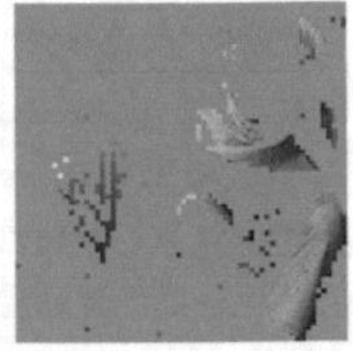

Fig. 9.5 The segmentation of the 4096 image blocks from the Barbara image. The image (left) is segmented into three groups (right three). Roughly speaking, the first subspace contains mostly image blocks with homogeneous textures; the second and third subspaces contain blocks with textures of different spatial orientations and frequencies.

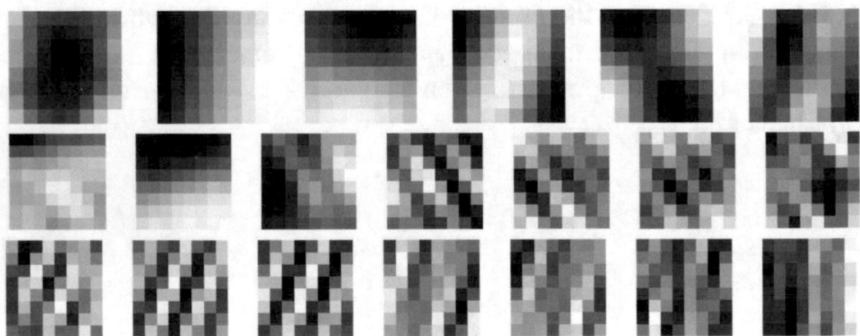

Fig. 9.6 The three sets of bases for the three subspaces (of blocks) shown in Figure 9.5, respectively. One row for one subspace and the number of base vectors (blocks) is the dimension of the subspace.

Example 9.1 (A Hybrid Linear Model for the Grayscale Barbara Image). Figure 9.5 and Figure 9.6 show intuitively a hybrid linear model identified for the 8×8 blocks of the standard 512×512 grayscale Barbara image. The total number of blocks is $N = 4096$. The algebraic subspace clustering algorithm identifies three subspaces for these blocks (for a given error tolerance), as shown in Figure 9.5. Figure 9.6 displays the three sets of bases for the three subspaces identified, respectively. It is worth noting that these bases are very consistent with the textures of the image blocks in the respective groups.

For this application, there are several good reasons why the algebraic subspace clustering algorithm is chosen over the statistical and the sparse subspace clustering algorithms:

- The dimension of the vectors is in the range that can be handled efficiently by the algebraic method, whereas the sparse clustering method is less effective, since the dimension is low and yet the number of samples is very large.
- To evaluate the complexity of the resulting model for compression purposes, we need to know explicitly the model parameters (number, dimensions, and bases of subspaces) through the identification process, which makes the nonparametric lossy-compression-based method less convenient to use.
- We normally do not know the number and dimensions of the subspaces in advance, nor we are given any reasonable initial guess of those subspaces, which makes the use of other parametric statistical methods such as K-subspace and EM difficult.

Nevertheless, be aware that the choice of the algebraic method here is primarily for simplicity and convenience (and it does serve the purpose well already). We do not rule out the possibility that one could modify other subspace clustering and estimation algorithms and obtain even better results for image compression in the future.

9.2.2 *Multiscale Hybrid Linear Models*

There are at least several reasons why the above hybrid linear model needs further improvement. Firstly, the hybrid linear model treats low-frequency/low-entropy regions of the image in the same way as the high-frequency/high-entropy regions, which is inefficient. Secondly, by treating all blocks the same, the hybrid linear model fails to exploit stronger correlations that typically exist among adjacent image blocks.[13] Finally, estimating the hybrid linear model is computationally expensive when the image is large. For example, if we use 2×2 blocks, a 512×512 color image will have $M = 65{,}536$ data vectors in $\mathbb{R}^{12}$. Estimating a hybrid linear model for such a huge number of vectors can be difficult (if not impossible) on a regular PC. In this section, we introduce a multiscale hybrid linear representation that is able to resolve the above issues.

The basic ideas of multiscale representations such as the Laplacian pyramid (Burt and Adelson 1983) have been exploited for image compression for decades

[13] For instance, if we take all the $b \times b$ blocks and scramble them arbitrarily, the scrambled image would be fit equally well by the same hybrid linear model for the original image.

(e.g., wavelets, subband coding). A multiscale method will give a more compact representation because it encodes low-frequency/low-entropy parts and high-frequency/high-entropy parts separately. The low-frequency/low-entropy parts are invariant after low-pass filtering and down-sampling, and can therefore be extracted from the much smaller down-sampled image. Only the high-frequency/high-entropy parts need to be represented at a level of higher resolution. Furthermore, the stronger correlations among adjacent image blocks will be captured in the down-sampled images, because every four image blocks are merged into one block in the down-sampled image. At each level, the number of imagery data vectors is one-fourth of that at one level above. Thus, the computational cost can also be reduced.

We now introduce a multiscale implementation of the hybrid linear model. We use the subscript l to indicate the level in the pyramid of down-sampled images $\mathcal{I}_l$.[14] The finest level (the original image) is indicated by $l = 0$. The larger the value of l, the coarser the down-sampled image. We define the highest level to be $l = L$.

Pyramid of Down-Sampled Images
First, the level-l image $\mathcal{I}_l$ passes a low-pass filter F_1 (averaging or Gaussian filter, etc.) and is down-sampled by 2 to get a coarser version image $\mathcal{I}_{l+1}$:

$$\mathcal{I}_{l+1} \doteq F_1(\mathcal{I}_l) \downarrow 2, \quad l = 0, \dots, L-1. \tag{9.16}$$

The coarsest level-L image $\mathcal{I}_L$ is approximated by $\hat{\mathcal{I}}_L$ using a hybrid linear model with the MSE ε_L^2. The number of coefficients needed for the approximation is Ω_L.

Pyramid of Residual Images
At all other levels l, $l = 0, \dots, L-1$, we do *not* need to approximate the down-sampled image $\mathcal{I}_l$, because it has been roughly approximated by the image at level-$(l+1)$ upsampled by 2. We only need to approximate the residual of this level, denoted by $\mathcal{I}_l'$:

$$\mathcal{I}_l' \doteq \mathcal{I}_l - F_2(\hat{\mathcal{I}}_{l+1}) \uparrow 2, \quad l = 0, \dots, L-1, \tag{9.17}$$

where F_2 is an interpolation filter. Each of these residual images $\mathcal{I}_l'$, $l = 0, \dots, L-1$ is approximated by $\hat{\mathcal{I}}_l'$ using a hybrid linear model with the MSE ε_l^2. The number of coefficients needed for the approximation is Ω_l, for each $l = 0, \dots, L-1$.

Pyramid of Approximated Images
The approximated image at the level l is denoted by $\hat{\mathcal{I}}_l$:

$$\hat{\mathcal{I}}_l \doteq \hat{\mathcal{I}}_l' + F_2(\hat{\mathcal{I}}_{l+1}) \uparrow 2, \quad l = 0, \dots, L-1. \tag{9.18}$$

Figure 9.7 shows the structure of a three-level ($L = 2$) approximation of the image $\mathcal{I}$. Only the hybrid linear models for $\hat{\mathcal{I}}_2$, $\hat{\mathcal{I}}_1'$, and $\hat{\mathcal{I}}_0'$, which are approximations

[14] This is not to be confused with the subscript i used to indicate different segments $\mathcal{I}_i$ of an image.

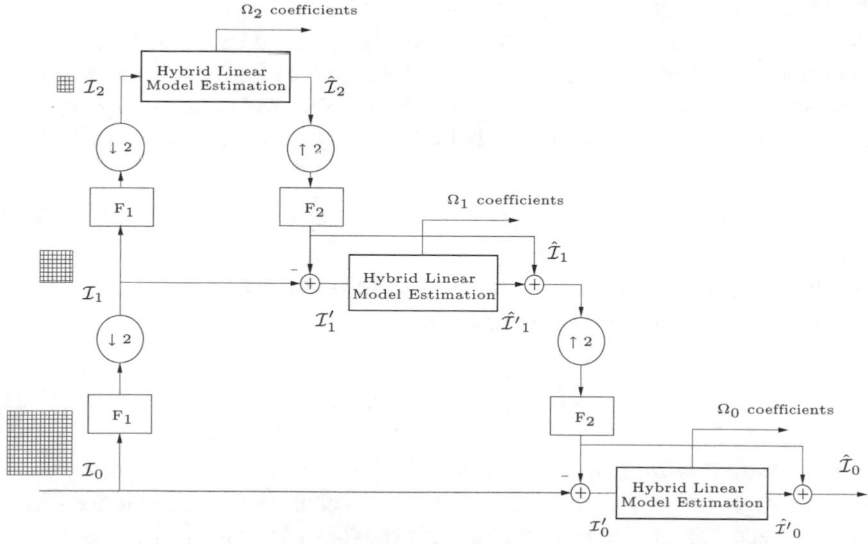

Fig. 9.7 Laplacian pyramid of the multiscale hybrid linear model.

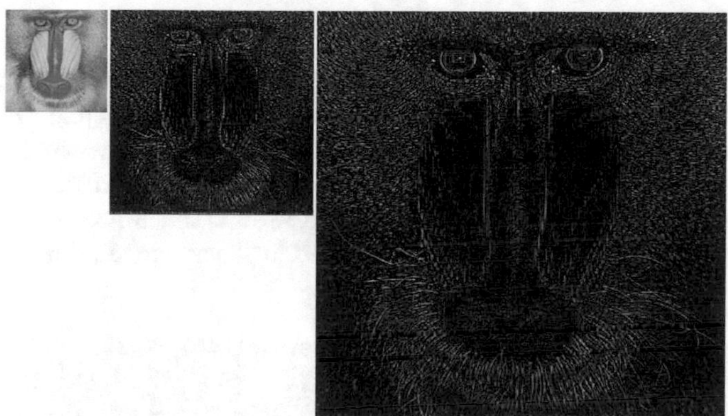

Fig. 9.8 Multiscale representation of the baboon image. Left: The coarsest level image $\mathcal{I}_2$. Middle: The residual image $\mathcal{I}'_1$. Right: The residual image $\mathcal{I}'_0$. The data at each level are modeled by the hybrid linear models. The contrast of the middle and right images has been adjusted so that they are visible.

for $\mathcal{I}_2$, $\mathcal{I}'_1$, and $\mathcal{I}'_0$, respectively, are needed for the final representation of the image. Figure 9.8 shows $\mathcal{I}_2$, $\mathcal{I}'_1$, and $\mathcal{I}'_0$ for the baboon image.

The total number of coefficients needed for the representation will be

$$\Omega = \sum_{l=0}^{L} \Omega_l. \tag{9.19}$$

MSE Threshold at Different Scale Levels
The MSE thresholds at different levels should be different but related, because the up-sampling by 2 will enlarge one pixel at level-$(l+1)$ into four pixels at level-l. If the MSE of the level $(l+1)$ is ε_{l+1}^2, the MSE of the level l after the up-sampling will become $4\varepsilon_{l+1}^2$. So the MSE thresholds of level $(l+1)$ and level l are related by

$$\varepsilon_{l+1}^2 = \frac{1}{4}\varepsilon_l^2, \quad l = 0, \ldots, L-1. \tag{9.20}$$

Usually, the user will give only the desired MSE for the approximation of the original image, which is ε^2. So we have

$$\varepsilon_l^2 = \frac{1}{4^l}\varepsilon^2, \quad l = 0, \ldots, L. \tag{9.21}$$

Vector Energy Constraint at Each Level
At each level-l, $l = 0, \ldots, L-1$, not all the vectors of the residual need to be approximated. We need to approximate only the (block) vectors $\{\boldsymbol{x}_j\}$ of the residual image $\mathcal{I}_l'$ that satisfy the following constraint:

$$\|\boldsymbol{x}_j'\|^2 > \varepsilon_l^2. \tag{9.22}$$

In practice, the energy of most of the residual vectors is close to zero. Only a small portion of the vectors at each level l need to be modeled (e.g., Figure 9.9). This property of the multiscale scheme not only significantly reduces the overall representation complexity Ω but also reduces the overall computational cost, since the number of data vectors processed at each level is much less than that of the original image. In addition, for a single hybrid linear model, when the image

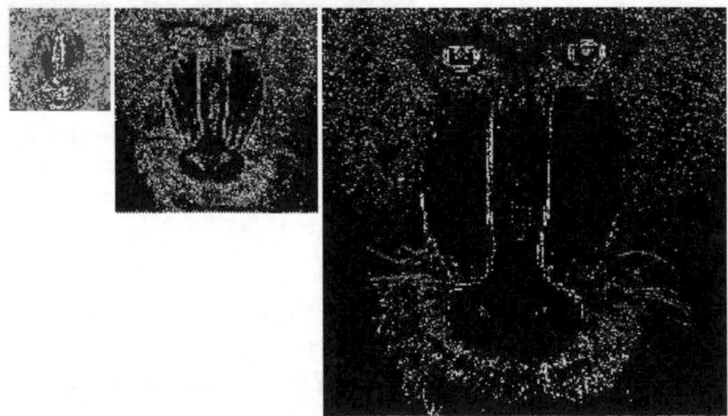

Fig. 9.9 The segmentation of (residual) vectors at the three levels—different subspaces are denoted by different colors. The black regions correspond to data vectors whose energy is below the MSE threshold ε_l^2 in equation (9.22).

Algorithm 9.2 (Multiscale Hybrid Linear Model Estimation).

function $\hat{\mathcal{I}}$ = MultiscaleModel($\mathcal{I}$, $level$, ε^2)
if $level <$ **MAXLEVEL then**
    $\mathcal{I}_{down}$ = Downsample($\mathrm{F}_1(\mathcal{I})$);
    $\hat{\mathcal{I}}_{nextlevel}$ = MultiscaleModel($\mathcal{I}_{down}$, $level + 1$, $\frac{1}{4}\varepsilon^2$);
end if
if $level =$ **MAXLEVEL then**
    $\mathcal{I}' = \mathcal{I}$;
else
    $\mathcal{I}_{up}$ = F_2(Upsample($\hat{\mathcal{I}}_{nextlevel}$));
    $\mathcal{I}' = \mathcal{I} - \mathcal{I}_{up}$;
end if
$\hat{\mathcal{I}'}$ = HybridLinearModel($\mathcal{I}'$, ε^2);
return $\mathcal{I}_{up} + \mathcal{I}'$.

Fig. 9.10 Testing images: the hill image (480 × 320) and the baboon image (512 × 512).

size increases, the computational cost will increase in proportion to the square of the image size. In the multiscale model, if the image size increases, we can correspondingly increase the number of levels, and the complexity increases only linearly in proportion to the image size.

The overall process of estimating the multiscale hybrid linear model can be written as the recursive pseudocode in Algorithm 9.2.

9.2.3 Experiments and Comparisons

Comparison of Different Lossy Representations

The first experiment is conducted on two standard images commonly used to compare image compression schemes: the 480 × 320 hill image and the 512 × 512 baboon image shown in Figure 9.10. We choose these two images because they are representative of two different types of images. The hill image contains large low-frequency/low-entropy regions, and the baboon image contains mostly high-

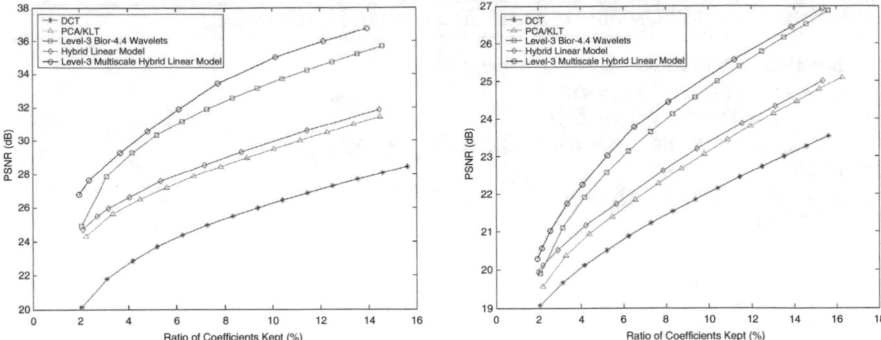

Fig. 9.11 Left: Comparison of several image representations for the hill image. Right: Comparison for the baboon image. The multiscale hybrid linear model achieves the best PSNR among all the methods for both images.

frequency/high-entropy regions. The size of the blocks b is chosen to be 2, and the level of the pyramid is 3; we will test the effect of changing these parameters in subsequent experiments. In Figure 9.11, the results of the multiscale hybrid linear model are compared with several other commonly used image representations including DCT, PCA/KLT, single-scale hybrid linear model, and Level-3 (Daubechies) biorthogonal 4.4 wavelets (adopted by JPEG-2000). The x-axis of the figures is the ratio of coefficients (including the overhead) kept for the representation, which is defined as

$$\eta = \frac{\Omega}{WHc}. \tag{9.23}$$

The y-axis is the PSNR of the approximated image defined in equation (9.8). The multiscale hybrid linear model achieves the best PSNR among all the methods for both images. Figure 9.12 shows the two recovered images using the same number of coefficients for the hybrid linear model and the wavelets. Notice that in the area around the whiskers of the baboon, the hybrid linear model preserves the detail of the textures better than the wavelets. But the multiscale hybrid linear model produces a slight block effect in the smooth regions.

Effect of the Number of Scale Levels

The second experiment shown in Figure 9.13 compares the multiscale hybrid linear representation with wavelets for different numbers of levels. It is conducted on the hill and baboon images with 2×2 blocks. The performance increases while the number of levels is increased from 3 to 4. But if we keep increasing the number of levels to 5, the level-5 curves of both wavelets and our method (which are not shown in the figures) coincide with the level-4 curves. The performance cannot improve any more, because the down-sampled images in the fifth level are so small that it is hard for them to be further compressed. Only when the image is large can we use more levels of down-sampling to achieve a more compressed representation.

Fig. 9.12 Left: The baboon image recovered from the multiscale hybrid linear model using 7.5% coefficients of the original image. (PSNR=24.64). Right: The baboon image recovered from wavelets using the same percentage of coefficients. (PSNR=23.94).

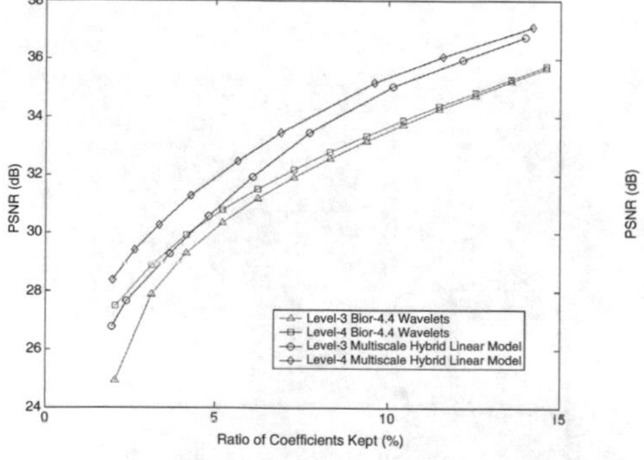

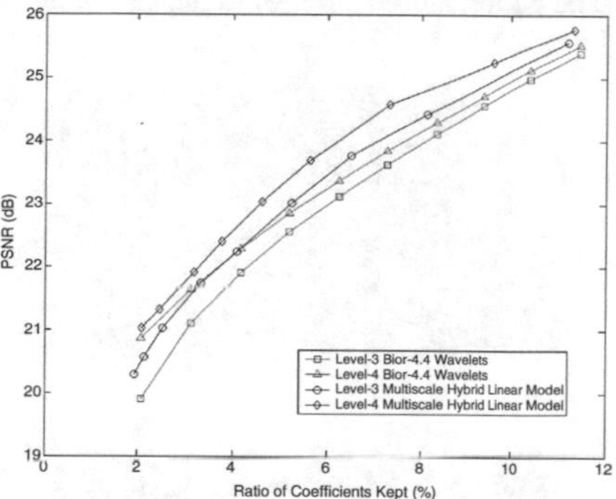

Fig. 9.13 Top: Comparison of the multiscale hybrid linear model with wavelets for level 3 and level 4 for the hill image. Bottom: The same comparison for the baboon image. The performance increases while the number of levels increases from 3 to 4.

Effect of the Block Size

The third experiment shown in Figure 9.14 compares the multiscale hybrid linear models with different block sizes from 2×2 to 16×16. The dimension of the ambient space of the data vectors $\boldsymbol{x}$ ranges from 12 to 192 accordingly. The testing image is the baboon image, and the number of down-sampling levels is 3. For large blocks, the number of data vectors is small, but the dimension of the subspaces is large. So the overhead will be large and seriously degrade the performance. Also the block effect will be more obvious when the block size is large. This experiment shows that 2 is the optimal block size, which also happens to be compatible with the simplest down-sampling scheme.

We have tested the multiscale hybrid linear model on a wide range of images, with some representative ones shown in Figure 9.15. From our experiments and

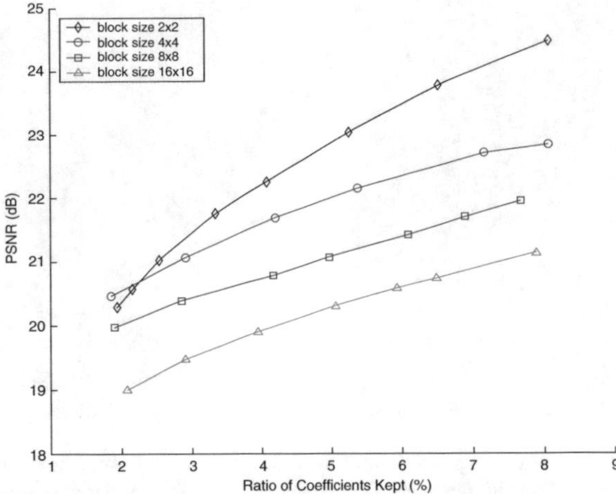

Fig. 9.14 Comparison of the multiscale hybrid linear model with different block sizes: 16, 8, 4, 2. The performance increases as the size of blocks decreases.

Fig. 9.15 A few standard testing images. From the top left to the bottom right: monarch (768 × 512), sail (768 × 512), canyon (752 × 512), tiger (480 × 320), street (480 × 320), tree (512 × 768), tissue (microscopic) (1408 × 1664), Lena (512 × 512), earth (satellite) (512 × 512), urban (aerial) (512 × 512), bricks (696 × 648). The multiscale hybrid linear model outperforms wavelets except for the Lena and monarch images.

experience, we observe that the multiscale hybrid linear model is more suitable than wavelets for representing images with multiple high-frequency/high-entropy regions, such as those with sharp 2D edges and rich textures. Wavelets are prone to blur sharp 2D edges but are better at representing low-frequency/low-entropy regions. This probably explains why the hybrid linear model performs slightly worse than wavelets for the Lena and the monarch images: the backgrounds of those two images are out of focus, so that they do not contain much high-frequency/high-entropy content.

Another limitation of the hybrid linear model is that it does not perform well on grayscale images (e.g., the Barbara image, Figure 9.5). For a grayscale image, the dimension D of a 2×2 block is only 4. Such a low dimension is inadequate for any further dimensional reduction. If we use a larger block size, say 8×8, the block effect will also degrade the performance.

Unlike pre-fixed transformations such as wavelets, our method involves identifying the subspaces and their bases. Computationally, it is more costly. With unoptimized MATLAB codes, the overall model estimation takes 30 seconds to 3 minutes on a Pentium 4 1.8-GHz PC depending on the image size and the desired PSNR. The smaller the PSNR, the shorter the running time, because the number of blocks needed to be coded in higher levels will be less.

9.3 Multiscale Hybrid Linear Models in Wavelet Domain

From the discussion in the previous section, we have noticed that wavelets can achieve a better representation for smooth regions and avoid the block artifacts. Therefore, in this section, we will combine the hybrid linear model with the wavelet approach to build multiscale hybrid linear models in the wavelet domain. For readers who are not familiar with wavelets, we recommend the books of (Vetterli and Kovacevic 1995).

9.3.1 Imagery Data Vectors in the Wavelet Domain

In the wavelet domain, an image is typically transformed into an octave tree of subbands by certain separable wavelets. At each level, the LH, HL, HH subbands contain the information about high-frequency edges, and the LL subband is further decomposed into subbands at the next level. Figure 9.16 shows the octave tree structure of a level-2 wavelet decomposition. As shown in Figure 9.17, the vectors $\{\boldsymbol{x}_j \in \mathbb{R}^D\}_{j=1}^N$ are constructed by stacking the corresponding wavelet coefficients in the LH, HL, HH subbands. The dimension of the vectors is $D = 3c$, because there are c color channels. One of the reasons for this choice of vectors is that for edges along the same direction, these coefficients are linearly related and reside in a lower-dimensional subspace. To see this, let us first assume that the color along an edge is

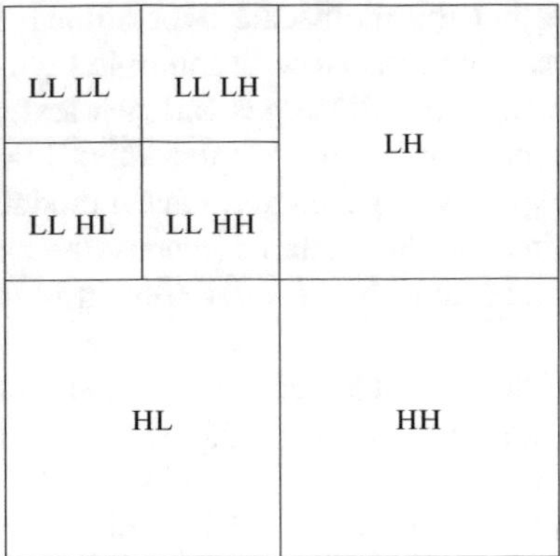

Fig. 9.16 The subbands of a level-2 wavelet decomposition.

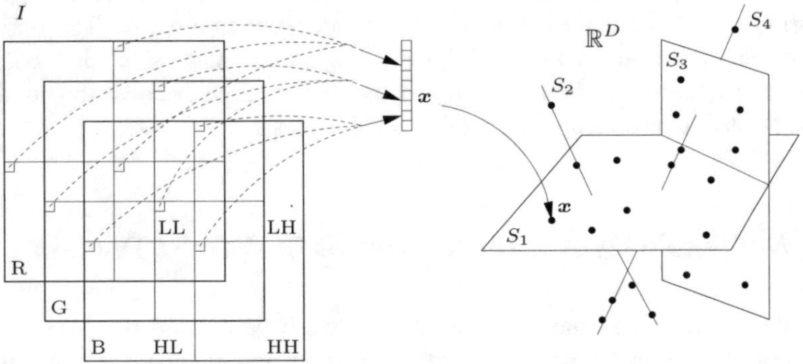

Fig. 9.17 The construction of imagery data vectors in the wavelet domain. These data vectors are assumed to reside in multiple (affine) subspaces that may have different dimensions.

constant. If the edge is along the horizontal, vertical, or diagonal direction, there will be an edge in the coefficients in the LH, HL, or HH subband, respectively. The other two subbands will be zero. So the dimension of the imagery data vectors associated with such an edge will be 1. If the edge is not exactly in one of these three directions, there will be an edge in the coefficients of all three subbands. For example, if the direction of the edge is between the horizontal and diagonal, the amplitude of the coefficients in the LH and HH subbands will be large. The coefficients in the HL subband will be insignificant relative to the coefficients in the other two subbands. So the dimension of the data vectors associated with this edge is approximately 2 (subject to a small error ε^2). If the color along an edge is changing, the dimension of the subspace will be higher, but generally lower than the ambient dimension $D = 3c$. Notice that the above scheme is only one of many possible ways in which one may construct the imagery data vector in the wavelet domain. For instance, one may construct the vector using coefficients across different scales. It remains

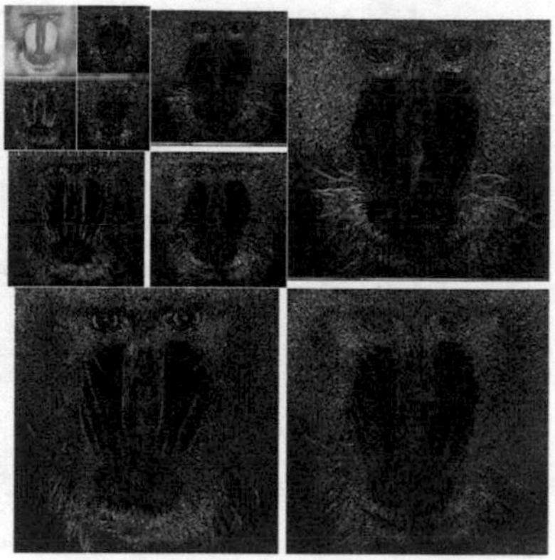

Fig. 9.18 The subbands of level-3 bior-4.4 wavelet decomposition of the baboon image.

an open question whether such new constructions may lead to even more efficient representations than the one presented here.

9.3.2 *Hybrid Linear Models in the Wavelet Domain*

In the wavelet domain, there is no need to build a down-sampling pyramid. The multilevel wavelet decomposition already gives a multiscale structure in the wavelet domain. For example, Figure 9.18 shows the octave tree structure of a level-3 bior-4.4 wavelet transformation of the baboon image. At each level, we may construct the imagery data vectors in the wavelet domain according to the previous section. A hybrid linear model will be identified for the so-obtained vectors at each level. Figure 9.19 shows the segmentation results using the hybrid linear model at three scale levels for the baboon image.

In the nonlinear wavelet approximation, the coefficients that are below an error threshold will be ignored. Similarly, in our model, not all the vectors of the imagery data vectors need to be modeled and approximated. We need to approximate only the (coefficient) vectors $\{\boldsymbol{x}_j\}$ that satisfy the following constraint:

$$\|\boldsymbol{x}_j\|^2 > \varepsilon^2. \tag{9.24}$$

Notice that here we do not need to scale the error tolerance at different levels, because the wavelet basis is orthonormal by construction. In practice, the energy of most of the vectors is close to zero. Only a small portion of the vectors at each level need to be modeled (e.g., Figure 9.19).

The overall process of estimating the multiscale hybrid linear model in the wavelet domain can be summarized as the pseudocode in Algorithm 9.3.

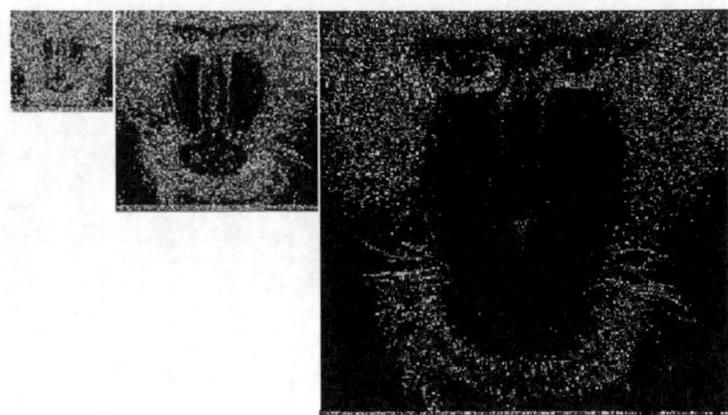

Fig. 9.19 The segmentation of data vectors constructed from the three subbands at each level—different subspaces are denoted by different colors. The black regions correspond to data vectors whose energy is below the MSE threshold ε^2 in equation (9.24).

Algorithm 9.3 (Multiscale Hybrid Linear Model: Wavelet Domain).

function $\hat{\mathcal{I}} = \text{MultiscaleModel}(\mathcal{I}, level, \varepsilon^2)$
$\tilde{\mathcal{I}} = \text{WaveletTransform}(\mathcal{I}, level)$;
for each $level$ **do**
    $\hat{\tilde{\mathcal{I}}}_{level} = \text{HybridLinearModel}(\tilde{\mathcal{I}}_{level}, \varepsilon^2)$;
end for
$\hat{\mathcal{I}} = \text{InverseWaveletTransform}(\hat{\tilde{\mathcal{I}}}, level)$;
return $\hat{\mathcal{I}}$.

9.3.3 Comparison with Other Lossy Representations

In this section, in order to obtain a fair comparison, the experimental setting is the same as that of the spatial domain in the previous section. The experiment is conducted on the same two standard images: the 480×320 hill image and the 512×512 baboon image shown in Figure 9.10.

The number of levels of the model is also chosen to be 3. In Figure 9.20, the results are compared with several other commonly used image representations including DCT, PCA/KLT, single-scale hybrid linear model, and Level-3 bi-orthogonal 4.4 wavelets (JPEG 2000) as well as the multiscale hybrid linear model in the spatial domain. The multiscale hybrid linear model in the wavelet domain achieves better PSNR than that in the spatial domain. Figure 9.21 shows the three recovered images using the same number of coefficients for wavelets, the hybrid linear model in the spatial domain, and that in the wavelet domain, respectively. Figure 9.22 shows a visual comparison with the enlarged bottom-right corners of the images in Figure 9.21.

Notice that in the area around the baboon's whiskers, the wavelets blur both the whiskers and the subtle details in the background. The multiscale hybrid linear

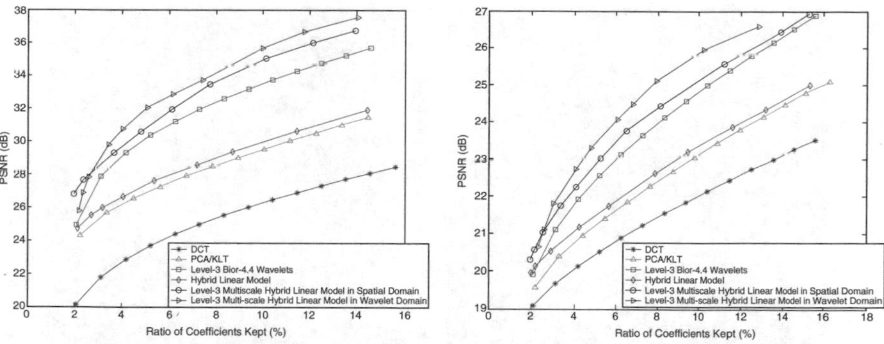

Fig. 9.20 Top: Comparison of several image representations for the hill image. Bottom: Comparison for the baboon image. The multiscale hybrid linear model in the wavelet domain achieves better PSNR than that in the spatial domain.

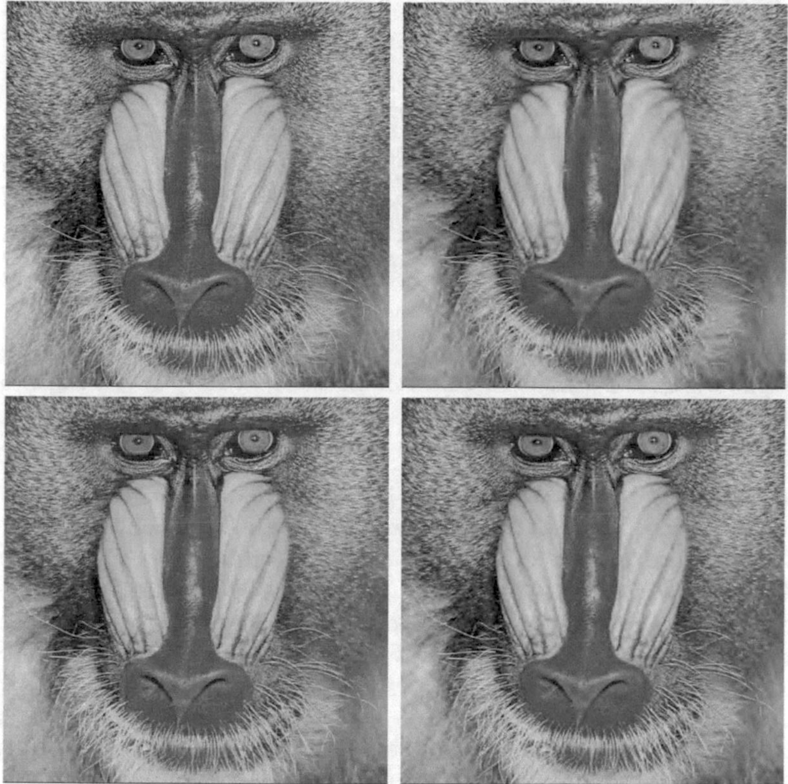

Fig. 9.21 Visual comparison of three representations for the baboon image approximated with 7.5% coefficients. Top left: The original image. Top right: The level-3 biorthogonal 4.4 wavelets (PSNR=23.94). Bottom left: The level-3 multiscale hybrid linear model in the spatial domain (PSNR=24.64). Bottom right: The level-3 multi scale hybrid linear model in the wavelet domain (PSNR=24.88).

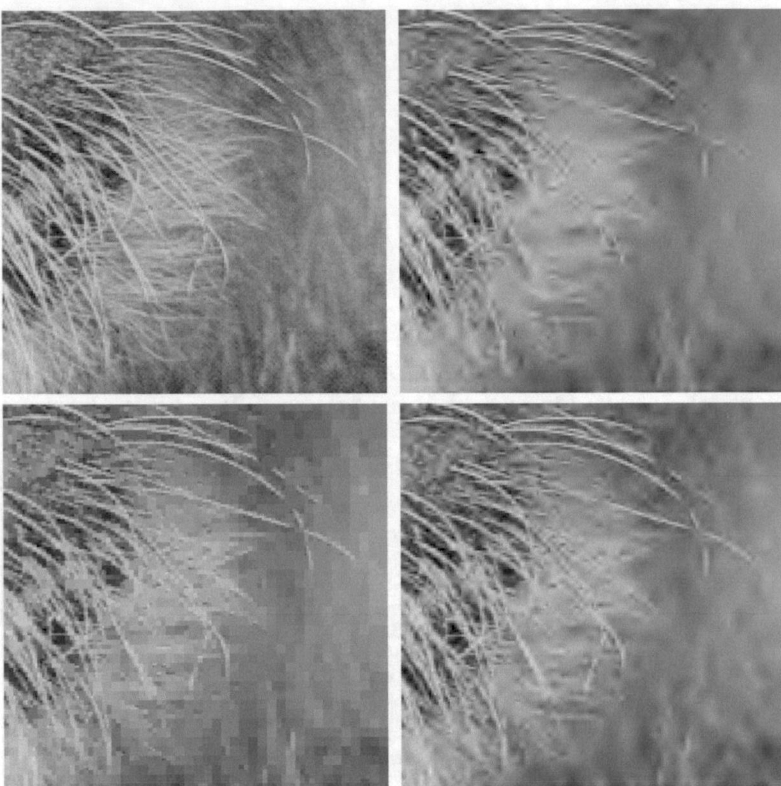

Fig. 9.22 Enlarged bottom-right corner of the images in Figure 9.21. Top left: The original image. Top right: The level-3 biorthogonal 4.4 wavelets. Bottom left: The level-3 multiscale hybrid linear model in the spatial domain. Bottom right: the level-3 multiscale hybrid linear model in the wavelet domain.

model (in the spatial domain) preserves the sharp edges around the whiskers but generates slight block artifacts in the relatively smooth background area. The multiscale hybrid linear model in the wavelet domain successfully eliminates the block artifacts, keeps the sharp edges around the whiskers, and preserves more details than the wavelets in the background. Among the three methods, the multiscale hybrid linear model in the wavelet domain achieves not only the highest PSNR, but also produces the best visual effect.

As we know from the previous section, the multiscale hybrid linear model in the spatial domain performs slightly worse than the wavelets for the Lena and monarch images (Figure 9.15). Nevertheless, in the wavelet domain, the multiscale hybrid linear model can generate very competitive results, as shown in Figure 9.23. The multiscale hybrid linear model in the wavelet domain achieves better PSNR than the wavelets for the monarch image. For the Lena image, the comparison is mixed and merits further investigation.

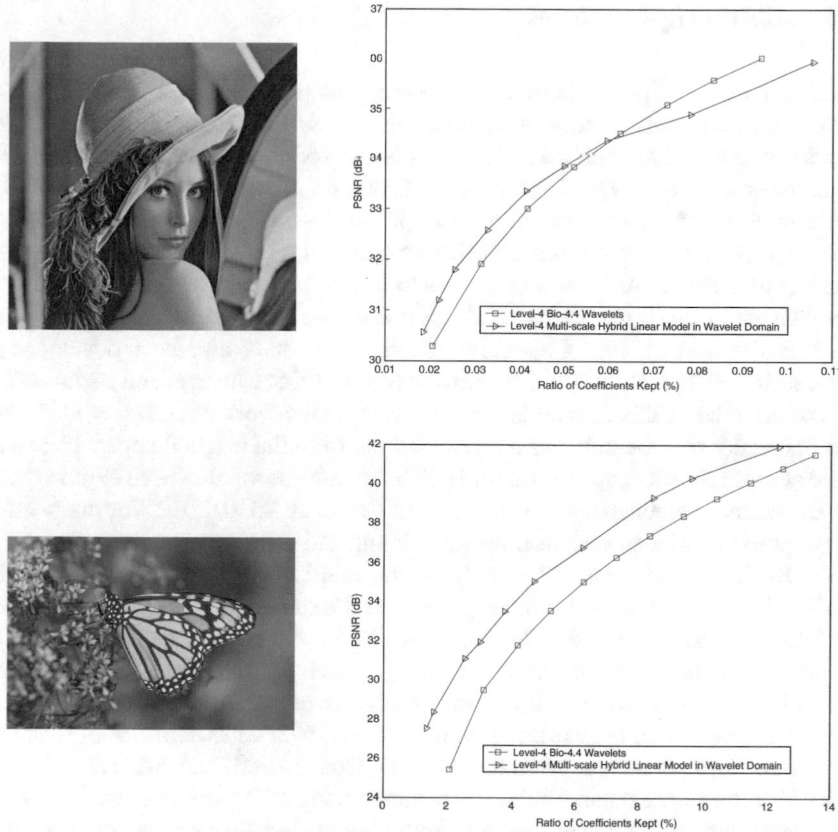

Fig. 9.23 Top: Comparison of multiscale hybrid linear model in the wavelet domain with wavelets for the Lena image. Bottom: Comparison of the multiscale hybrid linear model in the wavelet domain with wavelets for the monarch image. The multiscale hybrid linear model in the wavelet domain achieves better PSNR than wavelets for a wide range of PSNR for these two images.

The above hybrid linear model (in the wavelet domain) does not produce very competitive results for grayscale images, since the dimension of the vector is merely 3, and there is little room for further dimensional reduction. For grayscale images, one may have to choose a slightly larger window in the wavelet domain or construct the vector using wavelet coefficients across different scales. A thorough investigation of all the possible cases is beyond the scope of this book. The purpose here is simply to demonstrate (using arguably the simplest cases) the great potential of a new spectrum of image representations suggested by combining multisubspace methods with conventional image representation/approximation schemes. The quest for more efficient and more compact representations for natural images without doubt will continue as long as these new models and tools allow us to discover and exploit rich new structures in the imagery data.

9.4 Bibliographic Notes

There is a vast amount of literature on finding adaptive bases (or transforms) for more compact representations of signals. Adaptive wavelet transforms and adapted wavelet packets have been extensively studied (Coifman and Wickerhauser 1992; Ramchandran et al. 1996; Meyer 2002, 2000; Ramchandran and Vetterli 1993; Delsarte et al. 1992; Pavlovic et al. 1998). The idea is to search for an optimal transform (in terms of certain criteria) among a limited (although large) set of possible transforms. Another approach is to find some universal optimal transform based on the given signals (Effros and Chou 1995; Rabiee et al. 1996; Delsarte et al. 1992; Pavlovic et al. 1998). Spatially adapted bases have also been developed in work such as (Chen et al. 2003; Sikora and Makai 1995; Muresan and Parks 2003).

The material of this chapter is mainly based on the work of (Hong et al. 2006), and the reader may be able to find more details from the original paper. The same class of multiple-subspace (or multiple-PCA) models have also been explored and developed more extensively in later work of (Yu et al. 2010, 2012) for many other image processing tasks such as image inpainting and denoising.

As we have discussed in the chapter, the multisubspace hybrid linear model is also closely related to the more general and expressive *sparse representation* models. In (Olshausen and D.J.Field 1996), the authors have identified a set of nonorthogonal base vectors for natural images such that the representation of the image is sparse (i.e., only a few base vectors or atoms are needed to represent each image block). In the work of (Donoho 1995, 1998; Chen et al. 1998; Elad and Bruckstein 2002; Feuer and Nemirovski 2003; Donoho and Elad 2003; Starck et al. 2003; Elad and Bruckstein 2001), a systematic body of theory and algorithms has been developed for effectively and efficiently computing the sparsest representation for a signal with respect to a given (incoherent) dictionary.

This work have inspired many researchers to seek ever more compact and sparse representations through the identification of better, possibly overcomplete, dictionaries (Spielman et al. 2012; Sun et al. 2015), as well as to develop ever more scalable and efficient optimization techniques for computing sparse representations. Within this more powerful framework, researchers have significantly advanced the state of the art of many important image processing tasks such as image compression, image denoising, image deblurring, image inpainting, and image superresolution (Yang et al. 2010). It is well beyond the scope of this book to give complete and comprehensive coverage of all these wonderful applications. Nevertheless, we hope that through the simple and basic example introduced in this chapter, the reader will already be able to grasp the main ideas behind such new classes of models and methods for image (and data) processing.

Chapter 10
Image Segmentation

The whole is more than the sum of its parts.

—Aristotle

Image segmentation is the task of partitioning a natural image into multiple contiguous regions, also known as segments, whereby adjacent regions are separated by salient edges or contours, and each region consists of pixels with homogeneous color or texture. In computer vision, this is widely accepted as a crucial step for any high-level vision tasks such as object recognition and understanding image semantics.

To be precise, given an image I defined on a 2D grid of pixels $\Omega = \{u = (i,j) \mid 1 \leq i \leq m, 1 \leq j \leq n\}$, image segmentation seeks to assign each pixel $u \in \Omega$ to one of a number of segments $\mathcal{R} = \{R_1, \ldots, R_K\}$ with $\Omega = \cup_{i=1}^{k} R_i$. This is equivalent to assigning a label to each pixel:

$$l : u \mapsto l(u) \in \{1, \ldots, K\},$$

where the label $l(u)$ indicates to which segment pixel u belongs. For a segmentation to be useful, we often prefer that two pixels, say $u, v \in \Omega$, that belong to the same segment $l(u) = l(v)$ should have similar color or texture in their neighborhoods $I(u) \approx I(v)$; in addition, spatially close pixels u and v tend to have the same label unless separated by a clear edge. In other words, the label function $l(u)$ is a piecewise constant function on Ω. One benefit from segmentation is that it essentially establishes an "equivalence" relationship among pixels that belong to the same segment. Hence any higher-level vision tasks need to deal with only the K segments, which is dramatically smaller in number than the number of pixels $N = m \times n$.

Strictly speaking, natural image segmentation is an inherently ambiguous problem: it is difficult, if not impossible, to define what the "optimal" segmentation is

R. Vidal et al., *Generalized Principal Component Analysis*, Interdisciplinary Applied Mathematics 40, DOI 10.1007/978-0-387-87811-9_10

for any given image. Generally speaking, one desires that the resulting segments be highly correlated to objects in the image. That is, they should be "semantically meaningful": Ideally, each segment should correspond to an integral part of an object, and each object should consist of very few segments. Since most high-level vision tasks (such as object recognition) aim to emulate functionalities of human visual systems, evaluation of segmentation results is often based on qualitative and quantitative comparisons with segmentation results done by humans.[1]

10.1 Basic Models and Principles

Since it is nearly impossible to accurately model the human visual system, most practical image segmentation methods rely on a few tractable principles and models. Despite diverse goals for middle- or higher-level vision tasks, it is widely accepted that a good segmentation should group image pixels into regions whose statistical characteristics (of color, texture, or other feature) are homogeneous or stationary, and whose boundaries are "simple" and "spatially accurate" with respect to image edges (Haralick and Shapiro 1985). How can we translate such simple principles into a computable form?

10.1.1 Problem Formulation

The image segmentation problem, as stated above, is essentially to assign a segment label $l(u)$ to each pixel u based on its color, texture, and relationships with other (nearby) pixels. Let us represent the color or texture at a pixel u to be a feature vector $\boldsymbol{x}(u) \in \mathbb{R}^m$. We may assume that the feature vectors satisfy certain a distribution $\boldsymbol{x} \sim p(\boldsymbol{x}; \theta)$ that encodes statistics of natural images. For simplicity, we assume for now that this distribution (and its parameter θ) has already been learned or is known to us. In a probabilistic inference setting, given the field of feature vectors $\{\boldsymbol{x}(u), u \in \Omega\}$ for all pixels of the image I, we are interested in inferring the label values $\{l(u), u \in \Omega\}$ through maximizing the a posteriori probability:

$$\max_{l} p(l \mid \boldsymbol{x}; \theta) \propto p(l)p(\boldsymbol{x} \mid l; \theta). \tag{10.1}$$

Obviously, what form this a posteriori takes affects not only what kind of segmentation we get but also whether it allows trackable algorithms to compute the globally optimal segmentation l^*.

[1]Note that there is even ambiguity in segmentation done by different humans. In later sections, we will see how we could make such human-based evaluation somewhat meaningful.

Let us examine the above distributions more closely. It is natural to assume that the feature vector $\boldsymbol{x}(u)$ is conditionally independent given its label $l(u)$ and the prior distribution parameters θ. Hence the conditional probability $p(\boldsymbol{x} \mid l; \theta)$ factorizes:

$$p(\boldsymbol{x} \mid l; \theta) = \prod_{u \in \Omega} p(\boldsymbol{x}(u)|l(u); \theta) \doteq \exp\Big[-\sum_{u} \phi_1(u)\Big]. \tag{10.2}$$

Here we use $\phi_1(\cdot)$ to denote an energy function that depends on only a single pixel. For the case that $p(\boldsymbol{x}; \theta)$ is a mixture of Gaussian models and the feature vectors $\boldsymbol{x}$ associated with the kth segment follow a Gaussian distribution $\mathcal{N}(\mu_k, \Sigma_k)$, we have

$$p(\boldsymbol{x}(u)|l(u); \theta) = \exp\Big[-\frac{1}{Z}(\boldsymbol{x}(u) - \mu_k)^\top \Sigma_k^{-1}(\boldsymbol{x}(u) - \mu_k)\Big]. \tag{10.3}$$

The distribution $p(l)$ is in general not factorable, since the labels of pixels are not independent of each other: labels are more likely to be spatially contiguous, since adjacent pixels likely belong to the same segment. In addition, $p(l)$ could encode other (global or semiglobal) prior information about pixel labels that are not encoded in the features. For instance, if we can precompute the set of edges for the image, then two pixels are unlikely to have the same label if they are separated by a strong edge. More generally, this distribution could encode information about even higher orders of label dependencies among more than two pixels. In other words, we could assume that $p(l)$ is of the form

$$p(l) = p_2(l)p_3(l)\cdots, \tag{10.4}$$

where $p_i(l)$ encodes label dependency among i-tuples of pixels. In most applications, it suffices to consider dependency among pairwise pixels, and the probability $p(l)$ takes the form

$$p(l) = p_2(l) = \exp\Big[-\sum_{u,v} \phi_2(u, v)\Big], \tag{10.5}$$

where $\phi_2(\cdot, \cdot)$ is an energy function that depends on pairs of pixels. A very popular choice for $\phi_2(u, v)$ is

$$\phi_2(u, v) = w_{u,v}|l(u) - l(v)|^p, \tag{10.6}$$

where $w_{u,v}$ is a weight that indicates dependency of labels between two pixels. For example, in the simple Ising model, we have $w_{u,v} \neq 0$ only when u, v are adjacent pixels. The weight could also encode any pairwise "similarity" between two pixels: whether their features are similar to each other; or whether they are separated by an edge in between.

Given the above forms for $p(\boldsymbol{x} \mid l; \theta)$ and $p(l)$, to infer the labels from the maximum a posteriori estimate (10.1) is to optimize an objective function of the form

$$\arg\max_{l} p(l \mid \boldsymbol{x}; \theta) = \arg\max_{l} \exp\big[-E(u)\big], \tag{10.7}$$

where the energy function $E(u)$ takes the special form

$$E(u) \doteq \sum_{u} \phi_1(u) + \sum_{u,v} \phi_2(u, v). \tag{10.8}$$

There has been extensive study in the computer vision literature in the past three years about how to effectively and efficiently minimize an energy function of this form. In fact, it is known that if $K = 2$, then the global optimal solution to the above problem can be found via the graph cut algorithm. There has been much work that generalizes to the case with more than two segments and even to energy functions that consist of higher-order dependency terms.

10.1.2 *Image Segmentation as Subspace Clustering*

In this chapter, however, we are not so interested in the general image segmentation problem per se. We want to study how the image segmentation problem is closely related to the mixture of subspaces (or degenerate Gaussians) model that we have studied in this book. In addition, we will see how some of the optimization techniques that we have learned from this book for subspace clustering could offer a rather effective solution to the segmentation problem, at least certain parts of it.

To simplify the problem, we assume that the energy function contains only the unary term $\phi_1(u)$.[2] Then the image segmentation problem naturally reduces to a data clustering problem with respect to a given mixture model $p(\boldsymbol{x}; \theta)$: for any feature vector $\boldsymbol{y} = \boldsymbol{x}(u)$ with $k = l(u)$, it is of distribution $p(\boldsymbol{y}; \theta_k)$.

The simplest distributions for features of each segment include that $\boldsymbol{x}(u)$ is constant and that $\boldsymbol{x}(u)$ is an isotropic Gaussian distribution. However, such simple models have many limitations. The constant distribution essentially assumes that the image is piecewise constant. That does not capture the variabilities that we often see in natural image segments: change of color intensity, illumination, etc. In addition, in a region with a somewhat complex texture, windows centered at different pixels in the region may have different appearances although they are statistically similar. The isotropic Gaussian distribution is also problematic: features

---

[2]We will see how to incorporate pairwise information such as edges into such a simplified framework later. In particular, as we will see, such information can be incorporated through a special initialization to the segmentation algorithm.

from different textural regions differ not only in their means, but also in the shapes of their distributions. When we extract high-dimensional features from a large neighborhood around each pixel, the features of each segment normally do not span the whole feature space and more likely lie on a lower-dimensional manifold. As we have seen in previous chapters, strong empirical evidence has shown that it is more reasonable to assume that features or patches of an image segment (approximately) span a low-dimensional subspace in the feature space, and all features of the whole image then lie in a mixture of subspaces (or degenerate Gaussians).

Therefore, with the unary assumption for the energy function $E(u)$ and the additional subspace assumption for the distribution of features for each segment, the image segmentation problem is largely equivalent to the problem of subspace clustering. Notice that in the above formulation, we have assumed that the model of the feature distribution $p(\boldsymbol{x};\theta)$ is known. Obviously, the model parameters θ need to be estimated for the image before segmentation can be computed. In real image segmentation algorithms, this is often done by minimizing the energy function $E(u,\theta)$ with respect to the segmentation $l(u)$ and the model parameters θ in an alternating fashion, just as in the EM algorithm.

10.1.3 Minimum Coding Length Principle

In this chapter, we will show that the compression-based subspace clustering method studied in Chapter 6 is very pertinent to the image segmentation problem and offers a very effective and useful solution, for the following reasons:

- In image segmentation, we are mostly interested in the clustering result itself and do not care so much about the precise parameters for the mixture model. The compression-based method does not explicitly estimate the mixture model, and it does not need to know in advance how many segments there are.
- The agglomerative nature of the compression-based clustering algorithm can be easily modified to exploit the spatial adjacency between image pixels and regions, which not only improves efficiency through reduced search space but also can generate segments that respect spatial continuity.
- By controlling the level of quantization error, we can directly control the level of variability in the segments obtained. If we change the quantization error from small to large, we naturally obtain a hierarchy of image segmentation results of different levels of granularity (see Figure 10.5).
- In the end, as a byproduct, compression-based image segmentation gives an accurate upper bound of the image complexity in terms of binary coding length. Such a quantity can be used for other useful purposes such as image compression.

As we have mentioned above, by casting the image segmentation problem as a subspace clustering problem, we are unable to model and exploit higher-order label dependencies among pixels. As we will see, this can be naturally remedied in the compressed-based approach in two ways: we can initialize the compression-based algorithm with initial super-pixel segments that respect the local edges; and within the same compression framework, we can encodes the boundary information of each homogeneous texture region by counting the number of bits needed to encode the boundary (with an adaptive chain code).

Based on the *minimum description length* (MDL) principle (which is popular for model selection in statistics), the optimal segmentation of an image is defined to be the one that minimizes its total coding length, including codes needed for both the segments and their boundaries. At any fixed quantization level, the final coding length gives a purely objective measure for how good the segmentation is in terms of how compactly one can represent the resulting segmented image with binary codes.

We conduct extensive experiments to compare the results with human segmentation using the Berkeley segmentation data set (BSD) ((Martin et al. 2001)). Although the method is conceptually simple and the quantity optimized is purely objective, the segmentation results match extremely well those made by humans, exceeding or competing with the best segmentation algorithms.

10.2 Encoding Image Textures and Boundaries

10.2.1 Construction of Texture Features

We first introduce how to construct texture vectors that represent homogeneous textures in image segments. In order to capture the variation of texture, one can directly apply a $w \times w$ cutoff window around a pixel across the three color channels and stack the color values inside the window in a vector form as in (Yang et al. 2008).[3]

Figure 10.1 illustrates the process of constructing texture features. Let the w-neighborhood $\mathcal{W}_w(u)$ be the set of all pixels in a $w \times w$ window across three color channels (e.g., RGB or $L^*a^*b^*$) centered at pixel u. Define the set of features X by taking the w-neighborhood around each pixel in I, and then stacking the window as a column vector of the data matrix:

$$X \doteq [\boldsymbol{x}(u) \in \mathbb{R}^{3w^2} : \boldsymbol{x}(u) = \mathcal{W}_w(u)^S \text{ for } u \in I]. \tag{10.9}$$

For ease of computation, we further reduce the dimensionality of these features by projecting the set of all features in X onto their first D principal components. We denote the set of features with reduced dimensionality by $\hat{X}$. We have observed that

[3]Another popular approach for constructing texture vectors is to use multivariate responses of a fixed 2D texture filter bank. A previous study by (Varma and Zisserman 2003) has argued that the difference in segmentation results between the two approaches is small, and yet it is more expensive to compute 2D filter bank responses.

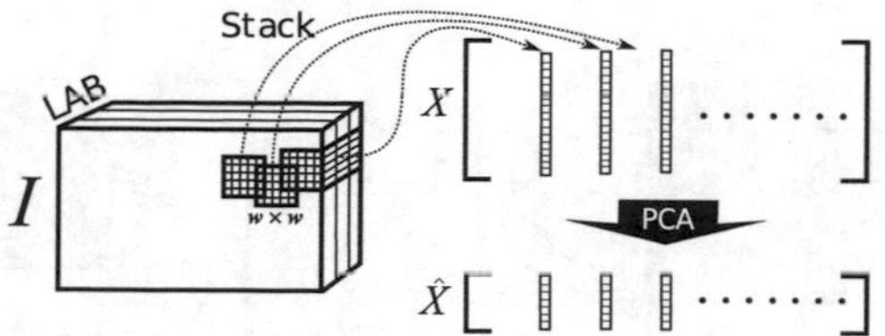

Fig. 10.1 Texture features are constructed by stacking the $w \times w$ windows around all pixels of a color image I into a data matrix X and then projecting to a low-dimensional space via principal component analysis (PCA).

for many natural images, the first eight principal components of X contain over 99% of the energy. In this chapter, we choose to assign $D = 8$.

Over all distributions with the same variance, it is known that the Gaussian distribution has the highest rate distortion, and is in this sense the worst-case distribution for compression. Thus using the rate-distortion for a Gaussian distribution, we obtain an *upper bound* for the coding length of the true distribution.

10.2.2 Texture Encoding

To describe encoding texture vectors, we first consider a single region R with N pixels. From Chapter 6, for a fixed quantization error ε, the expected number of bits needed to code the set of N feature windows $\hat{X}$ up to distortion ε^2 is given by

$$L_\varepsilon(\hat{X}) \doteq \underbrace{\frac{D}{2}\log_2\det\left(I + \frac{D}{\varepsilon^2}\Sigma\right)}_{\text{codebook}} + \underbrace{\frac{N}{2}\log_2\det\left(I + \frac{D}{\varepsilon^2}\Sigma\right)}_{\text{data}} + \underbrace{\frac{D}{2}\log_2\left(1 + \frac{\|\boldsymbol{\mu}\|^2}{\varepsilon^2}\right)}_{\text{mean}},$$

where $\boldsymbol{\mu}$ and Σ are the mean and covariance of the vectors in $\hat{X}$. The coding length function $L_\varepsilon(\hat{X})$ is the sum of three terms, for the D principal vectors as the codebook, the N vectors with respect to that codebook, and the mean of the Gaussian distribution, respectively.

The coding length function above is uniquely determined by the mean and covariance $(\boldsymbol{\mu}, \Sigma)$. To estimate them empirically, we need to exclude the windows that cross the boundary of R (as shown in Figure 10.2(a)). Such windows contain textures from the adjacent regions, which cannot be well modeled by a single Gaussian as the interior windows. Hence, the empirical mean $\hat{\boldsymbol{\mu}}_w$ and covariance $\hat{\Sigma}_w$ of R are estimated only from the *interior* of R:

$$\mathcal{I}_w(R) \doteq \{u \in R : \forall v \in \mathcal{W}_w(u), v \in R\}. \tag{10.10}$$

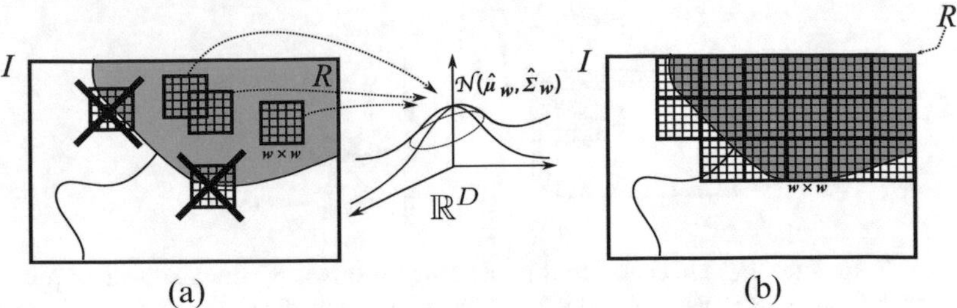

Fig. 10.2 (a) Only windows from the interior of a region are used to compute the empirical mean $\hat{\mu}_w$ and covariance $\hat{\Sigma}_w$. (b) Only nonoverlapping windows that can tile R as a grid are encoded.

The coding length function $L_\varepsilon(\hat{X})$ encodes all texture vectors in $\hat{X}$ to represent region R.[4] This is highly redundant, because the N windows in the region *overlap* with each other. Thus, to obtain a more accurate code of R that closely approximates its true entropy, we need to code only the *nonoverlapping* windows that can tile R as a grid, as in Figure 10.2 (b).

Ideally, if R is a rectangular region of size $mw \times nw$, where m and n are positive integers, then clearly we can tile R with exactly $mn = \frac{N}{w^2}$ windows. So for coding the region R, a more proper coding length measure is given by

$$L_{w,\varepsilon}(R) \doteq \left(\frac{D}{2} + \frac{N}{2w^2}\right) \log_2 \det\left(I + \frac{D}{\varepsilon^2}\hat{\Sigma}_w\right) + \frac{D}{2}\log_2\left(1 + \frac{\|\hat{\mu}_w\|^2}{\varepsilon^2}\right). \quad (10.11)$$

Real regions in natural images normally do not have such nice rectangular shapes. However, (10.11) remains a good approximation to the actual coding length of a region R with relatively smooth boundaries.[5]

10.2.3 Boundary Encoding

To code windows from multiple regions in an image, one must know to which region each window belongs, so that each window can be decoded with respect to the correct codebook. For generic samples from multiple classes, one can estimate the distribution of each class label and then code the membership of the samples using a scheme that is asymptotically optimal for that class distribution (such as

[4]The image segmentation example shown in Section 6.4.2 in Chapter 6 was done using such a coding length function.

[5]For a large region with a sufficiently smooth boundary, the number of boundary-crossing windows is significantly smaller than the number of those in the interior. For boundary-crossing windows, their average coding length is roughly proportional to the number of pixels inside the region if the Gaussian distribution is sufficiently isotropic.

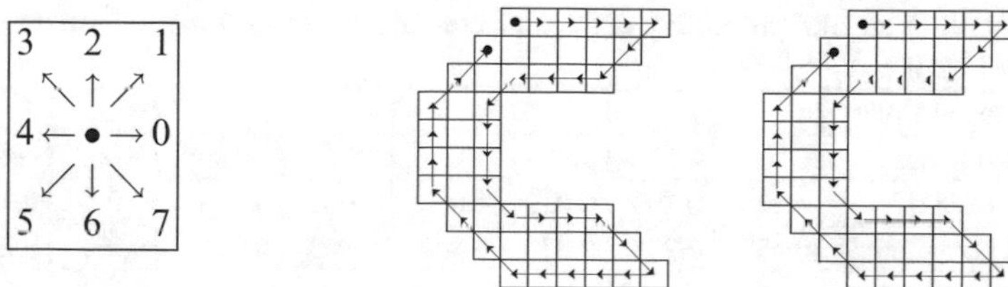

Fig. 10.3 Left: The Freeman chain code of an edge orientation along eight possible directions. **Middle:** Representation of the boundary of a region in an image with respect to the Freeman chain code. **Right:** Representation with respect to the difference chain code.

the Huffman code used in (Yang et al. 2008)). Such coding schemes are highly inefficient for natural image segmentation, since they do not leverage the spatial continuity of pixels in the same region. In fact, for this application, pixels from the same region form a connected component. Thus, the most efficient way of coding group membership for regions in images is to code the *boundary* of the region containing the pixels.

A well-known scheme for representing boundaries of image regions is the *Freeman chain code*. In this coding scheme, the orientation of an edge element is quantized along eight discrete directions, shown in Figure 10.3. Let $\{o_t\}_{t=1}^T$ denote the orientations of the T boundary pixels of R. Since each chain code can be encoded using three bits, the coding length of the boundary of R is

$$B(R) = 3\sum_{i=0}^{7} \#(o_t = i). \tag{10.12}$$

The coding length $B(R)$ can be further improved by using an adaptive Huffman code that leverages the prior distribution of the chain codes. Though the distribution of the chain codes is essentially uniform in most images, for regions with smooth boundaries, we expect that the orientations of consecutive edges are similar, and so consecutive chain codes will not differ by much. Given an initial orientation (expressed in chain code) o_t, the *difference chain code* of the following orientation o_{t+1} is $\Delta o_t \doteq \mod(o_t - o_{t+1}, 8)$. Figure 10.3 compares the original Freeman chain code with the difference chain code for representing the boundary of a region. Notice that for this region, the difference encoding uses only half of the possible codes, with most being zeros, while the Freeman encoding uses all eight chain codes. Given the prior distribution $P[\Delta o]$ of difference chain codes, $B(R)$ can be encoded more efficiently using a lossless Huffman coding scheme:

$$B(R) = -\sum_{i=0}^{7} \#(\Delta o_t = i) \log_2(P[\Delta o = i]). \tag{10.13}$$

For natural images, we estimate $P[\Delta o]$ using images from the BSD that were manually segmented by humans. We compare the distribution with the one estimated

Table 10.1 The prior probability of the difference chain codes estimated from the BSD and by (Liu and Zalik 2005).

Difference Code	0	1	2	3	4	5	6	7
Angle change	0°	45°	90°	135°	180°	−135°	−90°	−45°
Prob. (BSD)	0.585	0.190	0.020	0.000	0.002	0.003	0.031	0.169
Prob. (Liu-Zalik)	0.453	0.244	0.022	0.006	0.003	0.006	0.022	0.244

by (Liu and Zalik 2005), which used 1000 images of curves, contour patterns, and shapes obtained from the web. As the results in Table 10.1 show, the regions of natural images tend to have smoother boundaries when segmented by humans.

10.3 Compression-Based Image Segmentation

In this section, we discuss how to use the coding length functions to construct a better compression-based image segmentation algorithm. We first describe a basic approach. Then we propose a hierarchical scheme to deal with small regions using multiscale texture windows. Finally, we investigate a simple yet effective regression scheme to adaptively choose a proper distortion parameter ε based on a set of manually labeled segmentation examples.

10.3.1 Minimizing Total Coding Length

Suppose an image I can be segmented into nonoverlapping regions $\mathcal{R} = \{R_1, \dots, R_k\}$, $\cup_{i=1}^k R_i = I$. The total coding length of the image I is

$$L^s_{w,\varepsilon}(\mathcal{R}) \doteq \sum_{i=1}^{k} L_{w,\varepsilon}(R_i) + \tfrac{1}{2}B(R_i). \tag{10.14}$$

Here, the boundary term is scaled by a half, because we need to represent the boundary between any two regions only once. The optimal segmentation of I is the one that minimizes (10.14). Finding this optimal segmentation is, in general, a combinatorial task, but we can often do so using an *agglomerative* approximation similar to that introduced in Chapter 6.

To initialize the optimization process, one can assume that each image pixel (and its windowed texture vector) belongs to an individual group of its own. However, this presents a problem that the maximal size of the texture window can be one only without intersecting with other adjacent regions (i.e., other neighboring pixels). In this chapter, similar to (Yang et al. 2008), we utilize an oversegmentation step to initialize the optimization by *superpixels*. A superpixel is a small region in the image that does not contain strong edges in its interior. Superpixels provide a coarser quantization of an image than the underlying pixels while respecting strong edges between the adjacent homogeneous regions. There are several methods that can be used to obtain a superpixel initialization, including those of (Mori et al. 2004; Felzenszwalb and Huttenlocher 2004) and (Ren et al. 2005). We have found that (Mori et al. 2004)[6] works well for our purposes here.

Given an oversegmentation of the image, at each iteration we find the pair of regions R_i and R_j that will maximally decrease (10.14) if merged:

$$\begin{aligned}
(R_i^*, R_j^*) &= \arg\max R_i, R_j \in \mathcal{R} \Delta L_{w,\varepsilon}(R_i, R_j), \quad \text{where} \\
\Delta L_{w,\varepsilon}(R_i, R_j) &\doteq L^s_{w,\varepsilon}(\mathcal{R}) - L^s_{w,\varepsilon}((\mathcal{R} \backslash \{R_i, R_j\}) \cup \{R_i \cup R_j\}) \\
&= L_{w,\varepsilon}(R_i) + L_{w,\varepsilon}(R_j) - L_{w,\varepsilon}(R_i \cup R_j) \\
&\quad + \tfrac{1}{2}(B(R_i) + B(R_j) - B(R_i \cup R_j)).
\end{aligned} \tag{10.15}$$

Here $\Delta L_{w,\varepsilon}(R_i, R_j)$ essentially captures the difference in the lossy coding lengths of the texture regions R_i and R_j and their boundaries before and after the merging. If $\Delta L(R_i^*, R_j^*) > 0$, we merge R_i^* and R_j^* into one region and repeat the process until the coding length $L^s_{w,\varepsilon}(\mathcal{R})$ cannot be further reduced.

To model the spatial locality of textures, we further construct a *region adjacency graph* (RAG) $\mathcal{G} = (\mathcal{V}, \mathcal{E})$. Each vertex $v_i \in \mathcal{V}$ corresponds to region $R_i \in \mathcal{R}$, and an edge $e_{ij} \in \mathcal{E}$ indicates that regions R_i and R_j are adjacent in the image. To perform image segmentation, we simply apply a constrained version of the above agglomerative procedure, merging only regions that are adjacent in the image. The proposed region-merging method has been widely used by other image segmentation algorithms ((Haralick and Shapiro 1985; Tremeau and Borel 1997; Deng and Manjunath 2001)).

In terms of the computational complexity, one can show that the agglomerative clustering process that iteratively minimizes (10.14) is a polynomial-time algorithm. More specifically, let w be the window size, n the image size, and k the number of initial superpixel segments. One can show that the computational complexity of agglomerative clustering is bounded by $O(kw^6 + n^2w^2)$. Also note that the complexity bound has ignored the cost to sort and maintain the ordering of the coding length difference (10.15), since the algorithm can use a heap structure to efficiently implement the sorting and resorting algorithms ((Kurita 1995)).

10.3.2 Hierarchical Implementation

The above region-merging scheme is based on the assumption of a fixed texture window size, and clearly cannot effectively deal with regions or superpixels that are very small. In such cases, the majority of the texture windows will intersect with

[6]We use the publicly available code for this method available at http://www.cs.sfu.ca/~mori/research/superpixels/ with parameter `N_sp` = 200.

the boundary of the regions. We say that a region R is *degenerate* with respect to window size w if $\mathcal{I}_w(R) = \emptyset$. For such regions, the w-neighborhoods of all pixels will contain pixels from other regions, and so $\hat{\boldsymbol{\mu}}$ and $\hat{\Sigma}$ cannot be reliably estimated. These regions are degenerate precisely because of the window size; for any w-degenerate region R, there is $1 \leq w' < w$ such that $\mathcal{I}_{w'}(R) \neq \emptyset$. We say that R is *marginally nondegenerate* with respect to window size w if $\mathcal{I}_w(R) \neq \emptyset$ and $\mathcal{I}_{w+2}(R) = \emptyset$. To deal with these degenerate regions, we propose to use a hierarchy of window sizes. Starting from the largest window size, we recursively apply the above scheme with ever smaller window sizes till all degenerate regions have been merged with their adjacent ones. In particular, we start from 7×7 and reduce to 5×5, 3×3, and 1×1. Please refer to Figure 10.4 for an example of this hierarchical scheme.

Notice that at a fixed window size, the region-merging process is similar to the compression-based texture merging (CTM) approach proposed in (Yang et al. 2008). Nevertheless, the new coding length function and the hierarchical implementation give a much more accurate approximation to the true image entropy and hence lead to much better segmentation results. We summarize the overall algorithm for image segmentation in Algorithm 10.1, which we refer to as *texture and boundary encoding-based segmentation* (TBES).

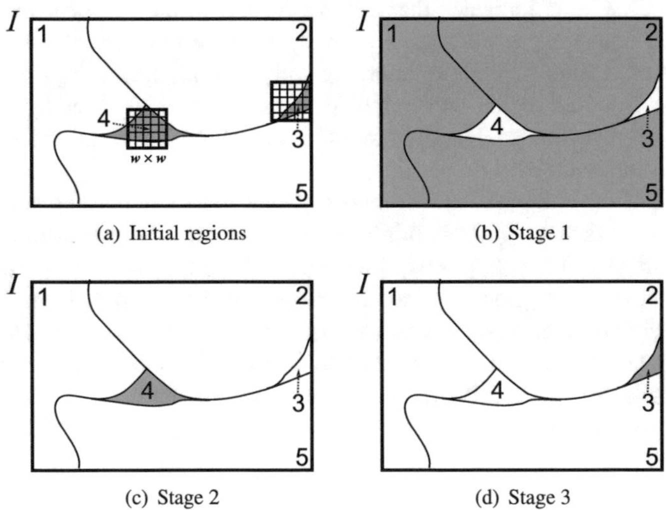

Fig. 10.4 An example illustrates the scheme for hierarchical image segmentation. (a) Initial set of regions. Note that regions 3 and 4 are degenerate with respect to the window size w. (b) In the first stage, only nondegenerate regions 1, 2, and 5 are considered for merging. (c) In the next stage, w is reduced, causing region 4 to be marginally nondegenerate. We consider merging region 4 with its nondegenerate neighbors. (d) In the last stage, w is reduced enough so that region 3 becomes nondegenerate. These stages are repeated until the overall coding length can no longer be reduced.

Algorithm 10.1 Texture and Boundary Encoding-based Segmentation (TBES)

Given image I, distortion ε, max window size w_M, superpixels $\mathcal{R} = \{R_1, \ldots, R_k\}$,

for $w = 1 : 2 : w_M$ **do**
  Construct $\hat{X}_w$ by stacking the $w \times w$ windows around each $u \in I$ as column vectors and applying PCA.
end for
Construct RAG $\mathcal{G} = (\mathcal{V}, \mathcal{E})$, where $\mathcal{V} \sim \mathcal{R}$ and $e_{ij} \in \mathcal{E}$ only if R_i and R_j are adjacent in I.
$w = w_M$
repeat
  if $w = w_M$ **then**
    Find R_i and R_j such that $e_{ij} \in \mathcal{E}, \mathcal{I}_w(R_i) \neq \emptyset, \mathcal{I}_w(R_j) \neq \emptyset$, and $\Delta L_{w,\varepsilon}(R_i, R_j)$ is maximal.
  else
    Find R_i and R_j such that $e_{ij} \in \mathcal{E}$, $\mathcal{I}_w(R_i) \neq \emptyset$, $\mathcal{I}_w(R_j) \neq \emptyset$, $\mathcal{I}_{w+2}(R_i) = \emptyset$ or $\mathcal{I}_{w+2}(R_j) = \emptyset$ and $\Delta L_{w,\varepsilon}(R_i, R_j)$ is maximal.
  end if
  if $\Delta L_{w,\varepsilon}(R_i, R_j) > 0$ **then**
    $\mathcal{R} := (\mathcal{R} \setminus \{R_i, R_j\}) \cup \{R_i \cup R_j\}$.
    Update $\mathcal{G}$ based on the newly merged region.
    $w = w_M$
  else if $w \neq 1$ **then**
    $w = w - 2$
  end if
until $\mathcal{I}_{w_M}(R) \neq \emptyset, \quad \forall R \in \mathcal{R}$ **and** $\Delta L_{w_M,\varepsilon}(R_i, R_j) \leq 0, \quad \forall R_i, R_j \in \mathcal{R}$
Output: The set of regions $\mathcal{R}$.

10.3.3 *Choosing the Proper Distortion Level*

Algorithm 10.1 requires a single parameter, the distortion level ε, which determines the granularity of the segmentation. The optimality of ε is measured by the segmentation that best matches with human perception. As shown in Figure 10.5, since natural images have different scales of resolution, no single choice of ε is optimal for all images. In this section, we propose a solution to adaptively selecting a proper distortion parameter such that the segmentation result better approximates human perception. The method assumes that a set of training images $\mathcal{I} = \{I_1, \cdots, I_K\}$ has been manually segmented by human users as the ground truth set $\mathcal{S}_g = \{\mathcal{R}_g(I_1), \cdots, \mathcal{R}_g(I_K)\}$.

To objectively quantify how well a given segmentation matches with human perception, we first need a measure for the discrepancy between two segmentations $\mathcal{R}_1$ and $\mathcal{R}_2$, denoted by $d(\mathcal{R}_1, \mathcal{R}_2)$. Intuitively, the discrepancy measure should be small when $\mathcal{R}_1$ and $\mathcal{R}_2$ are similar in some specific sense.[7] Given a measure d, the best ε for I_i, denoted by ε_i^*, can be obtained by

[7] We will discuss several discrepancy measures in Section 10.4.2, such as the probabilistic Rand index (PRI) and variation of information (VOI).

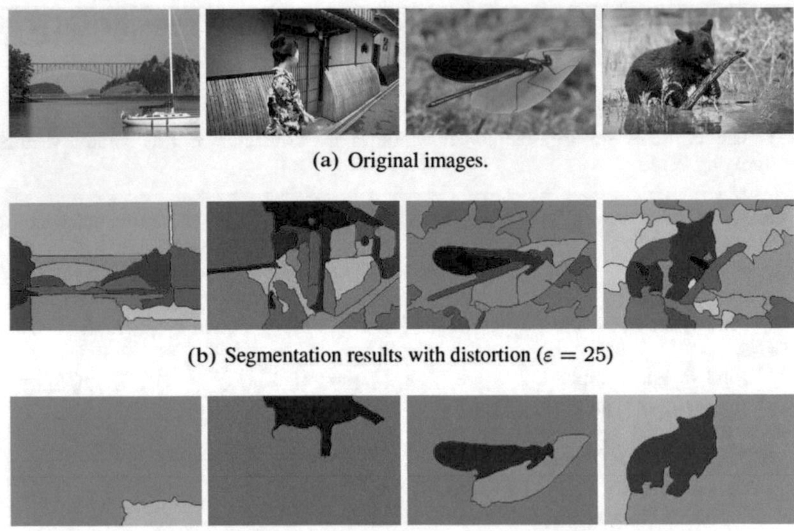

(a) Original images.

(b) Segmentation results with distortion ($\varepsilon = 25$)

(c) Segmentation results with distortion ($\varepsilon = 400$)

Fig. 10.5 A comparison of segmentation results with respect to different distortion levels. The low distortion generates better segmentations for the left two images, while the high distortion generates better results for the right two images.

$$\varepsilon_i^* = \arg\min_{\varepsilon} d(\mathcal{R}_\varepsilon(I_i), \mathcal{R}_g(I_i)), \quad \text{for each } I_i \in \mathcal{I}. \tag{10.16}$$

An example of the relationship between ε and a discrepancy measure d is shown in Figure 10.6.

Since ground truth segmentations are not available for nontraining images, we shall use the training images $\mathcal{S}_g = \{\mathcal{R}_g(I_i)\}$ to infer ε for a test image. A classical technique for estimating a continuous parameter, such as ε, from training data is *linear regression* ((Duda et al. 2000)). The method requires a pair $(\varepsilon_i, \boldsymbol{f}_i)$ per training image I_i, where ε_i is the "optimal" distortion for image I_i and $\boldsymbol{f}_i$ is a set of features extracted from I_i. Then the regression parameters $\boldsymbol{w}$ can be estimated by solving the following objective function:

$$\boldsymbol{w}^* = \arg\min \boldsymbol{w} \sum_i (\boldsymbol{w}^\top \boldsymbol{f}_i - \varepsilon_i^*)^2. \tag{10.17}$$

The distortion level ε with respect to a new test image I with its feature vector $\boldsymbol{f}$ is given by $\varepsilon(\boldsymbol{f}) \doteq \boldsymbol{w}^{*\top} \boldsymbol{f}$.

The features $\boldsymbol{f}_i$ in (10.17) should be chosen to effectively model the statistics of the image, so that the relationship between ε and $\boldsymbol{f}_i$ is well approximated by the linear function $\varepsilon_i \approx \boldsymbol{w}^\top \boldsymbol{f}_i$. A simple idea to define $\boldsymbol{f}_i$ could consider how contrastive the regions in I_i are. Intuitively, when the textures in I_i are similar, such as in camouflage images, stronger sensitivity to contrast in patterns is required. Since

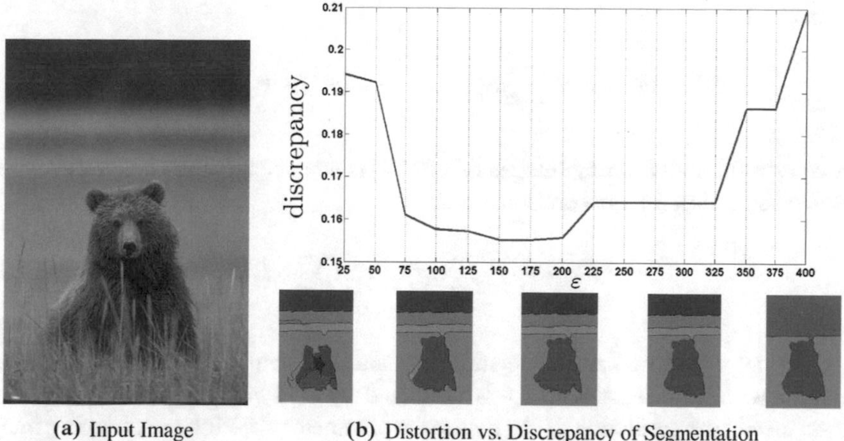

(a) Input Image (b) Distortion vs. Discrepancy of Segmentation

Fig. 10.6 The effect of distortion ε on the discrepancy $d(\mathcal{R}_\varepsilon(I_i), \mathcal{R}_g(I_i))$ on an example image. The discrepancy shown in the plot is the probability that an arbitrary pair of pixels do not have consistent labels in $\mathcal{R}_\varepsilon(I_i)$ and $\mathcal{R}_g(I_i)$, namely, PRI^C (please refer to Section 10.4.2).

computing the standard deviation of pixel intensities gives a measure of pattern contrast, we resize each I_i with multiple scales, and define the features $\boldsymbol{f}_i$ as the standard deviations of the pixel intensities at the multiple image resolutions.

Another issue in linear regression is that the classical model (10.17) is insufficient to accurately predict the distortion level for Algorithm 10.1. In particular, the discrepancy measure d is used only to determine the optimal ε^* for a training image. Segmentation results for other choices of ε are not used in the regression. However, it is possible to better estimate the distortion ε by taking into account the segmentation results around a neighborhood of the optimal distortion ε^* in the training set.

For agglomerative image segmentation, the discrepancy measures that we use in this chapter exhibit a simple behavior. Specifically, as ε deviates from ε^* in either direction, the discrepancy between the segmentation and the ground truth almost increases monotonically. This is because as ε deviates from ε^*, it leads to overseg-mentation or undersegmentation, both of which have larger discrepancies from the ground truth (see Figure 10.6). Motivated by this observation, we approximate the discrepancy function d by a convex quadratic form:

$$d(\mathcal{R}_\varepsilon(I_i), \mathcal{R}_g(I_i)) \approx a_i\varepsilon^2 + b_i\varepsilon + c_i, \quad \text{where } a_i > 0. \tag{10.18}$$

The parameters (a_i, b_i, c_i) are then estimated by least squares fitting with respect to the pairs (d, ε). The latter is attained by sampling the function $d(\mathcal{R}_\varepsilon(I_i), \mathcal{R}_g(I_i))$ at different ε's.

Once we substitute (10.18) in (10.16) in combination with the linear model $\varepsilon = \boldsymbol{w}^\top \boldsymbol{f}_i$, the objective function to recover the linear regression parameter $\boldsymbol{w}^*$ is given by

$$\boldsymbol{w}^* = \arg\min \boldsymbol{w} \sum_i a_i(\boldsymbol{w}^\top f_i)^2 + b_i(\boldsymbol{w}^\top f_i) + c_i. \tag{10.19}$$

Since $a_i > 0$ for all training images I_k, (10.19) is an unconstrained convex program. Thus it has a closed-form solution:

$$\boldsymbol{w}^* = -\frac{1}{2}(\sum_i a_i \boldsymbol{f}_i \boldsymbol{f}_i^\top)^{-1}(\sum_i b_i \boldsymbol{f}_i). \tag{10.20}$$

Once $\boldsymbol{w}^*$ is learned from the training data, the optimal distortion of the test image I with its feature vector $\boldsymbol{f}$ is predicted by $\varepsilon(\boldsymbol{f}) = \boldsymbol{w}^{*\top}\boldsymbol{f}$. We caution that based on $\boldsymbol{w}^*$, the prediction of the distortion parameter $\varepsilon(\boldsymbol{f}_i)$ for each training image I_i may not necessarily be the same as ε_i^* selected from the ground truth $\mathcal{R}_g(I_i)$. Nevertheless, the proposed solution ensures that the linear model minimizes the average discrepancy over the training data.

10.4 Experimental Evaluation

In this section, we conduct extensive evaluations to validate the performance of the TBES algorithm. The experiment is based on the publicly available Berkeley segmentation data set (BSD) ((Martin et al. 2001)). BSD consists of 300 natural images, which cover a variety of natural scene categories, such as portraits, animals, landscapes, and beaches. The database is partitioned into a training set of 200 images and a testing set of 100 images. It also provides ground-truth segmentation results of all the images obtained by several human subjects. On average, five segmentation maps are available per image. Multiple ground truth allows us to investigate how human subjects agree with each other.

The implementation of the TBES algorithm and the benchmark scripts are available online at http://perception.csl.illinois.edu/coding/image_segmentation/ or on request.

10.4.1 Color Spaces and Compressibility

The optimal coding length of textured regions of an image depends in part on the color space. We seek to determine the color space in which natural images are most compressible based on the proposed lossy compression scheme (10.14). It has been noted in the literature that the *Lab* color space (also known as $L^*a^*b^*$) better approximates the perceptually uniform color metric (Jain 1989). This has motivated some of the previous work (Yang et al. 2008; Rao et al. 2009) to utilize such a representation in methods for natural image segmentation. In order to check the validity of this assumption, particularly for the compression-based segmentation scheme, we perform a study on five color spaces that have been widely used in the literature, namely, Lab, YUV, RGB, XYZ, and HSV.

We use the manually segmented training images in the Berkeley data set to rank the compressibility of the five color spaces. Given a color space, for any image and corresponding segmentation, the number of bits required to encode texture information is computed as $L_\varepsilon(\hat{X})$, with features constructed as in Section 10.2.1. The average coding length of an image is computed as the one over all ground-truth segmentation maps for that image. Finally, the average coding length of the data set is computed over all the images in the data set.

We note that the volume of the pixel distribution (and thus the coding length) can change if the pixel values are rescaled. This means that one color space can look more compressible by merely producing numbers in a smaller range, say [0, 1] as opposed to another that is in the range [0, 255]. In order to achieve a fair comparison, we normalize the feature vectors by a scale factor c, which is constant across features from the same color space:

$$c = 1/\sqrt{\bar{\lambda}_{\max}}, \tag{10.21}$$

where $\bar{\lambda}_{\max}$ is the average of the maximum eigenvalues of the feature covariance matrix over all regions and all images in the data set.

The average (normalized) coding lengths of five representative color spaces are shown in Figure 10.7. Among all five color spaces examined, Lab has the shortest coding length. Therefore, in experiments for the rest of the chapter, input images are first converted to the Lab color space.

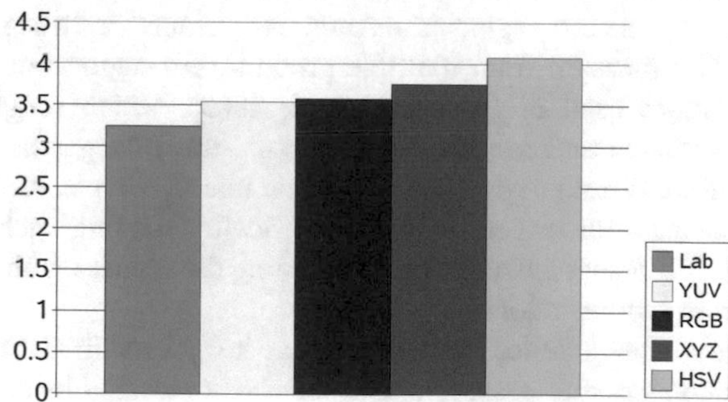

Fig. 10.7 Average coding length of an image in five representative color spaces.

10.4.2 Experimental Setup

To quantitatively evaluate the performance of this method, we use four metrics for comparing pairs of image segmentation: the *probabilistic Rand index* (PRI) ((Rand 1971)), *variation of information* (VOI) ((Meila 2005)), *boundary displacement error* (BDE) ((Freixenet et al. 2002)), and the *global F-measure* ((Arbelaez 2006)):

1. The *probabilistic Rand index* (PRI) is a classical metric that measures the probability that an arbitrary pair of samples have consistent labels in the two partitions. The PRI metric is in the range $[0, 1]$, with higher values indicating greater similarity between two partitions. When used to adaptively choose ε as described in Section 10.3.3, we use $\mathrm{PRI}^C \doteq (1 - \mathrm{PRI})$.
2. The *variation of information* (VOI) measures the sum of information loss and information gain between the two clusterings, and thus it roughly measures the extent to which one clustering can explain the other. The VOI metric is nonnegative, with lower values indicating greater similarity.
3. The *boundary displacement error* (BDE) measures the average displacement error of boundary pixels between two segmented images. Particularly, it defines the error of one boundary pixel as the distance between the pixel and the closest pixel in the other boundary image.
4. The *global F-measure* (GFM) is the harmonic mean of precision and recall, a pair of complementary metrics for measuring the accuracy of the boundaries in an image segmentation given the ground-truth boundaries. Precision measures the fraction of true boundary pixels in the test segmentation. Recall measures the fraction of ground-truth boundary pixels in the test segmentation. When used to adaptively choose ε, we use $\mathrm{GFM}^C \doteq (1 - \mathrm{GFM})$.

In cases in which we have multiple ground-truth segmentations, to compute the PRI, VOI, or BDE measure for a test segmentation, we simply average the results of the metrics between the test segmentation and each ground-truth segmentation. To compute the GFM measure from multiple ground-truth segmentations, we apply the same techniques used in (Arbelaez et al. 2009), which roughly aggregate the boundary precision and recall over all ground-truth images as an ensemble. With multiple ground-truth segmentations for an image, we can also estimate the human performance with respect to these metrics by treating each ground-truth segmentation as a test segmentation and computing the metrics with respect to the other ground-truth segmentations.

The adaptive ε scheme relies on the feature vector $\boldsymbol{f}$ used in (10.19) as follows. The image I is converted to grayscale, and its size is rescaled by a set of specific factors. The standard deviation of pixel intensity of each rescaled image constitutes a component of the feature vector. Empirically, we have observed that using four scale factors, i.e., $\boldsymbol{f} \in \mathbb{R}^4$, produces good segmentation results for the algorithm on the BSD database.

The parameters (a_k, b_k, c_k) in the quadratic form in (10.18) are estimated as follows. We sample $25 \leq \varepsilon \leq 400$ uniformly, in steps of 25, and compute the corresponding $d(S_\varepsilon(I_k), S_g(I_k))$ for each sample. This gives a set $\{(d_{k,n}, \varepsilon_{k,n})\}_{n=1}^{16}$ for an image I_k. We use this set to estimate (a_k, b_k, c_k) by the least squares method.

10.4.3 Results and Discussions

We quantitatively compare the performance of the TBES method with seven *publicly available* image segmentation methods, namely, *mean shift* (MS) by (Comanicu and Meer 2002), *Markov Chain Monte Carlo* (MCMC) by (Tu and Zhu 2002), *F&H* by (Felzenszwalb and Huttenlocher 2004), *multiscale NCut* (MNC) by (Cour et al. 2005), *compression-based texture merging* (CTM) by (Yang et al. 2008), *ultrametric contour maps* (UCM) by (Arbelaez et al. 2009), and *saliency driven total variation* (SDTV) by (Donoser et al. 2009), respectively. The user-defined parameters of these methods have been tuned by the training subset of each data set to achieve the best performance with respect to each segmentation index. Then the performance of each method is evaluated based on the test subset.

Table 10.2 shows the segmentation accuracy of TBES compared to the human ground truth and the other seven algorithms.[8] In addition to the evaluation of the algorithms, multiple ground-truth segmentations in BSD allow us to estimate the human performance with respect to these metrics. This was achieved by treating each ground-truth segmentation as a test segmentation and computing the metrics with respect to the other ground-truth segmentations. To qualitatively inspect the segmentation, Figure 10.8 illustrates some representative results.

Among all the algorithms in Table 10.2, TBES achieves the best performance with respect to PRI and VOI. It is also worth noting that there seems to be a large gap in terms of VOI between all the algorithm indices and the human index (e.g., 1.705 for TBES versus 1.163 for human). With respect to BDE and GFM, UCM achieves the best performance, which is mainly due to the fact that UCM was designed to construct texture regions from the hierarchies of (strong) image contours and edges. In this category, TBES still achieves the second-best performance, largely exceeding the indices posted by the rest of the algorithms in the literature.

Note that in Table 10.2, TBES consistently outperforms CTM, on which the fundamental lossy-coding framework of TBES is based. To clarify the contribution of each new TBES component, we further provide an analysis of the efficacy of the components of TBES in a "leave-one-out" comparison. In Table 10.3, the performance of TBES with certain functions individually disabled is shown. The variations of the code include disabling adaptive choice of epsilon, discounting overlapping windows, hierarchical window sizes, and boundary coding, respectively.

[8]The quantitative performance of several existing algorithms was also evaluated in a recent work ((Arbelaez et al. 2009)), which was published roughly at the same time as this work. The reported results therein generally agree with our findings.

Table 10.2 Comparison on the BSD using the PRI, VOI, BDE, and GFM indices. For PRI and GFM, higher values indicate better segmentation; for VOI and BDE, lower values indicate better segmentation.

BSD	PRI	VOI	BDE	GFM
Human	0.868	1.163	7.983	0.787
TBES	**0.807**	**1.705**	12.681	0.647
MS	0.772	2.004	13.976	0.600
MCMC	0.768	2.261	13.897	0.467
F&H	0.770	2.188	14.057	0.579
MNC	0.742	2.651	13.461	0.590
CTM	0.755	1.897	14.066	0.595
UCM	0.796	1.715	**10.954**	**0.706**
SDTV	0.801	1.790	15.513	0.593

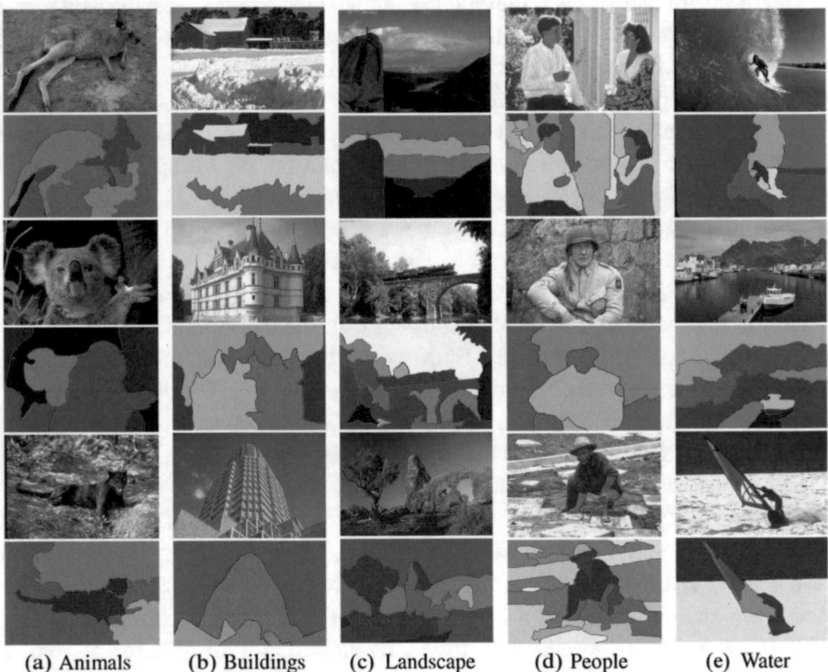

Fig. 10.8 Representative segmentation results (in color) of the TBES algorithm on various image categories from BSD. For each image pair, the top is the original input image, and the bottom is the segmentation result, where each texture region is rendered by its mean color. The distortion ε was chosen adaptively to optimize PRI.

Table 10.3 A comparison of the efficacy of the individual components of the TBES algorithm. The first row shows the performance of TBES, and each following row corresponds to disabling one component of TBES. $\mathrm{TBES}_{(\varepsilon)}$, $\mathrm{TBSE}_{(w)}$, $\mathrm{TBSE}_{(h)}$, and $\mathrm{TBSE}_{(b)}$ correspond to disabling the code for adaptive choice of epsilon, discounting overlapping windows, hierarchical window sizes, and boundary coding, respectively. For $\mathrm{TBES}_{(\varepsilon)}$, a fixed $\varepsilon = 150$ is chosen. For $\mathrm{TBES}_{(h)}$ that disables hierarchical resolutions, a fixed window size $w = 7$ is chosen. The best performance values are highlighted in boldface.

BSD	PRI	VOI	BDE	GFM
TBES	**0.807**	**1.705**	**12.681**	**0.647**
$\mathrm{TBES}_{(\varepsilon)}$	0.793	1.792	15.020	0.545
$\mathrm{TBES}_{(w)}$	0.790	1.788	13.972	0.597
$\mathrm{TBES}_{(h)}$	0.794	1.743	13.335	0.613
$\mathrm{TBES}_{(b)}$	0.796	1.775	13.659	0.638

Clearly, since TBES retains the best performance over all four segmentation metrics in Table 10.3, it shows that disabling any segmentation criterion would worsen its performance. Since $\mathrm{TBES}_{(\varepsilon)}$ gives the overall worst performance in Table 10.3, one can conclude that adaptively choosing the distortion level ε is the single most important heuristic in TBES, which justifies our argument in Section 10.3.3 that since natural images represent different scene categories with different scales of resolution, no single choice of ε is optimal for all images. Furthermore, it is interesting to observe that all the variations of TBES in Table 10.3 still achieve better segmentation metrics than the original CTM algorithm in Table 10.2.

Finally, we briefly discuss a few images on which this method fails to achieve a good segmentation. The examples are shown in Figure 10.9. The main causes for visually inferior segmentation are camouflage, shadows, non-Gaussian textures, and thin regions:

1. It is easy to see that the texture of animal camouflages is deliberately chosen to be similar to the background texture. The algorithm falls behind humans in this situation, arguably because human vision can *recognize* the holistic shape and texture of the animals based on experience.
2. Since shades of the same texture may appear very different in images, TBES may break up the regions into more or less the same level of shade.
3. Some patterns in natural images do not follow the Gaussian texture assumption. Examples include geometric patterns such as lines or curves.
4. Thin regions, such as spiders' legs, are problematic for TBES for two reasons. First, it has trouble to properly form low-level superpixels used as the

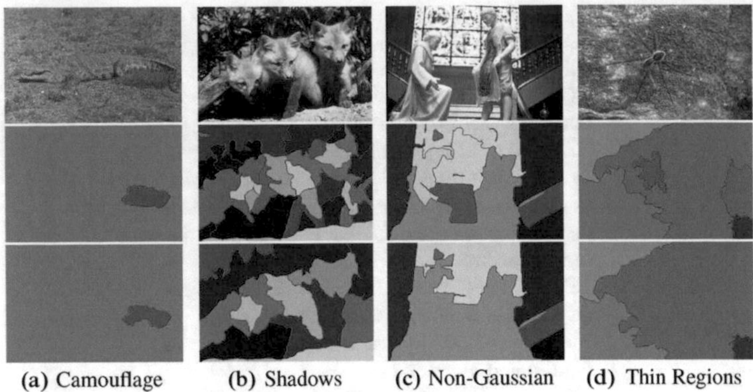

Fig. 10.9 Examples from BSD (in color) where the TBES algorithm failed to obtain a reasonable segmentation. **Top:** Original input images. **Middle:** Segmentation with respect to PRI. **Bottom:** Segmentation w.r.t VOI.

initialization. Second, large enough windows that can better capture the statistics of the texture can barely fit into such thin regions. Consequently, texture estimation at these regions is ill conditioned and unstable.

To realize whether these problems are unique to the TBES method or are more universal, we have investigated similar problematic cases with the other methods reported here ((Comanicu and Meer 2002; Tu and Zhu 2002; Felzenszwalb and Huttenlocher 2004; Cour et al. 2005; Yang et al. 2008; Arbelaez et al. 2009; Donoser et al. 2009)). None of the methods were able to handle camouflage very well. Shadows are challenging for these methods as well. However, we observe that UCM performs relatively better in this case. For geometric patterns, CTM seems to be slightly better than others, but still is an oversegmentation. In the category of thin regions, all algorithms performed very poorly, but mean-shift is better by, for example, roughly picking up some of the spider's legs. It is further worth pointing out an interesting observation about PRI versus VOI that the former prefers oversegmentation and the latter prefers undersegmentation (as shown in Figure 10.9).

To summarize, in this chapter, we have studied natural image segmentation as a subspace clustering problem. The proposed segmentation algorithm is based on the information-theoretic approach introduced in Chapter 6. We have shown how to customize that method to take into account boundary information that is special and important in image segmentation. In particular, the texture and boundary information of each texture region is encoded using a Gaussian distribution and adaptive chain code, respectively. The partitioning of an image is achieved by an agglomerative clustering process applied to a hierarchy of decreasing window sizes. Based on the MDL principle, the optimal segmentation of the image is defined as the one that minimizes its total coding length. Since the lossy coding length function also depends on a distortion parameter that determines the granularity of the segmentation, one can use a simple linear regression to learn the optimal distortion parameter from a set of training images when it is provided by the user. Our experiments have validated that this simple method outperforms other existing methods in terms of region-based segmentation indices (i.e., PRI and VOI), and is among the top solutions in terms of contour-based segmentation indices (i.e., BDE and GFM).

10.5 Bibliographic Notes

Image segmentation has been a very active research area in computer vision over the past 20 years, and there has been a vast amount of literature on this topic. Since it is impossible to give a comprehensive survey, we list here only some recent work that is related to the ideas and assumptions behind the method introduced in this chapter.

Mixture Models for Image Segmentation
It has been widely accepted in the image segmentation community that one should use a mixture model to model texture (features) of different regions. For example, normalized cuts (NC) by (Shi and Malik 2000), multiscale normalized cuts (MNC) by (Cour et al. 2005), F&H by (Felzenszwalb and Huttenlocher 2004), normalized tree partitioning by (Wang et al. 2008a), and multilayer spectral segmentation by (Kim et al. 2010) formulate the segmentation of a mixture model as a graph minimum-cut problem, while mean shift (MS) by (Comanicu and Meer 2002) seeks a partition of a color image based on different modes within the estimated empirical distribution.

Statistical Models for Texture
Over the years, there have been many proposed methods to model the representation of image textures in natural images. One model that has been shown to be successful in encoding textures both empirically and theoretically is the Gaussian *mesh Markov model* (MMM) (Levina and Bickel 2006). Particularly in texture synthesis, the Gaussian MMM can provide consistent estimates of the joint distribution of the pixels in a window, which then can be used to fill in missing texture patches via a simple nonparametric scheme (Efros and Leung 1999). Independent image patches that are sampled from such a Gaussian mesh model follow a Gaussian distribution. This to some extent justifies the (degenerate) Gaussian model that we have assumed in this chapter for the window-based feature vectors for each image segment.

Edges and Contours for Image Segmentation.
Region contours/edges convey important information about the saliency of the objects in the image and their shapes (see (Elder and Zucker 1996; Gevers and Smeulders 1997; Arbelaez 2006; Zhu et al. 2007; Ren et al. 2008)). Several recent methods have been proposed to combine the cues of homogeneous color and texture with the cue of contours in the segmentation process, including (Malik et al. 2001; Tu and Zhu 2002; Kim et al. 2005).

Hierarchical Image Segmentation.
The properties of local features (including texture and edges) usually do not share the same level of homogeneity at the same spatial scale. Thus, salient image regions can only be extracted from a hierarchy of image features under multiple scales (see (Yu 2005; Ren et al. 2005; Yang et al. 2008; Donoser et al. 2009)). The method introduced in this chapter utilizes the notion of hierarchy in two places; one is in the construction of feature vectors (in Section 10.3.2), and one is through varying the level of quantization error (in Section 10.3.3).

Compression-Based Image Segmentation.
As we have discussed in Chapter 6, the notion of lossy *minimum description length* (MDL) has been shown to be very effective for evaluating clustering results of general mixed data ((Ma et al. 2007)). The MDL principle and the compression-based clustering method were first applied to the segmentation of natural images by (Yang et al. 2008). The method is known as *compression-based texture merging* (CTM). The method presented in this chapter is based on the work of (Rao et al. 2009; Mobahi et al. 2011), which can be considered a greatly improved version of CTM. It remedies the issue of redundant encoding of the overlapping image windows and replaces the simple Huffman coding of the membership with a more accurate boundary coding scheme. As result, not only is the overall coding length more accurate, but also the segmentation results are much better.

Chapter 11
Motion Segmentation

I can calculate the motion of heavenly bodies, but not the madness of people.

—Isaac Newton

The previous two chapters have shown how to use a mixture of subspaces to represent and segment static images. In those cases, different subspaces were used to account for multiple characteristics of natural images, e.g., different textures. In this chapter, we will show how to use a mixture of subspaces to represent and segment time series, e.g., video and motion capture data. In particular, we will use different subspaces to account for multiple characteristics of the dynamics of a time series, such as multiple moving objects or multiple temporal events.

The first few sections of this chapter will focus on the problem of segmenting a video into multiple regions corresponding to different moving objects, i.e., the *motion segmentation problem*. Depending on whether one is interested in segmenting the 2D motion field in the image plane (optical flow) or the 3D motion (rotation and translation) of multiple objects in 3D space, the motion segmentation problem can be divided into two main categories: *2D motion segmentation* and *3D motion segmentation*, respectively. Moreover, depending on the type of image measurements (image derivatives, optical flows, feature point trajectories) and camera projection models (orthographic, spherical, perspective), different types of motion models can be used to describe each moving object. In this chapter, we will focus on the 3D motion segmentation problem from 2D point trajectories. We will first study the problem of segmenting multiple motions from consecutive multiple orthographic views. This problem corresponds to the very important practical setting of a surveillance camera that is monitoring moving vehicles or pedestrians from a far distance. We show that this problem is equivalent to clustering point trajectories into multiple affine subspaces of dimension one to three. We will then study how to segment multiple motions from two separate perspective views. While in this case

R. Vidal et al., *Generalized Principal Component Analysis*, Interdisciplinary Applied Mathematics 40, DOI 10.1007/978-0-387-87811-9_11

the point trajectories do not live in a linear or affine subspace, we will show that by using a nonlinear embedding, the problem can still be reduced to the problem of clustering multiple subspaces in $\mathbb{R}^9$.

The last section of this chapter will focus on the problem of segmenting a video into multiple temporal segments corresponding to different temporal events, i.e., the *temporal segmentation problem*. For example, segmenting a news video into multiple video shots, segmenting motion capture data of humans into a sequence of actions (walking, running), or segmenting robotic surgery data of a certain task (suturing) into a sequence of surgical gestures (grabbing the needle, inserting the needle, etc.). By assuming that the data within each segment are the output of a linear autoregressive dynamical model, we show that the data points within each segment live in a low-dimensional subspace; hence the temporal segmentation problem can be tackled using the subspace clustering algorithms discussed in Part II.

11.1 The 3D Motion Segmentation Problem

Consider a video sequence taken by a moving camera observing n rigidly moving objects. For example, Figure 11.1(a) shows the first frame of a video of a moving car taken by a moving camera. In this case, there are two motions in the scene: the motion of the red car and the motion of the background induced by the camera. Notice that from image measurements alone, it is impossible to distinguish the case in which the camera is stationary from that in which the camera itself is moving: all that can be inferred from image measurement is the relative motion of the objects in the scene with respect to the camera. Therefore, we assume for simplicity that the camera is stationary. Under this assumption, the static background becomes one of the moving objects, and its hypothetical motion is induced by the motion of the camera.

To formulate the motion segmentation problem, we first need to describe the motion of objects relative to a camera and the geometry of the camera. In this book, we largely follow the convention and terminology of (Ma et al. 2003). We describe the pose of each object by an element of the special Euclidean group $SE(3)$, $g = (R, \boldsymbol{T}) \in SE(3)$, where $R \in SO(3)$ is a rotation matrix and $\boldsymbol{T} \in \mathbb{R}^3$ is a translation vector.[1] The pose will be expressed relative to a world coordinate system W, which we assume to coincide with the camera coordinate system, as illustrated in Figure 11.2. More specifically, let the pose of object $i = 1, \dots, n$ at frame $f = 1, \dots, F$ be denoted by $g_{fi} \in SE(3)$. The motion of the ith object between the first and the fth frames is given by $(R_{fi}, \boldsymbol{T}_{fi}) = g_{fi} g_{1i}^{-1} \in SE(3)$.[2]

[1]The special Euclidean group is defined as $SE(3) = \{(R, \boldsymbol{T}) : R \in SO(3), \boldsymbol{T} \in \mathbb{R}^3\}$, where $SO(3) = \{R \in \mathbb{R}^{3\times 3} : R^\top R = I \text{ and } \det(R) = 1\}$ is the special orthogonal group.

[2]The inverse of $g \in SE(3)$ is $g^{-1} = (R^\top, -R^\top \boldsymbol{T}) \in SE(3)$, and the product of two transformations $g_1 = (R_1, \boldsymbol{T}_1)$ and $g_2 = (R_2, \boldsymbol{T}_2)$ is defined as $g_1 g_2 = (R_1 R_2, R_1 \boldsymbol{T}_1 + \boldsymbol{T}_2)$.

(a) First frame of the video

(b) First frame with feature points superimposed

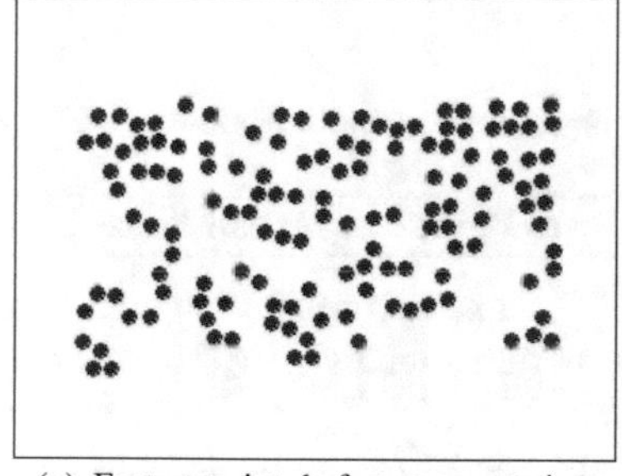
(c) Feature points before segmentation

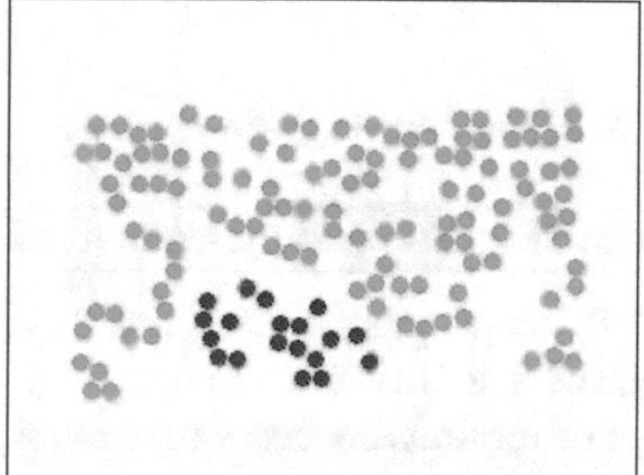
(d) Feature points after segmentation

Fig. 11.1 An image with multiple moving objects.

Let us now consider a collection of P_i 3D points lying on the ith object and let $\{\boldsymbol{X}_p \in \mathbb{R}^3\}_{p \in G_i}$ be the coordinates of these points relative to the world frame, where $G_i \subset \{1, \ldots, P\}$ is the set of points corresponding to the ith motion. As illustrated in Figure 11.2, the coordinates $\boldsymbol{X}_{fp}$ of point p relative to the fth camera frame are obtained by applying the rigid-body transformation $(R_{fi}, \boldsymbol{T}_{fi})$ to $\boldsymbol{X}_p$:

$$\boldsymbol{X}_{fp} = R_{fi}\boldsymbol{X}_p + \boldsymbol{T}_{fi} \in \mathbb{R}^3 \quad \forall p \in G_i. \tag{11.1}$$

The projection of $\boldsymbol{X}_{fp}$ onto the camera plane is thus given by

$$\boldsymbol{x}_{fp} = \pi_f(R_{fi}\boldsymbol{X}_p + \boldsymbol{T}_{fi}) \quad \forall p \in G_i, \tag{11.2}$$

where $\pi_f : \mathbb{R}^3 \mapsto \mathbb{R}^2$ is the camera projection model. We will discuss the specific form of the map π (orthographic, perspective, etc.) in a moment. As an example, Figure 11.1(b) illustrates the set of image points $\{\boldsymbol{x}_{1p} \in \mathbb{R}^2\}_{p=1}^P$ extracted for the first frame of the video shown in Figure 11.1(a).

With the above notation, the 3D motion segmentation problem can be formulated as the problem of segmenting the point trajectories $\{\boldsymbol{x}_{fp}\}_{p=1,\ldots,P}^{f=1,\ldots,F}$ into n groups corresponding to the n different motions in the video. Figure 11.1 provides an example, where Figure 11.1(c) shows the first frame of the point trajectories to be

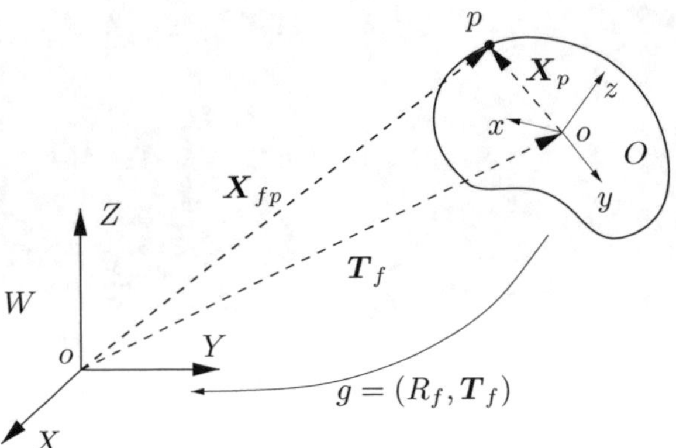

Fig. 11.2 A rigid-body motion between a moving object frame O and a world (camera) frame W.

Problem 11.1 (3D Motion Segmentation from Point Correspondences)

Given P image point trajectories $\{\boldsymbol{x}_{fp}\}_{p=1,\ldots,P}^{f=1,\ldots,F}$ taken from F views of a motion sequence related by a collection of n 3D motion models $\{\mathcal{M}_i\}_{i=1}^n$, estimate the number of motion models n, the model parameters $\{\mathcal{M}_i\}_{i=1}^n$, and the segmentation of the image point trajectories, i.e., the motion model i that corresponds to the pth trajectory $\{\boldsymbol{x}_{fp}\}_{f=1,\ldots,F}$.

---

segmented, while Figure 11.1(d) shows the same point trajectories segmented into two groups corresponding to the red car and the background, respectively. More formally, we state our goal as Problem 11.1.

As we will see in the remainder of this chapter, in some cases the camera model is such that the 3D motions impose linear constraints on the image measurements; thus Problem 11.1 is a direct application of the subspace clustering algorithms discussed earlier. In other cases, the motion models are more complex, e.g., bilinear in the image measurements, and the subspace clustering algorithms need to be extended to deal with such classes of motion segmentation problems.

In practice, the motion segmentation problem is more challenging than what is stated in Problem 11.1, because the point trajectories may be corrupted. For example, objects may appear and disappear from the field of view, or parts of the object may become occluded. In such cases, the feature point trajectories will be incomplete, and we will need to resort to subspace clustering algorithms that can handle missing entries. Moreover, the feature point trajectories need to be extracted using some tracking or matching algorithm, and such algorithms could make mistakes that may lead to erroneous trajectories. In such cases, we will need to resort to subspace clustering algorithms that can handle corrupted entries.

The next two sections show how to solve important special cases of the motion segmentation problem in Problem 11.1 using the tools from Part II of this book.

11.2 Motion Segmentation from Multiple Affine Views

In this section, we study a special but important case of the 3D motion segmentation problem in which the camera project is *affine*. We will first review the geometry of this problem and show that it is equivalent to clustering multiple low-dimensional affine subspaces of a high-dimensional space. We will then show how different subspace clustering algorithms can be used to solve this problem.

11.2.1 Affine Projection of a Rigid-Body Motion

Consider a video consisting of a single rigid-body motion first. Let the pose of the objet relative to the camera at frame $f = 1, \dots, F$ be denoted by $(R_f, \boldsymbol{T}_f) \in SE(3)$. Let $\boldsymbol{X}_p$ be the coordinates of a point in 3D space, let $\boldsymbol{X}_{fp} = R_f\boldsymbol{X}_p + \boldsymbol{T}_f$ be its coordinates at frame $f = 1, \dots, F$, and let $\boldsymbol{x}_{fp} = \pi_f(R_f\boldsymbol{X}_p + \boldsymbol{T}_f)$ be its projection onto the camera plane at frame f (see Section 11.1 for details). Adopting the orthographic projection model, we obtain

$$\boldsymbol{x}_{fp} \doteq \begin{bmatrix} 1 & 0 & 0 \\ 0 & 1 & 0 \end{bmatrix} \boldsymbol{X}_{fp} = \begin{bmatrix} 1 & 0 & 0 \\ 0 & 1 & 0 \end{bmatrix} \begin{bmatrix} R_f & \boldsymbol{T}_f \end{bmatrix} \begin{bmatrix} \boldsymbol{X}_p \\ 1 \end{bmatrix}. \tag{11.3}$$

This projection equation is specified relative to a very particular reference frame centered at the optical center with one axis aligned with the optical axis. In practice, when one captures digital images, the measurements are obtained in pixel coordinates, which are related to the image coordinates by the transformation

$$\tilde{\boldsymbol{x}}_{fp} \doteq \begin{bmatrix} s_x & s_\theta \\ 0 & s_y \end{bmatrix} \boldsymbol{x}_{fp} + \begin{bmatrix} o_x \\ o_y \end{bmatrix} = \underbrace{\begin{bmatrix} s_x & s_\theta & o_x \\ 0 & s_y & o_y \end{bmatrix}}_{K_f} \begin{bmatrix} \boldsymbol{x}_{fp} \\ 1 \end{bmatrix}, \tag{11.4}$$

where $K_f \in \mathbb{R}^{2\times 3}$ is the camera calibration matrix, s_x and s_y are scale factors in the x and y coordinates of the image, s_θ is a skew factor, and (o_x, o_y) is a translation of the image coordinates that moves the origin from the center of the image to its upper-left corner.

Combining the motion model (11.1), the orthographic projection model (11.3), and the calibration model (11.4) leads to the following *affine camera model*:

$$\tilde{\boldsymbol{x}}_{fp} = K_f \begin{bmatrix} 1 & 0 & 0 & 0 \\ 0 & 1 & 0 & 0 \\ 0 & 0 & 0 & 1 \end{bmatrix} \begin{bmatrix} R_f & \boldsymbol{T}_f \\ \boldsymbol{0}^\top & 1 \end{bmatrix} \begin{bmatrix} \boldsymbol{X}_p \\ 1 \end{bmatrix} = \underbrace{\begin{bmatrix} M_f & \boldsymbol{t}_f \end{bmatrix}}_{A_f} \begin{bmatrix} \boldsymbol{X}_p \\ 1 \end{bmatrix}, \tag{11.5}$$

where $A_f \in \mathbb{R}^{2\times 4}$ is the *affine camera matrix* at frame f, which depends on the camera calibration parameters $K_f \in \mathbb{R}^{2\times 3}$ and the object pose relative to the camera $(R_f, \boldsymbol{T}_f) \in SE(3)$. Notice that the rows of each A_f involve linear combinations of the first two rows of the rotation matrix R_f; hence A_f is of rank 2. With an abuse of notation, we will drop the tilde from $\tilde{\boldsymbol{x}}_{fp}$ and write it as $\boldsymbol{x}_{fp}$.

11.2.2 Motion Subspace of a Rigid-Body Motion

Let $\boldsymbol{w}_p \in \mathbb{R}^{2F}$ be the 2D trajectory of point p. Then

$$\boldsymbol{w}_p \doteq \begin{bmatrix} \boldsymbol{x}_{1p} \\ \vdots \\ \boldsymbol{x}_{Fp} \end{bmatrix} = \begin{bmatrix} M_1 \\ \vdots \\ M_F \end{bmatrix} \boldsymbol{X}_p + \begin{bmatrix} \boldsymbol{t}_1 \\ \vdots \\ \boldsymbol{t}_F \end{bmatrix} = M\boldsymbol{X}_p + \boldsymbol{t}, \tag{11.6}$$

where $M \in \mathbb{R}^{2F\times 3}$ and $\boldsymbol{t} \in \mathbb{R}^{2F}$. It follows that the matrix of all 2D trajectories $W \doteq \begin{bmatrix} \boldsymbol{w}_1 \cdots \boldsymbol{w}_P \end{bmatrix} \in \mathbb{R}^{2F\times P}$ can be decomposed as

$$W = MS + \boldsymbol{t}\mathbf{1}^\top, \tag{11.7}$$

where $S = \begin{bmatrix} \boldsymbol{X}_1 \cdots \boldsymbol{X}_P \end{bmatrix} \in \mathbb{R}^{3\times P}$. Therefore, the 2D trajectories associated with a single rigid-body motion observed by an affine camera live in an affine subspace of $\mathbb{R}^{2F}$ spanned by the columns of M. This affine subspace is of dimension at most three, and we call it the *motion subspace* of the rigid-body motion.

In the motion segmentation literature, this affine motion subspace is usually interpreted as a linear subspace. Specifically, notice that we can rewrite (11.7) as

$$W = \begin{bmatrix} M\, \boldsymbol{t} \end{bmatrix} \begin{bmatrix} S \\ \mathbf{1}^\top \end{bmatrix}. \tag{11.8}$$

Since $\begin{bmatrix} M\, \boldsymbol{t} \end{bmatrix} \in \mathbb{R}^{2F\times 4}$, we have $\operatorname{rank}(W) \leq 4$; hence the span of the columns of W can be interpreted as a linear subspace of $\mathbb{R}^{2F}$ of dimension at most four. This observation is important, since it will enable us to apply subspace clustering algorithms designed for *linear* subspaces to the motion segmentation problem.

11.2.3 Segmentation of Multiple Rigid-Body Motions

Let us now consider a stationary camera observing n different rigid-body motions. The ith rigid-body motion is represented by the transformation

$$\begin{bmatrix} R_{if} & \boldsymbol{T}_{if} \\ \mathbf{0}^\top & 1 \end{bmatrix}, \tag{11.9}$$

which defines the pose of the ith body during frame f. The 2×4 projection matrix associated with the ith body is then given by

$$A_{if} = K_f \begin{bmatrix} 1 & 0 & 0 & 0 \\ 0 & 1 & 0 & 0 \\ 0 & 0 & 0 & 1 \end{bmatrix} \begin{bmatrix} R_{if} & \boldsymbol{T}_{if} \\ \boldsymbol{0}^\top & 1 \end{bmatrix}. \tag{11.10}$$

A point $\boldsymbol{X}_p$ belonging to the ith body is projected to the point

$$\boldsymbol{x}_{fp} = A_{if} \begin{bmatrix} \boldsymbol{X}_p \\ 1 \end{bmatrix}. \tag{11.11}$$

Let $W_i \in \mathbb{R}^{2F \times P_i}$ denote the matrix of 2D trajectories associated with the ith body, where P_i is the number of points belonging to the ith moving object and $P = \sum_{i=1}^{n} P_i$ is the total number of trajectories. Then the data matrix consisting of the trajectories of all points can be written as

$$W = \big[W_1, W_2, \ldots, W_n\big]\,\Gamma \quad \in \mathbb{R}^{2F \times P}, \tag{11.12}$$

where $\Gamma \in \mathbb{R}^{P \times P}$ is an unknown permutation matrix that specifies the segmentation of the points according to different rigid-body motions.

The *3D motion segmentation problem* is the task of clustering these P trajectories according to the n rigid-body motions. Since the trajectories associated with each object live in an affine subspace of $\mathbb{R}^{2F}$ of dimension $d_i \in \{1, 2, 3\}$, $i = 1, \ldots, n$, the 3D motion segmentation problem is equivalent to clustering a set of points into n affine subspaces of $\mathbb{R}^{2F}$ of unknown dimensions $d_i \in \{1, 2, 3\}$. Therefore, we can apply any of the existing subspace clustering algorithms for affine subspaces, e.g., the SLBF and SCC algorithms in Chapter 7 and the SSC algorithm in Chapter 8, to the columns of W to solve the 3D motion segmentation problem. Alternatively, since each affine subspace of dimension d can be interpreted as a linear subspace of dimension $d + 1$, we can also apply any of the existing subspace clustering algorithms for linear subspaces, e.g., the ASC, ALC, LSA, LLMC, SCC, SSC, and LRSC algorithms described in Chapters 5–8.

11.2.4 Experiments on Multiview Motion Segmentation

The Hopkins 155 Motion Segmentation Data Set

We evaluate a number of subspace clustering algorithms on the Hopkins 155 motion segmentation database (Tron and Vidal 2007), which is available online at http://www.vision.jhu.edu/data/hopkins155. The database consists of 155 sequences of two and three motions, which can be divided into three main categories: checkerboard, traffic, and articulated sequences.

1. The checkerboard sequences contain multiple objects moving independently and arbitrarily in 3D space; hence the motion trajectories are expected to lie in independent affine subspaces of dimension three.
2. The traffic sequences contain cars moving independently on the ground plane; hence the motion trajectories are expected to lie in independent affine subspaces of dimension two.
3. The articulated sequences contain motions of people, cranes, etc., where object parts do not move independently, and so the motion subspaces are expected to be dependent.

For each sequence, the trajectories are extracted automatically with a tracker, and outliers are manually removed. Therefore, the trajectories are corrupted by noise, but do not have missing entries or outliers. Figure 11.3 shows sample images from videos in the database with the feature points superimposed.

Figure 11.4 shows the singular values of the matrix of feature point trajectories of a single motion for several videos in the data set. Note that the singular value curve has a knee around 4, corroborating the approximate 4-dimensionality of the motion data in each video.

Experimental Setup

In order to make the experimental results comparable to those in the existing literature, for each method we apply the same preprocessing steps described in their respective papers. Specifically, we project the trajectories onto a subspace of dimension $r \leq 2F$ using PCA. Historically, there have been two choices for the

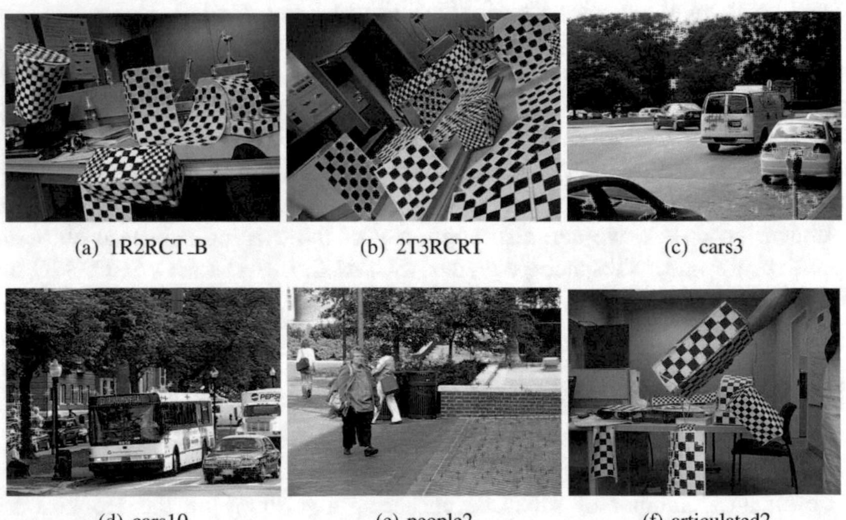

(a) 1R2RCT_B (b) 2T3RCRT (c) cars3

(d) cars10 (e) people2 (f) articulated2

Fig. 11.3 Sample images from some sequences of the Hopkins 155 database with tracked points superimposed. Figures (a) and (b) correspond to checkerboard sequences, figures (c) and (d) to traffic sequences, and figures (e) and (f) to articulated sequences.

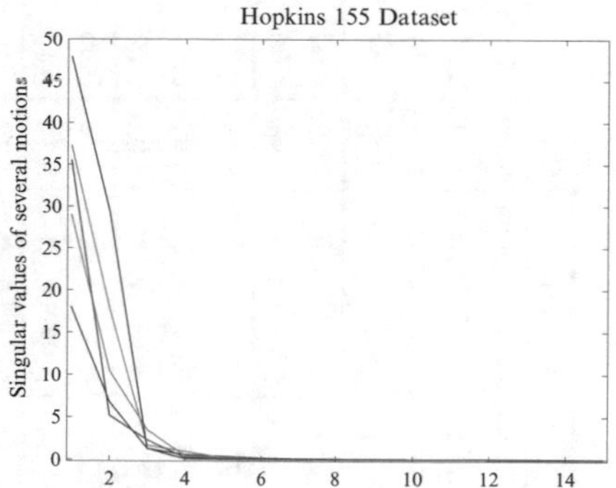

Fig. 11.4 Singular values of several motions in the Hopkins 155 data set. Each motion corresponds to a subspace of dimension at most 4.

dimension of the projection: $r = 5$ and $r = 4n$. The choice of $r = 5$ is motivated by the ASC algorithm, which models 3D affine subspaces as 4D linear subspaces. Since $d_{\max} = 4$, ASC chooses $r = d_{\max} + 1 = 5$. The choice of $d = 4n$ is motivated by factorization methods, which use the fact that for independent subspaces, $r = \text{rank}(X) = 4n$. In our experiments, we use $r = 5$ for ASC and RANSAC and $r = 4n$ for ASC, LLMC, LSA, SCC, and SSC. For LRSC, we use $r = 2F$, since the method searches for the low-rank representation. For ALC, r is chosen automatically for each sequence as the minimum r such that $r \geq 8\log(2F/r)$. We will refer to this choice as the *sparsity-preserving* (sp) projection. Also, for the algorithms that make use of K-means, either a single restart is used when the algorithm is initialized by another algorithm (LLMC, SCC), or 10 restarts are used when it is initialized at random (ASC, LLMC, LSA). SSC uses 20 restarts.

Clustering Errors

For each algorithm and each sequence, we record the clustering error, defined as

$$\text{Clustering error} = \frac{\text{\# of misclassified points}}{\text{total \# of points}} \times 100\%. \tag{11.13}$$

Table 11.1 reports the average and median clustering errors, and Figure 11.5 shows the percentage of sequences for which the clustering error is below a given percentage. More detailed statistics with the clustering errors of each algorithm on each of the 155 sequences can be found at http://www.vision.jhu.edu/data/hopkins155/.

Comparison and Conclusions

Examining the experimental results, we can draw the following conclusions about the performance of the algorithms tested.

Table 11.1 Classification errors of several subspace clustering algorithms on the Hopkins 155 motion segmentation database. All algorithms use two parameters (d, r), where d is the dimension of the subspaces and r is the dimension of the projection. Affine subspace clustering algorithms treat subspaces as 3-dimensional affine subspaces, i.e., $d = 3$, while linear subspace clustering algorithms treat subspaces as 4-dimensional linear subspaces, i.e., $d = 4$. The dimensions of the projections are $r = 5$, $r = 4n$, where n is the number of motions, and $r = 2F$, where F is the number of frames. ALC uses a sparsity-preserving (sp) dimension for the projection. All algorithms use PCA to perform the projection, except for SSC, which uses a random projection with entries drawn from a Bernoulli (SSC-B) or normal (SSC-N) distribution. The results for ASC correspond to the spectral clustering-based ASC algorithm. LLMC-ASC denotes LLMC initialized by the ASC algorithm.

	Two motions								Three motions								All	
	Check. (78)		Traffic (31)		Articul. (11)		All (120)		Check. (26)		Traffic (7)		Articul. (2)		All (35)		(155)	
	Mean	Median	Mean	Median	Mean	Median	Mean	Median	Mean	Median	Mean	Median	Mean	Median	Mean	Median	Mean	Median
ASC (4,5)	6.09	1.03	1.41	0.00	2.88	0.00	4.59	0.38	31.95	32.93	19.83	19.55	16.85	16.85	28.66	28.26	10.34	2.54
ASC (4n-1,4n)	4.78	0.51	1.63	0.00	6.18	3.20	4.10	0.44	36.99	36.26	39.68	40.92	29.62	29.62	37.11	37.18	11.55	1.36
RANSAC (4,5)	6.52	1.75	2.55	0.21	7.25	2.64	5.56	1.18	25.78	26.00	12.83	11.45	21.38	21.38	22.94	22.03	9.76	3.21
LSA (4,5)	8.84	3.43	2.15	1.00	4.66	1.28	6.73	1.99	30.37	31.98	27.02	34.01	23.11	23.11	29.28	31.63	11.82	4.00
LSA (4,4n)	2.57	0.27	5.43	1.48	4.10	1.22	3.45	0.59	5.80	1.77	25.07	23.79	7.25	7.25	9.73	2.33	4.94	0.90
LLMC (4,5)	4.85	0.00	1.96	0.00	6.16	1.37	4.22	0.00	9.06	7.09	6.45	0.00	5.26	5.26	8.33	3.19	5.15	0.00
LLMC (4,4n)	3.96	0.23	3.53	0.33	6.48	1.30	4.08	0.24	8.48	5.80	6.04	4.09	9.38	9.38	8.04	4.93	4.97	0.87
MSL	4.46	0.00	2.23	0.00	7.23	0.00	4.14	0.00	10.38	4.61	1.80	0.00	2.71	2.71	8.23	1.76	5.03	0.00
ALC (4,5)	2.56	0.00	2.83	0.30	6.90	0.89	3.03	0.00	6.78	0.92	4.01	1.35	7.25	7.25	6.26	1.02	3.76	0.26
ALC (4,sp)	1.49	0.27	1.75	1.51	10.70	0.95	2.40	0.43	5.00	0.66	8.86	0.51	21.08	21.08	6.69	0.67	3.37	0.49
SCC (4, 4n)	1.30	0.04	1.07	0.44	3.68	0.67	1.46	0.16	5.68	2.96	2.35	2.07	10.94	10.94	5.31	2.40	2.33	
SCC (4, 2F)	1.31	0.06	1.02	0.26	3.21	0.76	1.41	0.10	6.31	1.97	3.31	3.31	9.58	9.58	5.90	1.99	2.42	
SLBF (3, 2F)	1.59	0.00	0.20	0.00	0.80	0.00	1.16	0.00	4.57	0.94	**0.38**	0.00	2.66	2.66	3.63	0.64	1.66	
LRSC (4,4n)							2.58	0.00							6.68	1.76	3.49	0.09
SSC-B (4,4n)	**0.83**	0.00	0.23	0.00	1.63	0.00	**0.75**	0.00	4.49	0.54	0.61	0.00	1.60	1.60	3.55	0.25	1.45	0.00
SSC-N (4,4n)	1.12	0.00	**0.02**	0.00	**0.62**	0.00	0.82	0.00	**2.97**	0.27	0.58	0.00	**1.42**	0.00	**2.45**	0.20	**1.24**	0.00

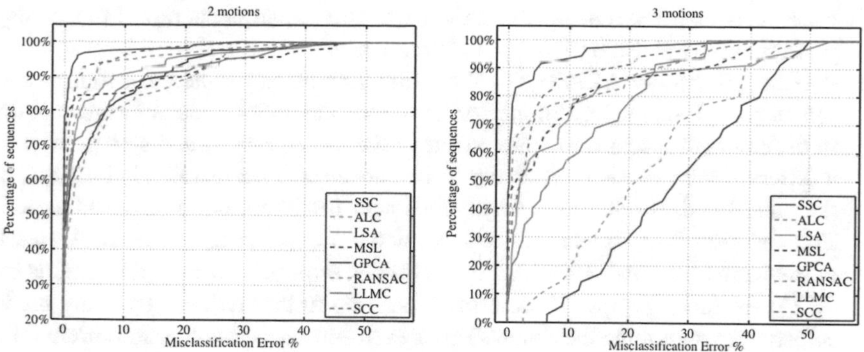

Fig. 11.5 Percentage of sequences for which the clustering error is less than or equal to a given percentage of error. The algorithms tested are ASC(4,5), RANSAC(4,5), LSA(4,4n), LLMC(4,4n), MSL, ALC(4,sp), SCC(4,4n), SSC-N(4,4n).

ASC (Algebraic Subspace Clustering): To avoid using multiple polynomials, we use an implementation of ASC based on hyperplanes in which the data are interpreted as a subspace of dimension $r-1$ in $\mathbb{R}^r$, where $r=5$ or $r=4n$. For two motions, ASC achieves a classification error of 4.59% for $r=5$ and 4.10% for $r=4n$. Notice that ASC is among the most accurate methods for the traffic and articulated sequences, which are sequences with dependent motion subspaces. However, ASC has higher errors on the checkerboard sequences, which constitute the majority of the database. This result is expected, because ASC is best designed for dependent subspaces. Notice also that increasing r from 5 to $4n$ improves the results for checkerboard sequences, but not for the traffic and articulated sequences. This is also expected, because the rank of the data matrix should be high for sequences with full-dimensional and independent motions (checkerboard), and low for sequences with degenerate (traffic) and dependent (articulated) motions. This suggests that using model selection to determine a different value of r for each sequence should improve the results. For three motions, the results are completely different, with a clustering error of 29%–37%. This is expected, because the number of coefficients fitted by ASC grows exponentially with the number of motions, while the number of feature points remains of the same order. Furthermore, ASC uses a least-squares method for fitting the polynomial, which neglects nonlinear constraints among the coefficients. The number of nonlinear constraints neglected also increases with the number of subspaces.

RANSAC (RANdom SAmple Consensus): The results for this purely statistical algorithm are similar to what we found for ASC. In the case of two motions, the results are a bit worse than those of ASC. In the case of three motions, the results are better than those of ASC, but still quite far from those of the best-performing algorithms. This is expected, because as the number of motions increases, the probability of drawing a set of points from the same group decreases significantly. Another drawback of RANSAC is that its performance varies between two runs on the same data. Our experiments report the average performance over 1000 trials for each sequence.

LSA (Local Subspace Affinity): When the dimension for the projection is chosen as $r = 5$, this algorithm performs worse than ASC. This is because points in different subspaces are closer to each other when $r = 5$, and so a point from a different subspace is more likely to be chosen as a nearest neighbor. ASC, on the other hand, is not affected by points near the intersection of the subspaces. The situation is completely different when $r = 4n$. In this case, LSA clearly outperforms ASC and RANSAC, achieving an error of 3.45% for two groups and 9.73% for three groups. These errors could be further reduced by using model selection to determine the dimension of each subspace. Another important thing to observe is that LSA performs better on the checkerboard sequences, but has larger errors than ASC on the traffic and articulated sequences. This confirms that LSA has difficulties with dependent subspaces.

LLMC (Locally Linear Manifold Clustering): The results of this algorithm also represent a clear improvement over ASC and RANSAC, especially for three motions. The only cases in which ASC outperforms LLMC are for traffic and articulated sequences. This is expected, because LLMC is not designed to handle dependent subspaces. Unlike LSA, LLMC is not significantly affected by the choice of r, with a clustering error of 5.15% for $r = 5$ and 4.97% for $r = 4n$. Notice also that the performance of LLMC improves when initialized with ASC to 4.87% for $r = 5$ and 4.37% for $r = 4n$. However, there are a few sequences for which LLMC performs worse than ASC even when LLMC is initialized by ASC. This happens for sequences with dependent motions, which are not well handled by LLMC.

MSL (Multistage Learning (Kanatani 2001)): By looking at the average clustering error, we can see that MSL, LSA, and LLMC have a similar accuracy. Furthermore, their segmentation results remain consistent in going from two to three motions. However, sometimes the MSL method gets stuck in a local minimum. This is reflected by high clustering errors for some sequences, as can be seen by the long tails in Figure 11.5.

ALC (Agglomerative Lossy Compression): The ALC algorithm represents a significant increase in performance with respect to all previous algorithms, especially for the checkerboard sequences, which constitute the majority of the database. However, ALC does not perform very well on the articulated sequences. This is because ALC typically needs the samples from a group to cover the subspace with sufficient density, while many of the articulated scenes have very few feature point trajectories. With regard to the projection dimension, the results indicate that overall, ALC performs better with an automatic choice of the projection, rather than with a fixed choice of $r = 5$. One drawback of ALC is that it needs to be run about 100 times for different choices of the distortion parameter ε in order to obtain the right number of motions and the best segmentation results.

SCC (Spectral Curvature Clustering): This algorithm performs even better than ALC in almost all motion categories. The only exception is for the articulated sequences with three motions. This is because these sequences contain few trajectories for the sampling strategy to operate correctly. Another advantage of SCC with respect to ALC is that it is not very sensitive to the choice of the parameter c (number of sampled subsets), while ALC needs to be run for several choices of the distortion parameter ε. Notice also that the performance of SCC is not significantly affected by the dimension of the projection $r = 5$, $r = 4n$, or $r = 2F$.

SSC (Sparse Subspace Clustering): This algorithm performs extremely well, not only for checkerboard sequences, which have independent and fully dimensional motion subspaces, but also for traffic and articulated sequences, which are the bottleneck of almost all existing methods, because they contain degenerate and dependent motion subspaces. Overall, SSC is not sensitive to the dimension of the projection ($r = 5$ vs. $r = 4n$ vs. $r = 2F$) or the parameter μ.

SLBF (Spectral Local Best-Fit Flats): This algorithm performs extremely well for all motion sequences. Its performance is essentially on a par with that of SSC. We refer the reader to (Zhang et al. 2010) for additional experiments.

LRSC (Low-Rank Subspace Clustering): Overall, LRSC compares favorably against LSA and SCC. However, LRSC does not perform as well as SSC. Also, the performance of LRSC depends on the choice of the parameters. In particular, notice that choosing τ that depends on the number of motions and size of each sequence gives better results than using a fixed τ. We also notice that LRSC has almost the same performance with or without projection.

The reader should be aware that the above comparison of all the algorithms is merely on the motion segmentation data set. The conclusions drawn above should not be blindly generalized to all other types of data or problems the user may encounter in practice. As we have discussed extensively in the previous chapters, the choice of the most suitable clustering algorithm for an application often depends on the particular settings and conditions. Each of the subspace clustering algorithms has its own strengths and limitations, and the user should exercise discretion carefully based on the special nature of the data and tasks at hand.

11.3 Motion Segmentation from Two Perspective Views

In this section, we consider the 3D motion segmentation problem in cases in which the projection model is the more general perspective projection. As a result, the resulting 3D motion model will be *bilinear* in the image measurements. In particular, we consider the segmentation of rigid-body motions from point correspondences in two perspective views of nonplanar (Section 11.3.3) and planar (Section 11.3.4) scenes. In both cases, we show that the motion segmentation problem can be solved using extensions of the subspace clustering algorithms presented earlier to certain classes of bilinear surfaces.

11.3.1 Perspective Projection of a Rigid-Body Motion

As before, we consider a video consisting of a single rigid-body motion and denote its pose at frame $f = 1, \dots, F$ by $(R_f, \boldsymbol{T}_f) \in SE(3)$. Let $\boldsymbol{X}_p$ be the coordinates of a point p in 3D space. At frame $f = 1, \dots, F$, the coordinates of this point are transformed to $\boldsymbol{X}_{fp} = R_f\boldsymbol{X}_p + \boldsymbol{T}_f$. Let $\boldsymbol{x}_{fp} = \pi_f(R_f\boldsymbol{X}_p + \boldsymbol{T}_f)$ be the projection of $\boldsymbol{X}_{fp}$ onto the camera plane at frame f (see Section 11.1 for details). Adopting the perspective projection model shown in Figure 11.6, we obtain

$$\boldsymbol{x}_{fp} \doteq \begin{bmatrix} x_{fp} \\ y_{fp} \\ 1 \end{bmatrix} = \frac{1}{Z_{fp}} \begin{bmatrix} \phi & 0 & 0 \\ 0 & \phi & 0 \\ 0 & 0 & 1 \end{bmatrix} \begin{bmatrix} X_{fp} \\ Y_{fp} \\ Z_{fp} \end{bmatrix} = \frac{1}{Z_{fp}} \begin{bmatrix} \phi & 0 & 0 \\ 0 & \phi & 0 \\ 0 & 0 & 1 \end{bmatrix} \boldsymbol{X}_{fp}, \tag{11.14}$$

where ϕ is the camera's focal length and Z_{fp} is the depth of the point.

Notice that we have written the image point $\boldsymbol{x}_{fp}$ in homogeneous coordinates by appending a one as its third coordinate. This allows us to rewrite the calibration model in (11.4) as a matrix multiplication

$$\tilde{\boldsymbol{x}}_{fp} = \begin{bmatrix} s_x & s_\theta & o_x \\ 0 & s_y & o_y \\ 0 & 0 & 1 \end{bmatrix} \boldsymbol{x}_{fp}. \tag{11.15}$$

Combining the motion model (11.1), the projection model (11.14), and the calibration model (11.4) leads to the following *perspective camera model*:

$$Z_{fp}\tilde{\boldsymbol{x}}_{fp} = \underbrace{\begin{bmatrix} s_x & s_\theta & o_x \\ 0 & s_y & o_y \\ 0 & 0 & 1 \end{bmatrix} \begin{bmatrix} \phi & 0 & 0 \\ 0 & \phi & 0 \\ 0 & 0 & 1 \end{bmatrix}}_{K_f \in \mathbb{R}^{3\times 3}} \begin{bmatrix} R_f & \boldsymbol{T}_f \end{bmatrix} \begin{bmatrix} \boldsymbol{X}_p \\ 1 \end{bmatrix}, \tag{11.16}$$

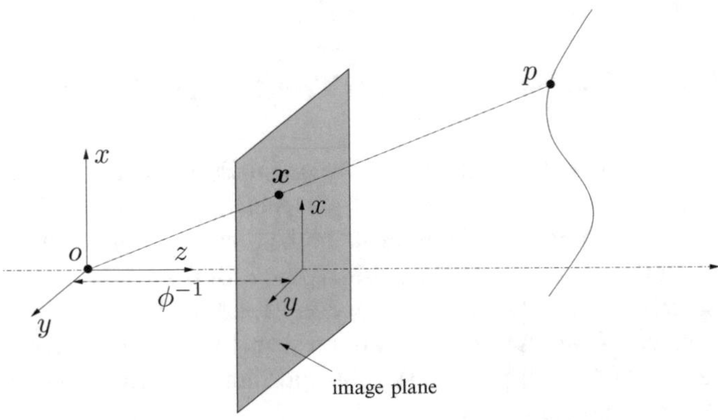

Fig. 11.6 Frontal pinhole imaging model: the image of a 3D point p is the point $\boldsymbol{x}$ at the intersection of the ray going through the optical center o and the image plane at a distance ϕ^{-1} in front of the optical center.

where $K_f \in \mathbb{R}^{3\times 3}$ is called the *camera calibration matrix* in the fth frame. When $K_f = I$, we say that the camera is calibrated.

If we also use homogeneous coordinates to represent $\boldsymbol{X}_p$ as $\tilde{\boldsymbol{X}}_p = \begin{bmatrix} \boldsymbol{X}_p \\ 1 \end{bmatrix}$, then we can write the projection model as a matrix multiplication

$$Z_{fp}\tilde{\boldsymbol{x}}_{fp} = \Pi_f \tilde{\boldsymbol{X}}_p, \tag{11.17}$$

where $\Pi_f = K_f \begin{bmatrix} R_f & \boldsymbol{T}_f \end{bmatrix} \in \mathbb{R}^{3\times 4}$ is called the *perspective projection matrix*. Finally, notice that if we define $\tilde{R}_f = K_f R_f$ and $\tilde{\boldsymbol{T}}_f = K_f \boldsymbol{T}_f$, we can rewrite the projection equation as $Z_{fp}\tilde{\boldsymbol{x}}_{fp} = \tilde{R}_f \boldsymbol{X}_p + \tilde{\boldsymbol{T}}_f$. For the sake of simplicity, we will drop the tildes from the notation and write the projection model simply as

$$Z_{fp}\boldsymbol{x}_{fp} = R_f \boldsymbol{X}_p + \boldsymbol{T}_f. \tag{11.18}$$

11.3.2 Segmentation of 3D Translational Motions

In this subsection, we consider the problem of segmenting n 3D translational motions from multiple point correspondences in two perspective views. We assume that the 3D scene is nonplanar and that the translation of each object between the two views is nonzero. As we shall see, in this case the image measurements are related by a bilinear constraint. However, we will show that this bilinear constraint can be converted into a linear constraint on a nonlinear embedding of the point correspondences. As a result, the motion segmentation problem from two perspective views will be reduced to a problem of clustering planes in $\mathbb{R}^3$.

Let us first consider the case of a single 3D translational object. Let $\boldsymbol{x}_1$ and $\boldsymbol{x}_2$ be images of point $\boldsymbol{X}$ in the first and second frames of a video sequence consisting of $F = 2$ frames. Then the projection equation reduces to

$$Z_2\boldsymbol{x}_2 = \boldsymbol{X} + \boldsymbol{T}_2 = Z_1\boldsymbol{x}_1 + \boldsymbol{T}_2. \tag{11.19}$$

As illustrated in Figure 11.7, the vectors $\boldsymbol{x}_2$, $\boldsymbol{T}_2$ and $\boldsymbol{x}_1$ must be coplanar; hence their triple product must be zero, i.e.,

$$(\boldsymbol{x}_1 \times \boldsymbol{x}_2) \cdot \boldsymbol{T}_2 = 0. \tag{11.20}$$

While this equation is nonlinear in $(\boldsymbol{x}_1, \boldsymbol{x}_2)$, it is linear in

$$\boldsymbol{y} = \boldsymbol{x}_1 \times \boldsymbol{x}_2. \tag{11.21}$$

Indeed, $\boldsymbol{T}_2^\top \boldsymbol{y} = 0$ is simply the equation of a plane in $\boldsymbol{y}$ with normal vector $\boldsymbol{T}_2$.

Consider now the case of n moving objects and let $\boldsymbol{T}_{i2}$, $i = 1, \ldots, n$, denote their translations between the two views. Let $\{(\boldsymbol{x}_{1p}, \boldsymbol{x}_{2p})\}_{p=1}^{P}$ denote P point correspondences between the two views (in homogeneous coordinates). Then the set of points $\{\boldsymbol{y}_p = \boldsymbol{x}_{1p} \times \boldsymbol{x}_{2p}\}_{p=1}^{P}$ lies in a union of n planes in $\mathbb{R}^3$, and we can cluster the point correspondences using a subspace clustering algorithm for linear subspaces.

11.3.3 Segmentation of Rigid-Body Motions

In this subsection, we consider the problem of segmenting n 3D rigid-body motions from point correspondences in two perspective views. We assume that the 3D scene is nonplanar and that the translation of each object between the two views is nonzero. As we shall see, in this case the image measurements are related by a bilinear constraint. However, we will show that this bilinear constraint can be converted into a linear constraint on a nonlinear embedding of the point correspondences. As a result, the motion segmentation problem from two perspective views will be reduced to a problem of clustering 8-dimensional subspaces of $\mathbb{R}^9$.

Let us first consider the case of a single 3D rigid-body motion. Let $\boldsymbol{x}_1$ and $\boldsymbol{x}_2$ be, respectively, images of a point $\boldsymbol{X}$ in the first and second frames of a video sequence consisting of $F = 2$ frames. Then the projection equation reduces to

$$Z_2\boldsymbol{x}_2 = R_2\boldsymbol{X} + \boldsymbol{T}_2 = Z_1R_2\boldsymbol{x}_1 + \boldsymbol{T}_2. \tag{11.22}$$

As illustrated in Figure 11.7, the vectors $\boldsymbol{x}_2$, $\boldsymbol{T}_2$, and $R\boldsymbol{x}_1$ must be coplanar; hence their triple product must be zero, i.e.,

$$\boldsymbol{x}_2 \cdot (\boldsymbol{T}_2 \times R_2\boldsymbol{x}_1) = 0 \iff \boldsymbol{x}_2^\top \widehat{\boldsymbol{T}}_2 R_2 \boldsymbol{x}_1 = 0, \tag{11.23}$$

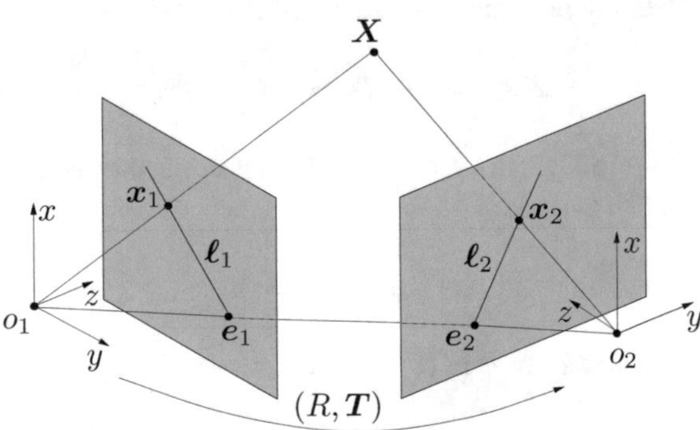

Fig. 11.7 Epipolar geometry: Two projections $\boldsymbol{x}_1, \boldsymbol{x}_2 \in \mathbb{R}^3$ of a 3D point $\boldsymbol{X}$ from two vantage points. The relative Euclidean transformation between the two vantage points is given by $(R, \boldsymbol{T}) \in SE(3)$. The intersections of the line (o_1, o_2) with each image plane are called *epipoles* and are denoted by $\boldsymbol{e}_1$ and $\boldsymbol{e}_2$. The intersections of the plane $(o_1, o_2, \boldsymbol{X})$ with the two image planes are called *epipolar lines* and are denoted by $\boldsymbol{\ell}_1$ and $\boldsymbol{\ell}_2$.

where for a vector $\boldsymbol{T} = (T_1, T_2, T_3)^\top \in \mathbb{R}^3$, the matrix $\widehat{\boldsymbol{T}}$ is a skew-symmetric matrix generating the cross product by $\boldsymbol{T}$, i.e., $\widehat{\boldsymbol{T}}\boldsymbol{x} = \boldsymbol{T} \times \boldsymbol{x}$, and is defined as

$$\widehat{\boldsymbol{T}} = \begin{bmatrix} 0 & -T_3 & T_2 \\ T_3 & 0 & -T_1 \\ -T_2 & T_1 & 0 \end{bmatrix}. \tag{11.24}$$

Let $E = \widehat{\boldsymbol{T}}_2 R_2 \in \mathbb{R}^{3\times 3}$. This matrix is known in the computer vision literature as the *fundamental matrix*. With this notation, (11.23) becomes

$$\boldsymbol{x}_2^\top E \boldsymbol{x}_1 = 0. \tag{11.25}$$

This equation is known as the *epipolar constraint* and establishes a fundamental relationship between two perspective images of a rigid-body motion. While this equation is nonlinear in $(\boldsymbol{x}_1, \boldsymbol{x}_2)$, notice that it is linear in the Kronecker product of $\boldsymbol{x}_1$ and $\boldsymbol{x}_2$, $\boldsymbol{y} = \boldsymbol{x}_1 \otimes \boldsymbol{x}_2 \in \mathbb{R}^9$. Specifically, we have

$$\boldsymbol{x}_2^\top E \boldsymbol{x}_1 = \sum_{ij} x_{2i} e_{ij} x_{1j} = \boldsymbol{e}^\top \boldsymbol{y} = 0, \tag{11.26}$$

where

$$\boldsymbol{e} = \left[e_{11}, e_{21}, e_{31}, e_{12}, e_{22}, e_{32}, e_{13}, e_{23}, e_{33}\right]^\top, \tag{11.27}$$

$$\boldsymbol{y} = \left[x_{11}x_{21}, x_{11}x_{22}, x_{11}x_{23}, x_{12}x_{21}, x_{12}x_{22}, x_{12}x_{23}, \ldots, x_{13}x_{23}\right]^\top. \tag{11.28}$$

This is simply the equation of a hyperplane in $\boldsymbol{y}$ with normal vector $\boldsymbol{e} \in \mathbb{R}^9$.

Consider now the case of n moving objects and let $\{E_i\}_{i=1}^n$ denote their fundamental matrices between the two views. Let $\{(\boldsymbol{x}_{1p}, \boldsymbol{x}_{2p})\}_{p=1}^P$ denote P point correspondences between the two views (in homogeneous coordinates). Then the set of points $\boldsymbol{y}_p = \boldsymbol{x}_{1p} \otimes \boldsymbol{x}_{2p}$ lies in a union of n hyperplanes in $\mathbb{R}^9$, and we can cluster the point correspondences using a subspace clustering algorithm for linear subspaces.

11.3.4 Segmentation of Rotational Motions or Planar Scenes

The motion segmentation scheme described in the previous subsection assumes that the displacement of each object between the two views relative to the camera is nonzero, i.e., $\boldsymbol{T} \neq \boldsymbol{0}$, for otherwise, the individual fundamental matrices $E = \widehat{\boldsymbol{T}}R$ would be zero. Furthermore, it also requires that the 3D points be in general configuration; otherwise, one could not uniquely recover each fundamental matrix from its epipolar constraint. The latter case occurs, for example, in the case of planar objects.

Both in the case of a purely rotating object (relative to the camera) and in the case of a planar object, the motion $(R, \boldsymbol{T})$ between the two views $\boldsymbol{x}_1$ and $\boldsymbol{x}_2$ can be described by a homography matrix $H \in \mathbb{R}^{3\times 3}$ as

$$\boldsymbol{x}_2 \sim H\boldsymbol{x}_1 = \begin{bmatrix} h_{11} & h_{12} & h_{13} \\ h_{21} & h_{22} & h_{23} \\ h_{31} & h_{32} & h_{33} \end{bmatrix} \boldsymbol{x}_1, \tag{11.29}$$

where $\sim$ means equality up to a scale factor. In the case of a purely rotating object, we have $H = R_2$, because the motion equations in (11.22) reduce to

$$Z_2\boldsymbol{x}_2 = Z_1 R_2 \boldsymbol{x}_1 \iff \boldsymbol{x}_2 \sim R_2\boldsymbol{x}_1. \tag{11.30}$$

In the case of a planar object, we have $\boldsymbol{n}^\top \boldsymbol{X} = d$, where $\boldsymbol{n} \in \mathbb{S}^2$ is the normal to the plane and d is the distance from the plane to the origin of the first view. It follows from the motion equations in (11.22) that

$$Z_2\boldsymbol{x}_2 = R_2\boldsymbol{X} + \boldsymbol{T}_2\frac{\boldsymbol{n}^\top \boldsymbol{X}}{d} = Z_1(R_2 + \frac{1}{d}\boldsymbol{T}_2\boldsymbol{n}^\top)\boldsymbol{x}_1. \tag{11.31}$$

Therefore, $\boldsymbol{x}_2 \sim H\boldsymbol{x}_1$ with $H = R_2 + \frac{1}{d}\boldsymbol{T}_2\boldsymbol{n}^\top$, as claimed.

The equation $\boldsymbol{x}_2 \sim H\boldsymbol{x}_1$ is known as the *homographic constraint*. While this equation is nonlinear in $(\boldsymbol{x}_1, \boldsymbol{x}_2)$, notice that it is linear in the Kronecker product of $\boldsymbol{x}_1$ and $\boldsymbol{x}_2$, $\boldsymbol{y} = \boldsymbol{x}_1 \otimes \boldsymbol{x}_2 \in \mathbb{R}^9$. Specifically, we have

$$\begin{bmatrix} 0 & h_{31} & -h_{21} & 0 & h_{32} & -h_{22} & 0 & h_{33} & -h_{23} \\ -h_{31} & 0 & h_{11} & -h_{32} & 0 & h_{12} & -h_{33} & 0 & h_{13} \end{bmatrix} \boldsymbol{y} = 0. \tag{11.32}$$

This is simply the equation of a linear subspace of dimension seven in $\mathbb{R}^9$.

Consider now the case of n moving objects and let $\{H_i\}_{i=1}^n$ denote their homographs between the two views. Let $\{(\boldsymbol{x}_{1p}, \boldsymbol{x}_{2p})\}_{p=1}^P$ denote P point correspondences between the two views (in homogeneous coordinates). Then the set of points $\boldsymbol{y}_p = \boldsymbol{x}_{1p} \otimes \boldsymbol{x}_{2p}$ lies in a union of n seven-dimensional subspaces in $\mathbb{R}^9$, and we can cluster the point correspondences using a subspace clustering algorithm for linear subspaces.

11.3.5 Experiments on Two-View Motion Segmentation

In this subsection we evaluate the performance of the motion segmentation algorithms from two perspective views on the Hopkins 155 data set.

Experimental Setup
Since this data set provides trajectories of data points over several frames, we use the following three settings to form the data points to be clustered:

Setting 1: Use corresponding feature points between the first and the last frames to form the embedded vector $\boldsymbol{y}$ in $\mathbb{R}^3$ or $\mathbb{R}^9$, as in (11.21) or (11.28).

Setting 2: Use corresponding features between the first and the middle frames, and those between the middle and the last frames. For example, if there are 10 frames, we use frame pairs $(1, 5)$ and $(6, 10)$. A data point is then formed by simply concatenating the two embedded vectors as in (11.21) or (11.28). Thus, data points in this setting have dimension 6 or 18.

Setting 3: Use corresponding features from multiple pairs of frames that have the same frame interval (which is to be half of the total number of frames). For instance, for a sequence of 10 frames, we use feature points from the pairs $(1, 6), (2, 7), \ldots, (5, 10)$. Again, all the embedded vectors formed as in (11.21) or (11.28) are concatenated as a single data point.

The so-obtained data points are then clustered using the ASC algorithm described in Chapter 5, the SASC algorithm described in Chapter 7, and the SSC algorithm described in Chapter 8. For the ASC algorithm, we consider the following two variants:

1. ASC by line intersection (ASC-LI), i.e., Algorithm 5.3 with a fixed number of groups ($n = 2$ or $n = 3$) and the best out of 10 randomly chosen lines.
2. ASC by minimum distance (ASC-MD), i.e., Algorithm 5.4 with the heuristic distance in (5.45) being used to select a point closest to each subspace with $\delta = 10^{-5}$.

For the SASC algorithm, we consider the following two variants:

1. SASC with the angle-based affinity (SASC-angle), i.e., Algorithm 7.5 with the affinity in (7.24) with parameter $q = 20$ for the 3D translational motion model and $q = 1$ for the 3D rigid-body motion model.
2. SASC with the distance-based affinity (SASC-distance), i.e., Algorithm 7.5 with the affinity in (7.26) with parameter $q = 20$ for the 3D translational motion model and $q = 1$ for the 3D rigid-body motion model.

For the SSC algorithm, we solve the optimization problem in (8.95) with parameter $\tau = \alpha\tau_{\min,2}$, where $\alpha = 10^4$ for settings 1 and 2 and $\alpha = 10^5$ for setting 3, and $\tau_{\min,2}$ is defined in (8.96).

The resulting affinity matrix C is postprocessed columnwise in two steps. In the first step, the absolute value of the entries of each column are sorted in descending order, and the top k entries are preserved with others set to zero, where k is the smallest number such that the ℓ_1 norm of the top k entries is larger than 0.7 times the ℓ_1 norm of the entire vector. Then in the second step, the columns are normalized to unit ℓ_∞-norm.

Table 11.2 Clustering error (%) of various subspace clustering methods on the Hopkins 155 data set using two-view 3D translational and two-view 3D rigid-body motion models for each one of the three settings. The methods tested are: ASC by Line Intersection (ASC-LI), ASC by Minimum Distance (ASC-MD), SASC with angle affinity (SASC-angle), SASC with distance affinity (SASC-distance), and SSC. For the sake of comparison, we also report the motion segmentation results using the multiview affine motion model from Table 11.1.

		2 motions		3 motions		All	
Model	Method	Mean	Median	Mean	Median	Mean	Median
Trans 1	ASC-LI	26.95	29.79	41.73	44.56	30.29	33.09
	ASC-MD	27.50	28.99	39.82	43.29	30.28	33.60
	SASC-angle	27.21	30.88	45.47	49.90	31.33	33.33
	SASC-distance	32.95	35.76	44.35	44.05	35.52	37.19
	SSC	18.19	16.00	33.99	35.04	21.76	20.52
Trans 2	ASC-LI	30.30	31.62	43.44	44.33	33.27	34.76
	ASC-MD	18.19	16.21	31.78	32.98	21.26	19.05
	SASC-angle	27.52	28.31	43.33	44.66	31.09	34.11
	SASC-distance	34.96	38.00	47.36	47.33	37.76	39.88
	SSC	9.79	0.80	21.04	19.82	12.33	1.59
Trans 3	SSC	8.26	0.91	12.07	0.67	9.12	0.89
Rigid 1	ASC-LI	32.52	34.19	44.89	47.10	35.31	36.51
	ASC-MD	28.04	30.88	38.79	39.16	30.47	32.14
	SASC-angle	21.09	20.66	33.30	32.99	23.85	23.11
	SASC-distance	21.28	20.17	33.31	34.21	23.99	22.94
	SSC	3.30	0.00	14.58	10.34	5.84	0.39
Rigid 2	ASC-LI	31.25	33.81	43.24	42.20	33.96	35.76
	ASC-MD	26.48	31.30	43.91	45.96	30.42	34.06
	SASC-angle	23.18	23.20	39.72	43.15	26.91	28.45
	SASC-distance	23.80	24.39	33.95	37.02	26.10	27.46
	SSC	2.06	0.00	5.56	1.49	2.85	0.00
Rigid 3	SSC	1.69	0.00	4.61	0.44	2.35	0.00
Affine	ASC(4,5)	4.59	0.38	28.66	28.26	10.34	2.54
	ASC(4n-1,4n)	4.10	0.44	37.11	37.18	11.55	1.36
	SSC-B	0.75	0.00	3.55	0.25	1.45	0.00
	SSC-N	0.82	0.00	2.45	0.20	1.24	0.00

Results and Conclusions
For each algorithm and each sequence, we record the clustering error as defined in (11.13). Table 11.2 reports the average and median clustering errors of ASC, SASC, and SSC for different settings. Comparing all the experimental results, we can draw the following conclusions:

Number of Views. Notice that most methods perform the worst when using only two views (Setting 1), and that performance generally improves as more frames are used. In particular, the best results are obtained by the multiview affine motion model, which uses all the frames.

Translational Motion Model. Notice that none of the methods performs well when using the 3D translational motion model. This is to be expected, since one of the motions in each video (the background motion) is due to the camera motion, which cannot be well approximated by a purely translational motion. Nonetheless, it is interesting to see that SSC is the only method that performs reasonably well, in spite of the simplicity of the motion model.

Rigid-Body Motion Model. Notice also that with the exception of SSC, none of the methods performs well when using the 3D rigid-body motion model. This is to be expected for settings 1 and 2, which use only a few frames, but setting 3 uses sufficiently many frames, and hence we would have expected the performance of perspective algorithms to match that of affine algorithms. This is indeed the case for SSC. However, notice that affine algorithms still do better. This is expected, since the data in the Hopkins 155 database can be well approximated by the affine model, as reported in (Tron and Vidal 2007).

Clustering Methods. Observe that the spectral ASC methods perform generally better than the purely algebraic ASC methods. This is expected, since spectral methods are more robust to noise than purely algebraic methods. Observe also that in general, SSC gives the best results for all motion models and settings. This is surprising, because most of the two-view models represent the motion of each object with a hyperplane, which is the case for which algebraic methods are best suited. SSC, on the other hand, is best suited for low-dimensional subspaces of a higher-dimensional space, which is the case for the affine motion model. Nonetheless, we see that SSC performs reasonably well already with the rigid-body motion model for setting 1, which involves only two views.

11.4 Temporal Motion Segmentation

The previous two sections showed how a mixture of subspaces model can be used to segment time-series data in the *spatial domain*. In particular, given a video sequence, we focused on the problem of segmenting it into multiple spatial regions corresponding to different rigid-body motions. In this section, we will show that the mixture of subspaces model can also be used to segment time-series data in the *temporal domain*. In particular, we will study the problem of segmenting a time series into multiple temporal segments of homogeneous dynamics. We will also present applications to the segmentation of a video sequence into multiple video shots corresponding to different events, and the segmentation of human motion data into multiple actions or gestures.

11.4.1 Dynamical Models of Time-Series Data

Let $\{\boldsymbol{x}_t \in \mathbb{R}^D\}_{t=0}^T$ be a time series, e.g., a video sequence. We assume that this time series can be decomposed into multiple temporal segments $[t_i, t_{i+1}]$ for $i = 1, \dots, n$, where $0 = t_0 < t_1 < \cdots < t_n = T$, each of homogeneous dynamics. In particular, we assume that the dynamics of the time series within each segment can be described by a linear autoregressive (AR) dynamical model

$$\boldsymbol{x}_t = a_1 \boldsymbol{x}_{t-1} + a_2 \boldsymbol{x}_{t-2} + \cdots + a_d \boldsymbol{x}_{t-d} + \boldsymbol{w}_t, \quad \boldsymbol{x}_t, \boldsymbol{w}_t \in \mathbb{R}^D, \tag{11.33}$$

where $\boldsymbol{w}_t$ is i.i.d. zero-mean Gaussian noise with covariance $\sigma^2 I$.

Given a time series $\{\boldsymbol{x}_t\}_{t=0}^T$, the *temporal segmentation problem* is the problem of finding the number of segments n and the switching times $\{t_i\}$. In addition, we may also be interested in finding the model parameters for each segment.

Linear dynamical models such as (11.33) have been used extensively for modeling and recognition of dynamic visual phenomena, including the recognition of human gestures, actions, and activities from motion capture and video data, and the recognition of surgical gestures from robotic surgery and video data (see Section 11.5 for a brief overview). In such applications, the dimension D of the data at each time instance is much higher than the order d of the dynamical system. In this case, it is easy to show that all the data points $\{\boldsymbol{x}_t\}$ associated with the same dynamical system belong to a d-dimensional subspace of $\mathbb{R}^D$. In particular, notice that when $\boldsymbol{w}_t = \boldsymbol{0}$, we have that $\boldsymbol{x}_{d+1}$ is a linear combination of $\boldsymbol{x}_1, \dots, \boldsymbol{x}_d$, and $\boldsymbol{x}_{d+2}$ is a linear combination of $\boldsymbol{x}_2, \dots, \boldsymbol{x}_{d+1}$, hence of $\boldsymbol{x}_1, \dots, \boldsymbol{x}_d$. By induction, it follows that $\boldsymbol{x}_t$ is a linear combination of $\boldsymbol{x}_1, \dots, \boldsymbol{x}_d$ for all $t \geq d+1$. Assuming that $\boldsymbol{x}_1, \dots, \boldsymbol{x}_d$ are linearly independent, we can see that there exist a full-rank matrix $U \in \mathbb{R}^{D \times d}$ and a vector $\boldsymbol{y}_t \in \mathbb{R}^d$ such that $\boldsymbol{x}_t = U\boldsymbol{y}_t$ for all $t \geq 1$. More generally, when $\boldsymbol{w}_t \neq 0$, we have

$$\boldsymbol{x}_t = U\boldsymbol{y}_t + \tilde{\boldsymbol{w}}_t, \tag{11.34}$$

where $\tilde{\boldsymbol{w}}_t = \sum_{i=1}^t \beta_i \boldsymbol{w}_i$ is zero-mean Gaussian noise. Therefore, when σ is small, the set of $\{\boldsymbol{x}_t\}$ lies approximately in a linear subspace of $\mathbb{R}^D$ of dimension d.

Consider now a time series with multiple segments, each one modeled with an AR model. In this case, we assume that the time series follows a switched linear autoregressive (SAR) dynamical model

$$\boldsymbol{x}_t = a_1(\lambda_t)\boldsymbol{x}_{t-1} + a_2(\lambda_t)\boldsymbol{x}_{t-2} + \cdots + a_{d_i}(\lambda_t)\boldsymbol{x}_{t-d_{\lambda_t}} + \boldsymbol{w}_t, \tag{11.35}$$

where $\lambda_t \in \{1, 2, \dots, n\}$ determines which of the n AR models is active at time t. We assume that λ_t is unknown. In fact, our goal is to determine which data point comes from which model (segmentation problem) without knowing the model parameters $\{a_j(i)\}_{i=1,\dots,n}^{j=1,\dots,d}$. To that end, let $G_i = \{t : \lambda_t = i\}$ be the set of time instants associated with the ith model. It follows that the data points $\{\boldsymbol{x}_t\}_{t \in G_i}$ live approximately in a subspace of dimension d_i. Therefore, the set of all data points $\{\boldsymbol{x}_t\}$ lives in a union of n subspaces of dimensions $\{d_i\}_{i=1}^n$. As a consequence, we can use any of the generic subspace clustering algorithms described in Part II to segment the time series into multiple temporal segments.

In some applications, such as dynamical system identification, we are required not only to cluster the temporal observations, but also to identify the orders and the parameters of all ARX models. In such cases, we need to understand more precisely the conditions under which the systems are identifiable and how to recover such parameters accordingly. We leave a more careful study of such conditions to Chapter 12, and concentrate here only on the clustering problem.

11.4.2 Experiments on Temporal Video Segmentation

In this section, we consider the problem of segmenting a video sequence into multiple temporal segments. In this case, a data point $\boldsymbol{x}_t \in \mathbb{R}^D$ at time t corresponds to a video frame (an image) that has been reshaped into a D-dimensional vector, where D is the number of pixels. Since D is too large for most of the subspace clustering algorithms discussed in this book, we usually reduce the dimension of the data either by downsampling each image or by applying PCA to the collection of all frames. In this experiment, we first use PCA to project the frames to a space of lower dimension 6. The projected data are then modeled as the output of a switched AR model as above.

We test two subspace clustering algorithms, namely SASC and SSC, on two videos of news. The first video (CNNObamaHillary) contains $F = 402$ frames, and each frame is an RGB image of size 215×321. The second video (CNNWorldCup) contains $F = 118$ frames of RGB images of size 325×577. Sample frames of the two test videos with ground truth segmentation are shown in Figure 11.8.

For SASC, we test its two variants described in Section 11.3.5. The first one is SASC-angle, where we use the angle-based affinity with $q = 5$. The second one is SASC-distance, where we use the distance-based affinity with $q = 0.5$. For SSC, we use the same model as in Section 11.3.5, with parameter $\tau = \alpha\tau_{\min,2}$, where $\alpha = 10^6$ and $\tau_{\min,2}$ is defined in (8.96).

Table 11.3 gives the clustering errors, defined in equation (11.13), by applying spectral clustering to the affinity matrices produced by these methods, which are plotted in Figure 11.9. Observe that SASC-angle fails to capture the multiple events in the video and gives large clustering errors. In contrast, SASC-distance gives a much better affinity with reasonably good clustering errors. Nonetheless, the off-diagonal entries in the affinity are still high. Overall, SSC gives a much better affinity matrix, which subsequently leads to a perfect segmentation of the video.

Now, it is important to note that in our model, each segment is assumed to be the output from a (linear) dynamical system. As such, it need not correspond to a

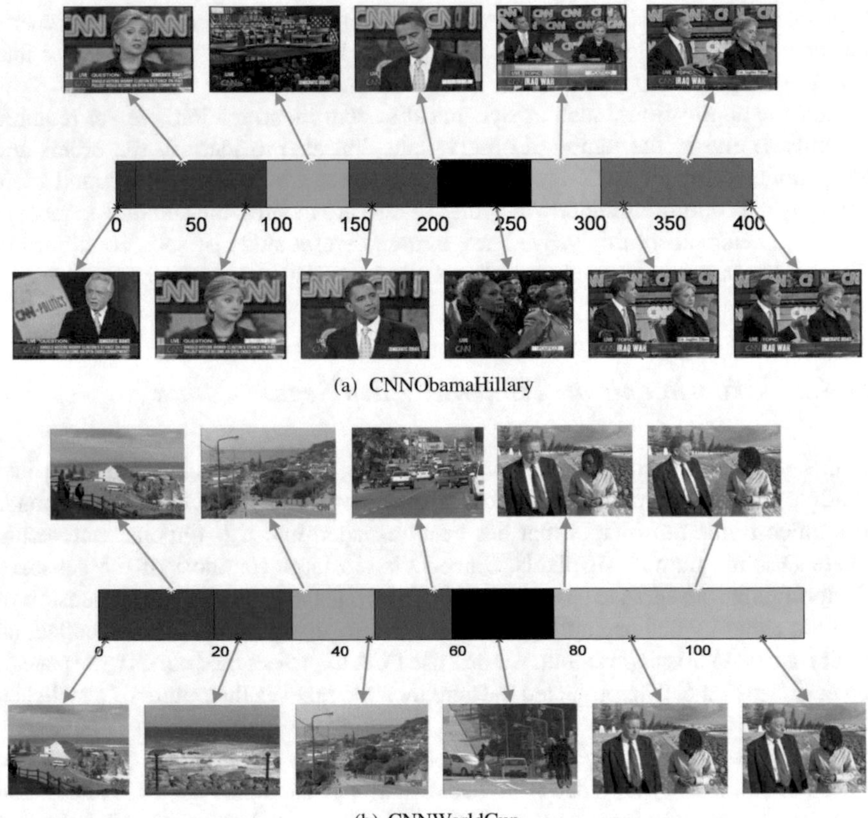

(a) CNNObamaHillary

(b) CNNWorldCup

Fig. 11.8 An illustration of the two videos we use in the experiments. Different colors represent different segments of the videos.

Table 11.3 Clustering error (%) of SSC and SASC for video segmentation obtained by applying spectral clustering to the affinity matrices in Figure 11.9.

Video /Method	SSC	SASC-angle	SASC-distance
CNNObamaHillary	0.00	64.94	0.00
CNNWorldCup	0.00	49.15	5.08

semantically cohesive event in a video. Moreover, as discussed before, our model does not enforce any temporal coherence, and so the segmentation need not to be continuous in time. Nonetheless, we observe in the experiments that such a grossly simplified model is able to capture temporally coherent and semantically meaningful events in simple videos.

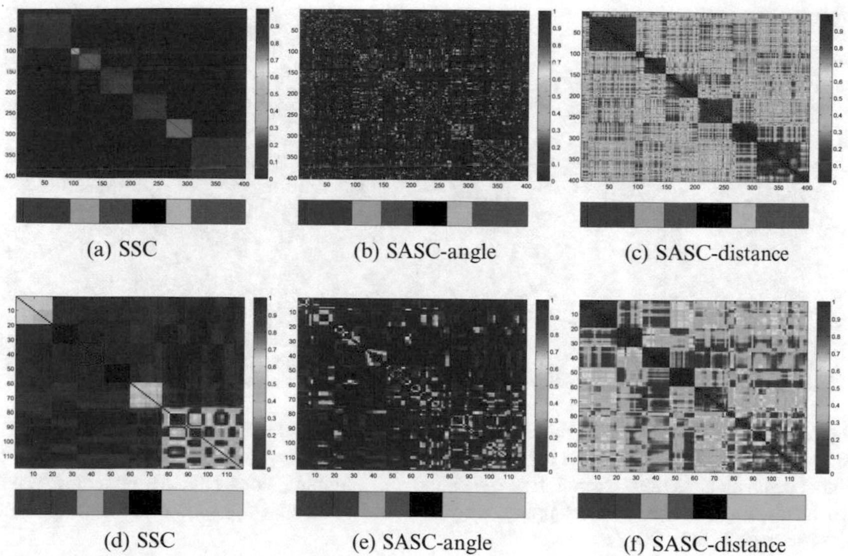

(a) SSC (b) SASC-angle (c) SASC-distance

(d) SSC (e) SASC-angle (f) SASC-distance

Fig. 11.9 Affinity matrices given by SSC and SASC for two news videos. Top row: CNNObamaHillary video. Bottom row: World cup video. The color bar beneath each subplot is the ground truth segmentation. The affinity matrix is obtained by taking the absolute value, symmetrizing, normalizing to (0, 1), and then setting diagonal entries to zero.

11.4.3 Experiments on Segmentation of Human Motion Data

In this section, we consider the problem of segmenting human motion data into multiple segments corresponding to different actions or gestures. Human motion data typically consist of time series data for the position, velocity, or acceleration of a collection of 3D points on a moving human body. Such data could be obtained by a motion capture system, by accelerometers installed in the human body, or using the Microsoft®Kinect sensor. When the points in 3D space coincide with human joints, the time series data could also include joint angles. Figure 11.10 shows an example from the Berkeley Multimodal Human Action Database (MHAD) taken from (Ofli et al. 2013), which illustrates multiple points tracked in the human body as well as 3D reconstructions obtained from the Kinect sensor for different human actions.

Given time series data $\boldsymbol{x}_t \in \mathbb{R}^D$, where $\boldsymbol{x}_t$ corresponds to motion information (position, velocity, acceleration, joint angles) from all points in the human body, the task is to segment this time series into a collection of segments corresponding to different actions. Since in this case, the dimension of the data D is typically in the range 10–100, we can apply most subspace clustering methods in this book without the need for reducing the dimension of the data.

In this experiment, we apply two subspace clustering algorithms introduced earlier in Chapter 8, namely SSC and LRSC, to segment human motion data from

Fig. 11.10 Snapshots from all the actions available in the Berkeley Multimodal Human Action Database (MHAD) are displayed together with the corresponding point clouds obtained from the Kinect depth data. Actions (from left to right): jumping, jumping jacks, bending, punching, waving two hands, waving one hand, clapping, throwing, sit down/stand up, sit down, stand up. Image taken from (Ofli et al. 2013).

the Carnegie Mellon University Motion Capture (MOCAP) database (CMU 2003). Following the experimental setup in (Barbic et al. 2004; Zhou et al. 2008), we use 14 sequences from subject 86. Each sequence consists of approximately 8000 frames of about 10 human actions, such as walk, squat, run, stand, jump, drink, punch. We apply SSC and LRSC to every other frame from each sequence. For SSC, we use the noisy model in (8.95) and set the parameter $\tau = \alpha\tau_{\min,2}$, where $\alpha = 30$ and $\tau_{\min,2}$ is defined in (8.96). For LRSC, we use the noisy model in (8.31) with parameter $\tau = 10{,}000$. Then the same postprocessing steps described in Section 11.3.5 are applied to the affinity matrices obtained by both methods. In addition, given that the segments are expected to be temporally continuous, we can improve the results by conducting a certain temporal smoothing. To be more precise, if a motion segment given by the algorithm has length less than a threshold, then we divide the segment into two halves and merge each into its adjacent segment. Methods with this postprocessing step are denoted by "-H."

Table 11.4 shows the average clustering performance on the 14 sequences, while Figure 11.11 visualizes segmentation results on one representative sequence. The metric for evaluating the performance is the average per cluster accuracy, which is given by

$$\text{Segmentation error (\%)} = \frac{1}{n}\sum_{i=1}^{n}\frac{\#\text{ of misclassified points for group } i}{\text{total }\#\text{ of points in group } i}. \tag{11.36}$$

As we see from the results, SSC generally does a better job than LRSC in segmenting the motion data sequences. In addition, the simple temporal postprocessing seems to be able to further improve the segmentation results, especially

Table 11.4 Segmentation error (%) of applying SSC/LRSC to 14 sequences of subject 86 from the MOCAP data set

Seq.	ACA	SSC	SSC-H	LRSC	LRSC-H
1	26.82	6.54	6.54	14.59	3.67
2	25.19	6.09	3.55	26.84	4.59
3	7.74	4.69	2.18	35.83	31.69
4	9.88	42.57	41.27	46.19	29.86
5	28.92	8.16	3.88	49.77	38.58
6	14.85	8.98	3.92	51.49	31.61
7	13.04	22.01	24.96	19.45	21.31
8	5.07	20.73	18.74	37.41	34.82
9	13.91	6.68	6.68	11.90	9.84
10	4.07	1.45	1.33	43.30	28.48
11	7.20	5.24	3.19	41.87	35.73
12	8.99	23.15	30.65	48.18	34.86
13	41.99	33.76	33.76	47.50	38.76
14	9.35	63.48	54.78	43.98	16.37
Avg.	15.50	18.11	16.82	37.02	25.73

Fig. 11.11 Comparison of different clustering methods on the sequence 2 of subject 86 from the motion capture data. The first row illustrates the ground truth segmentation of the sequence. The following rows are results given by different algorithms. Different colors correspond to different actions.

effective for LRSC. For comparison, in both the table and figure, we have also shown the results of the state-of-the-art human motion segmentation method aligned cluster analysis (ACA) (Zhou et al. 2008). The ACA method relies on using hidden Markov models to explicitly incorporate dynamical and temporal constraints for segmentation, whereas SSC and LRSC here treat each frame independently. Therefore, the performance of ACA is understandably better.

11.5 Bibliographical Notes

Bibliographical Notes on 3D Motion Segmentation

Three-dimensional motion estimation and segmentation has been an active topic of research in the computer vision community over the past few years. Earlier work (Feng and Perona 1998) solves this problem by first clustering the features corresponding to the same motion using, e.g., K-means or spectral clustering, and then estimating a single motion model for each group. This can also be done in a probabilistic framework (Torr 1998) in which a maximum-likelihood estimate of the parameters of each motion model is sought by alternating between feature clustering and single-body motion estimation using the expectation maximization (EM) algorithm. However, the convergence of EM to the global maximum depends strongly on initialization (Torr et al. 2001).

In order to deal with the initialization problem of EM-like approaches, recent work has concentrated on the study of the geometry of dynamic scenes, including the analysis of multiple points moving linearly with constant speed (Han and Kanade 2000; Shashua and Levin 2001) or in a conic section (Avidan and Shashua 2000), multiple points moving in a plane (Sturm 2002), multiple translating planes (Wolf and Shashua 2001a), self-calibration from multiple motions (Fitzgibbon and Zisserman 2000; Han and Kanade 2001), multiple moving objects seen by an affine camera (Boult and Brown 1991; Costeira and Kanade 1998; Kanatani 2001; Wu et al. 2001; Kanatani and Matsunaga 2002; Zelnik-Manor and Irani 2003; Kanatani and Sugaya 2003; Vidal and Hartley 2004), and two-object segmentation from two perspective views (Wolf and Shashua 2001b). The case of multiple moving objects seen by two perspective views was studied in (Vidal et al. 2002b; Vidal and Sastry 2003; Vidal and Ma 2004; Vidal et al. 2006; Ma et al. 2003), and has been extended to three perspective views via the so-called *multibody trifocal tensor* (Hartley and Vidal 2004). Such works have been the basis for the material presented in this chapter. Extensions to omnidirectional cameras can be found in (Shakernia et al. 2003; Vidal 2005).

Bibliographical Notes on Time Series Analysis

Over the past decade, there has been an increasing interest in the application of system-theoretic techniques to the modeling of high-dimensional time-series data using linear dynamical systems (LDSs) such as (11.33) and their extensions. For instance, (Béjar et al. 2012) uses LDSs to model surgical gestures in kinematic and video data from the DaVinci robot; (Ravichandran et al. 2006; Ghoreyshi and Vidal 2007) use LDSs to model the appearance of a deforming heart in a magnetic resonance image sequence; (Doretto et al. 2003; Yuan et al. 2004; Vidal and Ravichandran 2005; Woolfe and Fitzgibbon 2006; Doretto and Soatto 2006; Hyndman et al. 2007; Ravichandran and Vidal 2008; Ravichandran et al. 2009; Ravichandran and Vidal 2011; Ravichandran et al. 2013) use LDSs to model the appearance of *dynamic textures*, such as water and fire, in a video sequence; the same idea is exploited in (Chan and Vasconcelos 2005a,b) but for different applications such as classification and retrieval of traffic video scenes and for motion segmentation; (Nunez and Cipriano 2009) uses ARMA models to characterize different froth types in the flotation process (mineral processing); (Bissacco et al. 2001; Chaudhry et al. 2009; Li et al. 2011) use LDSs to model human gaits, such as walking and running, in motion capture and video data; (Kim et al. 2009) uses LDSs to detect salient motions in video by detecting mode changes in video patches; (Aggarwal et al. 2004) uses LDSs to model the appearance of moving faces; (Saisan et al. 2004) uses LDSs to model audio-visual lip articulations; (Rahimi et al. 2005) presents a semisupervised regression model that can be used for rigid pose estimation and tracking problems even in case of articulated or nonrigid (such as lips) objects; (Xiong et al. 2011, 2012) use dynamical systems to perform different computer vision tasks including tracking in the presence of occlusions; continuing in the tracking domain, (Ayazoglu et al. 2011) uses dynamical systems to perform multicamera tracking; LDSs have been extensively used also for action recognition. A few of the most recents publications are (Nascimento et al. 2005; Ali et al. 2007; Wang et al. 2008b; Li et al. 2011; Turaga et al. 2011). Given a high-dimensional time series, one can use standard system identification techniques, e.g., subspace identification (Overschee and Moor 1993), to learn the parameters of an LDS model. Given a model, novel time series can be synthesized by simulating the model forward. For example, impressive synthesis of dynamic textures has been demonstrated by a number of papers (Doretto et al. 2003; Doretto and Soatto 2003; Yuan et al. 2004; Szummer and Picard 1996). The same ideas have also been used for the synthesis of lip articulations using speech as the driving input (Saisan et al. 2004).

Chapter 12
Hybrid System Identification

From the earliest traceable cosmical changes down to the latest results of civilization, we shall find that the transformation of the homogeneous into the heterogeneous is that in which Progress essentially consists.

—Herbert Spencer

Hybrid systems are mathematical models that are used to describe continuous processes that occasionally exhibit discontinuous behaviors due to sudden changes of dynamics. For instance, the continuous trajectory of a bouncing ball results from alternating between free fall and elastic contact with the ground. However, hybrid systems can also be used to describe a complex process or time series that does not itself exhibit discontinuous behaviors, by approximating the process or series with a simpler class of dynamical models. For example, a nonlinear dynamical system can be approximated by switching among a set of linear systems, each approximating the nonlinear system in a subset of its state space. As another example, a video sequence can be segmented to different scenes by fitting a piecewise linear dynamical model to the entire sequence.

In recent years, there has been significant interest and progress in the study of the analysis, stability, and control of hybrid systems. When the system parameters are known, many successful theories have been developed to characterize the behaviors of hybrid systems under different switching mechanisms. However, in practice, the parameters and the switching mechanism of a hybrid system are often not known or derivable from first principles. We are faced with the task of identifying the system from its input and output measurements.

In this chapter, we show how to apply the algebraic subspace clustering (ASC) method (see Chapter 5) to the problem of identifying a class of discrete-time hybrid

R. Vidal et al., *Generalized Principal Component Analysis*, Interdisciplinary Applied Mathematics 40, DOI 10.1007/978-0-387-87811-9_12

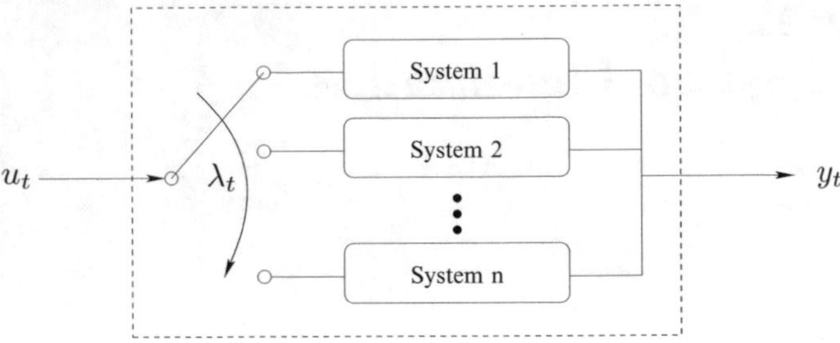

Fig. 12.1 The input/output diagram of a hybrid system switching among n constituent systems. The identification problem requires inferring what is in the black box (including the n systems and the switching mechanism λ_t) from its input u_t and output y_t.

systems known as hybrid auto regressive exogenous (ARX) systems.[1] We know from classic identification theory of linear systems that the configuration space of the input/output data generated by a single ARX system, say

$$y_t = \sum_{j=1}^{n_a} a_j y_{t-j} + \sum_{j=1}^{n_c} c_j u_{t-j} + w_t, \quad y_t, u_t, w_t \in \mathbb{R}, \tag{12.1}$$

is a linear subspace. The problem of identifying the system is equivalent to identifying this subspace from a finite number of (noisy) samples on the subspace (as we will review briefly in Section 12.2). Unfortunately, for a hybrid system that switches among multiple ARX systems, as shown in Figure 12.1, when the orders of the constituent systems are different, depending on the switching sequence λ_t, the configuration space of the hybrid ARX system might *not* simply be a union of the configuration spaces of the constituent ARX systems. Therefore, the problem of identifying the hybrid ARX system is *not* a trivial subspace segmentation problem.

In this chapter, we show how to incorporate some special (algebraic and dynamical) structures of a hybrid ARX system so that the identification problem can still be solved by a special version of the algebraic subspace clustering method. In particular, we will show that a hybrid ARX system can still be correctly identified from a special polynomial p that fits the input/output data of the hybrid ARX system; the last nonzero term of p has the lowest degree-lexicographic order in the ideal $\mathfrak{a}$ of polynomials. This polynomial is unique, factorable, and independent of the switching sequence. The nonrepeated factors of this polynomial correspond to the constituent ARX systems; hence the number of systems is given by the number of nonrepeated factors (Section 12.3).

[1] ARX systems are an extremely popular class of dynamical models that are widely used in control, signal processing, communications, and economics. In image/video processing, they can be used to model videos of dynamical scenes.

Although the analysis and algorithm will be developed primarily in a noise-free algebraic setting, the algebraic subspace clustering and identification algorithm is numerically stable and works with moderate noise. Simulation and experimental results show that the algorithm performs extremely well for both synthetic and real data, in comparison with the existing iterative (e.g., EM-based) identification algorithms (Section 12.4).

12.1 Problem Statement

Now let us consider a hybrid ARX system—a system that switches among multiple, say n, ARX systems of the type (12.1). Mathematically, such a system can be described as

$$y_t = \sum_{j=1}^{n_a(\lambda_t)} a_j(\lambda_t) y_{t-j} + \sum_{j=1}^{n_c(\lambda_t)} c_j(\lambda_t) u_{t-j} \quad (+\, w_t), \tag{12.2}$$

where $u_t \in \mathbb{R}$ is the *input*, $y_t \in \mathbb{R}$ is the *output*, $\lambda_t \in \{1, 2, \ldots, n\}$ is the *discrete state*, and $n_a(i)$, $n_c(i)$, $\{a_j(i)\}_{j=1}^{n_a(i)}$ and $\{c_j(i)\}_{j=1}^{n_c(i)}$ are, respectively, the orders and the system parameters of the ith ARX system for $i = 1, \ldots, n$. The last term w_t is zero for a deterministic ARX system and a white-noise random process for a stochastic system. The purpose of this chapter is to provide an analytic solution to the deterministic case, which approximates the stochastic case when w_t is small.

The discrete state λ_t, also called the *mode* of the system, can evolve due to a variety of mechanisms. In the least-restrictive case, $\{\lambda_t\}$ is a deterministic but unknown sequence that can take a finite number of possible values, which we can assume to coincide with a collection of integers:

$$\lambda : t \in \mathbb{Z} \mapsto \lambda_t \in \{1,\ 2,\ \ldots,\ n\}.$$

One can further restrict the set of switching sequences by assuming that λ_t is a realization of an irreducible Markov chain, governed by transition probabilities

$$\pi(i,j) \doteq P(\lambda_{t+1} = j | \lambda_t = i).$$

In this case, the system (12.2) is often called a "jump-Markov linear system" (JMLS). Alternatively, one can assume that λ_t is a piecewise constant function of the "continuous states" of the system (12.2),

$$\lambda : (y_{t-1}, \ldots, y_{t-n_a}) \in \mathbb{R}^{n_a} \mapsto \lambda_t \in \{1,\ 2,\ \ldots,\ n\}.$$

In this case, the system (12.2) is often called a "piecewise ARX" (PWARX) system.

Problem 12.1 (Identification of Hybrid Auto Regressive eXogenous Systems). Given input/output data $\{u_t, y_t\}_{t=0}^T$ generated by a hybrid ARX system such as (12.2), identify the number of constituent systems n, the orders of each ARX system $\{n_a(i), n_c(i)\}_{i=1}^n$, the system parameters $\{a_j(i)\}_{j=1}^{n_a(i)}$ and $\{c_j(i)\}_{j=1}^{n_c(i)}$, and the discrete states $\{\lambda_t\}$.

In this chapter, we will consider the first scenario, so that our results also apply to other switching mechanisms if that information becomes available. Therefore, our method does not depend on any particular switching mechanism. Once the switching sequence has been identified, the switching mechanism can be further retrieved.

The following problem summarizes the goal of this chapter. In the sequel, we characterize a set of (sufficient) conditions that allow one to solve the above problem as well as develop an efficient algorithm for it.

12.2 Identification of a Single ARX System

For the sake of completeness and comparison, let us first review some classical results for the identification of a single discrete-time ARX system

$$y_t = a_1 y_{t-1} + \cdots + a_{n_a} y_{t-n_a} + c_1 u_{t-1} + \cdots + c_{n_c} u_{t-n_c}. \tag{12.3}$$

From the theory of signals and systems, given the infinite sequences of the input $\{y_t\}$ and the output $\{u_t\}$, we can compute their Z-transforms $\hat{y}(z)$ and $\hat{u}(z)$, respectively. The transfer function $\hat{H}(z) \doteq \hat{y}(z)/\hat{u}(z)$ of the system (12.3) is given by

$$\begin{aligned}\hat{H}(z) &= z^{\max(n_a-n_c,0)}\tilde{H}(z)\\ &= \frac{z^{\max(n_a-n_c,0)}(z^{n_c-1}c_1 + z^{n_c-2}c_2 + \cdots + c_{n_c})}{z^{\max(n_c-n_a,0)}(z^{n_a} - z^{n_a-1}a_1 - z^{n_a-2}a_2 - \cdots - a_{n_a})}.\end{aligned} \tag{12.4}$$

Then we can identify the parameters of the ARX model by directly computing $\hat{H}(z)$ as $\hat{y}(z)/\hat{u}(z)$.[2] This requires the ARX model to be *identifiable*, i.e., $\tilde{H}(z)$ must have no pole-zero cancellation,[3] and $\hat{u}(z)$ to have no zero in common with a pole of $\hat{H}(z)$ and vice versa.

[2]Notice that this scheme is impractical, since it requires one to obtain the typically infinitely long output sequence $\{y_t\}$.

[3]That is, the polynomials $z^{\max(n_c-n_a,0)}(z^{n_a} - z^{n_a-1}a_1 - z^{n_a-2}a_2 - \cdots - a_{n_a})$ and $z^{n_c-1}c_1 + z^{n_c-2}c_2 + \cdots + c_{n_c}$ are coprime.

Alternatively, we may identify the system via the identification of a *subspace* associated with the input/output data. Let us define $D \doteq n_a + n_c + 1$ and the vector of *regressors* to be

$$\boldsymbol{x}_t \doteq \left[y_t, y_{t-1}, \ldots, y_{t-n_a}, u_{t-1}, u_{t-2}, \ldots, u_{t-n_c}\right]^\top \quad \in \mathbb{R}^D. \tag{12.5}$$

Thus, for all time t, the so-defined $\boldsymbol{x}_t$ is orthogonal to the vector that consists of the parameters of the ARX system:

$$\boldsymbol{b} \doteq \left[1, -a_1, -a_2, \ldots, -a_{n_a}, -c_1, -c_2, \ldots, -c_{n_c}\right]^\top \quad \in \mathbb{R}^D. \tag{12.6}$$

That is, $\forall t$, $\boldsymbol{x}_t$ and $\boldsymbol{b}$ satisfy the equation $\boldsymbol{b}^\top \boldsymbol{x}_t = 0$. In other words, $\boldsymbol{b}$ is the normal vector to the hyperplane spanned by (the rows of) the following *data matrix*:

$$L(n_a, n_c) \doteq [\boldsymbol{x}_{\max(n_a,n_c)}, \ldots, \boldsymbol{x}_{t-1}, \boldsymbol{x}_t, \boldsymbol{x}_{t+1}, \ldots]^\top \quad \in \mathbb{R}^{\infty \times D}. \tag{12.7}$$

When the model orders n_a, n_c are known, we can readily solve for the model parameters $\boldsymbol{b}$ from the null space of $L(n_a, n_c)$ via SVD.

In practice, however, the model orders may be unknown, and only upper bounds $\bar{n}_a$ and $\bar{n}_c$ may be available. Thus, the vector of regressors $\boldsymbol{x}_t$ can be defined as

$$\boldsymbol{x}_t \doteq \left[y_t, y_{t-1}, y_{t-2}, \ldots, y_{t-\bar{n}_a}, u_{t-1}, u_{t-2}, \ldots, u_{t-\bar{n}_c}\right]^\top \quad \in \mathbb{R}^D, \tag{12.8}$$

where $D = \bar{n}_a + \bar{n}_c + 1$. Obviously, the vector

$$\boldsymbol{b} \doteq \left[1, -a_1, \ldots, -a_{n_a}, \boldsymbol{0}_{1\times(\bar{n}_a - n_a)}, -c_1, \ldots, -c_{n_c}, \boldsymbol{0}_{1\times(\bar{n}_c - n_c)}\right]^\top \tag{12.9}$$

satisfies the equation $\boldsymbol{x}_t^\top \boldsymbol{b} = 0$ for all t. Notice that here the vector $\boldsymbol{b}$ is the one in (12.6) with additional $\bar{n}_a - n_a$ and $\bar{n}_c - n_c$ zeros filled in after the terms $-a_{n_a}$ and $-c_{n_c}$, respectively.

Let us define the data matrix $L(\bar{n}_a, \bar{n}_c)$ in the same way as in equation (12.7). Because of the possibly redundant embedding (12.8), the vector $\boldsymbol{b}$ is no longer the only one in the null space of L. It is easy to verify that all the following vectors are also in the null space of L:

$$\begin{aligned} \boldsymbol{b}^1 &\doteq \left[\boldsymbol{0}_1, 1, -a_1, \ldots, -a_{n_a}, \boldsymbol{0}_{\bar{n}_a - n_a - 1}, \boldsymbol{0}_1, -c_1, \ldots, -c_{n_c}, \boldsymbol{0}_{\bar{n}_c - n_c - 1}\right]^\top, \\ \boldsymbol{b}^2 &\doteq \left[\boldsymbol{0}_2, 1, -a_1, \ldots, -a_{n_a}, \boldsymbol{0}_{\bar{n}_a - n_a - 2}, \boldsymbol{0}_2, -c_1, \ldots, -c_{n_c}, \boldsymbol{0}_{\bar{n}_c - n_c - 2}\right]^\top, \\ &\vdots \qquad\qquad\qquad\qquad\qquad\qquad \vdots \end{aligned} \tag{12.10}$$

Therefore, the data $\{\boldsymbol{x}_t\}$ span a low-dimensional linear subspace $\mathcal{S}$ in the ambient space $\mathbb{R}^D$.[4] Each of the vectors defined above uniquely determines the original

[4]Only when the initial conditions $\{y_{t_0-1}, \ldots, y_{t_0-\bar{n}_a}\}$ are arbitrary do the data span a hyperplane in $\mathbb{R}^D$ with $\boldsymbol{b}$ as the only normal vector.

system (12.3), including its order and coefficients. However, a vector in the null space of L is in general a linear combination of all such vectors, and it is not necessarily one of the above. Thus, in order to identify the original system from the data matrix L, we need to seek a vector in its null space that has certain desired structure.

Notice that the last $\bar{n}_c - n_c$ entries of $\boldsymbol{b}$ in (12.9) are zero; hence the last nonzero entry of $\boldsymbol{b}$ has the lowest order—in terms of the ordering of the entries of $\boldsymbol{x}_t$—among all vectors that are in the null space of L. Therefore, we can obtain the first $\bar{n}_a + n_c + 1$ entries of $\boldsymbol{b}$ from the null space of the submatrix of L defined by its first $\bar{n}_a + n_c + 1$ columns. Since n_c is unknown, we can incrementally take the first $j = 1, 2, \ldots$ columns of the matrix L from left to right:

$$L^1 \doteq L(:, 1:1), \quad L^2 \doteq L(:, 1:2), \quad \ldots, \quad L^j \doteq L(:, 1:j), \tag{12.11}$$

until the rank of the submatrix L^j stops increasing for the first time for some $j = m$.[5]

Remark 12.1 (Identifying $\boldsymbol{b}$ and m in the Stochastic Case)**.** *In the stochastic case (i.e., $w_t \neq 0$), the ultimate goal is to minimize the (squared) modeling error $\sum_t w_t^2 = \sum_t (\boldsymbol{b}^\top \boldsymbol{x}_t)^2$, which corresponds to the maximum-likelihood estimate when w_t is white noise. Then the optimal solution $\boldsymbol{b}^*$ can be found in a least-squares sense as the singular vector that corresponds to the smallest singular value of L^m. However, in the noisy case, we cannot directly estimate m from the rank of L^j, since it might be of full rank for all j. Based on model selection techniques, m can be estimated from a noisy L^j as*

$$m = \underset{j=1,\ldots,D}{\operatorname{argmin}} \left\{ \frac{\sigma_j^2(L^j)}{\sum_{k=1}^{j-1} \sigma_k^2(L^j)} + \kappa \cdot j \right\}, \tag{12.12}$$

where $\sigma_k(L^j)$ is the kth singular value of L^j, and $\kappa \in \mathbb{R}$ is a parameter weighting the two terms. The above criterion minimizes a cost function that consists of a data fitting term and a model complexity term. The data fitting term measures how well the data are approximated by the model, in this case how close the matrix L^j is to dropping rank. The model complexity term penalizes choosing models of high complexity, in this case choosing a large rank.

There is, however, a much more direct way of dealing with the case of unknown orders. The following lemma shows that the system orders n_a and n_c together with the system parameters $\boldsymbol{b}$ can all be simultaneously and uniquely computed from the data.

Lemma 12.2 (Identifying the Orders of an ARX System)**.** *Suppose we are given data generated by an identifiable ARX model whose input $\hat{u}(z)$ shares no poles or zeros with the zeros or poles, respectively, of the model transfer function $\hat{H}(z)$. If $\bar{n}_a + \bar{n}_c + 1 \leq n_a + n_c + 1$, then*

[5]If n_c was known, then we would have $m = \bar{n}_a + n_c + 1$.

$$\operatorname{rank}\big(L(\bar{n}_a,\bar{n}_c)\big) = \begin{cases} \bar{n}_a+\bar{n}_c & \text{if and only if} \quad \bar{n}_a = n_a \text{ and } \bar{n}_c = n_c, \\ \bar{n}_a+\bar{n}_c+1 & \text{otherwise.} \end{cases} \tag{12.13}$$

Therefore, the systems' orders can be computed uniquely as

$$(n_a,n_c) = \underset{(\bar{n}_a,\bar{n}_c)\in\mathbb{Z}^2}{\operatorname{argmin}} \{\bar{n}_a+\bar{n}_c : \operatorname{rank}(L(\bar{n}_a,\bar{n}_c)) = \bar{n}_a+\bar{n}_c\}. \tag{12.14}$$

The parameter vector $\boldsymbol{b}$ *is the unique vector in the null space of* $L(n_a,n_c)$*.*

Proof. Suppose $\operatorname{rank}\big(L(\bar{n}_a,\bar{n}_c)\big) \leq \bar{n}_a+\bar{n}_c$ and $\boldsymbol{b}' = [1, b'_1, b'_2, \ldots, b'_{\bar{n}_a+\bar{n}_c}] \in \mathbb{R}^{\bar{n}_a+\bar{n}_c+1}$ is a nonzero vector such that $L\boldsymbol{b}' = 0$. Consider the Z-transform of $L\boldsymbol{b}' = 0$:

$$\hat{y}(z) + b'_1 z^{-1}\hat{y}(z) + \cdots + b'_{\bar{n}_a} z^{-\bar{n}_a}\hat{y}(z)$$
$$+\, b'_{\bar{n}_a+1} z^{-1}\hat{u}(z) + \cdots + b'_{\bar{n}_a+\bar{n}_c} z^{-\bar{n}_a-\bar{n}_c}\hat{u}(z) = 0.$$

Since $\hat{u}(z)$ does not have any of the poles or zeros of the transfer function $\hat{H}(z)$ in (12.4), the ratio $\hat{y}(z)/\hat{u}(z)$ derived from the above equation should be a rational function whose numerator and denominator contain those of $\hat{H}(z)$ as factors, respectively. Since $\bar{n}_a+\bar{n}_c \leq n_a+n_c$, this happens only if $\bar{n}_a = n_a$ and $\bar{n}_c = n_c$ and the vector $\boldsymbol{b}'$ is exactly the same as $\boldsymbol{b}$ in (12.6). □

Remark 12.3 (Identifying n_a, n_c in the Stochastic Case)**.** *In the stochastic case (i.e.,* $w_t \neq 0$*), we cannot directly estimate* n_a, n_c *from the rank of* $L(\bar{n}_a,\bar{n}_c)$*, since it might be of full rank for all* $\bar{n}_a, \bar{n}_c$*. From model selection methods,* n_a, n_c *can be estimated from a noisy L as*

$$(n_a,n_c) = \underset{(\bar{n}_a,\bar{n}_c)\in\mathbb{Z}^2}{\operatorname{argmin}} \left\{ \frac{\sigma^2_{\bar{n}_a+\bar{n}_c+1}(L(\bar{n}_a,\bar{n}_c))}{\sum_{k=1}^{\bar{n}_a+\bar{n}_c}\sigma_k^2(L(\bar{n}_a,\bar{n}_c))} + \kappa\cdot(\bar{n}_a+\bar{n}_c) \right\}, \tag{12.15}$$

where $\sigma_k(L)$ *is the kth singular value of L, and* $\kappa \in \mathbb{R}$ *is a parameter weighting the two terms, the first for the model fitting error and the second for the model complexity.*

In principle, the above lemma allows us to identify the precise orders n_a, n_c and the vector $\boldsymbol{b}$ of the ARX system from the (infinite) sequences of input $\{u_t\}$ and output $\{y_t\}$. In practice, we are usually given a finite input/output sequence. In such cases, we need to assume that the sequence of regressors is *sufficiently exciting*, i.e., the $T\times(n_a+n_c+1)$ submatrix

$$L \doteq [\boldsymbol{x}_{\max(n_a,n_c)}, \ldots, \boldsymbol{x}_{\max(n_a,n_c)+T-1}]^\top$$

has the same rank $n_a + n_c$ as the "full" L matrix defined in (12.7). Then the maximum-likelihood estimate for $\boldsymbol{b} \in \mathbb{R}^{n_a+n_c+1}$ can be identified as the singular vector that corresponds to the smallest singular value of L.

This condition for sufficient exciting for finite data can also be expressed in terms of only the input sequence. As shown in (Anderson and Johnson 1982), the regressors are sufficiently exciting if the input sequence $\{u_t\}$ is, i.e., if the vectors

$$\boldsymbol{u}_t \doteq [u_t, u_{t-1}, \ldots, u_{t-n_a-n_c+1}]^\top \in \mathbb{R}^{n_a+n_c}, \qquad n_a + n_c - 1 \leq t \leq T,$$

span an $(n_a + n_c)$-dimensional subspace.

12.3 Identification of Hybrid ARX Systems

From our discussion in the previous section, we know that the regressors generated by an identifiable ARX system with sufficiently exciting input live in a linear subspace in $\mathbb{R}^D$, where $D = \bar{n}_a + \bar{n}_c + 1$ and $\bar{n}_a, \bar{n}_c$ are upper bounds on the orders of the system. The problem of identifying the ARX system becomes one of seeking a vector in the orthogonal complement to this subspace that has a certain desired structure. We show in this section how to generalize these concepts to the more challenging problem of identifying a hybrid ARX system (Problem 12.1). Most of our development will focus on the case of single-input single-output (SISO) systems.

Consider an input/output sequence $\{u_t, y_t\}$ generated by a hybrid ARX system switching among a set of n ARX systems with parameters $\{\boldsymbol{b}_i\}_{i=1}^n$ and possibly different orders $\{n_a(i), n_c(i)\}_{i=1}^n$. We assume that the hybrid ARX system is *identifiable*, i.e., for all $i = 1, \ldots, n$, the rational function $\tilde{H}_i(z)$ associated with the ith ARX model has no zero-pole cancellation, and the configuration subspaces of all the ARX models do not contain one another.[6] In general, we also assume that we do not know the exact orders of the systems but know only certain upper bounds on them, i.e.,

$$\bar{n}_a \geq \max\{n_a(1), \ldots, n_a(n)\}, \quad \bar{n}_c \geq \max\{n_c(1), \ldots, n_c(n)\}.$$

[6]One way to ensure this is to assume that for all $i \neq j = 1, \ldots, n$, $\tilde{H}_i(z)$ and $\tilde{H}_j(z)$ do not have all their zeros and poles in common. That is, there is no ARX system that can simulate another ARX system with a smaller order. However, this is unnecessary, because two ARX systems can have different configuration spaces even if one system's zeros and poles are a subset of the other's.

Very often, we do not know the exact number of systems involved either, but know only an upper bound on it, i.e., $\bar{n} \geq n$.[7] In this section, we study how to identify the hybrid ARX system despite these uncertainties.

12.3.1 The Hybrid Decoupling Polynomial

One of the difficulties in identifying hybrid ARX systems is that we do not know the switching sequence λ_t; hence we cannot directly apply the subspace identification technique described in the previous section to each of the n ARX systems. As we will soon see, in fact both the number of subspaces and their dimensions depend not only on the number of systems and their orders but also on the switching sequence. This motivates us to look for relationships between the data $\{\boldsymbol{x}_t \in \mathbb{R}^D\}$ and the system parameters $\{\boldsymbol{b}_i \in \mathbb{R}^D\}$ that do not depend on the switching sequence. To this end, recall that for every t, there exists a state $\lambda_t = i \in \{1, 2, \ldots, n\}$ such that $\boldsymbol{b}_i^\top \boldsymbol{x}_t = 0$. Therefore, the following polynomial equation must be satisfied by the system parameters and the input/output data for any switching sequence and mechanism (JMLS or PWARX):

$$p_n(\boldsymbol{x}_t) \doteq \prod_{i=1}^{n} (\boldsymbol{b}_i^\top \boldsymbol{x}_t) = 0. \tag{12.16}$$

We call this polynomial equation the *hybrid decoupling polynomial* (HDP). In the absence of knowledge about the switching mechanism, the HDP encodes all the information about the system parameters that we can obtain from the input/output data.

The HDP eliminates the discrete state by taking the product of the equations defining each of the ARX systems. While taking the product is not the only way of algebraically eliminating the discrete state, this leads to an algebraic equation with a very nice algebraic structure. The HDP is simply a homogeneous multivariate polynomial of degree n in D variables:

$$p_n(z) \doteq \prod_{i=1}^{n} (\boldsymbol{b}_i^\top z) = 0, \tag{12.17}$$

which can be written linearly in terms of its coefficients as

$$p_n(z) \doteq \sum h_{n_1,\ldots,n_D} z_1^{n_1} \cdots z_D^{n_D} = \boldsymbol{h}_n^\top \nu_n(z) = 0. \tag{12.18}$$

[7] This is the case when a particular switching sequence visits only a subset of all the discrete states.

In equation (12.18), $h_{n_1,\dots,n_D} \in \mathbb{R}$ is the coefficient of the monomial $z_1^{n_1} z_2^{n_2} \cdots z_D^{n_D}$. Obviously, the vector $\boldsymbol{h}_n = (h_{n_1,\dots,n_D})$ encodes the parameters of all the constituent ARX systems. We will show in the sequel how this vector can be correctly recovered from the data and how the parameters of each individual ARX system can be further retrieved from it.

12.3.2 *Identifying the Hybrid Decoupling Polynomial*

Let us assume for now that we know the number of systems n. We will show later how to relax this assumption. Since the HDP (12.16)–(12.18) is satisfied by all the data points $\{\boldsymbol{x}_t\}_{t=1}^T$, we can use it to derive the following linear system on the vector $\boldsymbol{h}_n$:

$$L_n(\bar{n}_a, \bar{n}_c)\, \boldsymbol{h}_n \doteq \begin{bmatrix} \nu_n(\boldsymbol{x}_{\max\{\bar{n}_a,\bar{n}_c\}})^\top \\ \nu_n(\boldsymbol{x}_{\max\{\bar{n}_a,\bar{n}_c\}+1})^\top \\ \vdots \\ \nu_n(\boldsymbol{x}_{\max\{\bar{n}_a,\bar{n}_c\}+T-1})^\top \end{bmatrix} \boldsymbol{h}_n = \boldsymbol{0}_{T\times 1}, \tag{12.19}$$

where $L_n(\bar{n}_a, \bar{n}_c) \in \mathbb{R}^{T\times M_n(D)}$ is the matrix of the input/output data embedded via the Veronese map.

Definition 12.4 (Sufficiently Exciting Switching and Input Sequences)**.** *A switching and input sequence $\{\lambda_t, u_t\}$ is called* sufficiently exciting *for a hybrid ARX system if the data points $\{\boldsymbol{x}_t\}$ generated by $\{\lambda_t, u_t\}$ are sufficient to determine the union of the subspaces associated with the constituent ARX systems as an algebraic variety, in the sense of Theorem C.10 of Appendix C.*

Given the data matrix $L_n(\bar{n}_a, \bar{n}_c)$ from a sufficiently exciting switching and input sequence, we would like to retrieve the coefficient vector $\boldsymbol{h}_n$ from its null space. There are two potential difficulties. First, since the maximum orders $\bar{n}_a, \bar{n}_c$ may not be tight for every constituent ARX system, the null space of $L_n(\bar{n}_a, \bar{n}_c)$ may be more than one-dimensional, as we have known from a single ARX system. Second, even if we know the discrete state for each time, the structure of the data associated with each state is not exactly the same as that of the ARX system itself: Suppose we switch to the ith system at time t_0. Then we have $\boldsymbol{b}_i^\top \boldsymbol{x}_{t_0} = 0$. However, the vectors $\boldsymbol{b}$ given in equation (12.10) are no longer orthogonal to $\boldsymbol{x}_{t_0}$ even if the embedding is redundant for the ith system. In a sense, the regressor at a switching time usually lives in a subspace whose dimension is higher than that of the subspace associated with the ARX model generating the regressor. Therefore, the configuration space of the data $\{\boldsymbol{x}_t\}$ of a hybrid ARX system will *not* be exactly the union of all the subspaces associated with the constituent ARX systems. Let us denote the former by an algebraic variety Z' and the latter by Z. Then in general, we have $Z' \supseteq Z$.

In order to retrieve $\boldsymbol{h}_n$ uniquely from the data matrix L_n, we need to utilize its additional structure.

Lemma 12.5 (Structure of the Hybrid Decoupling Polynomial). *The monomial associated with the last nonzero entry of the coefficient vector $\boldsymbol{h}_n$ of the hybrid decoupling polynomial $p_n(z) = \boldsymbol{h}_n^\top \nu_n(z)$ has the lowest degree-lexicographic order in all the polynomials in $\mathfrak{a}(Z) \cap R_n$.*[8]

Proof. Every polynomial of degree n in the ideal $\mathfrak{a}(Z)$ is a superposition of the polynomials $\prod_{i=1}^n (\boldsymbol{b}_{\sigma(i)}^\top z)$, where $\boldsymbol{b}_{\sigma(i)}$ is a normal vector to the subspace associated with the ith ARX system.[9] Notice that $\boldsymbol{h}_n$ is the symmetric tensor of $\boldsymbol{b}_1, \boldsymbol{b}_2, \ldots, \boldsymbol{b}_n$ defined in (12.9). For the ith ARX system, the last nonzero entry of the vector $\boldsymbol{b}_i$ always has the lowest degree-lexicographic order among all normal vectors that are orthogonal to the regressors $z = \boldsymbol{x}_t$ associated with the ith system. Therefore, the last nonzero entry of $\boldsymbol{h}_n$ must have the lowest degree-lexicographic order. □

Theorem 12.6 (Identifying the Hybrid Decoupling Polynomial). *Suppose that $\{u_t, y_t\}_{t=0}^T$ are the input/output data generated by an identifiable hybrid ARX system. Let $L_n^j \in \mathbb{R}^{T \times j}$ be the first j columns of the embedded data matrix $L_n(\bar{n}_a, \bar{n}_c)$, and let*

$$m \doteq \min\left\{j : \operatorname{rank}\left(L_n^j\right) = j - 1\right\}. \tag{12.20}$$

If T is sufficiently large and the input and switching sequences are sufficiently exciting, then the coefficient vector $\boldsymbol{h}_n$ of the hybrid decoupling polynomial is given by

$$\boldsymbol{h}_n = \left[\left(\boldsymbol{h}_n^m\right)^\top, \ \boldsymbol{0}_{1 \times (M_n(D) - m)}\right]^\top \in \mathbb{R}^{M_n(D)}, \tag{12.21}$$

where $\boldsymbol{h}_n^m \in \mathbb{R}^m$ is the unique vector that satisfies

$$L_n^m \boldsymbol{h}_n^m = \boldsymbol{0} \quad \text{and} \quad \boldsymbol{h}_n^m(1) = 1. \tag{12.22}$$

Proof. Let Z be the union of the subspaces associated with the n constituent ARX systems. Since the input and switching sequence are sufficiently exciting in the sense of Definition 12.4, according to Theorem C.10 of Appendix C, every polynomial of degree less than or equal to n that vanishes on all the data points must be in the set $\mathfrak{a}(Z) \cap R_{\leq n}$.[10]

---

[8] R_n is the set of homogeneous polynomials of degree n; see Appendix C.

[9] This is easily verifiable from the fact that the derivatives of the polynomials in $\mathfrak{a}(Z)$ are exactly the normal vectors of the subspaces.

[10] $R_{\leq n}$ is the set of polynomials of degree up to n; see Appendix C.

From our discussion before the theorem, the configuration space Z' of the data $\{\boldsymbol{x}_t\}$ associated with the hybrid ARX system is in general a superset of Z. The ideal $\mathfrak{a}'(Z')$ of polynomials that vanish on the configuration space Z' is then a subideal of the ideal $\mathfrak{a}(Z)$ associated with the union of the subspaces. Furthermore, regardless of the switching sequence, the hybrid decoupling polynomial $p_n(z)$ is always in $\mathfrak{a}' \cap R_n \subseteq \mathfrak{a} \cap R_n$. According to Lemma 12.5, the last nonzero term of $p_n(z)$ has the lowest degree-lexicographic order among all polynomials of degree n in $\mathfrak{a}$, and so it has the lowest degree-lexicographic order among all polynomials of degree n in $\mathfrak{a}'$. Since every solution $L_n\tilde{\boldsymbol{h}} = 0$ gives a polynomial $\tilde{p}_n(z) = \tilde{\boldsymbol{h}}_n^\top \nu_n(z) \in \mathfrak{a} \cap R_n$ of degree n that vanishes on all data points, the last nonzero entry of $\boldsymbol{h}_n$ given by (12.21) has the lowest degree-lexicographic order. Therefore, we have $p_n(z) = \boldsymbol{h}_n^\top \nu_n(z)$.

□

In fact, to compute the coefficients $\boldsymbol{h}_n$ of the hybrid decoupling polynomial, we can do better than checking the rank of the submatrix L_n^j for every $j = 1, 2, \ldots$. The following corollary provides one alternative scheme.

Corollary 12.7 (Zero Coefficients of the Decoupling Polynomial). *Consider a set of vectors $\boldsymbol{b}_i \in \mathbb{R}^D, i = 1, \ldots, n$. Suppose that one of the $\boldsymbol{b}_i$ has a maximal number of zeros on its right, and without loss of generality, assume that it is*

$$\boldsymbol{b}_1 = [b_{11}, b_{12}, \ldots, b_{1n_1}, 0, \ldots, 0]^\top, \quad \text{with} \quad b_{1n_1} \neq 0.$$

The multivariate polynomial $p_n(z) \doteq (\boldsymbol{b}_1^\top z)(\boldsymbol{b}_2^\top z) \cdots (\boldsymbol{b}_n^\top z)$ has zero coefficients for all the monomials of $\nu_n\big([z_{n_1+1}, z_{n_1+2}, \ldots, z_D]\big)$; but the coefficients cannot all be zeros for the monomials of $\nu_n\big([z_{n_1}, z_{n_1+1}, \ldots, z_D]\big)$.

This corollary allows us to narrow down the range for m (where L_n^j first drops rank) because m must fall between two consecutive values of the following:

$$1, \quad M_n(D) - M_n(D-1), \quad M_n(D) - M_n(D-2), \quad \ldots, \quad M_n(D) - 1.$$

Remark 12.8 (Suboptimality in the Stochastic Case). *In the stochastic case (i.e., $w_t \neq 0$), we can still solve for $\boldsymbol{h}_n^m$ in (12.22) in a least-squares sense as the singular vector of L_n^m associated with its smallest singular value, using a similar model selection criterion for m as in Remark 12.1. However, in contrast to the single-system case, the so-found $\boldsymbol{h}_n$ no longer minimizes the sum of least-square errors $\sum_t w_t^2 = \sum_t (\boldsymbol{b}_{\lambda_t}^\top \boldsymbol{x}_t)^2$. Instead, it minimizes (in a least-squares sense) a "weighted version" of this objective:*

$$\sum_t \alpha_t (\boldsymbol{b}_{\lambda_t}^\top \boldsymbol{x}_t)^2 \doteq \sum_t \prod_{i \neq \lambda_t} (\boldsymbol{b}_i^\top \boldsymbol{x}_t)^2 (\boldsymbol{b}_{\lambda_t}^\top \boldsymbol{x}_t)^2, \tag{12.23}$$

where the weight α_t is conveniently chosen to be $\prod_{i\neq\lambda_t}(\boldsymbol{b}_i^\top \boldsymbol{x}_t)^2$. Such a "softening" of the objective function allows a global algebraic solution. It offers a suboptimal approximation to the original stochastic objective when the variance of w_t is small. One can use the solution as the initialization for any other (local) nonlinear optimization scheme (such as expectation maximization) to further minimize the original stochastic objective.

Notice that in the above theorem, we have assumed that the switching sequence is such that all the ARX systems are sufficiently visited. What if only a subset of the n systems are sufficiently visited? Furthermore, in practice, we sometimes do not even know the correct number of systems involved and know only an upper bound for it. The question is whether the above theorem still applies when the degree n we choose for the Veronese embedding is strictly larger than the actual number of systems. This is answered by the following corollary, whose proof is straightforward.

Corollary 12.9 (Identifying the Number of ARX Systems). *Let $\{u_t, y_t\}_{t=0}^T$ be the input/output data generated by a hybrid ARX system with $n < \bar{n}$ discrete states. If T is sufficiently large and the input and switching sequences are sufficiently exciting, then the vector $\boldsymbol{h}_{\bar{n}}$ found by Theorem 12.6 is the symmetric tensor product*

$$\boldsymbol{h}_{\bar{n}} = \mathrm{Sym}\big(\boldsymbol{b}_1 \otimes \boldsymbol{b}_2 \cdots \otimes \boldsymbol{b}_n \otimes \underbrace{\boldsymbol{e}_1 \otimes \cdots \otimes \boldsymbol{e}_1}_{\bar{n}-n}\big), \tag{12.24}$$

where $\boldsymbol{e}_1 \doteq [1, 0, \ldots, 0]^\top \in \mathbb{R}^D$, i.e., $\boldsymbol{h}_{\bar{n}}$ is the coefficients of the polynomial

$$p_{\bar{n}}(z) = \boldsymbol{h}_{\bar{n}}^\top \nu_{\bar{n}}(z) = \big(\boldsymbol{b}_1^\top z\big)\big(\boldsymbol{b}_2^\top z\big) \cdots \big(\boldsymbol{b}_n^\top z\big) z_1^{\bar{n}-n}. \tag{12.25}$$

Therefore, even if we may overestimate the number of constituent systems or the switching sequence does not visit all the systems, the solution given by Theorem 12.6 will simply treat the nonexistent (or not visited) systems as if they had zero order,[11] and the information about the rest of the systems will be conveniently recovered.

12.3.3 *Identifying System Parameters and Discrete States*

Theorem 12.6 allows us to determine the hybrid decoupling polynomial $p_n(z) = \boldsymbol{h}_n^\top \nu_n(z)$ from input/output data $\{u_t, y_t\}_{t=0}^T$. The rest of the problem is to recover the

[11] That is, the coefficient vector $\boldsymbol{b} = \boldsymbol{e}_1$ corresponds to the "system" $y_t = 0$ with $n_a = n_c = 0$, which is a trivial ARX system.

system parameters $\{\boldsymbol{b}_i\}_{i=1}^n$ from $\boldsymbol{h}_n$. To this end, recall from Chapter 5 that given $\boldsymbol{h}_n$, one can recover the model parameters by looking at the partial derivative of $p_n(z)$ given in (12.17),

$$\nabla p_n(z) \doteq \frac{\partial p_n(z)}{\partial z} = \sum_{i=1}^{n} \prod_{\ell \neq i} (\boldsymbol{b}_\ell^\top z)\boldsymbol{b}_i. \tag{12.26}$$

If z belongs to the hyperplane $\mathcal{H}_i = \{z : \boldsymbol{b}_i^\top z = 0\}$, then since the first entry of $\boldsymbol{b}_i$ by definition is equal to one, after replacing $\boldsymbol{b}_i^\top z = 0$ in (12.26), we obtain

$$\boldsymbol{b}_i = \left.\frac{\nabla p_n(z)}{\boldsymbol{e}_1^\top \nabla p_n(z)}\right|_{z \in \mathcal{H}_i} \in \mathbb{R}^D, \tag{12.27}$$

where $\boldsymbol{e}_1 = [1, 0, \ldots, 0]^\top \in \mathbb{R}^D$. Therefore, we can estimate the system parameters directly from the derivatives of $p_n(z)$ at a collection of n points $\{z_i \in \mathcal{H}_i\}_{i=1}^n$ lying in the n hyperplanes, respectively.

In order to find the set of points $\{z_i \in \mathcal{H}_i\}_{i=1}^n$, let us consider a line with base point z_0 and direction $\boldsymbol{v}$, $\mathcal{L} = \{z_0 + \alpha \boldsymbol{v}, \alpha \in \mathbb{R}\}$. If $z_0 \neq 0$, z_0 is not parallel to $\boldsymbol{v}$, and $\boldsymbol{b}_i^\top \boldsymbol{v} \neq 0$, then the line $\mathcal{L}$ in general intersects the n hyperplanes $\cup_{i=1}^n \mathcal{H}_i = \{z : p_n(z) = 0\}$ at n distinct points

$$z_i = z_0 + \alpha_i \boldsymbol{v} \in \mathcal{H}_i \cap \mathcal{L}, \qquad i = 1, \ldots, n, \tag{12.28}$$

where $\{\alpha_i\}$ are the roots of the univariate polynomial

$$q_n(\alpha) = p_n(z_0 + \alpha \boldsymbol{v}). \tag{12.29}$$

We are left with choosing the parameters $\boldsymbol{x}_0$ and $\boldsymbol{v}$ for the line $\mathcal{L}$. The base point $\boldsymbol{x}_0$ can be chosen as any nonzero vector in $\mathbb{R}^D$. Given z_0, the direction $\boldsymbol{v}$ must be chosen not parallel to z_0 and such that $\boldsymbol{b}_i^\top \boldsymbol{v} \neq 0$, for all $i = 1, \ldots, n$. Since the latter constraint is equivalent to $p_n(\boldsymbol{v}) \neq 0$, and p_n is known, we can immediately choose $\boldsymbol{v}$ even though we do not know the system parameters $\{\boldsymbol{b}_i\}_{i=1}^n$.

Be aware that if we have chosen for the Veronese embedding a number $\bar{n}$ that is strictly larger than n, the polynomial $p_{\bar{n}}(z)$ will be of the form $(\boldsymbol{b}_1^\top z)(\boldsymbol{b}_2^\top z) \cdots (\boldsymbol{b}_n^\top z) z_1^{\bar{n}-n}$. Then the line $\mathcal{L}$ will have only $n+1$ intersections with the n hyperplanes $\mathcal{H}_1, \ldots, \mathcal{H}_n$ and the hyperplane $\mathcal{H}_0 \doteq \{z : \boldsymbol{e}_1^\top z = z_1 = 0\}$. The intersection $z_0 = \mathcal{H}_0 \cap \mathcal{L}$ has a multiplicity of $\bar{n} - n$; and $\nabla p_{\bar{n}}(z_0) \sim \boldsymbol{e}_1$ if $\bar{n} - n = 1$ or $\nabla p_{\bar{n}}(z_0) = 0$ if $\bar{n} - n > 1$. We have essentially proven the following theorem.

Theorem 12.10 (Identifying the Constituent System Parameters). *Given the input/output data $\{u_t, y_t\}_{t=0}^T$ generated by a hybrid ARX system with n discrete states, the system parameters $\{\boldsymbol{b}_i\}_{i=1}^n$ can be computed from the hybrid decoupling polynomial $p_{\bar{n}}(z) = \boldsymbol{h}_{\bar{n}}^\top \nu_{\bar{n}}(z)$ for any $\bar{n} \geq n$ as follows:*

1. *Choose* $z_0 \neq 0$ *and* $\boldsymbol{v}$ *such that* $\boldsymbol{v} \neq \gamma z_0$ *and* $p_{\bar{n}}(\boldsymbol{v}) \neq 0$.
2. *Solve for the* $\bar{n}$ *roots* $\{\alpha_i\}_{i=1}^{\bar{n}}$ *of* $q_{\bar{n}}(\alpha) = p_{\bar{n}}(z_0 + \alpha\boldsymbol{v}) = 0$.
3. *For all the roots* $z_i = z_0 + \alpha_i\boldsymbol{v}$ *with* $z_1 \neq 0$, *compute the system parameters* $\{\boldsymbol{b}_i\}_{i=1}^n$ *as*

$$\boldsymbol{b}_i = \frac{\nabla p_{\bar{n}}(z_i)}{\boldsymbol{e}_1^\top \nabla p_{\bar{n}}(z_i)} \quad \in \mathbb{R}^D, \qquad i = 1, 2, \ldots, n. \tag{12.30}$$

Remark 12.11 (Alternative Ways of Identifying $\{\boldsymbol{b}_i\}_{i=1}^n$ from Noisy Data)**.** *In the presence of noise, we can still estimate the normal vectors* $\{\boldsymbol{b}_i\}_{i=1}^n$ *as in Theorem 12.10. However, the quality of the estimates will depend on the choice of the parameters* z_0 *and* $\boldsymbol{v}$. *In this case, one can choose multiple* $(z_0, \boldsymbol{v})$ *satisfying the above conditions, obtain the system parameters for each choice, and let* $\{\boldsymbol{b}_i\}_{i=1}^n$ *be the parameters that better reconstruct* $\boldsymbol{h}_n$. *Alternatively, one can directly choose* $\{z_i\}_{i=1}^n$ *from points in the data set that fit the decoupling polynomial in a certain optimal way, as discussed Chapter 5. That allows us to bypass the problem of solving for the (real) roots of the real polynomial* $q_{\bar{n}}(\alpha)$.

Once the system parameters $\{\boldsymbol{b}_i\}_{i=1}^n$ are recovered, we can then reconstruct the orders $n_a(i), n_c(i)$ of each constituent ARX system as well as the discrete state trajectory $\{\lambda_t\}$ from the input/output data $\{\boldsymbol{x}_t\}_{t=0}^T$. Notice that for each time t, there exists a generally unique i such that $\boldsymbol{b}_i^\top \boldsymbol{x}_t = 0$. Therefore, the discrete state λ_t can be easily identified as

$$\lambda_t = \operatorname*{argmin}_{i=1,\ldots,n} \left(\boldsymbol{b}_i^\top \boldsymbol{x}_t\right)^2. \tag{12.31}$$

There will be ambiguity in the value of λ_t only if $\boldsymbol{x}_t$ happens to be at (or close to) the intersection of more than one subspace associated to the constituent ARX systems. However, the set of all such points is a zero-measure set of the variety $Z \subseteq \{z : p_n(z) = 0\}$.

12.3.4 The Basic Algorithm and Its Extensions

Based on the results that we have derived so far, we summarize the main steps for solving the identification of a hybrid ARX system (Problem 12.1) as the following Algorithm 12.1. Notice that the algorithm is different from the general-purpose ASC algorithm given in Chapter 5. By utilizing the structure in the system parameters $\{\boldsymbol{b}_i\}$ and subsequently in their symmetric tensor product $\boldsymbol{h}_n$, the algorithm guarantees that the so-found polynomial p_n is the desired hybrid decoupling polynomial.

Algorithm 12.1 (Identification of an SISO Hybrid ARX System).

Given the input/output data $\{y_t, u_t\}$ from a sufficiently excited hybrid ARX system, and the upper bound on the number $\bar{n}$ and maximum orders $(\bar{n}_a, \bar{n}_c)$ of its constituent ARX systems:

1. **Veronese Embedding.** Construct the data matrix $L_{\bar{n}}(\bar{n}_a, \bar{n}_c)$ via the Veronese map based on the given number $\bar{n}$ of systems and the maximum orders $(\bar{n}_a, \bar{n}_c)$.
2. **Hybrid Decoupling Polynomial.** Compute the coefficients of the polynomial $p_{\bar{n}}(z) \doteq \boldsymbol{h}_{\bar{n}}^\top \nu_{\bar{n}}(z) = \prod_{i=1}^{n} (\boldsymbol{b}_i^\top z) z_1^{\bar{n}-n} = 0$ from the data matrix $L_{\bar{n}}$ according to Theorem 12.6 and Corollary 12.9. In the stochastic case, comply with Remarks 12.1 and 12.8.
3. **Constituent System Parameters.** Retrieve the parameters $\{\boldsymbol{b}_i\}_{i=1}^{n}$ of each constituent ARX system from $p_{\bar{n}}(z)$ according to Theorem 12.10. In the noisy case, comply with Remark 12.11.
4. **Key System Parameters.** The correct number of system n is the number of $\boldsymbol{b}_i \neq \boldsymbol{e}_1$; The correct orders $n_a(i), n_c(i)$ are determined from such $\boldsymbol{b}_i$ according to their definition (12.9). The discrete state λ_t for each time t is given by equation (12.31).

Different Embedding Orders
The order of stacking $\{y_t\}$ and $\{u_t\}$ in the vector $\boldsymbol{x}_t$ in (12.8) is more efficient for the algorithm when the values of $n_a(i)$ are approximately the same for all the constituent systems and the $n_c(i)$ are much smaller than $n_a(i)$. However, if the $n_a(i)$ are rather different for different systems and the $n_c(i)$ and $n_a(i)$ are roughly the same, the following ordering in time t,

$$\boldsymbol{x}_t \doteq \left[y_t, y_{t-1}, u_{t-1}, y_{t-2}, u_{t-2}, \dots, y_{t-\bar{n}_a}, u_{t-\bar{n}_a}\right]^\top \quad \in \mathbb{R}^D, \tag{12.32}$$

results in fewer nonzero leading coefficients in $\boldsymbol{h}_n$. Thus the above algorithm becomes more efficient. Nevertheless, if all the systems have the same $n_a = n_c$, then both embeddings have the same efficiency.

Inferring the Switching Mechanisms
Once the system parameters and the discrete state have been identified, the problem of estimating the switching mechanisms, e.g., the partition of the state space for PWARX or the parameters of the jump Markov process for JMLS, becomes simpler. We refer the interested reader to (Bemporad et al. 2003; Ferrari-Trecate et al. 2003) for specific algorithms.

12.4 Simulations and Experiments

In this section, we evaluate the performance of the proposed algorithm with respect to the model orders and the amount of noise. We also present experiments on real data from a component placement process in a pick-and-place machine.

12.4.1 Error in the Estimation of the Model Parameters

Consider the following PWARX system taken from (Niessen and A.Juloski 2004):

$$y_t = \begin{cases} 0.5u_{t-1} + 0.5 + w_{t-1} & \text{if} \quad u_{t-1} \in [-2.5, 0], \\ -u_{t-1} + 2 + w_{t-1} & \text{if} \quad u_{t-1} \in (0, 2.5]. \end{cases} \tag{12.33}$$

The input sequence u_t consists of 100 points, 80% uniformly distributed in $[-2.5, 2.5]$ and 20% uniformly distributed in $[0.85, 1.15]$. The noise is $w_t \overset{\text{i.i.d.}}{\sim} \mathcal{N}(0, 0.005)$. The error between the estimated parameters $\hat{\boldsymbol{b}}$ and the true parameters $\boldsymbol{b}$ is defined as

$$\text{error} = \max_{i=1,\dots,m.} \min_{j=1,\dots,n.} \frac{\|\hat{\boldsymbol{b}}_i - \boldsymbol{b}_j\|}{\|[\boldsymbol{0}_{(D-1)\times 1} \; I_{D-1}]\boldsymbol{b}_j\|}.$$

We applied our algorithm with known parameters $n = 2$, $n_a = 0$, and $n_c = 1$. Our algorithm gives an estimate for the ARX model parameters of $[0.5047, 0.5102]^\top$ and $[-0.9646, 1.9496]^\top$, which corresponds to an error of 0.0276. Table 12.1 compares our results with those reported in (Niessen and A.Juloski 2004) for the algorithms of (Ferrari-Trecate et al. 2003) and (Bemporad et al. 2003). Notice that our algorithm provides a purely algebraic solution to the problem that does not perform iterative refinement. Nevertheless, it provides an error comparable to the errors of the other algorithms that are based on iterative refinement.

12.4.2 Error as a Function of the Model Orders

Consider the following PWAR system taken from (Niessen and A.Juloski 2004):

$$y_t = \begin{cases} 2y_{t-1} + 0u_{t-1} + 10 + w_t & \text{if} \quad y_{t-1} \in [-10, 0], \\ -1.5y_{t-1} + 0u_{t-1} + 10 + w_t & \text{if} \quad y_{t-1} \in (0, 10], \end{cases} \tag{12.34}$$

with initial condition $y_0 = -10$, input $u_t \overset{\text{i.i.d.}}{\sim} \mathcal{U}(-10, 10)$, and noise $w_t \overset{\text{i.i.d.}}{\sim} \mathcal{N}(0, 0.01)$.

Table 12.1 Comparison of error in the estimation of the model parameters.

Algorithms	Errors
Ferrari-Trecate et. al.	0.0045
Bemporad et. al.	0.0334
Algorithm 12.1	0.0276

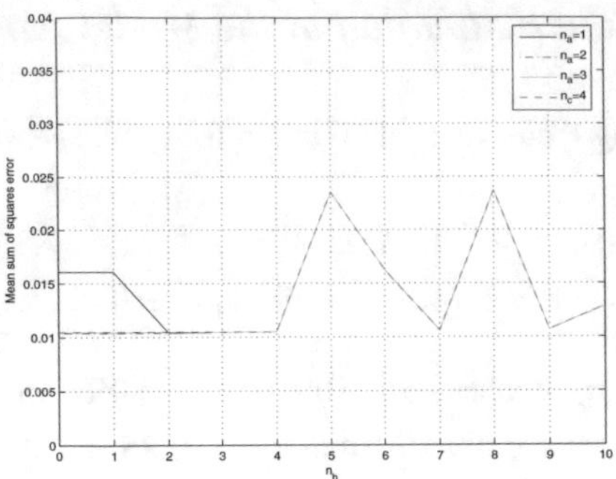

Fig. 12.2 Mean sum of squares error for various orders of the ARX models.

We applied our algorithm[12] with known number of models $n = 2$, but unknown model orders (n_a, n_c). We evaluated the performance of our algorithm as a function of the orders (n_a, n_c). We used a fixed value for (n_a, n_c) and searched for the polynomial in the null space of $L_n(n_a, n_c)$ with the smallest degree-lexicographic order. We repeated the experiment for multiple values of $n_a = 1, \ldots, 4$ and $n_c = 1, \ldots, 10$, to evaluate the effectiveness of equation (12.12) at finding the "correct" null space of $L_n(n_a, n_c)$. Figure 12.2 shows the results for $\kappa = 10^{-5}$. Notice that for the entire range of values of n_a and n_c, the algorithm gives an error that is very close to the theoretical bound of 0.01 (the noise variance).

For the correct system orders $n_a = 1$ and $n_c = 0$, the estimates of the ARX model parameters from our algorithm are $[1.9878, 0, 10.0161]^\top$ and $[-1.4810, 0, 10.0052]^\top$, which have an error of 0.0020. These results are significantly better than those reported in (Niessen and A.Juloski 2004) for the Ferrari-Trecate and Bemporad algorithms.

12.4.3 Error as a Function of Noise

Consider the following PWAR model taken from (Niessen and A.Juloski 2004):

$$y_t = \begin{cases} 2u_{t-1} + 10 + w_t & \text{if} \quad u_{t-1} \in [-10, 0], \\ -1.5u_{t-1} + 10 + w_t & \text{if} \quad u_{t-1} \in (0, 10], \end{cases} \tag{12.35}$$

with input $u_t \overset{\text{i.i.d.}}{\sim} \mathcal{U}(-10, 10)$ and noise $w_t \overset{\text{i.i.d.}}{\sim} \mathcal{N}(0, \sigma_\eta^2)$. We run our algorithm with $n = 2$, $n_a = 0$, and $n_c = 1$ for 10 different values of σ_η and compute the mean and the variance of the error in the estimated model parameters, as shown in

[12]Since here the system is an affine ARX model with a constant input, we need to slightly modify our algorithm by using the homogeneous representation for the regressor $\boldsymbol{x}_t$, i.e., appending an entry of "1."

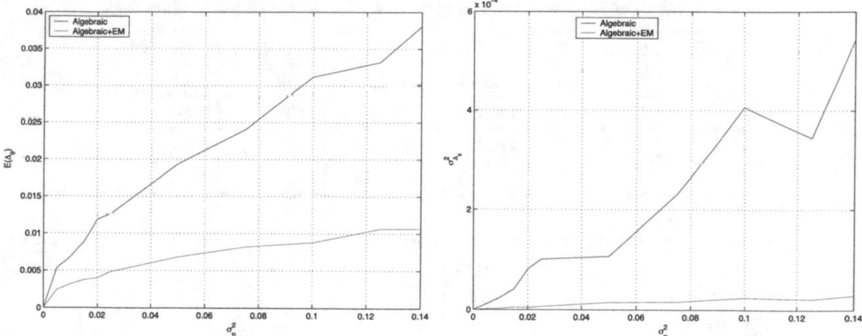

Fig. 12.3 Means (left) and variances (right) of the error in the estimation of the model parameters for different levels of noise. Blue curves are for the purely algebraic Algorithm 12.1; Green curves are for the EM algorithm initialized with the solutions from Algorithm 12.1.

Figure 12.3. The algorithm estimates the parameters with an error of less than 3.7% for the levels of noise considered. Again, the errors provided by the purely algebraic algorithm (Algorithm 12.1) without any iterative refinement are comparable to those of the Ferrari-Trecate and Bemporad algorithms reported in (Niessen and A.Juloski 2004), which are about 2 $\sim$ 3%. Furthermore, if we use the solutions offered by our algebraic algorithm to initialize other iterative refinement algorithms such as the expectation and maximization (EM) algorithm, then the error is reduced significantly to about 1% (see Figure 12.3 left).

12.4.4 Experimental Results on Test Data Sets

We applied our algorithm with $n = n_a = n_c = 2$ to four data sets of $T = 60{,}000$ measurements from a component placement process in a pick-and-place machine (Juloski et al. 2004).[13]

Since the methods of (Ferrari-Trecate et al. 2003) and (Bemporad et al. 2003) cannot handle large data sets, for comparison purposes we first report results on a down-sampled data set of 750 points.[14] The 750 points are separated into two overlapping groups of points. The first 500 points are used for identification, and the last 500 points are used for validation. Table 12.2 shows the average sum of squared residuals (SSR): one-step-ahead prediction errors, and the average sum of squared simulation errors (SSE) obtained by our method for all four data sets, as well as the SSE of the Ferrari-Trecate and Bemporad algorithms for the first data set as reported in (Niessen and A.Juloski 2004). Figure 12.4 shows the true and simulated outputs for data set 1.

[13]We thank Prof. A. Juloski for providing us with the data sets.

[14]We take one out of every 80 samples.

Table 12.2 Training and simulation errors for down-sampled data sets.

Data Set	n	n_a	n_c	ASC SSR	SSE	F-T SSE	Bem. SSE
1	2	2	2	0.0803	0.1195	1.98	2.15
2	2	2	2	0.4765	0.4678	N/A	N/A
3	2	2	2	0.6692	0.7368	N/A	N/A
4	2	2	2	3.1004	3.8430	N/A	N/A

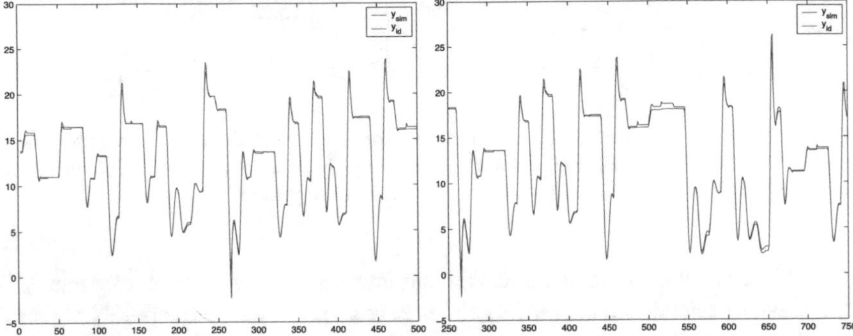

Fig. 12.4 Training and simulation sequences for down-sampled data set 1.

Table 12.3 Training and simulation errors for complete data sets.

Data Set	n	n_a	n_c	SSR	SSE
1 with all points	2	2	2	$4.9696 \cdot 10^{-6}$	$5.3426 \cdot 10^{-6}$
2 with all points	2	2	2	$9.2464 \cdot 10^{-6}$	$7.9081 \cdot 10^{-6}$
3 with all points	2	2	2	$2.3010 \cdot 10^{-5}$	$2.5290 \cdot 10^{-5}$
4 with all points	2	2	2	$7.5906 \cdot 10^{-6}$	$9.6362 \cdot 10^{-6}$

We now report the results of our algorithm tested on the entire data sets. We split the 60,000 measurements into two groups of 30,000 points each. The first 30,000 are used for identification and the last 30,000 for simulation. Table 12.3 shows the average sum of squared residual error (SSR) and the average sum of squared simulation error (SSE) obtained by our method for all four data sets. Figure 12.5 shows the true and simulated outputs for data set 1.

Overall, the algorithm demonstrates a very good performance in all four data sets. The running time of a MATLAB implementation of our algorithm is 0.15 seconds for the 500 data points and 0.841 seconds for 30,000 data points.

12.5 Bibliographic Notes

Work on identification (and filtering) of hybrid systems first appeared in the 1970s; a review of the state of the art as of 1982 can be found in (Tugnait 1982). After a decade-long hiatus, the problem has recently been enjoying considerable interest

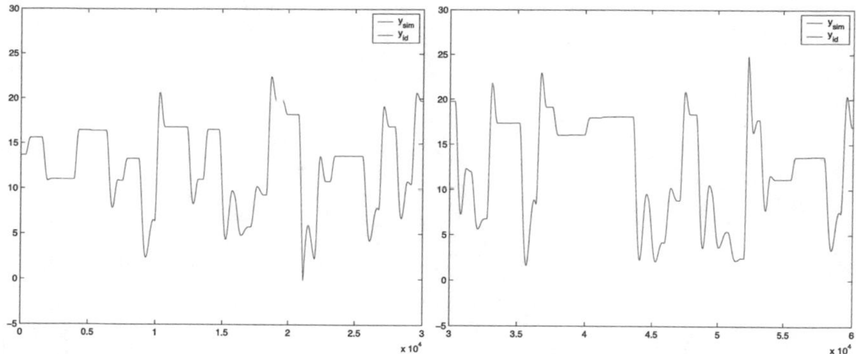

Fig. 12.5 Training and simulation sequences for complete data sets: the simulated and the identified sequences overlap almost exactly.

(Bemporad et al. 2000; Ezzine and Haddad 1989; Sun et al. 2002; Szigeti 1992; Vidal et al. 2002a, 2003a). Much related work has also appeared in the machine-learning community (Billio et al. 1999; Blake et al. 1999; Doucet et al. 2000; Ghahramani and Hinton 1998; Murphy 1998; Pavlovic et al. 1999).

When the model parameters and the switching mechanism are *known*, the identification problem reduces to the design of observers for the hybrid state (Alessandri and Coletta 2001; Balluchi et al. 2002; Ferrari-Trecate et al. 2002; Vecchio and Murray 2004), together with the study of observability conditions under which hybrid observers operate correctly (Babaali and Egerstedt 2004; Bemporad et al. 2000; Collins and Schuppen 2004; Vidal et al. 2002a, 2003a; Hwang et al. 2003; Santis et al. 2003).

When the model parameters and the switching mechanism are both *unknown*, the identification problem becomes much more challenging. Existing work has concentrated on the class of piecewise affine and piecewise ARX systems, i.e., models in which the regressor space is partitioned into polyhedra with affine or ARX submodels for each polyhedron. For instance, (Ferrari-Trecate et al. 2003) assumes that the number of systems is known, and it proposes an identification algorithm that combines clustering, regression, and classification techniques; (Bemporad et al. 2001) solves for the model parameters and the partition of the state space using mixed-integer linear and quadratic programming; (Bemporad et al. 2003) uses a greedy approach for partitioning a set of infeasible inequalities into a minimum number of feasible subsystems, and then iterates between assigning data points to models and computing the model parameters.

The connection between algebraic subspace clustering and identification of hybrid ARX systems was first noticed in (Vidal et al. 2003c; Vidal 2004). Material presented in this chapter follows that in (Ma and Vidal 2005).

Chapter 13
Final Words

Regarding the fundamental investigations of mathematics, there is no final ending … no first beginning.

—Felix Klein

As we have stated from the very beginning of this book, the ultimate goal of our quest is to be able to effectively and efficiently extract low-dimensional structures in high-dimensional data. Our intention is for this book to serve as an introductory textbook for readers who are interested in modern data science and engineering, including both its mathematical and computational foundations as well as its applications. By using what is arguably the most basic and useful class of structures, i.e., linear subspaces, this book introduces some of the most fundamental geometrical, statistical, and optimization principles for data analysis. While these mathematical models and principles are classical and timeless, the problems and results presented in this book are rather modern and timely. Compared with classical methods for learning low-dimensional subspaces (such as PCA (Jolliffe 1986)), the methods discussed in this book significantly enrich our data analysis arsenal with modern methods that are robust to imperfect data (due to uncontrolled data acquisition processes) and can handle mixed heterogenous structures in the data.

In this final chapter, we discuss a few related topics that are not explicitly covered in this book because many of them are still open and active research areas. Nonetheless, we believe that these topics are all very crucial for the future development of modern data science and engineering, and the topics covered in this book serve as a good foundation for readers to venture into these more advanced topics.

R. Vidal et al., *Generalized Principal Component Analysis*, Interdisciplinary Applied Mathematics 40, DOI 10.1007/978-0-387-87811-9_13

13.1 Unbalanced and Multimodal Data

In practical applications, data are often highly imbalanced across classes. In the face clustering example we have used throughout the book, the number of face images from each individual may vary from individual to individual. As a result, when subspace clustering methods are applied to such imbalanced data, they may introduce a bias toward the class that has more samples and may have low performance on the minority class. Resampling methods—whereby underrepresented classes are oversampled and overrepresented classes are undersampled (He and Garcia 2009; He and Ma 2013)—can be applied to make the samples balanced. However, these methods can fail, because they completely ignore low-dimensional structures that are common in multiclass data. A promising approach to handling this issue is to automatically select a small subset of representatives for a large data set using sparse representation techniques (Elhamifar et al. 2012b,a). Interestingly, such methods are able to exploit multisubspace structures and guarantee that sufficiently many samples from each subspace are selected as representatives.

Another limitation of the techniques described in this book is that they require the data to come from the same modality. In practice, the desired information is often buried in various complementary types of data, say a combination of texts, audios, and images. How can we convert these different types of data into a common representation so that we can apply some of the methods in this book? So far, most of the techniques for handling multimodal data (say those popular in the multimedia literature) first extract features from each data type and then simply concatenate the features for analysis. It seems that there is increasing need to put information fusion from different data types on a firm theoretical and algorithmic foundation. Recent work in the area of domain adaptation aims to address this challenge (Patel et al. 2014; Jhuo et al. 2012; Qiu et al. 2012; Shekhar et al. 2013).

13.2 Unsupervised and Semisupervised Learning

According to the ontology of machine learning, all methods introduced in this book belong to the category of *unsupervised learning*. That is, they try to automatically learn the subspace structures of the data set without any manual labeling of the data classes[1] or manual setting of the model parameters. The reason for favoring unsupervised learning in the modern era of Big Data era is obvious: it is cost- and time-prohibitive to manually label massive data sets.

Nevertheless, in many practical situations and tasks, it is reasonable to assume that a small portion of the data set can be properly labeled in advance. Mathematically, the difficulty of the learning task can be dramatically alleviated even if a

[1]For instance, label whether a data point is an outlier; or label which subspace a data point belongs to in advance.

tiny subset of the data are labeled. We have seen some concrete examples in this book that support this view. In the algebraic subspace clustering method described in Chapter 5, we saw that although identifying individual subspaces through factorizing the vanishing polynomials (the vanishing ideal) is computationally intractable, the task can be significantly simplified once we are able to identify a sample point that belongs to one of the subspaces. Similar situations may naturally arise in many practical tasks. For instance, each Facebook user may have a few of the photos in his or her album labeled properly, yet it is desirable to use all the images in the Facebook repository (labeled or unlabeled) to build an effective face recognition or face labeling system.

Naturally, improving the effectiveness and efficiency of the subspace learning algorithms described in this book in the semisupervised learning setting will be a very meaningful and useful direction for future investigation. In particular, there is a need to develop principles that provide good guidelines for data sampling and labeling in such new settings, which, to the best of our knowledge, is still lacking.

13.3 Data Acquisition and Online Data Analysis

In this book, we have assumed that the data have all been collected in advance and have already been converted to a vector or matrix form ready for analysis. This may not be the case in many practical situations. For many demands of data analysis on the Internet or sensor networks, new data are accumulated on a daily basis and need to be stored, processed, and analyzed together with all the data that have been collected before. One natural example is how to analyze video streams from a network of cameras in a metropolitan area, either for traffic violations, security surveillance, or crime investigations. There is an obvious need for developing a real-time or online version of all the data analysis algorithms so that we can learn structures of the data adaptively as new data arrive and as the data structures evolve in time.

Toward the end of the book, in Chapters 11 and 12, we touched on applications of analyzing dynamical data such as videos and hybrid linear dynamical systems. However, to apply the methods in this book, we typically have to process such data in a *batch fashion*. To our knowledge, in the literature, there have already been good progress made toward developing online versions of some of the algorithms featured in this book, e.g., robust principal component analysis (Feng et al. 2013). There has also been good success in applying sparse representation and data clustering to real-time tasks such as object tracking in videos (Zhang et al. 2014). There has also been good progress on developing online versions of the algebraic subspace clustering algorithm for applications in online hybrid system identification (Vidal 2008). Nevertheless, how to develop online data analysis methods in a systematic and principled fashion remains an active research area.

One important issue associated with online data processing is how to control the data acquisition process so that we can more effectively collect the most informative samples for the task at hand. If we could have some control over what data to collect and how to collect them, the subsequent data analysis tasks could potentially be dramatically simplified. This is one of the main messages advocated and supported by compressive sensing theory (Candès 2006; Baraniuk 2007).

13.4 Other Low-Dimensional Models

The class of models studied in this book, although very fundamental, can become inadequate for practical data sets that exhibit more sophisticated structures. As we have studied in Chapter 4, linear subspaces are no longer effective for data sets that have significant nonlinear structures. In such cases, the linear subspace model of PCA needs to be replaced with a low-dimensional surface or submanifold. However, although we have seen in Chapter 4 how such a nonlinear manifold can be learned through parametric or nonparametric techniques, we never dealt with data that may lie on a mixture of nonlinear manifolds.

Union of Manifolds
Note that a union of manifolds is a much more general (and expressive) class of models, which is also known in the literature as *stratifications* (Haro et al. 2008, 2006). Learning manifolds and stratifications remains an active research area, and many effective algorithms have been proposed so far. However, the theory and algorithms for manifold and stratification learning are still far from having reached the same level of maturity as those for subspace models covered in this book. Existing methods for clustering data in a union of manifolds include generalizations of the manifold learning algorithms discussed in Chapter 4, such as (Souvenir and Pless 2005), which is based on alternating minimization, and the locally linear manifold clustering (LLMC) algorithm (Polito and Perona 2002; Goh and Vidal 2007), which we discussed in Chapter 7 in the context of affine subspaces, but which generalizes to nonlinear manifolds. Another algorithm is sparse manifold clustering and embedding (SMCE) (Elhamifar and Vidal 2011), which generalizes the sparse subspace clustering algorithm discussed in Chapter 8. However, as stated before, a theoretical analysis of the conditions under which these methods give the correct clustering is still missing. Finally, there are also extensions of both LLMC and SSC to Riemannian manifolds, which have appeared in (Goh and Vidal 2008) and (Cetingül et al. 2014), respectively.

Compressive Sensing and Decomposable Structures
The rise of compressive sensing (Candès 2006; Baraniuk 2007) has brought to our attention a large family of low-dimensional structures in high-dimensional spaces, the so-called *decomposable structures* (Negahban et al. 2010; Candès and Recht 2011). In a sense, sparse signals, low-rank matrices (low-dimensional subspaces), and mixture of subspaces are all special cases of such structures, as we have seen in Chapters 3 and 8. All decomposable structures have similarly nice geometric and statistical properties as sparse signals and low-rank matrices:

they all can be recovered from nearly minimum samples via tractable means (say convex optimization). In addition, those structures can be arbitrarily combined (sum, union, and intersection) to generate an even broader family of low-dimensional structures. Nevertheless, beyond sparse and low-rank models, our understanding of and practice with structures in this broad family remains rather limited to this day. There is already evidence indicating that many such low-dimensional models and structures will play important roles in future data analysis.

Deep Learning and Deep Neural Networks
If the rise of compressive sensing is due to a series of mathematical breakthroughs, the revival of deep learning (Hinton et al. 2006) is largely attributed to some empirical successes of deep neural networks in classifying practical data such as speeches and images (Jarret et al. 2009). Since low-dimensional linear maps (such as the auto-encoders or the convolutional neural networks) are the key building blocks for each layer of a deep neural network, knowledge given in this book about low-dimensional linear models serves as a good foundation for thoroughly studying properties of hierarchical linear models such as deep neural networks and the treelike graphical models we used in Chapter 9. Recent theoretical advances in the analysis of deep neural networks have indicated strong connections of learning deep neural networks with dictionary learning (Spielman et al. 2012; Sun et al. 2015), sparse regularization (Arora et al. 2014), and matrix/tensor factorization (Haeffele et al. 2014; Haeffele and Vidal 2015). There are good reasons to believe that such advances will eventually lead to a rigorous and profound mathematical theory for deep networks and deep learning, similar to what has been established for sparse models in compressive sensing.

13.5 Computability and Scalability

According to the 2014 Big Data report from the White House, "*We are only in the very nascent stage of the so-called 'Internet of Things'.*" Our government, society, industry, and scientific community have been suddenly inundated with unprecedentedly massive data sets from the Internet (texts, audios, images, and videos, etc.) that contain important information about our daily lives and businesses. This has presented tremendous opportunities and challenges for the information technology industry and community, which require *correct mathematical algorithms and computing technologies to effectively and efficiently analyze those massive data sets and extract useful information from them.*

From Intractable to Tractable
While this book has taken only a few baby steps toward meeting the grand challenge of big data analysis, we have touched on a number of of promising and significant areas of progress in that direction. As we may recall, the problem of generalizing PCA to data with incomplete or corrupted entries or to data from multiple subspaces

is in general a highly combinatorial problem that is computationally intractable.[2] For instance, we saw in Chapter 5 and Appendix C that a precise characterization of the geometric structures of general subspace arrangements requires sophisticated (algebraic) geometric techniques whose computational complexity explodes as the dimension or the size of the data set increases. Now, there has been a very long history of research attempting to tackle instances of the GPCA problem with greedy, heuristic, brute force, or ad hoc algorithms. Although some of these algorithms have produced good results for many practical instances of the problem, one must realize that such algorithms do not provide any strong guarantee of success for general cases.[3] Because of this, at the beginning phase of our study of GPCA, we were wondering ourselves whether we would have to live with the fact that there will never be tractable algorithms for solving these GPCA problems with both correctness and efficiency guarantees. Fortunately, this did not turn out to be the case. With the help of more advanced statistical and computational tools from compressive sensing and convex optimization, researchers were able to develop tractable and efficient algorithms that provide provably correct solutions to the GPCA problems under broad conditions (see Chapters 3 and 8). Along the way, we have begun to realize how limited our understanding of high-dimensional data sets was and how surprisingly optimistic the situation has turned out to be.

From Tractable to Practical
However, having tractable solutions does not mean that the existing algorithms can already meet the modern challenge of big data analysis. Most of the algorithms introduced in this book are capable of handling data size or dimension up to the order of 10^4–10^5 on a typical computer. There has been tremendous effort in the computational community to speed and scale up core computational components heavily utilized by algorithms introduced in this book, including SVD for robust PCA or spectral clustering and ℓ_1 minimization for SSC. Many of the Internet-size data sets and problems require the scaling up of those algorithms by at least a few orders of magnitude. Hence, it is extremely important to investigate alternative optimization techniques that are more suitable for parallel and distributed computing and require less communication and memory. The drive for ever more scalable methods has become the source of inspiration for many ingenious new results in modern high-dimensional statistics and parallel optimization. For instance, the new factorization method mentioned in Chapter 3 has resulted from the effort to try to scale up the matrix completion or matrix recovery problem, instead of relying on the relatively expensive SVD. Recent promising generalizations of this approach have appeared in (Bach et al. 2008; Bach 2013; Haeffele et al. 2014; Udell et al. 2015). The search for ever more efficient and scalable sparse recovery algorithms has revolutionized optimization in the past few years with many new

[2]Strictly speaking, both problems are NP-hard in their general cases.

[3]One must be aware that success on instances can never be used as justification for the correctness of a proposed method.

parallel and distributed algorithms that are able to be implemented on commercial cloud computing platforms (Deng et al. 2013; Peng et al. 2013). Hence, we have sufficient reasons to be optimistic that for most methods and algorithms introduced in this book, researchers will be able to implement them and make them available to everyone on typical cloud computing platforms (such as the Hadoop MapReduce and the Spark systems) in the near future.

13.6 Theory, Algorithms, Systems, and Applications

As the demand for big data analysis is driven by many Internet-scale or world-scale applications, the ever more popular and powerful cloud computing platforms can be viewed as necessary technological infrastructures to support such tasks. However, big data and cloud computing would not have generated so much excitement in the scientific and research communities if they had required nothing more than scaling up what we used to do in the past. As this book has demonstrated, the challenges of analyzing massive high-dimensional data sets under uncontrolled engineering conditions has pushed researchers into the new realm of high-dimensional geometry, statistics, and optimization. We have begun to understand phenomena that were never imagined in classical low-dimensional settings or for tasks with small data sets.

The rise of compressive sensing and sparse representation has begun to provide researchers with a solid theoretical foundation for understanding the geometric and statistical properties of large high-dimensional data sets, whereas the revival of deep learning has begun to provide researchers with efficient computational platforms for handling practical (reinforced) learning tasks with large-scale, high-dimensional input data sets. Almost around the same time, the quest to seek tractable and efficient algorithms is revolutionizing optimization tools needed to learn such complex models and analyze such massive data sets. As we have mentioned before, many such optimization and learning algorithms can be easily implemented on modern cloud computing platforms, and hence can be scaled up to arbitrary sizes.

All these exciting developments make us believe that we are witnessing a *perfect storm* that takes place only occasionally in the history of science and engineering, whereby fundamental mathematical theories and significant engineering endeavors are fueling each other's explosive development. Never before have we seen long-isolated research fields in mathematics, statistics, optimization algorithms, computer systems, and industrial applications work so closely together on a common set of challenges. As a result, every field is making progress at an unprecedented rate, feeding on or fueling the progress and success of other fields. We anticipate that this trend will continue for quite some time, until a new body of scientific and engineering knowledge is fully developed. We hope that this book helps scientists and researchers move one step closer toward that grand goal.

Appendix A
Basic Facts from Optimization

Since the fabric of the universe is most perfect and the work of a most wise Creator, nothing at all takes place in the universe in which some rule of maximum or minimum does not appear.

—L. Euler

In engineering practice, there are often many possible or feasible solutions to a given problem. For instance, there might be multiple models that can explain the same observed data. In such situations, it is desirable to find a solution that is better than others in the sense that it optimizes certain objective function, e.g., it maximizes a likelihood function. To make this book more self-contained, we review in this appendix some of the key facts and tools from optimization. This appendix is by no means meant to be a complete tutorial in optimization. The reader is referred to (Bertsekas 1999; Boyd and Vandenberghe 2004) for details.

A.1 Unconstrained Optimization

The goal of *unconstrained optimization* is to find the minimum value of a function $f : \mathbb{R}^n \to \mathbb{R}$ as well as the point $\boldsymbol{x}^* \in \arg\min_{\boldsymbol{x}} f(\boldsymbol{x})$ at which the function achieves its minimum value, i.e., a point $\boldsymbol{x}^*$ such that $f(\boldsymbol{x}^*) \leq f(\boldsymbol{x}), \forall \boldsymbol{x} \in \mathbb{R}^n$. Notice that in general, the minimum value or optimal solution of a function may not exist, and even if it does exist, it may not be unique. For simplicity and convenience, unless otherwise stated, we will always assume that the function f is twice differentiable. We denote the gradient and Hessian of the function f by ∇f and $\nabla^2 f$, respectively. Notice that $\nabla f(\boldsymbol{x})$ is an n-dimensional vector and $\nabla^2 f(\boldsymbol{x})$ is an $n \times n$ matrix. More precisely, they are defined to be

R. Vidal et al., *Generalized Principal Component Analysis*, Interdisciplinary Applied Mathematics 40, DOI 10.1007/978-0-387-87811-9

$$\nabla f(\boldsymbol{x}) \doteq \begin{bmatrix} \frac{\partial f(\boldsymbol{x})}{\partial x_1} \\ \frac{\partial f(\boldsymbol{x})}{\partial x_2} \\ \vdots \\ \frac{\partial f(\boldsymbol{x})}{\partial x_n} \end{bmatrix}, \quad \nabla^2 f(\boldsymbol{x}) \doteq \begin{bmatrix} \frac{\partial^2 f(\boldsymbol{x})}{\partial x_1 \partial x_1} & \frac{\partial^2 f(\boldsymbol{x})}{\partial x_1 \partial x_2} & \cdots & \frac{\partial^2 f(\boldsymbol{x})}{\partial x_1 \partial x_n} \\ \frac{\partial^2 f(\boldsymbol{x})}{\partial x_2 \partial x_1} & \frac{\partial^2 f(\boldsymbol{x})}{\partial x_2 \partial x_2} & \cdots & \frac{\partial^2 f(\boldsymbol{x})}{\partial x_2 \partial x_n} \\ \vdots & \vdots & \ddots & \vdots \\ \frac{\partial^2 f(\boldsymbol{x})}{\partial x_n \partial x_1} & \frac{\partial^2 f(\boldsymbol{x})}{\partial x_n \partial x_2} & \cdots & \frac{\partial^2 f(\boldsymbol{x})}{\partial x_n \partial x_n} \end{bmatrix}, \tag{A.1}$$

where $\boldsymbol{x} = \left[x_1, \ldots, x_n\right]^\top$. Sometimes, we use $\nabla_{\boldsymbol{x}} f(\boldsymbol{x}, \boldsymbol{y})$ to indicate the gradient with respect to $\boldsymbol{x}$ only, and similarly for the Hessian.

A.1.1 *Optimality Conditions*

We use $N(\boldsymbol{x}, \varepsilon)$ to denote an ε-ball around the point $\boldsymbol{x}$. We say that a point $\boldsymbol{x}^*$ is a local minimum of f if there exists an $\varepsilon > 0$ such that $f(\boldsymbol{x}^*) \le f(\boldsymbol{x})$ for all $\boldsymbol{x} \in N(\boldsymbol{x}^*, \varepsilon)$. We say that $\boldsymbol{x}^*$ is a strict minimum if equality holds only if $\boldsymbol{x} = \boldsymbol{x}^*$. If the size of the neighborhood can be arbitrarily large, we say that $\boldsymbol{x}^*$ is the global minimum of the function.

It is not difficult to prove by contradiction (see Exercise A.1) that a necessary condition for a point $\boldsymbol{x}^*$ to be a local minimum is that the gradient $\nabla f(\boldsymbol{x})$ vanish at $\boldsymbol{x}^*$, or more precisely,

$$\nabla f(\boldsymbol{x}^*) = \boldsymbol{0}. \tag{A.2}$$

The following proposition gives sufficient conditions for a point $\boldsymbol{x}^*$ to be a local minimum in terms of its gradient and Hessian.

Proposition A.1 (Second-Order Sufficient Optimality Conditions). *If a point* $\boldsymbol{x}^* \in \mathbb{R}^n$ *satisfies the conditions*

$$\nabla f(\boldsymbol{x}^*) = 0, \quad \nabla^2 f(\boldsymbol{x}^*) \succ \boldsymbol{0}, \tag{A.3}$$

then $\boldsymbol{x}^*$ *is a (strict) local minimum of* $f(x)$.

In practice, the above conditions can be used to find all possible local minima of a given function. Of course, in general, local minima of a function are often not unique and do not have to be the global minimum. However, if f is convex defined on a convex domain, then every local minimum must be the global minimum.

A.1.2 *Convex Set and Convex Function*

Definition A.2 (Convex Set). *A set* $\mathcal{X} \subseteq \mathbb{R}^n$ *is said to be* convex *if for every* $\boldsymbol{x}, \boldsymbol{y} \in \mathcal{X}$ *and* $\lambda \in [0, 1]$, *we have* $\lambda \boldsymbol{x} + (1 - \lambda)\boldsymbol{y} \in \mathcal{X}$.

For convenience, the empty set is considered a special convex set. In most optimization problems that we consider in this book, we are searching for the minimum of a function over a convex domain. It is easy to verify many useful properties of convex sets. For example, the intersection of any two convex sets is also a convex set (see Exercise A.2).

Given any set (convex or not), we can associate with it a convex set as follows:

Definition A.3 (Convex Hull). *Given a set $\mathcal{X} = \{\boldsymbol{x}_i\} \subseteq \mathbb{R}^n$, we define its* convex hull, *denoted by conv($\mathcal{X}$), to be*

$$conv(\mathcal{X}) \doteq \left\{ \boldsymbol{y} : \boldsymbol{y} = \sum_{i=1}^{k} \lambda_i \boldsymbol{x}_i, \text{ where } k \in \mathbb{N}, \lambda_i \geq 0 \text{ and } \sum_{i=1}^{k} \lambda_i = 1 \right\}. \tag{A.4}$$

It is easy to show that a convex hull must be a convex set and that the convex hull of a convex set is the convex set itself (see Exercise A.2).

Definition A.4 (Convex Function). *A function $f : \mathcal{X} \to \mathbb{R}$ defined on a convex domain $\mathcal{X} \subseteq \mathbb{R}^n$ is said to be* convex *if for all $\boldsymbol{x}, \boldsymbol{y} \in \mathcal{X}$ and $\lambda \in [0, 1]$, we have*

$$f(\lambda \boldsymbol{x} + (1 - \lambda)\boldsymbol{y}) \leq \lambda f(\boldsymbol{x}) + (1 - \lambda) f(\boldsymbol{y}). \tag{A.5}$$

We say that f is strictly convex if the inequality is strict for $\boldsymbol{x} \neq \boldsymbol{y}$ and $\lambda \in (0, 1)$.

Convex functions are extremely important for optimization largely because their minima and maxima have some very useful properties.

Theorem A.5 (Minima of Convex Function). *If a convex function f defined over a convex domain $\mathcal{X} \subseteq \mathbb{R}^n$ has a minimum, then it has the following properties:*

1. *Every local minimum of f is also a global minimum.*
2. *The set of all minima of f is a convex set.*
3. *If the function f is strictly convex, it has a unique minimum $\boldsymbol{x}^*$.*

Proof. Let f^* denote the global minimum value of f over $\mathcal{X}$, and choose a point $\boldsymbol{x}^*$ where f reaches the global minimum value, i.e., $f(\boldsymbol{x}^*) = f^*$.

1. To prove the first statement, let us assume for the sake of contradiction that f has a local minimum at $\boldsymbol{y}^*$. Then, due to the convexity of f, we have that for all $\lambda \in [0, 1]$,

$$f(\lambda \boldsymbol{x}^* + (1 - \lambda)\boldsymbol{y}^*) = f(\boldsymbol{y}^* + \lambda(\boldsymbol{x}^* - \boldsymbol{y}^*)) \leq \lambda f(\boldsymbol{x}^*) + (1 - \lambda) f(\boldsymbol{y}^*).$$

If $f(\boldsymbol{x}^*) < f(\boldsymbol{y}^*)$, then $\lambda f(\boldsymbol{x}^*) + (1 - \lambda) f(\boldsymbol{y}^*) < f(\boldsymbol{y}^*)$ for every $\lambda \in (0, 1]$. Therefore, we have $f(\boldsymbol{y}^* + \lambda(\boldsymbol{x}^* - \boldsymbol{y}^*)) < f(\boldsymbol{y}^*)$ for all $\lambda \in (0, 1]$. This contradicts the assumption that $\boldsymbol{y}^*$ is a local minimum of f.
2. To prove the second statement, we need to show that for every $c \in \mathbb{R}$, the set $\{x : f(x) \leq c\}$ is convex. We leave this as an exercise to the reader (see Exercise A.3). The claim then follows by choosing $c = f^*$.

3. To prove the third statement, let us assume for the sake of contradiction that f has two different local minima $\boldsymbol{x}^*$ and $\boldsymbol{y}^* \neq \boldsymbol{x}^*$. Due to the first statement, we have $f(\boldsymbol{x}^*) = f(\boldsymbol{y}^*)$. Since f is strictly convex, we further have

$$f(\lambda \boldsymbol{x}^* + (1-\lambda)\boldsymbol{y}^*) < \lambda f(\boldsymbol{x}^*) + (1-\lambda) f(\boldsymbol{y}^*) = f(\boldsymbol{x}^*)$$

for all $\lambda \in (0, 1)$. Since the domain $\mathcal{X}$ is convex, $\lambda \boldsymbol{x}^* + (1-\lambda)\boldsymbol{y}^* \in \mathcal{X}$. This contradicts that $\boldsymbol{x}^*$ is the global minimum of f over $\mathcal{X}$. Therefore, the minimum $\boldsymbol{x}^*$ must be unique.

□

Sometimes, we are also interested in the maximum value of a convex function f over a convex set $\mathcal{X}$. We have the following statement.

Theorem A.6 (Maxima of Convex Function over Compact Convex Domain). *Let f be a convex function defined on a compact convex domain $\mathcal{X}$. Then f reaches its maximum value at the boundary of $\mathcal{X}$. More precisely, we have*

$$\max_{\boldsymbol{x} \in \mathcal{X}} f(\boldsymbol{x}) = \max_{\boldsymbol{x} \in \partial \mathcal{X}} f(\boldsymbol{x}),$$

where $\partial \mathcal{X}$ denotes the boundary of the set $\mathcal{X}$.

We leave the proof as an exercise for the reader to become familiar with the properties of convex functions (see Exercise A.3).

Besides the above notion of (strict) convexity, the following two relaxed notions of convexity are also often used.

Definition A.7 (Quasiconvex). *A function $f : \mathcal{X} \to \mathbb{R}$ defined on a convex domain $\mathcal{X} \subseteq \mathbb{R}^n$ is said to be* quasiconvex *if for all $\boldsymbol{x}, \boldsymbol{y} \in \mathcal{X}$ and $\lambda \in [0, 1]$, we have*

$$f(\lambda \boldsymbol{x} + (1-\lambda)\boldsymbol{y}) \le \max\{f(\boldsymbol{x}), f(\boldsymbol{y})\}.$$

Definition A.8 (Pseudoconvex). *A function $f : \mathcal{X} \to \mathbb{R}$ defined on a convex domain $\mathcal{X} \subseteq \mathbb{R}^n$ is said to be* pseudoconvex *if for all $\boldsymbol{y} \in \mathbb{R}^n$ such that $\nabla f(\boldsymbol{x})^\top (\boldsymbol{y} - \boldsymbol{x}) \ge 0$, we have*

$$f(\boldsymbol{x}) \le f(\boldsymbol{y}).$$

A.1.3 Subgradient

Sometimes, the function we are trying to minimize is not necessarily smooth everywhere. In this case, the "gradient" of the function cannot be evaluated at every point. This leads to a generalized notion of the gradient called a *subgradient*.

Definition A.9 (Subgradient of a Convex Function). *The subgradient of a convex function $f : \mathcal{X} \rightarrow \mathbb{R}$ at a point $\boldsymbol{x} \in \mathcal{X}$, where $\mathcal{X}$ is convex, is defined to be the set*

$$\partial f(\boldsymbol{x}) \doteq \{\boldsymbol{v} \in \mathbb{R}^n : f(\boldsymbol{y}) \geq f(\boldsymbol{x}) + \boldsymbol{v}^\top(\boldsymbol{y} - \boldsymbol{x}),\ \forall \boldsymbol{y} \in \mathcal{X}\}. \tag{A.6}$$

Most conditions and results for minimizing a smooth convex function generalize to a nonsmooth convex function if one replaces gradient with subgradient. For instance, a point $\boldsymbol{x}^*$ is a minimum of a convex function f if and only if $\boldsymbol{0} \in \partial f(\boldsymbol{x}^*)$.

A.1.4 Gradient Descent Algorithm

There is an extremely rich history and literature on how to optimize a function. For many of the problems in this book, we are mostly interested in a simple method that can be easily implemented to obtain the optimal solution. Hence, in this section, we introduce a few simple methods that are pertinent to these problems, even though they do not necessarily represent the most advanced optimization techniques.

Almost all methods for minimizing a function f are based on a very simple idea. We begin with an initial guess $\boldsymbol{x} = \boldsymbol{x}^0$, and successively update $\boldsymbol{x}$ to $\boldsymbol{x}^1, \boldsymbol{x}^2, \ldots$, such that the value $f(\boldsymbol{x})$ decreases at each iteration; that is, $f(\boldsymbol{x}^{i+1}) \leq f(\boldsymbol{x}^i)$. Of course, the safest way to ensure a decrease of the value of the objective function is to follow the "direction of descent," which in our case would be the opposite direction to the gradient vector $\nabla f(\boldsymbol{x}^i)$. This idea gives rise to the classic *steepest descent method* for searching for the minimum. At each iteration, the variables are updated as

$$\boldsymbol{x}^{i+1} = \boldsymbol{x}^i - \alpha^i \nabla f(\boldsymbol{x}^i), \tag{A.7}$$

for some scalar $\alpha^i > 0$, called the *step size*.

There exist many different choices for the step size α^i, and the simplest one is of course to set it to be a small constant, but that does not always result in a decrease in the value of $f(x)$ at each iteration. Instead, α^i is often chosen to be the value α^* that is given by solving a one-dimensional minimization problem:

$$\alpha^* = \arg\min_{\alpha \geq 0} f(\boldsymbol{x}^i - \alpha \nabla f(\boldsymbol{x}^i)). \tag{A.8}$$

This is called the *minimization rule*.

Although the vector $-\nabla f(\boldsymbol{x}^i)$ points to the steepest descent direction locally around $\boldsymbol{x}^i$, it is not necessarily the best choice for searching for the minimum at a larger scale. For instance, if $f(\boldsymbol{x})$ can be approximated by a quadratic function $f(\boldsymbol{x}) \approx \frac{1}{2}(\boldsymbol{x} - \boldsymbol{x}^*)^\top K(\boldsymbol{x} - \boldsymbol{x}^*) + c$ and the matrix K has very large condition number,[1]

[1] That is, the ratio $\mathrm{cond}(K) = \frac{\lambda_{\max}}{\lambda_{\min}}$ between the largest and smallest eigenvalues of K is large.

then the simple gradient descent method typically has very poor convergence. In general, it is easy to establish that the error $f(\boldsymbol{x}^i) - f(\boldsymbol{x}^*)$ of gradient-based methods necessarily drops on the order of $o(i^{-1})$. Further, for a general class of objective functions, one can show that the optimal rate of convergence for gradient-based methods does not exceed $o(i^{-2})$ (Nemirovskii and Yudin 1979).

To improve the convergence of the gradient descent method, one can generalize the variable update equation to the form

$$\boldsymbol{x}^{i+1} = \boldsymbol{x}^i - \alpha^i D^i \nabla f(\boldsymbol{x}^i), \tag{A.9}$$

where $D^i \in \mathbb{R}^{n \times n}$ is a positive definite symmetric matrix to be determined in each particular algorithm. The steepest descent method in (A.7) becomes a particular case of (A.9), where $D^i \equiv I$. In general, D^i can be viewed as a weight matrix that adjusts the descent direction according to more sophisticated local information about the function f than the gradient alone. A simple choice for D^i would be a diagonal matrix that scales the descent speed differently in each axial direction. A more principled choice for D^i would be the inverse of the Hessian $D^i = [\nabla^2 f(x^i)]^{-1}$, which gives the classical Newton's method. This method typically has a much faster convergence rate than simple gradient-based methods. For example, it finds the minimum of a quadratic function in one step. In general, it can also be established that under fairly general conditions, optimization schemes based on Newton-type iterations often have a linear convergence rate, that is, the error $f(\boldsymbol{x}^i) - f(\boldsymbol{x}^*)$ reduces on the order of $o(\rho^i)$ for some $\rho \in (0, 1)$.

Despite the fast convergence of Newton's method, in many modern high-dimensional optimization problems that we encounter in this book, this choice is not very practical, because it is extremely costly to compute and store the Hessian matrix and its inverse. Hence, most modern optimization methods for large-scale optimization rely on smart modifications to the classical gradient descent method that are based on only first-order derivatives of the objective function. For interested readers, we point to the seminal work of (Nesterov 1983; Beck and Teboulle 2009) on *accelerated proximal gradient* algorithms that achieve a convergence rate of $o(i^{-2})$ for a large class of convex objective functions.

A.1.5 Alternating Direction Minimization

In many optimization problems that we encounter in this book, we are required to minimize an objective function f that has special structures. For example, if we partition the variables $\boldsymbol{x} \in \mathbb{R}^n$ into, say, N blocks $\boldsymbol{x} = (x_1, \ldots, x_N)$, it may be very convenient to minimize f with respect to one block of variables at a time.

Such methods are also known in the optimization literature as *block coordinate descent* (BCD) methods (Tseng 2001) or alternating direction minimization (ADM) methods, especially when $N = 2$. For example, in the matrix factorization problem discussed in Section 2.1.2, our goal is to obtain a factorization (U, V) that best approximates a given matrix M by minimizing the objective function

$$\|M - UV^\top\|_F^2. \tag{A.10}$$

If we fix one factor, say U, then finding the best V that minimizes the error is a simple quadratic problem and has a closed-form solution. Hence, it is rather natural to minimize such an objective function by iteratively minimizing with respect one factor at a time. As another example, in some of the convex optimization problems that we utilize for recovering low-rank matrices or sparse vectors, the objective function is often of the special form

$$f(\boldsymbol{x}) = f_0(\boldsymbol{x}) + \sum_{i=1}^{N} f_i(x_i), \tag{A.11}$$

where $f_i(\cdot)$ is a component that depends only on the ith block variables x_i, and $f_0(x)$ typically is a simple function with nice properties. Such a function is said to have a *separable* form. Again, it is natural to minimize such a function in a (block) coordinate descent fashion especially if $f_0(\boldsymbol{x}) + f_i(x_i)$ is much easier to minimize with respect to each coordinate block x_i.

Below, we formally describe the block coordinate descent (BCD) method, as a special version of what was described in (Tseng 2001):

- Initialization. Choose any $\boldsymbol{x}^0 = (x_1^0, \dots, x_N^0) \in \mathbb{R}^n$.
- For the $(i+1)$th iteration, $i \geq 0$, given $\boldsymbol{x}^i = (x_1^i, \dots, x_N^i) \in \mathbb{R}^n$ from the previous iteration, choose $s = i \,(\mathrm{mod}\, N)$ and compute
  - $x_s^{i+1} = \arg\min_{x_s} f(x_1^i, \dots, x_{s-1}^i, x_s, x_{s+1}^i, \dots, x_N^i)$;
  - $x_j^{i+1} = x_j^i, \quad \forall j \neq s.$
- Repeat the process till convergence or the maximum number of iterations has been reached.

Although the above alternating minimization scheme is very widely used in engineering solutions for real-world optimization problems, theoretically it is important to know about when it is guaranteed to converge, at least to a local minimum of the objective function f. There has been a vast amount of classical literature that characterizes the convergence of the BCD method for various classes of objective functions. We here summarize some of the well-known convergence results, which are helpful in justifying the optimization techniques used for problems in this book. For detailed and rigorous proofs of these results, we refer the reader to the references given below.

Proposition A.10 (Convergence of Block Coordinate Descent). *Given a function $f : \mathbb{R}^n \to \mathbb{R}$ bounded from below, the BCD method converges to a stationary point of f under each of the following conditions:*

- *The function f is strictly convex (Warga 1963).*
- *The function f is pseudoconvex (Zadeh 1970).*
- *The function f is quadratic (Luo and Tseng 1993).*
- *The function f is pseudoconvex in each pair of blocks (x_j, x_k) for every $j, k \in \{1, \dots N\}$ (Tseng 2001).*
- *The function f has unique minimum in each coordinate block (Luenberger 1973).*

In fact, if the function f is not pseudoconvex, a counterexample (Powell 1973) exists in which the method may cycle without approaching any stationary point of f. The last result suggests that the alternating minimization scheme for the matrix factorization problem is guaranteed to converge to a stationary point.

A.2 Constrained Optimization

In this section, we consider the problem of minimizing a function $f : \mathbb{R}^n \to \mathbb{R}$ subject to equality constraints on the variable $\boldsymbol{x} \in \mathbb{R}^n$, i.e.,

$$\boldsymbol{x}^* = \arg\min f(\boldsymbol{x}) \quad \text{subject to} \quad h(\boldsymbol{x}) = 0, \tag{A.12}$$

where $h = [h_1, h_2, \dots, h_m]^\top$ is a smooth (multidimensional) function (or map) from $\mathbb{R}^n$ to $\mathbb{R}^m$. For each constraint $h_i(\boldsymbol{x}) = 0$ to be independently effective at the minimum $\boldsymbol{x}^*$, we often assume that their gradients

$$\nabla h_1(\boldsymbol{x}^*),\ \nabla h_2(\boldsymbol{x}^*),\ \dots,\ \nabla h_m(\boldsymbol{x}^*) \quad \in \mathbb{R}^n \tag{A.13}$$

are linearly independent. If so, the constraints are called *regular*.

A.2.1 *Optimality Conditions and Lagrangian Multipliers*

For simplicity, we always assume that the functions f and h are at least twice continuously differentiable. Then the main theorem of Lagrange is as follows.

Theorem A.11 (Lagrange multiplier theorem; necessary conditions). *Let $\boldsymbol{x}^*$ be a local minimum of a function f subject to regular constraints $h(\boldsymbol{x}^*) = 0$. Then there exists a unique vector $\lambda^* = [\lambda_1^*, \lambda_2^*, \dots, \lambda_m^*]^\top \in \mathbb{R}^m$, called* Lagrange multipliers, *such that*

$$\nabla f(\boldsymbol{x}^*) + \sum_{i=1}^{m} \lambda_i^* \nabla h_i(\boldsymbol{x}^*) = \mathbf{0}. \tag{A.14}$$

Furthermore, we have

$$v^\top\Big(\nabla^2 f(x^*) + \sum_{i=1}^{m} \lambda_i^* \nabla^2 h_i(x^*)\Big)v \geq \mathbf{0} \tag{A.15}$$

for all vectors $v \in \mathbb{R}^n$ *that satisfy* $\nabla h_i(x^*)^\top v = 0$, *for* $i = 1, 2, \ldots, m$.

Theorem A.12 (Lagrange multiplier theorem; sufficient conditions). *Assume that* $x^* \in \mathbb{R}^n$ *and* $\lambda^* = [\lambda_1^*, \lambda_2^*, \ldots, \lambda_m^*]^\top \in \mathbb{R}^m$ *satisfy*

$$\nabla f(x^*) + \sum_{i=1}^{m} \lambda_i^* \nabla h_i(x^*) = 0, \quad h_i(x^*) = 0,\ i = 1, 2, \ldots, m, \tag{A.16}$$

and furthermore, we have

$$v^\top\Big(\nabla^2 f(x^*) + \sum_{i=1}^{m} \lambda_i^* \nabla^2 h_i(x^*)\Big)v \geq \mathbf{0}, \tag{A.17}$$

for all vectors $v \in \mathbb{R}^n$ *that satisfy* $\nabla h_i(x^*)^\top v = 0$, *for* $i = 1, 2, \ldots, m$. *Then* x^* *is a strict local minimum of* f *subject to* $h(x) = 0$.

The Lagrangian function

If we define for convenience the *Lagrangian function* $\mathcal{L} : \mathbb{R}^{n+m} \to \mathbb{R}$ as

$$\mathcal{L}(x, \lambda) \doteq f(x) + \lambda^\top h(x), \tag{A.18}$$

then the necessary conditions in Theorem A.11 can be written as

$$\nabla_x \mathcal{L}(x^*, \lambda^*) = \mathbf{0}, \qquad \nabla_\lambda \mathcal{L}(x^*, \lambda^*) = \mathbf{0}, \tag{A.19}$$

$$v^\top \nabla_x^2 \mathcal{L}(x^*, \lambda^*) v \geq \mathbf{0},\ \forall v : v^\top \nabla h(x^*) = \mathbf{0}. \tag{A.20}$$

The conditions (A.19) give a system of $n + m$ equations with $n + m$ unknowns: the entries of x^* and λ^*. If the constraint $h(x) = 0$ is regular, then in principle, this system of equations is independent, and we should be able to solve for x^* and λ^*. The solutions will contain all the (local) minima, but it is possible that some of them need not be minima at all. Nevertheless, whether we are able to solve these equations or not, they usually provide rich information about the minima of the constrained optimization. We illustrate how we can utilize the necessary conditions for Lagrange multipliers to find the optimal solution to a constrained optimization problem with the following example.

Example A.13 [Matrix Lagrange Multipliers] Consider the problem of projecting a given matrix $M \in \mathbb{R}^{n\times n}$ onto the space of orthogonal matrices $O(n) = \{U \in \mathbb{R}^{n\times n} : U^\top U = I\}$. That is, we want to find a matrix $U \in \mathbb{R}^{n\times n}$ that minimizes

$$\min_{U} \|M - U\|_F^2 \quad \text{subject to} \quad U^\top U = I. \tag{A.21}$$

Notice that there are n^2 constraints in $U^\top U = I$. This suggests using n^2 Lagrange multipliers, which can be conveniently represented as the entries of a matrix $\Lambda \in \mathbb{R}^{n \times n}$. However, since the matrix $U^\top U$ is symmetric, there are only $n(n+1)/2$ independent constraints. Therefore, the matrix Λ needs to be chosen to be symmetric. Now, since the inner product between the two matrices A and B can be conveniently written as $\langle A, B\rangle = \text{trace}(A^\top B)$, the Lagrangian function can be written as

$$\mathcal{L}(U, \Lambda) = \|M - U\|_F^2 + \text{trace}(\Lambda(U^\top U - I)). \tag{A.22}$$

The necessary condition $\frac{\partial \mathcal{L}}{\partial U} = \mathbf{0}$ in Theorem A.11 gives

$$(U - M) + U\Lambda = \mathbf{0}. \tag{A.23}$$

This gives $\Lambda = U^\top M - I$. Since Λ is symmetric, so is $U^\top M$. Let $M = W\Sigma V^\top$ be the singular value decomposition of M. Both W, V are orthogonal matrices. If the singular values of M are all different, then in order for $U^\top M = U^\top W\Sigma V^\top$ to be symmetric, we must have $U^\top W = V$; hence $U = WV^\top$.

As we see from the above example, for some constrained optimization problems, the necessary conditions of the Lagrangian alone allow us to solve for the optimal solution. In general, of course, this is not always possible, and we have to resort to numerical solutions to find the optimal solution.

A.2.2 Augmented Lagrange Multipler Methods

If we are not able to solve for the minima from the equations given by the necessary conditions, we must resort to a numerical optimization scheme. The basic idea is to try to convert the original constrained optimization to an unconstrained one by introducing extra *penalty* terms to the objective function. A typical choice is the *augmented Lagrangian function* $\mathcal{L}_c : \mathbb{R}^{n+m} \to \mathbb{R}$, defined as

$$\mathcal{L}_c(\boldsymbol{x}, \lambda) \doteq f(\boldsymbol{x}) + \lambda^\top h(\boldsymbol{x}) + \frac{c}{2}\|h(\boldsymbol{x})\|^2, \tag{A.24}$$

where $c > 0$ is a positive penalty parameter. It is reasonable to expect that for very large c, the location $\boldsymbol{x}^*$ of the global minimum of the unconstrained minimization

$$(\boldsymbol{x}^*, \lambda^*) = \arg\max_{\lambda} \min_{\boldsymbol{x}} \mathcal{L}_c(\boldsymbol{x}\lambda) \tag{A.25}$$

should be very close to the global minimum of the original constrained minimization.

Proposition A.14 (Convergence of ALM (Bertsekas 1999)). *For* $i = 0, 1, \ldots$, *let* $\boldsymbol{x}^i$ *be a global minimum of the unconstrained optimization problem*

$$\min_{\boldsymbol{x}} \mathscr{L}_{c^i}(\boldsymbol{x}, \lambda^i), \tag{A.26}$$

where $\{\lambda^i\}$ *is bounded,* $0 < c^i < c^{i+1}$ *for all i, and* $c^i \to \infty$. *Then the limit of the sequence* $\{\boldsymbol{x}^i\}$ *is a global minimum of the original constrained optimization problem.*

This result leads to the classical augmented Lagrangian algorithm for solving the constrained optimization problem (A.12) via the following iteration:

$$\begin{aligned} \boldsymbol{x}^{i+1} &= \arg\min_{\boldsymbol{x}} \mathscr{L}_{c_i}(\boldsymbol{x}, \lambda^i), \\ \lambda^{i+1} &= \lambda^i + c^{i+1} h(\boldsymbol{x}^{i+1}). \end{aligned} \tag{A.27}$$

It is easy to see that if $\{\lambda^i\}$ is a bounded sequence and $c^i \to \infty$, then we must have $h(\boldsymbol{x}^i) \to \boldsymbol{0}$; hence the constraint will be enforced at the point $\boldsymbol{x}^*$ to which the algorithm converges. Moreover, the limit point λ^* of the bounded sequence $\{\lambda^i\}$ will be the desired Lagrange multiplier in Theorem A.11.

A.2.3 Alternating Direction Method of Multipliers

A very common class of optimization problems that one encounters in practice is to optimize some convex objective function subject to a set of linear constraints. Very often, including some cases we have encountered in this book, the objective function f has a separable form that makes it amenable to simpler optimization schemes such as alternating minimization, discussed earlier. For example, consider the following optimization problem:

$$\min_{\boldsymbol{x}} f(\boldsymbol{x}) = \sum_{i=1}^{N} f_i(x_i), \quad \text{subject to} \quad \sum_{i=1}^{N} A_i x_i = \boldsymbol{b}, \tag{A.28}$$

where each f_i is a convex function. In some cases, the component functions f_i need not be smooth. For instance, in the robust PCA problem discussed in Section 3.2, we aim to solve

$$\min_{L,S} \|L\|_* + \lambda \|S\|_1, \quad \text{subject to} \quad D = L + S, \tag{A.29}$$

where the nuclear norm $\|\cdot\|_*$ and the ℓ^1-norm $\|\cdot\|_1$ are not smooth.

In this subsection, we show how to use the augmented Lagrangian method to solve this class of optimization problems in an effective and efficient way. Notice that the augmented Lagrangian function for this class of problems precisely resembles the separable form (A.11) studied earlier. We are particularly interested in simple and scalable algorithms that utilize only first-order information of the objective function and do not involve any expensive computations.

Most of the cases in which we are interested in this book involve (or can be reduced to) only two terms, say

$$\min_{x,y} f(x) + g(y) \quad \text{subject to} \quad Ax + By = b, \tag{A.30}$$

where $f : \mathbb{R}^n \to \mathbb{R}$ and $g : \mathbb{R}^p \to \mathbb{R}$ are two convex functions, and $A \in \mathbb{R}^{m\times n}$, $B \in \mathbb{R}^{m\times p}$, and $\boldsymbol{b} \in \mathbb{R}^m$ together specify m linear constraints. For simplicity, we will first illustrate the basic algorithm and results using the two-term problem, and we will later discuss how to generalize to multiple terms.

Let us define the augmented Lagrangian function for problem (A.30):

$$\mathcal{L}_\mu(x, y; \lambda) \doteq f(x) + g(y) + \langle \lambda, Ax + By - b \rangle + \frac{1}{2\mu}\|Ax + By - b\|_2^2, \tag{A.31}$$

where $\mu > 0$ is a penalty parameter. According to the augmented Lagrangian method, μ should be a decreasing sequence converging to 0. Then, following the classical augmented Lagrangian method (Bertsekas 1999), we can solve problem (A.30) via the following iteration:

$$\begin{aligned} (x^{i+1}, y^{i+1}) &= \arg\min_{x,y} \mathcal{L}_\mu(x, y; \lambda^i), \\ \lambda^{i+1} &= \lambda^i + (Ax^{i+1} + By^{i+1} - b)/\mu. \end{aligned} \tag{A.32}$$

However, the joint minimization over both x and y can be very difficult. Fortunately, as in the case of the robust PCA problem, the minimization over x or y with the other variables fixed is often much simpler. This leads to the *alternating direction method of multipliers* (ADMM), which follows the following iteration scheme:

$$\begin{aligned} x^{i+1} &= \arg\min_{x} \mathcal{L}_\mu(x, y^i; \lambda^i), \\ y^{i+1} &= \arg\min_{y} \mathcal{L}_\mu(x^{i+1}, y; \lambda^i), \\ \lambda^{i+1} &= \lambda^i + (Ax^{i+1} + By^{i+1} - b)/\mu. \end{aligned} \tag{A.33}$$

This alternating direction technique is known as the Douglas–Rachford operator splitting method and is known to converge to the global optimal solution (see (Ma 2012) and references therein).

In the robust PCA problem, both A and B are identity operators, and the associated optimization problems for the two alternating minimizations are both very simple to solve. For instance, the minimization with respect to the sparse term is $\min_S \|S\|_1 + \alpha\|S - M\|$ for some fixed matrix M and constant α. The solution is given by a simple entrywise soft thresholding. The minimization with respect to the low-rank term is a simple singular-value soft thresholding.

However, in many other problems, the operators A and B are not necessarily the identities, and the problem of minimizing each component may no longer be so simple, even for a case such as the ℓ^1-norm. Note that by completing the squares, we can write the iteration scheme (A.33) explicitly as

$$\begin{aligned} \boldsymbol{x}^{i+1} &= \arg\min_{\boldsymbol{x}} f(\boldsymbol{x}) + \frac{1}{2\mu}\|A\boldsymbol{x} + B\boldsymbol{y}^i - \boldsymbol{b} + \mu\boldsymbol{\lambda}^i\|_2^2, \\ \boldsymbol{y}^{i+1} &= \arg\min_{\boldsymbol{y}} g(\boldsymbol{y}) + \frac{1}{2\mu}\|A\boldsymbol{x}^{i+1} + B\boldsymbol{y} - \boldsymbol{b} + \mu\boldsymbol{\lambda}^i\|_2^2, \\ \boldsymbol{\lambda}^{i+1} &= \boldsymbol{\lambda}^i + (A\boldsymbol{x}^{i+1} + B\boldsymbol{y}^{i+1} - \boldsymbol{b})/\mu. \end{aligned} \tag{A.34}$$

Although one can always resort to some iterative scheme to find the minimum solutions to the above two subproblems, the computational cost can be very high. One technique proposed to simplify the above minimization approximates the quadratic penalty term $\frac{1}{2}\|A\boldsymbol{x} + B\boldsymbol{y}^i - \boldsymbol{b} + \mu\boldsymbol{\lambda}^k\|_2^2$ with another proximal quadratic term

$$\begin{aligned} &\frac{1}{2\tau_1}\|\boldsymbol{x} - (\boldsymbol{x}^i - \tau_1 A^\top(A\boldsymbol{x}^i + B\boldsymbol{y}^i - \boldsymbol{b} + \mu\boldsymbol{\lambda}^i))\|_2^2 \\ &= \langle \boldsymbol{x} - \boldsymbol{x}^i, A^\top(A\boldsymbol{x}^i + B\boldsymbol{y}^i - \boldsymbol{b} + \mu\boldsymbol{\lambda}^i)\rangle + \frac{1}{2\tau_1}\|\boldsymbol{x} - \boldsymbol{x}^i\|_2^2 + c, \end{aligned} \tag{A.35}$$

where c is a constant. Notice that this term can be interpreted to approximate the original quadratic term with its Taylor expansion at the previous iteration point x^i up to the second-order term, where $A^\top(A\boldsymbol{x}^i + B\boldsymbol{y}^i - \boldsymbol{b} + \mu\boldsymbol{\lambda}^i)$ is the gradient of the quadratic term at $\boldsymbol{x}^i$, but the Hessian $A^\top A$ is approximated with a constant $1/\tau_1$. To ensure that the approximation is an upper bound of the original function, we want $\tau_1 < 1/\lambda_{\max}(A^\top A)$. If we do the same for the subproblem for updating $\boldsymbol{y}$, then the ADMM iteration scheme can be replaced by the so-called *alternating proximal gradient minimization* (APGM) scheme:

$$\begin{aligned} \boldsymbol{x}^{i+1} &= \arg\min_{\boldsymbol{x}} f(\boldsymbol{x}) + \frac{1}{2\mu\tau_1}\|\boldsymbol{x} - (\boldsymbol{x}^i - \tau_1 A^\top(A\boldsymbol{x}^i + B\boldsymbol{y}^i - \boldsymbol{b} + \mu\boldsymbol{\lambda}^i))\|_2^2, \\ \boldsymbol{y}^{i+1} &= \arg\min_{\boldsymbol{y}} g(\boldsymbol{y}) + \frac{1}{2\mu\tau_2}\|\boldsymbol{y} - (\boldsymbol{y}^i - \tau_2 B^\top(A\boldsymbol{x}^{i+1} + B\boldsymbol{y}^i - \boldsymbol{b} + \mu\boldsymbol{\lambda}^i))\|_2^2, \\ \boldsymbol{\lambda}^{i+1} &= \boldsymbol{\lambda}^i + (A\boldsymbol{x}^{i+1} + B\boldsymbol{y}^{i+1} - \boldsymbol{b})/\mu. \end{aligned} \tag{A.36}$$

Could the approximation affect the convergence of the ADMM method? The following result ensures that the global convergence remains intact if ADMM is replaced with APGM.

Proposition A.15 (Convergence of ADMM with Proximal Gradient (Ma 2012)). *For $\tau_1 < 1/\lambda_{\max}(A^\top A)$ and $\tau_2 < 1/\lambda_{\max}(B^\top B)$, the sequence $\{(\boldsymbol{x}^i, \boldsymbol{y}^i, \lambda^i)\}$ produced by the above APGM scheme* (A.36) *converges to the global optimal solution of problem* (A.30).

This result is very useful. Although it is established only for the two-term problem, it essentially offers an effective solution for the multiterm problem (A.28): we can always partition the N terms into two blocks and apply the APGM scheme. The convergence is ensured. Of course, in practice, the speed of convergence could be different for different partitions.

A.3 Exercises

Exercise A.1 Show that a necessary condition for a point $\boldsymbol{x}^*$ to be a local minimum of a differentiable function f is that the gradient $\nabla f(\boldsymbol{x})$ vanish at $\boldsymbol{x}^*$, i.e., $\nabla f(\boldsymbol{x}^*) = 0$.

Exercise A.2 Show that:

1. The intersection of two convex sets is convex.
2. The convex hull of a set is convex.
3. The convex hull of a convex set is the set itself.

Exercise A.3 Let $f : \mathcal{X} \to \mathbb{R}$ be a convex function defined over a convex domain $\mathcal{X} \subseteq \mathbb{R}^n$. Show that:

1. For every $c \in \mathbb{R}$, the set $\{\boldsymbol{x} \in \mathcal{X} : f(\boldsymbol{x}) \le c\}$ is convex.
2. If $\mathcal{X}$ is compact, then f reaches its maximum value at the boundary of $\mathcal{X}$, i.e., $\max_{\boldsymbol{x} \in \mathcal{X}} f(\boldsymbol{x}) = \max_{\boldsymbol{x} \in \partial \mathcal{X}} f(\boldsymbol{x})$.
3. f is pseudoconvex.
4. f is quasiconvex.

Appendix B
Basic Facts from Mathematical Statistics

A knowledge of statistics is like a knowledge of foreign languages or of algebra; it may prove of use at any time under any circumstances.

—A.L. Bowley

In the practice of science and engineering, data are often modeled as samples of a random variable (or vector) drawn from a certain probability distribution. Mathematical statistics deals with the problem of inferring the underlying distribution from the given samples. To render the problem tractable, we typically assume that the unknown distribution belongs to some parametric family (e.g., Gaussian), and formulate the problem as one of estimating the parameters of the distribution from the samples.

In this appendix, we provide a brief review of some of the most relevant concepts and results from mathematical statistics used in this book. The review is not meant to be exhaustive, but rather to make the book self-contained for readers who already have some basic knowledge in probability theory and statistics. For a more formal and thorough introduction to mathematical statistics, we refer the reader to the classic books (Wilks 1962) and (Bickel and Doksum 2000).

B.1 Estimation of Parametric Models

Assume that you are given independent and identically distributed (i.i.d.) samples from an unknown parametric distribution from which you wish to estimate some properties of the distribution. In this section, we show how to estimate the parameters of the distribution, such as the mean and variance, from the i.i.d. samples.

R. Vidal et al., *Generalized Principal Component Analysis*, Interdisciplinary Applied Mathematics 40, DOI 10.1007/978-0-387-87811-9

We study different types of estimators, such as minimum variance and maximum likelihood estimators, and their properties, such as unbiasedness, efficiency, and consistency.

B.1.1 Sufficient Statistics

Let $\boldsymbol{x}$ be a random variable or vector. For simplicity, we assume that the distribution of $\boldsymbol{x}$ has a density $p_\theta(\boldsymbol{x})$, where the parameter vector $\theta = [\theta_1, \theta_2, \ldots, \theta_d]^\top \in \Theta \subset \mathbb{R}^d$, once known, uniquely determines the density function $p_\theta(\cdot)$. Now suppose that $\mathcal{X} = \{\boldsymbol{x}_j\}_{j=1}^N$ is a set of i.i.d. samples of $\boldsymbol{x}$ drawn according to the density $p_\theta(\boldsymbol{x})$. Then $\mathcal{X}$ has the density

$$p_\theta(\mathcal{X}) = \prod_{j=1}^N p_\theta(\boldsymbol{x}_j). \tag{B.1}$$

We call any real or vector-valued function of $\mathcal{X}$ a *statistic* and denote it by $T(\mathcal{X})$. The goal is to choose a function $T(\cdot)$ that gives a "good" estimate of the true parameter θ. To that end, we introduce the concept of *sufficient statistics*.

Definition B.1 (Sufficient Statistic). *A statistic $T(\mathcal{X})$ is said to be* sufficient *for θ if the conditional distribution of $\mathcal{X}$ given $T(\mathcal{X})$, $p_\theta(\mathcal{X} \mid T(\mathcal{X}))$ is not a function of θ.*

Intuitively, a sufficient statistic $T(\mathcal{X})$ with respect to θ is a statistic that contains all the information that is useful to estimate θ. In other words, we can throw away the given samples and estimate θ from $T(\mathcal{X})$ without any loss of information. Unfortunately, the above definition is not very useful for finding sufficient statistics. Instead, one typically resorts to the following factorization theorem.

Theorem B.2 (Fisher–Neyman). *A statistic $T(\mathcal{X})$ is sufficient for θ if and only if there exist a function $g(t, \theta)$ and a function $h(\mathcal{X})$ such that*

$$p_\theta(\mathcal{X}) = g(T(\mathcal{X}), \theta)h(\mathcal{X}). \tag{B.2}$$

Example B.3 (Sufficient Statistic of a Gaussian Random Variable). For Gaussian data $\boldsymbol{x}_j \sim \mathcal{N}(\boldsymbol{\mu}, \Sigma)$, where $\boldsymbol{x}_j \in \mathbb{R}^D$, $\boldsymbol{\mu} \in \mathbb{R}^D$ and $\Sigma \in \mathbb{R}^{D\times D}$, the statistic $T(\mathcal{X}) = (\sum_{j=1}^N \boldsymbol{x}_j, \sum_{j=1}^N \boldsymbol{x}_j\boldsymbol{x}_j^\top)$ is a sufficient statistic for $\theta = (\boldsymbol{\mu}, \Sigma)$, because

$$\begin{aligned} p_\theta(\mathcal{X}) &= \prod_{j=1}^N \frac{1}{(2\pi)^{D/2}\det(\Sigma)^{1/2}} \exp(-\frac{(\boldsymbol{x}_j - \boldsymbol{\mu})^\top \Sigma^{-1}(\boldsymbol{x}_j - \boldsymbol{\mu})}{2}) \\ &= \frac{\exp\big(-\frac{1}{2}(\operatorname{trace}\big(\Sigma^{-1}\sum_{j=1}^N \boldsymbol{x}_j\boldsymbol{x}_j^\top\big) - 2\boldsymbol{\mu}^\top\Sigma^{-1}\sum_{j=1}^N \boldsymbol{x}_j + \boldsymbol{\mu}^\top\boldsymbol{\mu})\big)}{(2\pi)^{ND/2}\det(\Sigma)^{N/2}} \\ &= g(T(\mathcal{X}), \theta) \cdot 1. \end{aligned} \tag{B.3}$$

B.1.2 Mean Square Error, Efficiency, and Fisher Information

Notice that sufficient statistics are not unique. For instance, $T(\mathcal{X}) = \mathcal{X}$ is a sufficient statistic, and every one-to-one function of a sufficient statistic is also a sufficient statistic. Therefore, it is important to devise some criteria for choosing a "good" sufficient statistic.

A popular measure of "goodness" of a statistic $T(\mathcal{X}) \in \mathbb{R}^d$ as an estimate of $\theta \in \mathbb{R}^d$ is the *mean squared error* (MSE) between $T(\mathcal{X})$ and θ:

$$R(\theta, T) = \mathbb{E}_\theta[\|T(\mathcal{X}) - \theta\|^2]. \tag{B.4}$$

In some literature, such a function is also referred to as the "risk function," whence the capital letter R. Notice that the expression $R(\theta, T)$ can be rewritten as

$$\begin{aligned} R(\theta, T) &= \mathbb{E}_\theta[\|T(\mathcal{X}) - \mathbb{E}_\theta[T(\mathcal{X})] + \mathbb{E}_\theta[T(\mathcal{X})] - \theta\|^2] \\ &= \mathbb{E}_\theta[\|T(\mathcal{X}) - \mathbb{E}_\theta[T(\mathcal{X})]\|^2] + \|\mathbb{E}_\theta[T(\mathcal{X})] - \theta\|^2 \\ &\doteq \mathrm{Var}_\theta(T(\mathcal{X})) + \|\boldsymbol{b}_\theta(T(\mathcal{X}))\|^2, \end{aligned} \tag{B.5}$$

where $\boldsymbol{b}_\theta(T(\mathcal{X})) = \mathbb{E}_\theta[T(\mathcal{X})] - \theta$ is called the *bias* of the estimate $T(\mathcal{X})$, and $\mathrm{Var}_\theta(T(\mathcal{X})) \in \mathbb{R}$ is the trace of the covariance matrix

$$\mathrm{Cov}_\theta(T(\mathcal{X})) \doteq \mathbb{E}_\theta[(T(\mathcal{X}) - \mathbb{E}_\theta[T(\mathcal{X})])(T(\mathcal{X}) - \mathbb{E}_\theta[T(\mathcal{X})])^\top] \in \mathbb{R}^{d\times d}. \tag{B.6}$$

We refer to $\mathrm{Var}_\theta(T(\mathcal{X}))$ as the "variance" of $T(\mathcal{X})$. Thus, a good estimate is one that has both small bias and small variance.

Example B.4 For Gaussian data $\boldsymbol{x}_j \sim \mathcal{N}(\boldsymbol{\mu}, \Sigma)$, where $\boldsymbol{x}_j \in \mathbb{R}^D$, $\boldsymbol{\mu} \in \mathbb{R}^D$ and $\Sigma \in \mathbb{R}^{D\times D}$, the statistic $T(\mathcal{X}) = \frac{1}{N}\sum_{j=1}^N \boldsymbol{x}_j$ is an unbiased estimator of $\boldsymbol{\mu}$, because

$$\mathbb{E}_\theta[T(\mathcal{X})] = \frac{1}{N}\sum_{j=1}^N \mathbb{E}_\theta[\boldsymbol{x}_j] = \frac{1}{N}N\boldsymbol{\mu} = \boldsymbol{\mu}. \tag{B.7}$$

We can use the MSE to compare two estimators. We define the *relative efficiency* of two estimators T_1 and T_2 as the ratio

$$\nu_{1,2}(\theta) \doteq \frac{R(\theta, T_2)}{R(\theta, T_1)}. \tag{B.8}$$

The larger the relative efficiency ν_{12}, the smaller the MSE of T_1 relative to that of T_2. Thus, T_1 gives a more accurate, or "sharper," estimate for θ.

Notice that in general, the relative efficiency is a function of θ. Therefore, one estimator could have lower MSE for some values of θ, and another estimator could have lower MSE for other values of θ. In fact, there is no such thing as a universally optimal estimator that gives an error smaller than that of any other estimator for all θ. For instance, if the true parameter is θ_0, the estimator $S(\mathcal{X}) = \theta_0$ achieves the smallest possible error $R(\theta, S) = 0$. Thus, the universally optimal estimate, say T, would need to have $R(\theta_0, T) = 0$, too. Since θ_0 can be arbitrary, T would need to estimate every potential parameter θ perfectly, which is impossible except for trivial cases. One can view this as a manifestation of the so-called *no free lunch theorem* known in learning theory: without any prior knowledge about θ, we can expect a statistical estimate to be better than others most of the time, but we can never expect it to be the best *all the time*. Thus, in the future, whenever we claim that some estimate is "optimal," the claim will be in the restricted sense that it is optimal within a special class of estimates considered (e.g., unbiased estimates).

In the case of unbiased estimators, the MSE reduces to the variance. Therefore, we can compare two estimators by comparing their variances. Theorem B.5 gives a lower bound on the variance of an estimator, which allows us to evaluate the efficiency of an estimator by comparing its variance to this lower bound (see Definition B.6). Before stating the theorem, we need to introduce some notation.

Assume that $p_\theta(\boldsymbol{x})$ is differentiable with respect to θ and define the *Fisher information matrix* as

$$I(\theta) \doteq \mathbb{E}_\theta\Big[\big(\frac{\partial}{\partial\theta}\log p_\theta(\mathcal{X})\big)\big(\frac{\partial}{\partial\theta}\log p_\theta(\mathcal{X})\big)^\top\Big] \in \mathbb{R}^{d\times d}. \tag{B.9}$$

Also, assume that the function $\psi(\theta) \doteq \mathbb{E}_\theta[T(\mathcal{X})] = [\psi_1(\theta), \psi_2(\theta), \ldots, \psi_d(\theta)]^\top$ is differentiable with respect to θ and define

$$\frac{\partial\psi(\theta)}{\partial\theta} \doteq \begin{bmatrix} \frac{\partial\psi_1(\theta)}{\partial\theta_1} & \frac{\partial\psi_1(\theta)}{\partial\theta_2} & \cdots & \frac{\partial\psi_1(\theta)}{\partial\theta_d} \\ \frac{\partial\psi_2(\theta)}{\partial\theta_1} & \frac{\partial\psi_2(\theta)}{\partial\theta_2} & \cdots & \frac{\partial\psi_2(\theta)}{\partial\theta_d} \\ \vdots & \vdots & \ddots & \vdots \\ \frac{\partial\psi_d(\theta)}{\partial\theta_1} & \frac{\partial\psi_d(\theta)}{\partial\theta_2} & \cdots & \frac{\partial\psi_d(\theta)}{\partial\theta_d} \end{bmatrix} \in \mathbb{R}^{d\times d}. \tag{B.10}$$

We have the following result.

Theorem B.5 (Cramér–Rao Lower Bound). *Let $T(\mathcal{X})$ be an estimator for θ and assume that the following regularity conditions on the density p_θ and the estimator $T(\mathcal{X})$ hold:*

1. *The information matrix is well defined. That is, for all $\mathcal{X}$ such that $p_\theta(\mathcal{X}) > 0$, $\frac{\partial}{\partial\theta}\ln p_\theta(\mathcal{X})$ exists and is finite.*
2. *The operations of integration with respect to $\mathcal{X}$ and differentiation with respect to θ commute, i.e.,*

$$\frac{\partial}{\partial\theta}\int T(\mathcal{X})p_\theta(\mathcal{X})d\mathcal{X} = \int T(\mathcal{X})\frac{\partial}{\partial\theta}p_\theta(\mathcal{X})d\mathcal{X}. \tag{B.11}$$

3. *For all θ, $\psi(\theta)$ is differentiable.*

We have that for all θ,

$$Cov_\theta(T(\mathcal{X})) \geq \frac{\partial \psi(\theta)}{\partial \theta} I(\theta)^{-1} \Big(\frac{\partial \psi(\theta)}{\partial \theta}\Big)^\top, \tag{B.12}$$

where the inequality is between positive semidefinite symmetric matrices.

In the case of an unbiased estimator we have $\psi(\theta) = \theta$ and $\psi'(\theta) = I$. Therefore, the information inequality gives the following lower bound for the variance of an unbiased estimate: $\mathrm{Cov}(T(\mathcal{X})) \geq I(\theta)^{-1}$. This bound is often referred to as the *Cramér–Rao lower bound*. Since $\mathcal{X} = \{\boldsymbol{x}_j\}_{j=1}^N$ are i.i.d. samples from $p_\theta(\boldsymbol{x})$, if we define $I_1(\theta) \doteq \mathbb{E}_\theta\big[\frac{\partial}{\partial\theta}\log p_\theta(\boldsymbol{x}_1)(\frac{\partial}{\partial\theta}\log p_\theta(\boldsymbol{x}_1))^\top\big] \in \mathbb{R}^{d\times d}$, we obtain

$$I(\theta) = NI_1(\theta). \tag{B.13}$$

Thus, the Cramér–Rao lower bound can be rewritten as $\mathrm{Cov}_\theta(T(\mathcal{X})) \geq \frac{1}{N} I_1(\theta)^{-1}$.

Definition B.6 (Efficiency). *We define the efficiency of an unbiased estimator $T(\mathcal{X})$ as*

$$\nu(\theta) = \frac{\mathrm{trace}\left(I^{-1}(\theta)\right)}{Var_\theta(T(\mathcal{X}))} = \frac{1}{N}\frac{\mathrm{trace}\left(I_1^{-1}(\theta)\right)}{Var_\theta(T(\mathcal{X}))}. \tag{B.14}$$

An unbiased estimator $T(\mathcal{X})$ is called efficient *if it achieves the Cramér–Rao lower bound, i.e., if $\nu(\theta) = 1$ for all θ.*

Next, we describe a procedure for finding an efficient estimator whenever possible.

B.1.3 The Rao–Blackwell Theorem and Uniformly Minimum-Variance Unbiased Estimator

To find a good estimate for θ in the MSE sense, we can resort to the Rao–Blackwell theorem. This theorem allows us to take an arbitrary estimate $S(\mathcal{X})$ of θ and produce a new estimate $S^*(\mathcal{X})$ whose MSE is at least as good as that of $S(\mathcal{X})$.

Theorem B.7 (Rao–Blackwell). *If $T(\mathcal{X})$ is a sufficient statistic for θ and $S(\mathcal{X})$ is any estimate of θ, then $\tilde{S}(\mathcal{X}) = \mathbb{E}_\theta[S(\mathcal{X}) \mid T(\mathcal{X})]$ is such that*

$$\forall \theta \; R(\theta, \tilde{S}) \leq R(\theta, S). \tag{B.15}$$

The above procedure for transforming an estimator using the Rao–Blackwell theorem is often called Rao–Blackwellization. This procedure can significantly improve the estimate of θ. However, it is not guaranteed to produce an optimal estimate of θ in the MSE sense.

As we mentioned earlier, to make the estimation problem well conditioned, we must restrict the class of estimates. For instance, we may require the estimate $S(\mathcal{X})$ to be unbiased, i.e., $b_\theta(S(\mathcal{X})) = 0$. Then the problem of finding the best unbiased estimate becomes

$$\min_{S(\cdot)} R(\theta, S) = \mathrm{Var}_\theta(S(\mathcal{X})) \quad \text{s.t.} \quad \mathbb{E}_\theta[S(\mathcal{X})] = \theta. \tag{B.16}$$

The optimal $S^*(\mathcal{X})$, if it exists, is called the *uniformly minimum variance unbiased* (UMVU) estimate. In general, an unbiased estimator of θ need not exist, and so $S^*(\mathcal{X})$ is not always well defined. However, if an unbiased estimator of θ does exist, then so does $S^*(\mathcal{X})$. Moreover, if the sufficient statistic $T(\mathcal{X})$ is complete, as defined next, then $S^*(\mathcal{X})$ is unique and can be found by Rao–Blackwellization.

Definition B.8 (Complete Statistic). *A statistic T is said to be complete if for every real-valued function $g(\cdot)$ such that $\mathbb{E}_\theta[g(T(\mathcal{X}))] = 0$ for all θ, we have that $p_\theta(g(T(\mathcal{X})) = 0) = 1$ for all θ.*

Starting with an unbiased estimate $S(\mathcal{X})$ and a sufficient and complete statistic $T(\mathcal{X})$, the following theorem simplifies the computation of the UMVU estimate.

Theorem B.9 (Lehmann–Scheffé). *If $T(\mathcal{X})$ is a complete sufficient statistic and $S(\mathcal{X})$ is any unbiased estimate of θ, then $S^*(\mathcal{X}) = \mathbb{E}_\theta[S(\mathcal{X}) \mid T(\mathcal{X})]$ is an UMVU estimate of θ. If further, $\mathrm{Var}_\theta(S^*(\mathcal{X})) < \infty$ for all θ, then $S^*(\mathcal{X})$ is the unique UMVU estimator.*

While the above procedure gives us an optimal unbiased estimate in the MSE sense, the UMVU estimate is often too difficult to compute in practice. Furthermore, the property of unbiasedness is not invariant under functional transformation: if $T(\mathcal{X})$ is an unbiased estimate for θ, then $g(T(\mathcal{X}))$ is in general not an unbiased estimate for $g(\theta)$. To have the functional invariant property, we often resort to the so-called maximum likelihood estimator, as described next.

B.1.4 Maximum Likelihood (ML) Estimator

Recall that the joint distribution of the N i.i.d. samples $\{\boldsymbol{x}_j\}_{j=1}^N$ has the density $p_\theta(\mathcal{X}) = \prod_{j=1}^N p_\theta(\boldsymbol{x}_j)$, and consider this density a function of θ with $\mathcal{X}$ fixed. We call this function the *likelihood function* and denote it by $L(\theta, \mathcal{X}) = p_\theta(\mathcal{X})$. The *maximum likelihood (ML) estimate* of θ, it if exists, is given by the solution to the following optimization problem:

$$\hat{\theta}_N = \arg\max_{\theta \in \Theta} \Big(L(\theta, \mathcal{X}) = p_\theta(\mathcal{X}) = \prod_{j=1}^N p_\theta(\boldsymbol{x}_j) \Big). \tag{B.17}$$

where Θ is the space of parameters. Since the logarithmic function is monotonic, we may choose to maximize the log-likelihood function instead:

$$\hat{\theta}_N = \arg\max_{\theta\in\Theta} \Big(\ell(\theta, \mathcal{X}) = \log(L(\theta, \mathcal{X})) = \sum_{j=1}^{N} \log p_\theta(\boldsymbol{x}_j)\Big), \tag{B.18}$$

which often turns out to be more convenient to use in practice. Thus, a necessary condition for the optimality of $\hat{\theta}_N$ is that

$$\frac{\partial \ell(\theta, \mathcal{X})}{\partial \theta}\Big|_{\hat{\theta}_N} = 0. \tag{B.19}$$

The ML estimate is a more popular choice than the UMVU estimate, because its existence is easier to establish, and it is usually easier to compute than the UMVU estimate. Moreover, the ML estimate is invariant under functional transformations. That is, if $\hat{\theta}_N$ is an ML estimate of θ, then $g(\hat{\theta}_N)$ is an ML estimate of $g(\theta)$. Furthermore, when the sample size is large, the ML estimate is asymptotically optimal for a wide variety of parametric models. Thus, both UMVU and ML estimates give essentially the same answer, as explained next in more detail.

B.1.5 Consistency and Asymptotic Efficiency of the ML Estimator

In general, we would like an estimate $\hat{\theta}_N$ obtained from N samples $\{\boldsymbol{x}_j\}_{j=1}^N$ to perform better and better as the number of samples increases. In this section, we characterize the asymptotic properties of an estimator. To do so, we need to make a number of technical assumptions.

Assumption B.10 *Assume that the space of parameters Θ is compact and that the density $p_\theta(\boldsymbol{x})$ is continuous and twice differentiable in θ for all $\boldsymbol{x}$ and identifiable, i.e., $p_\theta = p_{\theta_0} \iff \theta = \theta_0$. Assume also that there exists a function $K(\boldsymbol{x})$ such that $\mathbb{E}_{\theta_0}[K(\boldsymbol{x})] < \infty$ and $\log p_\theta(\boldsymbol{x}) - \log p_{\theta_0}(\boldsymbol{x}) \le K(\boldsymbol{x})$ for all $\boldsymbol{x}$ and θ.*

Given these assumptions, a first approach to characterizing the asymptotic behavior of an estimator is through the notion of *consistency*.

Definition B.11 (Consistency). *An estimate $\hat{\theta}_N$ of θ is said to be consistent if it converges in probability to θ ($\hat{\theta}_N \to \theta$), i.e.,*

$$\lim_{N\to\infty} P\big[\|\hat{\theta}_N - \theta\| \ge \varepsilon\big] = 0, \quad \forall \varepsilon > 0. \tag{B.20}$$

The following classical result from statistics characterizes the consistency of the ML estimator.

Proposition B.12. *Let $\{x_j\}_{j=1}^N$ be i.i.d. samples from $p_{\theta_0}(\mathbf{x})$. Under the regularity assumptions in B.10, every sequence of ML estimates $\hat{\theta}_N$ converges to θ_0 in probability. In other words, every maximum likelihood estimate is consistent.*

A second approach to characterizing the asymptotic behavior of an estimator is through the notion of *asymptotic unbiasedness.*

Definition B.13 (Asymptotic Unbiasedness). *Let $\boldsymbol{\mu}_N = \mathbb{E}_\theta[\hat{\theta}_N] \in \mathbb{R}^d$ and $\Sigma_N = Cov_\theta(\hat{\theta}_N) \in \mathbb{R}^{d\times d}$. We say that an estimate $\hat{\theta}_N$ of θ is asymptotically unbiased if*

$$\lim_{N\to\infty} \sqrt{N}(\boldsymbol{\mu}_N - \theta) = 0, \quad \textit{and} \quad \lim_{N\to\infty} N\Sigma_N = \Sigma > 0 \tag{B.21}$$

for some positive definite symmetric matrix $\Sigma \in \mathbb{R}^{d\times d}$.

It is easy to see that asymptotic unbiasedness is a stronger property than consistency. That is, an estimate can be consistent but asymptotically biased. In addition, most "reasonable" estimates $\hat{\theta}_N$ (e.g., the ML estimate) are often asymptotically normally distributed with mean $\boldsymbol{\mu}_N$ and covariance matrix Σ_N due to the central limit theorem. Therefore, the asymptotic distribution of an asymptotically unbiased estimate is uniquely characterized by the parameters θ and Σ.

A third way to characterize the asymptotic behavior of an estimator is through the notion of *asymptotic efficiency*. Given two asymptotically unbiased estimates, say $\hat{\theta}_N^{(1)}$ and $\hat{\theta}_N^{(2)}$, their relative *asymptotic efficiency* is defined as the ratio

$$\nu_{1,2}(\theta) \doteq \frac{\det(\Sigma^{(2)})}{\det(\Sigma^{(1)})}, \tag{B.22}$$

where $\Sigma^{(i)} = \lim_{N\to\infty} N\mathrm{Cov}_\theta\big(\hat{\theta}_N^{(i)}\big)$, for $i = 1, 2$. The larger the efficiency ratio $\nu_{1,2}$, the smaller the asymptotic variance of $\hat{\theta}^{(1)}$, relative to that of $\hat{\theta}^{(2)}$. Thus, $\hat{\theta}^{(1)}$ gives a more accurate or "sharper" estimate for θ, although both $\hat{\theta}^{(1)}$ and $\hat{\theta}^{(2)}$ are asymptotically unbiased. Nevertheless, according to Theorem B.5, an estimate cannot be arbitrarily more efficient than others. That is, for every asymptotically unbiased estimate $\hat{\theta}_N$, using (B.13) and (B.21), its covariance matrix is bounded asymptotically from below by the Cramér–Rao bound:

$$\lim_{N\to\infty} N\Sigma_N = \Sigma \geq I_1(\theta)^{-1}. \tag{B.23}$$

Definition B.14 (Asymptotic Efficiency). *An estimate $\hat{\theta}_N$ is said to be asymptotically efficient if it is asymptotically normal and achieves equality in the Cramér–Rao bound* (B.23).

Asymptotic efficiency is a desirable property for an estimate, and it is sometimes referred to as asymptotic optimality. It often can be shown that UMVU estimates are asymptotically efficient. We also have the following result.

Proposition B.15. *Let $\{x_j\}_{j=1}^N$ be i.i.d. samples from $p_{\theta_0}(x)$. Assume that the regularity conditions in B.10 hold and that the Fisher information matrix $I_1(\theta_0)$ is positive definite. Then there is a consistent sequence of ML estimators $\hat{\theta}_N$ such that $\sqrt{N}(\hat{\theta}_N - \theta_0)$ converges in distribution to $\mathcal{N}(0, I_1(\theta_0)^{-1})$. In other words, the sequence $\hat{\theta}_N$ is asymptotically unbiased and asymptotically efficient.*

Proof. We here outline the basic ideas for a "proof," which can also be used to establish for other estimates their asymptotic unbiasedness or efficiency with respect to the ML estimate. Define the function

$$\psi(x, \theta) \doteq \frac{\partial}{\partial \theta} \log p_\theta(x) \in \mathbb{R}^d. \tag{B.24}$$

If the maximum likelihood estimate $\hat{\theta}_N$ exists, it must satisfy the equation

$$\left.\frac{\partial \ell(\theta, \mathcal{X})}{\partial \theta}\right|_{\hat{\theta}_N} = \sum_{j=1}^N \psi(x_j, \hat{\theta}_N) = 0. \tag{B.25}$$

By the mean value theorem, we have

$$\sum_{j=1}^N \psi(x_j, \hat{\theta}_N) - \sum_{j=1}^N \psi(x_j, \theta) = \Big[\sum_{j=1}^N \frac{\partial \psi(x_j, \theta_N^*)}{\partial \theta}\Big]\Big(\hat{\theta}_N - \theta\Big), \tag{B.26}$$

where θ_N^* is a point between θ and $\hat{\theta}_N$. Using (B.25), we obtain

$$\sqrt{N}(\hat{\theta}_N - \theta) = \Big[\frac{1}{N}\sum_{j=1}^N \frac{\partial \psi(x_j, \theta_N^*)}{\partial \theta}\Big]^{-1}\Big(-N^{-\frac{1}{2}}\sum_{j=1}^N \psi(x_j, \theta)\Big). \tag{B.27}$$

Now, it follows from Proposition B.12 that $\hat{\theta}_N$ is consistent. This implies that $\lim_{N\to\infty} \frac{1}{N}\sum_{j=1}^N \frac{\partial \psi(x_j, \theta_N^*)}{\partial \theta} = \lim_{N\to\infty} \frac{1}{N}\sum_{j=1}^N \frac{\partial \psi(x_j, \theta)}{\partial \theta}$. By the law of large numbers, the last limit is equal to

$$\begin{aligned}
\mathbb{E}_\theta\Big[\frac{\partial \psi(x_1, \theta)}{\partial \theta}\Big] &= \mathbb{E}_\theta\Big[\frac{\partial^2}{\partial \theta^2} \log p_\theta(x_1)\Big] = \int \frac{\partial}{\partial \theta}\Big(\frac{\frac{\partial}{\partial \theta} p_\theta(x_1)}{p_\theta(x_1)}\Big) p_\theta(x_1) \\
&= \int \frac{p_\theta(x_1)\frac{\partial^2}{\partial \theta^2} p_\theta(x_1) - \frac{\partial}{\partial \theta} p_\theta(x_1)(\frac{\partial}{\partial \theta} p_\theta(x_1))^\top}{p_\theta(x_1)^2} p_\theta(x_1) \\
&= \frac{\partial^2}{\partial \theta^2}\int p_\theta(x_1) - \int \frac{\partial}{\partial \theta} \log p_\theta(x_1)(\frac{\partial}{\partial \theta} \log p_\theta(x_1))^\top p_\theta(x_1) \\
&= -\mathbb{E}_\theta\Big[\frac{\partial}{\partial \theta} \log p_\theta(x_1)\big(\frac{\partial}{\partial \theta} \log p_\theta(x_1)\big)^\top\Big] = -I_1(\theta).
\end{aligned}$$

The remaining term in (B.27) involves the sum of the random vectors $\frac{\partial}{\partial\theta}\log p_\theta(\boldsymbol{x}_j)$. These vectors are i.i.d. with mean $\mathbb{E}_\theta[\frac{\partial}{\partial\theta}\log p_\theta(\boldsymbol{x}_j)] = \int \frac{\partial}{\partial\theta} p_\theta(\boldsymbol{x}_j) = 0$ and covariance $\mathbb{E}_\theta[\frac{\partial}{\partial\theta}\log p_\theta(\boldsymbol{x}_j)(\frac{\partial}{\partial\theta}\log p_\theta(\boldsymbol{x}_j))^\top] = I_1(\theta)$. Thus, by the central limit theorem, the right-hand side of (B.27) converges in distribution to $\mathcal{N}(0, I_1(\theta)^{-1})$. That is, the ML estimate is asymptotically unbiased, and its asymptotic variance reaches the Cramér–Rao lower bound. □

When the sample size is large, one can appeal to the law of large numbers to derive an information-theoretic justification for the ML estimate, which can be somewhat more revealing. Notice that maximizing the log-likelihood function is equivalent to minimizing the following objective function:

$$\min_{\theta\in\Theta}\Big(H(\theta,N) \doteq \frac{1}{N}\sum_{j=1}^{N}\big(-\log p_\theta(\boldsymbol{x}_j)\big)\Big). \tag{B.28}$$

In information theory, the quantity $-\log p_\theta(\boldsymbol{x})$ is associated with the number of bits required to represent a random event $\boldsymbol{x}$ that has the probability $p_\theta(\boldsymbol{x})$ (Cover and Thomas 1991). When the sample size N is large, due to the law of large numbers, the quantity $H(\theta,N)$ converges to

$$\lim_{N\to\infty} H(\theta,N) = H(\theta) = \mathbb{E}_{\theta_0}[-\log p_\theta(\boldsymbol{x})] = \int \big(-\log p_\theta(\boldsymbol{x})\big)p_{\theta_0}(\boldsymbol{x})\,d\boldsymbol{x}, \tag{B.29}$$

where $p_{\theta_0}(\boldsymbol{x})$ is the true distribution. Notice that the above quantity is a measure similar to the notion of "entropy": $H(\theta)$ is asymptotically the average code length of the sample set $\{\boldsymbol{x}_j\}$ when we assume that it is of the distribution $p_\theta(\boldsymbol{x})$, while $\boldsymbol{x}$ is actually drawn according to $p_{\theta_0}(\boldsymbol{x})$. Thus, the goal of ML estimation is to find the $\hat{\theta}$ that minimizes the empirical entropy of the given sample set. This is obviously a smart thing to do, since such an estimate $\hat{\theta}$ gives the most compact representation of the given sample data if an optimal coding scheme is adopted (Cover and Thomas 1991). We refer to this as the "minimum entropy principle."

Notice also that the $\hat{\theta}$ that minimizes $\int\big(-\log p_\theta(\boldsymbol{x})\big)p_{\theta_0}(\boldsymbol{x})\,d\boldsymbol{x}$ is the same as that which minimizes the so-called *Kullback–Leibler (KL) divergence* between the two distributions $p_{\theta_0}(\boldsymbol{x})$ and $p_\theta(\boldsymbol{x})$, i.e.,

$$KL\big(p_{\theta_0}(\boldsymbol{x})\,||\,p_\theta(\boldsymbol{x})\big) \doteq \int \log\Big(\frac{p_{\theta_0}(\boldsymbol{x})}{p_\theta(\boldsymbol{x})}\Big)p_{\theta_0}(\boldsymbol{x})\,d\boldsymbol{x}. \tag{B.30}$$

One may show that under general conditions, the KL divergence is always nonnegative and becomes zero if and only if $\theta = \theta_0$. In essence, when the sample size is large, the ML objective is equivalent to minimizing the KL divergence.

However, the ML estimate is known to have very bad performance in some models even with a large number of samples. This is particularly the case when the models have many redundant parameters or the distributions are degenerate. Furthermore, both UMVU and ML estimates are not the optimal estimates in a

Bayesian[1] or minimax[2] sense. For instance, the ML estimate can be viewed as a special Bayesian estimate only when the parameter θ is uniformly distributed.

B.2 ML Estimation for Models with Latent Variables

In many practical situations, we need to estimate a statistical model in which only part of the random variables or vectors are observed, and the rest are "missing," or "hidden," or "latent," or "unobserved." For instance, suppose that two random vectors $(\boldsymbol{x}, z)$ have a joint distribution with density $p_\theta(\boldsymbol{x}, z)$, but only samples of $\boldsymbol{x}$, $\mathcal{X} = \{\boldsymbol{x}_j\}_{j=1}^N$, are observed, while the corresponding samples of z, $\mathcal{Z} = \{z_j\}_{j=1}^N$, are not available. As before, we wish to find an optimal estimate $\hat{\theta}$ for θ from the observations.

Since samples of z are not available, there is no way one can find the maximum likelihood estimate of θ from the *complete log-likelihood function*:

$$\ell_c(\theta, \mathcal{X}, \mathcal{Z}) = \sum_{j=1}^N \log p_\theta(\boldsymbol{x}_j, z_j). \tag{B.31}$$

Instead, it makes sense to use the marginal distribution of $\boldsymbol{x}$, $p_\theta(\boldsymbol{x})$, and find the maximum likelihood estimate from the *incomplete log-likelihood function*[3]

$$\ell(\theta, \mathcal{X}) = \sum_{j=1}^N \log(p_\theta(\boldsymbol{x}_j)) = \sum_{j=1}^N \log\Big(\int p_\theta(\boldsymbol{x}_j, z)dz\Big). \tag{B.32}$$

The problem is now reduced to a standard ML estimation problem, and one can adopt any appropriate optimization method (say conjugate gradient) to find the maximum. Thus, it seems that there is no need to involve z at all.

In practice, however, there are several reasons why marginalizing over z may not be the best approach. First, for some models $p_\theta(\boldsymbol{x}, z)$, computing the marginal $p_\theta(\boldsymbol{x})$ can be intractable (e.g., summing over a combinatorial number of values for z), or it can destroy good structures in the models. Second, directly maximizing $\ell(\theta, \mathcal{X})$ may turn out to be a very difficult optimization problem (e.g., high-dimensional, having many local minima). Third, in some applications, it is desirable to obtain an estimate of the unobservables z from the observables $\boldsymbol{x}$.

---

[1]A Bayesian estimate T^* is the solution to the problem $\min_T \int R(\theta, T)\pi(\theta)\, d\theta$ for a given prior distribution $\pi(\theta)$ of θ. That is, T^* is the best estimate in terms of its average risk.

[2]A minimax estimate T^* is the solution to the problem $\min_T \max_\theta R(\theta, T)$. That is, T^* is the best estimate according to its worst performance. Of course, such a T^* does not have to always exist or be easier to compute than the ML estimate.

[3]In this section, we assume that z is a continuous variable. Whenever z is discrete, we can simply replace the integrals by sums, as we will do in the next section when we cover mixture models.

B.2.1 Expectation Maximization (EM)

An alternative approach to marginalizing over the hidden variables is to take the expectation over the hidden variables. More specifically, instead of maximizing the incomplete log-likelihood $\ell(\theta, \mathcal{X})$, we can estimate the conditional density of the hidden variables given the observations $\mathcal{X}$ and an estimate θ^k for the parameters, i.e., $p_{\theta^k}(\mathcal{Z} \mid \mathcal{X})$, and maximize the expected value of the complete log-likelihood $\ell_c(\theta, \mathcal{X}, \mathcal{Z})$ with respect to the distribution $p_{\theta^k}(\mathcal{Z} \mid \mathcal{X})$.

This alternative approach has several potential advantages. First, it provides an estimate for the density of $z \mid \boldsymbol{x}$, if needed. Second, the computation of the expected complete log-likelihood is often much simpler than the computation of the incomplete log-likelihood, as we will see. Third, the maximization of the expected log-likelihood is often much simpler than the maximization of the incomplete log-likelihood, as we will see.

In order to derive this alternative approach, let us recall the following identities:

$$\forall z \;\; p_\theta(\boldsymbol{x}) = \frac{p_\theta(\boldsymbol{x}, z)}{p_\theta(z \mid \boldsymbol{x})} \quad \text{and} \quad \forall \boldsymbol{x} \;\; \int p_\theta(z \mid \boldsymbol{x})\, dz = 1. \tag{B.33}$$

Using these identities, we can rewrite the incomplete log-likelihood as

$$\ell(\theta, \mathcal{X}) = \sum_{j=1}^{N} \log p_\theta(\boldsymbol{x}_j) = \sum_{j=1}^{N} \int p_\theta(z \mid \boldsymbol{x}_j) \log\left(\frac{p_\theta(\boldsymbol{x}_j, z)}{p_\theta(z \mid \boldsymbol{x}_j)}\right) dz \tag{B.34}$$

$$= \max_{w_j} \sum_{j=1}^{N} \int w_j(z) \log\left(\frac{p_\theta(\boldsymbol{x}_j, z)}{w_j(z)}\right) dz, \tag{B.35}$$

where $w_j(z)$ is a density, i.e., $w_j(z) \geq 0 \; \forall z$ and $\int w_j(z) dz = 1 \; \forall j = 1, \ldots, N$. To see the last step, we use the method of Lagrange multipliers. The Lagrangian function is

$$\mathcal{L}(w_j, \lambda) = \int w_j(z) \log\left(\frac{p_\theta(\boldsymbol{x}_j, z)}{w_j(z)}\right) dz + \lambda(1 - \int w_j(z) dz). \tag{B.36}$$

Setting the variation of $\mathcal{L}$ with respect to w_j to zero, we obtain[4]

$$\frac{\partial \mathcal{L}}{\partial w_j} = \log\left(\frac{p_\theta(\boldsymbol{x}_j, z)}{w_j(z)}\right) - 1 - \lambda = 0 \implies w_j^*(z) = p_\theta(\boldsymbol{x}_j, z) e^{-\lambda - 1}. \tag{B.37}$$

[4]Here w_j is a function of z, which is in general a continuous random variable. Therefore, we use the variation with respect to w_j in lieu of the derivative with respect to w_j. We can use the derivative, instead, whenever z is a discrete random variable.

Enforcing $\int w_j^*(z)dz = 1$, we obtain

$$w_j^*(z) = \frac{p_\theta(\boldsymbol{x}_j, z)}{\int p_\theta(\boldsymbol{x}_j, z)dz} = \frac{p_\theta(\boldsymbol{x}_j, z)}{p_\theta(\boldsymbol{x}_j)} = p_\theta(z \mid \boldsymbol{x}_j). \tag{B.38}$$

Thus, it follows from (B.34)–(B.35) that the maximization of $\ell(\theta, \mathcal{X})$ is equivalent to the following optimization problem:

$$\max_{\theta\in\Theta} \ell(\theta, \mathcal{X}) = \max_{\theta\in\Theta} \max_{\{w_j\}} \sum_{j=1}^N \int w_j(z) \log\left(\frac{p_\theta(\boldsymbol{x}_j, z)}{w_j(z)}\right) dz. \tag{B.39}$$

We solve the optimization problem on the right-hand side using an alternating maximization strategy (see Appendix A.1.5). Given θ, the optimal density $w_j(z)$ is given by $w_j^*(z) \doteq p_\theta(z \mid \boldsymbol{x}_j)$, which is the *a posteriori* density of z given $\boldsymbol{x}_j$ and θ. Given $w_j(z)$, the optimal parameter θ is given by

$$\begin{aligned}\theta^* &= \arg\max_{\theta\in\Theta} \sum_{j=1}^N \int w_j(z_j) \log p_\theta(\boldsymbol{x}_j, z_j) dz_j \qquad \text{(B.40)}\\ &= \arg\max_{\theta\in\Theta} \sum_{j=1}^N \mathbb{E}_{w_j}[\log(p_\theta(\boldsymbol{x}_j, z_j) \mid \boldsymbol{x}_j)] = \arg\max_{\theta\in\Theta} \mathbb{E}_w[\ell_c(\theta, \mathcal{X}, \mathcal{Z}) \mid \mathcal{X}],\end{aligned}$$

where the last expectation is taken with respect to the density $w(\mathcal{Z}) = \prod_{j=1}^N w_j(z_j) = p_\theta(\mathcal{Z} \mid \mathcal{X})$. Therefore, θ^* maximizes the expected complete log-likelihood taken with respect to the a posteriori density of the hidden variables given the observed ones. By alternating between these two steps, we obtain the well-known expectation maximization (EM) algorithm (Dempster et al. 1977) for maximizing the incomplete log-likelihood $\ell(\theta, \mathcal{X})$, which we summarize in Algorithm B.1.

Each iteration of this coordinate ascent algorithm does not decrease the value of the objective function in (B.39). Moreover, each iteration does not decrease the value of the incomplete log-likelihood because

$$\ell(\theta^{k+1}, \mathcal{X}) = \sum_{j=1}^N \int p_{\theta^{k+1}}(z \mid \boldsymbol{x}_j) \log \frac{p_{\theta^{k+1}}(\boldsymbol{x}_j, z)}{p_{\theta^{k+1}}(z \mid \boldsymbol{x}_j)} dz \tag{B.43}$$

$$\geq \sum_{j=1}^N \int p_{\theta^k}(z \mid \boldsymbol{x}_j) \log \frac{p_{\theta^{k+1}}(\boldsymbol{x}_j, z)}{p_{\theta^k}(z \mid \boldsymbol{x}_j)} dz \tag{B.44}$$

$$\geq \sum_{j=1}^N \int p_{\theta^k}(z \mid \boldsymbol{x}_j) \log \frac{p_{\theta^k}(\boldsymbol{x}_j, z)}{p_{\theta^k}(z \mid \boldsymbol{x}_j)} dz = \ell(\theta^k, \mathcal{X}). \tag{B.45}$$

Algorithm B.1 (Expectation Maximization)

Input: Data points $\{x_j\}_{j=1}^N$ and initial parameter vector θ^0.

$k \leftarrow 0$.

while not converged **do**

  E-step: For fixed $\theta = \theta^k$, solve for each $w_j(z), j = 1, \ldots, N$, as

$$w_j^k(z) = p_{\theta^k}(z \mid x_j). \tag{B.41}$$

  M-step: For fixed w_j^k, solve for θ as

$$\theta^{k+1} = \arg\max_{\theta \in \Theta} \sum_{j=1}^{N} \int w_j^k(z) \log(p_\theta(x_j, z))\, dz. \tag{B.42}$$

end while

$k \leftarrow k + 1$.

Output: Converged parameter $\hat{\theta}$.

The first equality follows from (B.34), while the first inequality follows from (B.35) after replacing the optimal $w_j^*(z) = p_{\theta^{k+1}}(z \mid x_j)$ by $p_{\theta^k}(z \mid x_j)$. The second inequality follows from (B.42) by replacing the optimal θ^{k+1} by θ^k, while the second equality follows from (B.34). When the cost function no longer increases, the process reaches a (local) extremum θ^* of the function $\ell(\theta, \mathcal{X})$.

The following result establishes the convergence of the EM algorithm.

Proposition B.16. *The expectation maximization process converges to one of the stationary points (extrema) of the log-likelihood function $\ell(\theta, \mathcal{X})$.*

For a more thorough exposition and complete proof of the convergence of the EM algorithm, one may refer to the paper (Wu 1983) and the book (McLanchlan and Krishnan 1997). See also Appendix A.1.5 for a discussion on the convergence of the alternating maximization approach. However, for the EM algorithm to converge to the maximum likelihood estimate (usually the global maximum) of $L(\theta, \mathcal{X})$, a good initialization is crucial.

Notice also that each step of the EM algorithm is in general a much simpler optimization problem than directly maximizing the incomplete log-likelihood $\ell(\theta, \mathcal{X})$. For many popular models (e.g., mixtures of Gaussians), one might even be able to find closed-form formulas for both steps, as shown next.

B.2.2 Maximum a Posteriori Expectation Maximization (MAP-EM)

Another alternative approach to marginalizing over the hidden variables is to take the maximum over the hidden variables. More specifically, instead of maximizing the incomplete log-likelihood $\ell(\theta, \mathcal{X})$ with respect to θ, we maximize the complete log-likelihood $\ell_c(\theta, \mathcal{X}, \mathcal{Z})$ with respect to both θ and $\mathcal{Z}$, i.e.,

$$\max_{\theta \in \Theta} \max_{\{z_j\}} \prod_{j=1}^{N} p_\theta(\boldsymbol{x}_j, z_j) \equiv \max_{\theta \in \Theta} \max_{\{z_j\}} \sum_{j=1}^{N} \log p_\theta(\boldsymbol{x}_j, z_j). \tag{B.46}$$

Observe that this problem is equivalent to

$$\max_{\theta \in \Theta} \sum_{j=1}^{N} \max_{z_j} \log p_\theta(\boldsymbol{x}_j, z_j) \equiv \max_{\theta \in \Theta} \sum_{j=1}^{N} \log p_\theta(\boldsymbol{x}_j, \hat{z}_j), \tag{B.47}$$

where

$$\hat{z}_j = \arg\max_{z} p_\theta(\boldsymbol{x}_j, z) = \arg\max_{z} p_\theta(z \mid \boldsymbol{x}_j), \tag{B.48}$$

is the maximum a posteriori (MAP) estimate of the latent variable z_j given $\boldsymbol{x}_j$. Therefore, when θ is fixed, we can solve for each z_j independently. This observation motivates us to consider an alternating maximization strategy (see Appendix A.1.5) for estimating θ. Specifically, given $\theta = \theta^k$, we solve for each hidden variable as $\hat{z}_j^k = \arg\max_z p_{\theta^k}(z \mid \boldsymbol{x}_j)$. Then, given $\mathcal{Z}$, we find the parameter θ that maximizes the complete log-likelihood with the hidden variables replaced by their MAP values, i.e., we estimate θ as $\hat{\theta}^{k+1} = \arg\max_\theta \sum_{j=1}^{N} \log p_\theta(\boldsymbol{x}_j, \hat{z}_j^k)$.

For the sake of completeness, Algorithm B.2 summarizes this MAP-EM strategy. Notice that there is a clear connection with the EM algorithm: if in the EM algorithm wereplace $w_j^k(z)$ by the Dirac delta $\delta(z - \hat{z}_j^k)$, then the M-step of EM reduces to the

Algorithm B.2 (Maximum a Posteriori Expectation Maximization)

Input: Data points $\{\boldsymbol{x}_j\}_{j=1}^N$ and initial parameter vector θ^0.

$k \leftarrow 0$.

while not converged **do**

MAP-step: For fixed $\theta = \theta^k$, solve for each $z_j, j = 1, \ldots, N$, as

$$z_j^k = \arg\max_{z} p_{\theta^k}(z \mid \boldsymbol{x}_j). \tag{B.49}$$

M-step: For fixed z_j^k, solve for θ as

$$\theta^{k+1} = \arg\max_{\theta \in \Theta} \sum_{j=1}^{N} \log p_\theta(\boldsymbol{x}_j, z_j^k). \tag{B.50}$$

end while

$k \leftarrow k + 1$.

Output: Converged parameter $\hat{\theta}$.

M-step of MAP-EM. Thus, we can view the MAP-EM algorithm pretty much as an EM algorithm in which the E-step is replaced by a MAP-step. This, of course, results in an approximation, and the resulting MAP-EM algorithm no longer provides an ML estimator for θ. In spite of this drawback, the MAP-EM algorithm is used as an approximate EM method, especially for mixture models, as discussed in the next section.

B.3 Estimation of Mixture Models

Mixture models are an important class of probabilistic models in which the data $\{\boldsymbol{x}_j\}_{j=1}^N$ are sampled from a distribution $p_\theta(\boldsymbol{x})$ that is a superposition of multiple distributions $\{p_{\theta_i}(\boldsymbol{x})\}_{i=1}^n$. Specifically, the mixture distribution is given by

$$p_\theta(\boldsymbol{x}) = \pi_1 p_{\theta_1}(\boldsymbol{x}) + \pi_2 p_{\theta_2}(\boldsymbol{x}) + \cdots + \pi_n p_{\theta_n}(\boldsymbol{x}), \tag{B.51}$$

where θ_i denotes the parameters of the ith distribution, $\pi_i > 0$ denotes the prior probability of drawing a point from the ith model and is such that $\sum_{i=1}^n \pi_i = 1$, and $\theta = (\theta_1, \ldots, \theta_n, \pi_1, \ldots, \pi_n)$ denotes the parameters of the mixture model. Such a distribution can be interpreted as the marginal distribution of a model with a latent random variable $z \in \{1, 2, \ldots, n\}$ that indicates the model from which $\boldsymbol{x}$ was sampled. To see this, notice that the marginal distribution can be written as

$$\begin{aligned} p_\theta(\boldsymbol{x}) &= \sum_z p_\theta(\boldsymbol{x}, z) = \sum_z p_\theta(\boldsymbol{x} \mid z) p_\theta(z) \\ &= \sum_{i=1}^n p_\theta(\boldsymbol{x} \mid z = i) p_\theta(z = i) = \sum_{i=1}^n p_{\theta_i}(\boldsymbol{x}) \pi_i, \end{aligned} \tag{B.52}$$

where $p_{\theta_i}(\boldsymbol{x}) \doteq p_\theta(\boldsymbol{x} \mid z = i)$ and $\pi_i \doteq p_\theta(z = i) > 0$, $i = 1, 2, \ldots, n$. The variables $\{\pi_i\}_{i=1}^n$ are often called the *mixing proportions*.

B.3.1 EM for Mixture Models

The EM algorithm is often used to estimate the parameters of a mixture model. Unlike the general EM algorithm, where the latent variable z is real-valued, in the case of a mixture model the latent variable z is discrete. Specifically, let $z_j \in \{1, \ldots, n\}$ be the latent variable associated with data point $\boldsymbol{x}_j$. In the E-step, we assume that we know the parameters $\theta^k = (\theta_1^k, \ldots, \theta_n^k, \pi_1^k, \ldots, \pi_n^k)$ of the mixture model and use them to compute the a posteriori distribution of $z_j \mid \boldsymbol{x}_j$, i.e.,

$$w_{ij}^k = p_{\theta^k}(z_j = i \mid \boldsymbol{x}_j) = \frac{p_{\theta^k}(\boldsymbol{x}_j \mid z_j = i) p_{\theta^k}(z_j = i)}{p_{\theta^k}(\boldsymbol{x}_j)} = \frac{p_{\theta_i^k}(\boldsymbol{x}_j)\pi_i^k}{\sum_{i=1}^n p_{\theta_i^k}(\boldsymbol{x}_j)\pi_i^k}. \tag{B.53}$$

In the M-step, we maximize the expected log-likelihood in (B.42),

$$\sum_{j=1}^N \sum_{i=1}^n w_{ij}^k \log(p_\theta(\boldsymbol{x}_j, z_j = i)) = \sum_{j=1}^N \sum_{i=1}^n w_{ij}^k \log(p_{\theta_i}(\boldsymbol{x}_j)\pi_i), \tag{B.54}$$

with respect to θ, and we obtain (see Exercise B.3)

$$\pi_i^{k+1} = \arg\max_{\pi_i} \sum_{j=1}^N w_{ij}^k \log(\pi_i) = \frac{\sum_{j=1}^N w_{ij}^k}{\sum_{j=1}^N \sum_{i=1}^n w_{ij}^k}, \tag{B.55}$$

$$\theta_i^{k+1} = \arg\max_{\theta_i} \sum_{j=1}^N w_{ij}^k \log(p_{\theta_i}(\boldsymbol{x}_j)). \tag{B.56}$$

Therefore, the parameters $\{\pi_i\}$ can be obtained in closed form. Whether the parameters $\{\theta_i\}$ can also be obtained in closed form will depend on the specific form of $p_{\theta_i}(\boldsymbol{x})$. Example B.17 shows that this is so for a mixture of Gaussians.

Once the model parameters are estimated from the EM algorithm, the "membership" $c_j \in \{1, 2, \ldots, n\}$ for a given sample point $\boldsymbol{x}_j$, i.e., the component distribution from which $\boldsymbol{x}_j$ is most likely drawn, can be determined by the Bayesian rule from its a posteriori probability:

$$c_j = \arg\max_{i=1,\ldots,n} p_\theta(z_j = i \mid \boldsymbol{x}_j) = \arg\max_{i=1,\ldots,n} \hat{w}_{ij}. \tag{B.57}$$

Example B.17 (EM for a Mixture of Gaussians). In the case that each mixture component is a Gaussian model with parameter $\theta_i = (\boldsymbol{\mu}_i, \Sigma_i)$, we have

$$p_{\theta_i}(\boldsymbol{x}) = \frac{1}{(2\pi)^{D/2}\det(\Sigma_i)^{1/2}} \exp\Big(-\frac{(\boldsymbol{x}-\boldsymbol{\mu}_i)^\top \Sigma_i^{-1}(\boldsymbol{x}-\boldsymbol{\mu}_i)}{2}\Big). \tag{B.58}$$

In the E-step, w_{ij}^k can be computed in closed form from (B.53) as

$$w_{ij}^k = \frac{p_{\theta_i^k}(\boldsymbol{x}_j)\pi_i^k}{\sum_{i=1}^n p_{\theta_i^k}(\boldsymbol{x}_j)\pi_i^k}. \tag{B.59}$$

Then the M-step is given by

$$\pi_i^{k+1} = \arg\max_{\pi_i} \sum_{j=1}^N w_{ij}^k \log \pi_i = \frac{\sum_{j=1}^N w_{ij}^k}{\sum_{j=1}^N \sum_{i=1}^n w_{ij}^k}, \tag{B.60}$$

$$\theta_i^{k+1} = \arg\max_{\theta_i} \sum_{j=1}^{N} w_{ij}^k \Big(-\frac{1}{2}(\boldsymbol{x}_j - \boldsymbol{\mu}_i)^\top \Sigma_i^{-1} (\boldsymbol{x}_j - \boldsymbol{\mu}_i) - \frac{1}{2}\det(\Sigma_i) \Big). \tag{B.61}$$

The above solution for the mixing proportions π_i^{k+1} follows from Exercise B.3, while the solution for $\theta_i^{k+1} = (\boldsymbol{\mu}_i^{k+1}, \Sigma_i^{k+1})$ follows from Exercise B.4 and is given by

$$\boldsymbol{\mu}_i^{k+1} = \frac{\sum_{j=1}^{N} w_{ij}^k \boldsymbol{x}_j}{\sum_{j=1}^{N} w_{ij}^k} \quad \text{and} \quad \Sigma_i^{k+1} = \frac{\sum_{j=1}^{N} w_{ij}^k (\boldsymbol{x}_j - \boldsymbol{\mu}_i^{k+1})(\boldsymbol{x}_j - \boldsymbol{\mu}_i^{k+1})^\top}{\sum_{j=1}^{N} w_{ij}^k}. \tag{B.62}$$

B.3.2 MAP-EM for Mixture Models

The EM algorithm for mixture models is based on alternating between computing the expected log-likelihood (E-step), which involves taking the expectation with respect to the latent variables, and maximizing the expected log-likelihood (M-step). As discussed in Appendix B.2.2, the MAP-EM algorithm is an alternative approach in which instead of taking the expectation, we directly maximize over the latent variables. As we will see in this section, this results in an approximate EM algorithm in which, in the E-step, each data point is assigned to the model that maximizes the posterior of the latent variables, whence the name MAP-EM.

To see this, let $z_j \in \{1, 2, \dots, n\}$ be the latent variable denoting the model that generated $\boldsymbol{x}_j$. The MAP-EM algorithm finds the model parameters and latent variables that maximize the complete log likelihood, i.e.,

$$\max_{\theta \in \Theta} \sum_{j=1}^{N} \max_{z_j} \log p_\theta(\boldsymbol{x}_j, z_j) \equiv \max_{\theta \in \Theta} \sum_{j=1}^{N} \max_{i=1,\dots,n} \log(p_\theta(\boldsymbol{x}_j | z_j = i)\pi_i). \tag{B.63}$$

Observe that this problem can be rewritten as[5]

$$\max_{\theta \in \Theta} \max_{\{w_{ij}\}} \sum_{j=1}^{N} \sum_{i=1}^{n} w_{ij} \log(p_{\theta_i}(\boldsymbol{x}_j)\pi_i), \tag{B.64}$$

[5]One may interpret this objective as follows. For each sample, we find the component distribution that maximizes the posterior. Once we have decided to "assign" $\boldsymbol{x}_j$ to the distribution $p_{\theta_i}(\boldsymbol{x})$, it takes $-\log p_{\theta_i}(\boldsymbol{x}_j)$ bits to encode $\boldsymbol{x}_j$. Thus, the above objective function is equivalent to minimizing the sum of the coding lengths given the membership of all the samples.

where $w_{ij} \in \{0, 1\}$ is an auxiliary variable encoding the assignment of points to models, which is defined as

$$w_{ij} = \begin{cases} 1 & \text{if } i = \arg\max_{\ell=1,\ldots,n} p_{\theta_\ell}(\boldsymbol{x}_j \mid z = \ell)\pi_\ell \\ 0 & \text{otherwise,} \end{cases} \tag{B.65}$$

and is such that for all $j = 1, \ldots, N$, $\sum_{i=1}^n w_{ij} = 1$.

Notice the striking connection between the *hard assignment* of points to models in (B.65) and the *soft assignment* done in the E-step of the EM algorithm for mixture models in (B.53). Notice also that when w_{ij} is fixed, the objective function in (B.64) is the same as that in the M-step of the EM algorithm for mixture models in (B.54). Thus, if we apply an alternating maximization strategy (see Appendix A.1.5) to the problem in (B.64), we obtain an algorithm that alternates between the following two steps:

MAP-step: Given θ, solve for w_{ij} such that $\sum_{i=1}^n w_{ij} = 1$. The optimal solution is given by (B.65) and involves assigning each data point to the model that maximizes the posterior probability, whence the name MAP-EM.

M-step: Given w_{ij}, solve for $\theta \in \Theta$. This problem is identical to the M-step in (B.54), whose solution is given by (B.55) and (B.56).

Notice that this MAP-EM algorithm for mixture models is a particular case of the MAP-EM algorithm described in Appendix B.2.2. Notice also that this MAP-EM algorithm for mixture models is very similar to the EM algorithm for mixture models, except that the *soft assignments* in the E-step in (B.53) are replaced by the *hard assignments* in (B.65). Thus, the MAP-EM algorithm is effectively an approximate EM algorithm.

Example B.18 (MAP-EM for a Mixture of Gaussians and the K-means Algorithm). In the case that each mixture component is a Gaussian model with parameter $\theta_i = (\boldsymbol{\mu}_i, \Sigma_i)$, we have

$$p_{\theta_i}(\boldsymbol{x}) = \frac{1}{(2\pi)^{D/2}\det(\Sigma_i)^{1/2}} \exp\Big(-\frac{(\boldsymbol{x} - \boldsymbol{\mu}_i)^\top \Sigma_i^{-1}(\boldsymbol{x} - \boldsymbol{\mu}_i)}{2}\Big). \tag{B.66}$$

In the E-step, given $\theta^k = (\theta_1^k, \ldots \theta_n^k, \pi_1^k, \ldots, \pi_n^k)$, w_{ij}^k can be computed in closed form as

$$w_{ij}^k = \begin{cases} 1 & \text{if } i = \arg\max_{\ell=1,\ldots,n} p_{\theta_\ell^k}(\boldsymbol{x}_j)\pi_\ell^k \\ 0 & \text{otherwise.} \end{cases} \tag{B.67}$$

Then, in the M-step, given w_{ij}^k, the mixing proportions π_i and the Gaussian parameters θ_i are given by

$$\pi_i^{k+1} = \frac{\sum_{j=1}^N w_{ij}^k}{\sum_{j=1}^N \sum_{i=1}^n w_{ij}^k}, \tag{B.68}$$

$$\boldsymbol{\mu}_i^{k+1} = \frac{\sum\limits_{j=1}^N w_{ij}^k \boldsymbol{x}_j}{\sum\limits_{j=1}^N w_{ij}^k} \quad \text{and} \quad \Sigma_i^{k+1} = \frac{\sum\limits_{j=1}^N w_{ij}^k (\boldsymbol{x}_j - \boldsymbol{\mu}_i^{k+1})(\boldsymbol{x}_j - \boldsymbol{\mu}_i^{k+1})^\top}{\sum\limits_{j=1}^N w_{ij}^k}. \tag{B.69}$$

Therefore, the MAP-EM algorithm alternates between assigning points to models using the MAP rule and recomputing the model parameters for each cluster.

Assume further that the mixture of Gaussians model is such that all mixing proportions are equal, i.e., $\pi_i = 1/n$ for all $i = 1, \dots, n$, and all covariance matrices are equal to the identity matrix, i.e., $\Sigma_i = I$ for all $i = 1, \dots, n$. In this case, the quantity $(\boldsymbol{x} - \boldsymbol{\mu}_i)^\top \Sigma_i^{-1} (\boldsymbol{x} - \boldsymbol{\mu}_i)$ reduces to the Euclidean distance $\|\boldsymbol{x} - \boldsymbol{\mu}_i\|^2$ from point $\boldsymbol{x}$ to the mean for the ith cluster $\boldsymbol{\mu}_i$. Therefore, the MAP-EM algorithm for a mixture of isotropic Gaussians with equal mixing proportions alternates between the following two steps:

MAP-step Given $\theta^k = (\boldsymbol{\mu}_1^k, \dots \boldsymbol{\mu}_n^k)$, assign each point to its closest cluster center, i.e.,

$$w_{ij}^k = \begin{cases} 1 & \text{if } i = \underset{\ell=1,\dots,n}{\arg\min} \|\boldsymbol{x}_j - \boldsymbol{\mu}_\ell\|_2^2, \\ 0 & \text{otherwise.} \end{cases} \tag{B.70}$$

M-step Given w_{ij}^k, update each cluster center as the average of the points assigned to that cluster, i.e.,

$$\boldsymbol{\mu}_i^{k+1} = \frac{\sum\limits_{j=1}^N w_{ij}^k \boldsymbol{x}_j}{\sum\limits_{j=1}^N w_{ij}^k}. \tag{B.71}$$

This particular case of the MAP-EM algorithm gives rise to a very popular clustering algorithm called *K-means* (see (Lloyd 1957; Forgy 1965; Jancey 1966; MacQueen 1967)), where $-\log p_{\theta_i}(\boldsymbol{x})$ reduces to the simple Euclidean distance to a cluster center. This algorithm is discussed in more detail in Section 4.3.1.

B.3.3 A Case in Which EM Fails

One difficulty with the EM algorithm is that a stationary value θ^* to which the algorithm converges is not necessarily the global maximum. Furthermore, for distributions as simple as a mixture of Gaussians, the global maximum of a likelihood function may not even exist, especially when some component distributions may become nearly singular. We illustrate this caution via the following example.

Example B.19 (ML Estimate of Two Mixed Gaussians (Vapnik 1995)). Consider a distribution $p(x)$, $x \in \mathbb{R}$, that is a mixture of two Gaussian (normal) distributions:

$$p(x, \mu, \sigma) = \frac{1}{2\sigma\sqrt{2\pi}} \exp\left\{-\frac{(x-\mu)^2}{2\sigma^2}\right\} + \frac{1}{2\sqrt{2\pi}} \exp\left\{-\frac{x^2}{2}\right\}, \tag{B.72}$$

where $\theta = (\mu, \sigma)$ are unknown. Then for given data $\mathcal{X} = \{x_1, x_2, \ldots, x_N\}$ and constant $A > 0$, there exists a small σ_0 such that for $\mu = x_1$, the log-likelihood will exceed A (regardless of the true μ, σ):

$$l(\mathcal{X}, \theta)\big|_{\mu=x_1,\sigma=\sigma_0} = \sum_{j=1}^{N} \ln p(x_i \mid \mu = x_1, \sigma = \sigma_0) \tag{B.73}$$

$$> \ln\left(\frac{1}{2\sigma_0\sqrt{2\pi}}\right) + \sum_{j=2}^{N} \ln\left(\frac{1}{2\sqrt{2\pi}} \exp\left\{-\frac{x_j^2}{2}\right\}\right) \tag{B.74}$$

$$= -\ln\sigma_0 - \sum_{j=2}^{N} \frac{x_j^2}{2} - N \ln 2\sqrt{2\pi} > A. \tag{B.75}$$

Therefore, the maximum of the log-likelihood does not even exist, and the ML objective would not provide a valid solution to estimating the unknown parameters. In fact, in this case, the true parameter corresponds to the largest (finite) local maximum of the log-likelihood.

From this simple example, we can see that the ML method does not apply to all probability densities.[6] If we insist on using it for mixtures of Gaussians, we should try to avoid the situation in which the variance can be arbitrarily small, i.e., $\sigma \to 0$. Unfortunately, this is often the case with random variables in high-dimensional spaces, where their distributions typically concentrate on low-dimensional subspaces or manifolds.

[6]It generally applies well to a class of density functions that are bounded by a common finite value from above. Hence EM would work well for generic Gaussians.

B.4 Model-Selection Criteria

So far, we have studied the following problem: given N independent samples $\{\boldsymbol{x}_j\}_{j=1}^N$ drawn from a distribution $p_\theta(\boldsymbol{x})$, where $p_\theta(\boldsymbol{x})$ belongs to a family of distributions indexed by the model parameter θ, obtain an estimate θ^* of θ. In doing so, we have assumed that the parameter $\theta \in \mathbb{R}^d$ is of fixed dimension d.

In practice, however, we may not know exactly the family of distributions to which the model belongs. Instead, we might know only that the model belongs to one of several possible families of distributions $p_{\theta(m)}(\boldsymbol{x})$, where m is a (discrete) index for the model families, $\theta(m) \in \mathbb{R}^{d(m)}$ is the vector of parameters for model family m, and $d(m)$ is the number of independent model parameters for that family. For instance, in the mixture model (B.51), the number of mixture components n could be unknown and would need to be estimated together with the mixture model parameters. In this case, for each value of n we can define a parameter vector $\theta(n) = (\theta_1, \dots, \theta_n, \pi_1, \dots, \pi_n)$ of dimension[7] $d(n) = nd + n - 1$. Therefore, the challenge is to choose among different models of different dimensions.

The problem of determining both the model type m and its parameter $\theta(m)$ is conventionally referred to as a *model selection* problem (as opposed to parameter estimation). Many important model-selection criteria have been developed in the statistics community and the algorithmic complexity community for general classes of models. These criteria include:

- The Akaike information criterion (AIC) (Akaike 1977) (also known as the C_p statistics (Mallows 1973)) and geometric AIC (G-AIC) (Kanatani 2003);
- The Bayesian information criterion (BIC) (also known as the Schwartz criterion); and
- Minimum description length (MDL) (Rissanen 1978) and minimum message length (MML) (Wallace and Boulton 1968).

Although these criteria were originally motivated and derived from different viewpoints (or in different contexts), they all share a common characteristic: the optimal model should be one that strikes a good *balance* between the *model complexity*, which typically depends on the dimension of the parameter space, and the *data fidelity* to the chosen model, which is typically measured as the sum of squared errors from the data points to the model. In fact, some of the criteria are essentially equivalent to each other despite their different origins. For instance, to a large extent, the AIC is equivalent to the C_p statistics, and the BIC is equivalent to the MDL.

In what follows, we give a brief review of the AIC and the BIC to illustrate the key ideas behind model selection. However, we emphasize here that in general, no model-selection criterion is always better than others under all circumstances, and the best criterion depends on the purpose of the model. For a more detailed exposition of these and many other model-selection criteria, we refer the reader to (Burnham and Anderson 2002).

---

[7] We subtract one parameter because $\sum_{i=1}^n \pi_i = 1$.

B.4.1 Akaike Information Criterion

Given N independent sample points $\mathcal{X} = \{\boldsymbol{x}_j\}_{j=1}^N$ drawn from a distribution $p_{\theta_0}(\boldsymbol{x})$, recall that the maximum-likelihood estimate $\hat{\theta}_N$ of the parameter θ is the one that maximizes the log-likelihood function $\ell(\theta, \mathcal{X}) = \sum_{j=1}^N \log p_\theta(\boldsymbol{x}_j)$.

The *Akaike information criterion* (AIC) for model selection is motivated from an information-theoretic viewpoint. In this approach, the quality of the obtained model is measured by the average code length used by the optimal coding scheme of $p_{\hat{\theta}_N}(\boldsymbol{x})$ for a random variable with actual distribution $p_{\theta_0}(\boldsymbol{x})$, i.e.,

$$\mathbb{E}_{\theta_0}[-\log p_{\hat{\theta}_N}(\boldsymbol{x})] = \int -\log\big(p_{\hat{\theta}_N}(\boldsymbol{x})\big)p_{\theta_0}(\boldsymbol{x})\,d\boldsymbol{x}. \tag{B.76}$$

The AIC relies on an approximation to the above expected log-likelihood loss that holds asymptotically as $N \to \infty$:

$$2\mathbb{E}_{\theta_0}[-\log p_{\hat{\theta}_N}(\boldsymbol{x})] \approx -2\ell(\hat{\theta}_N, \mathcal{X}) + 2d \doteq \text{AIC}, \tag{B.77}$$

where d is the number of free parameters for the class of models of interest.

For Gaussian noise models with variance σ^2, we have

$$\ell(\hat{\theta}_N, \mathcal{X}) = -\frac{1}{2\sigma^2}\sum_{j=1}^N \|\boldsymbol{x}_j - \hat{\boldsymbol{x}}_j\|^2, \tag{B.78}$$

where $\hat{\boldsymbol{x}}_j = \mathbb{E}_{\hat{\theta}_N}[\boldsymbol{x}_j]$ is the best estimate of $\boldsymbol{x}_j$ given the model $p_{\hat{\theta}_N}(\boldsymbol{x})$. Thus, if σ^2 is known (or approximated by the empirical sample variance), minimizing the AIC is equivalent to minimizing the so-called C_p statistic:

$$C_p \doteq \frac{1}{\sigma^2}\sum_{j=1}^N \|\boldsymbol{x}_j - \hat{\boldsymbol{x}}_j\|^2 + 2d - N, \tag{B.79}$$

where the first term is obviously the mean squared error (a measure of data fidelity), and the second term is an affine function of the dimension of the parameter space (a measure of the complexity of the model).

Now consider multiple classes of models whose parameter spaces are of different dimensions and denote the dimension of model class m by $d(m)$. Then the AIC selects the model class m^* that minimizes the following objective function:

$$\text{AIC}(m) = \frac{1}{\sigma^2}\sum_{j=1}^N \|\boldsymbol{x}_j - \hat{\boldsymbol{x}}_j(m)\|^2 + 2d(m), \tag{B.80}$$

where $\hat{\boldsymbol{x}}_j(m) = \mathbb{E}_{\hat{\theta}_N(m)}[\boldsymbol{x}_j]$ is the best estimate of $\boldsymbol{x}_j$ given the model $p_{\hat{\theta}_N(m)}(\boldsymbol{x})$, and $\hat{\theta}_N(m)$ is the maximum-likelihood estimate of θ for model family m.

B.4.2 Bayesian Information Criterion

The *Bayesian information criterion* (BIC) for model selection is motivated from a Bayesian inference viewpoint. In this approach, we assume a *prior* distribution of the model $p(\theta \mid m)$ and wish to choose the model class m^* that maximizes the *posterior* probability $p(m \mid \mathcal{X})$. Using the Bayesian rule, this is equivalent to maximizing

$$p(m \mid \mathcal{X}) \propto p(m)p(\mathcal{X} \mid m) = p(m) \int p(\mathcal{X} \mid \theta, m)p(\theta \mid m)\, d\theta. \tag{B.81}$$

If we assume that each model class is equally probable, this further reduces to maximizing the likelihood $p(\mathcal{X} \mid m)$ among all the model classes. This is equivalent to minimizing the negative log-likelihood $-2\log p(\mathcal{X} \mid m)$. With certain approximations, one can show that for general distributions, the following relationship holds asymptotically as $N \to \infty$:

$$\mathrm{BIC}(m) \doteq -2\log p(\mathcal{X} \mid m) = -2\ell(\hat{\theta}_N(m), \mathcal{X}) + \log(N)d(m) \tag{B.82}$$

$$= \frac{1}{\sigma^2}\sum_{j=1}^{N} \|\boldsymbol{x}_j - \hat{\boldsymbol{x}}_j(m)\|^2 + \log(N)d(m). \tag{B.83}$$

As before, $\hat{\theta}_N(m)$ is the maximum-likelihood estimate of θ given m, $d(m)$ is the number of parameters for class m, and σ^2 is the variance of a Gaussian noise model. Notice that when N and σ are known, the BIC is very similar to the AIC, except that the factor 2 in front of the second term in the AIC is replaced by $\log(N)$ in the BIC. Because we normally have $N \gg e^2$, the BIC penalizes complex models much more than the AIC does. Thus, the BIC tends to choose simpler models.

B.5 Robust Statistical Methods

For all the model-estimation and selection techniques discussed above, we have always assumed that the given data samples $\{\boldsymbol{x}_j\}_{j=1}^N$ are independent samples drawn from the same distribution $p_{\theta_0}(\boldsymbol{x})$. By an appeal to the law of large numbers, the asymptotic optimality of the estimate normally does not depend the particular set of samples given.[8] However, in many practical situations, the validity of the given

[8]The fact that almost all sets of i.i.d. samples are "typical" or "representative" of the given distribution has been at the heart of the development of Shannon's information theory.

data as independent samples of the model becomes questionable. Sometimes, the given data can be corrupted by or mixed with samples of a different (probabilistic) nature; or it can simply be the case that the given data are not a typical set of i.i.d. samples from the distribution in question. For the purpose of model estimation, these seemingly different interpretations are actually equivalent: we need to somehow infer the correct model while *accommodating* an atypical set of samples of the distribution (or the model). Obviously, this is an impossible task unless we impose some restrictions on how atypical the samples are. It is customary to assume that only a portion of the samples are different from or inconsistent with the rest of the data. Those samples are often referred to as *outliers*, and they may have a significant effect on the model inferred from data.

Unfortunately, despite centuries of interest and study,[9] there is no universally agreed definition of what an outlier is, especially for multivariate data. Roughly speaking, most definitions (or tests) for an outlier are based on one of the following guidelines:

1. The outliers are a set of samples that have relatively *large influence* on the estimated model parameters. A measure of influence is normally the difference between the model estimated with or without the sample in question.
2. The outliers are a set of *small-probability* samples with respect to the distribution in question. The given data set is therefore an atypical set if such small-probability samples constitute a significant portion of the data.
3. The outliers are a set of samples that are *not consistent* with (the model inferred from) the remainder of the data. A measure of inconsistency is normally the error residual of the sample in question with respect to the model.

Nevertheless, as we will soon see, for popular distributions such as the Gaussian, they all lead to more or less equivalent ways of detecting or accommodating outliers. However, under different conditions, different approaches that follow each of the above guidelines may give rise to solutions that can be more convenient and efficient than others.

B.5.1 Influence-Based Outlier Detection

When we try to estimate the parameter of the distribution $p_\theta(\boldsymbol{x})$ from a set of samples $\{\boldsymbol{x}_j\}_{j=1}^N$, every sample $\boldsymbol{x}_j$ might have an uneven effect on the estimated parameter $\hat{\theta}_N$. The samples that have a relatively large effect are called *influential samples*, and they can be regarded as outliers.

[9] The earliest documented discussions among astronomers about outliers or "erroneous observations" date back to the mid-eighteenth century. See (Barnett and Lewis 1983; Huber 1981; Bickel 1976) for a more thorough exposition of the studies of outliers in statistics.

To measure the influence of a particular sample $\boldsymbol{x}_j$, we may compare the difference between the parameter $\hat{\theta}_N$ estimated from all the N samples and the parameter $\hat{\theta}_N^{(j)}$ estimated from all but the jth sample. Without loss of generality, we here consider the maximum-likelihood estimate of the model:

$$\hat{\theta}_N = \arg\max_{\theta\in\Theta} \sum_{i=1}^{N} \log p_\theta(\boldsymbol{x}_i), \tag{B.84}$$

$$\hat{\theta}_N^{(j)} = \arg\max_{\theta\in\Theta} \sum_{i\neq j} \log p_\theta(\boldsymbol{x}_i), \tag{B.85}$$

and measure the influence of $\boldsymbol{x}_j$ on the estimation of θ by the difference

$$\hat{\theta}_N - \hat{\theta}_N^{(j)}. \tag{B.86}$$

Assume that $p_\theta(\boldsymbol{x})$ is analytic in θ and define the gradients of the above objective functions as

$$f(\theta) \doteq \sum_{i=1}^{N} \frac{1}{p_\theta(\boldsymbol{x}_i)} \frac{\partial p_\theta(\boldsymbol{x}_i)}{\partial \theta} \tag{B.87}$$

$$f^{(j)}(\theta) \doteq \sum_{i\neq j} \frac{1}{p_\theta(\boldsymbol{x}_i)} \frac{\partial p_\theta(\boldsymbol{x}_i)}{\partial \theta}. \tag{B.88}$$

If we now evaluate the function $f(\theta)$ at $\theta = \hat{\theta}_N^{(j)}$ using the Taylor series of $f(\theta)$ at $\theta = \hat{\theta}_N$, we obtain

$$f(\hat{\theta}_N^{(j)}) = f(\hat{\theta}_N) + f'(\hat{\theta}_N)(\hat{\theta}_N^{(j)} - \hat{\theta}_N) + o(\|\hat{\theta}_N - \hat{\theta}_N^{(j)}\|). \tag{B.89}$$

Since we have $f(\hat{\theta}_N) = 0$ and $f^{(j)}(\hat{\theta}_N^{(j)}) = 0$, the difference in the estimate caused by the jth sample is

$$\hat{\theta}_N^{(j)} - \hat{\theta}_N \approx \left(f'(\hat{\theta}_N)\right)^{\dagger} \left[\frac{1}{p_{\hat{\theta}_N^{(j)}}(\boldsymbol{x}_j)} \frac{\partial p_{\hat{\theta}_N^{(j)}}(\boldsymbol{x}_j)}{\partial \theta}\right]. \tag{B.90}$$

Notice that in the expression on the right-hand side, the factor $\left(f'(\hat{\theta}_N)\right)^{\dagger}$ is common for all samples.

Proposition B.20 (Approximate Sample Influence). *The difference between the ML estimate $\hat{\theta}_N$ from N samples and the ML estimate $\hat{\theta}_N^{(j)}$ without the jth sample $\boldsymbol{x}_j$ depends approximately linearly on the quantity*

$$\frac{1}{p_{\hat{\theta}_N^{(j)}}(\boldsymbol{x}_j)} \frac{\partial p_{\hat{\theta}_N^{(j)}}(\boldsymbol{x}_j)}{\partial \theta}. \tag{B.91}$$

In the special case that $p_\theta(\boldsymbol{x})$ is the Gaussian distribution $\mathcal{N}(\boldsymbol{\mu}, \sigma^2 I)$ with σ^2 known, the above equation gives the influence of the jth sample on the estimate of $\boldsymbol{\mu}$:

$$\hat{\boldsymbol{\mu}}_N^{(j)} - \hat{\boldsymbol{\mu}}_N \approx \alpha(\boldsymbol{x}_j - \hat{\boldsymbol{\mu}}_N^{(j)}), \tag{B.92}$$

where α is some constant depending on σ. That is, the sample influence can be measured by the distance between the sample and the mean estimated without the sample; or equivalently, the smaller the probability of a sample with respect to the estimated (Gaussian) distribution, the larger its influence on the estimated mean. Therefore, the three guidelines for defining outliers become very much equivalent for a Gaussian distribution.

In general, to evaluate the influence of all the samples, one needs to estimate the model $N + 1$ times, which is reasonable only if each estimate is not too costly to compute. In light of this drawback, some first-order approximations of the influence values were developed in roughly the same period during which the sample influence function was proposed (Campbell 1978; Critchley 1985), when computational resources were scarcer than they are today. In robust statistics, formulas that approximate an influence function are referred to as *theoretical influence functions*.

B.5.2 Probability-Based Outlier Detection

Assume that the data are drawn from a zero-mean[10] multivariate Gaussian distribution $\mathcal{N}(\mathbf{0}, \Sigma_{\boldsymbol{x}})$. If there were no outliers, the maximum likelihood estimate of $\Sigma_{\boldsymbol{x}}$ would be given by $\hat{\Sigma}_N = \frac{1}{N}\sum_{j=1}^N \boldsymbol{x}_j\boldsymbol{x}_j^\top \in \mathbb{R}^{D\times D}$. Therefore, we could approximate the probability that a sample $\boldsymbol{x}_j$ comes from this Gaussian model by

$$p(\boldsymbol{x}_j; \hat{\Sigma}_N) = \frac{1}{(2\pi)^{D/2}\det(\hat{\Sigma}_N)^{1/2}} \exp\left(-\frac{1}{2}\boldsymbol{x}_j^\top \hat{\Sigma}_N^{-1}\boldsymbol{x}_j\right). \tag{B.93}$$

If we adopt the guideline that outliers are samples that have a small probability with respect to the estimated model, then the outliers are exactly those samples that have a relatively large residual:

[10]We here are interested only in how to robustly estimate the covariance, or "scale," of the distribution. In case the mean, or "location," of the distribution is not known, a separate robust procedure can be employed to determine the mean before the covariance; see (Barnett and Lewis 1983).

$$\varepsilon_j = \boldsymbol{x}_j^\top \hat{\Sigma}_N^{-1} \boldsymbol{x}_j, \quad j = 1, 2, \ldots, N, \tag{B.94}$$

also known as the *Mahalanobis distance*.[11]

In principle, we could use $p(\boldsymbol{x}_j, \hat{\Sigma}_N)$ or ε_j to determine whether $\boldsymbol{x}_j$ is an outlier. However, the above estimate of the covariance matrix $\Sigma_{\boldsymbol{x}}$ is obtained using all the samples, including the outliers themselves. Therefore, if $\hat{\Sigma}_N$ is very different from $\Sigma_{\boldsymbol{x}}$, the outliers could be incorrectly detected. In order to improve the estimate of $\Sigma_{\boldsymbol{x}}$, one can recompute $\hat{\Sigma}_N$ by discarding or downweighting samples that have low probability or large Mahalanobis distance. Let $w_j \in [0, 1]$ be a weight assigned to the jth point such that $w_j \approx 1$ if $\boldsymbol{x}_j$ is an inlier and $w_j \approx 0$ if $\boldsymbol{x}_j$ is an outlier. Then a new estimate of $\Sigma_{\boldsymbol{x}}$ can be obtained as

$$\hat{\Sigma}_N = \frac{\sum_{j=1}^N w_j \boldsymbol{x}_j \boldsymbol{x}_j^\top}{\sum_{j=1}^N w_j}. \tag{B.95}$$

Maximum-Likelihood-Type Estimators (M-Estimators)

If we choose $w(\varepsilon) \equiv \varepsilon$, the above expression gives the original estimate of the covariance matrix $\hat{\Sigma}_N = \frac{1}{N}\sum_{j=1}^N \boldsymbol{x}_j \boldsymbol{x}_j^\top$. Alternatively, if we simply want to discard all samples with a Mahalanobis distance larger than a certain threshold $\varepsilon_0 > 0$, we can choose the following weight function:

$$w(\varepsilon) = \begin{cases} \varepsilon, & \text{for } \varepsilon \le \varepsilon_0, \\ 0, & \text{for } \varepsilon > \varepsilon_0. \end{cases} \tag{B.96}$$

Nevertheless, under the assumption that the distribution is elliptically symmetric and is contaminated by an associated normal distribution, the following weight function gives a more robust estimate of the covariance matrix (Hampel 1974; Campbell 1980):

$$w(\varepsilon) = \begin{cases} \varepsilon, & \text{for } \varepsilon \le \varepsilon_0, \\ \varepsilon_0 \exp[-\frac{1}{2a}(\varepsilon - \varepsilon_0)^2] & \text{for } \varepsilon > \varepsilon_0, \end{cases} \tag{B.97}$$

with $\varepsilon_0 = \sqrt{D + b}$ for some suitable choice of positive values for a and b, and D denotes the dimension of the space. Many other weight functions have also been proposed in the statistics literature. They serve as the basis for a class of robust estimators, known as *M-estimators* (maximum-likelihood-type estimators) (Huber 1981; Barnett and Lewis 1983). Nevertheless, most M-estimators differ only in how the samples are downweighted, but no one of them seems to dominate the others in terms of performance in all circumstances.

[11] In fact, it can be shown (Ferguson 1961) that if the outliers have a Gaussian distribution of a different covariance matrix $a\Sigma$, then ε_j is a sufficient statistic for the test that maximizes the probability of correct decision about the outlier (in the class of tests that are invariant under linear transformations). The interested reader may want to find out how this distance is equivalent (or related) to the sample influence $\hat{\Sigma}_N^{(j)} - \hat{\Sigma}_N$ or the approximate sample influence given in (B.91).

Notice that calculating the robust estimate $\hat{\Sigma}_N$ as in (B.95) is not easy, because the weights w_j also depend on the resulting $\hat{\Sigma}_N$. There is no surprise that many known algorithms are based on Monte Carlo (Maronna 1976; Campbell 1980).

Multivariate Trimming (MVT)
One drawback of the M-estimators is that their "breakdown point" is inversely proportional to the dimension of the data space. The breakdown point is an important measure of robustness of any estimator. Roughly speaking, it is the largest proportion of contamination that the estimator can tolerate. Thus, the M-estimators become much less robust when the dimension of the data is high.

One way to resolve this problem is to modify the M-estimators by simply trimming out a percentage of the samples with relatively large Mahalanobis distance and then using the remaining samples to reestimate the covariance matrix. Then each time we have a new estimate of the covariance matrix, we can recalculate the Mahalanobis distance of every sample and reselect samples that need to be trimmed. We can repeat the above process until a stable estimate of the covariance matrix is obtained. This iterative scheme is known as *multivariate trimming* (MVT), another popular robust estimator. By construction, the breakdown point of MVT does not depend on the dimension of the problem and depends only on the chosen trimming percentage.

When the percentage of outliers is somehow known, it is relatively easy to determine how many samples need to be trimmed, and it usually takes only a few iterations for MTV to converge. However, if the percentage is wrongfully specified, MVT is known to have trouble converging, or it may converge to a wrong estimate of the covariance matrix.

B.5.3 Random-Sampling-Based Outlier Detection

When the outliers constitute a large portion (up to 50% or even more) of the data set, the (ML) estimate $\hat{\theta}_N$ obtained from all the samples can be so severely corrupted that the sample influence and the Mahalanobis distance computed based on it become useless in discriminating between outliers and valid samples.[12] This motivates estimating the model parameter θ using only a (randomly sampled) small subset of the samples to begin with. In this section, we describe two such methods: least median of squares (LMS) and random sample consensus (RANSAC).

[12]Thus, the iterative process is likely to converge to a local minimum other than the true model parameter. Sometimes, it can even be the case that the roles of inliers and outliers are exchanged with respect to the converged estimate.

Least Median Estimation

If we knew that fewer than half of the samples are potential outliers, we could use only half of the samples to estimate the model parameter. But which half of the samples should we use? We know that the maximum-likelihood estimate minimizes the sum of negative log-likelihoods:

$$\hat{\theta}_N = \arg\min_{\theta\in\Theta} \sum_{j=1}^{N} \Big(-\log(p_\theta(\boldsymbol{x}_j)) \Big). \tag{B.98}$$

Since outliers should have small probability, hence large negative log-likelihood, we can order the values of the negative log-likelihood and eliminate from the above objective half of the samples that have relatively larger values:

$$\hat{\theta}_{N/2} = \arg\min_{\theta\in\Theta} \sum_{j} \big(-\log(p_\theta(\boldsymbol{x}_j)) \big), \tag{B.99}$$

where the sum is over the points $\boldsymbol{x}_j$ such that

$$-\log(p_\theta(\boldsymbol{x}_j)) \leq \operatorname*{median}_{\boldsymbol{x}_\ell\in\mathcal{X}} \big(-\log(p_\theta(\boldsymbol{x}_\ell)) \big). \tag{B.100}$$

A popular approximation to the above objective is simply to minimize the median value of the negative log-likelihood:

$$\hat{\theta}_M \doteq \arg\min_{\theta\in\Theta} \operatorname*{median}_{\boldsymbol{x}_j\in\mathcal{X}} \big(-\log_\theta(p(\boldsymbol{x}_j)) \big). \tag{B.101}$$

We call $\hat{\theta}_M$ the *least median estimate*. In the case of a Gaussian noise model, $-\log p(\boldsymbol{x}_j, \theta)$ is proportional to the squared error:

$$-\log(p_\theta(\boldsymbol{x}_j)) \propto \|\boldsymbol{x}_j - \hat{\boldsymbol{x}}_j\|^2. \tag{B.102}$$

For this reason, the estimate $\hat{\theta}_M$ is often known as the *least median of squares* (LMS) estimate.[13]

However, without knowing θ, it is impossible to order the log-likelihoods or the squared errors, let alone compute the median. A typical method to resolve this difficulty is to *randomly sample* a number of small subsets of the data:

$$\mathcal{X}_1, \mathcal{X}_2, \ldots, \mathcal{X}_m \subset \mathcal{X}, \tag{B.103}$$

[13] The importance of the median for robust estimation was pointed out first in the article (Hampel 1974).

where each subset $\mathcal{X}_i$ is independently drawn and contains $k \ll N$ samples. If p is the fraction of valid samples (the "inliers"), one can show (see Exercise B.8) that with probability $q = 1 - (1 - p^k)^m$, one of the above subsets will contain only valid samples. In other words, if q is the probability that one of the selected subsets contains only valid samples, we need to randomly sample at least

$$m \geq \frac{\log(1-q)}{\log(1-p^k)} \tag{B.104}$$

subsets of k samples.

Using each subset $\mathcal{X}_i$, we can compute an estimate $\hat{\theta}_i$ of the model and use the estimate to compute the median for the remaining $N - k$ samples in $\mathcal{X} \setminus \mathcal{X}_i$:

$$\hat{M}_i \doteq \underset{x_j \in \mathcal{X} \setminus \mathcal{X}_i}{\text{median}} \big(-\log(p_{\hat{\theta}_i}(\boldsymbol{x}_j)) \big). \tag{B.105}$$

Then the least median estimate $\hat{\theta}_M$ is approximated by the $\hat{\theta}_{i^*}$ that gives the smallest median $\hat{M}_{i^*} = \min_i \hat{M}_i$.

In the case of a Gaussian noise model, based on the order statistics of squared errors, we can use the median statistic to obtain an (asymptotically unbiased) estimate of the variance, or scale, of the error as follows:

$$\hat{\sigma} = \frac{N+5}{N\Phi^{-1}(0.5+p/2)} \sqrt{\underset{x_j \in \mathcal{X}}{\text{median}} \|\boldsymbol{x}_j - \hat{\boldsymbol{x}}_j\|^2}, \tag{B.106}$$

where $p = 0.5$ for the median statistic. Then one can use $\hat{\sigma}$ to find "good" samples in $\mathcal{X}$ whose squared errors are less than $\lambda\sigma^2$ for some chosen constant λ (normally less than 5). Using such good samples, we can recompute a more efficient (ML) estimate $\hat{\theta}$ of the model.

Random Sample Consensus (RANSAC)

In theory, the breakdown point of the least median estimate is up to 50% outliers. In many practical situations, however, outlying samples may constitute more than half of the data. Random sample consensus (RANSAC) (Fischler and Bolles 1981) is a method that is designed to work for such highly contaminated data.

In many respects, RANSAC is actually very similar to LMS. The main difference is that instead of looking at the median statistic,[14] RANSAC tries to find, among all the estimates $\{\hat{\theta}_i\}$ obtained from the subsets $\{\mathcal{X}_i\}$, the one that maximizes the number of samples that have an error residual (measured either by the negative log-likelihood or the squared error) smaller than a prespecified error tolerance:

[14]Which becomes meaningless when the fraction of outliers is over 50%.

$$\hat{\theta}_{i*} \doteq \arg\max_{\hat{\theta}_i} \#\{x_j \in \mathcal{X} : -\log(x_j, \hat{\theta}_i) \leq \tau\}. \tag{B.107}$$

In other words, $\hat{\theta}_{i*}$ achieves the highest "consensus" among all the random sample estimates $\{\hat{\theta}_i\}$, whence the name "random sample consensus" (RANSAC). To improve the efficiency of the estimate, we can recompute an ML estimate $\hat{\theta}$ of the model from all the samples that are consistent with $\hat{\theta}_{i*}$.

Notice that for RANSAC, one needs to specify the error tolerance τ a priori. In other words, RANSAC requires knowing the variance σ^2 of the error a priori, while LMS normally does not. There have been a few variations of RANSAC in the literature that relax this requirement. We here do not elaborate on them, and interested readers may refer to (Steward 1999) and references therein.

However, when the dimension of the model is large or the model has a large number of mixture components, random sampling techniques have not been very effective. The reason is largely that in this case, the number of subsets needed in (B.104) grows prohibitively large. The reader may refer to (Yang et al. 2006) for an empirical study that extends RANSAC-type ideas to the case of a mixture of subspaces.

B.6 Exercises

Exercise B.1 (ML Estimates of the Parameters of a Gaussian) Let $\boldsymbol{x} \in \mathbb{R}^D$ be a random vector with distribution $\mathcal{N}(\boldsymbol{\mu}_x, \Sigma_x)$, where $\boldsymbol{\mu}_x = \mathbb{E}(\boldsymbol{x}) \in \mathbb{R}^D$ and $\Sigma_x = \mathbb{E}(\boldsymbol{x} - \boldsymbol{\mu})(\boldsymbol{x} - \boldsymbol{\mu})^\top \in \mathbb{R}^{D \times D}$ are, respectively, the mean and the covariance of $\boldsymbol{x}$. Show that the maximum likelihood estimates of $\boldsymbol{\mu}_x$ and Σ_x are, respectively, given by

$$\hat{\boldsymbol{\mu}}_N \doteq \frac{1}{N}\sum_{j=1}^N \boldsymbol{x}_j \quad \text{and} \quad \hat{\Sigma}_N \doteq \frac{1}{N}\sum_{j=1}^N (\boldsymbol{x}_j - \hat{\boldsymbol{\mu}}_N)(\boldsymbol{x}_j - \hat{\boldsymbol{\mu}}_N)^\top. \tag{B.108}$$

Exercise B.2 (Invariance of ML Estimator) Let $\hat{\theta}_N$ be the maximum likelihood (ML) estimate of θ obtained from N i.i.d. samples $\{\boldsymbol{x}_j\}_{j=1}^N$ from the distribution $p_\theta(\boldsymbol{x})$. Show that $g(\hat{\theta}_N)$ is an ML estimate of $g(\theta)$. What are the conditions that need to be imposed on $g(\theta)$ in order for $g(\hat{\theta}_N)$ to be an ML estimate of $g(\theta)$?

Exercise B.3 (ML Estimates of the Mixing Proportions) Let $W = [w_{ij}] \in \mathbb{R}^{n \times N}$ be a left stochastic matrix, i.e., $w_{ij} \geq 0$ and $\sum_{i=1}^n w_{ij} = 1$ for all $j = 1, \dots, N$. Let π be a stochastic vector, i.e., $\pi \in \Pi = \{(\pi_1, \dots, \pi_n) : \pi_i \geq 0$, and $\sum_{i=1}^n \pi_i = 1\}$. Show that

$$\arg\max_{\pi\in\Pi} \sum_{j=1}^{N} w_{ij}\log(\pi_i) = \frac{\sum_{j=1}^{N} w_{ij}}{\sum_{i=1}^{n}\sum_{j=1}^{N} w_{ij}}. \tag{B.109}$$

Exercise B.4 (ML Estimates of the Parameters of a Mixture of Gaussians) Let $W = [w_{ij}] \in \mathbb{R}^{n\times N}$ be a left stochastic matrix, i.e., $w_{ij} \geq 0$ and $\sum_{i=1}^{n} w_{ij} = 1$ for all $j = 1,\ldots,N$. Show that the solution to the optimization problem

$$\max_{\boldsymbol{\mu}_i,\Sigma_i} \sum_{j=1}^{N} w_{ij}\Big(-\frac{1}{2}(\boldsymbol{x}_j-\boldsymbol{\mu}_i)^\top \Sigma_i^{-1}(\boldsymbol{x}_j-\boldsymbol{\mu}_i) - \frac{1}{2}\det(\Sigma_i)\Big) \tag{B.110}$$

is

$$\boldsymbol{\mu}_i = \frac{\sum_{j=1}^{N} w_{ij}\boldsymbol{x}_j}{\sum_{j=1}^{N} w_{ij}} \quad \text{and} \quad \Sigma_i = \frac{\sum_{j=1}^{N} w_{ij}(\boldsymbol{x}_j-\boldsymbol{\mu}_i)(\boldsymbol{x}_j-\boldsymbol{\mu}_i)^\top}{\sum_{j=1}^{N} w_{ij}}. \tag{B.111}$$

Exercise B.5 Study MATLAB's `gmdistribution` class, which is described at http://www.mathworks.com/help/stats/gmdistribution-class.html, and reproduce the example on clustering using Gaussian mixture models described at http://www.mathworks.com/help/stats/gaussian-mixture-models.html. That is, use the function `mvnrnd` to generate data sampled from a mixture of two Gaussians and the function `fitgmdist` to estimate the parameters of the mixture model from the data. Then plot the isocontours of the estimated distribution, the clustering of the data, and the soft assignment weights.

Exercise B.6 Reproduce three of the examples described at http://www.mathworks.com/help/stats/fitgmdist.html. Specifically, reproduce the examples entitled *Cluster Data Using a Gaussian Mixture Model*, *Regularize Gaussian Mixture Model Estimation*, and *Determine the Best Gaussian Mixture Fit Using AIC*. In addition, add a new example called *Determine the Best Gaussian Mixture Fit Using BIC* and compare it to AIC.

Exercise B.7 Implement the EM and MAP-EM algorithms for a mixture of Gaussians. The format of your function should be as follows.

Function `[group,mu,Sigma,pi]=GMM(x,n,method,group0,restarts)`

Parameters

`x`	$D \times N$ matrix whose columns are the data points
`n`	number of groups
`method`	'EM', 'MAPEM'
`group0`	$1 \times N$ vector containing an initial soft or hard assignment of points to groups

Returned values

`group`	$1 \times N$ vector containing the soft or hard assignments of points to groups
`mu`	$D \times n$ matrix whose ith column is the mean for the ith group
`Sigma`	$D \times D \times n$ tensor whose ith slice is the covariance matrix of the ith group
`pi`	$n \times 1$ vector whose entries are the mixing proportions

Estimates the parameters of a Gaussian mixture model

Generate data from a mixture of two Gaussians in $\mathbb{R}^2$ with means $(-1,-1)$ and $(1,1)$, equal covariance matrices $\sigma^2 I$, and equal mixing proportions $\pi_1 = \pi_2 = 1/2$. Increase σ from 0.1 to 1 and plot the clustering error as a function of σ. Plot also the error in the estimated parameters as a function of σ. Compare your results with those produced by the MATLAB function `fitgmdist`.

Exercise B.8 (RANSAC) Suppose you are given N data points such that p% are inliers and $(1-p)$% are outliers. Suppose you wish to fit a model to the inliers and that $k \ll N$ is the minimum number of points needed to estimate the model.

1. Suppose that you sample k out of N data points with replacement. What is the probability that all k points are inliers?
2. Suppose that not all k points are inliers, and so you keep sampling k points m times. Show that the probability that after m trials all k points are inliers for the first time is $1-(1-p^k)^m$.
3. Show that the number of trials needed so that the probability that all k points are inliers is at least q is given by

$$m \geq \frac{\log(1-q)}{\log(1-p^k)}. \tag{B.112}$$

Appendix C
Basic Facts from Algebraic Geometry

Algebra is but written geometry; geometry is but drawn algebra.

—Sophie Germain

A centuries-old practice in science and engineering it to fit polynomials to a given set of data points. In this book, we often use the set of zeros of (multivariate) polynomials to model a given data set. In mathematics, polynomials and their zero sets are studied in algebraic geometry, with Hilbert's Nullstellensatz establishing the basic link between algebra (polynomials) and geometry (the zero set of polynomials, a geometric object). In order to make this book self-contained, we review in this appendix some of the basic algebraic notions and facts that are used in this book, especially in Chapter 5. In particular, we will introduce the special algebraic properties of multiple subspaces as algebraic sets. For a more systematic introduction to abstract algebra and algebraic geometry, the reader may refer to the classic texts of Lang (Lang 1993) and Eisenbud (Eisenbud 1996).

C.1 Abstract Algebra Basics

C.1.1 Polynomial Rings

Consider a D-dimensional vector space over a field R (of characteristic 0), denoted by R^D, where R is usually the field of real numbers $\mathbb{R}$ or the field of complex numbers $\mathbb{C}$.

Let $R[\boldsymbol{x}] = [x_1, x_2, \ldots, x_D]$ be the set of all polynomials in D variables $x_1, x_2, \ldots, x_D$. Then $R[\boldsymbol{x}]$ is a *commutative ring* with two basic operations: "summation" and "multiplication" of polynomials. The elements of R are called *scalars* or *constants*. A *monomial* is a product of the variables; its degree is the number

R. Vidal et al., *Generalized Principal Component Analysis*, Interdisciplinary Applied Mathematics 40, DOI 10.1007/978-0-387-87811-9

of the variables (counting repeats). A monomial of degree n is of the form $\boldsymbol{x}^{\boldsymbol{n}} = x_1^{n_1} x_2^{n_2} \cdots x_D^{n_D}$ with $0 \leq n_j \leq n$ and $n_1 + n_2 + \cdots + n_D = n$. Altogether, there are

$$M_n(D) \doteq \binom{D+n-1}{n} = \binom{D+n-1}{D-1}$$

different degree-n monomials.

Definition C.1 (Veronese Map). *For given n and D, the* Veronese map *of degree n, denoted by $\nu_n : R^D \to R^{M_n(D)}$, is defined as*

$$\nu_n : [x_1, \ldots, x_D]^\top \mapsto [\ldots, \boldsymbol{x}^{\boldsymbol{n}}, \ldots]^\top, \tag{C.1}$$

where $\boldsymbol{x}^{\boldsymbol{n}}$ are degree-n monomials of the form $x_1^{n_1} x_2^{n_2} \cdots x_D^{n_D}$ with $\boldsymbol{n} = (n_1, n_2, \ldots, n_D)$ chosen in the degree-lexicographic order.

Example C.2 (The Veronese Map of Degree 2 in Three Variables). If $\boldsymbol{x} = [x_1, x_2, x_3]^\top \in R^3$, the Veronese map of degree 2 is given by

$$\nu_2(\boldsymbol{x}) = [x_1^2, x_1x_2, x_1x_3, x_2^2, x_2x_3, x_3^2]^\top \in R^6.$$

In the context of kernel methods (Chapter 4), the Veronese map is usually referred to as the polynomial embedding, and the ambient space $R^{M_n(D)}$ is called the *feature* space.

A *term* is a scalar multiplying a monomial. A polynomial $p(\boldsymbol{x})$ is said to be *homogeneous* if all its terms have the same degree. Sometimes, the word *form* is used to mean a homogeneous polynomial. Every homogeneous polynomial $p(\boldsymbol{x})$ of degree n can be written as

$$p(\boldsymbol{x}) = \boldsymbol{c}_n^\top \nu_n(\boldsymbol{x}) = \sum c_{n_1,\ldots,n_D} x_1^{n_1} \cdots x_D^{n_D}, \tag{C.2}$$

where $c_{n_1,\ldots,n_D} \in R$ are the coefficients associated with the monomials $\boldsymbol{x}^{\boldsymbol{n}} = x_1^{n_1} \cdots x_D^{n_D}$.

In this book, we are primarily interested in the *algebra* of homogeneous polynomials in D variables.[1] Because of that, we view R^D as a projective space—the set of one-dimensional subspaces (meaning lines through the origin). Every one-dimensional subspace, say a line L, can be represented by a point $[a_1, a_2, \ldots, a_D]^\top \neq [0, 0, \ldots, 0]^\top$ on the line. The result is a projective $(D-1)$-space over R that can be regarded as the D-tuples $[a_1, a_2, \ldots, a_D]^\top$ of elements of R, modulo the equivalence relation $[a_1, a_2, \ldots, a_D]^\top \sim [ba_1, ba_2, \ldots, ba_D]^\top$ for all $b \neq 0$ in R.

If $p(x_1, x_2, \ldots, x_D)$ is a homogeneous polynomial of degree n, then for $b \in R$, we have

$$p(ba_1, ba_2, \ldots, ba_D) = b^n p(a_1, a_2, \ldots, a_D). \tag{C.3}$$

[1] For algebra of polynomials defined on R^D as an affine space, the reader may refer to (Lang 1993).

Therefore, whether $p(a_1, a_2, \ldots, a_D) = 0$ on a line L does not depend on the representative point chosen on the line L.

We may view $R[\boldsymbol{x}]$ as a *graded ring*, which can be decomposed as

$$R[\boldsymbol{x}] = \bigoplus_{i=0}^{\infty} R_i = R_0 \oplus R_1 \oplus \cdots \oplus R_n \oplus \cdots, \tag{C.4}$$

where R_i consists of all polynomials of degree i. In particular, $R_0 = R$ is the set of nonzero scalars (or constants). It is convention (and convenient) to define the degree of the zero element 0 in R to be infinite or -1. The set R_1 consists of all homogeneous polynomials of degree one, i.e., the set of 1-forms,

$$R_1 \doteq \{b_1x_1 + b_2x_2 + \cdots + b_Dx_D : \ [b_1, b_2, \ldots, b_D]^\top \in R^D\}. \tag{C.5}$$

Obviously, the dimension of R_1 as a vector space is also D; R_1 can also be viewed as the dual space $(R^D)^*$ of R^D. For convenience, we also define the following two sets:

$$R_{\leq m} \doteq \bigoplus_{i=0}^{m} R_i = R_0 \oplus R_1 \oplus \cdots \oplus R_m,$$

$$R_{\geq m} \doteq \bigoplus_{i=m}^{\infty} R_i = R_m \oplus R_{m+1} \oplus \cdots,$$

which are the set of polynomials of degree less than or equal to m and those of degree greater than or equal to m, respectively.

C.1.2 Ideals and Algebraic Sets

Definition C.3 (Ideal). *An* ideal *in the (commutative) polynomial ring $R[\boldsymbol{x}]$ is an additive subgroup I (with respect to the summation of polynomials) such that if $p(\boldsymbol{x}) \in I$ and $q(\boldsymbol{x}) \in R[\boldsymbol{x}]$, then $p(\boldsymbol{x})q(\boldsymbol{x}) \in I$.*

From the definition, one can verify that if I, J are two ideals of $R[\boldsymbol{x}]$, their intersection $K = I \cap J$ is also an ideal. The previously defined set $R_{\geq m}$ is an ideal for every m. In particular, $R_{\geq 1}$ is the so-called *irrelevant ideal*, sometimes denoted by R_+.

An ideal is said to be *generated* by a subset $\mathcal{G} \subset I$ if every element $p(\boldsymbol{x}) \in I$ can be written in the form

$$p(\boldsymbol{x}) = \sum_{i=1}^{k} q_i(\boldsymbol{x}) g_i(\boldsymbol{x}), \quad \text{with } q_i(\boldsymbol{x}) \in R[\boldsymbol{x}] \text{ and } g_i(\boldsymbol{x}) \in \mathcal{G}. \tag{C.6}$$

We write $(\mathcal{G})$ for the ideal generated by a subset $\mathcal{G} \subset R[\boldsymbol{x}]$; if $\mathcal{G}$ contains only a finite number of elements $\{g_1, \ldots, g_k\}$, we usually write $(g_1, \ldots, g_k)$ in place of $(\mathcal{G})$. An ideal I is *principal* if it can be generated by one element (i.e., $I = p(\boldsymbol{x})R[\boldsymbol{x}]$ for some polynomial $p(\boldsymbol{x})$). Given two ideals I and J, the ideal that is generated by the product of elements in I and J,

$$\{f(\boldsymbol{x})g(\boldsymbol{x}),\ f(\boldsymbol{x}) \in I, g(\boldsymbol{x}) \in J\}$$

is called the *product ideal*, denoted by IJ.

An ideal I of the polynomial ring $R[\boldsymbol{x}]$ is *prime* if $I \neq R[\boldsymbol{x}]$ and if $p(\boldsymbol{x}), q(\boldsymbol{x}) \in R[\boldsymbol{x}]$ and $p(\boldsymbol{x})q(\boldsymbol{x}) \in I$ implies that $p(\boldsymbol{x}) \in I$ or $q(\boldsymbol{x}) \in I$. If I is prime, then for any ideals J, K with $JK \subseteq I$, we have $J \subseteq I$ or $K \subseteq I$.

A polynomial $p(\boldsymbol{x})$ is said to be *prime* or irreducible if $p(\boldsymbol{x})$ generates a prime ideal. Equivalently, $p(\boldsymbol{x})$ is irreducible if $p(\boldsymbol{x})$ is not a nonzero scalar and whenever $p(\boldsymbol{x}) = f(\boldsymbol{x})g(\boldsymbol{x})$, then one of $f(\boldsymbol{x})$ and $g(\boldsymbol{x})$ is a nonzero scalar.

Definition C.4 (Homogeneous Ideal). *A homogeneous ideal of $R[\boldsymbol{x}]$ is an ideal that is generated by homogeneous polynomials.*

Note that the sum of two homogeneous polynomials of different degrees is no longer a homogeneous polynomial. Thus, a homogeneous ideal contains inhomogeneous polynomials too.

Definition C.5 (Algebraic Set). *Given a set of homogeneous polynomials $J \subset R[\boldsymbol{x}]$, we may define a corresponding (projective) algebraic set $Z(J)$ as a subset of R^D to be*

$$Z(J) \doteq \{[a_1, a_2, \ldots, a_D]^\top \in R^D | f(a_1, a_2, \ldots, a_D) = 0, \forall f \in J\}. \tag{C.7}$$

If we view algebraic sets as the closed sets of R^D, this assigns a topology to the space R^D, which is called the *Zariski topology*.[2]

If $X = Z(J)$ is an algebraic set, an algebraic subset $Y \subset X$ is a set of the form $Y = Z(K)$ (where K is a set of homogeneous polynomials) that happens to be contained in X. A nonempty algebraic set is said to be *irreducible* if it is not the union of two nonempty smaller algebraic subsets. We call irreducible algebraic sets *algebraic varieties*. For instance, every subspace of R^D is an irreducible algebraic variety.

There is an inverse construction of algebraic sets. Given any subset $X \subseteq R^D$, we define the *vanishing ideal of* X to be the set of all polynomials that vanish on X:

$$I(X) \doteq \{f(\boldsymbol{x}) \in R[\boldsymbol{x}] | f(a_1, a_2, \ldots, a_n) = 0, \forall [a_1, a_2, \ldots, a_n]^\top \in X\}. \tag{C.8}$$

[2]This is because the intersection of any algebraic sets is an algebraic set; and the union of finitely many algebraic sets is also an algebraic set.

One can verify that $I(X)$ is an ideal. Treating two polynomials as equivalent if they agree at all the points of X, we get the *coordinate ring* $A(X)$ of X as the quotient $R[\mathbf{x}]/I(X)$ (see (Eisenbud 1996) for details).

Now let us consider a set of homogeneous polynomials $J \subset R[\mathbf{x}]$ (which is not necessarily an ideal) and a subset $X \subset R^D$ (which is not necessarily an algebraic set).

Proposition C.6. *The following assertions are true:*

1. *$I(Z(J))$ is an ideal that contains J;*
2. *$Z(I(X))$ is an algebraic set that contains X.*

Proposition C.7. *If X is an algebraic set and $I(X)$ is the vanishing ideal of X, then X is irreducible if and only if I is a prime ideal.*

Proof. If X is irreducible and $f(\mathbf{x})g(\mathbf{x}) \in I$, since $Z(\{I,f(\mathbf{x})\}) \cup Z(\{I,g(\mathbf{x})\}) = X$, then either $X = Z(\{I,f(\mathbf{x})\})$ or $X = Z(\{I,g(\mathbf{x})\})$. That is, either $f(\mathbf{x})$ or $g(\mathbf{x})$ vanishes on X and is in I. Conversely, suppose $X = X_1 \cup X_2$. If both X_1 and X_2 are algebraic sets strictly smaller than X, then there exist polynomials $f_1(\mathbf{x})$ and $f_2(\mathbf{x})$ that vanish on X_1 and X_2 respectively, but not on X. Since the product $f_1(\mathbf{x})f_2(\mathbf{x})$ vanishes on X, we have $f_1(\mathbf{x})f_2(\mathbf{x}) \in I$, but neither $f_1(\mathbf{x})$ nor $f_2(\mathbf{x})$ is in I. So I is not prime. □

C.1.3 Algebra and Geometry: Hilbert's Nullstellensatz

In practice, we often use an algebraic set to model a given set of data points, and the (ideal of) polynomials that vanish on the set provides a natural parametric model for the data. One question that is of particular importance in this context is whether there is a one-to-one correspondence between ideals and algebraic sets. This is in general not true, since the ideals $I = (f^2(\mathbf{x}))$ and $J = (f(\mathbf{x}))$ both vanish on the same algebraic set as the zero set of the polynomial $f(\mathbf{x})$. Fortunately, this turns out to be essentially the only case that prevents the one-to-one correspondence between ideals and algebraic sets.

Definition C.8 (Radical Ideal)**.** *Given a (homogeneous) ideal I of $R[\mathbf{x}]$, the (homogeneous) radical ideal of I is defined to be*

$$rad(I) \doteq \{f(\mathbf{x}) \in R[\mathbf{x}] | f(\mathbf{x})^m \in I \text{ for some integer } m\}. \tag{C.9}$$

We leave it to the reader to verify that $\mathrm{rad}(I)$ is indeed an ideal and furthermore, that if I is homogeneous, then so is $\mathrm{rad}(I)$.

Hilbert proved in 1893 the following important theorem that establishes one of the fundamental results in algebraic geometry:

Theorem C.9 (Nullstellensatz)**.** *Let R be an algebraically closed field (e.g., $R = \mathbb{C}$). If $I \subset R[\mathbf{x}]$ is a (homogeneous) ideal, then*

$$I(Z(I)) = rad(I). \tag{C.10}$$

Thus, the maps $I \mapsto Z(I)$ and $X \mapsto I(X)$ induce a one-to-one correspondence between the collection of (projective) algebraic sets of R^D and (homogeneous) radical ideals of $R[\mathbf{x}]$.

One may find up to five different proofs for this theorem in (Eisenbud 1996).[3] The importance of the Nullstellensatz cannot be exaggerated. It is a natural extension of Gauss's fundamental theorem of algebra[4] to multivariate polynomials. One of the remarkable consequences of the Nullstellensatz is that it identifies a geometric object (algebraic sets) with an algebraic object (radical ideals).

In our context, we often assume that our data points are drawn from an algebraic set and use the set of vanishing polynomials as a parametric model for the data. Hilbert's Nullstellensatz guarantees that such a model for the data is well defined and unique. To some extent, when we fit vanishing polynomials to the data, we are essentially inferring the underlying algebraic set. In the next section, we will discuss how to extend Hilbert's Nullstellensatz to the practical situation in which we have only finitely many sample points from an algebraic set.

C.1.4 Algebraic Sampling Theory

We often face a common mathematical problem: how to identify a (projective) algebraic set $Z \subseteq R^D$ from a finite, though perhaps very large, number of sample points in Z. In general, the algebraic set Z is not necessarily irreducible,[5] and the ideal $I(Z)$ is not necessarily prime.

From an algebraic viewpoint, it is impossible to recover a continuous algebraic set Z from a finite number of discrete sample points. To see this, note that the set of all polynomials that vanish on one (projective) point z is a submaximal ideal[6] $\mathfrak{m}$ in the (homogeneous) polynomial ring $R[z]$. The set of polynomials that vanish on a set of sample points $\{z_1, z_2, \dots, z_i\} \subseteq Z$ is the intersection

$$\mathfrak{a}_i \doteq \mathfrak{m}_1 \cap \mathfrak{m}_2 \cap \cdots \cap \mathfrak{m}_i, \tag{C.11}$$

which is a radical ideal that is typically much larger than $I(Z)$.

[3] Strictly speaking, for homogeneous ideals, for the one-to-one correspondence to be exact, one should consider only proper radical ideals.

[4] Every degree-n polynomial in one variable has exactly n roots in an algebraically closed field such as $\mathbb{C}$ (counting repeats).

[5] For instance, it is often the case that Z is the union of many subspaces or algebraic surfaces.

[6] The ideal of a point in the affine space is a maximal ideal; and the ideal of a point in the projective space is called a submaximal ideal. They both are "maximal" in the sense that they cannot be a subideal of any other homogeneous ideal of the polynomial ring.

Thus, some additional assumptions must be imposed on the algebraic set in order to make the problem of inferring $I(Z)$ from the samples well defined. Typically, we assume that the ideal $I(Z)$ of the algebraic set Z in question is generated by a set of (homogeneous) polynomials whose degrees are bounded by a relatively small n. That is,

$$\begin{aligned} I(Z) &\doteq \big(f_1,f_2,\ldots,f_s\big) \quad \text{s.t.} \quad \deg(f_j) \leq n, \\ Z(I) &\doteq \big\{z \in R^D \mid f_i(z) = 0, i = 1,2,\ldots,s\big\}. \end{aligned}$$

We are interested in retrieving $I(Z)$ uniquely from a set of sample points $\{z_1,z_2,\ldots,z_i\} \subseteq Z$. In general, $I(Z)$ is always a proper subideal of $\mathfrak{a}_i$, regardless of how large i is. However, the information about $I(Z)$ can still be retrieved from $\mathfrak{a}_i$ in the following sense.

Theorem C.10 (Sampling of an Algebraic Set). *Consider a nonempty set $Z \subseteq R^D$ whose vanishing ideal $I(Z)$ is generated by polynomials in $R_{\leq n}$. Then there is a finite sequence $F_N = \{z_1,\ldots,z_N\}$ such that the subspace $I(F_N) \cap R_{\leq n}$ generates $I(Z)$.*

Proof. Let $I_{\leq n} = I(Z) \cap R_{\leq n}$. This vector space generates $I(Z)$. Let $\mathfrak{a}_0 = R[\boldsymbol{x}] = I(\emptyset)$. Let $\mathfrak{b}_0 = \mathfrak{a}_0 \cap R_{\leq n}$ and let $A_0 = (\mathfrak{b}_0)$, the ideal generated by polynomials in $\mathfrak{a}_0$ of degree less than or equal to n. Since $1 \in R[\boldsymbol{x}] \cap R_{\leq n}$ is the generator of this ideal, we have $A_0 = R[\boldsymbol{x}]$. Since $Z \neq \emptyset$, then $A_0 \neq I(Z)$. Set $N = 1$ and pick a point $z_1 \in Z$. Then $1(z_1) \neq 0$ (1 is the function that assigns 1 to every point of Z). Let $\mathfrak{a}_1$ be the ideal that vanishes on $\{z_1\}$ and define $\mathfrak{b}_1 = \mathfrak{a}_1 \cap R_{\leq n}$. Further, let $A_1 = (\mathfrak{b}_1)$.[7] Since $I(Z) \subseteq \mathfrak{a}_1$, it follows that $I_{\leq n} \subseteq \mathfrak{b}_1$. If $A_1 = I(Z)$, then we are done. Suppose then that $I(Z) \subset A_1$.

Let us do the induction at this point. Suppose we have found a finite sequence $F_N = \{z_1,z_2,\ldots,z_N\} \subset Z$ with

$$I(F_N) = \mathfrak{a}_N \tag{C.12}$$

$$\mathfrak{b}_N = \mathfrak{a}_N \cap R_{\leq n} \tag{C.13}$$

$$A_N = (\mathfrak{b}_N) \tag{C.14}$$

$$\mathfrak{b}_0 \supset \mathfrak{b}_1 \supset \cdots \supset \mathfrak{b}_N \supseteq I_{\leq n}. \tag{C.15}$$

It follows that $I_{\leq n} \subseteq \mathfrak{b}_N$ and that $I(Z) \subseteq A_N$. If equality holds here, then we are done. If not, then there exist a function $g \in \mathfrak{b}_N$ not in $I(Z)$ and an element $z_{N+1} \in Z$ for which $g(z_{N+1}) \neq 0$. Set $F_{N+1} = \{z_1,\ldots,z_N,z_{N+1}\}$. Then one gets $\mathfrak{a}_{N+1},\mathfrak{b}_{N+1},A_{N+1}$ as before with

$$\mathfrak{b}_0 \supset \mathfrak{b}_1 \supset \cdots \supset \mathfrak{b}_N \supset \mathfrak{b}_{N+1} \supseteq I_{\leq n}. \tag{C.16}$$

[7] Here we are using the convention that (S) is the ideal generated by the set S. Recall also that the ring $R[\boldsymbol{x}]$ is *Noetherian* by the Hilbert basis theorem, and so all ideals in the ring are finitely generated (Lang 1993).

We obtain a descending chain of subspaces of the vector space $R_{\leq n}$. This chain must stabilize, since the vector space is finite-dimensional. Hence there is an N for which $\mathfrak{b}_N = I_{\leq n}$, and we are done. □

We point out that in the above proof, no clear bound on the total number N of points needed is given.[8] Nevertheless, from the proof of the theorem, the set of finite sequences of samples that satisfy the theorem is an open set. This is of great practical importance: with probability one, the vanishing ideal of an algebraic set can be correctly determined from a randomly chosen sequence of samples.

Example C.11 (A Hyperplane in $\mathbb{R}^3$). Consider a plane $P = \{z \in R^3 : f(z) = az_1 + bz_2 + cz_3 = 0\}$. Given any two points in general position in the plane P, $f(\boldsymbol{x}) = ax_1 + bx_2 + cx_3$ will be the only (homogeneous) polynomial of degree 1 that fits the two points. In terms of the notation introduced earlier, we have $I(P) = \left(\mathfrak{a}_2 \cap R_{\leq 1}\right)$.

Example C.12 (Zero Polynomial). When $Z = R^D$, the only polynomial that vanishes on Z is the zero polynomial, i.e., $I(Z) = (0)$. Since the zero polynomial is considered to be of degree -1, we have $(\mathfrak{a}_N \cap R_{\leq n}) = \emptyset$ for any given n (and large enough N).

The above theorem can be viewed as a first step toward an algebraic analogy to the well-known Nyquist–Shannon sampling theorem in signal processing, which stipulates that a continuous signal with a limited frequency bandwidth Ω can be uniquely determined from a sequence of discrete samples with a sampling rate higher than 2Ω. Here a signal is replaced by an algebraic set, and the frequency bandwidth is replaced by the bound on the degree of polynomials. It has been widely practiced in engineering that a curve or surface described by polynomial equations can be recovered from a sufficient number of sample points in general configuration, a procedure often loosely referred to as "polynomial fitting." However, the algebraic basis for this is often not clarified, and the conditions for the uniqueness of the solution are usually not well characterized or specified. This problem certainly merits further investigation.

C.1.5 Decomposition of Ideals and Algebraic Sets

Modeling a data set as an algebraic set does not stop at obtaining its vanishing ideal (and polynomials). The ultimate goal is to extract all the internal geometric or algebraic structures of the algebraic set. For instance, if an algebraic set consists of

[8] However, loose bounds can be obtained from the dimension of $R_{\leq n}$ as a vector space. In fact, in the algorithm, we implicitly used the dimension of $R_{\leq n}$ as a bound for N.

multiple subspaces, called a subspace arrangement, we need to know how to derive from its vanishing ideal the number of subspaces, their dimensions, and a basis of each subspace.

Thus, given an algebraic set X or equivalently its vanishing ideal $I(X)$, we want to decompose or segment it into a union of subsets each of which can no longer be further decomposed. As we mentioned earlier, an algebraic set that cannot be decomposed into smaller algebraic sets is called irreducible. As one of the fundamental finiteness theorems of algebraic geometry, we have the following.

Theorem C.13. *An algebraic set can have only finitely many irreducible components. That is, for some n,*

$$X = X_1 \cup X_2 \cup \cdots \cup X_n, \tag{C.17}$$

where $X_1, X_2, \ldots, X_n$ are irreducible algebraic varieties.

Proof. The proof is essentially based on the fact that the polynomial ring $R[\boldsymbol{x}]$ is Noetherian (i.e., finitely generated), and there are only finitely many prime ideals containing $I(X)$ that are minimal with respect to inclusion (See (Eisenbud 1996)). □

The vanishing ideal $I(X_i)$ of each irreducible algebraic variety X_i must be a prime ideal that is minimal over the radical ideal $I(X)$ – there is no prime subideal of $I(X_i)$ that includes $I(X)$. The ideal $I(X)$ is precisely the intersection of all the minimal prime ideals:

$$I(X) = I(X_1) \cap I(X_2) \cap \cdots \cap I(X_n). \tag{C.18}$$

This intersection is called a *minimal primary decomposition* of the radical ideal $I(X)$. Thus the primary decomposition of a radical ideal is closely related to the notion of "segmenting" or "decomposing" an algebraic set into multiple irreducible algebraic varieties: if we know how to decompose the ideal, we can find the irreducible algebraic variety corresponding to each primary component.

We are particularly interested in a special class of algebraic sets known as subspace arrangements. One of the goals of subspace clustering and modeling is to decompose a subspace arrangement into individual (irreducible) subspaces (see Chapter 5). In later sections, we will further study the algebraic properties of subspace arrangements.

C.1.6 Hilbert Function, Polynomial, and Series

Finally, we introduce an important invariant of algebraic sets, given by the Hilbert function. Knowing the values of the Hilbert function can be very useful in the identification of subspace arrangements, especially the number of subspaces and their dimensions.

Given a (projective) algebraic set Z and its vanishing ideal $I(Z)$, we can grade the ideal by degree as

$$I(Z) = I_0(Z) \oplus I_1(Z) \oplus \cdots \oplus I_i(Z) \oplus \cdots . \tag{C.19}$$

The *Hilbert function* of Z is defined to be

$$h_I(i) \doteq \dim(I_i(Z)). \tag{C.20}$$

Notice that $h_I(i)$ is exactly the number of linearly independent polynomials of degree i that vanish on Z. In this book, we also refer to h_I as the Hilbert function of the algebraic set Z.[9]

The *Hilbert series*, also known as the Poincaré series, of the ideal I is defined to be the power series[10]

$$\mathcal{H}(I,t) \doteq \sum_{i=0}^{\infty} h_I(i)t^i = h_I(0) + h_I(1)t + h_I(2)t^2 + \cdots . \tag{C.21}$$

Thus, given $\mathcal{H}(I,t)$, we know all the values of the Hilbert function h_I from its coefficients.

Example C.14 (Hilbert Series of a Polynomial Ring). The Hilbert series of the polynomial ring $R[\boldsymbol{x}] = \mathbb{R}[x_1, x_2, \ldots, x_D]$ is

$$\mathcal{H}(R[\boldsymbol{x}],t) = \sum_{i=0}^{\infty} \dim(R_i)t^i = \sum_{i=0}^{\infty} \binom{D+i-1}{i} t^i = \frac{1}{(1-t)^D}. \tag{C.22}$$

One can verify the correctness of the formula with the special case $D = 1$. Obviously, the coefficients of the Hilbert series of any ideal (as a subset of $R[\boldsymbol{x}]$) are bounded by those of $\mathcal{H}(R[\boldsymbol{x}],t)$, and hence the Hilbert series converges.

Example C.15 (Hilbert Series of a Subspace). The above formula can be generalized to the vanishing ideal of a subspace S of dimension d in $\mathbb{R}^D$. Let the codimension of the subspace be $c = D - d$. We have

$$\mathcal{H}(I(S),t) = \left(\frac{1}{(1-t)^c} - 1\right) \cdot \left(\frac{1}{(1-t)^{D-c}}\right) = \frac{1-(1-t)^c}{(1-t)^D}. \tag{C.23}$$

[9] In the literature, however, the Hilbert function of an algebraic set Z is sometimes defined to be the dimension of the homogeneous components of the coordinate ring $A(Z) \doteq R[\boldsymbol{x}]/I(Z)$ of Z, which is the codimension of $I_i(Z)$ as a subspace in R_i.

[10] In general, the Hilbert series can be defined for any finitely generated graded module $E = \bigoplus_{i=1}^{\infty} E_i$ using any Euler–Poincaré $\mathbb{Z}$-valued function $h_E(\cdot)$ as $\mathcal{H}(E,t) \doteq \sum_{i=0}^{\infty} h_E(i)t^i$ (Lang 1993). Here, for $E = I$, we choose $h_I(i) = \dim(I_i)$.

The following theorem, also due to Hilbert, reveals that the values of the Hilbert function of an ideal have some remarkable properties:

Theorem C.16 (Hilbert Polynomial). *Let $I(Z)$ be the vanishing ideal of an algebraic set Z over $R[x_1, \ldots, x_D]$. Then the values of its Hilbert function $h_I(i)$ agree, for large i, with those of a polynomial of degree $\leq D$. This polynomial, denoted by $H_I(i)$, is called the* Hilbert polynomial *of* $I(Z)$.

Then in the above example, for the polynomial ring, the Hilbert function itself is a polynomial in i:

$$H_R(i) = h_R(i) = \binom{D+i-1}{i} = \frac{1}{(D-1)!}(D+i-1)(D+i-2)\cdots(i+1).$$

However, for a general ideal I (of an algebraic set), it is not necessarily true that all values of its Hilbert function h_I agree with those of its Hilbert polynomial H_I. They might agree only when i is large enough. Thus, for a given algebraic set (or ideal), it would be interesting to know how large i needs to be in order for the Hilbert function to coincide with a polynomial. As we will soon see, for subspace arrangements, there is a very elegant answer to this question. One can even derive closed-form formulas for the Hilbert polynomials. These results are very important and useful for the subspace clustering problem, both conceptually and computationally.

C.2 Ideals of Subspace Arrangements

In this book, the main problem that we study is how to cluster a collection of data points drawn from a subspace arrangement $\mathcal{A} = \{S_1, S_2, \ldots, S_n\}$, formally introduced in Chapter 5;[11] $Z_{\mathcal{A}} = S_1 \cup S_2 \cup \cdots \cup S_n$ is the union of all the subspaces, and $Z_{\mathcal{A}}$ can be naturally described as the zero set of a set of polynomials, which makes it an *algebraic set*. The solution to the above problem typically relies on inferring the subspace arrangement $Z_{\mathcal{A}}$ from the data points. Thus, knowing the algebraic properties of $Z_{\mathcal{A}}$ may significantly facilitate this task.

Although subspace arrangements seem to be a very simple class of algebraic sets, a full characterization of their algebraic properties is a surprisingly difficult, if not impossible, task. Subspace arrangements have been a centuries-old subject that still actively interweaves many mathematical fields: algebraic geometry and topology, combinatorics, and complexity theory, graph and lattice theory, etc. Although the results are extremely rich and deep, in fact only a few special classes of subspace arrangements have been well characterized.

[11]Unless stated otherwise, the subspace arrangement considered will always be a central arrangement, as in Definition 5.4.

In the remaining sections of this appendix, we examine some important concepts and properties of subspace arrangements that are closely related to the subspace-clustering problem. The purpose of these sections is twofold: 1. to provide a rigorous justification for the algebraic subspace clustering algorithms derived in Chapter 5; 2. to summarize some important in-depth properties of subspace arrangements, which may suggest potential improvements of the algorithms. For readers who are interested only in the basic subspace clustering algorithms and their applications, these sections can be skipped on a first reading.

Vanishing Ideal of a Subspace.
A d-dimensional subspace S can be defined by $k = D-d$ linearly independent linear forms $\{l_1, l_2, \ldots, l_k\}$:

$$S \doteq \{\boldsymbol{x} \in R^D : l_i(\boldsymbol{x}) = 0,\ i = 1, 2, \ldots, k = D - d\}, \tag{C.24}$$

where l_i is of the form $l_i(\boldsymbol{x}) = a_{i1}x_1 + a_{i2}x_2 + \cdots a_{iD}x_D$ with $a_{ij} \in R$. Let S^* denote the space of all linear forms that vanish on S. Then $\dim(S^*) \doteq k = D - d$. The subspace S is also called the zero set of S^*, i.e., points in the ambient space that vanish on all polynomials in S^*, which is denoted by $Z(S^*)$. We define

$$I(S) \doteq \{p \in R[\boldsymbol{x}] : p(\boldsymbol{x}) = 0,\ \forall \boldsymbol{x} \in S\}. \tag{C.25}$$

Clearly, $I(S)$ is an ideal generated by linear forms in S^*, and it contains polynomials of all degrees that vanish on the subspace S. Every polynomial $p(\boldsymbol{x})$ in $I(S)$ can be written as a superposition:

$$p = l_1 h_1 + l_2 h_2 + \cdots + l_k h_k \tag{C.26}$$

for some polynomials $h_1, h_2, \ldots, h_k \in R[\boldsymbol{x}]$. Furthermore, $I(S)$ is a prime ideal.[12]

Vanishing Ideal of a Subspace Arrangement
Given a subspace arrangement $Z_{\mathcal{A}} = S_1 \cup S_2 \cup \cdots \cup S_n$, its vanishing ideal is

$$I(Z_{\mathcal{A}}) = I(S_1) \cap I(S_2) \cap \cdots \cap I(S_n). \tag{C.27}$$

The ideal $I(Z_{\mathcal{A}})$ can be graded by the degree of its polynomials

$$I(Z_{\mathcal{A}}) = I_m(Z_{\mathcal{A}}) \oplus I_{m+1}(Z_{\mathcal{A}}) \oplus \cdots \oplus I_i(Z_{\mathcal{A}}) \oplus \cdots. \tag{C.28}$$

Each $I_i(Z_{\mathcal{A}})$ is a vector space that consists of forms of degree i in $I(Z_{\mathcal{A}})$, and $m \geq 1$ is the least degree of the polynomials in $I(Z_{\mathcal{A}})$. Notice that forms that vanish on $Z_{\mathcal{A}}$ may have degrees strictly less than n. One example is an arrangement of two lines and one plane in $\mathbb{R}^3$. Since any two lines lie on a plane, the arrangement can

[12] It is a prime ideal because for every product $p_1 p_2 \in I(S)$, either $p_1 \in I(S)$ or $p_2 \in I(S)$.

be embedded into a hyperplane arrangement of two planes, and there exist forms of second degree that vanish on the union of the three subspaces. The dimension of $I_i(Z_{\mathcal{A}})$ is known as the Hilbert function $h_I(i)$ of $Z_{\mathcal{A}}$.

Example C.17 (Boolean Arrangement). The Boolean arrangement is the collection of coordinate hyperplanes $H_j \doteq \{\boldsymbol{x} : x_j = 0\}, 1 \leq j \leq D$. The vanishing ideal of the Boolean arrangement is generated by a single polynomial $p(\boldsymbol{x}) = x_1 x_2 \cdots x_D$ of degree D.

Example C.18 (Braid Arrangement). The braid arrangement is the collection of hyperplanes $H_{jk} \doteq \{\boldsymbol{x} : x_j - x_k = 0\}, 1 \leq j \neq k \leq D$. Similarly, the vanishing ideal of the Braid arrangement is generated by a single polynomial $p(\boldsymbol{x}) = \prod_{1 \leq j < k \leq D}(x_j - x_k)$.

Theorem C.19 (Regularity of Subspace Arrangements). *The vanishing ideal $I(Z_{\mathcal{A}})$ of a subspace arrangement $Z_{\mathcal{A}} = S_1 \cup S_2 \cup \cdots \cup S_n$ is n-regular. This implies that $I(Z)$ has a set of generators with degree $\leq n$.*

Proof. For the concept of n-regularity and the proof of the above statement, please refer to (Derksen 2007) and references therein. □

Due to the above theorem, the subspace arrangement $Z_{\mathcal{A}}$ is uniquely determined as the zero set of all polynomials of degree up to n in its vanishing ideal, i.e., as the zero set of polynomials in

$$Z_{\mathcal{A}} = Z(I_{\leq n}),$$

where $I_{\leq n} \doteq I_0 \oplus I_1 \oplus \cdots \oplus I_n$.

Product Ideal of a Subspace Arrangement
Let $J(Z_{\mathcal{A}})$ be the ideal generated by the products of linear forms

$$\{l_1 \cdot l_2 \cdots l_n, \quad \forall l_j \in S_j^*, j = 1, \ldots, n\}.$$

Or equivalently, we can define $J(Z_{\mathcal{A}})$ to be the product of the n ideals $I(S_1), I(S_2), \ldots, I(S_n)$:

$$J(Z_{\mathcal{A}}) \doteq I(S_1) \cdot I(S_2) \cdots I(S_n).$$

Then the *product ideal* $J(Z_{\mathcal{A}})$ is a subideal of $I(Z_{\mathcal{A}})$. Nevertheless, the two ideals share the same zero set:

$$Z_{\mathcal{A}} = Z(J) = Z(I). \tag{C.29}$$

By definition, I is the largest ideal that vanishes on $Z_{\mathcal{A}}$. In fact, I is the *radical ideal* of the product ideal J, i.e., $I = \text{rad}(J)$. We may also grade the ideal $J(Z_{\mathcal{A}})$ by the degree

$$J(Z_{\mathcal{A}}) = J_n(Z_{\mathcal{A}}) \oplus J_{n+1}(Z_{\mathcal{A}}) \oplus \cdots \oplus J_i(Z_{\mathcal{A}}) \oplus \cdots. \tag{C.30}$$

Notice that unlike I, the lowest degree of polynomials in J always starts from n, the number of subspaces. The Hilbert function of J is denoted by $h_J(i) = \dim(J_i(Z_{\mathcal{A}}))$. As we will soon see, the Hilbert functions (or polynomials, or series) of the product ideal J and the vanishing ideal I have very interesting and important relationships.

C.3 Subspace Embedding and PL-Generated Ideals

Let $Z_{\mathcal{A}}$ be a central subspace arrangement $Z_{\mathcal{A}} = S_1 \cup S_2 \cup \cdots \cup S_n$. Let $Z_{\mathcal{A}'} = S'_1 \cup S'_2 \cup \cdots \cup S'_{n'}$ be another (central) subspace arrangement. If we have $Z_{\mathcal{A}} \subseteq Z_{\mathcal{A}'}$, then it is necessary that for all $S_j \subset Z_{\mathcal{A}}$, there exist $S'_{j'} \subset Z_{\mathcal{A}'}$ such that $S_j \subseteq S'_{j'}$. If so, we call

$$Z_{\mathcal{A}} \subseteq Z_{\mathcal{A}'}$$

a *subspace embedding*. Beware that it is possible that $n' < n$ for a subspace embedding, since more than one subspace S_j of $Z_{\mathcal{A}}$ may belong to the same subspace $S_{j'}$ of $Z_{\mathcal{A}'}$. The subspace arrangements in Theorem 5.14 are examples of subspace embeddings. If $Z_{\mathcal{A}'}$ happens to be a hyperplane arrangement, we call the embedding a *hyperplane embedding*.

Is the zero-set of each homogeneous component of $I(Z_{\mathcal{A}})$, in particular $I_m(Z_{\mathcal{A}})$, a subspace embedding of $Z_{\mathcal{A}}$? Unfortunately, this is not true, since counterexamples can be constructed.

Example C.20 (Five Lines in $\mathbb{R}^3$). Consider five points in $\mathbb{P}^2$ (or equivalently, five lines in $\mathbb{R}^3$). The Veronese embedding of order two of a point $\boldsymbol{x} = [x_1, x_2, x_3] \in \mathbb{R}^3$ is $[x_1^2, x_1x_2, x_1x_3, x_2^2, x_2x_3, x_3^2] \in \mathbb{R}^6$. For five points in general position, the matrix $\boldsymbol{V}_2 = [\nu_2(\boldsymbol{x}_1), \nu_2(\boldsymbol{x}_2), \ldots \nu_2(\boldsymbol{x}_5)]$ is of rank 5. Let $\boldsymbol{c}^\top$ be the only vector in the left null space of $\boldsymbol{V}_2$ such that $\boldsymbol{c}^\top \boldsymbol{V}_2 = 0$. Then $p(\boldsymbol{x}) = \boldsymbol{c}^\top \nu_2(\boldsymbol{x})$ is in general an irreducible quadratic polynomial. Thus, the zero set of $I_2(Z_{\mathcal{A}}) = p(\boldsymbol{x})$ is not a subspace arrangement but an (irreducible) cone in $\mathbb{R}^3$.

Nevertheless, the following statement allows us to retrieve a subspace embedding from any polynomials in the vanishing ideal $I(Z_{\mathcal{A}})$.

Theorem C.21 (Hyperplane Embedding via Differentiation)**.** *For every polynomial p in the vanishing ideal $I(Z_{\mathcal{A}})$ of a subspace arrangement $Z_{\mathcal{A}} = S_1 \cup S_2 \cup \cdots \cup S_n$ and n points $\{\boldsymbol{x}_i \in S_i\}_{i=1}^n$ in general position, the union of the hyperplanes $\cup_{i=1}^n H_i = \{\boldsymbol{x} : \nabla p(\boldsymbol{x}_i)^\top \boldsymbol{x} = 0\}$ is a hyperplane embedding of the subspace arrangement.*

Proof. The proof is based on the simple fact that the derivative (gradient) $\nabla f(\boldsymbol{x})$ of any smooth function $f(\boldsymbol{x})$ is orthogonal to (the tangent space of) its level set $f(\boldsymbol{x}) = c$. □

In the above statement, if we replace p with a collection of polynomials in the vanishing ideal, their derivatives give a subspace embedding in a similar fashion as the hyperplane embedding. When the collection contains all the generators of the vanishing ideal, the subspace embedding becomes tight: the resulting subspace arrangement coincides with the original one. This property has been used in the development of algebraic subspace clustering algorithms in Chapter 5.

Another concept that is closely related to subspace embedding is a *pl-generated ideal.*

Definition C.22 (pl-Generated Ideals). *An ideal is said to be* pl-generated *if it is generated by products of linear forms.*

If the ideal of a subspace arrangement $Z_{\mathcal{A}}$ is pl-generated, then the zero set of every generator gives a hyperplane embedding of $Z_{\mathcal{A}}$.

Example C.23 (Hyperplane Arrangements). If $Z_{\mathcal{A}}$ is a hyperplane arrangement, then $I(Z_{\mathcal{A}})$ is always pl-generated, since it is generated by a single polynomial of the form[13]

$$p(\boldsymbol{x}) = (\boldsymbol{b}_1^\top \boldsymbol{x})(\boldsymbol{b}_2^\top \boldsymbol{x}) \cdots (\boldsymbol{b}_n^\top \boldsymbol{x}), \tag{C.31}$$

where $\boldsymbol{b}_i \in R^D$ are the normal vectors to the hyperplanes.

Obviously, the vanishing ideal $I(S)$ of a single subspace S is always pl-generated. The following example shows that this is also true for an arrangement of two subspaces.

Example C.24 (Two Subspaces). Let us show that for an arrangement $Z_{\mathcal{A}}$ of two subspaces, $I(Z_{\mathcal{A}})$ is always pl-generated. Let $Z_{\mathcal{A}} = S_1 \cup S_2$ and define $U^* \doteq S_1^* \cap S_2^*$ and $V^* \doteq S_1^* \setminus U^*, W^* \doteq S_2^* \setminus U^*$. Let $(u_1, u_2, \ldots, u_k)$ be a basis for U^*, $(v_1, v_2, \ldots, v_l)$ a basis for V^*, and $(w_1, w_2, \ldots, w_m)$ a basis for W^*. Then $I(Z_{\mathcal{A}}) = I(S_1) \cap I(S_2)$ is generated by $(u_1, \ldots, u_k, v_1 w_1, v_1 w_2, \ldots, v_l w_m)$.

Now consider an arrangement of n subspaces $Z_{\mathcal{A}} = S_1 \cup S_2 \cup \cdots \cup S_n$. By its definition, the product ideal $J(Z_{\mathcal{A}})$ is always pl-generated. Now, is the vanishing ideal $I(Z_{\mathcal{A}})$ always pl-generated? Unfortunately, this is not true. Below are some counterexamples.

Example C.25 (Lines in $\mathbb{R}^3$ (Björner et al. 2005)). For a central arrangement $Z_{\mathcal{A}}$ of r lines in general position in $\mathbb{R}^3$, $I(Z_{\mathcal{A}})$ is not pl-generated when $r = 5$ or $r > 6$. Example C.20 gives a proof for the case with $r = 5$.

Example C.26 (Planes in $\mathbb{R}^4$ (Björner et al. 2005)). For a central arrangement $Z_{\mathcal{A}}$ of r planes in general position in $\mathbb{R}^4$, $I(Z_{\mathcal{A}})$ is not pl-generated for all $r > 2$.

[13] In algebra, an ideal that is generated by a single generator is called a principal ideal.

However, can each homogeneous component $I_i(Z_{\mathcal{A}})$ be "pl-generated" when i is large enough? For instance, can it be that $I_n = J_n = S_1^* \cdot S_2^* \cdots S_n^*$? This is in general not true for an arbitrary arrangement. Below is a counterexample.

Example C.27 (Three Subspaces in $\mathbb{R}^5$; due to R. Fossum). Consider $R[\boldsymbol{x}] = \mathbb{R}[x_1, \dots, x_5]$ and an arrangement $Z_{\mathcal{A}}$ of three three-dimensional subspaces in $\mathbb{R}^5$ whose vanishing ideals are given by, respectively,

$$I(S_1) = (x_1, x_2), \quad I(S_2) = (x_3, x_4), \quad I(S_3) = ((x_1 + x_3), (x_2 + x_4)).$$

Denote their intersection by $I = I(S_1) \cap I(S_2) \cap I(S_3)$. The intersection contains the element

$$x_1x_4 - x_2x_3 = (x_1 + x_3)x_4 - (x_2 + x_4)x_3 = x_1(x_2 + x_4) - x_2(x_1 + x_3).$$

Then every element $(x_1x_4 - x_2x_3)l(x_1, \dots, x_5)$ with l a linear form is in $I_3(Z_{\mathcal{A}})$, the homogeneous component of elements of degree three. In particular, $(x_1x_4 - x_2x_3)x_5$ is in $I_3(Z_{\mathcal{A}})$. However, one can check that this element cannot be written in the form

$$\sum_i (a_ix_1 + b_ix_2)(c_ix_2 + d_ix_4)(e_i(x_1 + x_3) + f_i(x_2 + x_4))$$

for any $a_i, b_i, c_i, d_i, e_i, f_i \in \mathbb{R}$. Thus, $I_3(Z_{\mathcal{A}})$ is not spanned by $S_1^* \cdot S_2^* \cdot S_3^*$.

However, notice that the subspaces in the above example are not in "general position": their intersections are not of the minimum possible dimension. Could $I_n = J_n = S_1^* \cdot S_2^* \cdots S_n^*$ be instead true for n subspaces if they are in general position? The answer is yes. In fact, we can say more than that. As we will see in the next section, from the Hilbert functions of I and J, we actually have

$$I_i = J_i, \quad \forall i \geq n$$

if $S_1, S_2, \dots, S_n$ are "transversal" (i.e., all intersections are of minimum possible dimension). In other words, J_i could differ from I_i only for $i < n$.

C.4 Hilbert Functions of Subspace Arrangements

In this section, we study the Hilbert functions of subspace arrangements defined in Section C.1.6. We first discuss a few reasons why in the context of generalized principal component analysis, it is very important to know the values of the Hilbert function for the vanishing ideal I or the product ideal J of a subspace arrangement. We then examine the values of the Hilbert function for a few special examples. Finally, we give a complete characterization of the Hilbert function, the Hilbert polynomial, and the Hilbert series of a general subspace arrangement. In particular, we give a closed-form formula for the Hilbert polynomial of the vanishing ideal and the product ideal of the subspace arrangement.

C.4.1 Hilbert Function and Algebraic Subspace Clustering

In general, for a subspace arrangement $Z_{\mathcal{A}} = S_1 \cup S_2 \cup \cdots \cup S_n$ in general position, the values of the Hilbert function $h_I(i)$ of its vanishing ideal $I(Z_{\mathcal{A}})$ are invariant under a continuous change of the positions of the subspaces. They depend only on the dimensions of the subspaces $d_1, d_2, \ldots, d_n$ or their codimensions $c_i = D - d_i$, $i = 1, 2, \ldots, n$. Thus, the Hilbert function gives a rich set of invariants of subspace arrangements. In the context of subspace clustering, such invariants can help to determine the type of the subspace arrangement, such as the number of subspaces and their individual dimensions from a given set of (possibly noisy) sample points.

To see this, consider a sufficiently large number of sample points in general position $\mathcal{X} = \{\boldsymbol{x}_1, \boldsymbol{x}_2, \ldots, \boldsymbol{x}_N\} \subset Z_{\mathcal{A}}$ that are drawn from the subspaces, and let the embedded data matrix (via the Veronese map of degree i) be

$$\boldsymbol{V}_i \doteq [\nu_i(\boldsymbol{x}_1), \nu_i(\boldsymbol{x}_2), \ldots, \nu_i(\boldsymbol{x}_N)]^\top. \tag{C.32}$$

According to the algebraic sampling theorem of Appendix C.1.4, the dimension of $\mathrm{Null}(\boldsymbol{V}_i)$ is exactly the number of linearly independent polynomials of degree i that vanish on $Z_{\mathcal{A}}$. That is, the following relation holds:

$$\dim(\mathrm{Null}(\boldsymbol{V}_i)) = h_I(i), \tag{C.33}$$

or equivalently,

$$\mathrm{rank}(\boldsymbol{V}_i) = \dim(R_i) - h_I(i). \tag{C.34}$$

Thus, if we know the Hilbert function for different subspace arrangements in advance, we can determine from the rank of the data matrix from which subspace arrangement the sample data points are drawn. The following example illustrates the basic idea.

Example C.28 (Three Subspaces in $\mathbb{R}^3$). Suppose that we know only that our data are drawn from an arrangement of three subspaces in $\mathbb{R}^3$. There are in total four different types of such arrangements, shown in Figure C.1. The values of their corresponding Hilbert functions are listed in Table C.1. Given a sufficiently large number N of sample points from one of the above subspace arrangements, the rank of the embedded data matrix $\boldsymbol{V}_3 \in \mathbb{R}^{N \times 10}$ can be, instead of any value between 1 and 10, only $10 - h_I(3) = 9, 8, 6, 3$, which correspond to the only four possible configurations of three subspaces in $\mathbb{R}^3$: three planes, two planes and one line, one plane and two lines, or three lines, respectively, as shown in Figure C.1.

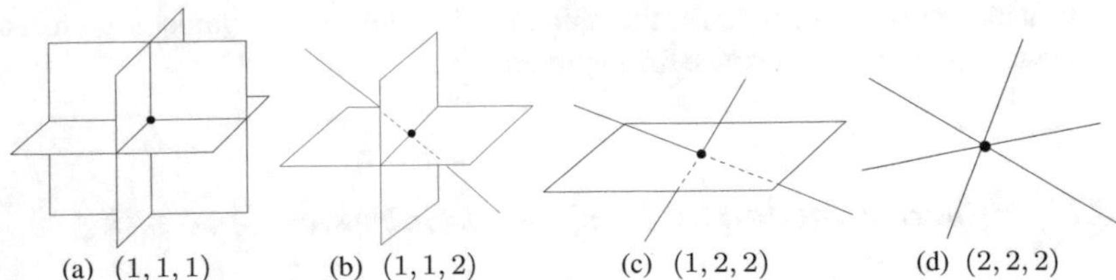

Fig. C.1 Four configurations of three subspaces in $\mathbb{R}^3$. The numbers are the codimensions (c_1, c_2, c_3) of the subspaces.

Table C.1 Values of the Hilbert functions of the four arrangements (assuming the subspaces are in general position).

c_1	c_2	c_3	$h_{I(Z_{\mathcal{A}})}(1)$	$h_{I(Z_{\mathcal{A}})}(2)$	$h_{I(Z_{\mathcal{A}})}(3)$
1	1	1	0	0	1
1	1	2	0	0	2
1	2	2	0	1	4
2	2	2	0	3	7

This suggests that given the dimensions of individual subspaces, we may know the rank of the embedded data matrix. Conversely, given the rank of the embedded data matrix, we can determine to a large extent the possible dimensions of the individual subspaces. Therefore, knowing the values of the Hilbert function will help us to at least rule out in advance impossible rank values for the embedded data matrix or the impossible subspace dimensions. This is particularly useful when the data are corrupted by noise, so that there is ambiguity in determining the rank of the embedded data matrix or the dimensions of the subspaces.

The next example illustrates how the values of the Hilbert function can help determine the correct number of subspaces.

Example C.29 (Overfit Hyperplane Arrangements in $\mathbb{R}^5$). Consider a data set sampled from a number of hyperplanes in general position in $\mathbb{R}^5$. Suppose we know only that the number of hyperplanes is at most 4, and we embed the data via the degree-4 Veronese map anyway. Table C.2 gives the possible values of the Hilbert function for an arrangement of 4, 3, 2, 1 hyperplanes in $\mathbb{R}^5$, respectively. Here we use the convention that the empty set has codimension 5 in $\mathbb{R}^5$.

The first row shows that if the number of hyperplanes is exactly equal to the degree of the Veronese map, then $h_I(4) = 1$, i.e., the data matrix V_4 has a rank-1 null space. The following rows show the values of $h_I(4)$ when the number of hyperplanes is $n = 3, 2, 1$, respectively. If the rank of the matrix V_4 matches any of these values, we know exactly the number of hyperplanes in the arrangement. Figure C.2 shows a superimposed plot of the singular values of V_4 for sample points drawn from $n = 1, 2, 3, 4$ hyperplanes in $\mathbb{R}^5$, respectively.

Table C.2 Values of the Hilbert function of (codimension-1) hyperplane arrangements in $\mathbb{R}^5$.

c_1	c_2	c_3	c_4	$h_{I(Z_{\mathcal{A}})}(4)$	rank$(\boldsymbol{V}_4)$
1	1	1	1	1	69
1	1	1	5	5	65
1	1	5	5	15	55
1	5	5	5	35	35

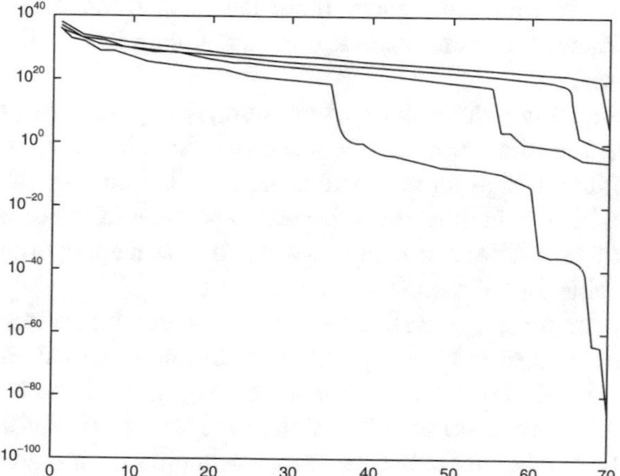

Fig. C.2 A superimposed semilog plot of the singular values of the embedded data matrix $\boldsymbol{V}_4$ for $n = 1, 2, 3, 4$ hyperplanes in $\mathbb{R}^5$, respectively. The rank drops at $35, 55, 65, 69$, which confirms the theoretical values of the Hilbert function.

Thus, in general, knowing the values of $h_I(i)$ even for $i > n$ may significantly help determine the correct number of subspaces in case the degree i of the Veronese map used for constructing the data matrix $\boldsymbol{V}_i$ is strictly higher than the number n of nontrivial subspaces in the arrangement.

The above examples show merely a few cases in which the values of the Hilbert function may facilitate solving the subspace clustering and modeling problem in Chapter 5, in particular the model-selection issue. It now remains as a question how to compute the values of the Hilbert function for arbitrary subspace arrangements.

Mathematically, we are interested in finding closed-form formulas, if they exist at all, for the Hilbert function (or the Hilbert polynomial, or the Hilbert series) of the subspace arrangements. As we will soon show, if the subspace arrangements are transversal (i.e., every intersection of subsets of the subspaces has the smallest possible dimension), we are able to show that the Hilbert function (of both I and J) agrees with the Hilbert polynomial (of both I and J) with $i \geq n$; and a closed-form formula for the Hilbert polynomial is known (and will be given later). However, no general formula is known for the Hilbert function (or series) of I, especially for the values $h_I(i)$ with $i < n$. For those values, one can still compute them in advance numerically based on the identity

$$h_I(i) = \dim(\text{Null}(V_i)) \tag{C.35}$$

from a sufficient set of samples on the subspace arrangements. The values for each type of arrangement needs to be computed only once, and the results can be stored in a table such as Table C.1 for each ambient space dimension D and number of subspaces n. We may later query these tables to retrieve information about the subspace arrangements and exploit relations among these values for different practical purposes.

However, computing the values of h_I numerically can be very expensive, especially when the dimension of the space (or the subspaces) is high. In order to densely sample the high-dimensional subspaces, the number of samples grows exponentially with the number of subspaces and their dimensions. Indeed, the MATLAB package that we are using runs out of the memory limit of 2 GB for computing the table for the case $D = 12$ and $n = 6$.

Fortunately, for most applications in image processing, computer vision, or systems identification, it is typically sufficient to know the values of $h_I(i)$ up to $n = 10$ and $D = 12$. For instance, for most images, the first $D = 12$ principal components already keep up to 99% of the total energy of the image, which is more than sufficient for any subsequent representation or compression purposes. Furthermore, if one chooses to use 2×2 blocks to represent a color image, then each block becomes one data point of dimension $2 \times 2 \times 3 = 12$. The number of segments sought for an image is typically less than ten. In system identification, the dimensions of the subspaces correspond to the orders of the systems, and they are typically less than 10.

C.4.2 Special Cases of the Hilbert Function

Before we study the Hilbert function for general subspace arrangements in the next section, we here give a few special cases for which we have computed certain values of the Hilbert function.

Example C.30 (Hyperplane Arrangements). Consider $Z_{\mathcal{A}} = S_1 \cup S_2 \cup \ldots \cup S_n \subset \mathbb{R}^D$ with each S_i a hyperplane. The subspaces S_i are of codimension 1, i.e., $c_1 = c_2 = \cdots = c_n = 1$. Then we have $h_I(n) = 1$, which is consistent with the fact that there is exactly one (factorable) polynomial of degree n that fits n hyperplanes. Furthermore, $h_I(i) = 0$ for all $i < n$, and

$$h_I(n+i) = \binom{D+i-1}{i}, \quad \forall i \geq 1.$$

We can generalize the case of hyperplanes to the following example.

Example C.31 (Subspaces Whose Duals Have No Intersection). Consider a subspace arrangement $Z_{\mathcal{A}} = S_1 \cup S_2 \cup \ldots \cup S_n \subset \mathbb{R}^D$ with $S_i^* \cap S_j^* = 0$ for all $i \neq j$. In other words, if the codimensions of $S_1, S_2, \ldots, S_n$ are $c_1, c_2, \ldots, c_n$, respectively, we have $c_1 + c_2 + \cdots + c_n \leq D$. Notice that hyperplane arrangements are a special case here. Generalizing the result in Example C.15, one can show that the Hilbert series of $I(Z_{\mathcal{A}})$ (and $J(Z_{\mathcal{A}})$) is

$$\mathcal{H}(I(Z_{\mathcal{A}}), t) = \mathcal{H}(J(Z_{\mathcal{A}}), t) = f(t) \doteq \frac{\prod_{i=1}^n \left(1 - (1-t)^{c_i}\right)}{(1-t)^D}. \tag{C.36}$$

The values of the Hilbert function $h_I(i)$ can be computed from the coefficients of the function $f(t)$ associated with t^i.

However, if the dual subspaces S_i^* have nontrivial intersections, the computation of Hilbert series and function becomes much more complicated. Below we give some special examples and leave the general study to the next section.

Example C.32 (Hilbert Function of Two Subspaces). We here derive a closed-form formula of $h_I(2)$ for an arrangement of $n = 2$ subspaces $Z_{\mathcal{A}} = S_1 \cup S_2$ in general position (see also Example C.24). Suppose their codimensions are c_1 and c_2, respectively. In $R_1 \sim \mathbb{R}^D$, the intersection of their dual subspaces S_1^* and S_2^* has the dimension

$$c \doteq \max\{c_1 + c_2 - D,\ 0\}. \tag{C.37}$$

Then we have

$$\begin{aligned} h_I(2) &= c \cdot (c+1)/2 + c \cdot (c_1 - c) + c \cdot (c_2 - c) + (c_1 - c) \cdot (c_2 - c) \\ &= c_1 \cdot c_2 - c \cdot (c-1)/2. \end{aligned} \tag{C.38}$$

Example C.33 (Three Subspaces in $\mathbb{R}^5$). Consider an arrangement of three subspaces $Z_{\mathcal{A}} = S_1 \cup S_2 \cup S_3 \subset \mathbb{R}^5$ in general position. After a change of coordinates, we may assume $S_1^* = \text{span}\{x_1, x_2, x_3\}$, $S_2^* = \text{span}\{x_1, x_4, x_5\}$, and $S_3^* = \text{span}\{x_2, x_3, x_4, x_5\}$. The value of $h_I(3)$ in this case is equal to $\dim(S_1^* \cdot S_2^* \cdot S_3^*)$. Firstly, we compute $S_1^* \cdot S_2^*$ and obtain a basis for it:

$$S_1^* \cdot S_2^* = \text{span}\{x_1^2, x_1x_4, x_1x_5, x_2x_1, x_2x_4, x_2x_5, x_3x_1, x_3x_4, x_3x_5\}.$$

From this, one can compute the basis for $S_1^* \cdot S_2^* \cdot S_3^*$:

$$\begin{aligned} S_1^* \cdot S_2^* \cdot S_3^* = \text{span}\{&x_1^2x_2, x_1x_2x_4, x_1x_2x_5, x_1x_2^2, x_2^2x_4, x_2^2x_5, x_1x_2x_3, x_2x_3x_4, \\ &x_2x_3x_5, x_1^2x_3, x_1x_3x_4, x_1x_3x_5, x_1x_3^2, x_3^2x_4, x_3^2x_5, x_1^2x_4, x_1x_4^2, \\ &x_1x_4x_5, x_2x_4^2, x_2x_4x_5, x_3x_4^2, x_3x_4x_5, x_1^2x_5, x_1x_5^2, x_2x_5^2, x_3x_5^2\}. \end{aligned}$$

Thus, we have $h_I(3) = 26$.

Table C.3 Values of the Hilbert function $h_I(5)$ for arrangements of five subspaces in $\mathbb{R}^3$.

c_1	c_2	c_3	c_4	c_5	$h_I(5)$
1	1	1	1	1	1
1	1	1	1	2	2
1	1	1	2	2	4
1	1	2	2	2	7
1	2	2	2	2	11
2	2	2	2	2	16

Example C.34 (Five Subspaces in $\mathbb{R}^3$). Consider an arrangement of five subspaces $S_1, S_2, \ldots, S_5$ in $\mathbb{R}^3$ of codimensions $c_1, c_2, \ldots, c_5$, respectively. We want to compute the value of $h_I(5)$, i.e., the dimension of homogeneous polynomials of degree five that vanish on the five subspaces $Z_{\mathcal{A}} = S_1 \cup S_2 \cup \cdots \cup S_5$. For all the possible values of $1 \leq c_1 \leq c_2 \leq \cdots \leq c_5 < 3$, we have computed the values of $\mathcal{D}_5^3$ and listed them in Table C.3. Notice that the values of $h_I(3)$ in the earlier Table C.1 form a subset of those of $h_I(5)$ in Table C.3. In fact, many relationships like this one exist among the values of the Hilbert function. If properly harnessed, they can significantly reduce the amount of work in computing the values of the Hilbert function.

Example C.35 (Five Subspaces in $\mathbb{R}^4$). Similar to the above example, we have computed the values of $h_I(5)$ for arrangements of five linear subspaces in $\mathbb{R}^4$. The results are given in Table C.4. In fact, using the numerical method described earlier, we have computed the values of $h_I(5)$ up to five subspaces in $\mathbb{R}^{12}$.

C.4.3 Formulas for the Hilbert Function

In this section, we give a general formula for the Hilbert polynomial of the subspace arrangement $Z_{\mathcal{A}} = S_1 \cup S_2 \cup \cdots \cup S_n$. However, due to limitations of space, we will not be able to give a detailed proof for all the results given here. Interested readers may refer to (Derksen 2007).

Let U be any subset of the set of indices $\underline{n} \doteq \{1, 2, \ldots, n\}$. We define the following ideals:

$$I_U \doteq \bigcap_{u \in U} I(S_u), \quad J_U \doteq \prod_{u \in U} I(S_u). \tag{C.39}$$

If U is empty, we use the convention $I_\emptyset = J_\emptyset = R$. We further define $V_U = \bigcap_{u \in U} S_u$, $d_U = \dim(V_U)$, and $c_U = D - d_U$.

Let us define polynomials $p_U(t)$ recursively as follows. First we define

Table C.4 Values of the Hilbert function $h_I(5)$ for arrangements of five subspaces in $\mathbb{R}^4$.

c_1	c_2	c_3	c_4	c_5	$h_I(5)$
1	1	1	1	1	1
1	1	1	1	2	2
1	1	1	1	3	3
1	1	1	2	2	4
1	1	1	2	3	6
1	1	1	3	3	8
1	1	2	2	2	8
1	1	2	2	3	11
1	1	2	3	3	14
1	1	3	3	3	17
1	2	2	2	2	15
1	2	2	2	3	19
1	2	2	3	3	23
1	2	3	3	3	27
1	3	3	3	3	31
2	2	2	2	2	26
2	2	2	2	3	31
2	2	2	3	3	36
2	2	3	3	3	41
2	3	3	3	3	46
3	3	3	3	3	51

$$p_{\emptyset}(t) = 1.$$

For $U \neq \emptyset$ and if $p_W(t)$ is already defined for all proper subsets W of U, then $p_U(t)$ is uniquely determined by the following equation:

$$\sum_{W \subseteq U} (-t)^{|W|} p_W(t) \equiv 0 \bmod (1-t)^{c_U}, \quad \deg(p_U(t)) < c_U. \tag{C.40}$$

Here $|W|$ is the number of indices in the set W.

With the above definitions, the Hilbert series of the product ideal J is given by

$$\mathcal{H}(J,t) = \frac{p_{\underline{n}}(t)t^n}{(1-t)^D}. \tag{C.41}$$

That is, the Hilbert series of the product ideal J depends only on the numbers $c_U, U \subseteq \underline{n}$. Thus, the values of the Hilbert function $h_J(i)$ are all combinatorial invariants—invariants that depend only on the values $\{c_U\}$ but not the particular position of the subspaces.

Definition C.36 (Transversal Subspaces). *The subspaces $S_1, S_2, \ldots, S_n$ are called* transversal *if $c_U = \min\left(D, \sum_{u \in U} c_u\right)$ for all $U \subseteq \underline{n}$. In other words, the intersection of any subset of the subspaces has the smallest possible dimension.*

Notice that the notion of "transversality" defined here is less strong than the typical notion of "general position." For instance, according to the above definition, three coplanar lines (through the origin) in $\mathbb{R}^3$ are transversal. However, they are not "in general position."

Theorem C.37 (Hilbert Function of a Transversal Subspace Arrangement). *Suppose that $S_1, S_2, \ldots, S_n$ are transversal. Then $\mathcal{H}(I,t) - f(t)$ and $\mathcal{H}(J,t) - f(t)$ are polynomials in t, where $f(t) = \frac{\prod_{i=1}^{n}\left(1-(1-t)^{c_i}\right)}{(1-t)^D}$.*

Thus, the difference between $\mathcal{H}(I,t)$ and $\mathcal{H}(J,t)$ is also a polynomial. We have the following corollary to the above theorem.

Corollary C.38. *If $S_1, S_2, \ldots, S_n$ are transversal, then $h_I(i) = H_I(i) = h_J(i) = H_J(i)$ for all $i \geq n$. That is, the Hilbert polynomials of both the vanishing ideal I and the product ideal J are the same, and the values of their Hilbert functions agree with the polynomial with $i \geq n$.*

One of the consequences of this corollary is that for transversal subspace arrangements, we must have $I_i = J_i$ for all $i \geq n$. This is a result that we have mentioned earlier, in Section C.3.

Example C.39 (Hilbert Series of Three Lines in $\mathbb{R}^3$). For example, suppose that Z_A is the union of three distinct lines (through the origin) in $\mathbb{R}^3$. Regardless of whether the three lines are coplanar, they are transversal. We have

$$\mathcal{H}(J(Z_A),t) = \frac{7t^3 - 9t^4 + 3t^5}{(1-t)^3} = 7t^3 + 12t^4 + 18t^5 + \cdots.$$

However, one has

$$\mathcal{H}(I(Z_A),t) = \frac{t + t^3 - t^4}{(1-t)^3} = t + 3t^2 + 7t^3 + 12t^4 + 18t^5 + \cdots$$

if the lines are coplanar, and

$$\mathcal{H}(I(Z_A),t) = \frac{3t^2 - 2t^3}{(1-t)^3} = 3t^2 + 7t^3 + 12t^4 + 18t^5 + \cdots$$

if the three lines are not coplanar. Notice that the coefficients of these Hilbert series become the same starting from the term t^3.

Then, using the recursive formula (C.41) of the Hilbert series $\mathcal{H}(J,t)$, we can derive a closed-form formula for the values of the Hilbert function $h_I(i)$ with $i \geq n$:

Corollary C.40 (A Closed-Form Formula for Hilbert Function). *If $S_1, S_2, \dots, S_n$ are transversal, then*

$$h_I(i) = h_J(i) = \sum_U (-1)^{|U|} \binom{D+i-1-c_U}{D-1-c_U}, \quad i \geq n, \tag{C.42}$$

where $c_U = \sum_{m \in U} c_m$ and the sum is over all index subsets U of $\underline{n}$ for which $c_U < D$.

Example C.41 (Three Subspaces in $\mathbb{R}^4$). Suppose that $Z_{\mathcal{A}} = S_1 \cup S_2 \cup S_3$ is a transversal arrangement in $\mathbb{R}^4$. Let d_1, d_2, d_3 (respectively c_1, c_2, c_3) be the dimensions (respectively codimensions) of S_1, S_2, S_3. We make a table of $h_I(n)$ for $n = 3, 4, 5$:

c_1, c_2, c_3	d_1, d_2, d_3	$h_I(3)$	$h_I(4)$	$h_I(5)$
1, 1, 1	3, 3, 3	1	4	10
1, 1, 2	3, 3, 2	2	7	16
1, 1, 3	3, 3, 1	3	9	19
1, 2, 2	3, 2, 2	4	12	25
1, 2, 3	3, 2, 1	6	15	29
1, 3, 3	3, 1, 1	8	18	33
2, 2, 2	2, 2, 2	8	20	38
2, 2, 3	2, 2, 1	11	24	43
2, 3, 3	2, 1, 1	14	28	48
3, 3, 3	1, 1, 1	17	32	53

Note that the codimensions c_1, c_2, c_3 are almost determined by $h_I(3)$. They are uniquely determined by $h_I(3)$ and $h_I(4)$.

The corollary below is a general result that explains why the codimensions of the subspaces c_1, c_2, c_3 can be uniquely determined by $h_I(3), h_I(4), h_I(5)$ in the above example. The corollary also reveals a strong theoretical connection between the Hilbert function and the algebraic subspace clustering problem.

Corollary C.42 (Subspace Dimensions from the Hilbert Function). *Consider a transversal arrangement of n subspaces. The codimensions $c_1, c_2, \dots, c_n$ are uniquely determined by the values of the Hilbert function $h_I(i)$ for $i = n, n+1, \dots, n+D-1$.*

As we have alluded to earlier, in the context of algebraic subspace clustering, these values of the Hilbert function are closely related to the ranks of the embedded data matrix $\boldsymbol{V}_i$ for $i = n, n+1, \dots, n+D-1$. Thus, knowing these ranks, we should in principle be able to uniquely determine the (co)dimensions of all the individual subspaces. These results suggest that knowing the values of the Hilbert function, one can potentially develop better algorithms for determining the correct subspace arrangement from a given set of data.

C.5 Bibliographic Notes

Subspace arrangements constitute a very special but important class of algebraic sets that have been studied in mathematics for centuries (Björner et al. 2005; Björner 1994; Orlik 1989). The importance as well as the difficulty of studying subspace arrangements can hardly be exaggerated. Different aspects of their properties have been and are still being investigated and exploited in many mathematical fields, including algebraic geometry and topology, combinatorics and complexity theory, and graph and lattice theory. See (Björner 1994) for a general review. Although the results about subspace arrangements are extremely rich and deep, only a few special classes of subspace arrangements have been fully characterized. Nevertheless, thanks to the work of (Derksen 2007), the Hilbert function, Hilbert polynomial, and Hilbert series of the vanishing ideal (and the product ideal) of transversal subspace arrangements have recently become well understood. This appendix gives a brief summary of these theoretical developments. These results have provided a sound theoretical foundation for many of the methods developed in this book for clustering and modeling multiple subspaces.

References

Agarwal, P., & Mustafa, N. (2004). k-means projective clustering. In *ACM Symposium on Principles of Database Systems*.

Agarwal, S., Lim, J., Zelnik-Manor, L., Perona, P., Kriegman, D., & Belongie, S. (2005). Beyond pairwise clustering. In *IEEE Conference on Computer Vision and Pattern Recognition* (Vol. 2, pp. 838–845).

Aggarwal, G., Roy-Chowdhury, A., & Chellappa, R. (2004). A system identification approach for video-based face recognition. In *Proceedings of International Conference on Pattern Recognition* (pp. 23–26).

Akaike, H. (1977). A new look at the statistical model selection. *IEEE Transactions on Automatic Control, 16*(6), 716–723.

Aldroubi, A., Cabrelli, C., & Molter, U. (2008). Optimal non-linear models for sparsity and sampling. *Journal of Fourier Analysis and Applications, 14*(5–6), 793–812.

Aldroubi, A., & Zaringhalam, K. (2009). Nonlinear least squares in $\mathbb{R}^N$. *Acta Applicandae Mathematicae, 107*(1–3), 325–337.

Alessandri, A., & Coletta, P. (2001). Design of Luenberger observers for a class of hybrid linear systems. In *Proceedings of Hybrid Systems: Computation and Control* (pp. 7–18). New York: Springer.

Ali, S., Basharat, A., & Shah, M. (2007). Chaotic invariants for human action recognition. In *Proceedings of International Conference on Computer Vision*.

Amaldi, E., & Kann, V. (1998). On the approximability of minimizing nonzero variables or unsatisfied relations in linear systems. *Theoretical Computer Science, 209*, 237–260.

Anderson, B., & Johnson, R. (1982). Exponential convergence of adaptive identification and control algorithms. *Automatica, 18*(1), 1–13.

Arbelaez, P. (2006). Boundary extraction in natural images using ultrametric contour maps. In *Workshop on Perceptual Organization in Computer Vision*.

Arbelaez, P., Maire, M., Fowlkes, C., & Malik, J. (2009). From contours to regions: An empirical evaluation. In *IEEE Conference on Computer Vision and Pattern Recognition*.

Arora, S., Bhaskara, A., Ge, R., & Ma, T. (2014). Provable bounds for learning some deep representations. In *Proceedings of International Conference on Machine Learning*.

Avidan, S., & Shashua, A. (2000). Trajectory triangulation: 3D reconstruction of moving points from a monocular image sequence. *IEEE Transactions on Pattern Analysis and Machine Intelligence, 22*(4), 348–357.

Ayazoglu, M., Li, B., Dicle, C., Sznaier, M., & Camps, O. (2011). Dynamic subspace-based coordinated multicamera tracking. In *IEEE International Conference on Computer Vision* (pp. 2462–2469)

R. Vidal et al., *Generalized Principal Component Analysis*, Interdisciplinary Applied Mathematics 40, DOI 10.1007/978-0-387-87811-9

Babaali, M., & Egerstedt, M. (2004). Observability of switched linear systems. In *Proceedings of Hybrid Systems: Computation and Control*. New York: Springer.

Bach, F. (2013). Convex relaxations of structured matrix factorizations. arXiv:1309.3117v1.

Bach, F., Mairal, J., & Ponce, J. (2008). *Convex sparse matrix factorizations*. http://arxiv.org/abs/0812.1869

Balluchi, A., Benvenuti, L., Benedetto, M. D., & Sangiovanni-Vincentelli, A. (2002). Design of observers for hybrid systems. In *Proceedings of Hybrid Systems: Computation and Control* (Vol. 2289, pp. 76–89). New York: Springer.

Baraniuk, R. (2007). Compressive sensing. *IEEE Signal Processing Magazine, 24*(4), 118–121.

Barbic, J., Safonova, A., Pan, J.-Y., Faloutsos, C., Hodgins, J. K., & Pollar, N. S. (2004). Segmenting motion capture data into distinct behaviors. In *Graphics Interface*.

Barnett, V., & Lewis, T. (1983). *Outliers in statistical data* (2nd ed.). New York: Wiley.

Basri, R., & Jacobs, D. (2003). Lambertian reflection and linear subspaces. *IEEE Transactions on Pattern Analysis and Machine Intelligence, 25*(2), 218–233.

Beck, A., & Teboulle, M. (2009). A fast iterative shrinkage-thresholding algorithm for linear inverse problems. *SIAM Journal on Imaging Sciences, 2*(1), 183–202.

Béjar, B., Zappella, L., & Vidal, R. (2012). Surgical gesture classification from video data. In *Medical Image Computing and Computer Assisted Intervention* (pp. 34–41).

Belhumeur, P., Hespanda, J., & Kriegeman, D. (1997). Eigenfaces vs. Fisherfaces: Recognition using class specific linear projection. *IEEE Transactions on Pattern Analysis and Machine Intelligence, 19*(7), 711–720.

Belhumeur, P., & Kriegman, D. (1998). What is the set of images of an object under all possible illumination conditions? *International Journal of Computer Vision, 28*(3), 1–16.

Belkin, M., & Niyogi, P. (2002). Laplacian eigenmaps and spectral techniques for embedding and clustering. In *Proceedings of Neural Information Processing Systems (NIPS)* (pp. 585–591).

Beltrami, E. (1873). Sulle funzioni bilineari. *Giornale di Mathematiche di Battaglini, 11*, 98–106.

Bemporad, A., Ferrari, G., & Morari, M. (2000). Observability and controllability of piecewise affine and hybrid systems. *IEEE Transactions on Automatic Control, 45*(10), 1864–1876.

Bemporad, A., Garulli, A., Paoletti, S., & Vicino, A. (2003). A greedy approach to identification of piecewise affine models. In *Hybrid systems: Computation and control. Lecture notes in computer science* (pp. 97–112). New York: Springer.

Bemporad, A., Roll, J., & Ljung, L. (2001). Identification of hybrid systems via mixed-integer programming. In *Proceedings of IEEE Conference on Decision & Control* (pp. 786–792).

Benson, H. (1994). Concave minimization: Theory, applications and algorithms. In R. Horst & P. M. Pardalos (Eds.), *Handbook of global optimization* (vol. 2, pp. 43–148), Springer Verlag.

Bertsekas, D. P. (1999). *Nonlinear programming* (2nd ed.). *Optimization and computation* (Vol. 2) Belmont: Athena Scientific.

Bickel, P. J. (1976). Another look at robustness: A review of reviews and some new developments. *Scandinavian Journal of Statistics, 3*(28), 145–168.

Bickel, P. J., & Doksum, K. A. (2000). *Mathematical statistics: Basic ideas and selected topics* (2nd ed.). Upper Saddle River: Prentice Hall.

Billio, M., Monfort, A., & Robert, C. (1999). Bayesian estimation of switching ARMA models. *Journal of Econometrics, 93*(2), 229–255.

Bissacco, A., Chiuso, A., Ma, Y., & Soatto, S. (2001). Recognition of human gaits. In *Proceedings of IEEE Conference on Computer Vision and Pattern Recognition* (Vol. 2, pp. 52–58).

Björner, A. (1994). Subspace arrangements. In *First European Congress of Mathematics, Vol. I (Paris, 1992). Progress in mathematics* (Vol. 119, pp. 321–370). Basel: Birkhäuser.

Björner, A., Peeva, I., & Sidman, J. (2005). Subspace arrangements defined by products of linear forms. *Journal of the London Mathematical Society, 71*(2), 273–288.

Blake, A., North, B., & Isard, M. (1999). Learning multi-class dynamics. *Advances in Neural Information Processing Systems, 11*, 389–395. Cambridge: MIT Press.

Bochnak, J., Coste, M., & Roy, M. F. (1998). *Real Algebraic Geometry*. New York: Springer.

Bottou, L., & Bengio, J. (1995). Convergence properties of the k-means algorithms. In *Neural Information Processing and Systems*.

Boult, T., & Brown, L. (1991). Factorization-based segmentation of motions. In *IEEE Workshop on Motion Understanding* (pp. 179–186).

Boyd, S., Parikh, N., Chu, E., Peleato, B., & Eckstein, J. (2010). Distributed optimization and statistical learning via the alternating direction method of multipliers. *Foundations and Trends in Machine Learning, 3*(1), 1–122.

Boyd, S., & Vandenberghe, L. (2004). *Convex Optimization*. Cambridge: Cambridge University Press.

Bradley, P. S., & Mangasarian, O. L. (2000). k-plane clustering. *Journal of Global Optimization, 16*(1), 23–32.

Brandt, S. (2002). Closed-form solutions for affine reconstruction under missing data. In *In Proceedings Statistical Methods for Video Processing (ECCV'02 Workshop)*.

Broomhead, D. S., & Kirby, M. (2000). A new approach to dimensionality reduction theory and algorithms. *SIAM Journal of Applied Mathematics, 60*(6), 2114–2142.

Bruckstein, A., Donoho, D., & Elad, M. (2009). From sparse solutions of systems of equations to sparse modeling of signals and images. *SIAM Review, 51*(1), 34–81.

Buchanan, A., & Fitzgibbon, A. (2005). Damped Newton algorithms for matrix factorization with missing data. In *IEEE Conference on Computer Vision and Pattern Recognition* (pp. 316–322).

Burer, S., & Monteiro, R. D. C. (2005). Local minima and convergence in low-rank semidefinite programming. *Mathematical Programming, Series A, 103*(3), 427–444.

Burges, C. (2005). Geometric methods for feature extraction and dimensional reduction - a guided tour. In *The data mining and knowledge discovery handbook* (pp. 59–92). Boston: Kluwer Academic.

Burges, C. J. C. (2010). Dimension reduction: A guided tour. *Foundations and Trends in Machine Learning, 2*(4), 275–365.

Burnham, K. P., & Anderson, D. R. (2002). *Model selection and multimodel inference: A practical information-theoretic approach*. New York: Springer.

Burt, P. J., & Adelson, E. H. (1983). The Laplacian pyramid as a compact image code. *IEEE Transactions on Communications, 31*(4), 532–540.

Cai, J.-F., Candés, E. J., & Shen, Z. (2008). A singular value thresholding algorithm for matrix completion. *SIAM Journal of Optimization, 20*(4), 1956–1982.

Campbell, N. (1978). The influence function as an aid in outlier detection in discriminant analysis. *Applied Statistics, 27*(3), 251–258.

Campbell, R. J. (1980). Robust procedures in multivariate analysis I: Robust covariance analysis. *Applied Statistics, 29*, 231–237.

Candès, E. (2006). Compressive sampling. In *Proceedings of the International Congress of Mathematics*.

Candès, E. (2008). The restricted isometry property and its implications for compressed sensing. *Comptes Rendus Mathematique, 346*(9–10), 589–592.

Candès, E., & Donoho, D. (2002). *New tight frames of curvelets and optimal representations of objects with smooth singularities*. Technical Report. Stanford University.

Candès, E., Li, X., Ma, Y., & Wright, J. (2011). Robust principal component analysis? *Journal of the ACM, 58*(3).

Candès, E., & Plan, Y. (2010). Matrix completion with noise. *Proceedings of the IEEE, 98*(6), 925–936.

Candès, E., & Recht, B. (2009). Exact matrix completion via convex optimization. *Foundations of Computational Mathematics, 9*, 717–772.

Candès, E., & Recht, B. (2011). Simple bounds for low-complexity model reconstruction. *Mathematical Programming Series A, 141*(1–2), 577–589.

Candès, E., & Tao, T. (2005). Decoding by linear programming. *IEEE Transactions on Information Theory, 51*(12), 4203–4215.

Candès, E., & Tao, T. (2010). The power of convex relaxation: Near-optimal matrix completion. *IEEE Transactions on Information Theory, 56*(5), 2053–2080.

Candès, E., & Wakin, M. (2008). An introduction to compressive sampling. *IEEE Signal Processing Magazine, 25*(2), 21–30.

Cattell, R. B. (1966). The scree test for the number of factors. *Multivariate Behavioral Research, 1*, 245–276.

Cetingül, H. E., Wright, M., Thompson, P., & Vidal, R. (2014). Segmentation of high angular resolution diffusion MRI using sparse riemannian manifold clustering. *IEEE Transactions on Medical Imaging, 33*(2), 301–317.

Chan, A., & Vasconcelos, N. (2005a). Classification and retrieval of traffic video using auto-regressive stochastic processes. In *Proceedings of 2005 IEEE Intelligent Vehicles Symposium* (pp. 771–776).

Chan, A., & Vasconcelos, N. (2005b). Mixtures of dynamic textures. In *IEEE International Conference on Computer Vision* (Vol. 1, pp. 641–647).

Chandrasekaran, V., Sanghavi, S., Parrilo, P., & Willsky, A. (2009). Sparse and low-rank matrix decompositions. In *IFAC Symposium on System Identification*.

Chaudhry, R., Ravichandran, A., Hager, G., & Vidal, R. (2009). Histograms of oriented optical flow and binet-cauchy kernels on nonlinear dynamical systems for the recognition of human actions. In *IEEE Conference on Computer Vision and Pattern Recognition*.

Chen, G., Atev, S., & Lerman, G. (2009). Kernel spectral curvature clustering (KSCC). In *Workshop on Dynamical Vision*.

Chen, G., & Lerman, G. (2009a). Foundations of a multi-way spectral clustering framework for hybrid linear modeling. *Foundations of Computational Mathematics, 9*(5), 517–558.

Chen, G., & Lerman, G. (2009b). Spectral curvature clustering (SCC). *International Journal of Computer Vision, 81*(3), 317–330.

Chen, J.-Q., Pappas, T. N., Mojsilovic, A., & Rogowitz, B. E. (2003). Image segmentation by spatially adaptive color and texture features. In *IEEE International Conference on Image Processing*.

Chen, S., Donoho, D., & Saunders, M. (1998). Atomic decomposition by basis pursuit. *SIAM Journal of Scientific Computing, 20*(1), 33–61.

Chung, F. (1997). *Spectral graph theory*. Washington: Conference Board of the Mathematical Sciences.

Cilibrasi, R., & Vitányi, P. M. (2005). Clustering by compression. *IEEE Transactions on Information Theory, 51*(4), 1523–1545.

CMU (2003). MOCAP database. http://mocap.cs.cmu.edu.

Coifman, R., & Wickerhauser, M. (1992). Entropy-based algorithms for best bases selection. *IEEE Transactions on Information Theory, 38*(2), 713–718.

Collins, M., Dasgupta, S., & Schapire, R. (2001). A generalization of principal component analysis to the exponential family. In *Neural Information Processing Systems* (Vol. 14)

Collins, P., & Schuppen, J. V. (2004). Observability of piecewise-affine hybrid systems. In *Proceedings of Hybrid Systems: Computation and Control*. New York: Springer.

Comanicu, D., & Meer, P. (2002). Mean shift: A robust approach toward feature space analysis. *IEEE Transactions on Pattern Analysis and Machine Intelligence (PAMI), 24*, 603–619.

Costeira, J., & Kanade, T. (1998). A multibody factorization method for independently moving objects. *International Journal of Computer Vision, 29*(3), 159–179.

Cour, T., Benezit, F., & Shi, J. (2005). Spectral segmentation with multiscale graph decomposition. In *Proceedings of the IEEE Conference on Computer Vision and Pattern Recognition (CVPR)*.

Cover, T., & Thomas, J. (1991). *Elements of information theory*. Wiley.

Cox, T. F., & Cox, M. A. A. (1994). *Multidimensional scaling*. London: Chapman and Hall.

Critchley, F. (1985). Influence in principal components analysis. *Biometrika, 72*(3), 627–636.

Davis, C., & Cahan, W. (1970). The rotation of eigenvectors by a pertubation. *SIAM Journal on Numerical Analysis, 7*(1), 1–46.

Davison, M. (1983). *Multidimensional Scaling*. New York: Wiley.

De la Torre, F., & Black, M. J. (2004). A framework for robust subspace learning. *International Journal of Computer Vision, 54*(1), 117–142.

Delsarte, P., Macq, B., & Slock, D. (1992). Signal-adapted multiresolution transform for image coding. *IEEE Transactions on Information Theory, 38*, 897–903.

Dempster, A., Laird, N., & Rubin, D. (1977). Maximum likelihood from incomplete data via the EM algorithm. *Journal of the Royal Statistical Society B, 39*(1), 1–38.

Deng, W., Lai, M.-J., Peng, Z., & Yin, W. (2013). Parallel multi-block admm with o(1/k) convergence. *UCLA CAM.*

Deng, Y., & Manjunath, B. (2001). Unsupervised segmentation of color-texture regions in images and video. *IEEE Transactions on Pattern Analysis and Machine Intelligence, 23*(8), 800–810.

Derksen, H. (2007). Hilbert series of subspace arrangements. *Journal of Pure and Applied Algebra, 209*(1), 91–98.

DeVore, R. (1998). Nonlinear approximation. *Acta Numerica, 7*, 51–150.

DeVore, R., Jawerth, B., & Lucier, B. (1992). Image compression through wavelet transform coding. *IEEE Transactions on Information Theory, 38*(2), 719–746.

Ding, C., Zha, H., He, X., Husbands, P., & Simon, H. D. (2004). Link analysis: Hubs and authoraties on the world wide web. *SIAM Review, 46*(2), 256–268.

Do, M. N., & Vetterli, M. (2002). Contourlets: A directional multiresolution image representation. In *IEEE International Conference on Image Processing.*

Donoho, D. (1995). Cart and best-ortho-basis: A connection. Manuscript.

Donoho, D. (1998). Sparse components analysis and optimal atomic decomposition. *Technical Report, Department of Statistics, Stanford University.*

Donoho, D., & Gavish, M. (2014). The optimal hard threshold for singular values is $4/\sqrt{3}$. *IEEE Transactions on Information Theory, 60*(8), 5040–5053.

Donoho, D., & Grimes, C. (2003). Hessian eigenmaps: Locally linear embedding techniques for high-dimensional data. *National Academy of Sciences, 100*(10), 5591–5596.

Donoho, D. L. (1999). Wedgelets: Nearly-minimax estimation of edges. *Annals of Statistics, 27*, 859–897.

Donoho, D. L. (2005). *Neighborly polytopes and sparse solution of underdetermined linear equations.* Technical Report. Stanford University.

Donoho, D. L. (2006). For most large underdetermined systems of linear equations the minimal ℓ^1-norm solution is also the sparsest solution. *Communications on Pure and Applied Mathematics, 59*(6), 797–829.

Donoho, D. L., & Elad, M. (2003). Optimally sparse representation in general (nonorthogonal) dictionaries via ℓ_1 minimization. *Proceedings of National Academy of Sciences, 100*(5), 2197–2202.

Donoho, D. L., Vetterli, M., DeVore, R., & Daubechies, I. (1998). Data compression and harmonic analysis. *IEEE Transactions on Information Theory, 44*(6), 2435–2476.

Donoser, M., Urschler, M., Hirzer, M., & Bischof, H. (2009). Saliency driven total variation segmentation. In *Proceedings of the International Conference on Computer Vision (ICCV).*

Doretto, G., Chiuso, A., Wu, Y., & Soatto, S. (2003). Dynamic textures. *International Journal of Computer Vision, 51*(2), 91–109.

Doretto, G., & Soatto, S. (2003). Editable dynamic textures. In *IEEE Conference on Computer Vision and Pattern Recognition* (Vol. II, pp. 137–142).

Doretto, G., & Soatto, S. (2006). Dynamic shape and appearance models. *IEEE Transactions on Pattern Analysis and Machine Intelligence, 28*(12), 2006–2019.

Doucet, A., Logothetis, A., & Krishnamurthy, V. (2000). Stochastic sampling algorithms for state estimation of jump Markov linear systems. *IEEE Transactions on Automatic Control, 45*(1), 188–202.

Duda, R., Hart, P., & Stork, D. (2000). *Pattern Classification* (2nd ed.). Wiley, New York.

Eckart, C., & Young, G. (1936). The approximation of one matrix by another of lower rank. *Psychometrika, 1*, 211–218.

Effros, M., & Chou, P. (1995). Weighted universal transform coding: Universal image compression with the Karhunen-Loéve transform. In *IEEE International Conference on Image Processing* (Vol. 2, pp. 61–64).

Efros, A. A., & Leung, T. K. (1999). Texture synthesis by non-parametric sampling. In *IEEE International Conference on Computer Vision* (pp. 1033–1038). Corfu, Greece.

Eisenbud, D. (1996). *Commutative algebra: With a view towards algebraic geometry. Graduate texts in mathematics*. New York: Springer.

Elad, M., & Bruckstein, A. (2001). On sparse signal representations. In *IEEE International Conference on Image Processing*.

Elad, M., & Bruckstein, A. (2002). A generalized uncertainty principle and sparse representation in pairs of bases. *IEEE Transactions on Information Theory, 48*(9), 2558–2567.

Elad, M., Figueiredo, M. A. T., & Ma, Y. (2010). On the role of sparse and redundant representations in image processing. *Proceedings of the IEEE, 98*(6), 972–982.

Elder, J., & Zucker, S. (1996). Computing contour closures. In *Proceedings of the European Conference on Computer Vision (ECCV)*.

Elhamifar, E., Sapiro, G., & Vidal, R. (2012a). Finding exemplars from pairwise dissimilarities via simultaneous sparse recovery. In *Neural Information Processing and Systems*.

Elhamifar, E., Sapiro, G., & Vidal, R. (2012b). See all by looking at a few: Sparse modeling for finding representative objects. In *IEEE Conference on Computer Vision and Pattern Recognition*.

Elhamifar, E., & Vidal, R. (2009). Sparse subspace clustering. In *IEEE Conference on Computer Vision and Pattern Recognition*.

Elhamifar, E., & Vidal, R. (2010). Clustering disjoint subspaces via sparse representation. In *IEEE International Conference on Acoustics, Speech, and Signal Processing*.

Elhamifar, E., & Vidal, R. (2011). Sparse manifold clustering and embedding. In *Neural Information Processing and Systems*.

Elhamifar, E., & Vidal, R. (2013). Sparse subspace clustering: Algorithm, theory, and applications. *IEEE Transactions on Pattern Analysis and Machine Intelligence, 35*(11), 2765–2781.

Ezzine, J., & Haddad, A. H. (1989). Controllability and observability of hybrid systems. *International Journal of Control, 49*(6), 2045–2055.

Favaro, P., Vidal, R., & Ravichandran, A. (2011). A closed form solution to robust subspace estimation and clustering. In *IEEE Conference on Computer Vision and Pattern Recognition*.

Fazel, M., Hindi, H., & Boyd, S. (2003). Log-det heuristic for matrix rank minimization with applications to Hankel and Euclidean distance matrices. In *Proceedings of the American Control Conference* (pp. 2156–2162).

Fei-Fei, L., Fergus, R., & Perona, P. (2004). Learning generative visual models from few training examples: An incremental bayesian approach tested on 101 object categories. In *Workshop on Generative Model Based Vision*.

Felzenszwalb, P. F., & Huttenlocher, D. P. (2004). Efficient graph-based image segmentation. *International Journal of Computer Vision (IJCV), 59*(2), 167–181.

Feng, J., Xu, H., Mannor, S., & Yang, S. (2013). Online PCA for contaminated data. In *NIPS*.

Feng, X., & Perona, P. (1998). Scene segmentation from 3D motion. In *IEEE Conference on Computer Vision and Pattern Recognition* (pp. 225–231).

Ferguson, T. (1961). On the rejection of outliers. In *Proceedings of the Fourth Berkeley Symposium on Mathematical Statistics and Probability*.

Ferrari-Trecate, G., Mignone, D., & Morari, M. (2002). Moving horizon estimation for hybrid systems. *IEEE Transactions on Automatic Control, 47*(10), 1663–1676.

Ferrari-Trecate, G., Muselli, M., Liberati, D., & Morari, M. (2003). A clustering technique for the identification of piecewise affine systems. *Automatica, 39*(2), 205–217.

Feuer, A., Nemirovski, A. (2003). On sparse representation in pairs of bases. *IEEE Transactions on Information Theory, 49*(6), 1579–1581.

Figueiredo, M. A. T., & Jain, A. K. (2002). Unsupervised learning of finite mixture models. *IEEE Transactions on Pattern Analysis and Machine Intelligence, 24*(3), 381–396.

Fischler, M. A., & Bolles, R. C. (1981). RANSAC random sample consensus: A paradigm for model fitting with applications to image analysis and automated cartography. *Communications of the ACM, 26*, 381–395.

Fisher, Y. (1995). *Fractal Image Compression: Theory and Application*. Springer-Verlag Telos.

Fitzgibbon, A., & Zisserman, A. (2000). Multibody structure and motion: 3D reconstruction of independently moving objects. In *European Conference on Computer Vision* (pp. 891–906).

Forgy, E. (1965). Cluster analysis of multivariate data: Efficiency vs. interpretability of classifications (abstract). *Biometrics, 21*, 768–769.

Freixenet, J., Munoz, X., Raba, D., Marti, J., & Cuff, X. (2002). Yet another survey on image segmentation. In *Proceedings of the European Conference on Computer Vision (ECCV)*.

Frey, B., Colmenarez, A., & Huang, T. (1998). Mixtures of local linear subspaces for face recognition. In *IEEE Conference on Computer Vision and Pattern Recognition*.

Gabriel, K. R. (1978). Least squares approximation of matrices by additive and multiplicative models. *Journal of the Royal Statistical Society B, 40*, 186–196.

Ganesh, A., Wright, J., Li, X., Candès, E., & Ma, Y. (2010). Dense error correction for low-rank matrices via principal component pursuit. In *International Symposium on Information Theory*.

Geman, S., & McClure, D. (1987). Statistical methods for tomographic image reconstruction. In *Proceedings of the 46th Session of the ISI, Bulletin of the ISI* (Vol. 52, pp. 5–21).

Georghiades, A., Belhumeur, P., & Kriegman, D. (2001). From few to many: Illumination cone models for face recognition under variable lighting and pose. *IEEE Transactions on Pattern Analysis and Machine Intelligence, 23*(6), 643–660.

Gersho, A., & Gray, R. M. (1992). *Vector Quantization and Signal Compression*. Boston: Kluwer Academic.

Gevers, T., & Smeulders, A. (1997). Combining region splitting and edge detection through guided Delaunay image subdivision. In *Proceedings of the IEEE Conference on Computer Vision and Pattern Recognition (CVPR)*.

Ghahramani, Z., & Beal, M. (2000). Variational inference for Bayesian mixtures of factor analysers. *Advances in Neural Information Processing Systems, 12*, 449–455.

Ghahramani, Z., & Hinton, G. (1996). The EM algorithm for mixtures of factor analyzers. *Technical Report CRG-TR-96-1, University of Toronto, Canada*.

Ghahramani, Z., & Hinton, G. E. (1998). Variational learning for switching state-space models. *Neural Computation, 12*(4), 963–996.

Ghoreyshi, A., & Vidal, R. (2007). Epicardial segmentation in dynamic cardiac MR sequences using priors on shape, intensity, and dynamics, in a level set framework. In *IEEE International Symposium on Biomedical Imaging* (pp. 860–863).

Gnanadesikan, R., & Kettenring, J. (1972). Robust estimates, residuals, and outlier detection with multiresponse data. *Biometrics, 28*(1), 81–124.

Goh, A., & Vidal, R. (2007). Segmenting motions of different types by unsupervised manifold clustering. In *IEEE Conference on Computer Vision and Pattern Recognition*.

Goh, A., & Vidal, R. (2008). Unsupervised Riemannian clustering of probability density functions. In *European Conference on Machine Learning*.

Goldfarb, D., & Ma, S. (2009). Convergence of fixed point continuation algorithms for matrix rank minimization. *Preprint*.

Golub, H., & Loan, C. V. (1996). *Matrix Computations* (2nd ed.). Baltimore: Johns Hopkins University Press.

Govindu, V. (2005). A tensor decomposition for geometric grouping and segmentation. In *IEEE Conference on Computer Vision and Pattern Recognition* (pp. 1150–1157).

Gower, J. (1966). Some distance properties of latent root and vector methods used in multivariate analysis. *Biometrika, 53*, 325–338.

Gross, D. (2011). Recovering low-rank matrices from few coefficients in any basis. *IEEE Trans on Information Theory, 57*(3), 1548–1566.

Gruber, A., & Weiss, Y. (2004). Multibody factorization with uncertainty and missing data using the EM algorithm. In *IEEE Conference on Computer Vision and Pattern Recognition* (Vol. I, pp. 707–714).

H.Aanaes, Fisker, R., Astrom, K., & Carstensen, J. M. (2002). Robust factorization. *IEEE Transactions on Pattern Analysis and Machine Intelligence, 24*(9), 1215–1225.

Haeffele, B., & Vidal, R. (2015). Global optimality in tensor factorization, deep learning, and beyond. *Preprint, http://arxiv.org/abs/1506.07540*.

Haeffele, B., Young, E., & Vidal, R. (2014). Structured low-rank matrix factorization: Optimality, algorithm, and applications to image processing. In *International Conference on Machine Learning*.

Hamkins, J., & Zeger, K. (2002). Gaussian source coding with spherical codes. *IEEE Transactions on Information Theory, 48*(11), 2980–2989.

Hampel, F., Ronchetti, E., Rousseeuw, P., & Stahel, W. (1986). *Robust statistics: The approach based on influence functions*. New York: Wiley.

Hampel, F. R. (1974). The influence curve and its role in robust estiamtion. *Journal of the American Statistical Association, 69*, 383–393.

Han, M., & Kanade, T. (2000). Reconstruction of a scene with multiple linearly moving objects. In *Proceedings of IEEE Conference on Computer Vision and Pattern Recognition* (Vol. 2, pp. 542–549).

Han, M., & Kanade, T. (2001). Multiple motion scene reconstruction from uncalibrated views. In *Proceedings of IEEE International Conference on Computer Vision* (Vol. 1, pp. 163–170).

Hansen, M., & Yu, B. (2001). Model selection and the principle of minimum description length. *Journal of American Statistical Association, 96*, 746–774.

Haralick, R., & Shapiro, L. (1985). Image segmentation techniques. *Computer Vision, Graphics, and Image Processing, 29*(1), 100–132.

Hardt, M. (2014). Understanding alternating minimization for matrix completion. In *Symposium on Foundations of Computer Science*.

Haro, G., Randall, G., & Sapiro, G. (2006). Stratification learning: Detecting mixed density and dimensionality in high dimensional point clouds. In *Neural Information Processing and Systems*.

Haro, G., Randall, G., & Sapiro, G. (2008). Translated poisson mixture model for stratification learning. *International Journal of Computer Vision, 80*(3), 358–374.

Harris, J. (1992). *Algebraic Geometry: A First Course*. New York: Springer.

Hartley, R., & Schaffalitzky, F. (2003). Powerfactorization: An approach to affine reconstruction with missing and uncertain data. In *Proceedings of Australia-Japan Advanced Workshop on Computer Vision*.

Hartley, R., & Vidal, R. (2004). The multibody trifocal tensor: Motion segmentation from 3 perspective views. In *IEEE Conference on Computer Vision and Pattern Recognition* (Vol. I, pp. 769–775).

Hartley, R., & Zisserman, A. (2004). *Multiple view geometry in computer vision* (2nd ed.). Cambridge: Cambridge University Press.

Hastie, T. (1984). Principal curves and surfaces. *Technical Report, Stanford University*.

Hastie, T., & Stuetzle, W. (1989). Principal curves. *Journal of the American Statistical Association, 84*(406), 502–516.

Hastie, T., Tibshirani, R., & Friedman, J. (2001). *The Elements of Statistical Learning*. New York: Springer.

He, H., & Garcia, E. (2009). Learning from imbalanced data. *IEEE Transactions on Knowledge and Data Engineering, 21*(9), 1263–1284.

He, H., & Ma, Y. (2013). *Imbalanced Learning: Foundations, Algorithms, and Applications*. New York: Wiley.

Hinton, G., Osindero, S., & Teh, Y.-W. (2006). A fast learning algorithm for deep belief nets. *Neural Computation, 18*(7), 1527–1554.

Hirsch, M. (1976). *Differential Topology*. New York: Springer.

Ho, J., Yang, M., Lim, J., Lee, K., & Kriegman, D. (2003). Clustering appearances of objects under varying illumination conditions. In *Proceedings of International Conference on Computer Vision and Pattern Recognition*.

Hong, W., Wright, J., Huang, K., & Ma, Y. (2006). Multi-scale hybrid linear models for lossy image representation. *IEEE Transactions on Image Processing, 15*(12), 3655–3671.

Horn, R. A., & Johnson, C. R. (1985). *Matrix Analysis*. Cambridge: Cambridge University Press.

Hotelling, H. (1933). Analysis of a complex of statistical variables into principal components. *Journal of Educational Psychology, 24*, 417–441.

Householder, A. S., & Young, G. (1938). Matrix approximation and latent roots. *American Mathematical Monthly, 45*, 165–171.

Huang, K., Ma, Y., & Vidal, R. (2004). Minimum effective dimension for mixtures of subspaces: A robust GPCA algorithm and its applications. In *IEEE Conference on Computer Vision and Pattern Recognition* (Vol. II, pp. 631–638).

Huber, P. (1981). *Robust Statistics*. New York: Wiley.

Hubert, L., Meulman, J., & Heiser, W. (2000). Two purposes for matrix factorization: A historical appraisal. *SIAM Review, 42*(1), 68–82.

Hwang, I., Balakrishnan, H., & Tomlin, C. (2003). Observability criteria and estimator design for stochastic linear hybrid systems. In *Proceedings of European Control Conference*.

Hyndman, M., Jepson, A., & Fleet, D. J. (2007). Higher-order autoregressive models for dynamic textures. In *British Machine Vision Conference* (pp. 76.1–76.10). doi:10.5244/C.21.76.

Jacobs, D. (2001). Linear fitting with missing data: Applications to structure-from-motion. *Computer Vision and Image Understanding, 82*, 57–81.

Jain, A. (1989). *Fundamentals of Digital Image Processing*. Upper Saddle River: Prentice Hall.

Jain, P., Meka, R., & Dhillon, I. (2010). Guaranteed rank minimization via singular value projection. In *Neural Information Processing Systems* (pp. 937–945).

Jain, P., & Netrapalli, P. (2014). Fast exact matrix completion with finite samples. In http://arxiv.org/pdf/1411.1087v1.pdf.

Jain, P., Netrapalli, P., & Sanghavi, S. (2012). Low-rank matrix completion using alternating minimization. In http://arxiv.org/pdf/1411.1087v1.pdf.

Janccy, R. (1966). Multidimensional group analysis. *Australian Journal of Botany, 14*, 127–130.

Jarret, K., Kavukcuoglu, K., Ranzato, M., & LeCun, Y. (2009). What is the best multi-stage architecture for object recognition. In *International Conference on Computer Vision*.

Jhuo, I.-H., Liu, D., Lee, D., & Chang, S.-F. (2012). Robust visual domain adaptation with low-rank reconstruction. In *IEEE Conference on Computer Vision and Pattern Recognition* (pp. 2168–2175).

Johnson, C. (1990). Matrix completion problems: A survey. In *Proceedings of Symposia in Applied Mathematics*.

Jolliffe, I. (1986). *Principal Component Analysis*. New York: Springer.

Jolliffe, I. (2002). *Principal Component Analysis* (2nd ed.). New York: Springer.

Jordan, M. (1874). Mémoire sur les formes bilinéaires. *Journal de Mathématiques Pures et Appliqués, 19*, 35–54.

Juloski, A., Heemels, W., & Ferrari-Trecate, G. (2004). Data-based hybrid modelling of the component placement process in pick-and-place machines. In *Control Engineering Practice*. Amsterdam: Elsevier.

Kamvar, S., Klein, D., & Manning, C. (2002). Interpreting and extending classical agglomerative clustering methods using a model-based approach. *Technical Report 2002-11, Stanford University Department of Computer Science*.

Kanatani, K. (1998). Geometric information criterion for model selection. *International Journal of Computer Vision* (pp. 171–189).

Kanatani, K. (2001). Motion segmentation by subspace separation and model selection. In *IEEE International Conference on Computer Vision* (Vol. 2, pp. 586–591).

Kanatani, K. (2002). Evaluation and selection of models for motion segmentation. In *Asian Conference on Computer Vision* (pp. 7–12).

Kanatani, K. (2003). How are statistical methods for geometric inference justified? In *Workshop on Statistical and Computational Theories of Vision, IEEE International Conference on Computer Vision*.

Kanatani, K., & Matsunaga, C. (2002). Estimating the number of independent motions for multibody motion segmentation. In *European Conference on Computer Vision* (pp. 25–31).

Kanatani, K., & Sugaya, Y. (2003). Multi-stage optimization for multi-body motion segmentation. In *Australia-Japan Advanced Workshop on Computer Vision* (pp. 335–349).

Ke, Q., & Kanade, T. (2005). Robust ℓ^1-norm factorization in the presence of outliers and missing data. In *IEEE Conference on Computer Vision and Pattern Recognition*.

Keshavan, R., Montanari, A., & Oh, S. (2010a). Matrix completion from a few entries. *IEEE Transactions on Information Theory*.

Keshavan, R., Montanari, A., & Oh, S. (2010b). Matrix completion from noisy entries. *Journal of Machine Learning Research, 11*, 2057–2078.

Keshavan, R. H. (2012). *Efficient algorithms for collaborative filtering*. Ph.D. Thesis. Stanford University.

Kim, J., Fisher, J., Yezzi, A., Cetin, M., & Willsky, A. (2005). A nonparametric statistical method for image segmentation using information theory and curve evolution. *PAMI, 14*(10), 1486–1502.

Kim, S. J., Doretto, G., Rittscher, J., Tu, P., Krahnstoever, N., & Pollefeys, M. (2009). A model change detection approach to dynamic scene modeling. In *Sixth IEEE International Conference on Advanced Video and Signal Based Surveillance, 2009 (AVSS '09)* (pp. 490–495).

Kim, S. J., Koh, K., Lustig, M., Boyd, S., & Gorinevsky, D. (2007). An interior-point method for large-scale l1-regularized least squares. *IEEE Journal on Selected Topics in Signal Processing, 1*(4), 606–617.

Kim, T., Lee, K., & Lee, S. (2010). Learning full pairwise affinities for spectral segmentation. In *Proceedings of the IEEE Conference on Computer Vision and Pattern Recognition (CVPR)*.

Kleinberg, J. M. (1999). Authorative sources in a hyberlinked environment. *Journal of the ACM, 48*, 604–632.

Kontogiorgis, S., & Meyer, R. (1989). A variable-penalty alternating direction method for convex optimization. *Mathematical Programming, 83*, 29–53.

Kruskal, J. (1964). Nonmetric multidimensional scaling: A numerical method. *Psychometrika*.

Kurita, T. (1995). An efficient clustering algorithm for region merging. *IEICE Transactions of Information and Systems, E78-D*(12), 1546–1551.

Lanczos, C. (1950). An iteration method for the solution of the eigenvalue problem of linear differential and integral operators. *Journal of Research of the National Bureau of Standards, 45*, 255–282.

Lang, S. (1993). *Algebra* (3rd ed.). Reading: Addison-Wesley.

Lee, J. A., & Verleysen, M. (2007). *Nonlinear Dimensionality Reduction* (1st ed.). New York: Springer.

Lee, K.-C., Ho, J., & Kriegman, D. (2005). Acquiring linear subspaces for face recognition under variable lighting. *IEEE Transactions on Pattern Analysis and Machine Intelligence, 27*(5), 684–698.

Leonardis, A., Bischof, H., & Maver, J. (2002). Multiple eigenspaces. *Pattern Recognition, 35*(11), 2613–2627.

LePennec, E., & Mallat, S. (2005). Sparse geometric image representation with bandelets. *IEEE Transactions on Image Processing, 14*(4), 423–438.

Levina, E., & Bickel, P. J. (2006). Texture synthesis and non-parametric resampling of random fields. *Annals of Statistics, 34*(4), 1751–1773.

Li, B., Ayazoglu, M., Mao, T., Camps, O. I., & Sznaier, M. (2011). Activity recognition using dynamic subspace angles. In *Proceedings of IEEE Conference on Computer Vision and Pattern Recognition* (pp. 3193–3200). New York: IEEE.

Lin, Z., Chen, M., Wu, L., & Ma, Y. (2011). The augmented Lagrange multiplier method for exact recovery of corrupted low-rank matrices. arXiv:1009.5055v2.

Lions, P., & Mercier, B. (1979). Splitting algorithms for the sum of two nonlinear operators. *SIAM Journal on Numerical Analysis, 16*(6), 964–979.

Liu, G., Lin, Z., Yan, S., Sun, J., & Ma, Y. (2013). Robust recovery of subspace structures by low-rank representation. *IEEE Trans. Pattern Analysis and Machine Intelligence, 35*(1), 171–184.

Liu, G., Lin, Z., & Yu, Y. (2010). Robust subspace segmentation by low-rank representation. In *International Conference on Machine Learning*.

Liu, Y. K., & Zalik, B. (2005). Efficient chain code with Huffman coding. *Pattern Recognition, 38*(4), 553–557.

Lloyd, S. (1957). *Least squares quantization in PCM*. Technical Report. Bell Laboratories. Published in 1982 in *IEEE Transactions on Information Theory, 28*, 128–137.

Luenberger, D. G. (1973). *Linear and Nonlinear Programming*. Reading: Addison-Wesley.

Luo, Z. Q., & Tseng, P. (1993). One the convergence rate of dual ascent methods for strictly convex minimization. *Mathematics of Operations Research, 18*, 846–867.

Ma, S. (2012). *Alternating proximal gradient method for convex minimization*. Technical Report.

Ma, Y., Derksen, H., Hong, W., & Wright, J. (2007). Segmentation of multivariate mixed data via lossy coding and compression. *IEEE Transactions on Pattern Analysis and Machine Intelligence, 29*(9), 1546–1562.

Ma, Y., Soatto, S., Kosecka, J., & Sastry, S. (2003). *An Invitation to 3D Vision: From Images to Geometric Models*. New York: Springer.

Ma, Y., & Vidal, R. (2005). Identification of deterministic switched ARX systems via identification of algebraic varieties. In *Hybrid Systems: Computation and Control* (pp. 449–465). New York: Springer.

Ma, Y., Yang, A. Y., Derksen, H., & Fossum, R. (2008). Estimation of subspace arrangements with applications in modeling and segmenting mixed data. *SIAM Review, 50*(3), 413–458.

MacQueen, J. (1967). Some methods for classification and analysis of multivariate observations. In *Proceedings of the Fifth Berkeley Symposium on Mathematical Statistics and Probability* (pp. 281–297).

Madiman, M., Harrison, M., & Kontoyiannis, I. (2004). Minimum description length vs. maximum likelihood in lossy data compression. In *Proceedings of the 2004 IEEE International Symposium on Information Theory*.

Malik, J., Belongie, S., Leung, T., & Shi, J. (2001). Contour and texture analysis for image segmentation. *International Journal of Computer Vision, 43*(1), 7–27.

Mallat, S. (1999). *A Wavelet Tour of Signal Processing* (2nd ed.). London: Academic.

Mallows, C. (1973). Some comments on C_p. *Technometrics, 15*, 661–675.

Maronna, R. A. (1976). Robust M-estimators of multivariate location and scatter. *Annals of Statistics, 4*, 51–67.

Martin, D., Fowlkes, C., Tal, D., & Malik, J. (2001). A Database of Human Segmented Natural Images and its Application to Evaluating Segmentation Algorithms and Measuring Ecological Statistics. In *IEEE International Conference on Computer Vision*.

McLanchlan, G. J., & Krishnan, T. (1997). *The EM Algorithms and Extentions*. Wiley Series in Probability and Statistics. John Wiley & Sons, Inc.

Meila, M. (2005). Comparing clusterings: An axiomatic view. In *Proceedings of the International Conference on Machine Learning*.

Mercer, J. (1909). Functions of positive and negative types and their connection with the theory of integral equations. *Philosophical Transactions, Royal Society London, A, 209*(1909), 415–446.

Meyer, F. (2000). Fast adaptive wavelet packet image compression. *IEEE Transactions on Image Processing, 9*(5), 792–800.

Meyer, F. (2002). Image compression with adaptive local cosines. *IEEE Transactions on Image Processing, 11*(6), 616–629.

Minka, T. (2000). Automatic choice of dimensionality for PCA. In *Neural Information Processing Systems* (Vol. 13, pp. 598–604).

Mirsky, L. (1975). A trace inequality of John von Neumann. *Monatshefte für Mathematic, 79*, 303–306.

Mobahi, H., Rao, S., Yang, A., & Sastry, S. (2011). Segmentation of natural images by texture and boundary compression. *International Journal of Computer Vision, 95*(1), 86–98.

Mori, G., Ren, X., Efros, A., & Malik, J. (2004). Recovering human body configurations: Combining segmentation and recognition. In *IEEE Conference on Computer Vision and Pattern Recognition*.

Muresan, D., & Parks, T. (2003). Adaptive principal components and image denoising. In *IEEE International Conference on Image Processing*.

Murphy, K. (1998). *Switching Kalman filters*. Technical Report. U.C. Berkeley.

Nascimento, J. C., Figueiredo, M. A. T., & Marques, J. S. (2005). Recognition of human activities using space dependent switched dynamical models. In *IEEE International Conference on Image Processing* (pp. 852–855).

Neal, R., & Hinton, G. (1998). A view of the EM algorithm that justifies incremental, sparse, and other variants. In M. Jordan (Ed.), *Learning in graphical models* (pp. 355–368). Boston: Kluwer Academic.

Negahban, S., Ravikumar, P., Wainwright, M., & Yu, B. (2010). A unified framework for analyzing *m*-estimators with decomposible regularizers. Available at http://arxiv.org/abs/1010.2731v1.

Nemirovskii, A. S., & Yudin, D. B. (1979). *Complexity of problems and efficiency of optimization methods* (in Russian). Moscow: Nauka.

Nesterov, Y. (1983). A method of solving a convex programming problem with convergence rate $O(1/k^2)$. *Soviet Mathematics Doklady, 27*(2), 372–376.

Ng, A., Weiss, Y., & Jordan, M. (2001). On spectral clustering: Analysis and an algorithm. In *Proceedings of Neural Information Processing Systems (NIPS)* (pp. 849–856).

Niessen, H., & A.Juloski (2004). Comparison of three procedures for identification of hybrid systems. In *Conference on Control Applications*.

Nunez, F., & Cipriano, A. (2009). Visual information model based predictor for froth speed control in flotation process. *Minerals Engineering, 22*(4), 366–371.

Ofli, F., Chaudhry, R., Kurillo, G., Vidal, R., & Bajcsy, R. (2013). Berkeley MHAD: A comprehensive multimodal human action database. In *IEEE Workshop on Applications of Computer Vision*.

Olshausen, B., & D.J.Field (1996). Emergence of simple-cell receptive field properties by learning a sparse code for natural images. *Nature, 381*(6583), 607–609.

Orlik, P. (1989). *Introduction to Arrangements. Conference Board of the Mathematical Sciences Regional Conference Series in Mathematics* (Vol. 72). Providence: American Mathematics Society.

Overschee, P. V., & Moor, B. D. (1993). Subspace algorithms for the stochastic identification problem. *Automatica, 29*(3), 649–660.

Patel, V. M., Gopalan, R., Li, R., & Chellappa, R. (2014). Visual domain adaptation: A survey of recent advances. *IEEE Signal Processing Magazine, 32*(3), 53–69.

Pavlovic, V., Moulin, P., & Ramchandran, K. (1998). An integrated framework for adaptive subband image coding. *IEEE Transactions on Signal Processing, 47*(4), 1024–1038.

Pavlovic, V., Rehg, J. M., Cham, T. J., & Murphy, K. P. (1999). A dynamic Bayesian network approach to figure tracking using learned dynamic models. In *Proceedings of the International Conference on Computer Vision* (pp. 94–101).

Pearson, K. (1901). On lines and planes of closest fit to systems of points in space. *The London, Edinburgh and Dublin Philosphical Magazine and Journal of Science, 2*, 559–572.

Peng, Z., Yan, M., & Yin, W. (2013). Parallel and distributed sparse optimization. In *Asilomar*.

Polito, M., & Perona, P. (2002). Grouping and dimensionality reduction by locally linear embedding. In *Proceedings of Neural Information Processing Systems (NIPS)*.

Powell, M. J. D. (1973). On search directions for minimization algorithms. *Mathematical Programming, 4*, 193–201.

Qiu, Q., Patel, V. M., Turaga, P., & Chellappa, R. (2012). Domain adaptive dictionary learning. In *European Conference on Computer Vision* (Vol. 7575, pp. 631–645).

Rabiee, H., Kashyap, R., & Safavian, S. (1996). Adaptive multiresolution image coding with matching and basis pursuits. In *IEEE International Conference on Image Processing*.

Rahimi, A., Darrell, T., & Recht, B. (2005). Learning appearance manifolds from video. In *IEEE Conference on Computer Vision and Pattern Recognition* (Vol. 1, pp. 868–875).

Ramchandran, K., & Vetterli, M. (1993). Best wavelet packets bases in a rate-distortion sense. *IEEE Transactions on Image Processing, 2*, 160–175.

Ramchandran, K., Vetterli, M., & Herley, C. (1996). Wavelets, subband coding, and best basis. *Proceedings of the IEEE, 84*(4), 541–560.

Rand, W. M. (1971). Objective criteria for the evaluation of clustering methods. *Journal of the American Statistical Association, 66*(336), 846–850.

Rao, S., Mobahi, H., Yang, A., & Sastry, S. (2009). Natural image segmentation with adaptive texture and boundary encoding. In *Asian Conference on Computer Vision, 1* (pp. 135–146).

Rao, S., Tron, R., Ma, Y., & Vidal, R. (2008). Motion segmentation via robust subspace separation in the presence of outlying, incomplete, or corrupted trajectories. In *IEEE Conference on Computer Vision and Pattern Recognition*.

Rao, S., Tron, R., Vidal, R., & Ma, Y. (2010). Motion segmentation in the presence of outlying, incomplete, or corrupted trajectories. *IEEE Transactions on Pattern Analysis and Machine Intelligence, 32*(10), 1832–1845.

Rao, S., Yang, A. Y., Wagner, A., & Ma, Y. (2005). Segmentation of hybrid motions via hybrid quadratic surface analysis. In *IEEE International Conference on Computer Vision* (pp. 2–9).

Ravichandran, A., Chaudhry, R., & Vidal, R. (2009). View-invariant dynamic texture recognition using a bag of dynamical systems. In *IEEE Conference on Computer Vision and Pattern Recognition*.

Ravichandran, A., Chaudhry, R., & Vidal, R. (2013). Categorizing dynamic textures using a bag of dynamical systems. *IEEE Transactions on Pattern Analysis and Machine Intelligence, 35*(2), 342–353.

Ravichandran, A., & Vidal, R. (2008). Video registration using dynamic textures. In *European Conference on Computer Vision*.

Ravichandran, A., & Vidal, R. (2011). Video registration using dynamic textures. *IEEE Transactions on Pattern Analysis and Machine Intelligence, 33*(1), 158–171.

Ravichandran, A., Vidal, R., & Halperin, H. (2006). Segmenting a beating heart using polysegment and spatial GPCA. In *IEEE International Symposium on Biomedical Imaging* (pp. 634–637).

Recht, B., Fazel, M., & Parrilo, P. (2010). Guaranteed minimum-rank solutions of linear matrix equations via nuclear norm minimization. *SIAM Review, 52*(3), 471–501.

Ren, X., Fowlkes, C., & Malik, J. (2005). Scale-invariant contour completion using condition random fields. In *IEEE International Conference on Computer Vision*.

Ren, X., Fowlkes, C., & Malik, J. (2008). Learning probabilistic models for contour completion in natural images. *International Journal of Computer Vision, 77*, 47–63.

Rissanen, J. (1978). Modeling by shortest data description. *Automatica, 14*, 465–471.

Rose, K. (1998). Deterministic annealing for clustering, compression, classification, regression, and related optimization problems. *Proceedings of the IEEE, 86*(11), 2210–2239.

Rousseeuw, P. (1984). Least median of squares regression. *Journal of American Statistics Association, 79*, 871–880.

Roweis, S., & Saul, L. (2000). Nonlinear dimensionality reduction by locally linear embedding. *Science, 290*(5500), 2323–2326.

Roweis, S., & Saul, L. (2003). Think globally, fit locally: Unsupervised learning of low dimensional manifolds. *Journal of Machine Learning Research, 4*, 119–155.

Saisan, P., Bissacco, A., Chiuso, A., & Soatto, S. (2004). Modeling and synthesis of facial motion driven by speech. In *European Conference on Computer Vision* (Vol. 3, pp. 456–467).

Santis, E., Benedetto, M. D., & Giordano, P. (2003). On observability and detectability of continuous-time linear switching systems. In *Proceedings of IEEE Conference on Decision & Control* (pp. 5777–5782).

Schindler, K., & Suter, D. (2005). Two-view multibody structure-and-motion with outliers. In *IEEE Conference on Computer Vision and Pattern Recognition*.

Schölkopf, B., & Smola, A. (2002). *Learning with kernels*. Cambridge: MIT Press.

Schölkopf, B., Smola, A., & Muller, K. R. (1998). Nonlinear component analysis as a kernel eigenvalue problem. *Neural Computation, 10*, 1299–1319.

Selim, S., & Ismail, M. A. (1984). K-means-type algorithms: A generalized convergence theorem and characterization of local optimality. *IEEE Transaction on Pattern Analysis and Machine Intelligence, 6*(1), 81–87.

Sha, F., & Saul, L. (2005). Analysis and extension of spectral methods for nonlinear dimensionality reduction. In *Proceedings of International Conference on Machine Learning* (pp. 784–791).

Shabalin, A., & Nobel, A. (2010). *Reconstruction of a low-rank matrix in the presence of gaussian noise* (pp. 1–34). arXiv preprint 1007.4148

Shakernia, O., Vidal, R., & Sastry, S. (2003). Multi-body motion estimation and segmentation from multiple central panoramic views. In *IEEE International Conference on Robotics and Automation* (Vol. 1, pp. 571–576).

Shapiro, J. M. (1993). Embedded image coding using zerotrees of wavelet coefficients. *IEEE Transactions on Signal Processing, 41*(12), 3445–3463.

Shashua, A., & Levin, A. (2001). Multi-frame infinitesimal motion model for the reconstruction of (dynamic) scenes with multiple linearly moving objects. In *Proceedings of IEEE International Conference on Computer Vision* (Vol. 2, pp. 592–599).

Shekhar, S., Patel, V. M., Nguyen, H. V., & Chellappa, R. (2013). Generalized domain-adaptive dictionaries. In *IEEE Conference on Computer Vision and Pattern Recognition.*

Shi, J., & Malik, J. (1998). Motion segmentation and tracking using normalized cuts. In *IEEE International Conference on Computer Vision* (pp. 1154–1160).

Shi, J., & Malik, J. (2000). Normalized cuts and image segmentation. *IEEE Transactions on Pattern Analysis and Machine Intelligence, 22*(8), 888–905.

Shi, T., Belkin, M., & Yin, B. (2008). Data spectroscopy: Eigenspace of convolution operators and clustering. *arXiv:0807.3719v1.*

Shizawa, M., & Mase, K. (1991). A unified computational theory for motion transparency and motion boundaries based on eigenenergy analysis. In *IEEE Conference on Computer Vision and Pattern Recognition* (pp. 289–295).

Shum, H.-Y., Ikeuchi, K., & Reddy, R. (1995). Principal component analysis with missing data and its application to polyhedral object modeling. *IEEE Transactions on Pattern Analysis and Machine Intelligence, 17*(9), 854–867.

Sikora, T., & Makai, B. (1995). Shape-adaptive DCT for generic coding of video. *IEEE Transactions on Circuits and Systems For Video Technology, 5*, 59–62.

Soltanolkotabi, M., & Candès, E. J. (2013). A geometric analysis of subspace clustering with outliers. *Annals of Statistics, 40*(4), 2195–2238.

Soltanolkotabi, M., Elhamifar, E., & Candès, E. J. (2014). Robust subspace clustering. *Annals of Statistics, 42*(2), 669–699.

Souvenir, R., & Pless, R. (2005). Manifold clustering. In *Proceedings of International Conference on Computer Vision* (Vol. I, pp. 648–653).

Spielman, D., Wang, H., & Wright, J. (2012). Exact recovery of sparsity-used dictionaries. *Conference on Learning Theory (COLT).*

Starck, J.-L., Elad, M., & Donoho, D. (2003). Image decomposition: Separation of texture from piecewise smooth content. In *Proceedings of the SPIE* (Vol. 5207, pp. 571–582).

Steward, C. V. (1999). Robust parameter estimation in computer vision. *SIAM Review, 41*(3), 513–537.

Sturm, P. (2002). Structure and motion for dynamic scenes - the case of points moving in planes. In *Proceedings of European Conference on Computer Vision* (pp. 867–882).

Sun, A., Ge, S. S., & Lee, T. H. (2002). Controllability and reachability criteria for switched linear systems. *Automatica, 38*, 775–786.

Sun, J., Qu, Q., & Wright, J. (2015). Complete dictionary recovery over the sphere. Preprint. http://arxiv.org/abs/1504.06785

Szigeti, F. (1992). A differential algebraic condition for controllability and observability of time varying linear systems. In *Proceedings of IEEE Conference on Decision and Control* (pp. 3088–3090).

Szummer, M., & Picard, R. W. (1996). Temporal texture modeling. In *IEEE International Conference on Image Processing* (Vol. 3, pp. 823–826).

Taubin, G. (1991). Estimation of planar curves, surfaces, and nonplanar space curves defined by implicit equations with applications to edge and range image segmentation. *IEEE Transactions on Pattern Analysis and Machine Intelligence, 13*(11), 1115–1138.

Tenenbaum, J. B., de Silva, V., & Langford, J. C. (2000). A global geometric framework for nonlinear dimensionality reduction. *Science, 290*(5500), 2319–2323.

Tibshirani, R. (1996). Regression shrinkage and selection via the LASSO. *Journal of the Royal Statistical Society B, 58*(1), 267–288.

Tipping, M., & Bishop, C. (1999a). Mixtures of probabilistic principal component analyzers. *Neural Computation, 11*(2), 443–482.

Tipping, M., & Bishop, C. (1999b). Probabilistic principal component analysis. *Journal of the Royal Statistical Society, 61*(3), 611–622.

Torgerson, W. (1958). *Theory and Methods of Scaling*. New York: Wiley.

Torr, P., & Davidson, C. (2003). IMPSAC: Synthesis of importance sampling and random sample consensus. *IEEE Transactions on Pattern Analysis and Machine Intelligence, 25*(3), 354–364.

Torr, P., Szcliski, R., & Anandan, P. (2001). An integrated Bayesian approach to layer extraction from image sequences. *IEEE Transactions on Pattern Analysis and Machine Intelligence, 23*(3), 297–303.

Torr, P. H. S. (1998). Geometric motion segmentation and model selection. *Philosophical Transactions of the Royal Society of London, 356*(1740), 1321–1340.

Tremeau, A., & Borel, N. (1997). A region growing and merging algorithm to color segmentation. *Pattern Recognition, 30*(7), 1191–1204.

Tron, R., & Vidal, R. (2007). A benchmark for the comparison of 3-D motion segmentation algorithms. In *IEEE Conference on Computer Vision and Pattern Recognition*.

Tse, D., & Viswanath, P. (2005). *Fundamentals of Wireless Communications*. Cambridge: Cambridge University Press.

Tseng, P. (2000). Nearest q-flat to m points. *Journal of Optimization Theory and Applications, 105*(1), 249–252.

Tseng, P. (2001). Convergence of a block coordinate descent method for nondifferentiable minimization. *Journal of Optimization Theory and Applications, 109*(3), 475–494.

Tu, Z., & Zhu, S. (2002). Image segmentation by data-driven Markov Chain Monte Carlo. *PAMI, 24*(5), 657–673.

Tugnait, J. K. (1982). Detection and estimation for abruptly changing systems. *Automatica, 18*(5), 607–615.

Turaga, P., Veeraraghavan, A., Srivastava, A., & Chellappa, R. (2011). Statistical computations on special manifolds for image and video-based recognition. *IEEE Transactions on Pattern Analysis and Machine Intelligence, 33*(11), 2273–2286.

Turk, M., & Pentland, A. (1991). Face recognition using eigenfaces. In *Proceedings of IEEE Conference on Computer Vision and Pattern Recognition* (pp. 586–591).

Udell, M., Horn, C., Zadeh, R., & Boyd, S. (2015). *Generalized low rank models*. Working manuscript.

Ueda, N., Nakan, R., & Ghahramani, Z. (2000). SMEM algorithm for mixture models. *Neural Computation, 12*, 2109–2128.

Vapnik, V. (1995). *The nature of statistical learning theory*. New York: Springer.

Varma, M., & Zisserman, A. (2003). Texture classification: Are filter banks necessary? In *Proceedings of the IEEE Conference on Computer Vision and Pattern Recognition*.

Vasilescu, M., & Terzopoulos, D. (2002). Multilinear analysis of image ensembles: Tensorfaces. In *Proceedings of European Conference on Computer Vision* (pp. 447–460).

Vecchio, D. D., & Murray, R. (2004). Observers for a class of hybrid systems on a lattice. In *Proceedings of Hybrid Systems: Computation and Control*. New York: Springer.

Vetterli, M., & Kovacevic, J. (1995). *Wavelets and subband coding*. Upper Saddle River: Prentice-Hall.

Vidal, R. (2004). Identification of PWARX hybrid models with unknown and possibly different orders. In *American Control Conference* (pp. 547–552).

Vidal, R. (2005). Multi-subspace methods for motion segmentation from affine, perspective and central panoramic cameras. In *IEEE Conference on Robotics and Automation* (pp. 1753–1758).

Vidal, R. (2008). Recursive identification of switched ARX systems. *Automatica, 44*(9), 2274–2287.

Vidal, R., Chiuso, A., & Soatto, S. (2002a). Observability and identifiability of jump linear systems. In *IEEE Conference on Decision and Control* (pp. 3614–3619).

Vidal, R., Chiuso, A., Soatto, S., & Sastry, S. (2003a). Observability of linear hybrid systems. In *Hybrid Systems: Computation and Control* (pp. 526–539). New York: Springer.

Vidal, R., & Favaro, P. (2014). Low rank subspace clustering (LRSC). *Pattern Recognition Letters, 43*, 47–61.

Vidal, R., & Hartley, R. (2004). Motion segmentation with missing data by PowerFactorization and Generalized PCA. In *IEEE Conference on Computer Vision and Pattern Recognition* (Vol. II, pp. 310–316).

Vidal, R., & Ma, Y. (2004). A unified algebraic approach to 2-D and 3-D motion segmentation. In *European Conference on Computer Vision* (pp. 1–15).

Vidal, R., Ma, Y., & Piazzi, J. (2004). A new GPCA algorithm for clustering subspaces by fitting, differentiating and dividing polynomials. In *IEEE Conference on Computer Vision and Pattern Recognition* (Vol. I, pp. 510–517).

Vidal, R., Ma, Y., & Sastry, S. (2003b). Generalized Principal Component Analysis (GPCA). In *IEEE Conference on Computer Vision and Pattern Recognition* (Vol. I, pp. 621–628).

Vidal, R., Ma, Y., Soatto, S., & Sastry, S. (2006). Two-view multibody structure from motion. *International Journal of Computer Vision, 68*(1), 7–25.

Vidal, R., & Ravichandran, A. (2005). Optical flow estimation and segmentation of multiple moving dynamic textures. In *IEEE Conference on Computer Vision and Pattern Recognition* (Vol. II, pp. 516–521).

Vidal, R., & Sastry, S. (2003). Optimal segmentation of dynamic scenes from two perspective views. In *IEEE Conference on Computer Vision and Pattern Recognition* (Vol. 2, pp. 281–286).

Vidal, R., Soatto, S., Ma, Y., & Sastry, S. (2002b). Segmentation of dynamic scenes from the multibody fundamental matrix. In *ECCV Workshop on Visual Modeling of Dynamic Scenes*.

Vidal, R., Soatto, S., Ma, Y., & Sastry, S. (2003c). An algebraic geometric approach to the identification of a class of linear hybrid systems. In *IEEE Conference on Decision and Control* (pp. 167–172).

Vidal, R., Tron, R., & Hartley, R. (2008). Multiframe motion segmentation with missing data using PowerFactorization and GPCA. *International Journal of Computer Vision, 79*(1), 85–105.

von Luxburg, U. (2007). A tutorial on spectral clustering. *Statistics and Computing, 17*(4), 395–416.

Wallace, C., & Boulton, D. (1968). An information measure for classification. *The Computer Journal, 11*, 185–194.

Wallace, C., & Dowe, D. (1999). Minimum message length and Kolmogrov complexity. *The Computer Journal, 42*(4), 270–283.

Wallace, G. K. (1991). The JPEG still picture compression standard. *Communications of the ACM. Special issue on digital multimedia systems, 34*(4), 30–44.

Wang, J., Jia, Y., Hua, X., Zhang, C., & Quan, L. (2008a). Normalized tree partitioning for image segmentation. In *IEEE Conference on Computer Vision and Pattern Recognition*.

Wang, J. M., Fleet, D. J., & Hertzmann, A. (2008b). Gaussian process dynamical models for human motion. *IEEE Transactions on Pattern Analysis and Machine Intelligence, 30*(2), 283–298.

Wang, Y.-X., & Xu, H. (2013). Noisy sparse subspace clustering. In *International Conference on Machine learning*.

Ward, J. (1963). Hierarchical grouping to optimize and objective function. *Journal of the American Statistical Association, 58*, 236–244.

Warga, J. (1963). Minimizing certain convex functions. *SIAM Journal on Applied Mathematics, 11*, 588–593.

Wei, S., & Lin, Z. (2010). Analysis and improvement of low rank representation for subspace segmentation. Technical Report MSR-TR-2010-177, Microsoft Research Asia.

Weinberger, K. Q., & Saul, L. (2004). Unsupervised learning of image manifolds by semidefinite programming. In *Proceedings of IEEE Conference on Computer Vision and Pattern Recognition* (pp. 988–955).

Wiberg, T. (1976). Computation of principal components when data are missing. In *Symposium on Computational Statistics* (pp. 229–326).

Wilks, S. S. (1962). *Mathematical Staistics*. New York: Wiley.

Williams, C. (2002). On a connection between kernel PCA and metric multidimensional scaling. *Machine Learning, 46*, 11–19.

Wolf, L., & Shashua, A. (2001a). Affine 3-D reconstruction from two projective images of independently translating planes. In *Proceedings of IEEE International Conference on Computer Vision* (pp. 238–244).

Wolf, L., & Shashua, A. (2001b). Two-body segmentation from two perspective views. In *IEEE Conference on Computer Vision and Pattern Recognition* (pp. 263–270).

Wolf, L., & Shashua, A. (2003). Learning over sets using kernel principal angles. *Journal of Machine Learning Research, 4*(10), 913–931.

Woolfe, F., & Fitzgibbon, A. (2006). Shift-invariant dynamic texture recognition. In *Proceedings of European Conference on Computer Vision*, pages II: 549–562.

Wright, J., Ganesh, A., Kerui, M., & Ma, Y. (2013). Compressive principal component analysis. *IMA Journal on Information and Inference, 2*(1), 32–68.

Wright, J., Ganesh, A., Rao, S., Peng, Y., & Ma, Y. (2009a). Robust principal component analysis: Exact recovery of corrupted low-rank matrices via convex optimization. In *NIPS*.

Wright, J., Ma, Y., Tao, Y., Lin, Z., & Shum, H.-Y. (2009b). Classification via minimum incremental coding length (MICL). *SIAM Journal on Imahing Sciences, 2*(2), 367–395.

Wu, J. (1983). On the convergence properties of the EM algorithm. *Annals of Statistics, 11*(1), 95–103.

Wu, Y., Zhang, Z., Huang, T., & Lin, J. (2001). Multibody grouping via orthogonal subspace decomposition. In *IEEE Conference on Computer Vision and Pattern Recognition* (Vol. 2, pp. 252–257).

Xiong, F., Camps, O., & Sznaier, M. (2011). Low order dynamics embedding for high dimensional time series. In *IEEE International Conference on Computer Vision* (pp. 2368–2374).

Xiong, F., Camps, O., & Sznaier, M. (2012). Dynamic context for tracking behind occlusions. In *European Conference on Computer Vision. Lecture notes in computer science* (Vol. 7576, pp. 580–593). Berlin/Heidelberg: Springer.

Xu, H., Caramanis, C., & Sanghavi, S. (2010). Robust pca via outlier pursuit. In *Neural Information Processing Systems (NIPS)*.

Yan, J., & Pollefeys, M. (2006). A general framework for motion segmentation: Independent, articulated, rigid, non-rigid, degenerate and non-degenerate. In *European Conference on Computer Vision* (pp. 94–106).

Yang, A., Wright, J., Ma, Y., & Sastry, S. (2008). Unsupervised segmentation of natural images via lossy data compression. *Computer Vision and Image Understanding, 110*(2), 212–225.

Yang, A. Y., Rao, S. R., & Ma, Y. (2006). Robust statistical estimation and segmentation of multiple subspaces. In *CVPR workshop on 25 years of RANSAC*.

Yang, J., Wright, J., Huang, T., & Ma, Y. (2010). Image super-resolution via sparse representation. *IEEE Transactions on Image Processing, 19*(11), 2861–2873.

Yang, M. H., Ahuja, N., & Kriegman, D. (2000). Face detection using mixtures of linear subspaces. In *IEEE International Conference on Automatic Face and Gesture Recognition*.

Yu, G., Sapiro, G., & Mallat, S. (2010). Image modeling and enhancement via structured sparse model selection. In *International Conference on Image Processing*.

Yu, G., Sapiro, G., & Mallat, S. (2012). Solving inverse problems with piecewise linear estimators: From gaussian mixture models to structured sparsity. *IEEE Transactions on Image Processing, 21*(5), 2481–2499.

Yu, S. (2005). Segmentation induced by scale invariance. In *IEEE Conference on Computer Vision and Pattern Recognition*.

Yuan, L., Wen, F., Liu, C., & Shum, H. (2004). Synthesizing dynamic texture with closed-loop linear dynamic system. In *European Conference on Computer Vision* (pp. 603–616).

Yuan, X., & Yang, J. (2009). Sparse and low-rank matrix decomposition via alternating direction methods. *Preprint*.

Zadeh, N. (1970). A note on the cyclic coordinate ascent method. *Management Science, 16*, 642–644.

Zelnik-Manor, L., & Irani, M. (2003). Degeneracies, dependencies and their implications in multi-body and multi-sequence factorization. In *IEEE Conference on Computer Vision and Pattern Recognition* (Vol. 2, pp. 287–293).

Zhang, K., Zhang, L., & Yang, M. (2014). Fast compressive tracking. *IEEE Transactions on Pattern Analysis and Machine Intelligence, 36*(10).

Zhang, T., Szlam, A., & Lerman, G. (2009). Median *k*-flats for hybrid linear modeling with many outliers. In *Workshop on Subspace Methods*.

Zhang, T., Szlam, A., Wang, Y., & Lerman, G. (2010). Randomized hybrid linear modeling via local best-fit flats. In *IEEE Conference on Computer Vision and Pattern Recognition* (pp. 1927–1934).

Zhang, Z., & Zha, H. (2005). Principal manifolds and nonlinear dimensionality reduction via tangent space alignment. *SIAM Journal on Scientific Computing, 26*(1), 313–338.

Zhou, F., la Torre, F. D., & Hodgins, J. K. (2008). Aligned cluster analysis for temporal segmentation of human motion. In *International Conference on Automatic Face and Gesture Recognition*.

Zhou, M., Wang, C., Chen, M., Paisley, J., Dunson, D., & Carin, L. (2010a). Nonparametric bayesian matrix completion. In *Sensor Array and Multichannel Signal Processing Workshop*.

Zhou, Z., Wright, J., Li, X., Candès, E., & Ma, Y. (2010b). Stable principal component pursuit. In *International Symposium on Information Theory*.

Zhu, Q., Song, G., & Shi, J. (2007). Untangling cycles for contour grouping. In *Proceedings of the International Conference on Computer Vision (ICCV)*.

Index

R. Vidal et al., *Generalized Principal Component Analysis*, Interdisciplinary Applied Mathematics 40, DOI 10.1007/978-0-387-87811-9

B

F

G

M